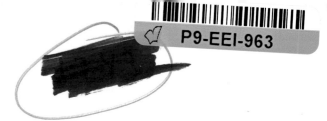

The Electromagnetic Spectrum

The Visible Spectrum

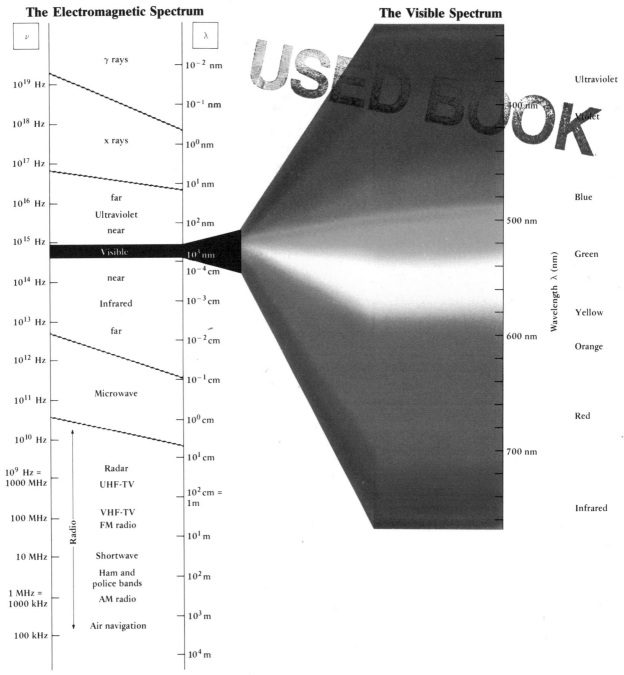

ν

γ rays

10^{19} Hz

10^{18} Hz   x rays

10^{17} Hz

10^{16} Hz   far
Ultraviolet
near

10^{15} Hz   Visible

10^{14} Hz   near

Infrared

10^{13} Hz   far

10^{12} Hz

Microwave

10^{11} Hz

10^{10} Hz

10^9 Hz =
1000 MHz   Radar
UHF-TV

100 MHz   VHF-TV
FM radio

10 MHz   Shortwave

Ham and
police bands

1 MHz =
1000 kHz   AM radio

100 kHz   Air navigation

Radio

λ

10^{-2} nm

10^{-1} nm

10^0 nm

10^1 nm

10^2 nm

10^3 nm

10^{-4} cm

10^{-3} cm

10^{-2} cm

10^{-1} cm

10^0 cm

10^1 cm

10^2 cm =
1 m

10^1 m

10^2 m

10^3 m

10^4 m

Ultraviolet

400 nm

Violet

Blue

500 nm

Green

Yellow

600 nm

Orange

Red

700 nm

Infrared

Wavelength λ (nm)

ANALYTICAL CHEMISTRY
An Introduction

SIXTH EDITION

Douglas A. Skoog
Stanford University

Donald M. West
San Jose State University

F. James Holler
University of Kentucky

SAUNDERS GOLDEN SUNBURST SERIES

SAUNDERS COLLEGE PUBLISHING
Harcourt Brace College Publishers

Philadelphia   Ft. Worth   Chicago   San Francisco
Montreal   Toronto   London   Sydney   Tokyo

Text Typeface: 10/12 Times Roman
Compositor: Bi-Comp, Inc.
Publisher: John Vondeling
Developmental Editor: Jennifer Bortel
Managing Editor: Carol Field
Project Editor: Maureen Iannuzzi
Copy Editor: Debbie Hardin
Manager of Art and Design: Carol Bleistine
Art Director: Anne Muldrow, Susan Blaker
Art Assistant: Caroline McGowan, Sue Kinney
Text Designer: LMD Services for Publishers
Cover Designer: Lawrence R. Didona
Text Artwork: Larry Ward
Director of EDP: Tim Frelick
Production Manager: Charlene Squibb
Marketing Manager: Marjorie Waldron

Cover Credit: Seldom has any event in the history of chemistry stimulated as much research activity and public interest as the discovery in 1985 of C_{60}, known as Buckminsterfullerene, or Buckyballs. In Feature 27-1, we describe the chromatographic separation and purification of Buckyballs. The photograph on the cover is a computer-generated electron density map of a Buckyball. The photograph was provided by SPL/Photo Researchers.

Printed in the United States of America

Analytical Chemistry, Sixth Edition

ISBN: 0-03-097285-X
Library of Congress Catalog Card Number:
    4 032 98765432

PREFACE

The sixth edition of *Analytical Chemistry: An Introduction*, like earlier editions, is an abbreviated version of the authors' other text, *Fundamentals of Analytical Chemistry*.[1] A primary audience of earlier editions of this text has been students in fields such as medicine, biology, geology, engineering, and physical sciences whose only exposure to analytical chemistry is a one-semester or one-quarter introductory course in quantitative analysis. The text has also found use in a sophomore-level course for chemistry majors where fewer laboratory experiments and less detailed and more descriptive treatment of topics is desired. Finally, the text has been used in freshman laboratory courses. This edition is designed to serve these same audiences.

OBJECTIVES

A major objective of this text is to provide a rigorous background in those chemical principles that are of particular importance to analytical chemistry. A second goal is to develop in the student an appreciation of the difficult task of judging the accuracy and precision of experimental data and to show how these judgments can be sharpened by the application of statistical methods. A third aim is to introduce the student to a wide range of techniques of modern analytical chemistry. A final goal is to teach those laboratory skills that will give students confidence in their ability to obtain high-quality analytical data.

COVERAGE

The material in this text covers both fundamental and practical aspects of chemical analysis. The first three chapters provide an introduction to the field of analytical chemistry and a review of elementary stoichiometry and equilibrium chemistry. Chapters 4 and 5 then deal with accuracy, precision, and the applications of statistics to data treatment. Chapter 6 provides a brief overview of the theory and applications of gravimetric methods of analysis. A systematic method for attacking multiple equilibrium problems and the effect of electrolytes on equilibria is then presented in Chapters 7 and 8. In Chapters 9 through 17 we consider the theory and practice of various titrimetric methods of analysis. The next two chapters are devoted to electroanalytical methods. Chapters 20 through 24 provide material on various spectroscopic methods of analysis. Chapters 25 through 27 deal with analytical separations, with Chapter 25 being devoted to extraction and ion exchange and Chapters 26 and 27 to chromatography. Finally, Chapters 28 and 29 provide discussions of the equipment and practice of chemical analysis and many detailed procedures for a variety of analyses.

FEATURES

We have included many features in this text to enhance learning and to provide a versatile teaching tool.

Organization

Analytical Chemistry: An Introduction provides a gradual progression from theoretical to practical topics in analytical chemistry. Each chapter is organized as a learning unit; most chapters can stand alone, thus affording the instructor the option of including or omitting topics as time and resources permit.

Mathematical Level

Generally, the theoretical presentations are based upon college algebra and assume no knowledge of calculus.

Important Equations

Equations that we consider most important are highlighted with a color screen for emphasis and ease of review.

Worked Examples

A large number of carefully worked out examples serve as an aid in understanding the principles of analytical chemistry. The examples also serve as models for the questions and problems found at the end of each chapter.

Questions and Problems

An extensive set of questions and problems is included at the end of all chapters except the first and last two. Answers to about half the questions and problems are given at the end of the book.

Features

A series of boxed and highlighted *Features* are found throughout the text. These *Features* contain derivation of equations, explanations of more difficult theoretical points, historical notes, and interesting applications of analytical chemistry in the modern world.

Appendixes and Endpapers

Included in the appendixes are tables of equilibrium constants and standard potentials. We have also included here material on the use of exponential numbers and logarithms, the least-squares method for deriving calibration curves, and volumetric calculations using normality and equivalent weights (terms that are not used in the book itself). The endpapers provide a full-color chart of chemical indicators, a color display of the electromagnetic spectrum, a table of atomic masses, a table of molar masses of compounds of particular interest in analytical chemistry, and a periodic table.

CHANGES IN THE SIXTH EDITION

Readers of the fifth edition will find numerous changes in the sixth edition in content as well as in style and format.

Content

- We have strengthened our statistics chapter by adding new sections that deal with comparison of two sample means and a sample mean with a population mean.
- We have placed several derivations of equations in the *Feature* sections, thus giving the instructor the option of assigning or not assigning the material.
- We have added brief sections on electrothermal atomic atomizers and plasma sources.
- We have deleted the sections in the fifth edition that were devoted to computer applications because a supplemental book by one of the authors is now available that allows students to perform many of the calculations of analytical chemistry quickly and accurately with Mathcad™, which is a commercially available mathematical notebook, computational tool, and equation solver.[2] This book introduces students to the program syntax needed to perform statistical calculations, solve systems of equilibrium equations, carry out least-squares analysis, analyze multicomponent mixtures using multiple linear regression, and accomplish many other computational and graphical tasks.
- We have also added a number of new *Features*, most of which have biomedical or ecological flavors. For example, in Chapter 1, we provide a survey of the steps in a typical analysis that takes the form of a case study of the accidental poisoning of deer by a herbicide. The principles of immunoassay detailed in another *Feature* affords an excellent example of the intelligent use of equilibrium methods for determining many medically important analytes. In still another *Feature*, we discuss acid rain and the buffer capacity of lakes. These *Features* and others show students the importance of analytical chemistry in such fields as biology, biochemistry, medicine, medical technology, geology, and environmental science.

Style and Format

- We have continued our efforts to make this text more readable and student friendly. Throughout, we have endeavored to use short sentences, simpler words, and the active voice.
- Wherever possible we have avoided the use of jargon and acronyms. We have expanded the use of marginal notes to highlight important definitions and concepts. Such notes are also used for line drawings and photographs of equipment, historical notes, and thought-provoking questions.
- We have also expanded the set of color plates that demonstrate color changes and other chemical phenomena of importance in analytical chemistry.

ACKNOWLEDGMENTS

We wish to acknowledge with thanks the helpful comments and suggestions of the following who have reviewed the manuscript in various stages of its production: Professor Thomas A. Ballintine, University of North Dakota; Professor Kenneth C. Brooks, University of Illinois at Urbana–Champaign; Professor William E. Curtin, Trinity University; Professor Edward T. Gray Jr., University of Hartford; Professor Barry Gump, California State University–Fresno; Professor Thomas R. Herrington, University of San Diego; Professor Edward Koubek, United States Naval Academy; Professor James W. Lang, University of Oregon; Professor Craig Lunte, University of Kansas; Professor Karen B. Sentell, University of Vermont; Professor Frank A. Settle Jr., Virginia Military Institute; and Professor John P. Walters, St. Olaf College.

Our thanks also to Professors C. Marvin Lang and Gary Shulfer of the University of Wisconsin, Stevens Point for providing color slides of the postage stamps that appear in the color plates and in the marginal notes. In addition we would like to thank Professor Mike McClure of Hopkinsville Community College, Hopkinsville, KY for providing details of the analysis described in the *Feature* in Chapter 1 and Ms. Susan Spencer of Autodesk, Inc. for providing Hyper-Chem™, which was used to generate a number of the molecular models depicted in the book.

Finally, we wish to thank the various people at Saunders College Publishing for their friendly assistance in completing this project in record time. Among these are Project Editor Maureen Iannuzzi, Copy Editor Debbie Hardin, Art Directors Anne Muldrow and Susan Blaker, and Production Manager Charlene Squibb. Our thanks also to our Publisher John Vondeling, Developmental Editor Jennifer Bortel, and Editorial Assistant Chiara Puffer.

Douglas A. Skoog
Stanford University

Donald M. West
San Jose State University

F. James Holler
University of Kentucky

May 31, 1993

CONTENTS OVERVIEW

CONTENTS

x   Contents

CHAPTER 1

INTRODUCTION

Analytical chemistry involves separating, identifying, and determining the relative amounts of the components making up a sample of matter. Qualitative analysis reveals the chemical identity of the *analytes*. Quantitative analysis gives the amount of one or more of these analytes in numerical terms. Therefore, qualitative information is required before a quantitative analysis can be undertaken. A separation step is usually a necessary part of both a qualitative and a quantitative analysis.

The components of a sample that are to be determined are often referred to as the *analytes*.

In this book we are concerned principally with quantitative methods of analysis and methods of analytical separations. However, we refer occasionally to qualitative methods.

1A   ROLE OF ANALYTICAL CHEMISTRY IN THE SCIENCES

Analytical chemistry has played a vital role in the development of science. For example, in 1894 Wilhelm Ostwald wrote,

Analytical chemistry, or the art of recognizing different substances and determining their constituents, takes a prominent position among the applications of science, since the questions which it enables us to answer arise wherever chemical processes are employed for scientific or technical purposes. Its supreme importance has caused it to be assiduously cultivated from a very early period in the history of chemistry, and its records comprise a large part of the quantitative work which is spread over the whole domain of science.

Since 1894, analytical chemistry has evolved from an art into a science with applications throughout industry, medicine, and all the sciences. To illustrate, consider a few examples. Parts per million of hydrocarbons, nitrogen oxides,

1

and carbon monoxide in the exhaust gases of automobiles determine the effectiveness of smog-control devices. Quantitative measurements for ionized calcium in blood serum help diagnose parathyroid disease in human patients. Quantitative determination of the nitrogen content of foods establishes their protein content and thus their nutritional value. Analysis of steel during its production permits adjustment in the concentrations of such elements as carbon, nickel, and chromium to give a product that has a desired strength, hardness, corrosion resistance, and ductility. Household gas supplies are continuously monitored for their mercaptan content to assure that the gas has a sufficiently obnoxious odor to act as a warning of dangerous leaks. Modern farmers tailor their fertilization and irrigation schedules to meet changing plant needs during the growing season. They gauge needs from quantitative analyses of the plants themselves and the soil in which they grow.

Quantitative analytical measurements also play a vital role in many research areas in chemistry, biochemistry, biology, geology, and the other sciences. For example, chemists unravel mechanisms of chemical reactions through reaction rate studies. The rate of consumption of reactants or formation of products in a chemical reaction is calculated from quantitative measurements made at various time intervals. Quantitative analyses for potassium, calcium, and sodium ions in the body fluids of animals permit physiologists to study the role these ions play in the conduction of nerve signals and the contraction or relaxation of muscles. Materials scientists rely heavily on quantitative analyses of crystalline germanium and silicon in their studies of the behavior of semiconductor devices. Impurities in these devices are in the concentration range of 1×10^{-6} to 1×10^{-10} percent. Archaeologists identify sources of volcanic glasses (obsidian) from the concentrations of minor elements in samples taken from various locations. This knowledge in turn makes it possible to trace prehistoric trade routes for tools and weapons fashioned from obsidian.

Many chemists and biochemists devote a significant part of their time in the laboratory gathering quantitative information about systems of interest to them. For such investigators, analytical chemistry serves as a tool to their scholarly efforts.

1B   CLASSIFICATION OF QUANTITATIVE METHODS OF ANALYSIS

We use a wide variety of measurements to complete analyses.

We compute the results of a typical quantitative analysis from two measurements. The first is the mass or volume of sample to be analyzed. The second is the measurement of some quantity that is proportional to the amount of analyte in that sample and normally completes the analysis. Chemists classify analytical methods according to the nature of this final measurement. In a *gravimetric* method the mass of the analyte or some compound that is chemically related to the analyte is determined. In a *volumetric* method, the volume of a solution containing sufficient reagent to react completely with the analyte is measured. *Electroanalytical* methods involve the measurement of such electrical properties as potential, current, resistance, and quantity of charge. *Spectroscopic* methods are based on measurements of the interaction between electromagnetic radiation and analyte atoms or molecules or on the production of such radiation by analytes. Finally, there is a group of miscellaneous methods for completing analyses. These include the measurement of such properties as mass-to-charge ratio (*mass*

Electromagnetic radiation includes X-ray, ultraviolet, visible, infrared, and microwave radiation.

spectrometry), rate of radioactive decay, heat of reaction, rate of reaction, thermal conductivity, optical activity, and refractive index.

1C   STEPS IN A TYPICAL QUANTITATIVE ANALYSIS

A typical quantitative analysis involves the sequence of steps in the flow chart of Figure 1-1. In some instances, we can leave out one or more of these steps. Ordinarily, however, all play an important role in the success of an analysis. The first twenty-four chapters of this book focus on the last three steps in Figure 1-1. In the measurement step, we determine one of the physical properties mentioned in Section 1B. In the calculation step, we compute the relative amount of the analyte present in the samples. In the estimation of reliability step, we evaluate the quality of the results.

The paragraphs that follow will give you an overview of a typical quantitative analysis.

1C-1   Selecting a Method of Analysis

Selecting the method to solve an analytical problem is a vital first step in any quantitative analysis. The choice is sometimes difficult, requiring experience as well as intuition. One important consideration in selection is the accuracy required. Unfortunately, high reliability nearly always requires a large investment of time and money. The chosen method is most often a compromise between accuracy and economics.

Another consideration that is related to economic factors is the number of samples to be analyzed. If there are many samples, we can afford to spend a good deal of time in preliminary operations, such as assembling and calibrating instruments and equipment and preparing standard solutions. On the other hand, with only a single sample or a few samples, we may find it more expedient to select a procedure that avoids or minimizes such preliminary steps.

A final consideration is that the method chosen will always be governed by the complexity of the sample being analyzed and the number of components in the *sample matrix*.

1C-2   Sampling

To produce meaningful information, an analysis must be performed on a sample whose composition faithfully represents that of the bulk of material from which it was taken. Where the bulk is large and inhomogeneous, great effort is required to get a representative sample. Consider, for example, a railroad car containing 25 tons of silver ore. Buyer and seller must agree on the value of the shipment based primarily on its silver content. The ore itself is inherently heterogeneous, consisting of lumps of varying size as well as varying silver content. The actual *assay* of this shipment will be performed on a sample that weighs about one gram. For the analysis to have significance, this small sample must have a composition that is representative of 25 tons (or approximately 22,700,000 g) of ore in the shipment. The task of sampling is to isolate 1 g of material that accurately reflects the average composition of the nearly 23,000,000 g of bulk sample. Obtaining such a representative sample is a difficult undertaking that requires a careful, systematic manipulation of the entire shipment.

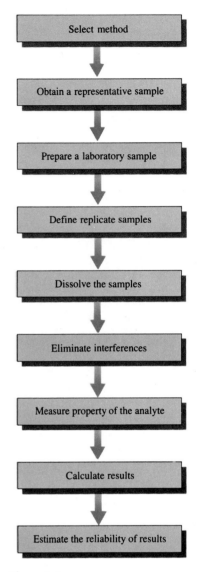

Figure 1-1
Steps in a quantitative analysis.

The *matrix*, or *sample matrix*, is the medium containing the analyte.

An *assay* is the process of determining how much of a given sample is the material indicated by its name. For example, a zinc alloy is assayed for its zinc content.

Many sampling problems are easier to solve than the one just described. Nevertheless, the chemist must have some assurance that the laboratory sample is representative of the whole before proceeding with an analysis.

1C-3 Preparing a Laboratory Sample

Frequently after sampling, solid materials are ground to decrease particle size, are mixed to ensure homogeneity, and are stored for various lengths of time before the analysis begins. During each of these steps, absorption or desorption of water may occur, depending on the humidity of the environment. A loss or gain of water changes the exact chemical composition of solids. Therefore, it is a good idea to carefully dry samples at the outset of an analysis. Alternatively, the moisture content of the sample can be determined at the time of the analysis in a separate analytical procedure.

1C-4 Defining Replicate Samples

Replicate samples are portions of a material of approximately the same size that are carried through an analytical procedure at the same time and in the same way.

Most chemical analyses are performed on *replicate samples* whose masses or volumes have been defined by careful measurements with an analytical balance or with a precise volumetric device. Obtaining replicate data on samples improves the quality of the results and provides a measure of their reliability.

Detailed instructions on techniques for measuring the mass or volume of samples appear in Sections 28B and 28G.

1C-5 Preparing Solutions of the Sample

Most analyses are performed on solutions of the sample. Ideally, the solvent should dissolve the entire sample (not just the analyte) rapidly. The conditions of dissolution should be sufficiently mild so that loss of the analyte cannot occur. Unfortunately, many materials that must be analyzed are insoluble in common solvents. Examples include silicate minerals, high-molecular-weight polymers, or specimens of animal tissue. Conversion of the analyte in such materials into a soluble form can be a difficult and time-consuming task. Methods for decomposing and dissolving samples appear in various parts of Chapter 29.

1C-6 Eliminating Interferences

Techniques or reactions that work well for only a single species or analyte are said to be *specific*. Techniques or reactions that work well for only a few analytes are *selective*.

Few chemical or physical properties of importance in chemical analysis are unique to a single chemical species. Instead, the reactions used and the properties measured are characteristic of a group of elements or compounds. This lack of truly specific reactions and properties greatly increases the complexity of an analysis. The analyst must devise a scheme that effectively isolates the species of interest from all others in the sample that can influence the final measurement. Substances that prevent the direct measurement of the analyte concentration are called *interferences*. It is important to separate interferences prior to the final measurement step in most analyses. No hard and fast rules can be given for the elimination of interferences. The resolution of this problem can be the most demanding aspect of an analysis. Chapters 25 through 27 describe separation methods.

1C-7 Calibration and Measurement

All analytical results depend on a final measurement X of a physical property of the analyte. This property must vary in a known and reproducible way with the concentration c_A of the analyte. Often, the physical property measured is directly proportional to the concentration. That is,

$$c_A = kX$$

where k is a proportionality constant. With two exceptions, analytical methods require the empirical determination of k with chemical standards for which c_A is known. The exceptions are gravimetric methods discussed in Chapter 6 and coulometric methods discussed in Chapter 19. The process of determining k is an important step in most analyses and is termed a *calibration*.

> Whenever possible standards for calibration should be made up in the same matrix as the samples to be analyzed.
>
> Only gravimetric and coulometric methods require no calibration step.

1C-8 Calculating Results

The computation of analyte concentrations from experimental data is ordinarily a simple and straightforward task, particularly with modern calculators or computers. Such computations are based on the raw experimental data collected in the measurement step, the stoichiometry of the chemical reaction of interest, and instrumental factors. These calculations are illustrated throughout this book.

1C-9 Evaluating Results and Estimating Their Reliability

Analytical results are incomplete without an estimate of their reliability. The experimenter must provide some measure of the uncertainties associated with computed results if the data are to have any value. Chapters 4 and 5 present detailed methods for carrying out this important final step in the analytical process.

Feature 1-1
DEER KILL: A CASE STUDY ILLUSTRATING THE USE OF ANALYTICAL CHEMISTRY IN THE SOLUTION OF TOXICOLOGICAL PROBLEMS

The tools of modern analytical chemistry serve in a wide variety of tasks in environmental and agricultural investigations. In this feature we describe a case study in which quantitative analysis was used to determine the agent that caused the deaths in a number of whitetail deer in a wildlife area of a Kentucky state park. We begin with a description of the problem and then show how the steps illustrated in Figure 1-1 were used to solve the analytical problem.

The Problem

The incident began when a park ranger found a dead whitetail deer near a pond in the Land Between the Lakes State Park in south central Kentucky. The park ranger enlisted the help of a chemist from the state veterinary diagnostic laboratory to help find the cause of death of the deer so that further deer kills might be prevented.

The ranger and the chemist investigated the site where the badly decomposed carcass of the deer had been found. Because of the advanced state of decomposition, no fresh organ tissue samples could be gathered. A few days after the investigation, the ranger found two more dead deer at essentially the same location. The chemist came to the site of the kill, loaded the deer on a truck for transport to his laboratory, and he and the ranger conducted a careful examination of the surrounding area. Perhaps some clue could be found to help determine the cause of the deaths.

The search encompassed about two acres surrounding the pond. The investigators noticed that grass around nearby power-line poles was wilted and discolored. They speculated that a herbicide had been used on the grass. A common ingredient in herbicides is arsenic in any one of a variety of forms such as arsenic trioxide, sodium arsenite, monosodium methanearsenate, or disodium methanearsenate. The latter compound is the disodium salt of methanearsenic acid, $CH_3AsO(OH)_2$, which is very soluble in water and thus is used as the active ingredient in many herbicides. The herbicidal activity of disodium methanearsenate is due to its reactivity with the sulfydryl (S—H) groups in the amino acid cysteine. When cysteine in plant enzymes reacts with arsenical compounds, the enzyme function is inhibited and the plant eventually dies. Unfortunately, similar chemical effects occur in animals as well. The investigators therefore collected samples of the discolored dead grass so that they could be tested along with samples from the organs of the deer. They hoped to analyze the samples to confirm the presence of arsenic and, if present, to determine its concentration in the samples.

Selecting a Method

A common method for the quantitative determination of arsenic in biological samples is found in the publication of the Association of Official Analytical Chemists (AOAC).[1] This method involves the distillation of arsenic as arsine, which is then determined by colorimetric measurements.

Obtaining Representative Samples

Back at the laboratory, the deer were dissected and the kidneys were removed for analysis. The kidneys were chosen because the suspected pathogen (arsenic) is rapidly eliminated from the animal body through the urinary tract where the arsenic tends to be concentrated.

Preparing a Laboratory Sample

Each kidney was cut into pieces and blended in a high-speed blender. This step served to reduce the size of the pieces of tissue and to homogenize the resulting laboratory sample.

---

[1] *Official Methods of Analysis*, 15th ed., p. 626. Washington, DC, 1990.

Defining Replicate Samples

Three 10-g samples of the homogenized tissue from each deer were placed in porcelain crucibles. Each crucible was heated cautiously over an open flame until the sample stopped smoking. The crucible was then cooled, placed in a furnace at room temperature, and then heated to 555°C in the furnace where it remained for two hours. This process is known as *dry ashing*. Dry ashing serves to free the analyte from organic material and convert any arsenic present to As_2O_5. Samples of the discolored grass were treated in a similar way to prepare them for dissolution.

Dissolving the Samples

The dry solid in each of the sample crucibles was dissolved in dilute HCl, which converted the As_2O_5 to soluble H_3AsO_4.

Eliminating Interferences

Arsenic can be separated from other substances that might interfere in the analysis by converting it to arsine, AsH_3, a toxic, colorless gas that evolves when a solution of H_3AsO_4 is treated with zinc. The solutions resulting from the deer and grass samples were combined with Sn^{2+} and a small amount of iodide ion was added to catalyze the reduction of H_3AsO_4 to H_3AsO_3 according to the following reaction:

$$H_3AsO_4 + SnCl_2 + 2HCl \rightarrow H_3AsO_3 + SnCl_4 + H_2O$$

The H_3AsO_3 was then converted to AsH_3 by the addition of zinc metal as follows:

$$H_3AsO_3 + 3Zn + 6HCl \rightarrow AsH_3(g) + 3ZnCl_2 + 3H_2O$$

The entire reaction was carried out in flasks equipped with a stopper and delivery tube so that the arsine gas could be collected in the absorber solution as shown in Figure 1-A. The arrangement ensures that interferences are left in the reaction flask and only the arsine is collected in the absorber in special transparent containers called *cuvettes*.

As the arsine bubbles into the solution in the cuvette, it reacts with silver diethyldithiocarbamate to form a colored complex compound according to the following equation.

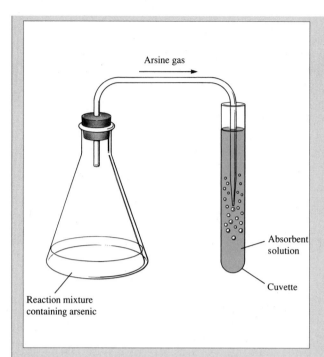

Figure 1-A
An arsine generator.

Besides the flasks containing the deer and grass samples, several flasks were prepared containing known concentrations of arsenic as well as a flask containing only the reagents and no arsenic. Such a sample is called a *blank*. The quantities of arsine collected from these flasks were used as standards for comparison with the unknown samples from the deer and grass.

Measuring the Amount of the Analyte

The highly colored complex of arsenic, which formed as the arsine bubbled into the pyridine solution of diethyldithiocarbamate, absorbed light at a wavelength of 535 nm. The extent of absorption of light can be used to measure the concentration of the absorbing species using a device called a spectrophotometer. The characteristics of these devices and their use in making quantitative measurements of concentration are discussed in detail in Chapter 22.

The cuvettes containing the absorbing colored complex of arsenic are shown in Figure 1-Ba. As you can see, the intensity of the color of the solutions is proportional to the quantity of arsenic in the samples. The analyst inserts the cuvettes into a spectrophotometer (see Figure 22-1) and reads a number (the absorbance) representing the color intensity of each of the solutions.

Calculating the Concentration

The absorbances for the solutions containing known concentrations of arsenic are plotted to produce a *calibration curve* of absorbance versus concentration as shown in Figure 1-Bb. The vertical lines between Figure 1-Ba and Figure 1-Bb show the correspondence between each solution and the corresponding point plotted on the graph. The intensity of the color of each solution is represented by the number plotted on the vertical axis of the calibration curve. The concentration of arsenic in parts per million in each standard solution corresponds to the vertical grid lines of the calibration curve as shown. The curve is then used to determine the concentration of the unknown solutions by finding the absorbances of the unknowns on the absorbance axis of the plot and reading the corresponding concentrations on the concentration axis. The lines leading from the cuvettes to the calibration curve show how this was done for the samples from the kidneys of the deer. The concentrations of arsenic in the deer were found on the concentration axis to be 16 ppm and 22 ppm, respectively.

Arsenic in the kidney tissue of animals is toxic at levels above about 10

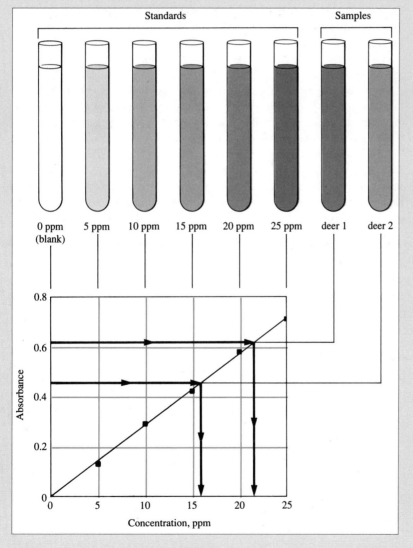

Figure 1-B

ppm, so it was probable that the deer were killed by ingesting an arsenic compound. The tests also showed that the samples of grass contained about 600 ppm arsenic. This very high level of arsenic suggested that the grass had been sprayed with an arsenical herbicide. The investigators concluded that the deer had probably died as a result of eating the poisoned grass.

Reliability of the Data

The data from these experiments were analyzed using the statistical methods described in Chapter 4 and Chapter 5. For each of the standard arsenic solutions and the deer samples, the average of the three absorbance measurements was calculated. The average absorbance for the replicates is a more reliable measure of the concentration of arsenic than a single measurement. Least-squares analysis of the standard data (see Appendix 6) was used to find the best straight line among the points (see Figure 1-Bb) and to calculate the concentrations of the unknown samples along with their uncertainties and confidence limits.

In this analysis, the formation of the highly colored product of the reaction served both to confirm the probable presence of arsenic and to provide a reliable estimate of its concentration in the deer and in the grass. Based on their results, the investigators recommended that the use of arsenical herbicides be suspended in the wildlife area to protect the deer and other animals that might eat plants in the area.

This case study illustrates how chemical analysis is used in the identification and determination of quantities of hazardous chemicals in the environment. Many of the methods and instruments of analytical chemistry are used routinely to provide vital information in environmental and toxicological studies of this type.

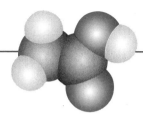

CHAPTER 2

SOME BASIC CONCEPTS

This chapter reviews several topics that you may have encountered before and that are of particular importance in analytical chemistry.

2A SOME IMPORTANT UNITS OF MEASUREMENT

2A-1 SI Units

Scientists throughout the world are adopting a standardized system of units known as the *International System of Units* (SI). This system is based on the seven fundamental base units shown in Table 2-1. Numerous other useful units, such as volts, hertz, joules, and coulombs, are derived from these base units.

In order to express small or large measured quantities in terms of a few simple digits, a variety of prefixes are used with these base units as well as other derived units. As shown in Table 2-2, these prefixes multiply the unit by various powers of ten. For example, the wavelength of yellow radiation used for determining sodium by flame photometry is about 5.9×10^{-7} m, which can be expressed more compactly as 590 nm (nanometers); the volume of a liquid injected onto a chromatographic column is often roughly 50×10^{-6} L or 50 μL (microliters); or the amount of memory on some computer hard disks is about 200,000,000 bytes or 200 Mbytes (megabytes).

In performing analyses, we often measure the mass of chemical species with a balance. For such measurements, metric units of kilograms (kg), grams (g), milligrams (mg), or micrograms (μg) are used. Volumes of liquids are measured in units of liters (L), milliliters (mL), and sometimes microliters (μL). The liter is an SI unit that is *defined* as exactly 10^{-3} m³. The milliliter is defined as 10^{-6} m³ or 1 cm³.

SI is the acronym for the French ''Système International d'Unités.''

The ångström unit Å is a non-SI unit of length that is widely used to express the wavelength of very short radiation such as X-rays (1 Å = 0.1 nm = 10^{-10} m). Thus, typical X-radiation lies in the range of 0.1 to 10 Å.

Table 2-1
SI BASE UNITS

Physical Quantity	Name of Unit	Abbreviation
Mass	kilogram	kg
Length	meter	m
Time	second	s
Temperature	kelvin	K
Amount of substance	mole	mol
Electric current	ampere	A
Luminous intensity	candela	cd

Table 2-2
PREFIXES FOR UNITS

Prefix	Abbreviation	Multiplier
giga-	G	10^9
mega-	M	10^6
kilo-	k	10^3
deci-	d	10^{-1}
centi-	c	10^{-2}
milli-	m	10^{-3}
micro-	μ	10^{-6}
nano-	n	10^{-9}
pico-	p	10^{-12}
femto-	f	10^{-15}
atto-	a	10^{-18}

Mass is an invariant measure of the amount of matter. *Weight* is the force of gravitational attraction between that matter and the earth.

This French postage stamp commemorates the Meter Convention of 1875. The stamp shows some signatures from the Treaty of the Meter, the seven SI units, and the definition of the meter as 1,650,763.73 wavelengths of the orange light given off by Kr-86. Since the stamp was issued in 1975, the meter has been redefined as the distance that light travels in a vacuum during 1/299,792,458 of a second.

Feature 2-1
THE DISTINCTION BETWEEN MASS AND WEIGHT

It is important to understand the difference between mass and weight. *Mass* is an invariant measure of the amount of matter in an object. *Weight* is the force of attraction between an object and the earth. Because gravitational attraction varies with geographic location, the weight of an object depends on where you weigh it. For example, a crucible weighs less in Denver than in Atlantic City (both cities are at approximately the same latitude) because the attractive force between crucible and earth is smaller at the higher altitude of Denver. Similarly, the crucible weighs more in Seattle than in Panama (both cities are at sea level) because the earth is somewhat flattened at the poles, and the force of attraction increases measurably with latitude. The mass of the crucible, however, remains constant regardless of where you measure it.

Weight and mass are related by the familiar expression

$$W = mg$$

where W is the weight of an object, m is its mass, and g is the acceleration due to gravity.

A chemical analysis is always based on mass so that the results will not depend on locality. A balance is used to compare the weight of an object with the weight of one or more standard masses. Because g affects both unknown and known in the same way at a given location, an equality in their weights indicates an equality in mass.

The distinction between mass and weight is often lost in common usage, and the process of comparing masses is ordinarily called *weighing*. In addition, the objects of known mass as well as the results of weighing are called *weights*. However, always bear in mind that analytical data are based on mass rather than weight, and throughout this book we will use mass rather than weight to describe the amounts of substances. On the other hand, for lack of a better term, we will use the verb *weigh* to describe the act of determining the mass of an object.

2A-2 The Mole

The *mole* (abbreviated mol) is the SI base unit for the amount of a chemical species. It is always associated with a chemical formula and refers to Avogadro's number (6.022×10^{23}) of particles represented by that formula. The *molar mass* ($\mathcal{M}$) of a substance is the mass in grams of 1 mol of that substance. Molar masses of compounds are derived by summing the masses of all the atoms appearing in a chemical formula. For example, the molar mass of formaldehyde CH_2O is

A *mole* of a chemical species is 6.022×10^{23} atoms, molecules, ions, electrons, or ion pairs.

In the older literature the terms *formula weight* and *molecular weight* are often used as synonyms for molar mass.

$$\mathcal{M}_{CH_2O} = \frac{1 \text{ mol } C}{\text{mol } CH_2O} \times \frac{12.0 \text{ g}}{\text{mol } C} + \frac{2 \text{ mol } H}{\text{mol } CH_2O} \times \frac{1.0 \text{ g}}{\text{mol } H} + \frac{1 \text{ mol } O}{\text{mol } CH_2O} \times \frac{16.0 \text{ g}}{\text{mol } O}$$

$$= 30.0 \text{ g/mol } CH_2O$$

and that of glucose ($C_6H_{12}O_6$) is

$$\mathcal{M}_{CH_6H_{12}O_6} = \frac{6 \text{ mol } C}{\text{mol } C_6H_{12}O_6} \times \frac{12.0 \text{ g}}{\text{mol } C} + \frac{12 \text{ mol } H}{\text{mol } C_6H_{12}O_6} \times \frac{1.0 \text{ g}}{\text{mol } H} + \frac{6 \text{ mol } O}{\text{mol } C_6H_{12}O_6}$$

$$\times \frac{16.0 \text{ g}}{\text{mol } O} = 180.0 \text{ g/mol } C_6H_{12}O_6$$

Thus, 1 mol of formaldehyde has a mass of 30.0 g and 1 mol of glucose has a mass of 180.0 g.

Feature 2-2

You should keep in mind that the masses for the elements listed in the table inside the back cover of this book are *relative masses* in terms of *atomic mass units* (amu), or *daltons*. The atomic mass unit is based on a relative scale in which the reference is the ^{12}C carbon isotope that is *assigned* a mass of exactly 12 amu. Thus, the amu is by definition 1/12 of the mass of one neutral ^{12}C atom. The *molar mass* of ^{12}C is then defined as the mass in *grams* of 6.022×10^{23} atoms of the carbon-12 isotope, or exactly 12 g. Likewise, the molar mass of any other element is the mass in grams of 6.022×10^{23} atoms of that element and is numerically equal to the atomic mass of the element in amu units. Thus, the atomic mass of naturally occurring oxygen is 15.9994 amu; its molar mass is 15.9994 g.

2A-3 The Millimole

1 mmol $= 10^{-3}$ mol

Sometimes it is more convenient to make calculations with millimoles (mmol) rather than moles, where the millimole is 1/1000 of a mole. The mass in grams of a millimole is likewise 1/1000 of the molar mass.

2A-4 Calculating the Amount of a Substance in Moles or Millimoles

The two examples that follow illustrate how the number of moles or millimoles of a species can be derived from its mass in grams or from the mass of a chemically related species.

Example 2-1

How many moles and millimoles of benzoic acid (122.1 g/mol) are contained in 2.00 g of the pure acid?

If we use HBz to represent benzoic acid, we can write that 1 mol HBz has a mass of 122.1 g. Thus,

$$\text{amount of HBz} = 2.00 \text{ g HBz} \times \frac{1 \text{ mol HBz}}{122.1 \text{ g HBz}} = 0.0164 \text{ mol HBz}$$

To obtain the number of millimoles, we divide by the millimolar mass (0.1221 g/mol). That is,

The number of moles of a species X is given by

$$\text{amount HBz} = 2.00 \text{ g HBz} \times \frac{1 \text{ mmol HBz}}{0.1221 \text{ g HBz}} = 16.4 \text{ mmol HBz}$$

$$\text{no. mol X} = \frac{gX}{g \text{ X/mol X}} = g \text{ X} \times \frac{\text{mol X}}{g \text{ X}}$$

The number of millimoles is given by

Example 2-2

$$\text{no. mmol X} = \frac{g \text{ X}}{g \text{ X/mmol X}}$$

$$= g \text{ X} \times \frac{\text{mmol X}}{g \text{ X}}$$

How many grams of Na^+ (22.99 g/mol) are contained in 25.0 g of Na_2SO_4 (142.0 g/mol)?

The chemical formula tells us that 1 mol of Na_2SO_4 contains 2 mol of Na^+. That is,

$$\text{no. mol Na}^+ = \text{no. mol Na}_2\text{SO}_4 \times \frac{2 \text{ mol Na}^+}{\text{mol Na}_2\text{SO}_4}$$

To obtain the number of moles Na_2SO_4 we proceed as in Example 2-1.

$$\text{no. mol } Na_2SO_4 = 25.0 \text{ g } Na_2SO_4 \times \frac{1 \text{ mol } Na_2SO_4}{142.0 \text{ g } Na_2SO_4}$$

Combining this equation with the first leads to

$$\text{no. mol } Na^+ = 25.0 \text{ g } Na_2SO_4 \times \frac{1 \text{ mol } Na_2SO_4}{142.0 \text{ g } Na_2SO_4} \times \frac{2 \text{ mol } Na^+}{\text{mol } Na_2SO_4}$$

To obtain the mass of sodium in 25.0 g Na_2SO_4, we multiply the number of moles Na^+ by the molar mass of Na^+, or 22.99 g. That is,

$$\text{mass } Na^+ = \text{no. mol } Na^+ \times \frac{22.99 \text{ g } Na^+}{\text{mol } Na^+}$$

Substituting the previous equation gives the number of grams of Na^+.

$$\text{mass } Na^+ = 25.0 \text{ g } Na_2SO_4 \times \frac{1 \text{ mol } Na_2SO_4}{142.0 \text{ g } Na_2SO_4} \times \frac{2 \text{ mol } Na^+}{\text{mol } Na_2SO_4} \times \frac{22.99 \text{ g } Na^+}{\text{mol } Na^+}$$

$$= 8.10 \text{ g } Na^+$$

> When you make calculations of this kind, include all units as we do throughout this chapter. This practice often reveals errors you have made in setting up equations.

2B   SOLUTIONS AND THEIR CONCENTRATIONS

2B-1   Concentration of Solutions

Chemists express the concentration of solids in solution in several ways. The most important of these are described in this section.

Molar Concentration

The molar concentration (c_X) of the solution of the chemical species X is the number of moles of that species that is contained in one liter of the solution (*not one liter of the solvent*). The unit of molar concentration is *molarity*, M, which has the dimensions of mol L^{-1}. Molarity also expresses the number of millimoles of a solute per milliliter of solution.

$$c_X = \frac{\text{no. mol solute}}{\text{no. L solution}} = \frac{\text{no. mmol solute}}{\text{no. mL solution}} \qquad \textbf{(2-1)}$$

Example 2-3

Calculate the molar concentration of ethanol in an aqueous solution that contains 2.30 g of C_2H_5OH (46.07 g/mol) in 3.50 L of solution.

Because the unit of molarity is the number of moles of solute per liter of solution, both of these quantities will be needed. The number of liters is given as 3.50, so all we need to do is to convert the number of grams of ethanol to the corresponding number of moles of ethanol and divide this number by the volume.

$$\text{no. mol C}_2\text{H}_5\text{OH} = 2.30 \text{ g C}_2\text{H}_5\text{OH} \times \frac{1 \text{ mol C}_2\text{H}_5\text{OH}}{46.07 \text{ g C}_2\text{H}_5\text{OH}} = 0.04992 \text{ mol C}_2\text{H}_5\text{OH}$$

To obtain the molar concentration, $c_{C_2H_5OH}$, we divide by the volume. Thus,

$$c_{C_2H_5OH} = \frac{0.0492 \text{ mol C}_2\text{H}_5\text{OH}}{3.50 \text{ L}}$$

$$= 0.0143 \text{ mol C}_2\text{H}_5\text{OH/L} = 0.0143 \text{ M}$$

Analytical molarity describes how a solution of a given molarity can be prepared.

The molarity c_X defined by Equation 2-1 is an analytical molarity.

Some chemists prefer to distinguish between species and analytical concentrations in a different way. They use *molar concentration* for species concentration and *formal concentration* (F) for analytical concentration. Applying this convention to our example, we can say that the formal concentration of H_2SO_4 is 1.0 F, whereas its molar concentration is 0.0 M because no undissociated H_2SO_4 exists in the solution.

Here, the *analytical molarity* of H_2SO_4 is given by $c_{H_2SO_4} = [SO_4^{2-}] + [HSO_4^-]$ because these are the only two sulfate-containing species in the solution.

Analytical Molarity. The *analytical molarity of a solution gives the total* number of moles of a solute in one liter of the solution (or the total number of millimoles in one milliliter). That is, the analytical molarity specifies a recipe by which the solution can be prepared. For example, a sulfuric acid solution that has an analytical concentration of 1.0 M can be prepared by dissolving 1.0 mol, or 98 g, of H_2SO_4 in water and diluting to exactly 1.0 L.

Equilibrium, or Species, Molarity. The *equilibrium,* or *species, molarity* expresses the molar concentration of a particular species in a solution at equilibrium. In order to state the species molarity, it is necessary to know how the solute behaves when it is dissolved in a solvent. For example, the species molarity of H_2SO_4 in a solution with an analytical concentration of 1.0 M is 0.0 M because the sulfuric acid is entirely dissociated into a mixture of H_3O^+, HSO_4^-, and SO_4^{2-} ions. Essentially no H_2SO_4 molecules as such are present in this solution. The equilibrium concentrations and thus the species molarity of these three ions are 1.01, 0.99, and 0.01 M, respectively.

Equilibrium molar concentrations are often symbolized by placing square brackets around the chemical formula for the species. Thus, for our solution of H_2SO_4 with an analytical concentration of 1.0 M, we can write

$$[H_2SO_4] = 0.00 \text{ M}$$

$$[HSO_4^-] = 0.99 \text{ M}$$

$$[H_3O^+] = 1.01 \text{ M}$$

$$[SO_4^{2-}] = 0.01 \text{ M}$$

Example 2-4

Calculate the analytical and equilibrium molar concentrations of the solute species in an aqueous solution that contains 285 mg of trichloroacetic acid, Cl_3CCOOH (163.4 g/mol), in 10.0 mL (the acid is 73% ionized in water).

As in Example 2-3, we calculate the number of moles of Cl_3CCOOH, which we designate as HA, and divide by the volume of the solution, 10.0 mL or 0.01000 L. Thus,

$$\text{no. mol HA} = 285 \text{ mg HA} \times \frac{1 \text{ g HA}}{1000 \text{ mg HA}} \times \frac{1 \text{ mol HA}}{163.4 \text{ g HA}}$$

$$= 1.744 \times 10^{-3} \text{ mol HA}$$

The analytical molar concentration, c_{HA}, is then

$$c_{HA} = \frac{1.744 \times 10^{-3} \text{ mol HA}}{10.0 \text{ mL}} \times \frac{1000 \text{ mL}}{L} = 0.174 \frac{\text{mol HA}}{L} = 0.174 \text{ M}$$

In this solution, 73% of the HA dissociates giving H^+ and A^-:

$$HA \rightleftharpoons H^+ + A^-$$

The species concentration of A^- is equal to 73% of the analytical concentration of HA. That is,

$$[A^-] = 0.174 \frac{\text{mol HA}}{L} \times \frac{73 \text{ mol A}^-}{100 \text{ mol HA}} = 0.127 \text{ M}$$

The species concentration of HA is equal to the analytical concentration of the acid (0.174 M) minus the concentration of A^-. That is,

$$[HA] = (0.174 - 0.127) = 0.047 \text{ M}$$

Because one mole H^+ is found formed for each mole A^-, we can also write

$$[H^+] = [A^-] = 0.127 \text{ M}$$

Example 2-5

Describe the preparation of 2.00 L of 0.108 M $BaCl_2$ from $BaCl_2 \cdot 2H_2O$ (244.3 g/mol).

In order to determine the number of grams of solute to be dissolved and diluted to 2.00 L, we note that 1 mol of the dihydrate yields 1 mol of $BaCl_2$. Therefore, to produce this solution we will need

$$2.00 \text{ L} \times \frac{0.108 \text{ mol BaCl}_2 \cdot 2 \text{ H}_2\text{O}}{L} = 0.216 \text{ mol BaCl}_2 \cdot 2 \text{ H}_2\text{O}$$

The mass of $BaCl_2 \cdot 2 H_2O$ is then

$$0.216 \text{ mol BaCl}_2 \cdot 2 \text{ H}_2\text{O} \times \frac{244.3 \text{ g BaCl}_2 \cdot 2 \text{ H}_2\text{O}}{\text{mol BaCl}_2 \cdot 2 \text{ H}_2\text{O}} = 52.8 \text{ g BaCl}_2 \cdot 2 \text{ H}_2\text{O}$$

Dissolve 52.8 g of $BaCl_2 \cdot 2 H_2O$ in water and dilute to 2.00 L.

Example 2-6

Describe the preparation of 500 mL of 0.0740 M Cl^- solution from solid $BaCl_2 \cdot 2H_2O$ (244.3 g/mol).

$$\text{mass } BaCl_2 \cdot 2H_2O = \frac{0.0740 \text{ mol } Cl^-}{L} \times 0.500 \text{ L} \times \frac{1 \text{ mol } BaCl_2 \cdot 2H_2O}{2 \text{ mol } Cl^-}$$

$$\times \frac{244.3 \text{ g } BaCl_2 \cdot 2H_2O}{\text{mol } BaCl_2 \cdot 2H_2O}$$

$$= 4.52 \text{ g } BaCl_2 \cdot 2H_2O$$

Dissolve 4.52 g of $BaCl_2 \cdot 2H_2O$ in water and dilute to 0.500 L or 500 mL.

Percent Concentration

We frequently express concentrations in terms of percent (parts per hundred). Unfortunately, this practice can be a source of ambiguity because percent composition of a solution can be expressed in several ways. Three common methods are

Weight percent should be more properly called mass percent and abbreviated m/m. The term weight percent is so widely used in the chemical literature, however, that we will use it throughout this book.

(1) $$\text{weight percent (w/w)} = \frac{\text{mass solute}}{\text{mass solution}} \times 100\%$$

(2) $$\text{volume percent (v/v)} = \frac{\text{volume solute}}{\text{volume soln}} \times 100\%$$

(3) $$\text{weight/volume percent (w/v)} = \frac{\text{mass solute, g}}{\text{volume soln, mL}} \times 100\%$$

Note that the denominator in each of these expressions refers to the *solution* rather than to the solvent. Note also that the first two expressions do not depend on the units employed, provided, of course, that there is consistency between numerator and denominator. The units in the third expression do not cancel and therefore must be specified. Of the three expressions, only mass percent has the virtue of being temperature-independent.

Weight percent is frequently employed to express the concentration of commercial aqueous reagents. For example, nitric acid is sold as a 70% solution, which means that the reagent contains 70 g of HNO_3 per 100 g of solution (see Example 2-10).

Volume percent is commonly used to specify the concentration of a solution prepared by diluting a pure liquid with another liquid. For example, a 5% aqueous solution of methanol *usually* means a solution prepared by diluting 5.0 mL of pure methanol with enough water to give 100 mL.

Weight/volume percent is often employed to indicate the composition of dilute aqueous solutions of solid reagents. For example, 5% aqueous silver nitrate *often* refers to a solution prepared by dissolving 5 g of silver nitrate in sufifcient water to give 100 mL of solution.

To avoid uncertainty, always explicitly specify the type of percent composition being discussed. If this information is missing, the user must decide intuitively which of the several types is involved. The potential error resulting from a wrong choice is considerable. For example, commercial 50% (w/w) sodium hydroxide contains 763 g of the reagent per liter, which corresponds to 76.3% (w/v) sodium hydroxide.

Always specify the type of percent when reporting concentrations in this way.

Parts Per Million and Parts Per Billion

For very dilute solutions, parts per million (ppm) is a convenient way to express concentration:

$$c_{ppm} = \frac{\text{mass of solute}}{\text{mass of solution}} \times 10^6 \text{ ppm}$$

where c_{ppm} is the concentration in parts per million. The units of mass in the numerator and denominator must agree. For even more dilute solutions, 10^9 ppb rather than 10^6 ppm is employed in the foregoing equation to give the results in parts per billion (ppb). The term *parts per thousand* (ppt) is also encountered, especially in oceanography.

A handy rule in calculating parts per million is to remember that for dilute aqueous solutions whose densities are approximately 1.00 g/mL, 1 ppm = 1.00 mg/L. That is,

$$c_{ppm} = \frac{\text{mass solute (mg)}}{\text{volume soln (L)}} \qquad (2\text{-}2)$$

Example 2-7

What is the molarity of K^+ in an aqueous solution that contains 63.3 ppm of $K_3Fe(CN)_6$ (329.2 g/mol)?

We wish to produce a solution that contains 63.3 g solute per 10^6 g solution. The density of such a dilute aqueous solution will be essentially that of pure water or about 1.00 g/mL or 1000 g/L. Thus, we may write

$$c_{K^+} = \frac{63.3 \text{ g } K_3Fe(CN)_6}{10^6 \text{ g soln}} \times \frac{10^3 \text{ g soln}}{\text{L soln}} \times \frac{1 \text{ mol } K_3Fe(CN)_6}{329.2 \text{ g } K_3Fe(CN)_6} \times \frac{3 \text{ mol } K^+}{\text{mol } K_3Fe(CN)_6}$$

$$c_{K^+} = 5.77 \times 10^{-4} \frac{\text{mol } K^+}{\text{L}}$$

Solution-Diluent Volume Ratios

The composition of a dilute solution is sometimes specified in terms of the volume of a concentrated solution and the volume of solvent to be used in diluting it. The volume of the concentrated solution is separated from that of the dilute solution by a colon. Thus, a 1:4 HCl solution contains four volumes of water for each volume of concentrated hydrochloric acid. This method of notation is frequently ambiguous in that the concentration of the original solution is not always obvious to the reader. Moreover, under some circumstances 1:4 means dilute one volume with three volumes. In view of such possibilities for misunderstanding, you should be alert to uncertainties associated with solution-diluent ratios when you encounter them.

p-Functions

The most well-known p-function is pH, which is the negative logarithm of $[H^+]$. That is, $pH = -\log [H^+]$. Chemists use pH, which is just a mathematical concept for convenience; it is not a chemical concept.

Scientists frequently express the concentration of a species in terms of its *p-function,* or *p-value.* The p-value is the negative logarithm (to the base 10) of the molar concentration of that species. So, for the species X,

$$pX = -\log [X]$$

As shown by the following examples, p-values offer the advantage of allowing concentrations that vary over ten or more orders of magnitude to be expressed in terms of small positive numbers.

Example 2-8

Calculate the p-value for each ion in a solution that is 2.00×10^{-3} M in NaCl and 5.4×10^{-4} M in HCl.

$$pH = -\log [H^+] = -\log (5.4 \times 10^{-4})$$
$$= -\log 5.4 - \log 10^{-4} = -0.73 - (-4) = 3.27$$

To obtain pNa, we write

$$pNa = -\log (2.00 \times 10^{-3}) = -\log 2.00 - \log 10^{-3}$$
$$= -0.301 - (-3.00) = 2.699$$

The total Cl^- concentration is given by the sum of the concentrations of the two solutes:

$$[Cl^-] = 2.00 \times 10^{-3}\,M + 5.4 \times 10^{-4}\,M = 2.00 \times 10^{-3}\,M + 0.54 \times 10^{-3}\,M$$
$$= 2.54 \times 10^{-3}\,M$$
$$pCl = -\log 2.54 \times 10^{-3} = 2.595$$

Note that in Examples 2-8 and 2-9 the results are rounded according to the rules listed on page 73.

Example 2-9

Calculate the molar concentration of Ag^+ in a solution that has a pAg of 6.372.

$$pAg = -\log [Ag^+] = 6.372$$
$$\log [Ag^+] = -6.372 = 0.628 - 7.000$$
$$[Ag^+] = \text{antilog} (0.628) \times \text{antilog} (-7.000) = 4.246 \times 10^{-7} = 4.25 \times 10^{-7}$$

The antilog key on some hand-held calculators is labeled 10^x. On some others you can obtain the antilog by pressing **INV** and then **LOG** .

2B-2 Density and Specific Gravity of Solutions

Density and specific gravity are terms often encountered in the analytical literature. The *density* of a substance is its mass per unit volume, whereas its *specific gravity* is the ratio of its mass to the mass of an equal volume of water at a specified temperature (ordinarily 4°C). Density has units of kilograms per liter or grams per milliliter in the metric system. Specific gravity is dimensionless and so is not tied to any particular system of units. For this reason, specific gravity is widely used in describing items of commerce. Since the density of water is approximately 1.00 g/mL and since we employ the metric system throughout this book, density and specific gravity are used interchangeably.

SPECIFIC GRAVITIES OF COMMERCIALLY AVAILABLE CONCENTRATED ACIDS AND BASES

Reagent	Typical Concentration, % (w/w)	Typical Specific Gravity
·Acetic acid	99.7	1.05
Ammonia	29.0	0.90
Hydrochloric acid	37.2	1.19
Hydrofluoric acid	49.5	1.15
	70.5	1.42
Nitric acid	71.0	1.67
Perchloric acid	86.0	1.71
Phosphoric acid	96.5	1.84
Sulfuric acid		

The density of a gas is usually expressed in g/L. The density of a liquid or a solid is generally expressed in g/mL.

Example 2-10

Calculate the molar concentration of HNO_3 (63.0 g/mol) in a solution that has a specific gravity of 1.42 and is 70% HNO_3 (w/w).

In order to obtain the molarity of the acid, we will first calculate the mass in grams of acid in one liter of solution. We will then convert the mass of acid per liter to moles of acid per liter.

To calculate the mass of acid per liter of concentrated solution we write

$$\frac{\text{g } HNO_3}{\text{L reagent}} = \frac{1.42 \text{ g reagent}}{\text{mL reagent}} \times \frac{10^3 \text{ mL reagent}}{\text{L reagent}} \times \frac{70 \text{ g } HNO_3}{100 \text{ g reagent}} = \frac{994 \text{ g } HNO_3}{\text{L reagent}}$$

Then to convert to moles per liter, we proceed as follows:

$$c_{HNO_3} = \frac{994 \text{ g } HNO_3}{\text{L reagent}} \times \frac{1 \text{ mol } HNO_3}{63.0 \text{ g } HNO_3} = \frac{15.8 \text{ mol } HNO_3}{\text{L reagent}} = 16 \text{ M}$$

Example 2-11

Describe the preparation of 100 mL of 6.0 M HCl from a concentrated solution that has a specific gravity of 1.18 and is 37% (w/w) HCl (36.5 g/mol).

Proceeding as in Example 2-10, we first calculate the molarity of the concentrated reagent. We then calculate the number of moles of acid that we need for the diluted solution. Finally, we divide the second figure by the first to obtain the volume of concentrated acid required. Thus, to obtain the molarity of the concentrated reagent, we write

$$c_{HCl} = \frac{1.18 \times 10^3 \text{ g reagent}}{\text{L reagent}} \times \frac{37 \text{ g HCl}}{100 \text{ g reagent}} \times \frac{1 \text{ mol HCl}}{36.5 \text{ g HCl}} = 12.0 \text{ M}$$

The number of moles HCl required is given by

$$\text{no. mol HCl} = 100 \text{ mL} \times \frac{1 \text{ L}}{1000 \text{ mL}} \times \frac{6.0 \text{ mol HCl}}{\text{L}} = 0.600 \text{ mol}$$

Finally, to obtain the volume of concentrated reagent, we write

$$\text{vol concd reagent} = 0.600 \ \text{mol HCl} \times \frac{1 \ \text{L reagent}}{12.0 \ \text{mol HCl}} = 0.0500 \ \text{L or } 50.0 \ \text{mL}$$

Thus dilute 50 mL of the concentrated reagent to 100 mL.

The solution to Example 2-11 is based on the following useful relationship, which we will be using countless times:

Equation 2-3 can be used with L and mol/L or mL and mmol/mL. Thus,

$$L_{concd} \times \frac{mol_{concd}}{L_{concd}} = L_{dil} \times \frac{mol_{dil}}{L_{dil}}$$

$$mL_{concd} \times \frac{mmol_{concd}}{mL_{concd}} = mL_{dil} \times \frac{mmol_{dil}}{mL_{dil}}$$

$$V_{concd} \times c_{concd} = V_{dil} \times c_{dil} \qquad \textbf{(2-3)}$$

where the two terms on the left are the volume and molar concentration of a concentrated solution that is being used to prepare a diluted solution having the volume and concentration given by the corresponding terms on the right. This equation is based on the fact that the number of moles of solute in the diluted solution must equal the number of moles in the concentrated reagent. Note that the volumes can be in milliliters or liters as long as the same units are used for both solutions.

2C   STOICHIOMETRIC CALCULATIONS

Stoichiometry is defined as the mass relationships among reacting chemical species. This section provides a brief review of stoichiometry and its applications to chemical calculations.

The *stoichiometry* of a reaction is the relationship among the number of moles of reactants and products as shown by a balanced equation.

A balanced chemical equation is a statement of the combining ratios, or stoichiometry, between reacting substances and their products. Thus, the equation

We often indicate the physical state of substances appearing in equations by the letters (g), (l), (s), or (aq), which refer to gaseous, liquid, solid, and aqueous solution states, respectively.

$$2NaI(aq) + Pb(NO_3)_2(aq) \rightleftharpoons PbI_2(s) + 2NaNO_3(aq)$$

indicates that 2 mol of aqueous sodium iodide combine with 1 mol of aqueous lead nitrate to produce 1 mol of solid lead iodide and 2 mol of aqueous sodium nitrate.[1]

Example 2-12 demonstrates how the masses in grams of reactants and products in a chemical reaction are related. As shown in Figure 2-1, a calculation of this type is a three-step process involving: (1) transformation of the known mass of a substance in grams to a corresponding number of moles; (2) multiplication by a factor (the *stoichiometric factor*) that accounts for the stoichiometry; and (3) reconversion of the data in moles back to the metric units called for in the answer.

---

[1] Here it is advantageous to show the reaction in terms of chemical compounds. If we wish to focus on reacting species, the net ionic equation is preferable:

$$2I^-(aq) + Pb^{2+}(aq) \rightleftharpoons PbI_2(s)$$

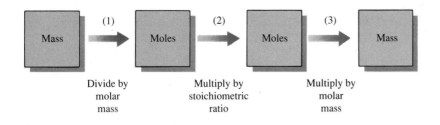

Figure 2-1

Flow diagram for making stoichiometric calculations.

Example 2-12

(a) What mass of $AgNO_3$ (169.9 g/mol) is needed to convert 2.33 g of Na_2CO_3 (106.0 g/mol) to Ag_2CO_3? (b) What mass of Ag_2CO_3 (275.7 g/mol) will be formed?

(a) $\qquad Na_2CO_3(aq) + 2AgNO_3(aq) \rightarrow Ag_2CO_3(s) + 2NaNO_3(aq)$

Step 1: no. mol Na_2CO_3 $= 2.33 \text{ g Na}_2\text{CO}_3 \times \dfrac{1 \text{ mol Na}_2\text{CO}_3}{106.0 \text{ g Na}_2\text{CO}_3}$

$$= 0.02198 \text{ mol Na}_2CO_3$$

Step 2: The balanced equation reveals that

$$\text{no. mol AgNO}_3 = 0.02198 \text{ mol Na}_2\text{CO}_3 \times \frac{2 \text{ mol AgNO}_3}{1 \text{ mol Na}_2\text{CO}_3}$$

$$= 0.04396 \text{ mol AgNO}_3$$

Here the stoichiometric factor is 2 mol $AgNO_3$/1 mol Na_2CO_3.

Step 3: mass $AgNO_3$ $= 0.04396 \text{ mol AgNO}_3 \times \dfrac{169.9 \text{ g AgNO}_3}{\text{mol AgNO}_3} = 7.47 \text{ g AgNO}_3$

(b) $\qquad$ no. mol Ag_2CO_3 = no. mol Na_2CO_3 = 0.02198 mol

$$\text{mass Ag}_2\text{CO}_3 = 0.02198 \text{ mol Ag}_2\text{CO}_3 \times \frac{275.7 \text{ g Ag}_2\text{CO}_3}{\text{mol Ag}_2\text{CO}_3} = 6.06 \text{ g Ag}_2CO_3$$

Example 2-13

What mass of Ag_2CO_3 (275.7 g/mol) is formed when 25.0 mL of 0.200 M $AgNO_3$ are mixed with 50.0 mL of 0.0800 M Na_2CO_3?

Mixing these two solutions will result in one (and only one) of three possible outcomes:

(1) an excess of $AgNO_3$ will remain after reaction is complete.
(2) an excess of Na_2CO_3 will remain after reaction is complete.
(3) neither reagent will be in excess (that is, the number of millimoles of Na_2CO_3 is exactly equal to one half the number of millimoles of $AgNO_3$).

As a first step, we must establish which of these situations applies by calculating the amounts of reactants (in chemical units) available at the outset.

Initial amounts are

$$\text{amount AgNO}_3 = 25.0 \text{ mL } \cancel{\text{AgNO}_3} \times \frac{0.200 \text{ mmol AgNO}_3}{\text{mL } \cancel{\text{AgNO}_3}}$$

$$= 5.00 \text{ mmol AgNO}_3$$

$$\text{amount Na}_2\text{CO}_3 = 50.0 \text{ mL } \cancel{\text{Na}_2\text{CO}_3} \times \frac{0.0800 \text{ mmol Na}_2\text{CO}_3}{\text{mL } \cancel{\text{Na}_2\text{CO}_3}}$$

$$= 4.00 \text{ mmol Na}_2\text{CO}_3$$

Because each CO_3^{2-} ion reacts with two Ag^+ ions, $2 \times 4.00 = 8.00$ mmol $AgNO_3$ would be required to react with the Na_2CO_3. Since too little $AgNO_3$ is present, situation (2) prevails and the amount of Ag_2CO_3 produced will be limited by the amount of $AgNO_3$ available. Thus,

$$\text{mass Ag}_2\text{CO}_3 = 5.00 \text{ } \cancel{\text{mmol AgNO}_3} \times \frac{1 \text{ mmol } \cancel{\text{Ag}_2\text{CO}_3}}{2 \text{ } \cancel{\text{mmol AgNO}_3}} \times \frac{0.2757 \text{ g Ag}_2\text{CO}_3}{\cancel{\text{mmol Ag}_2\text{CO}_3}}$$

$$= 0.689 \text{ g Ag}_2\text{CO}_3$$

Example 2-14

What will be the molar analytical Na_2CO_3 concentration in the solution produced when 25.0 mL of 0.200 M $AgNO_3$ are mixed with 50.0 mL of 0.0800 M Na_2CO_3? Formation of 5.00 mmol of $AgNO_3$ will require 2.50 mmol of Na_2CO_3. The number of millimoles of unreacted Na_2CO_3 is then given by

$$\text{amount Na}_2\text{CO}_3 = 4.00 \text{ mmol Na}_2\text{CO}_3 - 5.00 \text{ } \cancel{\text{mmol AgNO}_3} \times \frac{1 \text{ mmol Na}_2\text{CO}_3}{2 \text{ } \cancel{\text{mmol AgNO}_3}}$$

$$= 1.50 \text{ mmol Na}_2\text{CO}_3$$

By definition the molarity is the number of millimoles Na_2CO_3 per milliliter. Thus,

$$c_{\text{Na}_2\text{CO}_3} = \frac{1.50 \text{ mmol Na}_2\text{CO}_3}{(50.0 + 25.0) \text{ mL}} = 0.0200 \text{ M Na}_2\text{CO}_3$$

2D   QUESTIONS AND PROBLEMS

2-1. Define
*(a) dalton.
 (b) mole.
*(c) p-value.
 (d) stoichiometry.

2-2. Make a clear distinction between
*(a) the mass and the weight.
 (b) the atomic mass of an element and its molar mass.
*(c) density and specific gravity.
 (d) molar analytical concentration and molar equilibrium concentration.

---

* Answers to the asterisked problems are given in the answer section at the back of the book.

*2-3. What is the atomic mass of Fe? What is the molar mass of Fe?

2-4. The rest mass of an electron is 9.109×10^{-31} kg. Calculate the molar mass of the electron.

*2-5. Give two examples of units derived from the SI fundamental base units.

2-6. Simplify the following quantities using a unit with an appropriate prefix:
  *(a) 1.5×10^6 Hz.
  (b) 2.62×10^{-9} g.
  *(c) 6.23×10^4 μmol.
  (d) 4.0×10^9 s.
  *(e) 96,495 C.
  (f) 47,000 kg.

*2-7. How many moles of sodium ion are contained in 4.13 g of
  (a) NaBr?
  (b) $Na_2C_2O_4$?
  (c) $Na_2SO_4 \cdot 10H_2O$?
  (d) Na_3AsO_4?

2-8. How many potassium ions are contained in 0.561 g of
  (a) KCl?
  (b) K_2SO_4?
  (c) K_3PO_4?
  (d) $K_4Fe(CN)_6$?

*2-9. How many moles are contained in
  (a) 6.84 g of B_2O_3?
  (b) 296 mg of $Na_2B_4O_7 \cdot 10H_2O$?
  (c) 8.75 g of Mn_3O_4?
  (d) 67.4 mg of CaC_2O_4?

2-10. How many millimoles are contained in
  (a) 79.8 mg of H_2?
  (b) 8.43 g of SO_2?
  (c) 64.4 g of Na_2CO_3?
  (d) 411 mg of $KMnO_4$?

*2-11. How many millimoles of solute are contained in
  (a) 2.00 L of 2.76×10^{-3} M $KMnO_4$?
  (b) 750 mL of 0.0416 M KSCN?
  (c) 250 mL of a solution that contains 4.20 ppm of $CuSO_4$?
  (d) 3.50 L of 0.276 M KCl?

2-12. How many millimoles of solute are contained in
  (a) 4.25 mL of 0.0917 M KH_2PO_4?
  (b) 0.1020 L of 0.0643 M $HgCl_2$?
  (c) 2.81 L of a 49.0 ppm solution of $Mg(NO_3)_2$?
  (d) 79.8 mL of 0.1379 M NH_4VO_3?

*2-13. How many milligrams are contained in
  (a) 0.666 mol of HNO_3?
  (b) 300 mmol of MgO?
  (c) 19.0 mol of NH_4NO_3?
  (d) 5.32 mol of $(NH_4)_2Ce(NO_3)_6$ (548.22 g/mol)?

2-14. How many grams are contained in
  (a) 32.1 mmol of H_2O_2?
  (b) 0.466 mol of NH_4VO_3 (117.0 g/mol)?
  (c) 5.38 mol of $MgNH_4PO_4$?
  (d) 26.7 mmol of $KH(IO_3)_2$?

2-15. How many milligrams of solute are contained in
  *(a) 26.0 mL of 0.150 M sucrose (342 g/mol)?

  *(b) 2.92 L of 5.23×10^{-3} M H_2O_2?
  (c) 737 mL of a solution that contains 6.38 ppm of $Pb(NO_3)_2$?
  (d) 6.75 mL of 0.0619 M KNO_3?

2-16. How many grams of solute are contained in
  *(a) 450 mL of 0.164 M H_2O_2?
  *(b) 27.0 mL of 8.75×10^{-4} M benzoic acid (122 g/mol)?
  (c) 3.50 L of a solution that contains 21.7 ppm of $SnCl_2$?
  (d) 21.7 mL of 0.0125 M $KBrO_3$?

2-17. Calculate the p-value for each of the indicated ions in the following:
  *(a) Na^+, Cl^-, and OH^- in a solution that is 0.116 M in NaCl and 0.125 M in NaOH.
  (b) Ba^{2+}, Mn^{2+}, and Cl^- in a solution that is 3.80×10^{-3} M in $BaCl_2$ and 2.22 M in $MnCl_2$.
  *(c) H^+, Cl^-, and Zn^{2+} in a solution that is 1.50 M in HCl and 0.120 M in $ZnCl_2$.
  (d) Cu^{2+}, Zn^{2+}, and NO_3^- in a solution that is 4.32×10^{-2} M in $Cu(NO_3)_2$ and 0.101 M in $Zn(NO_3)_2$.
  *(e) K^+, OH^-, and $Fe(CN)_6^{4-}$ in a solution that is 3.79×10^{-6} M in $K_4Fe(CN)_6$ and 4.12×10^{-5} M in KOH.
  (f) H^+, Ba^{2+}, and ClO_4^- in a solution that is 2.75×10^{-4} M in $Ba(ClO_4)_2$ and 4.44×10^{-4} M in $HClO_4$.

2-18. Calculate the p-functions for each ion in a solution that is
  *(a) 0.0100 M in NaBr.
  (b) 0.0100 M in $BaBr_2$.
  *(c) 3.5×10^{-3} M in $Ba(OH)_2$.
  (d) 0.040 M in HCl and 0.020 M in NaCl.
  *(e) 5.2×10^{-3} M in $CaCl_2$ and 3.6×10^{-3} M in $BaCl_2$.
  (f) 4.8×10^{-8} M in $Zn(NO_3)_2$ and 5.6×10^{-7} M $Cd(NO_3)_2$.

2-19. Calculate the hydrogen ion concentration of a solution that has a pH of
  *(a) 9.19.   (e) 5.37.
  *(b) 4.83.   (f) 7.88.
  *(c) 2.17.   (g) -1.14.
  *(d) -0.033.   (h) 3.43.

2-20. Convert the following p-functions to molar concentrations:
  *(a) pCl = 7.14.   (e) pSO_4 = 6.10.
  *(b) pSCN = 0.033.   (f) pZn = 4.27.
  *(c) pOH = 9.61.   (g) pOCl = 2.81.
  *(d) pH = -0.93.   (h) pNO_3 = 4.08.

*2-21. Seawater contains an average of 1.08×10^3 ppm of Na^+ and 270 ppm SO_4^{2-}. Calculate
  (a) the molar concentrations of Na^+ and SO_4^{2-} given that the average density of seawater is 1.02 g/mL.
  (b) the pNa and pSO_4 for seawater.

2-22. Average human blood serum contains 18 mg of K^+ and 365 mg of Cl^- per 100 mL. Calculate
  (a) the molar concentration for each of these species; use 1.00 g/mL for the density of serum.
  (b) pK and pCl for human serum.

*2-23. A solution was prepared by dissolving 10.12 g of KCl $\cdot$ $MgCl_2 \cdot 6H_2O$ (277.85 g/mol) in sufficient water to give 2.000 L. Calculate
  (a) the molar analytical concentration of KCl $\cdot$ $MgCl_2$ in this solution.
  (b) the molar concentration of Mg^{2+}.

(c) the molar concentration of Cl^-.

(d) the weight/volume percentage of $KCl \cdot MgCl_2 \cdot 6H_2O$.

(e) the millimoles of Cl^- in 25.0 mL of this solution.

(f) ppm K^+.

(g) pMg for the solution.

(h) pCl for the solution.

2-24. A solution was prepared by dissolving 367 mg of $K_3Fe(CN)_6$ (329.2 g/mol) in sufficient water to give 750 mL. Calculate

(a) the molar analytical concentration of $K_3Fe(CN)_6$.

(b) the molar concentration of K^+.

(c) the molar concentration of $Fe(CN)_6^{3-}$.

(d) the weight/volume percentage of $K_3Fe(CN)_6$.

(e) the millimoles of K^+ in 50.0 mL of this solution.

(f) ppm $Fe(CN)_6^{3-}$.

(g) pK for the solution.

(h) $pFe(CN)_6$ for the solution.

*2-25. A 7.88% (w/w) $Fe(NO_3)_3$ (241.81 g/mol) solution has a density of 1.062 g/mL. Calculate

(a) the molar analytical concentration of $Fe(NO_3)_3$ in this solution.

(b) the molar NO_3^- concentration in the solution.

(c) the grams of $Fe(NO_3)_3$ contained in each liter of this solution.

2-26. A saturated aqueous solution of sodium chromate is 40.6% (w/w) Na_2CrO_4 (162.0 g/mol) and has a specific gravity of 1.430. Calculate

(a) the molar analytical concentration of Na_2CrO_4.

(b) pNa for this solution.

(c) the grams of solute in 1.00 L of saturated Na_2CrO_4.

*2-27. Describe the preparation of

(a) 500 mL of 6.50% (w/v) aqueous ethanol (C_2H_5OH, 46.1 g/mol).

(b) 500 g of 6.50% (w/w) aqueous ethanol.

(c) 500 mL of 6.50% (v/v) aqueous ethanol.

2-28. How would you prepare 600 mL of an aqueous solution that is

(a) 24.0% (w/v) acetone (58.08 g/mol)?

(b) 24.0% (w/w) acetone?

(c) 24.0% (v/v) acetone?

*2-29. Describe how you would prepare 2.00 L of 0.150 M perchloric acid from a concentrated solution that has a specific gravity of 1.66 and is 70% $HClO_4$ (w/w).

2-30. Describe how you would prepare 800 mL of 0.400 M aqueous ammonia from a concentrated solution that has a specific gravity of 0.90 and is 27% NH_3 (w/w).

2-31. Use chemical symbols (and small whole numbers) to express the stoichiometric factor, or ratio, needed to calculate the amount of substance in column B from a known amount of substance in column A.

Reaction	A	B
*(a) $Ag^+ + Cl^- \rightarrow AgCl(s)$	$AgNO_3$	$AgCl$
(b) $2Ag^+ + CrO_4^{2-} \rightarrow$ $Ag_2CrO_4(s)$	K_2CrO_4	Ag_2CrO_4
*(c) $PO_4^{3-} + 2H_3O^+ \rightarrow$ $H_2PO_4^- + 2H_2O$	Na_3PO_4	H_3O^+

Reaction	A	B
(d) $H_3PO_4 + OH^- \rightarrow H_2PO_4^-$	H_3PO_4	OH^-
*(e) $2Ag^+ + Cu(s) \rightarrow$ $2Ag(s) + Cu^{2+}$	Cu	Ag
(f) $I_2 + 2S_2O_3^{2-} \rightarrow$ $2I^- + S_4O_6^{2-}$	$Na_2S_2O_3$	I_2
*(g) $2CN^- + Ag^+ \rightarrow Ag(CN)_2^-$	$NaCN$	$NaAg(CN)_2$
(h) $Cu^{2+} + 4NH_3 \rightarrow Cu(NH_3)_4^{2+}$	NH_3	$Cu(NH_3)_4Cl_2$

2-32. Calculate the mass of silver bromide (187.8 g/mol) produced upon mixing 25.00 mL of 0.0661 M $AgNO_3$ (169.9 g/mol) with 40.0 mL of 0.0397 M KBr (119.0 g/mol).

*2-33. Describe the preparation of

(a) 500 mL of 0.0750 M $AgNO_3$ from the solid reagent.

(b) 1.00 L of 0.315 M HCl, starting with a 6.00 M solution of the reagent.

(c) 600 mL of a solution that is 0.0825 M in K^+, starting with solid $K_4Fe(CN)_6$.

(d) 400 mL of 3.00% (w/v) aqueous $BaCl_2$ from a 0.400 M $BaCl_2$ solution.

(e) 2.00 L of 0.120 M $HClO_4$ from the commercial reagent [60% $HClO_4$ (w/w), sp gr 1.60].

(f) 9.00 L of a solution that is 60.0 ppm in Na^+, starting with solid Na_2SO_4.

2-34. Describe the preparation of

(a) 5.00 L of 0.150 M $KMnO_4$ from the solid reagent.

(b) 4.00 L of 0.175 M $HClO_4$, starting with an 8.00 M solution of the reagent.

(c) 400 mL of a solution that is 0.0500 M in I^-, starting with MgI_2.

(d) 200 mL of 1.00% (w/v) aqueous $CuSO_4$, from a 0.218 M $CuSO_4$ solution.

(e) 1.50 L of 0.215 M NaOH from the concentrated commercial reagent [50% NaOH (w/w), sp gr 1.525].

(f) 1.50 L of a solution that is 12.0 ppm in K^+, starting with solid $K_4Fe(CN)_6$.

*2-35. What mass of solid $La(IO_3)_3$ (663.6 g/mol) is formed when 50.0 mL of 0.150 M La^{3+} are mixed with 75.0 mL of 0.202 M IO_3^-?

2-36. What mass of solid $PbCl_2$ (278.11 g/mol) is formed when 100 mL of 0.125 M Pb^{2+} are mixed with 200 mL of 0.175 M Cl^-?

*2-37. Exactly 0.1120 g of pure Na_2CO_3 was dissolved in 100.0 mL of 0.0497 M HCl.

(a) What mass of CO_2 was evolved?

(b) What was the molarity of the excess reactant (HCl or Na_2CO_3)?

2-38. Exactly 50.00 mL of a 0.4230 M solution of Na_3PO_4 were mixed with 100.00 mL of 0.5151 M $AgNO_3$.

(a) What mass of solid Ag_3PO_4 was formed?

(b) What is the molarity of the unreacted species (Na_3PO_4 or $AgNO_3$) after the reaction was complete?

*2-39. Exactly 75.00 mL of a 0.3333 M solution of Na_2SO_3 were treated with 150.0 mL of 0.3912 M $HClO_4$ and boiled to remove the SO_2 formed.

(a) How many grams of SO_2 were evolved?

(b) What was the concentration of the unreacted reagent (Na_2SO_3 or HCl) after the reaction was complete?

2-40. What mass of $MgNH_4PO_4$ precipitated when 200.0 mL of a 1.000% (w/v) solution of $MgCl_2$ were treated with 40.0 mL of 0.1753 M Na_3PO_4 and an excess of NH_4^+? What was the molarity of the excess reagent (Na_3PO_4 or $MgCl_2$) after the precipitation was complete?

***2-41.** What volume of 0.01000 M $AgNO_3$ would be required to precipitate all of the I^- in 200.0 mL of a solution that contained 2.643 ppt KI?

2-42. Exactly 750.0 mL of a solution that contained 650.1 ppm of $Ba(NO_3)_2$ were mixed with 200.0 mL of a solution that was 0.04100 M in $Al_2(SO_4)_3$.
(a) What mass of solid $BaSO_4$ was formed?
(b) What was the molarity of the unreacted reagent [$Al_2(SO_4)_3$ or $Ba(NO_3)_2$]?

AQUEOUS-SOLUTION CHEMISTRY

Chapter 3 treats aqueous-solution chemistry including chemical equilibrium and basic equilibrium constant calculations. This material is treated in most general chemistry courses.

3A THE CHEMICAL COMPOSITION OF AQUEOUS SOLUTIONS

Water is the most plentiful solvent available on earth, is easily purified, and is not toxic. It therefore finds widespread use as a medium for carrying out chemical analyses.

3A-1 Solutions of Electrolytes

Most of the solutes we will discuss are *electrolytes*. These solutes form ions when dissolved in water or certain other solvents and conduct electricity. *Strong electrolytes* ionize almost completely in a solvent, whereas *weak electrolytes* ionize only partially. Therefore, weak electrolytes conduct electricity to a smaller extent than do strong. Table 3-1 is a compilation of solutes that act as strong and weak electrolytes in water. Among the strong electrolytes listed are acids, bases, and salts.

> The reaction of an acid with a base produces a salt. Examples include NaCl, Na_2SO_4, and $NaOOCCH_3$ (sodium acetate).

3A-2 Acids and Bases

In 1923, two chemists, J. N. Brønsted in Denmark and J. M. Lowry in England, independently proposed a theory of acid/base behavior that is particularly useful in analytical chemistry. According to the Brønsted–Lowry theory, *an acid is a proton donor* and *a base is a proton acceptor*. In order for a species to behave

Table 3-1
CLASSIFICATION OF ELECTROLYTES

Strong	Weak
1. The inorganic acids HNO_3, $HClO_4$, H_2SO_4*, HCl, HI, HBr, $HClO_3$, $HBrO_3$	1. Many inorganic acids, including H_2CO_3, H_3BO_3, H_3PO_4, H_2S, H_2SO_3
2. Alkali and alkaline-earth hydroxides	2. Most organic acids
3. Most salts	3. Ammonia and most organic bases
	4. Halides, cyanides, and thiocyanates of Hg, Zn, and Cd

* H_2SO_4 is completely dissociated into HSO_4^- and H_3O^+ ions and for this reason is classified as a strong electrolyte. However, it should be noted that the HSO_4^- ion is a weak electrolyte, being only partially dissociated.

as an acid, a proton acceptor (or base) must be present. It is also true that for a species to behave as a base, a proton donor (or acid) must be present.

Conjugate Acids and Bases

An important feature of the Brønsted–Lowry concept is the idea that the entity produced when an acid gives up a proton is a potential proton acceptor and is called a *conjugate base* of the parent acid. For example, when the species acid₁ gives up a proton, the species base₁ is formed as shown by the reaction

$$acid_1 \rightleftharpoons base_1 + proton$$

Here, acid₁ and base₁ are a conjugate acid/base pair. Similarly, every base produces a *conjugate acid* as a result of accepting a proton. That is,

$$base_2 + proton \rightleftharpoons acid_2$$

When these two processes are combined, the result is an acid/base, or neutralization, reaction:

$$acid_1 + base_2 \rightleftharpoons base_1 + acid_2$$

The extent to which this reaction proceeds depends on the relative tendencies of the two bases to accept a proton (or the two acids to donate a proton). Examples of conjugate acid/base relationships are shown in Equations 3-1 through 3-4.

Many solvents are proton donors or proton acceptors and can thus induce basic or acidic behavior in solutes dissolved in them. For example, in an aqueous solution of ammonia, water can donate a proton and thus acts as an acid with respect to the solute:

$$\underset{base_1}{NH_3} + \underset{acid_2}{H_2O} \rightleftharpoons \underset{conjugate\ acid_1}{NH_4^+} + \underset{conjugate\ base_2}{OH^-} \tag{3-1}$$

In this reaction ammonia (base₁) reacts with water, which is labeled acid₂, to give the conjugate acid ammonium ion (acid₁) and hydroxide ion, which is the

An *acid* is a substance that donates protons; a *base* is a substance that accepts protons.

An acid donates protons only in the presence of a proton acceptor (a base). Likewise, a base accepts protons only in the presence of a proton donor (an acid).

A *conjugate base* is the species formed when an acid loses a proton. For example, acetate ion is the conjugate base of acetic acid; similarly, ammonium ion is the conjugate acid of ammonia.

A *conjugate acid* is the species formed when a base accepts a proton.

A substance acts as an acid only in the presence of a base, and a substance acts as a base only in the presence of an acid.

The species $H_9O_4^+$ appears to be the predominant form of the hydrated proton. Its structure is

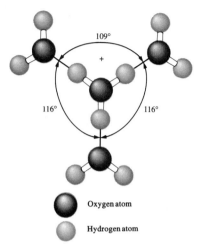

For convenience, in this book we use the simpler symbol H_3O^+ to indicate the hydrated proton in those chapters that deal with acid/base equilibria and acid/base equilibrium calculations. In the other chapters we simplify to the more convenient H^+. Note that H^+, H_3O^+, and $H_9O_4^+$ all represent the solvated proton.

Svante Arrhenius (1859–1927), Swedish chemist, formulated many of the early ideas regarding ionic dissociation in solution. His ideas were not accepted at first; in fact, he was given the lowest possible passing grade for his Ph.D. examination. In 1903 Arrhenius was awarded the Nobel Prize in chemistry for these revolutionary ideas. He was one of the first scientists to suggest the relationship between the amount of carbon dioxide in the atmosphere and global temperature, a phenomenon that has come to be known as the greenhouse effect.

conjugate base (base$_2$) of the acid water. In contrast, water acts as a proton acceptor, or base, in an aqueous solution of nitrous acid:

$$\underset{\text{base}_1}{H_2O} + \underset{\text{acid}_2}{HNO_2} \rightleftharpoons \underset{\substack{\text{conjugate} \\ \text{acid}_1}}{H_3O^+} + \underset{\substack{\text{conjugate} \\ \text{base}_2}}{NO_2^-} \qquad \textbf{(3-2)}$$

Nitrite ion is the conjugate base of the acid HNO_2. The conjugate acid in this case is the hydrated proton written as H_3O^+. This species is called the *hydronium ion* and consists of a proton covalently bonded to one water molecule. Higher hydrates such as $H_5O_2^+$ and $H_9O_4^+$ also exist in aqueous solution of protons. For convenience, however, chemists generally use the notations H_3O^+, or more simply H^+, in writing chemical equations in which the proton is involved.

Feature 3-1
AMPHIPROTIC SPECIES

Some compounds behave as both an acid and a base. An example is dihydrogen phosphate ion, $H_2PO_4^-$, which behaves as a base in the presence of a proton donor such as H_3O^+.

$$\underset{\text{base}_1}{H_2PO_4^-} + \underset{\text{acid}_2}{H_3O^+} \rightleftharpoons \underset{\text{acid}_1}{H_3PO_4} + \underset{\text{base}_2}{H_2O}$$

Here, H_3PO_4 is the conjugate acid of the original base. In the presence of a proton acceptor such as water, however, $H_2PO_4^-$ behaves as an acid and forms the conjugate base HPO_4^{2-}.

$$\underset{\text{acid}_1}{H_2PO_4^-} + \underset{\text{base}_2}{H_2O} \rightleftharpoons \underset{\text{base}_1}{HPO_4^{2-}} + \underset{\text{acid}_2}{H_3O^+}$$

The simple amino acids are an important class of amphiprotic compounds that contain both a weak acid and a weak base functional group. When dissolved in water, an amino acid such as glycine undergoes a kind of internal acid/base reaction to produce a *zwitterion*—a species that bears both a positive and a negative charge. Thus,

$$\underset{\text{glycine}}{NH_2CH_2COOH} \rightleftharpoons \underset{\text{zwitterion}}{NH_3^+CH_2COO^-}$$

This reaction is analogous to the acid/base reaction we observe between a carboxylic acid and an amine:

$$\underset{\text{acid}_1}{R_1COOH} + \underset{\text{base}_2}{R_2NH_2} \rightleftharpoons \underset{\text{base}_1}{R_1COO^-} + \underset{\text{acid}_2}{RNH_3^+}$$

Amphiprotic Solvents

Water is the classic example of an *amphiprotic* solvent—that is, a solvent that can act either as an acid (Equation 3-1) or as a base (Equation 3-2), depending on the solute. Other common amphiprotic solvents are methanol, ethanol, and anhydrous acetic acid. In methanol, for example, the equilibria analogous to those shown in Equations 3-1 and 3-2 are

$$\underset{\text{base}_1}{NH_3} + \underset{\text{acid}_2}{CH_3OH} \rightleftharpoons \underset{\substack{\text{conjugate} \\ \text{acid}_1}}{NH_4^+} + \underset{\substack{\text{conjugate} \\ \text{base}_2}}{CH_3O^-} \qquad (3\text{-}3)$$

$$\underset{\text{base}_1}{CH_3OH} + \underset{\text{acid}_2}{HNO_2} \rightleftharpoons \underset{\substack{\text{conjugate} \\ \text{acid}_1}}{CH_3OH_2^+} + \underset{\substack{\text{conjugate} \\ \text{base}_2}}{NO_2^-} \qquad (3\text{-}4)$$

It is important to emphasize that an acid that has donated a proton becomes a conjugate base capable of accepting a proton to reform the original acid; the converse holds equally well. Thus nitrite ion, the species produced by the loss of a proton from nitrous acid, is a potential acceptor of a proton from a suitable donor. It is this reaction that causes an aqueous solution of sodium nitrite to be slightly basic:

$$\underset{\text{base}_1}{NO_2^-} + \underset{\text{acid}_2}{H_2O} \rightleftharpoons \underset{\substack{\text{conjugate} \\ \text{acid}_1}}{HNO_2} + \underset{\substack{\text{conjugate} \\ \text{base}_2}}{OH^-}$$

3A-3 Autoprotolysis

Amphiprotic solvents undergo self-ionization, or *autoprotolysis* to form a pair of ionic species. Autoprotolysis is yet another example of acid/base behavior, as illustrated by the following equations.

$$\begin{array}{llll}
\text{base}_1 & + \ \text{acid}_2 & \rightleftharpoons \text{acid}_1 & + \ \text{base}_2 \\
H_2O & + \ H_2O & \rightleftharpoons H_3O^+ & + \ OH^- \\
CH_3OH & + \ CH_3OH & \rightleftharpoons CH_3OH_2^+ & + \ CH_3O^- \\
HCOOH & + \ HCOOH & \rightleftharpoons HCOOH_2^+ & + \ HCOO^- \\
NH_3 & + \ NH_3 & \rightleftharpoons NH_4^+ & + \ NH_2^-
\end{array}$$

The extent to which water undergoes autoprotolysis is slight at room temperature. Thus, the hydronium and hydroxide ion concentrations in pure water are only about 10^{-7} M. Despite the small values of these concentrations, this dissociation reaction is of utmost importance in understanding the behavior of aqueous solutions.

3A-4 Strengths of Acids and Bases

Figure 3-1 shows the dissociation reaction of a few common acids in water. The first two are *strong acids* because reaction with the solvent is sufficiently complete as to leave no undissociated solute molecules in aqueous solution. The remainder are *weak acids*, which react incompletely with water to give solutions that contain significant quantities of both the parent acid and its conjugate base. Note that acids can be cationic, anionic, or electrically neutral.

The acids in Figure 3-1 become progressively weaker from top to bottom. Perchloric acid and hydrochloric acid are completely dissociated. In contrast, only about 1% of acetic acid ($HC_2H_3O_2$) is dissociated. Ammonium ion is an even weaker acid; only about 0.01% of this ion is dissociated into hydronium

Figure 3-1

Dissociation reactions and relative strengths of some common acids and their conjugate bases.

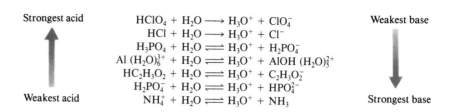

The common strong bases include NaOH, KOH, Ba(OH)$_2$, and the quaternary ammonium hydroxides, R$_4$NOH, where R is an alkyl group such as CH$_3$ or C$_2$H$_5$.

Of all the strong acids listed in the earlier marginal note, only perchloric acid is a strong acid in methanol and ethanol. Thus, these two alcohols are also differentiating solvents.

In a differentiating solvent, various acids dissociate to different degrees and thus have different strengths. In a leveling solvent, several acids are completely dissociated and are thus of the same strength.

ions and ammonia molecules. Another generality illustrated in Figure 3-1 is that the weakest acid forms the strongest conjugate base; that is, ammonia has a much stronger affinity for protons than any base above it. Perchlorate and chloride ions have no affinity for protons.

The tendency of a solvent to accept or donate protons determines the strength of a solute acid or base dissolved in it. For example, perchloric and hydrochloric acids are strong acids in water. If anhydrous acetic acid, a weaker proton acceptor than water, is substituted *as the solvent*, neither of these acids undergoes complete dissociation; instead, equilibria such as the following are established:

$$\underset{\text{base}_1}{CH_3COOH} + \underset{\text{acid}_2}{HClO_4} \rightleftharpoons \underset{\text{acid}_1}{CH_3COOH_2^+} + \underset{\text{base}_2}{ClO_4^-}$$

Perchloric acid is, however, considerably stronger than hydrochloric acid in this solvent, its dissociation being about 5000 times greater. Acetic acid thus acts as a *differentiating* solvent toward the two acids by revealing the inherent differences in their acidities. Water, on the other hand, is a *leveling* solvent for perchloric, hydrochloric, nitric, and sulfuric acids because all four are completely ionized in this solvent and thus exhibit no differences in strength. There are also differentiating and leveling solvents for bases.

3B   CHEMICAL EQUILIBRIUM

The reactions used in analytical chemistry never result in complete conversion of reactants to products. Instead, they proceed to a state of *chemical equilibrium* in which the ratio of concentrations of reactants and products is constant. *Equilibrium-constant expressions* are *algebraic equations* that describe the concentration relationships among reactants and products at equilibrium. Among other things, equilibrium-constant expressions permit calculation of the error resulting from the quantity of unreacted analyte that remains when a steady state has been reached.

The discussion that follows deals with use of equilibrium-constant expressions to gain information about analytical systems in which no more than one or two equilibria are present. Chapter 7 extends these methods to systems containing several simultaneous equilibria. Such complex systems are often encountered in analytical chemistry.

3B-1   The Equilibrium State

Consider the chemical equilibrium

$$H_3AsO_4 + 3I^- + 2H^+ \rightleftharpoons H_3AsO_3 + I_3^- + H_2O \qquad (3\text{-}5)$$

The rate of this reaction and the extent to which it proceeds to the right can be readily judged by observing the orange-red color of the triiodide ion I_3^- (the other participants in the reaction are colorless). If, for example, 1 mmol of arsenic acid H_3AsO_4 is added to 100 mL of a solution containing 3 mmol of potassium iodide, the red color of the triiodide ion appears almost immediately, and within a few seconds, the intensity of the color becomes constant, which shows that the triiodide concentration has become constant (see color plate 6a).

When equilibrium is achieved, properties of a system, such as color, stop changing.

A solution of identical color intensity (and hence identical triiodide concentration) can also be produced by adding 1 mmol of arsenous acid H_3AsO_3 to 100 mL of a solution containing 1 mmol of triiodide ion (see color plate 6b). Here, the color intensity is initially greater than in the first solution but rapidly decreases as a result of the reaction

$$H_3AsO_3 + I_3^- + H_2O \rightleftharpoons H_3AsO_4 + 3I^- + 2H^+$$

Ultimately the color of the two solutions is identical. Many other combinations of the four reactants yield solutions that are indistinguishable from the two just described.

The foregoing discussion illustrates that the concentration relationship at chemical equilibrium (that is, the *position of equilibrium*) is independent of the route by which the equilibrium state is achieved. However, this relationship is altered by the application of stress to the system. Such stresses include changes in temperature, in pressure (if one of the reactants or products is a gas), or in total concentration of a reactant or a product. These effects can be predicted qualitatively from the *principle of Le Châtelier*, which states that the position of chemical equilibrium always shifts in a direction that tends to relieve the effect of an applied stress. Thus, an increase in temperature alters the concentration relationship in the direction that tends to absorb heat, and an increase in pressure favors those participants that occupy a smaller total volume.

The position of a chemical equilibrium is independent of the route by which equilibrium is reached.

The Le Châtelier principle states that the position of an equilibrium always shifts in such a direction as to relieve a stress that is applied to the system.

In an analysis, the effect of introducing an additional amount of a participating species to the reaction mixture is particularly important. Here, the resulting stress is relieved by a shift in equilibrium in the direction that partially uses up the added substance. Thus, for the equilibrium we have been considering (Equation 3-5), the addition of arsenic acid (H_3AsO_4) or hydrogen ions causes an increase in color as more triiodide ion and arsenous acid are formed; the addition of arsenous acid has the reverse effect. An equilibrium shift brought about by changing the amount of one of the participating species is called a *mass-action effect*.

If it were possible to examine the system under discussion at the molecular level, we would find that reactions among the participating species continue even after equilibrium is achieved. The constant concentration ratio of reactants and products results from the equality in the rates of the forward and reverse reactions. In other words, chemical equilibrium is a dynamic state in which the rates of the forward and reverse reactions are identical.

Chemical reactions do not cease at equilibrium. Instead, the amounts of reactants and products are constant because the rates of the forward and reverse processes are identical.

3B-2 Equilibrium-Constant Expressions

The influence of concentration (or pressure if the species are gases) on the position of a chemical equilibrium is conveniently described in quantitative terms by means of an equilibrium-constant expression. Such expressions are readily derived

Thermodynamics is a branch of chemical science that deals with energy changes in chemical reactions. The position of a chemical equilibrium can be related to these energy changes.

Equilibrium-constant expressions provide *no* information as to whether a chemical reaction is fast enough to be used for an analysis.

Cato Guldberg (1836–1902) and Peter Waage (1833–1900) were Norwegian chemists whose primary interests were in the field of thermodynamics. In 1864, these workers were the first to propose the law of mass action, which is expressed in Equation 3-7.

from *thermodynamic theory*. They are of great practical importance because they allow us to predict the direction and completeness of a chemical reaction. We must emphasize, however, that an equilibrium-constant expression yields no information concerning the *rate* at which equilibrium is approached. In fact, we sometimes encounter reactions that have highly favorable equilibrium constants but are of little analytical use because their rates are low. This limitation can often be overcome by the use of a catalyst, which speeds the attainment of equilibrium without changing its position.

Let us consider a generalized equation for a chemical equilibrium

$$w\mathrm{W} + x\mathrm{X} \rightleftharpoons y\mathrm{Y} + z\mathrm{Z} \qquad (3\text{-}6)$$

where the capital letters represent the formulas of participating chemical species and the lower-case italic letters are the small whole numbers required to balance the equation. Thus, the equation states that w mol of W reacts with x mol of X to form y mol of Y and z mol of Z. The equilibrium-constant expression for this reaction is

$$K = \frac{[\mathrm{Y}]^y[\mathrm{Z}]^z}{[\mathrm{W}]^w[\mathrm{X}]^x} \qquad (3\text{-}7)$$

where the square-bracketed terms have the following meanings:

(1) molar concentration if the species is a dissolved solute,
(2) partial pressure in atmospheres if the species is a gas; in fact, we will often replace the square bracketed term (say [Z] in Equation 3-7) with the symbol p_z which stands for the partial pressure of the gas Z in atmospheres.

[Z] in Equation 3-7 is replaced with p_z in atmospheres if Z is a gas. If Z is a pure liquid, a pure solid, or the solvent present in excess, no term for this species is included in the equation.

If one (or more) of the species in Equation 3-7 is a pure liquid, a pure solid, or the solvent present in excess, no term for this species appears in the equilibrium-constant expression. For example, if Z in Equation 3-6 is the solvent H_2O, the equilibrium-constant expression simplifies to

$$K = \frac{[\mathrm{Y}]^y}{[\mathrm{W}]^w[\mathrm{X}]^x}$$

The reason for this simplification will become apparent in the section that follows.

The constant K in Equation 3-7 is a temperature-dependent numerical quantity called the *equilibrium constant*. Note that by convention, the concentrations of the products of the equilibrium always appear in the numerator of an equilibrium-constant expression and the concentrations of the reactants in the denominator.

Remember: Equation 3-7 is only an approximate form of an equilibrium-constant expression. The exact expression takes the form

$$K = \frac{a_\mathrm{Y}^y \times a_\mathrm{Z}^z}{a_\mathrm{W}^w \times a_\mathrm{X}^x} \qquad (3\text{-}8)$$

where a_Y, a_Z, a_W, and a_X are the *activities* of species Y, Z, W, and X (see Section 8B).

Equation 3-7 is only an approximate form of a thermodynamic equilibrium-constant expression. The exact form is given by Equation 3-8 in the margin. Generally we will use the approximate form of this equation because it is less tedious and time consuming to use. In Section 8B, we show when the use of Equation 3-7 is likely to lead to serious errors in equilibrium calculations and how Equation 3-8 is applied in these cases.

3B-3 Types of Equilibrium Constants Encountered in Analytical Chemistry

Table 3-2 summarizes the types of chemical equilibria and equilibrium constants that are of importance in analytical chemistry. Basic applications of some of these constants are illustrated in the paragraphs that follow.

Feature 3-2
STEPWISE AND OVERALL FORMATION CONSTANTS FOR COMPLEX IONS

Stepwise-formation constants are symbolized by K_i. For example,

$$Ni^{2+} + CN^- \rightleftharpoons Ni(CN)^+ \qquad K_1 = \frac{[Ni(CN)^+]}{[Ni^{2+}][CN^-]}$$

$$Ni(CN)^+ + CN^- \rightleftharpoons Ni(CN)_2 \qquad K_2 = \frac{[Ni(CN)_2]}{[Ni(CN)^+][CN^-]}$$

$$Ni(CN)_2 + CN^- \rightleftharpoons Ni(CN)_3^- \qquad K_3 = \frac{[Ni(CN)_3^-]}{[Ni(CN)_2][CN^-]}$$

$$Ni(CN)_3^- + CN^- \rightleftharpoons Ni(CN)_4^{2-} \qquad K_4 = \frac{[Ni(CN)_4^{2-}]}{[Ni(CN)_3^-][CN^-]}$$

Overall constants are designated by the symbol β_n. Thus,

$$Ni^{2+} + 2CN^- \rightleftharpoons Ni(CN)_2 \qquad \beta_2 = K_1 K_2 = \frac{[Ni(CN)_2]}{[Ni^{2+}][CN^-]^2}$$

$$Ni^{2+} + 3CN^- \rightleftharpoons Ni(CN)_3^- \qquad \beta_3 = K_1 K_2 K_3 = \frac{[Ni(CN)_3^-]}{[Ni^{2+}][CN^-]^3}$$

$$Ni^{2+} + 4CN^- \rightleftharpoons Ni(CN)_4^{2-} \qquad \beta_4 = K_1 K_2 K_3 K_4 = \frac{[Ni(CN)_4^{2-}]}{[Ni^{2+}][CN^-]^4}$$

Ion-Product Constant for Water

Aqueous solutions contain small amounts of hydronium and hydroxide ions as a consequence of the dissociation reaction

$$2H_2O \rightleftharpoons H_3O^+ + OH^- \tag{3-9}$$

An equilibrium constant for this reaction can be formulated as shown in Equation 3-7:

$$K = \frac{[H_3O^+][OH^-]}{[H_2O]^2} \tag{3-10}$$

The concentration of water in dilute aqueous solutions is enormous, however,

Table 3-2

EQUILIBRIA AND EQUILIBRIUM CONSTANTS OF IMPORTANCE TO ANALYTICAL CHEMISTRY

Type of Equilibrium	Name and Symbol of Equilibrium Constant	Typical Example	Equilibrium-Constant Expression
Dissociation of water	Ion-product constant, K_w	$2H_2O \rightleftharpoons H_3O^+ + OH^-$	$K_w = [H_3O^+][OH^-]$
Heterogeneous equilibrium between a slightly soluble substance and its ions in a saturated solution	Solubility product, K_{sp}	$BaSO_4(s) \rightleftharpoons Ba^{2+} + SO_4^{2-}$	$K_{sp} = [Ba^{2+}][SO_4^{2-}]$
Dissociation of a weak acid or base	Dissociation constant, K_a or K_b	$CH_3COOH + H_2O \rightleftharpoons$ $H_3O^+ + CH_3COO^-$ $CH_3COO^- + H_2O \rightleftharpoons$ $OH^- + CH_3COOH$	$K_a = \dfrac{[H_3O^+][CH_3COO^-]}{[CH_3COOH]}$ $K_b = \dfrac{[OH^-][CH_3COOH]}{[CH_3COO^-]}$
Formation of a complex ion	Formation constant, β_n	$Ni^{2+} + 4CN^- \rightleftharpoons Ni(CN)_4^{2-}$	$\beta_4 = \dfrac{[Ni(CN)_4^{2-}]}{[Ni^{2+}][CN^-]^4}$
Oxidation/reduction equilibrium	K_{redox}	$MnO_4^- + 5Fe^{2+} + 8H^+ \rightleftharpoons$ $Mn^{2+} + 5Fe^{3+} + 4H_2O$	$K_{redox} = \dfrac{[Mn^{2+}][Fe^{3+}]^5}{[MnO_4^-][Fe^{2+}]^5[H^+]^8}$
Distribution equilibrium between immiscible solvents	K_d	$I_2(aq) \rightleftharpoons I_2(org)$	$K_d = \dfrac{[I_2]_{org}}{[I_2]_{aq}}$

when compared with the concentration of hydrogen and hydroxide ions. As a consequence, $[H_2O]$ in Equation 3-10 can be taken as constant, and we write

$$K[H_2O]^2 = K_w = [H_3O^+][OH^-] \qquad \text{(3-11)}$$

where the new constant K_w is given a special name, the *ion-product constant for water*.

Feature 3-3

WHY $[H_2O]$ DOES NOT APPEAR IN EQUILIBRIUM-CONSTANT EXPRESSIONS FOR AQUEOUS SOLUTIONS

In a dilute aqueous solution the molar concentration of water is

$$[H_2O] = \frac{1000 \text{ g } H_2O}{L \ H_2O} \times \frac{1 \text{ mol } H_2O}{18.0 \text{ g } H_2O} = 55.6 \text{ M}$$

Let us suppose we have 0.1 mol of HCl in 1 L of water. The presence of this acid will shift to the left the equilibrium shown in Equation 3-9. Originally, however, there were only 10^{-7} mol/L OH$^-$ to consume the added protons.

Thus, even if all the OH^- ions are converted to H_2O, the water concentration will increase only to

$$[H_2O] = 55.6 \ \frac{\text{mol } H_2O}{\text{L } H_2O} + 1 \times 10^{-7} \ \frac{\text{mol } OH^-}{\text{L } H_2O} \times \frac{1 \text{ mol } H_2O}{\text{mol } OH^-} \cong 55.6 \text{ M}$$

The percent change in water concentration is

$$\frac{10^{-7} \text{ M}}{55.6 \text{ M}} \times 100\% = 2 \times 10^{-7}\%$$

which is certainly negligibly small. Thus, $K[H_2O]^2$ in Equation 3-10 is, for all practical purposes, a constant. That is,

$$K(55.6)^2 = K_w = 1.00 \times 10^{-14}$$

At 25°C, the ion-product constant for water is 1.008×10^{-14}. For convenience, we shall use the approximation that at room temperature $K_w \cong 1.00 \times 10^{-14}$. Table 3-3 shows the dependence of this constant on temperature.

The ion-product constant for water permits the ready calculation of the hydronium and hydroxide ion concentrations of aqueous solutions.

Example 3-1

Calculate the hydronium and hydroxide ion concentrations of pure water at 25° and 100°C.

Because OH^- and H_3O^+ are formed only from the dissociation of water, their concentrations must be equal:

$$[H_3O^+] = [OH^-]$$

Substitution into Equation 3-11 gives

$$[H_3O^+]^2 = [OH^-]^2 = K_w$$
$$[H_3O^+] = [OH^-] = \sqrt{K_w}$$

At 25°C,

$$[H_3O^+] = [OH^-] = \sqrt{1.00 \times 10^{-14}} = 1.00 \times 10^{-7}$$

At 100°C, from Table 3-3,

$$[H_3O^+] = [OH^-] = \sqrt{49 \times 10^{-14}} = 7.0 \times 10^{-7}$$

Example 3-2

Calculate the hydronium and hydroxide ion concentrations in 0.200 M aqueous NaOH.

Sodium hydroxide is a strong electrolyte, and its contribution to the hydroxide

Table 3-3
VARIATION OF K_w WITH TEMPERATURE

Temperature, °C	K_w
0	0.114×10^{-14}
25	1.01×10^{-14}
50	5.47×10^{-14}
100	49×10^{-14}

ion concentration in this solution is 0.200 mol/L. As in Example 3-1, hydroxide ions and hydronium ions are formed *in equal amounts* from dissociation of water. Therefore, we write

$$[OH^-] = 0.200 + [H_3O^+]$$

where $[H_3O^+]$ accounts for the hydroxide ions contributed by the solvent. The concentration of OH^- from the water is insignificant, however, when compared with 0.200, so we can write

$$[OH^-] \cong 0.200$$

Equation 3-11 is then used to calculate the hydronium ion concentration:

$$[H_3O^+] = \frac{K_w}{[OH^-]} = \frac{1.00 \times 10^{-14}}{0.200} = 5.00 \times 10^{-14}$$

Note that the approximation

$$[OH^-] = 0.200 + 5.00 \times 10^{-14} \cong 0.200$$

causes no significant error.

Solubility-Product Constants

When we say that a sparingly soluble salt is completely dissociated, *we do not imply that all of the salt dissolves.* Instead, the very small amount that *does* go into solution dissociates completely.

Most sparingly soluble salts are essentially completely dissociated in saturated aqueous solution. For example, when an excess of barium iodate is equilibrated with water, the dissociation process is adequately described by the equation

$$Ba(IO_3)_2(s) \rightleftharpoons Ba^{2+}(aq) + 2IO_3^-(aq)$$

Application of Equation 3-7 leads to

$$K = \frac{[Ba^{2+}][IO_3^-]^2}{[Ba(IO_3)_2(s)]}$$

The denominator represents the molar concentration of $Ba(IO_3)_2$ *in the solid*, which is a phase that is separate from but in contact with the saturated solution. The concentration of a compound in its solid state is, however, constant. In other words, the number of moles of $Ba(IO_3)_2$ divided by the *volume* of the solid $Ba(IO_3)_2$ is constant no matter how much excess solid is present. Therefore, the foregoing equation can be rewritten in the form

For Equation 3-12 to apply, it is necessary only that *some solid be present.* Always keep in mind that in the absence of *$Ba(IO_3)(s)$, Equation 3-12 is not valid.*

$$K[Ba(IO_3)_2(s)] = K_{sp} = [Ba^{2+}][IO_3^-]^2 \qquad \textbf{(3-12)}$$

where the new constant is called the *solubility-product constant* or the *solubility product.* It is important to appreciate that Equation 3-12 shows that the position of this equilibrium is independent of the *amount* of $Ba(IO_3)_2$, so long as some solid is present; that is, it does not matter whether the amount is a few milligrams or several grams.

A table of solubility-product constants for numerous inorganic salts is found in Appendix 1. The examples that follow demonstrate some typical uses of solubility-product expressions. Further applications are considered in Chapters 7 and 8.

The Solubility of a Precipitate in Pure Water

We can use the solubility-product expression to calculate the solubility of a sparingly soluble substance that ionizes in water.

Example 3-3

How many grams of $Ba(IO_3)_2$ (487 g/mol) can be dissolved in 500 mL of water at 25°C?

The solubility-product constant for $Ba(IO_3)_2$ is 1.57×10^{-9} (Appendix 1). The equilibrium between the solid and its ions in solution is described by the equation

$$Ba(IO_3)_2(s) \rightleftharpoons Ba^{2+} + 2IO_3^-$$

and so

$$K_{sp} = [Ba^{2+}][IO_3^-]^2 = 1.57 \times 10^{-9}$$

The equation describing the equilibrium reveals that 1 mol of Ba^{2+} is formed for each mole of $Ba(IO_3)_2$ that dissolves. Therefore,

$$\text{molar solubility of } Ba(IO_3)_2 = [Ba^{2+}]$$

Note that the molar solubility is equal to $[Ba^{2+}]$ or to $\frac{1}{2}[IO_3^-]$.

The iodate concentration is clearly twice that for barium ion:

$$[IO_3^-] = 2[Ba^{2+}]$$

Substituting this last equation into the equilibrium-constant expression gives

$$[Ba^{2+}](2[Ba^{2+}])^2 = 1.57 \times 10^{-9}$$
$$[Ba^{2+}] = \left(\frac{1.57 \times 10^{-9}}{4}\right)^{1/3} = 7.32 \times 10^{-4} \text{ M}$$

Since 1 mol Ba^{2+} is produced for every mole of $Ba(IO_3)_2$

$$\text{solubility} = 7.32 \times 10^{-4} \text{ M}$$

To compute the number of millimoles of $Ba(IO_3)_2$ dissolved in 500 mL of solution, we write

$$\text{no. mmol } Ba(IO_3)_2 = 7.32 \times 10^{-4} \frac{\text{mmol } Ba(IO_3)_2}{\text{mL}} \times 500 \text{ mL}$$

The mass of $Ba(IO_3)_2$ in 500 mL is given by

$$\text{mass } Ba(IO_3)_2 = (7.32 \times 10^{-4} \times 500) \ \text{mmol } \cancel{Ba(IO_3)_2}$$

$$\times \ 0.487 \ \frac{\text{g } Ba(IO_3)_2}{\text{mmol } \cancel{Ba(IO_3)_2}}$$

$$= 0.178 \text{ g}$$

The Effect of a Common Ion on the Solubility of a Precipitate. The common-ion effect, a mass-action effect predicted from the Le Châtelier principle, is demonstrated by the following examples.

Another example of the common-ion effect is shown in color plate 1.

Example 3-4

Calculate the molar solubility of $Ba(IO_3)_2$ in a solution that is 0.0200 M in $Ba(NO_3)_2$.

The solubility is no longer equal to $[Ba^{2+}]$ because $Ba(NO_3)_2$ is also a source of barium ions. We know, however, that solubility is related to $[IO_3^-]$:

$$\text{molar solubility of } Ba(IO_3)_2 = \tfrac{1}{2}[IO_3^-]$$

There are two sources of barium ions: $Ba(NO_3)_2$ and $Ba(IO_3)_2$. The contribution from the former is 0.0200 M, and that from the latter is equal to the molar solubility, or $\tfrac{1}{2}[IO_3^-]$. Thus,

$$[Ba^{2+}] = 0.0200 + \tfrac{1}{2}[IO_3^-]$$

Substitution of these quantities into the solubility-product expression yields

$$(0.0200 + \tfrac{1}{2}[IO_3^-])[IO_3^-]^2 = 1.57 \times 10^{-9}$$

Since the exact solution for $[IO_3^-]$ requires solving a cubic equation, we seek an approximation that simplifies the algebra. The small numerical value of K_{sp} suggests that the solubility of $Ba(IO_3)_2$ is not large, and this is confirmed by the result obtained in Example 3-3. Moreover, barium ion from $Ba(NO_3)_2$ will further repress the limited solubility of $Ba(IO_3)_2$. Thus, it is reasonable to seek a provisional answer to the problem by assuming that 0.0200 is large with respect to $\tfrac{1}{2}[IO_3^-]$. That is, $\tfrac{1}{2}[IO_3^-] \ll 0.0200$, and

$$[Ba^{2+}] = 0.0200 + \tfrac{1}{2}[IO_3^-] \cong 0.0200$$

The original equation then simplifies to

$$0.0200[IO_3^-]^2 = 1.57 \times 10^{-9}$$
$$[IO_3^-] = \sqrt{7.85 \times 10^{-8}} = 2.80 \times 10^{-4} \text{ M}$$

The assumption that $(0.0200 + \tfrac{1}{2} \times 2.80 \times 10^{-4}) \cong 0.0200$ does not appear to cause serious error because the second term, representing the amount of Ba^{2+}

arising from the dissociation of $Ba(IO_3)_2$, is only about 0.7% of 0.0200. Ordinarily, we consider an assumption of this type to be satisfactory if the discrepancy is less than 10%. Finally,

$$\text{solubility of } Ba(IO_3)_2 = \tfrac{1}{2}[IO_3^-] = \tfrac{1}{2} \times 2.80 \times 10^{-4} = 1.40 \times 10^{-4} \text{ M}$$

If we compare this result with the solubility of barium iodate in pure water (Example 3-3), we see that the presence of a small concentration of the common ion has lowered the molar solubility of $Ba(IO_3)_2$ by a factor of about five.

Example 3-5

Calculate the solubility of $Ba(IO_3)_2$ in a solution prepared by mixing 200 mL of 0.0100 M $Ba(NO_3)_2$ with 100 mL of 0.100 M $NaIO_3$.

We must first establish whether either reactant is present in excess at equilibrium. The amounts taken are

$$\text{no. mmol } Ba^{2+} = 200 \; \cancel{mL} \times 0.0100 \text{ mmol}/\cancel{mL} = 2.00$$
$$\text{no. mmol } IO_3^- = 100 \; \cancel{mL} \times 0.100 \text{ mmol}/\cancel{mL} = 10.0$$

If formation of $Ba(IO_3)_2$ is complete

$$\text{no. mmol excess } NaIO_3 = 10.0 - 2(2.00) = 6.0$$

Thus,

$$[IO_3^-] = \frac{6.0 \text{ mmol}}{300 \text{ mL}} = 0.0200 \text{ M}$$

As in Example 3-3,

$$\text{molar solubility of } Ba(IO_3)_2 = [Ba^{2+}]$$

Here, however,

$$[IO_3^-] = 0.0200 + 2[Ba^{2+}]$$

A 0.02 M excess of Ba^{2+} decreases the solubility of $Ba(IO_3)_2$ by a factor of about 5; this same excess of IO_3^- lowers the solubility by about 200.

where $2[Ba^{2+}]$ represents the iodate contributed by the sparingly soluble $Ba(IO_3)_2$. We can obtain a provisional answer after making the assumption that $[IO_3^-] \cong 0.0200$; thus

$$\text{solubility of } Ba(IO_3)_2 = [Ba^{2+}] = \frac{K_{sp}}{[IO_3^-]^2} = \frac{1.57 \times 10^{-9}}{(0.0200)^2}$$
$$= 3.93 \times 10^{-6} \text{ mol/L}$$

Since $0.0200 \gg 3.93 \times 10^{-6}$, the approximation is reasonable.

Note that the results from the last two examples demonstrate that an excess of iodate ions is more effective in decreasing the solubility of $Ba(IO_3)_2$ than is the same excess of barium ions.

Acid- and Base-Dissociation Constants

When a weak acid or a weak base is dissolved in water, partial dissociation occurs. Thus, for nitrous acid we can write

$$HNO_2 + H_2O \rightleftharpoons H_3O^+ + NO_2^- \qquad K_a = \frac{[H_3O^+][NO_2^-]}{[HNO_2]}$$

where K_a is the *acid-dissociation constant* for nitrous acid. In an analogous way, the *base-dissociation constant* for ammonia is

$$NH_3 + H_2O \rightleftharpoons NH_4^+ + OH^- \qquad K_b = \frac{[NH_4^+][OH^-]}{[NH_3]}$$

Note that $[H_2O]$ does not appear in the denominator of either equation because the concentration of water is so large relative to the concentration of the weak acid or base that the dissociation does not alter $[H_2O]$ appreciably (see Feature 3-3). Just as in the derivation of the ion-product constant for water, $[H_2O]$ is incorporated in the equilibrium constants K_a and K_b. Dissociation constants for weak acids are found in Appendix 2.

Dissociation Constants for Conjugate Acid/Base Pairs

Consider the dissociation-constant expressions for ammonia and its conjugate acid, ammonium ion:

$$NH_3 + H_2O \rightleftharpoons NH_4^+ + OH^- \qquad K_b = \frac{[NH_4^+][OH^-]}{[NH_3]}$$

$$NH_4^+ + H_2O \rightleftharpoons NH_3 + H_3O^+ \qquad K_a = \frac{[NH_3][H_3O^+]}{[NH_4^+]}$$

Multiplication of one equilibrium-constant expression by the other gives

$$K_a K_b = \frac{[NH_3][H_3O^+]}{[NH_4^+]} \times \frac{[NH_4^+][OH^-]}{[NH_3]} = [H_3O^+][OH^-]$$

but

$$K_w = [H_3O^+][OH^-]$$

and therefore

$$K_w = K_a K_b \qquad\qquad (3\text{-}13)$$

This relationship is general for all conjugate acid/base pairs. Many compilations of equilibrium-constant data list only acid-dissociation constants since it is so easy to calculate basic dissociation constants with Equation 3-13. For example, in Appendix 2 we find no data on the basic dissociation of ammonia (nor for any other bases). Instead we find the acid-dissociation constant for the conjugate acid, ammonium ion. That is,

$$NH_4^+ + H_2O \rightleftharpoons H_3O^+ + NH_3 \qquad K_a = \frac{[H_3O^+][NH_3]}{[NH_4^+]} = 5.70 \times 10^{-10}$$

and we can write

$$NH_3 + H_2O \rightleftharpoons NH_4^+ + OH^- \qquad K_b = \frac{[NH_4^+][OH^-]}{[NH_3]} = \frac{1.00 \times 10^{-14}}{5.70 \times 10^{-10}} = 1.75 \times 10^{-5}$$

Feature 3-4
RELATIVE STRENGTHS OF CONJUGATE ACID/BASE PAIRS

Equation 3-13 confirms the observation in Figure 3-1 that as the acid of a conjugate acid/base pair becomes weaker, its conjugate base becomes stronger and vice versa. Thus the conjugate base of an acid with a dissociation constant of 10^{-2} will have a basic dissociation constant of 10^{-12}, and an acid with a dissociation constant of 10^{-9} has a conjugate base with a dissociation constant of 10^{-5}.

Example 3-6

What is K_b for the equilibrium

$$CN^- + H_2O \rightleftharpoons HCN + OH^-$$

Appendix 2 lists a K_a value of 6.2×10^{-10} for HCN. Thus,

$$K_b = \frac{[HCN][OH^-]}{[CN^-]} = \frac{K_w}{K_a}$$

$$K_b = \frac{1.00 \times 10^{-14}}{6.2 \times 10^{-10}} = 1.6 \times 10^{-5}$$

Hydronium Ion Concentration in Solutions of Weak Acids. When the weak acid HA is dissolved in water, two equilibria are established that yield hydronium ions:

$$HA + H_2O \rightleftharpoons H_3O^+ + A^- \qquad K_a = \frac{[H_3O^+][A^-]}{[HA]}$$

$$2H_2O \rightleftharpoons H_3O^+ + OH^- \qquad K_w = [H_3O^+][OH^-]$$

Ordinarily, the hydronium ions produced from the first reaction suppress the dissociation of water to such an extent that the contribution of hydronium ions from the second equilibrium is negligible. Under these circumstances, one H_3O^+ ion is formed for each A^- ion and we write

$$[A^-] \cong [H_3O^+] \tag{3-14}$$

Furthermore, the sum of the molar concentrations of the weak acid and its conjugate base must equal the analytical concentration of the acid c_{HA} because the solution contains no other source of A^- ions. Thus,

$$c_{HA} = [A^-] + [HA] \tag{3-15}$$

Substituting $[H_3O^+]$ for $[A^-]$ (Equation 3-14) into Equation 3-15 yields

$$c_{HA} = [H_3O^+] + [HA]$$

which rearranges to

$$[HA] = c_{HA} - [H_3O^+] \tag{3-16}$$

When $[A^-]$ and $[HA]$ are replaced by their equivalent terms from Equations 3-14 and 3-16, the equilibrium-constant expression becomes

$$K_a = \frac{[H_3O^+]^2}{c_{HA} - [H_3O^+]} \tag{3-17}$$

which rearranges to

$$[H_3O^+]^2 + K_a[H_3O^+] - K_a c_{HA} = 0 \tag{3-18}$$

The positive solution to this quadratic equation is

$$[H_3O^+] = \frac{-K_a + \sqrt{K_a^2 + 4 K_a c_{HA}}}{2} \tag{3-19}$$

As an alternative to using Equation 3-19, Equation 3-18 may be solved by successive approximations as shown in Feature 3-5.

Equation 3-16 frequently can be simplified by making the assumption that dissociation does not appreciably decrease the molar concentration of HA. Thus, provided $[H_3O^+] \ll c_{HA}$, $c_{HA} - [H_3O^+] \cong c_{HA}$ and Equation 3-17 reduces to

$$K_a = \frac{[H_3O^+]^2}{c_{HA}}$$

$$[H_3O^+] = \sqrt{K_a c_{HA}} \tag{3-20}$$

Table 3-4

ERROR INTRODUCED BY ASSUMING H_3O^+ CONCENTRATION IS SMALL RELATIVE TO c_{HA} IN EQUATION 3-15

K_a	c_{HA}	$[H_3O^+]$ Using Assumption	$[H_3O^+]$ Using More Exact Equation	Percent Error
1.00×10^{-2}	1.00×10^{-3}	3.16×10^{-3}	0.92×10^{-3}	244
	1.00×10^{-2}	1.00×10^{-2}	0.62×10^{-2}	61
	1.00×10^{-1}	3.16×10^{-2}	2.70×10^{-2}	17
1.00×10^{-4}	1.00×10^{-4}	1.00×10^{-4}	0.62×10^{-4}	61
	1.00×10^{-3}	3.16×10^{-4}	2.70×10^{-4}	17
	1.00×10^{-2}	1.00×10^{-3}	0.95×10^{-3}	5.3
	1.00×10^{-1}	3.16×10^{-3}	3.11×10^{-3}	1.6
1.00×10^{-6}	1.00×10^{-5}	3.16×10^{-6}	2.70×10^{-6}	17
	1.00×10^{-4}	1.00×10^{-5}	0.95×10^{-5}	5.3
	1.00×10^{-3}	3.16×10^{-5}	3.11×10^{-5}	1.6
	1.00×10^{-2}	1.00×10^{-4}	9.95×10^{-5}	0.5
	1.00×10^{-1}	3.16×10^{-4}	3.16×10^{-4}	0.0

The magnitude of the error introduced by the assumption that $[H_3O^+] \ll c_{HA}$ increases as the molar concentration of acid becomes smaller and as the dissociation constant for the acid becomes larger. This statement is supported by the data in Table 3-4. Note that the error introduced by the assumption is about 0.5% when the ratio c_{HA}/K_a is 10^4. The error increases to about 1.6% when the ratio is 10^3, to about 5% when it is 10^2, and to about 17% when it is 10. Figure 3-2 illustrates the effect graphically. Note that the hydronium ion concentration computed with the approximation becomes equal to or greater than the molar concentration of the acid when the ratio is unity or smaller, which is a meaningless result.

It is usually best to make the simplifying assumption and obtain a trial value for $[H_3O^+]$ that can be compared with c_{HA} in Equation 3-16. If the trial value alters [HA] by an amount smaller than the allowable error in the calculation, the solution is satisfactory. Otherwise, the quadratic equation must be solved to obtain a better value for $[H_3O^+]$. Alternatively, the method of successive approximations (Feature 3-5) may be used.

Example 3-7

Calculate the hydronium ion concentration in 0.120 M nitrous acid.
   The principal equilibrium is

$$HNO_2 + H_2O \rightleftharpoons H_3O^+ + NO_2^-$$

for which (Appendix 2)

$$K_a = \frac{[H_3O^+][NO_2^-]}{[HNO_2]} = 7.1 \times 10^{-4}$$

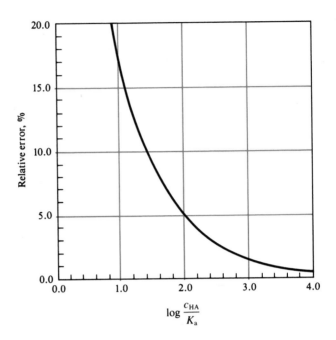

Figure 3-2
Relative error resulting from the assumption that $[H_3O^+] \ll c_{HA}$ in Equation 3-15.

Substitution into Equations 3-14 and 3-16 gives

$$[H_3O^+] = [NO_2^-]$$

$$[HNO_2] = 0.120 - [H_3O^+]$$

When these relationships are introduced into the expression for K_a, we obtain

$$\frac{[H_3O^+]^2}{0.120 - [H_3O^+]} = 7.1 \times 10^{-4}$$

If we now assume $[H_3O^+] \ll 0.120$, we find

$$[H_3O^+] = \sqrt{0.120 \times 7.1 \times 10^{-4}} = 9.2 \times 10^{-3}\ M$$

We now examine the assumption that $0.120 - 0.0092 \cong 0.120$ and see that the error is about 8%. The relative error in $[H_3O^+]$ is actually smaller than this figure, however, as we can see by calculating $\log (c_{HA}/K_a) = 2.2$, which suggests an error of about 4% (see Figure 3-2). If a more accurate figure is needed, solution of the quadratic equation yields $8.9 \times 10^{-3}\ M$ for the hydronium ion concentration.

Example 3-8

Calculate the hydronium ion concentration in a solution that is $2.0 \times 10^{-4}\ M$ in aniline hydrochloride, $C_6H_5NH_3Cl$.

In aqueous solution, dissociation of the salt to Cl^- and $C_6H_5NH_3^+$ is complete.

The weak acid $C_6H_5NH_3^+$ dissociates as follows:

$$C_6H_5NH_3^+ + H_2O \rightleftharpoons C_6H_5NH_2 + H_3O^+ \qquad K_a = \frac{[H_3O^+][C_6H_5NH_2]}{[C_6H_5NH_3^+]}$$

In Appendix 2 we find $K_a = 2.51 \times 10^{-5}$ for $C_6H_5NH_3^+$. Proceeding as in Example 3-7, we have

$$[H_3O^+] = [C_6H_5NH_2]$$
$$[C_6H_5NH_3^+] = 2.0 \times 10^{-4} - [H_3O^+]$$

Let us assume that $[H_3O^+] \ll 2.0 \times 10^{-4}$ and substitute the simplified value for $[C_6H_5NH_3^+]$ into the dissociation-constant expression:

$$\frac{[H_3O^+]^2}{2.0 \times 10^{-4}} = 2.51 \times 10^{-5}$$
$$[H_3O^+] = \sqrt{5.02 \times 10^{-9}} = 7.09 \times 10^{-5}\ M$$

Comparison of 7.09×10^{-5} with 2.0×10^{-4} suggests that a significant error has been introduced by the assumption that $[H_3O^+] \ll c_{C_6H_5NH_3^+}$ (Figure 3-2 indicates that this error is about 20%). Thus, unless only an approximate value for $[H_3O^+]$ is needed, we must use the exact expression

$$\frac{[H_3O^+]^2}{2.0 \times 10^{-4} - [H_3O^+]} = 2.51 \times 10^{-5}$$

which rearranges to

$$[H_3O^+]^2 + 2.51 \times 10^{-5}[H_3O^+] - 5.02 \times 10^{-9} = 0$$
$$[H_3O^+] = \frac{-2.51 \times 10^{-5} + \sqrt{(2.51 \times 10^{-5})^2 + 4 \times 5.02 \times 10^{-9}}}{2} = 5.94 \times 10^{-5}$$

The quadratic equation can also be solved by the iterative method shown in Feature 3-5.

Feature 3-5
THE METHOD OF SUCCESSIVE APPROXIMATIONS

For convenience let us write the quadratic equation in Example 3-8 in the form

$$x^2 + 2.51 \times 10^{-5}x - 5.02 \times 10^{-9} = 0$$

where $x = [H_3O^+]$.

As a first step, let us rearrange the equation to the form

$$x = \sqrt{5.02 \times 10^{-9} - 2.51 \times 10^{-5}x}$$

We then assume that x on the right-hand side of the equation is zero and calculate a first value, x_1.

$$x_1 = \sqrt{5.02 \times 10^{-9} - 2.51 \times 10^{-5} \times 0} = 7.09 \times 10^{-5}$$

We then substitute this value into the original equation and compute a second value, x_2. That is,

$$x_2 = \sqrt{5.02 \times 10^{-9} - 2.51 \times 10^{-5} \times 7.09 \times 10^{-5}} = 5.69 \times 10^{-5}$$

Repeating this calculation gives

$$x_3 = \sqrt{5.02 \times 10^{-9} - 2.51 \times 10^{-5} \times 5.69 \times 10^{-5}} = 5.99 \times 10^{-5}$$

Continuing in the same way we obtain

$$x_4 = 5.93 \times 10^{-5}$$
$$x_5 = 5.94 \times 10^{-5}$$
$$x_6 = 5.94 \times 10^{-5}$$

Note that after three iterations x_3 is 5.99×10^{-5}, which is within about 0.8% of the final value of 5.94×10^{-5} M.

Occasionally, the final result will oscillate between a high and a low value. In this case you may have to solve the quadratic equation.

The method of successive approximations is particularly useful when cubic- or higher-power equations need to be solved.

Hydronium Ion Concentration in Solutions of Weak Bases

The techniques discussed in previous sections are easily adapted to the calculation of the hydroxide or hydronium ion concentration in solutions of weak bases.

Aqueous ammonia is basic because of the reaction

$$NH_3 + H_2O \rightleftharpoons NH_4^+ + OH^-$$

The predominant species in such solutions has been demonstrated to be NH_3. Nevertheless, solutions of ammonia are still called ammonium hydroxide sometimes because at one time it was thought that NH_4OH rather than NH_3 was the undissociated form of the base. We write the equilibrium constant for this reaction as

$$K_b = \frac{[NH_4^+][OH^-]}{[NH_3]}$$

Example 3-9

Calculate the hydroxide ion concentration of a 0.0750 M NH_3 solution. The predominant equilibrium is

$$NH_3 + H_2O \rightleftharpoons NH_4^+ + OH^-$$

As shown on page 43,

$$K_b = \frac{[NH_4^+][OH^-]}{[NH_3]} = \frac{1.00 \times 10^{-14}}{5.70 \times 10^{-10}} = 1.75 \times 10^{-5}$$

The chemical equation shows that

$$[NH_4^+] = [OH^-]$$

Both NH_4^+ and NH_3 come from the 0.0750 M solution. Thus

$$[NH_4^+] + [NH_3] = c_{NH_3} = 0.0750 \text{ M}$$

If we substitute $[OH^-]$ for $[NH_4^+]$ and rearrange, we find that

$$[NH_3] = 0.0750 - [OH^-]$$

Substituting these quantities into the dissociation-constant expression yields

$$\frac{[OH^-]^2}{7.50 \times 10^{-2} - [OH^-]} = 1.75 \times 10^{-5}$$

Provided that $[OH^-] \ll 7.50 \times 10^{-2}$, this equation simplifies to

$$[OH^-]^2 \cong 7.50 \times 10^{-2} \times 1.75 \times 10^{-5}$$
$$[OH^-] = 1.15 \times 10^{-3} \text{ M}$$

On comparing the calculated value for $[OH^-]$ with 7.50×10^{-2}, we see that the error in $[OH^-]$ is less than 2%. If needed, a better value for $[OH^-]$ can be obtained by solving the quadratic equation.

Example 3-10

Calculate the hydroxide ion concentration in a 0.0100 M sodium hypochlorite solution.

The equilibrium between OCl^- and water is

$$OCl^- + H_2O \rightleftharpoons HOCl + OH^-$$

for which

$$K_b = \frac{[HOCl][OH^-]}{[OCl^-]}$$

Appendix 2 reveals that the acid-dissociation constant for HOCl is 3.0×10^{-8}. Therefore, we rearrange Equation 3-13 and write

$$K_b = \frac{K_w}{K_a} = \frac{1.00 \times 10^{-14}}{3.0 \times 10^{-8}} = 3.33 \times 10^{-7}$$

Proceeding as in Example 3-9, we have

$$[OH^-] = [HOCl]$$

$$[OCl^-] + [HOCl] = 0.0100$$

$$[OCl^-] = 0.0100 - [OH^-] \cong 0.0100$$

Here we have assumed that $[OH^-] \ll 0.0100$. Substitution into the equilibrium-constant expression gives

$$\frac{[OH^-]^2}{0.0100} = 3.33 \times 10^{-7}$$

$$[OH^-] = 5.8 \times 10^{-5} \, M$$

Note that the error resulting from the approximation is small.

3C   QUESTIONS AND PROBLEMS

3-1. Briefly describe or define
  *(a) a weak electrolyte.
  (b) a Brønsted-Lowry acid.
  *(c) the conjugate base of a Brønsted-Lowry acid.
  (d) neutralization, in terms of the Brønsted-Lowry concepts.
  *(e) an amphiprotic solute.
  (f) a zwitterion.
  *(g) autoprotolysis.
  (h) a strong acid.
  *(i) the Le Châtelier principle.
  (j) the common-ion effect.

3-2. Briefly describe or define
  *(a) an amphiprotic solvent.
  (b) a differentiating solvent.
  *(c) a leveling solvent.
  (d) a mass-action effect.

*3-3. Briefly explain why there is no term in an equilibrium-constant expression for water or for a pure solid, even though one or both appear in the balanced net ionic equation for the equilibrium.

3-4. Identify the acid on the left and its conjugate base on the right in the following equations:
  *(a) $HCN + H_2O \rightleftharpoons H_3O^+ + CN^-$
  (b) $HONH_2 + H_2O \rightleftharpoons HONH_3^+ + OH^-$
  *(c) $NH_4^+ + H_2O \rightleftharpoons NH_3 + H_3O^+$
  (d) $2HCO_3^- \rightleftharpoons H_2CO_3 + CO_3^{2-}$
  *(e) $PO_4^{3-} + H_2PO_4^- \rightleftharpoons 2HPO_4^{2-}$

3-5. Identify the base on the left and its conjugate acid on the right in the equations for Problem 3-4.

3-6. Write expressions for the autoprotolysis of
  *(a) H_2O.
  (b) CH_3COOH.
  *(c) CH_3NH_2.
  (d) CH_3CH_2OH.

3-7. Write equilibrium-constant expressions for
  *(a) solid AgI and its ions in solution.
  (b) solid $PbCl_2$ and its ions in solution.
  *(c) solid Ag_2CrO_4 and its ions in solution.
  (d) solid $BaCrO_4$ and its ions in solution.

3-8. Write the equilibrium-constant expressions, and obtain numerical values for each constant in
  *(a) the basic dissociation of ethylamine, $C_2H_5NH_2$.
  (b) the acidic dissociation of hydrogen cyanide, HCN.
  *(c) the acidic dissociation of pyridine hydrochloride, C_5H_5NHCl.
  (d) the basic dissociation of NaCN.
  *(e) the dissociation of H_3AsO_4 to H_3O^+ and AsO_4^{3-}.
  (f) the reaction of CO_3^{2-} with H_2O to give H_2CO_3 and OH^-.

3-9. Calculate the solubility-product constant for each of the following substances, given that the molar concentrations of their saturated solutions are as indicated:

*(a) AgSeCN (2.0×10^{-8} mol/L; products are Ag^+ and $SeCN^-$).

(b) $RaSO_4$ (6.6×10^{-6} mol/L).

*(c) $Ba(BrO_3)_2$ (9.2×10^{-3} mol/L).

(d) PbF_2 (1.9×10^{-3} mol/L).

*(e) $Ce(IO_3)_3$ (1.9×10^{-3} mol/L).

(f) BiI_3 (1.3×10^{-5} mol/L).

3-10. Calculate the solubility of the solutes in Problem 3-9 for solutions in which the cation concentration is 0.050 M.

3-11. Calculate the solubility of the solutes in Problem 3-9 for solutions in which the anion concentration is 0.050 M.

*3-12. The solubility product for Tl_2CrO_4 is 9.8×10^{-13}. What CrO_4^{2-} concentration is required to

(a) initiate precipitation of Tl_2CrO_4 from a solution that is 2.12×10^{-3} M in Tl^+?

(b) lower the concentration of Tl^+ in a solution to 1.00×10^{-6} M?

3-13. What hydroxide concentration is required to

(a) initiate precipitation of Fe^{3+} from a 1.00×10^{-3} M solution of $Fe_2(SO_4)_3$?

(b) lower the Fe^{3+} concentration in the foregoing solution to 1.00×10^{-9} M?

*3-14. The solubility-product constant for $Ce(IO_3)_3$ is 3.2×10^{-10}. What is the Ce^{3+} concentration in a solution prepared by mixing 50.0 mL of 0.0500 M Ce^{3+} with 50.00 mL of

(a) water?

(b) 0.050 M IO_3^-?

(c) 0.150 M IO_3^-?

(d) 0.300 M IO_3^-?

3-15. The solubility-product constant for K_2PtCl_6 is 1.1×10^{-5} ($K_2PtCl_6 \rightleftharpoons 2K^+ + PtCl_6^{2-}$). What is the K^+ concentration of a solution prepared by mixing 50.0 mL of 0.400 M KCl with 50.0 mL of

(a) 0.100 M $PtCl_6^{2-}$?

(b) 0.200 M $PtCl_6^{2-}$?

(c) 0.400 M $PtCl_6^{2-}$?

*3-16. The solubility products for a series of iodides are

TlI $K_{sp} = 6.5 \times 10^{-8}$

AgI $K_{sp} = 8.3 \times 10^{-17}$

PbI_2 $K_{sp} = 7.1 \times 10^{-9}$

BiI_3 $K_{sp} = 8.1 \times 10^{-19}$

List these four compounds in order of decreasing molar solubility in

(a) water.

(b) 0.10 M NaI.

(c) a 0.010 M solution of the solute cation.

3-17. The solubility products for a series of hydroxides are

BiOOH $K_{sp} = 4.0 \times 10^{-10} = [BiO^+][OH^-]$

$Be(OH)_2$ $K_{sp} = 7.0 \times 10^{-22}$

$Tm(OH)_3$ $K_{sp} = 3.0 \times 10^{-24}$

$Hf(OH)_4$ $K_{sp} = 4.0 \times 10^{-26}$

Which hydroxide has

(a) the lowest molar solubility in H_2O?

(b) the lowest molar solubility in a solution that is 0.10 M in NaOH?

3-18. Calculate the hydronium ion concentration of water at 100°C.

3-19. At 25°C, what are the molar H_3O^+ and OH^- concentrations in

*(a) 0.0200 M HOCl?

(b) 0.0800 M propanoic acid?

*(c) 0.200 M methylamine?

(d) 0.100 M trimethylamine?

*(e) 0.120 M NaOCl?

(f) 0.0860 M CH_3COONa?

*(g) 0.100 M hydroxylamine hydrochloride?

(h) 0.0500 M ethanolamine hydrochloride?

3-20. At 25°C, what is the hydronium ion concentration in

*(a) 0.100 M chloroacetic acid?

*(b) 0.100 M sodium chloroacetate?

(c) 0.0100 M methylamine?

(d) 0.0100 M methylamine hydrochloride?

*(e) 1.00×10^{-3} M aniline hydrochloride?

(f) 0.200 M HIO_3?

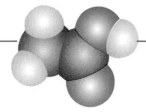

CHAPTER 4

ERRORS IN CHEMICAL ANALYSIS

Parts per million (ppm), that is, 20.00 parts of iron(III) per million parts of solution.

It is impossible to perform a chemical analysis in such a way that the results are totally free of errors, or uncertainties. All one can hope is to minimize these errors and to estimate their size with acceptable accuracy. In this and the next chapter, we explore the nature of experimental errors and their effects on the results of chemical analyses.

The effect of errors in analytical data is illustrated in Figure 4-1, which shows results for the quantitative determination of iron(III). Six equal portions of an aqueous solution known to contain exactly 20.00 ppm of iron(III) were analyzed in exactly the same way. Note that the results range from a low of 19.4 ppm to a high of 20.3 ppm of iron(III). The average $\bar{x}$ of the data is 19.8 ppm.

The true value of a measurement is never known exactly.

Every measurement is influenced by many uncertainties, which combine to produce a scatter of results like that shown in Figure 4-1. Measurement uncertainties can never be completely eliminated, so the true value for any quantity is always unknown. Often the probable magnitude of the error in a measurement can be evaluated, however. It is then possible to define limits within which the true value of a measured quantity lies at a given level of probability.

It is seldom easy to estimate the precision and accuracy of experimental data. Nevertheless, we must make such estimates whenever we collect laboratory results because *data of unknown precision and accuracy are worthless*. On the other hand, results that are not especially accurate may be of considerable value if the limits of uncertainty are known.

One of the first questions to answer before beginning an analysis is, ''What is the maximum error that I can tolerate in the result?'' The answer to this question determines the time required to do the work. For example, a tenfold increase in accuracy may take hours, days, or even weeks of added labor. *No one can afford to waste time generating data that are more reliable than is needed.*

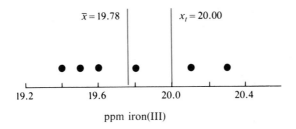

Figure 4-1
Results from six replicate determinations for iron in aqueous samples of a standard solution containing 20.00 ppm of iron(III).

4A DEFINITION OF TERMS

Chemists usually carry two to five portions (*replicates*) of a sample through an entire analytical procedure. Individual results from a set of measurements are seldom the same (Figure 4-1), so a central or ''best'' value is used for the set. We justify the extra effort required to analyze several samples in two ways. First, the central value of a set should be more reliable than any of the individual results. Second, variation in the data should provide a measure of the uncertainty associated with the central result. Either the *mean* or the *median* may serve as the central value for a set of replicated measurements.

4A-1 The Mean and Median

Mean, *arithmetic mean*, and *average* ($\bar{x}$) are synonyms for the quantity obtained by dividing the sum of replicate measurements by the number of measurements in the set:

$$\bar{x} = \frac{\sum_{i=1}^{N} x_i}{N}$$ (4-1)

The symbol Σx_i means to add all of the values x_i for the replicates.

where x_i represents the individual values of x making up a set of N replicate measurements.

The *median* is the middle result when replicate data are arranged in order of size. There are equal numbers of data that are larger and smaller than the median. For an odd number of results, the median can be evaluated directly by listing the results in increasing or decreasing order and choosing the central value. For an even number, the mean of the middle pair is used.

> The *median* is the middle value in a set of data that has been arranged in order of size. The median is used advantageously when a set of data contains an *outlier*—that is, a result that differs significantly from the rest of the data in the set. An outlier can have a significant effect on the mean of the set but has no effect on the median.

Example 4-1

Calculate the mean and median for the data shown in Figure 4-1.

$$\text{mean} = \bar{x} = \frac{19.4 + 19.5 + 19.6 + 19.8 + 20.1 + 20.3}{6} = 19.78 \approx 19.8 \text{ ppm Fe}$$

Because the set contains an even number of measurements, the median is the average of the central pair:

$$\text{median} = \frac{19.6 + 19.8}{2} = 19.7 \text{ ppm Fe}$$

Ideally, the mean and median are identical. Often they are not, however, particularly when the number of measurements in the set is small.

4A-2 Precision

Precision describes the reproducibility of measurements—that is, the closeness of results that have been obtained *in exactly the same way*. Generally, the precision of a measurement is determined by simply repeating the measurement.

Three terms are widely used to describe the precision of a set of replicate data: *standard deviation*, *variance*, and *coefficient of variation*. All of these terms are a function of the *deviation of the data from the mean*, which is defined as

$$\text{deviation from the mean} = d_i = |x_i - \bar{x}| \qquad \textbf{(4-2)}$$

> Deviations from the mean are always taken as positive.

The relationship between deviation from the mean and the three precision terms is given in Section 4C.

4A-3 Accuracy

Figure 4-2 illustrates the difference between accuracy and precision. *Accuracy* indicates the closeness of the measurement to its true or accepted value and is expressed by the *error*. Note the basic difference between accuracy and precision. Accuracy measures agreement between a result and its true value. Precision describes the agreement among several results that have been measured in the same way. Precision is determined by simply replicating a measurement. On the other hand, accuracy can never be determined exactly because the true value of a quantity can never be known exactly. An accepted value must be used instead. Accuracy is expressed in terms of either absolute or relative error.

Absolute Error

The *absolute error E* in the measurement of a quantity x_i is given by the equation

$$E = x_i - x_t \qquad \textbf{(4-3)}$$

> The term *absolute* has a different meaning here than it does in mathematics. An absolute value in mathematics means the magnitude of a number *ignoring its sign*. As we shall use it, the absolute error is the difference between an accepted value and an experimental result *including its sign*.

where x_t is the true, or accepted, value of the quantity. Returning to the data displayed in Figure 4-1, the absolute error of the result immediately to the left of the true value of 20.00 ppm is -0.2 ppm Fe; the result at 20.10 ppm is in error by $+0.1$ ppm Fe. Note that we retain the sign in stating the error. Thus, the negative sign in the first case shows that the experimental result is smaller than the accepted value.

Relative Error

Often, the *relative error* E_r is a more useful quantity than the absolute error. The percent relative error is given by the expression

$$E_r = \frac{x_i - x_t}{x_t} \times 100\% \qquad (4\text{-}4)$$

Relative error is also expressed in parts per thousand (ppt). So, the relative error for the mean of the data in Figure 4-1 is

$$E_r = \frac{19.8 - 20.00}{20.00} \times 100\% = -1\% \text{ or } -10 \text{ ppt}$$

4A-4 Types of Errors in Experimental Data

The precision of a measurement is readily determined by comparing data from carefully replicated experiments. Unfortunately, an estimate of the accuracy is not so easy to obtain. To determine the accuracy we have to know the true value, and ordinarily this is exactly what we are looking for.

It is tempting to assume that if we know the answer precisely, then we also know it accurately. The danger of this assumption is illustrated in Figure 4-3, which summarizes the results for the determination of nitrogen in two pure compounds. The dots show the absolute errors of replicate results obtained by four analysts. Note that Analyst 1 obtained relatively high precision and high accuracy. Analyst 2 had poor precision but good accuracy. The results of Analyst 3 are surprisingly common. The precision is excellent, but there is significant error in the numerical average for the data. Both the precision and the accuracy are poor for the results of Analyst 4.

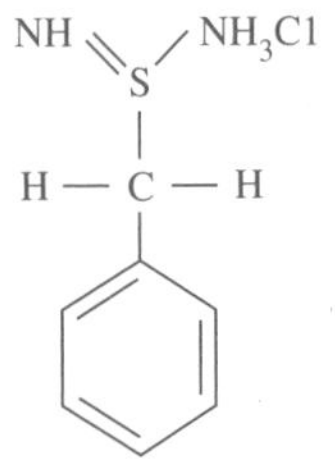

Figures 4-1 and 4-3 suggest that chemical analyses are affected by at least two types of errors. One type, called *random* (or *indeterminate*) *error*, causes data to be scattered more or less symmetrically around a mean value. Refer again to Figure 4-3, and notice that the scatter in the data, and thus the random error, for Analysts 1 and 3 is significantly less than that for Analysts 2 and 4. In general, then, the random error in a measurement is reflected by its precision.

A second type of error, called *systematic* (or *determinate*) *error*, causes the mean of a set of data to differ from the accepted value. For example, the mean of the data in Figure 4-1 has a systematic error of about -0.2 ppm Fe. The results of Analysts 1 and 2 in Figure 4-3 have little systematic error, but the data of Analysts 3 and 4 show determinate errors of about -0.7 and -1.2% nitrogen. In general, a systematic error causes the results in a series of replicate measurements to be all high or all low.

A third type of error is *gross error*. Gross errors differ from indeterminate and determinate errors. They usually occur only occasionally, are often large, and may cause a result to be either high or low. Gross errors lead to *outliers*—results that appear to differ markedly from all other data in a set of replicate measurements. There is no evidence of a gross error in Figures 4-1 and 4-3. Had

Low accuracy, low precision

High accuracy, low precision

Low accuracy, high precision

High accuracy, high precision

Figure 4-2
Accuracy and precision.

benzyl isothiourea hydrochloride

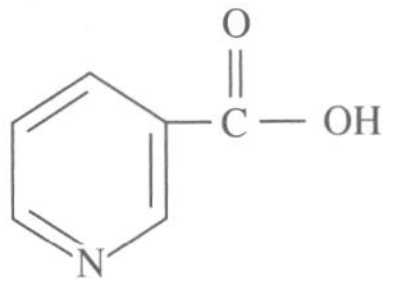

nicotinic acid

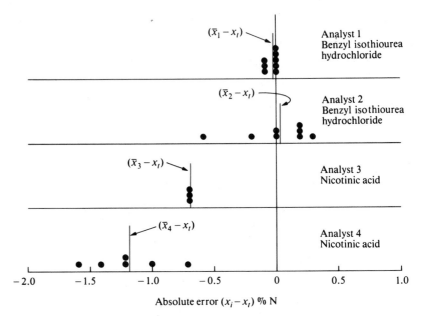

Figure 4-3
Absolute error in the micro-Kjeldahl determination of nitrogen. Each dot represents the error associated with a single determination. Each vertical line labeled $(\bar{x}_i - x_t)$ is the absolute average deviation of the set from the true value. (Data from C. O. Willits and C. L. Ogg, *J. Assoc. Anal. Chem.*, **1949**, *32*, 561. With permission.)

one of the results shown in Figure 4-1 occurred at 21.9 ppm Fe, it might have been an outlier.

4B   RANDOM ERRORS

Random, or *indeterminate*, errors arise when a system of measurement is extended to its maximum sensitivity. This type of error is caused by the many uncontrollable variables that are an inevitable part of every physical or chemical measurement. There are many contributors to random error, but none can be positively identified or measured because most are so small that they cannot be detected individually. The accumulated effect of the individual indeterminate uncertainties, however, causes the data from a set of replicate measurements to fluctuate randomly around the mean of the set. For example, the scatter of data in Figures 4-1 and 4-3 is a direct result of the accumulation of small random uncertainties. Notice that in Figure 4-3 the random error in the results of Analysts 2 and 4 is greater than in the results of Analysts 1 and 3.

4B-1   Sources of Random Errors

We can get a qualitative idea of the way small undetectable uncertainties produce a detectable random error in the following way. Imagine a situation in which just four small random errors combine to give an overall error. We will assume that each error has an equal probability of occurring and that each can cause the final result to be high or low by a fixed amount $\pm U$.

Table 4-1 shows all the possible ways the four errors can combine to give the indicated deviations from the mean value. Note that only one combination leads to a deviation of $+4U$, four combinations give a deviation of $+2U$, and six give

Table 4-1
POSSIBLE COMBINATIONS OF FOUR EQUAL-SIZED UNCERTAINTIES

Combinations of Uncertainties	Magnitude of Indeterminate Error	Number of Combinations	Relative Frequency
$+U_1 + U_2 + U_3 + U_4$	$+4U$	1	$1/16 = 0.0625$
$-U_1 + U_2 + U_3 + U_4$ $+U_1 - U_2 + U_3 + U_4$ $+U_1 + U_2 - U_3 + U_4$ $+U_1 + U_2 + U_3 - U_4$	$+2U$	4	$4/16 = 0.250$
$-U_1 - U_2 + U_3 + U_4$ $+U_1 + U_2 - U_3 - U_4$ $+U_1 - U_2 + U_3 - U_4$ $-U_1 + U_2 - U_3 + U_4$ $-U_1 + U_2 + U_3 - U_4$ $+U_1 - U_2 - U_3 + U_4$	0	6	$6/16 = 0.375$
$+U_1 - U_2 - U_3 - U_4$ $-U_1 + U_2 - U_3 - U_4$ $-U_1 - U_2 + U_3 - U_4$ $-U_1 - U_2 - U_3 + U_4$	$-2U$	4	$4/16 = 0.250$
$-U_1 - U_2 - U_3 - U_4$	$-4U$	1	$1/16 = 0.0625$

a deviation of $0U$. The negative errors have the same relationship. This ratio of $1:4:6:4:1$ is a measure of the probability of a deviation of each magnitude. Therefore, if we make a sufficiently large number of measurements, we can expect a frequency distribution like that shown in Figure 4-4a. Note that the ordinate in the plot is the relative frequency of occurrence of the five possible combinations.

Figure 4-4b shows the theoretical distribution for ten equal-sized uncertainties. Again we see that the most frequent occurrence is zero deviation from the mean. At the other extreme a maximum deviation of $10U$ occurs only about once in 500 measurements.

When the same procedure is applied to a very large number of individual errors, a bell-shaped curve like that shown in Figure 4-4c results. Such a plot is called a *Gaussian curve* or a *normal error curve*.

In our example, all the uncertainties have the same magnitude. This restriction is not necessary to derive the equation for a Gaussian curve.

4B-2 Distribution of Experimental Data

We find empirically that the distribution of replicate data from most quantitative analytical experiments approaches that of the Gaussian curve shown in Figure 4-4c. As an example, consider the data in Table 4-2 for the calibration of a 10-mL pipet. In this experiment a small flask and stopper are weighed. Ten milliliters of water are transferred to the flask with the pipet and the flask stoppered. Finally, the flask, the stopper, and the water are weighed again. The temperature of the water is also measured to establish its density. The mass of the water is then calculated by taking the difference between the two masses; this difference is

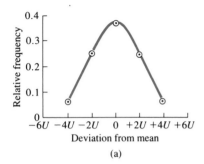

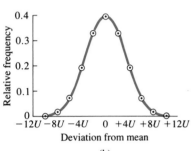

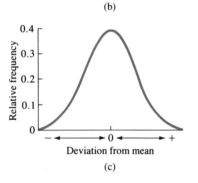

(a)

(b)

(c)

Figure 4-4

Frequency distribution for measurements containing (a) four indeterminate uncertainties; (b) ten indeterminate uncertainties; (c) a very large number of indeterminate uncertainties.

divided by the density of the water to find the volume delivered by the pipet. The experiment was repeated 50 times.

The data in Table 4-2 are typical of those obtained by an experienced worker weighing to the nearest milligram (which corresponds to 0.001 mL) on a top-loading balance and making every effort to avoid systematic error. Even so, the results vary from a low of 9.969 mL to a high of 9.994 mL. This 0.025 mL *spread*

Feature 4-1

FLIPPING COINS: A STUDENT ACTIVITY TO ILLUSTRATE A NORMAL DISTRIBUTION

If you flip a coin ten times, how many heads will you get? Try it, and record your result. Repeat the experiment. Are your results the same? Ask friends or members of your class to perform the same experiment and tabulate the results. The table below contains the results obtained by several classes of analytical chemistry students over the period from 1980 to 1993.

No. Heads	0	1	2	3	4	5	6	7	8	9	10
Frequency	1	1	20	40	93	91	78	43	20	7	1

Add your results to those in the table, and plot a histogram similar to the one shown below. Find the mean and standard deviation of your results and compare them to the values shown in the plot. The smooth curve in Figure 4-A is a normal curve for an infinite number of trials with the same mean and standard deviation as the data. Note that the mean of 5.04 is very close to the value of 5 that you would predict based on the laws of probability. As the number of trials increases, the histogram approaches the shape of the smooth curve and the mean approaches five.

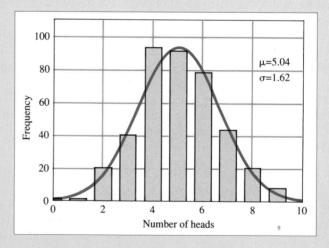

Figure 4-A

Results of a coin-flipping experiment by 395 students over a 13-year period.

Table 4-2

REPLICATE DATA ON THE CALIBRATION OF A
10-mL PIPET*

Trial	Volume, mL	Trial	Volume, mL	Trial	Volume, mL
1	9.988	18	9.975‡	35	9.976
2	9.973	19	9.980‡	36	9.990
3	9.986	20	9.994‡	37	9.988
4	9.980	21	9.992‡	38	9.971
5	9.975	22	9.984‡	39	9.986
6	9.982	23	9.981‡	40	9.978
7	9.986	24	9.987‡	41	9.986
8	9.982	25	9.978‡	42	9.982
9	9.981	26	9.983‡	43	9.977
10	9.990	27	9.982‡	44	9.977
11	9.980	28	9.991‡	45	9.986
12	9.989	29	9.981‡	46	9.978
13	9.978	30	9.969†	47	9.983
14	9.971	31	9.985‡	48	9.980
15	9.982	32	9.977‡	49	9.983
16	9.983	33	9.976‡	50	9.979
17	9.988	34	9.983‡		

Mean volume = 9.982 mL

Median volume = 9.982 mL

Spread = 0.025 mL

Standard deviation = 0.0056 mL

* Data listed in the order obtained.
† Minimum value.
‡ Maximum value.

of data results directly from an accumulation of all of the random uncertainties in
the experiment.

The information in Table 4-2 is easier to visualize when the data are rearranged
into frequency distribution groups, as in Table 4-3. Here, we tabulate the number
of data falling into a series of adjacent 0.003-mL *cells* and calculate the percentage
of measurements falling into each cell. Note that 26% of the data reside in the
cell containing the mean and median value of 9.982 mL and that more than half
the data are within ±0.004 mL of this mean.

The frequency distribution data in Table 4-3 are plotted as a bar graph, or
histogram (labeled *A* in Figure 4-5). We can imagine that as the number of
measurements increases, the histogram would approach the shape of the continuous curve shown as plot *B* in Figure 4-5. This curve is a Gaussian curve derived
for an infinite set of data. These data have the same mean (9.982 mL), the same
precision, and the same area under the curve as the histogram.

Variations in replicate results such as those in Table 4-2 result from numerous
small and individually undetectable random errors that are attributable to uncon-

A *histogram* is a bar graph such as that
shown by plot *A* in Figure 4-5.

Table 4-3
FREQUENCY DISTRIBUTION
OF DATA FROM TABLE 4-2

Volume Range, mL	Number in Range	% in Range
9.969 to 9.971	3	6
9.972 to 9.974	1	2
9.975 to 9.977	7	14
9.978 to 9.980	9	18
9.981 to 9.983	13	26
9.984 to 9.986	7	14
9.987 to 9.989	5	10
9.990 to 9.992	4	8
9.993 to 9.995	1	2

trollable variables in the experiment. Such small errors ordinarily tend to cancel one another and thus have a minimal effect. However, occasionally they occur in the same direction to produce a large positive or negative net error.

Sources of random uncertainties in the calibration of a pipet include: (1) visual judgments, such as the level of the water with respect to the marking on the pipet and the mercury level in the thermometer; (2) variations in the drainage time and in the angle of the pipet as it drains; (3) temperature fluctuations, which affect the volume of the pipet, the viscosity of the liquid, and the performance of the balance; and (4) vibrations and drafts that cause small variations in the balance readings. Undoubtedly, numerous other sources of random uncertainty also operate in this calibration process. Thus, many small and uncontrollable variables affect even as simple a process as calibrating a pipet. It is difficult or impossible to determine the influence of any one of the random errors arising from these variables, but their cumulative effect is responsible for the scatter of data around the mean.

4C THE STATISTICAL TREATMENT OF RANDOM ERROR

The random, or indeterminate, errors in the results of an analysis can be evaluated by the methods of statistics. Ordinarily, statistical analysis of analytical data is based on the assumption that random errors in an analysis follow a Gaussian, or normal, distribution such as that illustrated in curve *B* in Figure 4-5. Sometimes analytical data depart seriously from Gaussian behavior, but not often. Thus, we will base this discussion entirely on normally distributed random errors.

4C-1 The Sample and the Population

Do not confuse the *statistical sample* with the *analytical sample*. Four analytical samples analyzed in the laboratory represent a single statistical sample. This is an unfortunate duplication of the term *sample*.

In statistics, a finite number of experimental observations is called a *sample* of data. The sample is treated as a tiny fraction of an infinite number of observations that could in principle be made given infinite time. Statisticians call the theoretical infinite number of data a *population*, or a *universe*, of data. Statistical laws have been derived assuming a population of data; often they must be modified substantially when applied to a small sample because a few data may not be representative of the population. In the discussion that follows, we shall first

Figure 4-5

A histogram (*A*) showing distribution of the 50 results in Table 4-3 and a Gaussian curve (*B*) for data having the same mean and same standard deviation as the data in the histogram.

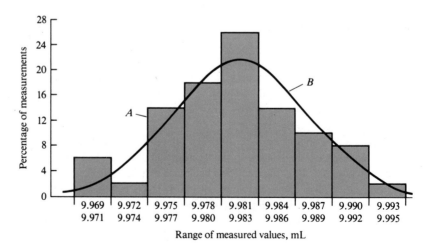

describe Gaussian statistics of populations. Then we will show how these relationships can be modified and applied to small samples of data.

4C-2 Properties of a Gaussian Curve

Figure 4-6a shows two Gaussian curves in which the relative frequency y of occurrence of various deviations from the mean is plotted as a function of deviation from the mean. It can be shown that curves such as these can be described by an equation that contains just two parameters, the *population mean* μ and the *population standard deviation* σ.

An equation for a Gaussian curve can take the form

$$y = \frac{1}{\sigma\sqrt{2\pi}} e^{-(x-\mu)^2/2\sigma^2}$$

The Population Mean μ and the Sample Mean $\bar{x}$

Statisticians find it useful to differentiate between a *sample mean* and a *population mean*. The former is the mean of a limited sample drawn from a population of data. It is defined by Equation 4-1, when N is a small number. In contrast, the population mean is the true mean for the population. It is also defined by Equation 4-1, with the added provision that N is so large that it approaches infinity. *In the absence of systematic error, the population mean is also the true value for the*

In the absence of determinate error, μ is the true value of a measured quantity.

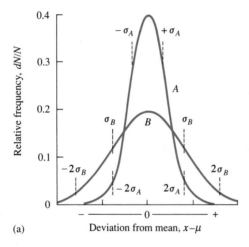

(a)

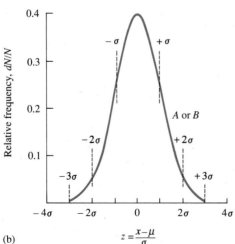

(b)

Figure 4-6

Normal error curves. The standard deviation for B is twice that for A, that is, $\sigma_B = 2\sigma_A$. (a) The abscissa is the deviation from the mean in the units of measurement. (b) The abscissa is the deviation from the mean in units of σ. Thus, A and B produce identical curves.

$$\bar{x} = \frac{\sum\limits_{i=1}^{N} x_i}{N} \text{ when } N \text{ is small}$$

$$\mu = \frac{\sum\limits_{i=1}^{N} x_i}{N} \text{ when } N \rightarrow \infty$$

The quantity $(x_i - \mu)$ in Equation 4-5 is the deviation of individual results from the mean of population of data; compare with Equation 4-2, which is for a sample of data.

measured quantity. To emphasize the difference between the two means, the sample mean is symbolized by $\bar{x}$ and the population mean by μ. More often than not, particularly when N is small, $\bar{x}$ differs from μ because a small sample of data does not exactly represent its population. The probable difference between $\bar{x}$ and μ decreases rapidly as the number of measurements making up the sample increases; ordinarily, by the time N reaches 20 to 30, this difference is negligible.

The Population Standard Deviation (σ)

The *population standard deviation* σ, which is a measure of the *precision* of a population of data, is given by the equation

$$\sigma = \sqrt{\frac{\sum\limits_{i=1}^{N} (x_i - \mu)^2}{N}} \tag{4-5}$$

where N is the number of replicate data making up the population.

The two curves in Figure 4-6a are for two populations of data that differ only in their standard deviations. The standard deviation for the data set yielding the broader but lower curve B is twice that for the measurements yielding curve A. The breadth of these curves is a measure of the precision of the two sets of data. Thus, the precision of the data leading to curve A is twice as good as that of the data that are represented by curve B.

Figure 4-6 shows another type of normal error curve in which the abscissa is now a new variable z, which is defined as

$$z = \frac{(x - \mu)}{\sigma} \tag{4-6}$$

Note that z is the deviation of the mean of a datum stated *in units of standard deviation.* That is, when $x - \mu = \sigma$, z is equal to one standard deviation; when $x - \mu = 2\sigma$, z is equal to two standard deviations; and so forth. Since z is the deviation of the mean in standard deviation units, a plot of relative frequency versus this parameter yields a single Gaussian curve that describes all populations of data regardless of standard deviation. Thus, Figure 4-6b is the normal error curve for both sets of data used to plot curves A and B in Figure 4-6a.

A normal error curve has several general properties: (1) The mean occurs at the central point of maximum frequency; (2) there is a symmetrical distribution of positive and negative deviations about the maximum; and (3) there is an exponential decrease in frequency as the magnitude of the deviations increases. Thus, small random uncertainties are observed much more often than very large ones.

Areas Under a Gaussian Curve

Regardless of its width, it can be shown that 68.3% of the area beneath a Gaussian curve for a population of data lies within one standard deviation ($\pm 1\sigma$) of the mean μ. Thus, 68.3% of the data making up the population lie within these bounds. Furthermore, approximately 95.5% of all data are within $\pm 2\sigma$ of the mean and 99.7% within $\pm 3\sigma$. In Figure 4-6 the vertical dashed lines show the areas bounded by $\pm 1\sigma$ and $\pm 2\sigma$.

Because of area relationships such as these, the standard deviation of a population of data is a useful predictive tool. For example, we can say that the chances are 68.3 in 100 that the random uncertainty of any single measurement is no more than $\pm 1\sigma$. Similarly, the chances are 95.5 in 100 that the error is less than $\pm 2\sigma$, and so forth.

4C-3 The Sample Standard Deviation as a Measure of Precision

Equation 4-5 must be modified when it is applied to a small sample of data. Thus, the *sample standard deviation s* is given by the equation

$$s = \sqrt{\frac{\sum_{i=1}^{N} (x_i - \bar{x})^2}{N - 1}} \qquad \textbf{(4-7)}$$

Note that Equation 4-7 differs from Equation 4-5 in two ways. First, the sample mean, $\bar{x}$, appears in the numerator of Equation 4-7 in place of the population mean, μ. Second, N in Equation 4-5 is replaced by the *number of degrees of freedom* $(N - 1)$. If this substitution is not used, the calculated s will on the average be less than the true standard deviation σ; that is, s will have a negative bias (see Feature 4-2).

Equation 4-7 applies to small sets of data. It says, "Find the deviations from the mean, square them, sum them, divide the sum by $N - 1$, and take the square root." The quantity $N - 1$ is called *the number of degrees of freedom*. Many scientific calculators have the standard deviation function built in.

An Alternative Expression for Sample Standard Deviation

To determine s with a calculator that does not have a standard deviation key, the following rearrangement of Equation 4-7 is easier to use:

$$s = \sqrt{\frac{\sum_{i=1}^{N} x_i^2 - \frac{\left(\sum_{i=1}^{N} x_i\right)^2}{N}}{N - 1}} \qquad \textbf{(4-8)}$$

Feature 4-2
THE SIGNIFICANCE OF NUMBER OF DEGREES OF FREEDOM

The number of degrees of freedom indicates the number of *independent* data that go into the computation of a standard deviation. Thus, when μ is unknown, two quantities must be extracted from a set of replicate data: $\bar{x}$ and s. One degree of freedom is used to calculate $\bar{x}$ because, with their signs retained, the sum of the individual deviations must add up to zero. Thus, when $N - 1$ deviations have been computed, the final one is known. Consequently, only $N - 1$ deviations provide an *independent* measure of the precision of the set. Failure to use $N - 1$ in calculating the standard deviation for small samples results in values of s that are on the average smaller than the true standard deviation σ.

Example 4-2

The following results were obtained in the replicate analysis of a blood sample for its lead content: 0.752, 0.756, 0.752, 0.751, and 0.760 ppm Pb. Calculate the mean and the standard deviation of this set of data.

To apply Equation 4-8, we calculate Σx_i^2 and $(\Sigma x_i)^2/N$.

Sample	x_i	x_i^2
1	0.752	0.565504
2	0.756	0.571536
3	0.752	0.565504
4	0.751	0.564001
5	0.760	0.577600

$$\Sigma x_i = 3.771 \quad \Sigma x_i^2 = 2.844145$$

$$\bar{x} = 3.771/5 = 0.7542 \approx 0.754$$

Note in Example 4-2 that $(\Sigma x_i)^2$ and Σx_i^2 are quite different numerically.

$$\frac{(\Sigma x_i)^2}{N} = \frac{(3.771)^2}{5} = 2.8440882$$

Substituting into Equation 4-8 leads to

$$s = \sqrt{\frac{2.844145 - 2.8440882}{5 - 1}} = \sqrt{\frac{0.0000568}{4}} = 0.00377 \approx 0.004$$

Note in Example 4-2 that the difference between Σx_i^2 and $(\Sigma x_i)^2/N$ is very small. If we had rounded these numbers off before subtracting them, a serious error would have appeared in the computed value of s. To avoid this source of error, *never round a standard deviation calculation until the very end.* Furthermore, for this same reason, never use Equation 4-8 to calculate the standard deviation of numbers containing five or more digits. Use Equation 4-7 instead.[1] Note also that hand-held calculators and small computers with a standard deviation function usually employ a version of Equation 4-8. So expect large errors in s when these devices are used to calculate the standard deviation of data that have five or more significant figures.

Any time we subtract two large, approximately equal numbers, the difference will usually have a relatively large uncertainty.

You should always keep in mind when you make statistical calculations that because of the uncertainty in $\bar{x}$, a sample standard deviation may differ significantly from the population standard deviation.

When $N \to \infty$, $\bar{x} \to \mu$ and $s \to \sigma$.

Standard Error of a Mean

The figures on percentage distribution just quoted refer to the probable error for a *single* measurement. If a series of replicate samples, each containing N data, are taken randomly from a population of data, the mean of each set will show

---

[1] In most cases, the first two or three digits in a set of data are identical to each other. As an alternative to using Equation 4-7, these identical digits can be dropped and the remaining digits used with Equation 4-8. For example, the standard deviation for the data in Example 4-2 could be based on 0.052, 0.056, 0.052, and so forth or even 52, 56, 52, and so forth.

less and less scatter as N increases. The standard deviation of each mean is known as the *standard error* of the mean and is given the symbol σ_m. It can be shown that the standard error is inversely proportional to the square root of the number of data N used to calculate the mean:

$$\sigma_m = \sigma/\sqrt{N} \qquad \textbf{(4-9)}$$

4C-4 The Reliability of s as a Measure of Precision

Most of the statistical tests we describe in Chapter 4 are based on sample standard deviations, and the probability of correctness of the results of these tests improves as the reliability of s becomes greater. Uncertainty in the calculated value of s decreases as N in Equation 4-8 increases. When N is greater than about 20, s and σ can be assumed to be approximately equal for all practical purposes. For example, if the 50 measurements in Table 4-2 are divided into 10 subgroups of 5 each, the value of s varies widely from one subgroup to another (0.0023 to 0.0079 mL) even though the average of the computed values of s is that of the entire set (0.0056 mL). In contrast, the computed values of s for two subsets of 25 each are nearly identical (0.0054 and 0.0058 mL).

> **Challenge:** Show that these statements are correct.
>
> When $N > 20$, $s \cong \sigma$.

The rapid improvement in the reliability of s with increases in N makes it feasible to obtain a good approximation of σ when the method of measurement is not excessively time-consuming and when an adequate supply of sample is available. For example, if the pH of numerous solutions is to be measured in the course of an investigation, it is useful to evaluate s in a series of preliminary experiments. This measurement is quick and requires only that a pair of rinsed and dried electrodes be immersed in the test solution and the pH read from a scale or a display. To determine s, 20 to 30 portions of a buffer solution of fixed pH can be measured with all steps of the procedure being followed exactly. Normally, it is safe to assume that the random error in this test is the same as that in subsequent measurements. The value of s calculated from Equation 4-5 is a good estimate of the theoretical σ.

Pooling Data to Improve the Reliability of s

For analyses that are time-consuming, the foregoing procedure is seldom practical. In this situation, data from a series of similar samples accumulated over time can often be pooled to provide an estimate of s that is superior to the value for any individual subset. Again, we must assume the same sources of random error in all the measurements. This assumption is usually valid if the samples have similar compositions and have been analyzed in exactly the same way.

To obtain a pooled estimate of the standard deviation, s_{pooled}, deviations from the mean for each subset are squared; the squares of all subsets are then summed and divided by an appropriate number of degrees of freedom. The pooled s is obtained by extracting the square root of the quotient. One degree of freedom is lost for each subset. Thus, the number of degrees of freedom for the pooled s is equal to the total number of measurements minus the number of subsets. Example 4-3 illustrates this type of computation.

Example 4-3

The mercury in samples of seven fish taken from Chesapeake Bay was determined by a method based on the absorption of radiation by gaseous elemental mercury (see Feature 24-1). Calculate a pooled estimate of the standard deviation for the method, based on the first three columns of data:

Specimen	Number of Samples Measured	Hg Content, ppm	Mean, ppm Hg	Sum of Squares of Deviations from Mean
1	3	1.80, 1.58, 1.64	1.673	0.0258
2	4	0.96, 0.98, 1.02, 1.10	1.015	0.0115
3	2	3.13, 3.35	3.240	0.0242
4	6	2.06, 1.93, 2.12, 2.16, 1.89, 1.95	2.018	0.0611
5	4	0.57, 0.58, 0.64, 0.49	0.570	0.0114
6	5	2.35, 2.44, 2.70, 2.48, 2.44	2.482	0.0685
7	4	1.11, 1.15, 1.22, 1.04	1.130	0.0170

Number of measurements = 28   Sum of squares = 0.2196

The values in the last two columns for specimen 1 were computed as follows:

x_i	$|(x_i - \bar{x})|$	$(x_i - \bar{x})^2$
1.80	0.127	0.0161
1.58	0.093	0.0086
1.64	0.033	0.0011

5.02 Sum of squares = 0.0258

$$\bar{x} = \frac{5.02}{3} = 1.673$$

The other data in columns 4 and 5 were obtained in a similar way. So

$$S_{\text{pooled}} = \sqrt{\frac{0.0258 + 0.0115 + 0.0242 + 0.0611 + 0.0114 + 0.0685 + 0.0170}{28 - 7}}$$

$$= 0.10 \text{ ppm Hg}$$

Note that one degree of freedom is lost for each of the seven samples. Because more than 20 degrees of freedom remain, however, the computed value of s is a good approximation of σ; that is, $s \rightarrow \sigma = 0.10$ ppm Hg.

Feature 4-3
EQUATION FOR CALCULATING POOLED STANDARD DEVIATIONS

The equation for computing a pooled standard deviation from several sets of data takes the form

$$s_{pooled} = \sqrt{\frac{\sum_{i=1}^{N_1} (x_i - \bar{x}_1)^2 + \sum_{j=1}^{N_2} (x_j - \bar{x}_2)^2 + \sum_{k=1}^{N_3} (x_k - \bar{x}_3)^2 + \cdots}{N_1 + N_2 + N_3 + \cdots - N_s}}$$

where N_1 is the number of data in set 1, N_2 is the number in set 2, and so forth. The term N_s is the number of data sets that are pooled.

4C-5 Alternative Terms for Expressing the Precision of Samples of Data

We ordinarily report the sample standard deviation as a measure of the precision of data. Three other terms are often encountered, however.

Variance (s^2)

The variance is the square of the standard deviation:

$$s^2 = \frac{\sum_{i=1}^{N} (x_i - \bar{x})^2}{N - 1} \qquad (4\text{-}10)$$

Note that the standard deviation has the same units as the data, whereas the variance has the units of the data squared. People who do scientific work tend to use standard deviation rather than variance as a measure of precision. It is easier to relate the precision of a measurement to the measurement itself if they both have the same units. The advantage of using variance is that variances are additive, as we will see later in this chapter.

Relative Standard Deviation (RSD) and Coefficient of Variation (CV)

We frequently quote standard deviations in relative rather than absolute terms. We calculate the relative standard deviation by dividing the standard deviation by the mean of the set of data. It is often expressed in parts per thousand (ppt) or in percent by multiplying this ratio by 1000 ppt or by 100%. For example,

$$\text{RSD} = (s/\bar{x}) \times 1000 \text{ ppt}$$

> The *coefficient of variation* is the percent relative standard deviation.

The relative standard deviation multiplied by 100% is called the *coefficient of variation* (CV).

$$CV = (s/\bar{x}) \times 100\% \qquad \text{(4-11)}$$

Relative standard deviations often give a clearer picture of data quality than do absolute standard deviations. As an example, suppose that a sample contains about 50 mg of copper and that the standard deviation of a copper determination is 2 mg. The CV for this sample is 4%. For a sample containing only 10 mg, the CV is 20%.

Spread or Range (*w*)

The *spread*, or *range*, is another term that is sometimes used to describe the precision of a set of replicate results. It is the difference between the largest value in the set and the smallest. Thus, the spread of the data in Figure 4-1 is 0.9 ppm Fe. The standard deviations for small sets of data can be quickly estimated by multiplying the range by a factor k. That is,

$$s = kw \qquad \text{(4-12)}$$

Values of k are given in Table 4-4.

> Prior to the invention of the hand-held scientific calculator in the 1960s, calculation of the standard deviation for a set of data was a tedious and time-consuming exercise that was often subject to arithmetic error. For this reason, scientists often resorted to the use of Equation 4-12 and Table 4-4 even though the standard deviation obtained is less reliable. With the availability of scientific computers, Equation 4-12 has become somewhat obsolete.

Example 4-4

For the set of data in Example 4-2 (page 64), calculate (a) the variance; (b) the relative standard deviation in parts per thousand; (c) the coefficient of variation; (d) the spread; and (e) an estimate of the standard deviation from the spread.
In Example 4-2, we found

$$\bar{x} = 0.754 \qquad \text{and} \qquad s = 0.0038$$

(a) $$s^2 = (0.0038)^2 = 1.4 \times 10^{-5}$$

(b) $$RSD = \frac{0.0038}{0.754} \times 1000 \text{ ppt} = 5.0 \text{ ppt}$$

(c) $$CV = \frac{0.0038}{0.754} \times 100\% = 0.50\%$$

(d) $$w = 0.760 - 0.751 = 0.009$$

(e) $$s = kw = 0.43 \times 0.009 = 0.0039 \approx 0.004$$

Table 4-4
FACTORS FOR ESTIMATING STANDARD DEVIATION FROM RANGE

N	k
2	0.89
3	0.59
4	0.49
5	0.43
6	0.39
7	0.37
8	0.35
9	0.34
10	0.32

4D   THE STANDARD DEVIATION OF COMPUTED RESULTS

We often need to estimate the standard deviation of a result that has been computed from two or more experimental data, each of which has a known sample standard deviation. The way such estimates are made depends on the type of arithmetic used.

4D-1 The Standard Deviation of Sums and Differences

Consider the summation:

$$
\begin{array}{ll}
+0.50 & (\pm 0.02) \\
+4.10 & (\pm 0.03) \\
\underline{-1.97} & (\pm 0.05) \\
-2.63 &
\end{array}
$$

where the numbers in parentheses are absolute standard deviations. If the signs of the three individual standard deviations happen to have the same sign, the standard deviation of the sum could be as large as $+0.02 + 0.03 + 0.05 = +0.10$ or $-0.02 - 0.03 - 0.05 = -0.10$. On the other hand, it is possible that the three standard deviations could combine to give an accumulated value of zero: $-0.02 - 0.03 + 0.05 = 0$ or $+0.02 + 0.03 - 0.05 = 0$. More likely, however, the standard deviation of the sum will lie between $+0.10$ and -0.10. It can be shown from statistical theory that the variances of a sum or difference can be obtained by adding the individual variances. Thus the most probable value for a standard deviation of a sum or difference can be found by taking the square root of the sum of the squares of the individual absolute standard deviations. So, for the computation

$$
y = a(\pm s_a) + b(\pm s_b) - c(\pm s_c)
$$

the standard deviation of the result s_y is given by

$$
s_y = \sqrt{s_a^2 + s_b^2 + s_c^2} \tag{4-13}
$$

where s_a, s_b, and s_c are the standard deviations of the three terms making up the result. Substituting the standard deviations from the example gives

$$
s_y = \sqrt{(\pm 0.02)^2 + (\pm 0.03)^2 + (\pm 0.05)^2} = \pm 0.06
$$

and the sum should be reported as 2.63 (± 0.06).

> The variance of a sum or difference is equal to the *sum* of the variances of the numbers making up that sum or difference.

> For a sum or a difference the *absolute standard deviation of the answer* is the square root of the sum of the squares of the *absolute standard deviations* of the numbers used to calculate the sum or difference.

4D-2 The Standard Deviation of Products and Quotients

Consider the following computation where the numbers in parentheses are again absolute standard deviations:

$$
\frac{4.10(\pm 0.02) \times 0.0050(\pm 0.0001)}{1.97(\pm 0.04)} = 0.010406(\pm ?)
$$

In this situation, the standard deviations of two of the numbers in the calculation are larger than the result itself. Evidently, we need a different approach for multiplication and division. The relative standard deviation of the product or quotient is determined by the *relative standard deviations* of the numbers forming the computed result. For example, in the case of

$$y = \frac{a \times b}{c} \qquad (4\text{-}14)$$

we obtain the relative standard deviation s_y/y of the result y by summing the squares of the relative standard deviations of a, b, and c and extracting the square root of the sum:

> For multiplication or division, the *relative standard deviation of the answer is* the square root of the sum of the squares of the *relative standard deviations* of the numbers that are multiplied or divided.

$$\frac{s_y}{y} = \sqrt{\left(\frac{s_a}{a}\right)^2 + \left(\frac{s_b}{b}\right)^2 + \left(\frac{s_c}{c}\right)^2} \qquad (4\text{-}15)$$

Applying this equation to the numerical example gives

$$\frac{s_y}{y} = \sqrt{\left(\frac{\pm 0.02}{4.10}\right)^2 + \left(\frac{\pm 0.0001}{0.005}\right)^2 + \left(\frac{\pm 0.04}{1.97}\right)^2}$$
$$= \sqrt{(0.0048)^2 + (0.0200)^2 + (0.0203)^2} = \pm 0.0289$$

In order to complete the calculation, we must find the absolute standard deviation of the result,

> To find the absolute standard deviation in a product or a quotient, first find the relative standard deviation in the result and then multiply it by the result.

$$s_y = y \times (\pm 0.0289) = 0.0104 \times (\pm 0.0289) = \pm 0.000301$$

and we can write the answer and its uncertainty as $0.0104(\pm 0.0003)$.

Example 4-5 demonstrates the calculation of the standard deviation of the result for a more complex calculation.

Example 4-5

Calculate the standard deviation of the result of

$$\frac{[14.3(\pm 0.2) - 11.6(\pm 0.2)] \times 0.050(\pm 0.001)}{[820(\pm 10) + 1030(\pm 5)] \times 42.3(\pm 0.4)} = 1.725(\pm ?) \times 10^{-6}$$

First, we must calculate the standard deviation of the sum and the difference. For the difference in the numerator,

$$s_a = \sqrt{(\pm 0.2)^2 + (\pm 0.2)^2} = \pm 0.283$$

and for the sum in the denominator,

$$s_b = \sqrt{(\pm 10)^2 + (\pm 5)^2} = 11.2$$

We may then rewrite the equation as

$$\frac{2.7(\pm 0.283) \times 0.050(\pm 0.001)}{1850(\pm 11.2) \times 42.3(\pm 0.4)} = 1.725 \times 10^{-6}$$

The equation now contains only products and quotients, and Equation 4-15 applies. Thus,

$$\frac{s_y}{y} = \sqrt{\left(\pm\frac{0.283}{2.7}\right)^2 + \left(\pm\frac{0.001}{0.050}\right)^2 + \left(\pm\frac{11.2}{1850}\right)^2 + \left(\pm\frac{0.4}{42.3}\right)^2} = 0.107$$

To obtain the absolute standard deviation, we write

$$s_y = y \times 0.107 = 1.725 \times 10^{-6} \times (\pm 0.107) = \pm 0.185 \times 10^{-6}$$

and round the answer to $1.7(\pm 0.2) \times 10^{-6}$.

4E   METHODS FOR REPORTING ANALYTICAL DATA

A numerical result is worthless unless something is known about its accuracy. Therefore, it is always essential to indicate the reliability of data. One of the best ways of indicating reliability is to give confidence limit at the 90% or 95% confidence level as we describe in Section 5B-2. Another method is to report the absolute standard deviation or the coefficient of variation of the data. In this case it is a good idea to indicate the number of data that were used to obtain the standard deviation so that the user of the data has some idea of the probable reliability of s. A less satisfactory but more common indicator of the quality of data is the *significant figure convention.*

4E-1   The Significant Figure Convention

A simple way of indicating the probable uncertainty associated with an experimental measurement is to round the result so that it contains only *significant figures.* By definition, the significant figures in a number are all of the certain digits *and the first uncertain digit.* For example, when you read the 50-mL buret section shown in Figure 4-7, you can tell that the liquid level is greater than 30.2 mL and less than 30.3 mL. You can also estimate the position of the liquid between the graduations to about ±0.02 mL. So, using the significant figure convention, you should report the volume delivered as 30.24 mL, which is four significant figures. In this example the first three digits are certain, and the last digit (4) is uncertain.

A zero may or may not be significant depending on its location in a number. A zero that is surrounded by other digits is always significant (such as in 30.24 mL) because it is read directly and with certainty from a scale or instrument readout. On the other hand, zeros that only locate the decimal point for us are not. If we write 30.24 mL as 0.03024 L, the number of significant figures is the same. The only function of the zero before the 3 is to locate the decimal point, so it is not significant. Terminal, or final, zeros may or may not be significant. For example, if the volume of a beaker is expressed as 2.0 L, the presence of the zero tells us that the volume is known to a few tenths of a liter. So, both the 2 and the zero are significant figures. If this same volume is reported as 2000 mL, the situation becomes confusing. The last two zeros are not significant because the uncertainty is still a few tenths of a liter or a few hundred milliliters.

Figure 4-7
Buret section showing the liquid level and meniscus.

Express data in scientific notation to avoid confusion in determining whether terminal zeros are significant.

Rules for determining the number of significant figures:

(1) Disregard all initial zeros.
(2) Disregard all final zeros *unless they follow a decimal point.*
(3) All remaining digits including zeros between non-zero digits are significant.

You have probably heard it said that a chain is only as strong as its weakest link. For addition and subtraction, the weak link is the *number of decimal places* in the number with the *smallest* number of decimal places.

When adding and subtracting numbers in scientific notation, express the numbers to the same power of ten. For example, to obtain the sum

$$2.432 \times 10^6 = \quad 2.432 \times 10^6$$
$$+6.512 \times 10^4 = +0.06512 \times 10^6$$
$$-1.227 \times 10^5 = -0.1227 \times 10^6$$
$$\qquad\qquad\qquad -2.37442 \times 10^6$$
$$\qquad\qquad\qquad = 2.374 \times 10^6$$

In order to follow the significant figure convention in a case such as this, use scientific notation and report the volume as 2.0×10^3 mL.

A limitation of the significant figure convention as an indicator of reliability of data is its ambiguity. For example, when a result is reported as 61.6, you have no way of knowing how large or small the uncertainty is in the last figure.

4E-2 Significant Figures in Numerical Computations

Care is required to determine the appropriate number of significant figures in the result of an arithmetic combination of two or more numbers.[2]

Sums and Differences

For addition and subtraction, the number of significant figures can be found by visual inspection. For example, in the expression

$$3.4 + 0.020 + 7.31 = 10.73 = 10.7$$

The second and third decimal places in the answer cannot be significant because 3.4 is uncertain in the first decimal place. Note that the result contains three significant digits even though two of the numbers involved have only two significant figures.

Products and Quotients

A rule of thumb suggested by some is that for multiplication and division the answer should be rounded so that it contains the same number of significant digits as the original number with the smallest number of significant digits. Unfortunately, this procedure often leads to incorrect rounding. For example, consider the two calculations

$$\frac{24 \times 4.52}{100.0} = 1.08 \quad \text{and} \quad \frac{24 \times 4.02}{100.0} = 0.965$$

By this rule, the first answer would be rounded to 1.1 and the second to 0.96. If, however, we assume a unit uncertainty in the last digit of each number in the first quotient, the relative uncertainties associated with each of these numbers are 1/24, 1/452, and 1/1000. Because the first relative uncertainty is much larger than the other two, the relative uncertainty in the result is also 1/24. The absolute uncertainty is then

$$1.08 \times 1/24 = 0.045 = 0.04$$

By the same argument the absolute uncertainty of the second answer is given by

$$0.965 \times 1/24 = 0.040 = 0.04$$

---

[2] For an extensive discussion of significant figures, see L. M. Schwartz, *J. Chem. Educ.*, **1985**, *62*, 693.

Therefore, the first result should be rounded to three significant figures, or 1.08, but the second should be rounded to only two, or 0.96.

Logarithms and Antilogarithms

Be especially careful in rounding the results of calculations involving logarithms. The following rules apply to most situations:[3]

1. In a logarithm of a number, keep as many digits to the right of the decimal point as there are significant figures in the original number.
2. In an antilogarithm of a number, keep as many digits as there are digits to the right of the decimal point in the original number.

The number of significant figures in the *mantissa*, or the digits to the right of the decimal point of a logarithm, is the same as the number of significant figures in the original number.

$$\log (\mathbf{9.57} \times 10^4) = 4.\mathbf{981}$$

Example 4-6

Round the following answers so that only significant digits are retained.

(a) $\log 4.000 \times 10^{-5} = -4.3979400$ and (b) antilog $12.5 = 3.162277 \times 10^{12}$.

(a) Following rule 1, we retain 4 digits to the right of the decimal point

$$\log \mathbf{4.000} \times 10^{-5} = -4.\mathbf{3979}$$

(b) Following rule 2, we may retain only 1 digit

$$\text{antilog } 12.\mathbf{5} = \mathbf{3} \times 10^{12}$$

4E-3 Rounding Data

Always round the computed results of a chemical analysis in an appropriate way. For example, consider the replicate results: 61.60, 61.46, 61.55, and 61.61. The mean of these data is 61.555, and the standard deviation is 0.069. When we round the mean, do we take 61.55 or 61.56? A good guide to follow when rounding a 5 is always to round to the nearest even number. In this way, we eliminate any tendency to round in a set direction. In other words, there is an equal likelihood that the nearest even number will be the higher or the lower in any given situation. Accordingly, we might choose to report the result as 61.56 ± 0.07. If we had reason to doubt the reliability of the estimated standard deviation, we might report the result as 61.6 ± 0.1.

In rounding a number ending in 5, always round so that the result ends with an even number.

4E-4 Rounding the Results from Chemical Computations

Throughout this book and others, the reader is asked to perform calculations with data whose precision is indicated only by the significant figure convention. In these circumstances, common sense assumptions must be made as to the uncertainty in each number. The uncertainty of the result is then estimated using

---

[3] D. E. Jones, *J. Chem. Educ.*, **1971**, *49*, 753.

the techniques presented in Section 4D. Finally, the result is rounded so that it contains only significant digits. *It is especially important to postpone rounding until the calculation is completed.* At least one extra digit beyond the significant digits should be carried through all of the computations in order to avoid a *rounding error.* This extra digit is sometimes called a ''guard'' digit. Modern calculators generally retain several extra digits that are not significant, and the user must be careful to round final results properly so that only significant figures are included. Example 4-7 illustrates this procedure.

Example 4-7

A 3.4842-g sample of a solid mixture containing benzoic acid, C_6H_5COOH (122.123 g/mol), was dissolved and titrated with base to a phenolphthalein end point. The acid consumed 41.36 mL of 0.2328 M NaOH. Calculate the percent benzoic acid (HBz) in the sample.

As will be shown in Section 9C-3, the computation takes the following form:

$$\% \text{ HBz} = \frac{41.36 \text{ mL} \times 0.2328 \frac{\text{mmol NaOH}}{\text{mL NaOH}} \times \frac{1 \text{ mmol HBz}}{\text{mmol NaOH}} \times \frac{122.123 \text{ g HBz}}{1000 \text{ mmol HBz}}}{3.4842 \text{ g sample}}$$
$$\times 100\%$$
$$= 33.749\%$$

Since all operations are either multiplication or division, the relative uncertainty of the answer is determined by the relative uncertainties of the experimental data. Let us estimate what these uncertainties are.

(1) The position of the liquid level in a buret can be estimated to ± 0.02 mL. Initial and final readings must be made, however, so that the standard deviation of the volume will be

$$\sqrt{(0.02)^2 + (0.02)^2} = \pm 0.028 \text{ mL} \qquad \text{(Equation 4-13)}$$

The relative uncertainty is then

$$\frac{\pm 0.028}{41.36} \times 1000 \text{ ppt} = \pm 0.68 \text{ ppt}$$

(2) Generally, the absolute uncertainty of a mass obtained with an analytical balance will be on the order of ± 0.0001 g. Thus the relative uncertainty of the denominator is

$$\frac{0.0001}{3.4842} \times 1000 \text{ ppt} = 0.029 \text{ ppt}$$

(3) We will assume that the absolute uncertainty in the molarity of the reagent solution is ± 0.0001, and so

$$\pm \frac{0.0001}{0.2328} \times 1000 \text{ ppt} = \pm 0.43 \text{ ppt}$$

(4) The relative uncertainty in the molar mass of HBz is several orders of magnitude smaller than that of the three experimental data and is of no

consequence. Note, however, that we should retain enough digits in the calculation so that the molar mass is given to at least one more digit (the guard digit) than any of the experimental data. Thus, in the calculation, we use 122.123 for the molar mass (here we are carrying two extra digits).

(5) No uncertainty is associated with 100% and the 1000 mmol HBz since these are exact numbers.

It is sufficient in cases like this to round the answer so that its relative uncertainty is of the *same order of magnitude* as the relative uncertainty of the number having the largest relative uncertainty. In practice, this means that the answer is rounded so that its relative uncertainty lies between 0.2 and 2 times the largest relative uncertainty of the input data.

We see that the largest relative uncertainty of the three input data is 0.68 ppt. The answer should then be rounded to the same order of magnitude as 0.68 or about 0.7 ppt. If the answer is rounded to 33.7, the suggested relative uncertainty is $(0.1/33.7) \times 1000$ ppt $= 3$ ppt, which is well over 2×0.7 ppt $= 1.4$ ppt. Rounding to 33.75 implies a relative uncertainty of $(0.01/33.75) \times 1000$ ppt $= 0.3$ ppt, which lies between 0.2×1 ppt and 2×1 ppt. Thus the answer is written as % HBz $= 33.75$.

With a little practice, you can make rounding decisions, such as that shown in Example 4-7, in your head. For example, looking again at the equation in this example, we estimate the relative uncertainty in the volume measurement to be somewhat less than 0.04 mL in about 40 mL or roughly 1 part in 1000. Similarly, the uncertainty in the molarity is roughly 1 part in 2000, and the uncertainty in the denominator is somewhat less than 1 in 34,000. So, the uncertainty in the result is determined by the uncertainty in the volumetric measurement, which, as we have said, is roughly 1 part in 1000. The uncertainty in the computed result is then 1/1000th of 33.75% or about 0.03%. Therefore we round to 33.75%.

It is important to emphasize that *rounding decisions are an important part of every calculation* and that such decisions *cannot* be based on the number of digits displayed on the readout of a calculator.

4F   QUESTIONS AND PROBLEMS

4-1. Explain the difference between
  *(a) accuracy and precision.
  (b) random and systematic error.
  *(c) mean and median.
  (d) absolute and relative error.
  *(e) variance and standard deviation.

4-2. Define
  *(a) range.
  (b) coefficient of variation.
  *(c) significant figures.
  (d) Gaussian distribution.

*4-3. Suggest some sources of random error in measuring the width of a 3-m table with a 1-m metal rule.

4-4. Consider the following sets of replicate measurements:

*A	B	*C	D	*E	F
2.4	69.94	0.0902	2.3	69.65	0.624
2.1	69.92	0.0884	2.6	69.63	0.613
2.1	69.80	0.0886	2.2	69.64	0.596
2.3		0.1000	2.4	69.21	0.607
1.5			2.9		0.582

_____

* Answers to the asterisked problems are given in the answer section at the back of the book.

For each set, calculate the
(a) mean.
(b) median.
(c) spread, or range.
(d) standard deviation.
(e) coefficient of variation.
(f) standard deviation estimate from the spread.

4-5. The accepted values for the sets of data in Problem 4-4 are: *set A, 2.0; set B, 69.75; *set C, 0.0930; set D, 3.0; *set E, 69.05; set F, 0.635. For each set, calculate
(a) the absolute error.
(b) the relative error in parts per thousand.

*4-6. Estimate the absolute standard deviation and the coefficient of variation for the results of the following calculations. Round each result so that it contains only significant digits. The numbers in parentheses are absolute standard deviations.
(a) $y = 6.75(\pm0.03) + 0.843(\pm0.001) - 7.021 (\pm0.001)$
$y = 0.572$
(b) $y = 19.97(\pm0.04) + 0.0030(\pm0.0001) +$
$y = 1.29 (\pm0.08) = 21.263$
(c) $y = 67.1(\pm0.3) \times 1.03(\pm0.02) \times 10^{-17}$
$y = 6.9113 \times 10^{-16}$
(d) $y = 243(\pm1) \times \dfrac{760(\pm2)}{1.006(\pm0.006)} = 183{,}578.5$
(e) $y = \dfrac{143(\pm6) - 64(\pm3)}{1249(\pm1) + 77(\pm8)} = 5.9578 \times 10^{-2}$
(f) $y = \dfrac{1.97(\pm0.01)}{243(\pm3)} = 8.106996 \times 10^{-3}$

4-7. Estimate the absolute standard deviation and the coefficient of variation for the results of the following calculations. Round each result to include only significant figures. The numbers in parentheses are absolute standard deviations.
(a) $y = -1.02(\pm0.02) \times 10^{-7} - 3.54(\pm0.2) \times 10^{-8}$
$= -1.374 \times 10^{-7}$
(b) $y = 100.20(\pm0.08) - 99.62(\pm0.06) +$
$0.200 (\pm0.004) = 0.780$
(c) $y = 0.0010(\pm0.0005) \times 18.10(\pm0.02) \times 200(\pm1)$
$= 3.62$
(d) $y = \dfrac{1.73(\pm0.03) \times 10^{-14}}{1.63(\pm0.04) \times 10^{-16}} = 106.1349693$
(e) $y = \dfrac{100(\pm1)}{2(\pm1)} = 50$
(f) $y = \dfrac{1.43(\pm0.02) \times 10^{-2} - 4.76(\pm0.06) \times 10^{-3}}{24.3(\pm0.7) + 8.06(\pm0.08)}$
$= 2.948 \times 10^{-4}$

4-8. Round each of the following results to include only significant figures.
*(a) $y = \log 1.73 = 0.238046$
(b) $y = \log 0.0432 = -1.364516$
*(c) $y = \log (6.022 \times 10^{23}) = 23.77960$
(d) $y = \log (4.213 \times 10^{-21}) = -20.375409$
*(e) $y = $ antilog $(-3.47) = 3.38844 \times 10^{-4}$
(f) $y = $ antilog $5.7 = 5.01187 \times 10^5$
*(g) $y = $ antilog $0.99 = 9.77237$
(h) $y = $ antilog $(-27.2424) = 5.722687 \times 10^{-28}$

*4-9. Analysis of several plant-food preparations for potassium ion yielded the following data:

Sample	Mean Percent K^+	Number of Observations	Deviation of Individual Results from Mean
1	4.80	5	0.13, 0.09, 0.07, 0.05, 0.06
2	8.04	3	0.09, 0.08, 0.12
3	3.77	4	0.02, 0.15, 0.07, 0.10
4	4.07	4	0.12, 0.06, 0.05, 0.11
5	6.84	5	0.06, 0.07, 0.13, 0.10, 0.09

(a) Evaluate the standard deviation s for each sample.
(b) Obtain a pooled estimate for s.

4-10. Six bottles of wine were analyzed for residual sugar, with the following results:

Bottle	Percent (w/v) Residual Sugar	Number of Observations	Deviation of Individual Results from Mean
1	0.94	3	0.050, 0.10, 0.08
2	1.08	4	0.060, 0.050, 0.090, 0.060
3	1.20	5	0.05, 0.12, 0.07, 0.00, 0.08
4	0.67	4	0.05, 0.10, 0.06, 0.09
5	0.83	3	0.07, 0.09, 0.10
6	0.76	4	0.06, 0.12, 0.04, 0.03

(a) Evaluate the standard deviation s for each set of data.
(b) Pool the data to establish an absolute standard deviation for the method.

*4-11. Nine samples of illicit heroin preparations were analyzed in duplicate by a gas chromatographic technique. Pool the following data to establish an absolute standard deviation for the procedure:

Sample	Heroin, %	Sample	Heroin, %
1	2.24, 2.27	6	1.07, 1.02
2	8.4, 8.7	7	14.4, 14.8
3	7.6, 7.5	8	21.9, 21.1
4	11.9, 12.6	9	8.8, 8.4
5	4.3, 4.2		

4-12. Calculate a pooled estimate of s from the following spectrophotometric analysis for NTA (nitrilotriacetic acid) in water from the Ohio River:

Sample	NTA, ppb
1	13, 16, 14, 9
2	38, 37, 38
3	25, 29, 23, 29, 26

4-13. The color change of a chemical indicator requires an overtitration of 0.03 mL. Calculate the percent relative error if the total volume of titrant is
*(a) 50.00 mL.   *(c) 25.0 mL.
 (b) 10.0 mL.   *(d) 40.0 mL.

4-14. A loss of 0.4 mg of Zn occurs in the course of an analysis for that element. Calculate the percent relative error due to this loss if the weight of Zn in the sample is
*(a) 40 mg.   *(c) 400 mg.
 (b) 175 mg.   *(d) 600 mg.

CHAPTER 5

EVALUATION OF ANALYTICAL DATA

In Chapter 4 we pointed out that experimental data of unknown reliability are useless. Thus, a vital part of performing an analysis is evaluating the probable uncertainty in the results. We also pointed out that three types of experimental errors can be recognized: systematic error, random error, and gross error. This chapter deals with how you can recognize these errors and how you can determine their effects.

5A SYSTEMATIC ERRORS

Systematic, or determinate, errors have a definite value, an assignable cause, and are of the same magnitude for replicate measurements made in the same way. Systematic errors lead to *bias* in measurement technique. Bias is illustrated by the two curves in Figure 5-1, which show the frequency distribution of replicate results in the analysis of identical samples by two methods that have random errors of identical size. Method A has no bias, so the population mean μ_A is the true value x_t. Method B has a bias, or systematic error, that is given by

$$\text{bias} = \mu_B - x_t = \mu_B - \mu_A \tag{5-1}$$

Note that bias affects all of the data in a set in the same way and that it can be either positive or negative.

5A-1 Sources of Systematic Errors

There are three types of systematic errors: (1) *Instrument errors* are caused by imperfections in measuring devices and instabilities in their power supplies. (2) *Method errors* arise from nonideal chemical or physical behavior of analytical

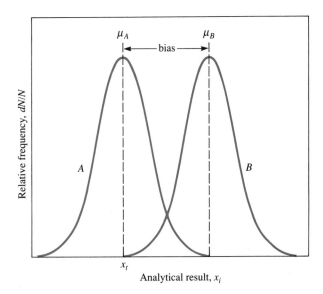

Figure 5-1
Illustration of bias: bias $= \mu_B - x_t$.

systems. (3) *Personal errors* result from the carelessness, inattention, or personal limitations of the experimenter.

Instrument Errors

All measuring devices are sources of systematic errors. For example, pipets, burets, and volumetric flasks may hold or deliver volumes slightly different from those indicated by their graduations. These differences result from using glassware at a temperature that differs significantly from the calibration temperature, from distortions in container walls due to heating while being dried, from errors in the original calibration, or from contaminants on the inner surfaces of the containers. Calibration eliminates most systematic errors of this type.

Electronic devices are subject to instrumental systematic errors. These uncertainties have many sources. For example, errors emerge as the voltage of a battery-operated power supply decreases with use. Errors result from increased resistance in circuits because of dirty electrical contacts. Temperature changes cause variation in resistors and standard voltage sources. Currents induced from 110-V power lines can affect electronic instruments. Errors from these and other sources are detectable and correctable.

Method Errors

The nonideal chemical or physical behaviors of the reagents and reactions used in an analysis often introduce systematic method errors. Such sources of nonideality include the slowness or incompleteness of some reactions, the instability of some species, the nonspecificity of most reagents, and the possible occurrence of side reactions that interfere with the measurement process. For example, a common method error in volumetric methods results from the small excess of reagent required to cause an indicator to undergo the color change that signals completion of the reaction. The accuracy of such an analysis is thus limited by the very phenomenon that makes the titration possible.

Another example of method error is illustrated by the data in Figures 4-3c and 4-3d. The results by Analysts 3 and 4 show a negative bias that can be traced to the chemical nature of the sample, nicotinic acid. The analytical method used involves the decomposition of the organic samples in hot concentrated sulfuric acid, which converts the nitrogen in the samples to ammonium sulfate. The amount of ammonia in the ammonium sulfate is then determined in the measurement step. Experiments have shown that compounds containing a pyridine ring such as nicotinic acid (see page 55) are incompletely decomposed by the sulfuric acid unless special precautions are taken. Without these precautions, low results are obtained. It is likely that the negative systematic errors $(\bar{x}_3 - x_t)$ and $(\bar{x}_4 - x_t)$ in Figure 4-3 are systematic errors that can be blamed on incomplete decomposition of the samples. Errors inherent in a method are difficult to detect and are thus the most serious of the three types of systematic error.

Personal Errors

Many measurements require personal judgments. Examples include estimating the position of a pointer between two scale divisions, the color of a solution at the end point in a titration, or the level of a liquid with respect to a graduation in a pipet or buret. Judgments of this type are often subject to systematic, unidirectional errors. For example, one person may read a pointer consistently high, another may be slightly slow in activating a timer, and a third may be less sensitive to color changes. An analyst who is insensitive to color changes tends to use excess reagent in a volumetric analysis.

A universal source of personal error is prejudice. Most of us, no matter how honest, have a natural tendency to estimate scale readings in a direction that improves the precision in a set of results. Or we may have a preconceived notion of the true value for the measurement. We then subconsciously cause the results to fall close to this value. Number bias is another source of personal error that varies considerably from person to person. The most common number bias encountered in estimating the position of a needle on a scale involves a preference for the digits 0 and 5. Also prevalent is a prejudice favoring small digits over large and even numbers over odd.

Color blindness is a good example of a handicap that amplifies personal errors in volumetric analysis. A famous color-blind analytical chemist enlisted his wife to come to the laboratory to help him detect color changes at end points of titrations.

Digital readouts on pH meters, laboratory balances, and other electronic instruments eliminate number bias because no judgment is involved in taking a reading.

Persons who make measurements must guard against personal bias to preserve the integrity of the data.

5A-2 The Effect of Systematic Errors on Analytical Results

Systematic errors may be either *constant* or *proportional*. The magnitude of an absolute constant error does not depend on the size of the quantity measured. Absolute proportional errors increase or decrease in proportion to the size of the sample taken for analysis.

Constant Errors

Constant errors become more serious as the size of the quantity measured decreases. The effect of solubility losses on the results of a gravimetric analysis illustrates this behavior.

Example 5-1

Suppose that 0.50 mg is lost as a result of washing a precipitate with 200 mL of wash liquid. If the precipitate weighs 500 mg, the relative error due to this solubility loss is $-(0.50/500) \times 100\% = -0.1\%$. Loss of the same quantity from 50 mg of precipitate results in a relative error of -1.0%. Thus the *relative error* resulting from a constant absolute error becomes greater as the quantity measured becomes smaller.

The excess of reagent required to bring about a color change during titration is another example of constant error. This volume, usually small, remains the same regardless of the total volume of reagent required for the titration. Again, the relative error from this source becomes more serious as the total volume decreases. One way of minimizing the effect of constant error is to use as large a sample as possible.

Proportional Errors

A common cause of proportional errors is the presence of interfering contaminants in the sample. For example, a widely used method for the determination of copper is based on the reaction of copper(II) ion with potassium iodide to give iodine. The quantity of iodine is then measured and is proportional to the amount of copper. Iron(III), if present, also liberates iodine from potassium iodide. Unless steps are taken to prevent this interference, high results are observed for the percentage of copper because the iodine produced will be a measure of the copper(II) *and* iron(III) in the sample. The size of this error is fixed by the *fraction* of iron contamination, which is independent of the size of the sample taken. If the sample size is doubled, for example, the amount of iodine liberated by both the copper and the iron contaminant is also doubled. Thus, the magnitude of the reported percentage of copper is independent of sample size.

5A-3 Detection of Systematic Instrument and Personal Errors

Systematic instrument errors are usually found and corrected by calibration. Periodic calibration of equipment is always desirable because the response of most instruments changes with time as a result of wear, corrosion, or mistreatment.

Most personal errors can be minimized by care and self-discipline. It is a good habit to check instrument readings, notebook entries, and calculations systematically. Errors that result from a known physical handicap can usually be avoided by a careful choice of method.

5A-4 Detection of Systematic Method Errors

Bias in an analytical method is particularly difficult to detect. We may take one or more of the following steps to recognize and adjust for a systematic error in an analysis.

Analysis of Standard Samples

The best way of estimating the bias of an analytical method is by the analysis of *standard reference materials*—materials that contain one or more analytes at known concentration levels. Standard reference materials are obtained in several ways.

Standard materials can sometimes be prepared by synthesis. Here, carefully measured quantities of the pure components of a material are measured out and

A standard reference material (SRM) contains one or more species of known concentration.

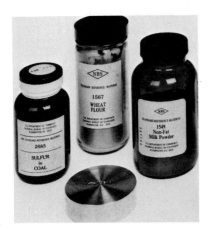

Standard reference materials from NIST. (Photo courtesy of the National Institute of Standards and Technology.)

In using SRMs it is often difficult to separate bias from ordinary random error.

mixed in such a way as to produce a homogeneous sample whose composition is known from the quantities taken. The overall composition of a synthetic standard material must approximate closely the composition of the samples to be analyzed. Great care must be taken to ensure that the concentration of analyte is known accurately. Unfortunately, the synthesis of such standard samples is often impossible or so difficult and time consuming that this approach is not practical.

Standard reference materials can be purchased from a number of governmental and industrial sources. For example, the National Institute of Standards and Technology (NIST) (formerly the National Bureau of Standards) offers more than 900 standard reference materials, including rocks and minerals, gas mixtures, glasses, hydrocarbon mixtures, polymers, urban dusts, rainwaters, and river sediments.[1] The concentration of one or more of the components in these materials has been determined in one of three ways: (1) by analysis using a previously validated reference method; (2) by analysis using two or more independent, reliable measurement methods; or (3) by analysis by a network of cooperating laboratories, technically competent and thoroughly knowledgeable with the material being tested.

Several commercial supply houses also offer analyzed materials for method testing.[2]

One of the problems you will encounter in using standard reference materials (SRMs) to establish the presence or absence of bias is that the mean of your analysis of replicates of the standard will ordinarily differ somewhat from the theoretical result. Then you are faced with the question of whether this difference is due to random error of your measurements or to bias in the method. In Section 5B-3, we demonstrate a statistical test that can be applied to aid your judgment in answering this question.

Independent Analysis

If standard samples are not available, a second independent and reliable analytical method can be used in parallel with the method being evaluated. The independent method should differ as much as possible from the one under study. This minimizes the possibility that some common factor in the sample has the same effect on both methods. Here again, a statistical test must be used to determine whether any difference is a result of random errors in the two methods or of bias in the method under study.

Blank Determinations

A *blank* solution contains the solvent and all of the reagents in an analysis, but none of the sample.

Blank determinations are useful for detecting certain types of constant errors. In a blank determination, or *blank*, all steps of the analysis are performed in the absence of a sample. The results from the blank are then applied as a correction

---

[1] See U.S. Department of Commerce, *NIST Standard Reference Materials Catalog*, 1992–93 ed., NIST Special Publication 260. Washington, D.C.: Government Printing Office, 1992. For a description of the reference material programs of the NIST, see R. A. Alvarez, S. D. Rasberry, and F. A. Uriano, *Anal. Chem.*, **1982**, *54*, 1226A; and F. A. Uriano, *ASTM Standardization News*, **1979**, *7*, 8.

[2] For sources of biological and environmental reference materials containing various elements, see C. Veillon, *Anal. Chem.*, **1986**, *58*, 851A.

to the sample measurements. Blank determinations reveal errors due to interfering contaminants from the reagents and vessels employed in analysis. Blanks also allow the analyst to correct titration data for the volume of reagent needed to cause an indicator to change color at the end point.

Variation in Sample Size

Example 5-1 demonstrates that the effect of a constant error decreases as the size of a measurement increases. Thus, constant errors can often be detected by varying the sample size.

5B   APPLICATION OF STATISTICS TO DATA EVALUATION

Experimentalists use statistical calculations to sharpen their judgments concerning the quality of experimental measurements. The most common applications of statistics to analytical chemistry include:

1. Deciding whether an outlying value in a set of replicate measurements is the result of a gross error and should be rejected or whether it is a legitimate part of the population that should be retained in calculating the mean for the set.
2. Defining the interval around the mean of a set within which the population mean can be expected to be found with a given probability.
3. Determining the number of replicate measurements required to assure that an experimental mean falls within a predetermined interval around the population mean (at a given probability).
4. Estimating the probability that (a) an experimental mean and a true value or (b) two experimental means are different—that is, whether the difference is real or is simply the result of random error. This test is particularly important in locating systematic errors in a method and determining whether two samples could come from the same source.
5. Treating calibration data.

We will examine each of these applications in the sections that follow.

5B-1   Gross Errors; Rejecting Outliers

When a set of data contains an outlying result that appears to differ excessively from the average, the decision must be made whether to retain or reject it. The choice of criteria for the rejection of a suspected result has its perils. If we set a stringent standard that makes the rejection of a questionable measurement difficult, we run the risk of retaining results that are spurious and have an inordinate effect on the mean of the data. If we set lenient limits on precision and make the rejection of a result easy, we are likely to discard measurements that rightfully belong in the set and thus introduce a bias to the data. It is an unfortunate fact that there is no universal rule to settle the question of retention or rejection.[3]

---

[3] J. Mandel, in *Treatise on Analytical Chemistry*, 2nd ed., I. M. Kolthoff and P. J. Elving, Eds., Part I, Vol. 1. pp. 282–89. New York: Wiley, 1978.

The Q test

As shown below, arrange the data in ascending order to select the outlier and its nearest neighbor.

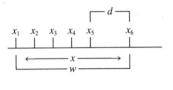

$$d = x_6 - x_5$$

$$w = x_6 - x_1$$

$$Q_{exp} = d/w$$

If $Q_{exp} > Q_{crit}$, reject x_6

Figure 5-2
The Q test for outliers.

Use caution when rejecting data for any reason.

The Q Test

The Q test is a simple, widely used statistical test.[4] In this test the absolute value of the difference d between the questionable result x_q and its nearest neighbor x_n is divided by the spread w of the entire set to give the quantity Q_{exp}:

$$Q_{exp} = \frac{d}{w} = \frac{|x_q - x_n|}{|x_1 - x_n|} \qquad (5\text{-}2)$$

This ratio is then compared with rejection values Q_{crit} found in Table 5-1. If Q_{exp} is greater than Q_{crit}, the questionable result can be rejected with the indicated degree of confidence (See Figure 5-2).

Example 5-2

The analysis of a calcite sample yielded CaO percentages of 55.95, 56.00, 56.04, 56.08, and 56.23. The last value appears anomalous; should it be retained or rejected?

The difference between 56.23 and 56.08 is 0.15%. The spread (56.23–55.95) is 0.28%. Thus,

$$Q_{exp} = \frac{56.23 - 56.08}{56.23 - 55.95} = \frac{0.15}{0.28} = 0.54$$

For five measurements, Q_{crit} at the 90% confidence level is 0.64. Because 0.54 < 0.64, we must retain the outlier.

Other Statistical Tests

Several other statistical tests have been developed as criteria for rejection or retention of outliers. Such tests, like the Q test, assume that the distribution of the population data is normal, or Gaussian. Unfortunately, this condition cannot be proved or disproved for samples that have fewer than about 50 results. Consequently, statistical rules, which are perfectly reliable for normal distributions of data, *should be used with extreme caution* when applied to samples containing only a few data. J. Mandel, in discussing treatment of small sets of data, writes, "Those who believe that they can discard observations with statistical sanction by using statistical rules for the rejection of outliers are simply deluding themselves."[5] Thus, statistical tests for rejection should be used only as aids to common sense when small samples are involved.

The blind application of statistical tests to retain or reject a suspect measurement in a small set of data is not likely to be much more reliable than an arbitrary decision. The application of good judgment based on broad experience with an analytical method is usually a better approach. In the end, the only valid reason for rejecting a result from a small set of data is the knowledge that a mistake

[4] R. B. Dean and W. J. Dixon, *Anal. Chem.*, **1951**, 23, 636.
[5] J. Mandel, in *Treatise on Analytical Chemistry*, 2nd ed., I. M. Kolthoff and P. J. Elving, Eds., Part I, Vol. 1. p. 282. New York: Wiley, 1978.

Table 5-1

CRITICAL VALUES FOR REJECTION QUOTIENT Q*

Number of Observations	Q_{crit} (Reject if $Q_{exp} > Q_{crit}$)		
	90% Confidence	**95% Confidence**	**99% Confidence**
3	0.941	0.970	0.994
4	0.765	0.829	0.926
5	0.642	0.710	0.821
6	0.560	0.625	0.740
7	0.507	0.568	0.680
8	0.468	0.526	0.634
9	0.437	0.493	0.598
10	0.412	0.466	0.568

* Reproduced from D. B. Rorabacher, *Anal. Chem.*, **1991**, *63*, 139. By courtesy of the American Chemical Society.

was made in the measurement process. Without this knowledge, *a cautious approach to rejection of an outlier is wise.*

Recommendations for the Treatment of Outliers

Recommendations for the treatment of a small set of results that contains a suspect value follow:

1. Reexamine carefully all data relating to the outlying result to see if a gross error could have affected its value. This recommendation demands *a properly kept laboratory notebook containing careful notations of all observations* (see Section 28I).
2. If possible, estimate the precision that you can reasonably expect from the procedure to be sure that the outlying result actually is questionable.
3. Repeat the analysis if sufficient sample and time are available. Agreement between the newly acquired data and those of the original set that appear to be valid will help justify the rejection of outliers. Furthermore, if retention is still indicated, the questionable result will have a smaller effect on the mean of the larger set of data.
4. If more data cannot be collected, apply the Q test to see if the doubtful result should be retained or rejected on statistical grounds.
5. If the Q test indicates retention, consider reporting the median of the set rather than the mean. The median has the great virtue of allowing inclusion of all data in a set without undue influence from an outlying value. In addition, the median of a normally distributed set containing three measurements provides a better estimate of the correct value than the mean of the set after the outlying value has been discarded.

5B-2 Random Errors; Confidence Limits

The exact value of the mean μ for a population of data can never be determined exactly because such a determination requires that an infinite number of measurements be made. However, statistical theory allows us to set limits around an

Confidence limits define an interval around $\bar{x}$ that probably contains μ.

experimentally determined mean $\bar{x}$ within which the true mean μ lies with a given degree of probability. These limits are called *confidence limits*, and the interval they define is known as the *confidence interval*.

The size of the confidence interval, which is derived from the sample standard deviation, depends on how accurately we know s—that is, how closely we think our sample standard deviation is to the true standard deviation. If we have reason to believe that s is a good approximation of σ, then the confidence interval can be significantly narrower than if the estimate of s is based on only two or three replicates.

The Confidence Interval When s Is a Good Approximation of σ

The confidence level is the probability expressed as a percent.

Figure 5-3 shows a series of five normal error curves. In each, the relative frequency is plotted as a function of the quantity z (Equation 4-6, page 62), which is the deviation from the mean *in units of the population standard deviation*. The shaded areas in each plot lie between the values of $-z$ and $+z$ that are indicated to the left and right of the curves. The numbers within the shaded areas are the percentages of the total area under the curve that are included within these values of z. For example, as shown in the top curve, 50% of the area under any Gaussian curve is located between -0.67σ and $+0.67\sigma$. Proceeding down, we see that 80% of the total area lies between -1.29σ and $+1.29\sigma$, and 90% lies between -1.64σ and $+1.64\sigma$. Relationships such as these allow us to define a range of values around a measurement within which the true mean is likely to lie with a certain probability, *provided we have a good estimate of σ*. For example, we may assume that 90 times out of 100, the true mean, μ, will be within $\pm1.64\sigma$ of any single measurement that we make (see Figure 5-3). Here, the *confidence level* is 90% and the *confidence interval* is $\pm z\sigma = \pm1.64\sigma$.

The confidence limits (CL) are the values above and below a measurement that bound its confidence interval.

We find a general expression for the *confidence limits* (CL) of a single measurement by rearranging Equation 4-6. (Remember that z can take positive or negative values.) Thus,

$$\text{CL for } \mu = x \pm z\sigma \tag{5-3}$$

Number of Measurements Averaged, N	Relative Size of Confidence Interval
1	1.00
2	0.71
3	0.58
4	0.50
5	0.45
6	0.41
10	0.32

For the mean of N measurements, the standard error of the mean, $\sigma/\sqrt{N}$ (Equation 4-9), is employed in place of σ. That is,

$$\text{CL for } \mu = \bar{x} \pm \frac{z\sigma}{\sqrt{N}} \tag{5-4}$$

Table 5-2
CONFIDENCE LEVELS FOR VARIOUS VALUES OF z

Confidence Levels, %	z
50	0.67
68	1.00
80	1.29
90	1.64
95	1.96
96	2.00
99	2.58
99.7	3.00
99.9	3.29

Values for z at various confidence levels are found in Table 5-2.

Example 5-3

Calculate the 80% and 95% confidence limits for (a) the first entry (1.80 ppm Hg) in Example 4-3, and (b) the mean value (1.67 ppm Hg) for specimen 1 in the same example. Assume that in each part, $s \rightarrow \sigma = 0.10$.

(a) From Table 5-2, we see that $z = 1.29$ and 1.96 for the two confidence levels.

Substituting into Equation 5-4

$$80\% \text{ CL} = 1.80 \pm \frac{1.29 \times 0.10}{\sqrt{1}} = 1.80 \pm 0.13$$

$$95\% \text{ CL} = 1.80 \pm \frac{1.96 \times 0.10}{\sqrt{1}} = 1.80 \pm 0.20$$

From these calculations, we conclude that the chances are 80 in 100 that μ, the population mean (and, *in the absence of systematic error*, the true value), lies in the interval between 1.67 and 1.93 ppm Hg. Furthermore, there is a 95% chance that it lies in the interval between 1.60 and 2.00 ppm Hg.

(b) For the three measurements,

$$80\% \text{ CL} = 1.67 \pm \frac{1.29 \times 0.10}{\sqrt{3}} = 1.67 \pm 0.07$$

$$95\% \text{ CL} = 1.67 \pm \frac{1.96 \times 0.10}{\sqrt{3}} = 1.67 \pm 0.11$$

Thus, the chances are 80 in 100 that the population mean is located in the interval from 1.60 to 1.74 ppm Hg and 95 in 100 that it lies between 1.56 and 1.78 ppm.

Example 5-4

How many replicate measurements of specimen 1 in Example 4-3 are needed to decrease the 95% confidence interval to ± 0.07 ppm Hg?

The confidence interval (CI) is given by the second term on the right-hand side of Equation 5-4 (CI $= \pm z\sigma/\sqrt{N}$). Thus

$$\text{CI} = \pm 0.07 = \pm \frac{z\sigma}{\sqrt{N}} = \pm \frac{1.96 \times 0.10}{\sqrt{N}}$$

$$\sqrt{N} = \pm \frac{1.96 \times 0.10}{0.07} = \pm 2.80$$

$$N = (\pm 2.8)^2 = 7.8$$

We conclude that eight measurements would provide a slightly better than 95% chance of the population mean lying within ± 0.07 ppm of the experimental mean.

Equation 5-4 tells us that the confidence interval for an analysis can be halved by carrying out four measurements. Sixteen measurements will narrow the interval by a factor of 4, and so on. We rapidly reach a point of diminishing returns in acquiring additional data. Ordinarily we take advantage of the relatively large gain attained by averaging two to four measurements but can seldom afford the time required for additional increases in confidence.

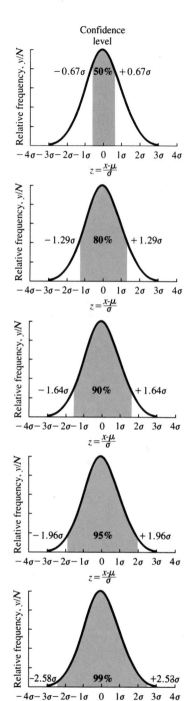

Figure 5-3

Areas under a Gaussian curve for various values of $\pm z$.

It is essential to keep in mind at all times that confidence intervals based on Equation 5-4 *apply only in the absence of bias and only if we can assume that* $s \cong \sigma$.

The Confidence Interval When σ Is Not Known

Often we are faced with limitations in time or amount of available sample that prevent us from accurately estimating σ. In such a case, a single set of replicate measurements must provide not only a mean but also an estimate of precision. As indicated earlier, s calculated from a small set of data may be quite uncertain. Thus, confidence limits are necessarily broader when a good estimate of σ is not available.

To account for the variability of s, we use the important statistical parameter t, which is defined in exactly the same way as z (Equation 4-6) except that s is substituted for σ:

The t statistic is often called *Student's t.* Student was the name used by W. S. Gossett when he wrote the classic paper on t that appeared in *Biometrika*, **1908**, *6*, 1. Gossett was employed by the Guinness Brewery to statistically analyze the results of determinations of the alcohol content of their products. As a result of this work, he discovered the now-famous statistical treatment of small sets of data. To avoid the disclosure of any trade secrets of his employer, Gossett published the paper under the name Student.

$$t = \frac{x - \mu}{s} \tag{5-5}$$

Like z in Equation 5-4, t depends on the desired confidence level. But t also depends on the number of degrees of freedom in the calculation of s. Table 5-3 provides values for t for a few degrees of freedom. More extensive tables are found in various mathematical and statistical handbooks. Note that $t \rightarrow z$ as the number of degrees of freedom becomes infinite.

Remember that the number of degrees of freedom is not equal to N.

The confidence limits for the mean $\bar{x}$ of N replicate measurements can be derived from t by an equation similar to Equation 5-4:

$$\text{CL for } \mu = \bar{x} \pm \frac{ts}{\sqrt{N}} \tag{5-6}$$

Example 5-5

A chemist obtained the following data for the alcohol content of a sample of blood: % C_2H_5OH: 0.084, 0.089, and 0.079. Calculate the 95% confidence limits for the mean assuming (a) there is no additional knowledge about the precision of the method, and (b) that on the basis of previous experience, it is known that $s \rightarrow \sigma = 0.005\%$ C_2H_5OH.

(a)

$$\Sigma x_i = 0.084 + 0.089 + 0.079 = 0.252$$

$$\Sigma x_i^2 = 0.007056 + 0.007921 + 0.006241 = 0.021218$$

$$s = \sqrt{\frac{0.021218 - (0.252)^2/3}{3 - 1}} = 0.0050\% \ C_2H_5OH$$

Here, $\bar{x} = 0.252/3 = 0.084$. Table 5-3 indicates that $t = 4.30$ for two degrees of freedom and 95% confidence. Thus,

$$95\% \text{ CL} = \bar{x} \pm \frac{ts}{\sqrt{N}} = 0.084 \pm \frac{4.30 \times 0.0050}{\sqrt{3}} = 0.084 \pm 0.012\% \ C_2H_5OH$$

(b) Because a good value of σ is available,

$$95\% \text{ CL} = \bar{x} \pm \frac{z\sigma}{\sqrt{N}} = 0.084 \pm \frac{1.96 \times 0.0050}{\sqrt{3}} = 0.084 \pm 0.006\% \text{ C}_2\text{H}_5\text{OH}$$

Note that a sure knowledge of σ decreases the confidence interval by a significant amount.

5B-3 Comparing a Mean with a True Value

As we noted in Section 5A-4, the best way to detect a systematic error in a method of analysis is to analyze standard reference material and compare the results with the accepted value for the concentration. If there is a difference between these two values, we must decide whether it results from bias in the method or from random error in the method. Statistics sharpen judgments of this kind.

Statistical tests involving comparison of two data are all based on a *null hypothesis*, which assumes the quantities being compared are the same. The probability of the observed differences appearing as a result of random error is then computed from statistical theory. Usually, if the observed difference is greater than or equal to the difference that would occur 5 times in 100 (the 5% probability level), the null hypothesis is considered questionable and the difference is judged to be significant. Other probability levels, such as 1 in 100 or 10 in 100, may also be adopted, depending on the certainty desired in the judgment.

To compare the results from an analytical method with what is thought to be a true value, we compute the difference $\bar{x} - \mu$, where $\bar{x}$ is the experimental mean and μ is the population mean, or the true value. This difference is then compared with the difference that could be caused by random error. If the observed difference is less than that computed for a chosen probability level, the null hypothesis that $\bar{x}$ and μ are the same cannot be rejected. That is, no significant systematic error has been demonstrated. It is important to realize, however, that this statement does not say that there is no systematic error. It says only that whatever bias is present is so small that it cannot be detected at the desired confidence level. If $\bar{x} - \mu$ is significantly larger than either the expected or the critical value, we may assume that the difference is real and that the method suffers from bias.

The critical value for the rejection of the null hypothesis is based on the t statistic described in the previous section. To obtain the desired critical value, we rearrange Equation 5-6 to give

$$\bar{x} - \mu = \pm \frac{ts}{\sqrt{N}} \tag{5-7}$$

where N is the number of replicate measurements used in the test. If a good estimate of σ is available, Equation 5-7 can be modified by replacing t with z and s with σ.

> In statistics, a null hypothesis postulates that two observed results are the same and any numerical differences between them is the consequence of random error.

> The only way to decide statistically that a mean is different from a true value is to reject the null hypothesis with a given probability.

Example 5-6

A new procedure for the rapid determination of sulfur in kerosenes was tested on a sample known from its method of preparation to contain 0.123% S. The results were % S = 0.112, 0.118, 0.115, and 0.119. Do the data indicate that there is a bias in the method?

$$\Sigma x_i = 0.112 + 0.118 + 0.115 + 0.119 = 0.464$$

$$\bar{x} = 0.464/4 = 0.116\% \text{ S}$$

$$\bar{x} - \mu = 0.116 - 0.123 = -0.007\% \text{ S}$$

To obtain s, we employ Equation 4-8. Thus,

$$\Sigma x_i^2 = 0.012544 + 0.013924 + 0.013255 + 0.014161 = 0.053854$$

$$s = \sqrt{\frac{0.053854 - (0.464)^2/4}{4 - 1}} = \sqrt{\frac{0.000030}{3}} = 0.0032$$

From Table 5-3, we find that at the 95% confidence level, t has a value of 3.18 for three degrees of freedom. Thus,

$$\frac{ts}{\sqrt{4}} = \frac{3.18 \times 0.0032}{\sqrt{4}} = \pm 0.0051$$

An experimental mean can be expected to deviate by ± 0.0051 or greater no more frequently than 5 times in 100. Thus, if we conclude that $\bar{x} - \mu = -0.007$ is a significant difference and that a systematic error is present, we will, on the average, be wrong fewer than 5 times in 100.

We must state the probability level when applying these tests.

If we make a similar calculation employing the value for t at the 99% confidence level, $ts/\sqrt{N}$ assumes a value of 0.0093. Thus, if we insist on being wrong no more often than 1 time in 100, we must conclude that no difference between the results has been *demonstrated*. Note that this statement is different from saying that there is no systematic error.

5B-4 Comparing Two Experimental Means

The results of chemical analyses are frequently used to determine whether two materials are identical. Here, the chemist must judge whether a difference in the means of two sets of identical analyses is real and constitutes evidence that the samples are different or whether the discrepancy is simply a consequence of random errors in the two sets. To illustrate, let us assume that N_1 replicate analyses of material 1 yielded a mean value of $\bar{x}_1$, and N_2 analyses of material 2 obtained by the same method gave a mean of $\bar{x}_2$. If the data were collected in an identical way, it is usually safe to assume that the standard deviations of the two sets of measurements are the same and modify Equation 5-7 to take into account that one set of results is being compared with a second rather than with the true mean of the data, μ.

Table 5-3
VALUES OF t FOR VARIOUS LEVELS OF PROBABILITY

Degrees of Freedom	Probability Level				
	80%	**90%**	**95%**	**99%**	**99.8%**
1	3.08	6.31	12.7	63.7	318.
2	1.89	2.92	4.30	9.92	22.3
3	1.64	2.35	3.18	5.84	10.2
4	1.53	2.13	2.78·	4.60	7.17
5	1.48	2.02	2.57	4.03	5.89
6	1.44	1.94	2.45	3.71	5.21
7	1.42	1.90	2.36	3.50	4.78
8	1.40	1.86	2.31	3.36	4.50
9	1.38	1.83	2.26	3.25	4.30
10	1.37	1.81	2.23	3.17	4.14
15	1.34	1.75	2.13	2.95	3.73
20	1.32	1.72	2.09	2.84	3.55
30	1.31	1.70	2.04	2.75	3.38
60	1.30	1.67	2.00	2.66	3.23
∞	1.29	1.64	1.96	2.58	3.09

In this case, as with the previous one, we invoke the null hypothesis that the samples are identical and that the observed difference in the results ($\bar{x}_1 - \bar{x}_2$) is the result of indeterminate errors. To test this hypothesis statistically, we employ a modification of Equation 5-7 that takes the form:

$$\bar{x}_1 - \bar{x}_2 = \pm t s_{\text{pooled}} \sqrt{\frac{N_1 + N_2}{N_1 N_2}} \tag{5-8}$$

where s_{pooled} is the pooled sample standard deviation obtained as in Example 4-3.[6]

The numerical value for the term on the right in Equation 5-8 is computed using t for the particular confidence level desired. The number of degrees of freedom for finding t in Table 5-3 is $N_1 + N_2 - 2$. If the experimental difference $\bar{x}_1 - \bar{x}_2$ is smaller than the computed value, the null hypothesis is not rejected and no significant difference between the means has been demonstrated. An experimental difference greater than the value computed from t indicates that there is a significant difference between the means.

If a good estimate of σ is available, Equation 5-8 can be modified by inserting z for t and σ for s.

---

[6] For a derivation of Equation 5-8, see D. A. Skoog, D. M. West, and F. J. Holler, *Fundamentals of Analytical Chemistry*, 6th ed., p. 50. Philadelphia: Saunders College Publishing, 1992.

Example 5-7

Two barrels of wine were analyzed for their alcohol content in order to determine whether they were from different sources. On the basis of six analyses, the average content of the first barrel was established to be 12.61% ethanol. Four analyses of the second barrel gave a mean of 12.53% alcohol. Together the ten analyses yielded a pooled value of $s = 0.070\%$. Do the data indicate a difference between the wines?

Here we employ Equation 5-8, using t for eight degrees of freedom $(10 - 2)$. At the 95% confidence level,

$$\pm ts \sqrt{\frac{N_1 + N_2}{N_1 N_2}} = \pm 2.31 \times 0.070 \sqrt{\frac{6 + 4}{6 \times 4}} = \pm 0.10\%$$

The observed difference is

$$\bar{x}_1 - \bar{x}_2 = 12.61 - 12.53 = 0.08\%$$

As often as 5 times in 100, random error will be responsible for a difference as great as 0.10%. At the 95% confidence level, then, no difference in the alcohol content of the wine has been established.

In Example 5-7, no significant difference between the two wines was detected at the 95% probability level. Note that this statement is not equivalent to saying that $\bar{x}_1$ is equal to $\bar{x}_2$, nor do the tests prove that the wines come from the same source. Indeed, it is even conceivable that one wine is a red and the other is a white. To establish with a reasonable probability that the two wines are from the same source would require extensive testing of other characteristics, such as taste, color, odor, and refractive index, as well as tartaric acid, sugar, and trace element content. If no significant differences are revealed by all of these tests and by others, then it might be possible to judge the two wines as having a common origin. In contrast, the finding of *one* significant difference in any test would show that the two wines are different. Thus, the establishment of a significant difference by a single test is much more revealing than the establishment of an absence of difference.

5B-5 The Least-Squares Method for Deriving Calibration Plots

Most analytical methods are based on a calibration curve in which a measured quantity y is plotted as a function of the known concentration x of a series of standards. Figure 5-4 shows a typical calibration curve to be used for the determination of isooctane in hydrocarbon samples. Here a series of solutions containing known concentrations of isooctane was prepared by mixing carefully measured quantities of the analyte and *n*-hexane. These mixtures were then passed through a chromatographic column, and the areas under the isooctane peak were recorded. A plot of the resulting data is shown in Figure 5-4. As is typical (and desirable), the plot approximates a straight line. Note, however, that not all the data fall exactly on the line because of the random errors in the measuring process. Thus, we must try to derive a ''best'' straight line from the points. A

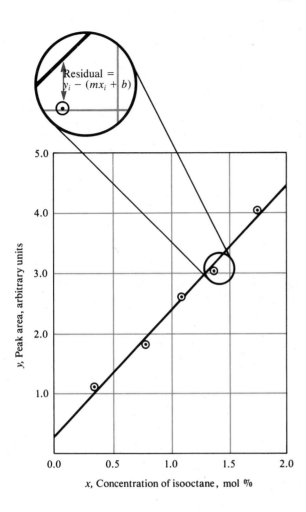

Figure 5-4
Calibration curve for the determination of isooctane in a hydrocarbon mixture.

statistical technique called *the method of least squares* provides the means for objectively obtaining an equation for such a line and also for specifying the uncertainties associated with its subsequent use.

In applying the method of least squares, we assume that there is a linear relationship between the peak areas (y) and the analyte concentration (x) as given by the equation

$$y = mx + b$$

where b is the intercept (the value of y when x is zero) and m is the slope of the line. We also assume that any deviation of individual points from the straight line results from error in the measurement. That is, we assume that we have prepared the standards carefully enough so that the random errors in the preparation process are negligible with respect to those in the measurement process. Usually it is possible to satisfy this assumption.

As illustrated in Figure 5-4, the vertical deviation of each point from the straight line is called a *residual*. The line generated by the least-squares method is the one that minimizes the sum of the squares of the residuals.

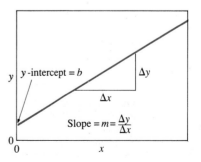

Figure 5-5
The slope-intercept form of a straight line.

Before carrying out a least-squares analysis, you should *always* plot the data to be sure that there is a linear relationship between the variables and to be sure there are no gross errors in the calibration data.

We will not give details here about the generation of a least-squares line because you probably will have available a hand-held calculator or small computer that will carry out this calculation with a touch of a few keys. For those who do not have such equipment, we have provided a procedure in Appendix 6 for determining a least-squares straight line and computing standard deviations of results based on the resulting calibration curve. We must add, however, that you should always plot your calibration data to be sure it is indeed linear and also to be sure that there are no outliers among the data that will distort the derived equation.

5C QUESTIONS AND PROBLEMS

*5-1. Name three types of systematic errors.

5-2. How are systematic method errors detected?

*5-3. What kind of systematic errors are detected by varying the sample size?

5-4. A method of analysis yields weights for gold that are low by 0.3 mg. Calculate the percent relative error caused by this uncertainty if the weight of gold in the sample is
*(a) 800 mg.   *(c) 100 mg.
(b) 500 mg.   (d) 25 mg.

5-5. The method described in Problem 5-4 is to be used for the analysis of ores that assay about 1.2% gold. What minimum sample weight should be taken if the relative error resulting from a 0.3-mg loss is not to exceed
*(a) −0.2%?   *(c) −0.8%?
(b) −0.5%?   (d) −1.2%?

5-6. Calculate the 95% confidence limit for each set of data in Problem 4-4. What do these confidence limits mean?

5-7. Calculate the 95% confidence limit for each set of data in Problem 4-4 if $s \rightarrow \sigma$ and has a value of
*(a) set A, 0.20.   (d) set D, 0.50.
(b) set B, 0.050.   *(e) set E, 0.15.
*(c) set C, 0.0070.   (f) set F, 0.015.

5-8. The last result in each set of data in Problem 4-4 may be an outlier. Apply the Q test (95% confidence level) to determine whether or not there is a statistical basis for rejection.

*5-9. An atomic absorption method for the determination of the amount of iron present in used jet engine oil was found from pooling 30 triplicate analyses to have a standard deviation $s \rightarrow \sigma = 2.4$ μg/mL. Calculate the 80% and 95% confidence limits for the result, 18.5 μg Fe/mL, if it was based on (a) a single analysis, (b) the mean of two analyses, (c) the mean of four analyses.

5-10. An atomic absorption method for determination of copper in fuels yielded a pooled standard deviation of $s \rightarrow \sigma = 0.32$ μg Cu/mL. The analysis of an oil from a reciprocating aircraft engine showed a copper content of 8.53 μg Cu/mL. Calculate the 90% and 99% confidence limit for the result if it was based on (a) a single analysis, (b) the mean of 4 analyses, (c) the mean of 16 analyses.

*5-11. How many replicate measurements are needed to decrease the 95% and 99% confidence intervals for the analysis described in Problem 5-9 to ±1.5 μg Fe/mL?

5-12. How many replicate measurements are necessary to decrease the 95% and 99% confidence intervals for the analysis described in Problem 5-10 to ±0.2 μg Cu/mL?

*5-13. A volumetric calcium analysis on triplicate samples of the blood serum of a patient believed to be suffering from a hyperparathyroid condition produced the following data: mmol Ca/L = 3.15, 3.25, 3.26. What is the 95% confidence limit for the mean of the data, assuming
(a) no prior information about the precision of the analysis?
(b) $s \rightarrow \sigma = 0.056$ mmol Ca/L?

5-14. A chemist obtained the following data for percent lindane in the triplicate analysis of an insecticide preparation: 7.47, 6.98, 7.27. Calculate the 90% confidence limit for the mean of the three data, assuming that
(a) the only information about the precision of the method is the precision for the three data.
(b) on the basis of long experience with the method, it is believed that $s \rightarrow \sigma = 0.28\%$ lindane.

5-15. A standard method for the determination of tetraethyl lead (TEL) in gasoline is reported to have a standard deviation of 0.040 mL TEL per gallon. If $s \rightarrow \sigma = 0.040$, how many replicate analyses should be made in order for the mean for the analysis of a sample to be within
*(a) ±0.03 mL/gal of the true mean 99% of the time?
(b) ±0.03 mL/gal of the true mean 95% of the time?
(c) ±0.02 mL/gal of the true mean 90% of the time?

*5-16. Apply the Q test to the following data sets to determine whether the outlying result should be retained or rejected at the 95% confidence level.
(a) 41.27, 41.61, 41.84, 41.70
(b) 7.295, 7.284, 7.388, 7.292

5-17. Apply the Q test to the following data sets to determine whether the outlying result should be retained or rejected at the 95% confidence level.
(a) 85.10, 84.62, 84.70
(b) 85.10, 84.62, 84.65, 84.70

*5-18. The sulfate ion concentration in natural water can be determined by measuring the turbidity that results when an excess of $BaCl_2$ is added to a measured quantity of the sample. A turbidimeter, the instrument used for this analysis, was calibrated with a series of standard Na_2SO_4 solutions. The following data were obtained in the calibration:

mg SO_4^{2-}/L, c_x	Turbidimeter Reading, R
0.00	0.06
5.00	1.48
10.00	2.28
15.0	3.98
20.0	4.61

Assume that there is a linear relationship between the instrument reading and concentration.

(a) Derive a least-squares equation for the relationship between the variables.

(b) Calculate the concentration of sulfate in a sample yielding a turbidimeter reading of 3.67. Calculate the absolute standard deviation of the result and the coefficient of variation.

(c) Repeat the calculations in (b) assuming that the 3.67 was a mean of six turbidimeter readings.

5-19. The following data were obtained in calibrating a calcium ion electrode for the determination of pCa. It is known that there is a linear relationship between the potential E and pCa.

pCa	E, mV
5.00	−53.8
4.00	−27.7
3.00	+2.7
2.00	+31.9
1.00	+65.1

(a) Derive a least-squares expression for the best straight line through the points.

(b) Calculate the pCa of a serum solution in which the electrode potential was 20.3 mV. Calculate the absolute and relative standard deviations for pCa if the result was from a single voltage measurement.

(c) Calculate the absolute and relative standard deviations for pCa if the millivolt reading in (b) was the mean of two replicate measurements. Repeat the calculation based on the mean of eight measurements.

*5-20. The following are relative peak areas for chromatograms of standard solutions of methyl vinyl ketone (MVK).

Concn MVK, mmol/L	Relative peak area
0.500	3.76
1.50	9.16
2.50	15.03
3.50	20.42
4.50	25.33
5.50	31.97

(a) Derive a least-squares expression assuming the variables bear a linear relationship to one another.

(b) A sample containing MVK yielded a relative peak area of 6.3. Calculate the concentration of MVK in the sample.

(c) Assume that the result in (b) represents a single measurement as well as the mean of four measurements. Calculate the respective absolute and relative standard deviations.

(d) Repeat the calculations in (b) and (c) for a sample that gave a peak area of 27.5.

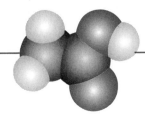

CHAPTER 6

GRAVIMETRIC METHODS OF ANALYSIS

Theodore W. Richards (1868–1928) and his graduate students at Harvard University developed or refined many of the techniques of gravimetric analysis involving silver and chlorine. These techniques were used to determine the atomic weights of 25 of the elements by preparing pure samples of the chlorides of the elements, decomposing known weights of these compounds, and determining their chloride content by gravimetric methods. For this work, Richards became the first American to receive the Nobel Prize in Chemistry in 1914.

Gravimetric methods, which are based on the measurement of mass, are of two major types.[1] In *precipitation methods*, the analyte is converted to a sparingly soluble precipitate. This precipitate is then filtered, washed free of impurities, and converted to a product of known composition by suitable heat treatment. The product is then weighed. For example, in a precipitation method for determining calcium in natural waters recommended by the Association of Official Analytical Chemists, an excess of oxalic acid, $H_2C_2O_4$, is added to a carefully measured volume of the sample. The addition of ammonia causes essentially all of the calcium in the sample to precipitate as calcium oxalate. The reaction is

$$Ca^{2+}(aq) + C_2O_4^{2-}(aq) \rightarrow CaC_2O_4(s)$$

The precipitate is collected in a weighed filtering crucible, dried, and ignited at a red heat. This process converts the precipitate quantitatively to calcium oxide. The reaction is

$$CaC_2O_4(s) \rightarrow CaO(s) + CO(g) + CO_2(g)$$

The crucible and precipitate are cooled, weighed, and the mass of calcium oxide is determined by difference. The calcium content of the sample is then computed as shown in Example 6-1, Section 6C.

In *volatilization methods*, the analyte or its decomposition products are volatilized at a suitable temperature. The volatile product is then collected and weighed,

---

[1] For an extensive treatment of gravimetric methods, see C. L. Rulfs, in *Treatise on Analytical Chemistry*, I. M. Kolthoff and P. J. Elving, Eds., Part I, Vol. 11, Chapter 13. New York: Wiley, 1975.

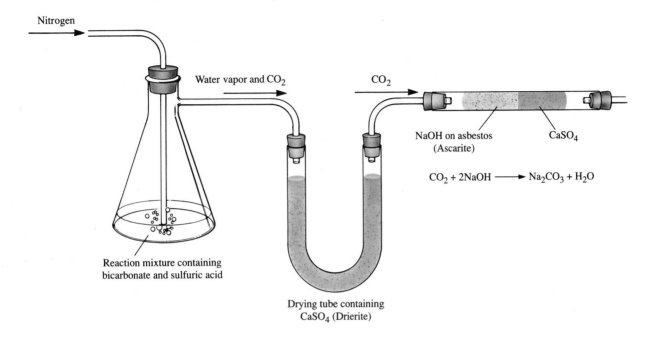

Figure 6-1

Apparatus for determining the sodium hydrogen carbonate content of antacid tablets by a gravimetric volatilization procedure.

or alternatively the mass of the product is determined indirectly from the loss in mass of the sample. An example of a gravimetric volatilization procedure is the determination of the sodium hydrogen carbonate content of antacid tablets. A weighed sample of the finely ground tablets is treated with dilute sulfuric acid to convert the sodium hydrogen carbonate to carbon dioxide:

$$NaHCO_3(aq) + H_2SO_4(aq) \rightarrow CO_2(g) + H_2O(l) + NaHSO_4(aq)$$

As shown in Figure 6-1, this reaction is carried out in a flask that is connected to a weighed absorption tube containing an absorbent (see Section 6D-4 for a description of the absorbent) that retains the carbon dioxide selectively as it is removed from the solution by heating. The difference in mass of the tube before and after absorption is used to calculate the amount of sodium hydrogen carbonate.

Gravimetric methods of analysis are based on mass measurements with an analytical balance, an instrument that yields highly accurate and precise data. If you perform a gravimetric chloride analysis in the laboratory, you may make some of the most accurate and precise measurements of your life.

6A   PROPERTIES OF PRECIPITATES AND PRECIPITATING REAGENTS

Ideally, a gravimetric precipitating agent should react *specifically* or, if not, *selectively* with the analyte. Specific reagents, which are rare, react only with a single chemical species. Selective reagents, which are more common, react with only a limited number of species. In addition to specificity or selectivity, the ideal precipitating reagent would react with the analyte to give a product that is

An example of a selective reagent is $AgNO_3$. The only common ions that it precipitates from acidic solution are Cl^-, Br^-, I^-, and SCN^-. Dimethylglyoxime, which is discussed in Section 6D-3, is a specific reagent that precipitates only Ni^{2+} from alkaline solutions.

1. easily filtered and washed free of contaminants;
2. of sufficiently low solubility so that no significant loss of the analyte occurs during filtration and washing;

Ignition in this context means to heat a solid at high temperature to convert organic matter to CO_2 and H_2O and to convert the analyte to its oxide or some other heat-stable form.

3. unreactive with constituents of the atmosphere;
4. of known composition after it is dried or, if necessary, ignited.

Few if any reagents produce precipitates that have all these desirable properties.

The variables that influence solubility (the second property in this list) are discussed in Chapter 7. In this section we consider methods for obtaining pure and easily filtered solids of known composition.

6A-1 Particle Size and Filterability of Precipitates[2]

Precipitates made up of large particles are generally desirable in gravimetric work because large particles are easy to filter and wash free of impurities. In addition, such precipitates are usually purer than are finely divided precipitates.

Factors That Determine the Particle Size of Precipitates

In diffuse light, colloidal suspensions may be perfectly clear and appear to contain no solid. The presence of the second phase can be detected, however, by shining the beam of a flashlight into the solution (see color plate 6). Because particles of colloidal dimensions scatter visible radiation, the path of the beam through the solution can be seen by the eye. This phenomenon is called the *Tyndall effect*.

The particles of colloidal suspension are not easily filtered. To trap these particles, the pore size of the filtering medium must be so small that filtrations take too long. With suitable treatment, however, the individual colloidal particles can be made to stick together to give a filterable mass.

Equation 6-1 is known as the Von Weimarn equation in recognition of the scientist who proposed it in 1925.

To increase the particle size of a precipitate, minimize the relative supersaturation during precipitate formation.

The particle size of solids formed by precipitation varies enormously. At one extreme are *colloidal suspensions*, whose tiny particles are invisible to the naked eye (10^{-7} to 10^{-4} cm in diameter). Colloidal particles show no tendency to settle from solution, nor are they easily filtered. At the other extreme are particles with dimensions on the order of tenths of a millimeter or greater. The temporary dispersion of such particles in the liquid phase is called a *crystalline suspension*. The particles of a crystalline suspension tend to settle spontaneously and are easily filtered.

Scientist have studied precipitate formation for many years, but the mechanism of the process is still not fully understood. It is certain, however, that the particle size of a precipitate is influenced by such experimental variables as precipitate solubility, temperature, reactant concentrations, and rate at which reactants are mixed. The net effect of these variables can be accounted for, at least qualitatively, by assuming that the particle size is related to a single property of the system called its *relative supersaturation*, where

$$\text{relative supersaturation} = \frac{Q - S}{S} \qquad \textbf{(6-1)}$$

In this equation Q is the concentration of the solute at any instant and S is its equilibrium solubility.

Generally precipitation reactions are slow so that even when a precipitating reagent is added drop by drop to a solution of an analyte, some supersaturation is likely. Experimental evidence indicates that the particle size of a precipitate varies inversely with the average relative supersaturation during the time the reagent is being introduced. Thus, when $(Q - S)/S$ is large, the precipitate tends to be colloidal; when $(Q - S)/S$ is small, a crystalline solid is more likely.

---

[2] For a more detailed treatment of precipitates, see H. A. Laitinen and W. E. Harris, *Chemical Analysis*, 2nd ed., Chapters 8 and 9. New York: McGraw-Hill, 1975; A. E. Nielsen, in *Treatise on Analytical Chemistry*, 2nd ed., I. M. Kolthoff and P. J. Elving, Eds., Part I, Vol. 3, Chapter 27. New York: Wiley, 1983.

Mechanism of Precipitate Formation

The effect of relative supersaturation on particle size can be explained if we assume that precipitates form by two different pathways: by *nucleation* and by *particle growth*. The particle size of a freshly formed precipitate is determined by which of these is favored.

In nucleation, a few ions, atoms, or molecules (perhaps as few as four or five) come together to form stable solid particles. Often, these nuclei form on the surface of suspended solid contaminants, such as dust particles. Further precipitation then involves a competition between additional nucleation and growth on existing nuclei (particle growth). If nucleation predominates, a precipitate containing a large number of small particles results; if growth predominates, a smaller number of large particles is produced.

The rate of nucleation is believed to increase enormously with increasing relative supersaturation. In contrast, the rate of particle growth is only moderately enhanced by high relative supersaturation. Thus, when a precipitate is formed at high relative supersaturation, nucleation is the major precipitation mechanism, and a large number of small particles are formed. On the other hand, at low relative supersaturations the rate of particle growth tends to predominate, and deposition of solid on existing particles occurs to the exclusion of further nucleation; a crystalline suspension results.

Nucleation is a process in which a minimum number of atoms, ions, or molecules join together to give a stable solid particle.

Precipitates form by nucleation and by particle growth. If nucleation predominates, a large number of very fine particles results; if particle growth predominates, a smaller number of larger particles are obtained.

Experimental Control of Particle Size

Experimental variables that minimize supersaturation and thus lead to crystalline precipitates include elevated temperatures to increase the solubility of the precipitate (S in Equation 6-1), dilute solutions (to minimize Q), and slow addition of the precipitating agent with good stirring. The last two measures also minimize the concentration of the solute (Q) at any given instant.

Larger particles can also be obtained by pH control, provided the solubility of the precipitate depends on pH. For example, large, easily filtered crystals of calcium oxalate are obtained by forming the bulk of the precipitate in a mildly acidic environment in which the salt is moderately soluble. The precipitation is then completed by slowly adding aqueous ammonia until the acidity is sufficiently low that essentially all of the calcium oxalate is precipitated. The additional precipitate produced during this step forms on the solid particles produced in the first step.

Unfortunately, many precipitates cannot be formed as crystals under practical laboratory conditions. A colloidal solid is generally encountered when a precipitate has such a low solubility that S in Equation 6-1 always remains negligible relative to Q. The relative supersaturation thus remains enormous throughout precipitate formation, and a colloidal suspension results. For example, under conditions feasible for an analysis, the hydrous oxides of iron(III), aluminum, and chromium(III), and the sulfides of most heavy-metal ions form only as colloids because of their very low solubilities.[3]

Precipitates having very low solubilities, such as many sulfides and hydrous oxides, generally form as colloids.

---

[3] Silver chloride illustrates that the relative supersaturation concept is imperfect. This compound ordinarily forms as a colloid, and yet its molar solubility is not significantly different from that of other compounds, such as $BaSO_4$, which generally form as crystals.

6A-2 Colloidal Precipitates

Colloidal suspensions are often stable for indefinite periods and are not directly usable for gravimetric analysis because their particles are too small to be readily filtered. Fortunately, the stability of most suspensions of this kind can be decreased by heating, by stirring, and by adding an electrolyte. These measures cause the individual colloidal particles to bind together to give an amorphous mass that settles out of solution and is filterable. The process of converting a colloidal suspension into a filterable solid is called *coagulation* or *agglomeration*.

Coagulation of Colloids

Colloidal suspensions are stable because the particles are either all positively charged or all negatively charged and thus repel one another. This charge results from cations or anions that are bound to the surface of the particles. The process by which ions are retained *on the surface of a solid* is known as *adsorption*. We can readily demonstrate that colloidal particles are charged by observing their migration when placed in an electrical field.

The tendency of ions to be adsorbed on an ionic solid surface originates in the normal bonding forces that are responsible for crystal growth. For example, a silver ion at the surface of a silver chloride particle has a partially unsatisfied bonding capacity for anions because of its surface location. Negative ions are attracted to this site by the same forces that hold chloride ions in the silver chloride lattice. Chloride ions at the surface of the solid exert an analogous attraction for cations dissolved in the solvent.

The kind and number of ions retained on the surface of a colloidal particle depends, in a complex way, on several variables. For a suspension produced in the course of a gravimetric analysis, however, the species adsorbed—and hence

Adsorption is a process in which a substance (gas, liquid, or solid) is held *on the surface* of a solid. In contrast, *absorption* involves retention of a substance *within* the pores of a solid.

The charge on a colloidal particle formed in a gravimetric analysis is determined by the charge of the lattice ion that is in excess when the precipitation is complete.

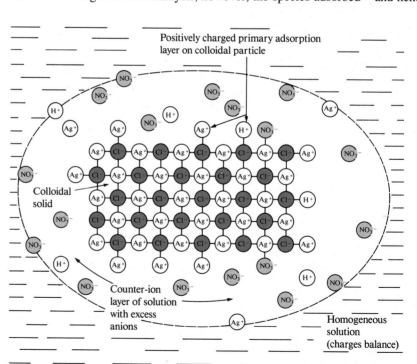

Figure 6-2
A colloidal silver chloride particle suspended in a solution of silver nitrate.

the charge on the particles—is readily predicted because lattice ions are generally more strongly held than any other ions. For example, when sodium chloride is first added to a solution containing silver nitrate, the colloidal particles of silver chloride formed are positively charged, as shown in Figure 6-2. This charge is due to adsorption of some of the excess silver ions in the medium. The charge on the particles becomes negative, however, when enough sodium chloride has been added to provide an excess of chloride ions. Now, the adsorbed species is primarily chloride ions. The surface charge is at a minimum when the supernatant liquid contains an excess of neither ion.

The extent of the adsorption, and thus the charge on a given particle, increases rapidly as the concentration of the lattice ion becomes greater. Eventually, however, the surface of the particles becomes covered with the adsorbed ions, and the charge then becomes constant and independent of concentration.

Figure 6-2 shows a colloidal silver chloride particle in a solution that contains an excess of silver nitrate. Attached directly to the solid surface is the *primary adsorption layer*, which consists mainly of adsorbed silver ions. Surrounding the charged particle is a layer of solution, called *the counter-ion layer*, which contains an excess of negative ions (principally nitrate). The primarily adsorbed silver ions and the negative counter-ion layer constitute an *electric double layer* that stabilizes a colloidal suspension by preventing individual particles from coming close enough to one another to agglomerate.

Figure 6-3a shows the effective charge on two silver chloride particles. The upper curve is for a particle in a solution that contains a reasonably large excess of silver nitrate, whereas the lower curve depicts a particle in a solution that has a much lower silver nitrate content. The effective charge can be thought of as a measure of the repulsive force the particle exerts on like particles in the solution. Note that the effective charge falls off rapidly as the distance from the surface increases and approaches zero at the points d_1 or d_2. These decreases in effective charge (in both cases positive) are caused by the negative charge of the excess counter ions in the double layer surrounding each particle. At points d_1 and d_2 the number of counter ions in the layers is approximately equal to the number of primarily adsorbed ions on the surface of the particles so that the effective charge of each particle approaches zero at this point.

The upper schematic in Figure 6-4 depicts two silver chloride particles and their counter-ion layers as they approach one another in the concentrated silver

> The electric double layer of a colloid consists of a layer of charge adsorbed on the surface of the particles and a layer with a net opposite charge in the solution surrounding the particles.

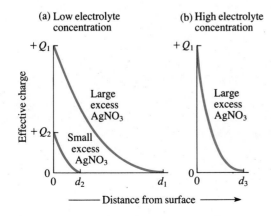

Figure 6-3

Effect of $AgNO_3$ and electrolyte concentration on the thickness of the double layer surrounding a colloidal AgCl particle in a solution containing excess of $AgNO_3$.

nitrate just considered. Note that the effective charge on the particles prevents them from approaching one another more closely than about $2d_1$—a distance that is too great for coagulation to occur. As shown in the lower part of Figure 6-4 in the more dilute silver nitrate solution, the two particles can approach within $2d_2$ of one another. Ultimately, as the concentration of silver nitrate is further decreased, the distance between particles becomes small enough so that the forces of agglomeration can take over, and a coagulated precipitate begins to appear.

Coagulation of a colloidal suspension can often be brought about by a short period of heating, particularly if accompanied by stirring. Heating decreases the number of adsorbed ions and thus the thickness d_i of the double layer. The particles may also gain enough kinetic energy at the higher temperature to overcome the barrier to close approach posed by the electrostatic repulsion of the double layers of particles.

An even more effective way to coagulate a colloid is to increase the electrolyte concentration of the solution. If we add a suitable ionic compound to a suspension, the concentration of counter ions increases in the vicinity of each particle. As a result, the volume of solution that contains sufficient counter ions to balance the charge of the primary adsorption layer decreases. The net effect of adding an electrolyte is thus a shrinkage of the counter-ion layer as shown in Figure 6-4. The particles can then approach one another more closely and agglomerate.

Peptization of Colloids

Peptization is the process by which a coagulated colloid reverts to its original dispersed state. When a coagulated colloid is washed, some of the electrolyte responsible for its coagulation is leached from the internal liquid in contact with the solid particles. Removal of this electrolyte has the effect of increasing the volume of the counter-ion layer. The repulsive forces responsible for the original colloidal state are then reestablished, and particles detach themselves from the coagulated mass. The washings become cloudy as the freshly dispersed particles pass through the filter.

We are thus faced with a dilemma in working with coagulated colloids. On the one hand, washing is needed to minimize contamination; on the other, there is the risk of losses resulting from peptization if pure water is used. The problem is commonly resolved by washing the precipitate with a solution containing an electrolyte that volatilizes during the subsequent drying or ignition step. For example, silver chloride is ordinarily washed with a dilute solution of nitric acid. While the precipitate undoubtedly becomes contaminated with the acid, no harm results since the nitric acid is volatilized during the ensuing drying step.

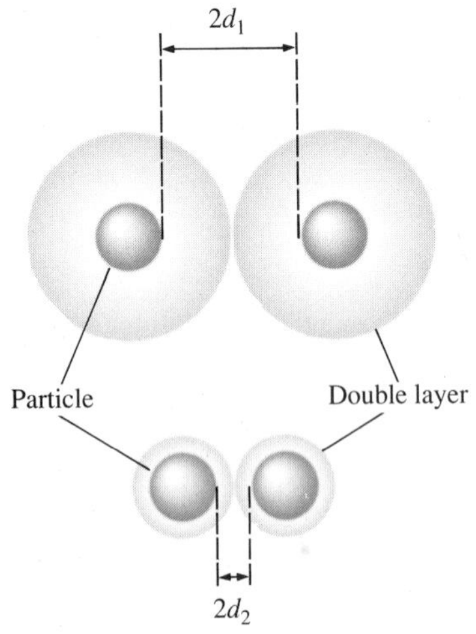

Practical Treatment of Colloidal Precipitates

Colloids are best precipitated from hot, stirred solutions containing sufficient electrolyte to ensure coagulation. The filterability of a coagulated colloid frequently improves if it is allowed to stand for an hour or more in contact with the hot solution from which it was formed. During this process, which is known as *digestion*, weakly bound water appears to be lost from the precipitate; the result is a denser mass that is easier to filter.

Colloidal suspensions can often be coagulated by heating, by stirring and by adding an electrolyte.

Figure 6-4
The electric double layer of a colloid consists of a layer of charge adsorbed on the surface of the particles and a layer of opposite charge in the solution surrounding the particles.

6A-3 Crystalline Precipitates

Crystalline precipitates are generally more easily filtered and purified than are coagulated colloids. In addition, the size of individual crystalline particles, and thus their filterability, can be controlled to a degree.

> Digestion is a process in which a precipitate is heated for an hour or more in the solution from which it was formed (known as the *mother liquor*).

Methods of Improving Particle Size and Filterability

The particle size of crystalline solids can often be improved significantly by minimizing Q and/or maximizing S in Equation 6-1. Minimization of Q is generally accomplished by using dilute solutions and adding the precipitating reagent slowly and with good mixing. Often S is increased by precipitating from hot solution or by adjusting the pH of the precipitation medium.

Digestion of crystalline precipitates (without stirring) for some time after formation frequently yields a purer, more filterable product. The improvement in filterability results from the dissolution and recrystallization that occur continuously and at an enhanced rate at elevated temperatures. Recrystallization apparently results in bridging between adjacent particles, a process that yields larger and more easily filtered crystalline aggregates. This view is supported by the observation that little improvement in filtering characteristics occurs if the mixture is stirred during digestion.

> Digestion improves the purity and filterability of both colloidal and crystalline precipitates.

6A-4 Coprecipitation

Coprecipitation is a phenomenon in which *otherwise soluble* compounds are removed from solution during precipitate formation. It is important to understand that the solution is *not* saturated with the coprecipitated species. Moreover, contamination of a precipitate by a second substance whose solubility product has been exceeded *does not constitute coprecipitation.*

There are four types of coprecipitation: *surface adsorption, mixed-crystal formation, occlusion,* and *mechanical entrapment.*[4] Surface adsorption and mixed-crystal formation are equilibrium processes, whereas occlusion and mechanical entrapment are controlled by the kinetics of crystal growth.

> Coprecipitation is a process in which *normally soluble* compounds are carried out of solution by a precipitate.

Surface Adsorption

Adsorption is a common source of coprecipitation that is likely to cause significant contamination of precipitates with large specific surface areas—that is, coagulated colloids (see Feature 6-1 for definition of specific area). Although adsorption does occur in crystalline solids, its effects on purity are usually undetectable because of the relatively small specific surface area of these solids.

Coagulation of a colloid does not significantly decrease the amount of adsorption because the coagulated solid still contains large internal surface areas that remain exposed to the solvent (Figure 6-5). The coprecipitated contaminant on the coagulated colloid consists of the lattice ion originally adsorbed on the surface

> Adsorption is often the major source of contamination in coagulated colloids but of no significance in crystalline precipitates.

---

[4] Several systems of classification of coprecipitation phenomena have been suggested. We follow the simple system proposed by A. E. Nielsen, in *Treatise on Analytical Chemistry*, 2nd ed., I. M. Kolthoff and P. J. Elving, Eds., Part I, Vol. 3, p. 333. New York: Wiley, 1983.

> The specific surface area of a solid is the ratio of its surface area (in cm^2) to its mass (g).

> In adsorption, a normally soluble compound is carried out of solution on the surface of a coagulated colloid. This compound consists of the primarily adsorbed ion and an ion of opposite charge from the counter-ion layer.

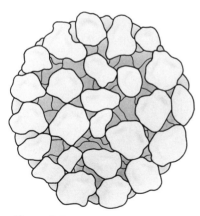

Figure 6-5
A coagulated colloid. Note the extensive *internal* surface area exposed to solvent.

Feature 6-1
SPECIFIC SURFACE AREA OF COLLOIDS

Specific surface area is defined as the surface area per unit mass of solid and is ordinarily expressed in terms of square centimeters per gram. For a given mass of solid, the specific surface area increases dramatically as particle size decreases and becomes enormous for colloids. For example, the solid cube shown in Figure 6-A, which has dimensions of 1 cm on a side, has a surface area of 6 cm^2. If this cube weighs 2 g, its specific surface area is 6 cm^2/2g = 3 cm^2/g. This cube could be divided into 1000 cubes each being 0.1 cm on a side. The surface area of each face of these cubes is now 0.1 cm × 0.1 cm = 0.01 cm^2 and the total area for the 6 faces of the cube is 0.06 cm^2. Because there are 1000 of these cubes, the total surface area for the 2 g of solid is now 60 cm^2; the specific surface area is then 30 cm^2/g. Continuing in this way we find that the specific surface area becomes 300 cm^2/g when we have 10^6 cubes that are 0.01 cm on a side. The particle size of a typical crystalline suspension lies in the region of 0.1 and 0.01 cm, so a typical crystalline precipitate has a specific surface area between 30 and 300 cm^2/g. Contrast these figures with that for 2 g of a colloid made up of 10^{18} particles having dimensions of 10^{-6} cm. Here, the specific area is 3 × 10^6 cm^2/g, which converts to something over 3000 ft^2/g. Thus 1 g of a colloidal suspension has a surface area that is equivalent to the floor area of a good-sized home.

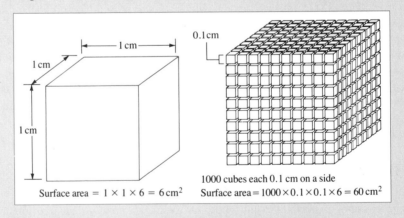

Surface area = 1 × 1 × 6 = 6 cm^2 Surface area = 1000 × 0.1 × 0.1 × 6 = 60 cm^2

Figure 6-A
Increase in surface area per unit mass with decrease in particle size.

before coagulation and the counter ion of opposite charge held in the film of solution immediately adjacent to the particle. *The net effect of surface adsorption is therefore the carrying down of an otherwise soluble compound as a surface contaminant.* For example, the coagulated silver chloride formed in the gravimetric determination of chloride ion is contaminated with primarily adsorbed silver ions along with nitrate or other anions in the counter-ion layer. As a consequence, silver nitrate, a normally soluble compound, is coprecipitated with the silver chloride.

Methods for Minimizing Adsorbed Impurities on Colloids. The purity of many coagulated colloids is improved by digestion. During this process, water is expelled from the solid to give a denser mass that has a smaller specific surface area for adsorption.

Washing a coagulated colloid with a solution containing a volatile electrolyte may also be helpful by displacing nonvolatile electrolyte species that may be present. Washing generally does not remove much of the primarily adsorbed ions because the attraction between these ions and the surface of the solid is too strong. Exchange occurs, however, between existing *counter ions* and ions in the wash liquid. For example, in the determination of silver by precipitation with chloride ion, the primarily adsorbed species is chloride. Washing with an acidic solution converts the counter-ion layer largely to hydrogen ions so that both chloride and hydrogen ions are retained by the solid. Volatile HCl is then given off when the precipitate is dried.

Regardless of the method of treatment, a coagulated colloid is always contaminated to some degree, even after extensive washing. The error introduced into the analysis from this source can be as low as 1 to 2 ppt, as in the coprecipitation of silver nitrate on silver chloride. On the other hand, coprecipitation of heavy-metal hydroxides on the hydrous oxides of trivalent iron or aluminum may introduce errors amounting to several percent, which is usually unacceptable.

Reprecipitation. A drastic but effective way to minimize the effects of adsorption is *reprecipitation*, or *double precipitation*. Here, the filtered solid is redissolved and reprecipitated. The first precipitate ordinarily carries down only a fraction of the contaminant present in the original solvent. Thus, the solution containing the redissolved precipitate has a significantly lower contaminant concentration than the original, and even less adsorption occurs during the second precipitation. Reprecipitation adds substantially to the time required for an analysis but is often necessary for such precipitates as the hydrous oxides of iron(III) and aluminum, which have extraordinary tendencies to adsorb the hydroxides of heavy-metal cations, such as zinc, cadmium, and manganese.

Mixed-Crystal Formation

In mixed-crystal formation, one of the ions in the crystal lattice of a solid is replaced by an ion of another element. For this exchange to occur, it is necessary that the two ions have the same charge and that their sizes differ by no more than about 5%. Furthermore, the two salts must belong to the same crystal class. For example, barium sulfate formed by adding barium chloride to a solution containing sulfate, lead, and acetate ions is found to be severely contaminated by lead sulfate even though acetate ions normally prevent precipitation of lead sulfate by complexing the lead. Here, lead ions replace some of the barium ions in the barium sulfate crystals. Other examples of coprecipitation by mixed-crystal formation include $MgKPO_4$ in $MgNH_4PO_4$, $SrSO_4$ in $BaSO_4$, and MnS in CdS.

The extent of mixed-crystal contamination is governed by the law of mass action and increases as the ratio of contaminant to analyte concentration increases. Mixed-crystal formation is a particularly troublesome type of coprecipitation

Surface adsorption does not usually affect the purity of a crystalline precipitate noticeably because the specific surface area of the solid is small.

Calcium oxalate is often freed from coprecipitated magnesium oxalate by reprecipitation. Here the precipitated calcium oxalate is dissolved in dilute sulfuric acid. Ammonium oxalate is added to the acid solution followed by ammonia, until a second precipitate of calcium oxalate forms. The new precipitate is largely free of magnesium because the concentration of this ion in the second solution is far less than it was in the first.

Mixed-crystal formation is a type of coprecipitation in which a contaminant ion replaces an ion in the lattice of a crystal.

because little can be done about it when certain combinations of ions are present in a sample matrix. This problem is encountered with both colloidal suspensions and crystalline precipitates. When mixed-crystal formation occurs, separation of the interfering ion may have to be carried out before the final precipitation step. Alternatively, a different precipitating reagent that does not give mixed crystals with the ions in question may be used.

Occlusion and Mechanical Entrapment

When a crystal is growing rapidly during precipitate formation, foreign ions in the counter-ion layer may become trapped, or *occluded*, within the growing crystal. Because supersaturation, and thus growth rate, decrease as a precipitation progresses, the amount of occluded material is greatest in that part of a crystal that forms first.

Mixed-crystal formation may occur in both colloidal and crystalline precipitates, whereas occlusion and mechanical entrapment are confined to crystalline precipitates.

Mechanical entrapment occurs when crystals lie close together during growth. Here, several crystals grow together and in so doing trap a portion of the solution in a tiny pocket.

Both occlusion and mechanical entrapment are minimal when the rate of precipitate formation is low—that is, under conditions of low supersaturation. In addition, digestion is often remarkably helpful in reducing these types of coprecipitation. Undoubtedly, the rapid solution and reprecipitation that goes on at the elevated temperature of digestion opens up the pockets and allows the impurities to escape into the solution.

Coprecipitation Errors

Coprecipitated impurities may cause either negative or positive errors in an analysis. If the contaminant is not a compound of the ion being determined, positive errors always result. Thus, a positive error is observed whenever colloidal silver chloride adsorbs silver nitrate during a chloride analysis. In contrast, when the contaminant does contain the ion being determined, either positive or negative errors may be observed. For example, in the determination of barium by precipitation as barium sulfate, occlusion of other barium salts occurs. If the occluded contaminant is barium nitrate, a positive error is observed because this compound has a larger molar mass than the barium sulfate that would have formed had no coprecipitation occurred. If barium chloride is the contaminant, the error is negative because its molar mass is less than that of the sulfate salt.

Coprecipitation can cause either negative or positive errors.

6A-5 Precipitation from Homogeneous Solution

Precipitation from homogeneous solution is a technique in which a precipitating agent is generated in a solution of the analyte by a slow chemical reaction.[5] Local reagent excesses do not occur because the precipitating agent appears gradually and homogeneously throughout the solution and reacts immediately with the analyte. As a result, the relative supersaturation is kept low during the entire precipitation. In general, homogeneously formed precipitates, both colloidal and crystalline, are better suited to analysis than are solids formed by direct addition of a precipitating reagent.

---

[5] For a general reference on this technique, see L. Gordon, M. L. Salutsky, and H. H. Willard, *Precipitation from Homogeneous Solution.* New York: Wiley, 1959.

Urea is often used for the homogeneous generation of hydroxide ion. The reaction can be expressed by the equation

$$(H_2N)_2CO + 3H_2O \rightarrow CO_2 + 2NH_4^+ + 2OH^-$$

This reaction proceeds slowly just below 100°C, and a 1- to 2-hour heating period is needed to complete a typical precipitation. Urea is particularly valuable for the precipitation of hydrous oxides or basic salts. For example, hydrous oxides of iron(III) and aluminum formed by direct addition of base are bulky and gelatinous masses that are heavily contaminated and difficult to filter. In contrast, when these same products are produced by homogeneous generation of hydroxide ion, they are dense, readily filtered, and have considerably higher purity. Figure 6-6 shows hydrous oxide precipitates of aluminum formed by direct addition of base and by homogeneous precipitation with urea. Similar pictures for hydrous oxides of iron(III) are shown in color plate 7. Homogeneous precipitation of crystalline precipitates also results in marked increases in crystal size as well as improvements in purity.

Representative methods based on precipitation by homogeneously generated reagents are given in Table 6-1.

6B   DRYING AND IGNITION OF PRECIPITATES

After filtration, a gravimetric precipitate is heated until its mass becomes constant. Heating removes the solvent and any volatile species carried down with the precipitate. Some precipitates are also ignited to decompose the solid and produce a compound of known composition. This new compound is often called the *weighing form.*

Jöns Jacob Berzelius (1779–1848), considered the leading chemist of his time, developed much of the apparatus and many of the techniques of nineteenth-century analytical chemistry. Examples include the use of ashless filter paper in gravimetry, the use of hydrofluoric acid to decompose silicates, and the use of the metric system in weight determinations. He performed thousands of analyses of pure compounds to determine the atomic weights of most of the elements known then. Berzelius also developed our present system of symbols for elements and compounds.

Figure 6-6

Hydrous aluminum oxide formed by (a) homogeneous generation of hydroxide ion, and (b) direct addition of base.

Table 6-1

METHODS FOR THE HOMOGENEOUS GENERATION OF PRECIPITATING AGENTS

Precipitating Agent	Reagent	Generation Reaction	Elements Precipitated
OH^-	Urea	$(NH_2)_2CO + 3H_2O \rightarrow$ $CO_2 + 2NH_4^+ + 2OH^-$	Al, Ga, Th, Bi, Fe, Sn
PO_4^{3-}	Trimethyl phosphate	$(CH_3O)_3PO + 3H_2O \rightarrow$ $3CH_3OH + H_3PO_4$	Zr, Hf
$C_2O_4^{2-}$	Ethyl oxalate	$(C_2H_5)_2C_2O_4 + 2H_2O \rightarrow$ $2C_2H_5OH + H_2C_2O_4$	Mg, Zn, Ca
SO_4^{2-}	Dimethyl sulfate	$(CH_3O)_2SO_2 + 4H_2O \rightarrow$ $2CH_3OH + SO_4^{2-} + 2H_3O^+$	Ba, Ca, Sr, Pb
CO_3^{2-}	Trichloroacetic acid	$Cl_3CCOOH + 2OH^- \rightarrow$ $CHCl_3 + CO_3^{2-} + H_2O$	La, Ba, Ra
H_2S	Thioacetamide*	$CH_3CSNH_2 + H_2O \rightarrow$ $CH_3CONH_2 + H_2S$	Sb, Mo, Cu, Cd
DMG†	Biacetyl + hydroxylamine	$CH_3COCOCH_3 +$ $2H_2NOH \rightarrow DMG + 2H_2O$	Ni
HOQ‡	8-Acetoxyquinoline§	$CH_3COOQ + H_2O \rightarrow$ $CH_3COOH + HOQ$	Al, U, Mg, Zn

* $CH_3-\overset{\overset{S}{\|}}{C}-NH_2$

† DMG = Dimethylglyoxime = $CH_3-\overset{\overset{OH}{\|}}{\underset{N}{C}}-\overset{\overset{OH}{\|}}{\underset{N}{C}}-CH_3$

‡ HOQ = 8-Hydroxyquinoline =

§ $CH_3-\overset{\overset{O}{\|}}{C}-O$

The temperature required to produce a suitable product varies from precipitate to precipitate. Figure 6-7 shows mass loss as a function of temperature for several common analytical precipitates. These data were obtained with an automatic thermobalance,[6] an instrument that records the mass of a substance continuously as its temperature is increased at a constant rate in a furnace (see Fig. 6-8). Heating three of the precipitates—silver chloride, barium sulfate, and aluminum oxide—simply causes removal of water and perhaps volatile contaminants. Note the vastly different temperatures required to produce an anhydrous precipitate of constant mass. Moisture is completely removed from silver chloride above 110°C, but dehydration of aluminum oxide is not complete until a temperature greater than 1000°C is achieved. It is of interest to note that aluminum oxide formed homogeneously with urea can be completely dehydrated at about 650°C.

---

[6]For descriptions of thermobalances, see W. W. Wendlandt, *Thermal Methods of Analysis*, 2nd ed. New York: Wiley, 1986.

The thermal curve for calcium oxalate is considerably more complex than the others in Figure 6-7. Below about 135°C, unbound water is eliminated to give the monohydrate $CaC_2O_4 \cdot H_2O$. This compound is then converted to the anhydrous oxalate CaC_2O_4 at 225°C. The abrupt change in mass at about 450°C signals the decomposition of calcium oxalate to calcium carbonate and carbon monoxide. The final step in the curve depicts the conversion of the carbonate to calcium oxide and carbon dioxide. It is evident that the compound finally weighed in a gravimetric calcium determination based on precipitation as oxalate is highly dependent on the ignition temperature.

6C   CALCULATION OF RESULTS FROM GRAVIMETRIC DATA

The results of a gravimetric analysis are generally computed from two experimental measurements: the mass of sample and the mass of a product of known composition formed from the analyte. The examples that follow illustrate how such computations are carried out.

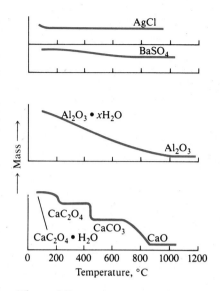

Figure 6-7

Effect of temperature on precipitate mass.

The temperature required to dehydrate a precipitate completely may be as low as 100°C or as high as 1000°C.

Example 6-1

The calcium in a 200.0-mL sample of a natural water was determined by precipitating the cation as CaC_2O_4. The precipitate was filtered, washed, and ignited in a crucible having an empty mass of 26.6002 g. The mass of the crucible plus CaO (56.08 g/mol) was 26.7134 g. Calculate the mass of Ca (40.08 g/mol) per 100 mL of the water.

The mass of CaO is

$$26.7134 \text{ g} - 26.6002 \text{ g} = 0.1132 \text{ g}$$

The number of moles Ca in the sample is equal to the number of moles CaO or

$$\text{amount Ca} = 0.1132 \text{ g CaO} \times \frac{1 \text{ mol CaO}}{56.08 \text{ g CaO}} \times \frac{1 \text{ mol Ca}}{\text{mol CaO}}$$

$$= 2.0185 \times 10^{-3} \text{ mol Ca}$$

$$\text{mass Ca/100 mL} = \frac{2.0185 \times 10^{-3} \text{ mol Ca} \times 40.08 \text{ g Ca/mol Ca}}{200 \text{ mL sample}} \times 100 \text{ mL}$$

$$= 0.04045 \text{ g Ca/100 mL}$$

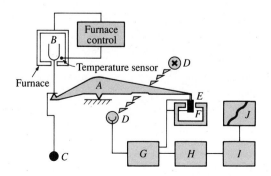

Figure 6-8

Schematic of a thermobalance: A: beam; B: sample cup and holder; C: counterweight; D: lamp and photodiodes; E: coil; F: magnet; G: control amplifier; H: tare calculator; I: amplifier; and J: recorder. (Courtesy of Mettler Instrument Corp., Hightstown, NJ.)

Example 6-2

An iron ore was analyzed by dissolving a 1.1324-g sample in concentrated HCl. The resulting solution was diluted with water, and the iron(III) was precipitated as the hydrous oxide $Fe_2O_3 \cdot xH_2O$ by the addition of NH_3. After filtration and washing, the residue was ignited at a high temperature to give 0.5394 g of pure Fe_2O_3 (159.69 g/mol). Calculate (a) the percent Fe (55.847 g/mol), and (b) the percent Fe_3O_4 (231.54 g/mol) in the sample.

For both parts of this problem we will need to calculate the number of moles of Fe_2O_3. Thus,

$$\text{amount } Fe_2O_3 = 0.5394 \text{ g } Fe_2O_3 \times \frac{1 \text{ mol } Fe_2O_3}{159.69 \text{ g } Fe_2O_3}$$

$$= 3.3778 \times 10^{-3} \text{ mol } Fe_2O_3$$

(a) The number of moles of Fe is twice the number of moles of Fe_2O_3. Thus

$$\text{mass Fe} = 3.3778 \times 10^{-3} \text{ mol } Fe_2O_3 \times \frac{2 \text{ mol Fe}}{\text{mol } Fe_2O_3} \times 55.847 \frac{\text{g Fe}}{\text{mol Fe}}$$

$$= 0.37728 \text{ g Fe}$$

$$\text{percent Fe} = \frac{0.37728 \text{ g Fe}}{1.1324 \text{ g sample}} \times 100\% = 33.317 = 33.32\%$$

(b) As shown by the following balanced equation, 3 mol of Fe_2O_3 are chemically equivalent to 2 mol of Fe_3O_4. That is,

$$3Fe_2O_3 \rightarrow 2Fe_3O_4 + \tfrac{1}{2}O_2$$

$$\text{mass } Fe_3O_4 = 3.3778 \times 10^{-3} \text{ mol } Fe_2O_3 \times \frac{2 \text{ mol } Fe_3O_4}{3 \text{ mol } Fe_2O_3} \times \frac{231.54 \text{ g } Fe_3O_4}{\text{mol } Fe_3O_4}$$

$$= 0.52140 \text{ g } Fe_3O_4$$

$$\text{percent } Fe_3O_4 = \frac{0.5140 \text{ g } Fe_3O_4}{1.1324 \text{ g sample}} \times 100\% = 46.044 = 46.04\%$$

Example 6-3

A 0.2356-g sample containing only NaCl (58.44 g/mol) and $BaCl_2$ (208.25 g/mol) yielded 0.4637 g of dried AgCl (143.32 g/mol). Calculate the percent of each halogen compound in the sample.

If we let x be the mass of NaCl in grams and y be the mass of $BaCl_2$ in grams, we can write as a first equation

$$x + y = 0.2356 \text{ g sample}$$

To obtain the mass of AgCl from the NaCl, we can write

$$\text{mass AgCl from NaCl} = x \text{ g NaCl} \times \frac{1 \text{ mol NaCl}}{58.442 \text{ g NaCl}} \times \frac{1 \text{ mol AgCl}}{\text{mol NaCl}}$$

$$\times \frac{143.32 \text{ g AgCl}}{\text{mol AgCl}}$$

$$= 2.45235 \, x \text{ g AgCl}$$

$$\text{mass AgCl from BaCl}_2 = y \text{ g } \cancel{\text{BaCl}_2} \times \frac{1 \text{ mol } \cancel{\text{BaCl}_2}}{208.23 \text{ g } \cancel{\text{BaCl}_2}} \times \frac{2 \text{ mol } \cancel{\text{AgCl}}}{\text{mol } \cancel{\text{BaCl}_2}}$$

$$\times \frac{143.32 \text{ g AgCl}}{\text{mol } \cancel{\text{AgCl}}}$$

$$= 1.37655 \, y \text{ g AgCl}$$

Thus

$$0.4637 \text{ g AgCl} = 2.45235 \, x \text{ g AgCl} + 1.37655 \, y \text{ g AgCl}$$

Substituting the first equation gives

$$0.4637 = 2.45235 \,(0.2356 - y) + 1.37655 \, y$$

This equation rearranges to

$$y = \frac{0.57777 - 0.4637}{2.45235 - 1.37655} = 0.10603 \text{ g BaCl}_2$$

and

$$\text{percent BaCl}_2 = \frac{0.10603 \text{ g BaCl}_2}{0.2356 \text{ g sample}} \times 100\% = 45.01\%$$

$$\text{percent NaCl} = 100.00 - 45.01 = 54.99\%$$

6D   APPLICATIONS OF GRAVIMETRIC METHODS

Gravimetric methods have been developed for most inorganic anions and cations, as well as for such neutral species as water, sulfur dioxide, carbon dioxide, and iodine. A variety of organic substances can also be determined gravimetrically. Examples include lactose in milk products, salicylates in drug preparations, phenolphthalein in laxatives, nicotine in pesticides, cholesterol in cereals, and benzaldehyde in almond extracts. Indeed, gravimetric methods are among the most widely applicable of all analytical procedures.

6D-1 Inorganic Precipitating Agents

Table 6-2 lists common inorganic precipitating agents. These reagents typically form slightly soluble salts or hydrous oxides with the analyte. As you can see from the many entries for each reagent, most inorganic reagents are not very selective. Table 6-3 lists several inorganic reagents that convert an analyte to its elemental form for weighing. Included in this list is an electric current, which deposits the element on the surface of an inert electrode. Electrogravimetric methods are discussed in Chapter 19.

6D-2 Organic Precipitating Agents

Numerous organic reagents have been developed for the gravimetric determination of inorganic species. Some of these reagents are significantly more selective in their reactions than are most of the inorganic reagents listed in Table 6-2.

Gravimetric methods do not require a calibration or standardization step (as do all other analytical procedures except coulometry) because the results are calculated directly from the experimental data and molar masses. Thus, when only one or two samples are to be analyzed, a gravimetric procedure may be the method of choice because it requires less time and effort than a procedure that requires preparation of standards and calibration.

Table 6-2
SOME INORGANIC PRECIPITATING AGENTS*

Precipitating Agent	Element Precipitated†
$NH_3(aq)$	**Be** (BeO), **Al** (Al_2O_3), **Sc** (Sc_2O_3), Cr (Cr_2O_3),‡ **Fe** (Fe_2O_3), Ga (Ga_2O_3), Zr (ZrO_2), **In** (In_2O_3), Sn (SnO_2), U (U_3O_8)
H_2S	Cu (CuO),‡ **Zn** (ZnO, or $ZnSO_4$), **Ge** (GeO_2), As ($\underline{As_2O_3}$, or As_2O_5), Mo (MoO_3), Sn (SnO_2),‡ Sb ($\underline{Sb_2O_3}$, or Sb_2O_5), Bi (Bi_2S_3)
$(NH_4)_2S$	Hg ($\underline{HgS}$), Co (Co_3O_4)
$(NH_4)_2HPO_4$	**Mg** ($Mg_2P_2O_7$), Al ($AlPO_4$), Mn ($Mn_2P_2O_7$), Zn ($Zn_2P_2O_7$), Zr ($Zr_2P_2O_7$), Cd ($Cd_2P_2O_7$), Bi ($BiPO_4$)
H_2SO_4	Li, Mn, **Sr, Cd, Pb, Ba** (all as sulfates)
H_2PtCl_6	K (K_2PtCl_6, or Pt), Rb ($\underline{Rb_2PtCl_6}$), Cs ($\underline{Cs_2PtCl_6}$)
$H_2C_2O_4$	Ca (CaO), Sr (SrO), **Th** (ThO_2)
$(NH_4)_2MoO_4$	Cd ($CdMoO_4$),‡ Pb ($\underline{PbMoO_4}$)
HCl	**Ag** (AgCl), Hg (Hg_2Cl_2), Na (as NaCl from butyl alcohol), Si (SiO_2)
$AgNO_3$	**Cl** (AgCl), Br ($\underline{AgBr}$), $I(\underline{AgI})$
$(NH_4)_2CO_3$	**Bi** (Bi_2O_3)
NH_4SCN	Cu [$Cu_2(SCN)_2$]
$NaHCO_3$	Ru, Os, Ir (precipitated as hydrous oxides; reduced with H_2 to metallic state)
HNO_3	Sn (SnO_2)
H_5IO_6	Hg [$Hg_5(IO_6)_2$]
NaCl, $Pb(NO_3)_2$	F (PbClF)
$BaCl_2$	SO_4^{2-} ($BaSO_4$)
$MgCl_2$, NH_4Cl	PO_4^{3-} ($Mg_2P_2O_7$)

* From W. F. Hillebrand, G. E. F. Lundell, H. A. Bright, and J. I. Hoffman, *Applied Inorganic Analysis*. New York: Wiley, 1953. By permission of John Wiley & Sons, Inc.
† Boldface type indicates that gravimetric analysis is the preferred method for the element or ion. The weighed form is indicated in parentheses.
‡ A double dagger indicates that the gravimetric method is seldom used. An underscore indicates the most reliable gravimetric method.

Chelates are cyclic metal-organic compounds in which the metal is a part of one or more five- or six-membered rings. The chelate pictured below is heme, which is a part of hemoglobin, the oxygen-carrying molecule in human blood.

Note the four six-membered rings that are formed with Fe^{2+}.

We encounter two types of organic reagents. One forms slightly soluble non-ionic products called *coordination compounds*; the other forms products in which the bonding between the inorganic species and the reagent is largely ionic.

Organic reagents that yield sparingly soluble coordination compounds typically contain at least two functional groups. Each of these groups is capable of bonding with a cation by donating a pair of electrons. The functional groups are located in the molecule such that a five- or six-membered ring results from the reaction. dReagents that form compounds of this type are called *chelating agents*, and their products are called *chelates*.

Metal chelates are relatively nonpolar and, as a consequence, have solubilities that are low in water but high in organic liquids. Usually these compounds have low densities and are often intensely colored. Because they are not wetted by water, coordination compounds are readily freed of moisture at low temperatures. Two widely used chelating reagents are described in the paragraphs that follow.

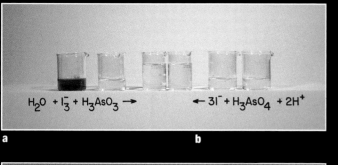

Color Plate 1

Chemical Equilibrium 1: Reaction between iodine and arsenic(III) at pH 1. (a) One mmol I_3^- added to one mmol H_3AsO_3. (b) Three mmol I^- added to one mmol H_3AsO_4 (Section 3B-1, page 33).

$$H_2O + I_3^- + H_3AsO_3 \longrightarrow \quad \longleftarrow 3I^- + H_3AsO_4 + 2H^+$$

a   b

Color Plate 2

Chemical Equilibrium 2: The same reaction as in color plate 6 carried out at pH 7 (Section 3B-1, page 33).

$$H_2O + I_3^- + H_2AsO_3^- \longrightarrow \quad \longleftarrow 3I^- + H_2AsO_4^- + 2H^+$$

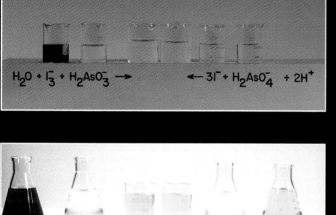

Color Plate 3

Chemical Equilibrium 3: Reaction between iodine and ferrocyanide. One mmol I_3^- added to two mmol $Fe(CN)_6^{4-}$. (b) Three mmol I^- added to two mmol $Fe(CN)_6^{3-}$ (Section 3B-1, page 33).

$$I_3^- + 2Fe(CN)_6^{4-} \longrightarrow \quad \longleftarrow 3I^- + 2Fe(CN)_6^{3-}$$

a   b

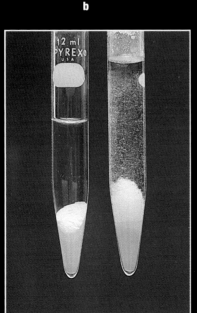

Color Plate 4

The common-ion effect. The test tube on the left contains a saturated solution of silver acetate, AgOAc. The following equilibrium is established in the test tube:

$$AgOAc(s) \rightleftharpoons Ag^+ (aq) + OAc^- (aq)$$

When $AgNO_3$ is added to the test tube, the equilibrium shifts to the left to form more AgOAc as shown in the test tube on the right (Section 3B-3, page 40).

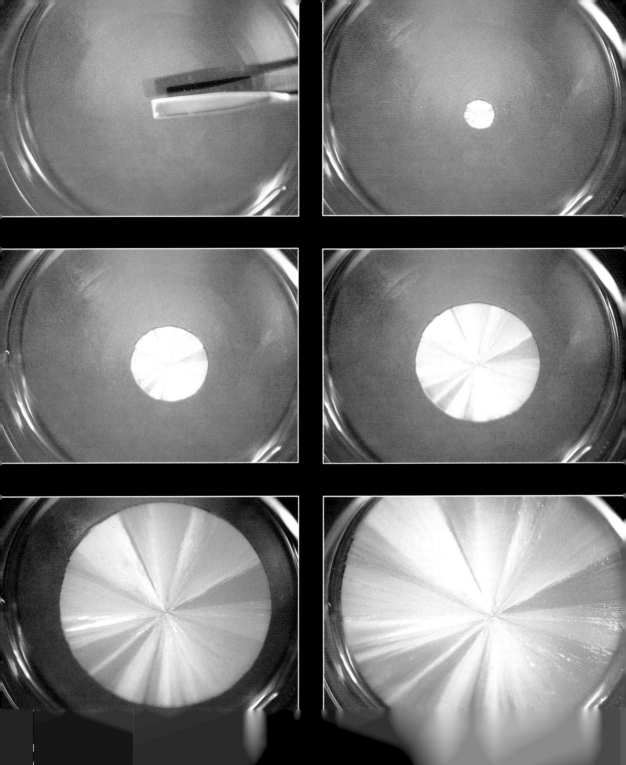

Color Plate 6
The Tyndall effect (see margin note, page 98).

Color Plate 7
Precipitation of 0.1041 g of Fe(III) with NH_3 (left), and homogeneous
precipitation of the same amount of iron with urea (right) (see Section

Color Plate 8
When dimethylglyoxim
to a slightly basic solut
Ni^{2+} *(aq)*, shown on th
bright red precipitate o
$Ni(C_4H_7N_2O_2)_2$ is forme

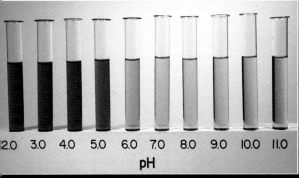

Methyl orange (3.1–4.4)

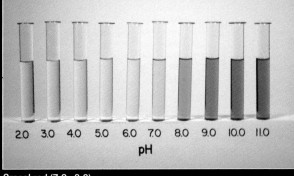

Bromocresol green (3.8–5.4)

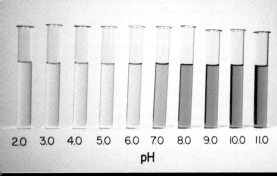

Methyl red (4.2–6.3)

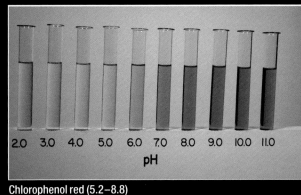

Chlorophenol red (5.2–8.8)

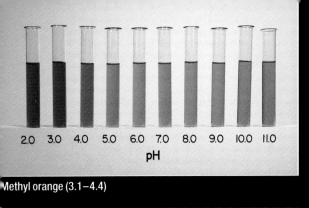

Bromothymol blue (6.0–7.6)

Cresol red (7.2–8.8)

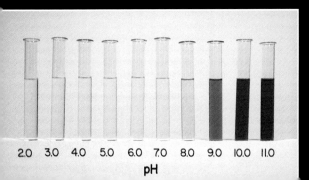

Color Plate 9
Acid-base indicators and their transition pH ranges
(Section 10A-2)

a   b   c   d

Color Plate 10
Argentometric determination of chloride:
Fajans method (Section 13B-1, page 230).
(a) aqueous 2′,7′-dichlorofluorescein;
(b) same, plus 1 mL 0.10 M Ag$^+$. Note the
absence of a precipitate; (c) same, plus
AgCl and an excess of Cl$^-$; (d) same,
plus AgCl and the first slight excess of Ag$^+$.

a   b   c   d

Color Plate 11
In (c) and (d) the AgCl has coagulated.
Note that in (d) the dye has been carried
down on the precipitate and that the
supernatant solution is clear, while in
(c) the dye remains in solution. (Section
13B-1, page 231).

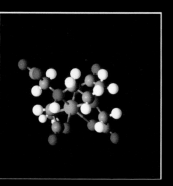

Color Plate 12
Model of an
EDTA/cation chelate
(Section 14B-2).

Color Plate 13
Reduction of
silver(I) by direct
reaction with
copper: the
"silver tree"
(Section 15A-2,
page 260).

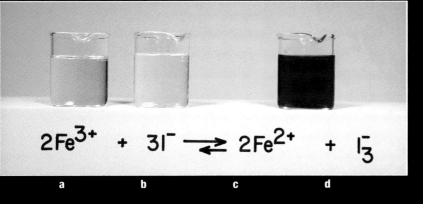

Reaction between iron(III)
and iodide (see margin note,
page 275).

$$2Fe^{3+} + 3I^- \rightleftharpoons 2Fe^{2+} + I_3^-$$

a b c d

Color Plate 16
Starch/iodine end point
(Section 17B-2):
(a) iodine solution;
(b) same, within a few drops
of the equivalence point;
(c) same as (b), with starch
added; (d) same as (c),
at the equivalence point.
(Section 17B-2, page 309).

a b c d

8-Hydroxyquinoline

Approximately two dozen cations form sparingly soluble chelates with 8-hydroxy-quinoline. The structure of magnesium 8-hydroxyquinolate is typical of these chelates:

8-Hydroxyquinoline is sometimes called *oxine*.

The solubilities of metal 8-hydroxyquinolates vary widely from cation to cation and are pH-dependent because 8-hydroxyquinoline is always deprotonated during the chelation reaction. Therefore, we can achieve a considerable degree of selectivity in the use of 8-hydroxyquinoline by controlling pH.

Dimethylglyoxime

Dimethylglyoxime is an organic precipitating agent of unparalleled specificity. Only nickel(II) is precipitated from a weakly alkaline solution. The reaction is

Nickel dimethylglyoxime is spectacular in appearance. It has a beautiful and vivid red color. A photograph of the precipitate is shown in color plate 8.

Creeping is the process by which a precipitate (usually a metal-organic chelate) moves up the walls of a wetted surface of a glass container or a filter paper.

Potassium tetraphenylboron is a salt rather than a chelate.

This precipitate is so bulky that only small amounts of nickel can be handled conveniently. It also has an exasperating tendency to creep up the sides of the container as it is filtered and washed. The solid is readily dried at 110°C and has the composition indicated by its formula.

Sodium Tetraphenylboron

Sodium tetraphenylboron, $(C_6H_5)_4B^-Na^+$, is an important example of an organic precipitating reagent that forms salt-like precipitates. In cold mineral-acid solutions, it is a near-specific precipitating agent for potassium and ammonium ions. The composition of the precipitates is stoichiometric and contains one mole of potassium or ammonium ion for each mole of tetraphenylboron ion; these ionic compounds are readily filtered and can be brought to constant mass at 105°C to 120°C. Only mercury(II), rubidium, and cesium interfere and must be removed by prior treatment.

Table 6-3
SOME REDUCING AGENTS EMPLOYED IN GRAVIMETRIC METHODS

Reducing Agent	Analyte
SO_2	Se, Au
$SO_2 + H_2NOH$	Te
H_2NOH	Se
$H_2C_2O_4$	Au
H_2	Re, Ir
HCOOH	Pt
$NaNO_2$	Au
$SnCl_2$	Hg
Electrolytic reduction	Co, Ni, Cu, Zn, Ag, In, Sn, Sb, Cd, Re, Bi

Table 6-4
GRAVIMETRIC METHODS FOR ORGANIC FUNCTIONAL GROUPS

Functional Group	Basis for Method	Reaction and Product Weighed*
Carbonyl	Mass of precipitate with 2,4-dinitrophenyl-hydrazine	$RCHO + H_2NNHC_6H_3(NO_2)_2 \rightarrow$ $\underline{R—CH = NNHC_6H_3(NO_2)_2}(s) + H_2O$ (RCOR′ reacts similarly)
Aromatic carbonyl	Mass of CO_2 formed at 230°C in quinoline; CO_2 distilled, absorbed, and weighed	$ArCHO \xrightarrow[CuCO_3]{230°C} Ar + \underline{CO_2}(g)$
Methoxyl and ethoxyl	Mass of AgI formed after distillation and decomposition of CH_3I or C_2H_5I	$ROCH_3 + HI \rightarrow ROH + CH_3I$ $RCOOCH_3 + HI \rightarrow RCOOH + CH_3I$ $ROC_2H_5 + HI \rightarrow ROH + C_2H_5I$ $\left.\right\}$ $CH_3I + Ag^+ + H_2O \rightarrow$ $\underline{AgI}(s) + CH_3OH$
Aromatic nitro	Mass loss of Sn	$RNO_2 + \frac{3}{2}\underline{Sn}(s) + 6 H^+ \rightarrow RNH_2 + \frac{3}{2}Sn^{4+} + 2 H_2O$
Azo	Mass loss of Cu	$RN = NR′ + 2\underline{Cu}(s) + 4 H^+ \rightarrow RNH_2 + R′NH_2 + 2 Cu^{2+}$
Phosphate	Mass of Ba salt	$RO\overset{\displaystyle O}{\overset{\|}{P}}(OH)_2 + Ba^{2+} \rightarrow \underline{ROPO_2Ba}(s) + 2 H^+$
Sulfamic acid	Mass of $BaSO_4$ after oxidation with HNO_2	$RNHSO_3H + HNO_2 + Ba^{2+} \rightarrow ROH + \underline{BaSO_4}(s) + N_2 + 2 H^+$
Sulfinic acid	Mass of Fe_2O_3 after ignition of Fe(III) sulfinate	$3 ROSOH + Fe^{3+} \rightarrow (ROSO)_3Fe(s) + 3 H^+$ $(ROSO)_3Fe \xrightarrow[O_2]{} CO_2 + H_2O + SO_2 + \underline{Fe_2O_3}(s)$

* The substance weighed is underlined.

6D-3 Organic Functional Group Analysis

Several reagents react selectively with certain organic functional groups and thus can be used for the determination of most compounds containing these groups. A list of gravimetric functional-group reagents is given in Table 6-4. Many of the reactions shown can also be used for volumetric and spectrophotometric applications.

6D-4 Volatilization Methods

The two most common gravimetric methods based on volatilization are those for water and carbon dioxide.

Water is quantitatively eliminated from many inorganic samples by ignition. In the direct determination, it is collected on any of several solid desiccants and its mass is determined from the mass gain of the desiccant.

The indirect method in which the amount of water is determined by the loss of mass of the sample during heating is less satisfactory because it must be assumed that water is the only component volatilized. This assumption is frequently unjustified because heating of many substances results in their decomposition and a consequent change in mass, irrespective of the presence of water. Nevertheless, the indirect method has found wide use for the determination of water in items of commerce. For example, a semiautomated instrument for the determination of moisture in cereal grains can be purchased. It consists of a platform balance on which a 10-g sample is heated with an infrared lamp. The percent residue is read directly.

Carbonates are ordinarily decomposed by acids to give carbon dioxide, which is readily evolved from solution by heat (see Example 6-2). As in the direct analysis for water, the mass of carbon dioxide is established from the increase in the mass of a solid absorbent. Ascarite II,[7] which consists of sodium hydroxide on a nonfibrous silicate, retains carbon dioxide by the reaction

$$2NaOH + CO_2 \rightarrow Na_2CO_3 + H_2O$$

The absorption tube must also contain a desiccant to prevent loss of the evolved water.

Sulfides and sulfites can also be determined by volatilization. Hydrogen sulfide or sulfur dioxide evolved from the sample after treatment with acid is collected in a suitable absorbent.

Finally, the classical method for the determination of carbon and hydrogen in organic compounds is a gravimetric procedure in which the combustion products (H_2O and CO_2) are collected selectively on weighed absorbents. The increase in mass serves as the analytical parameter.

> Automatic instruments for the routine determination of water in various products of agriculture and commerce are marketed by several instrument manufacturers.

6E   QUESTIONS AND PROBLEMS

6-1. Explain the difference between
 *(a) a colloidal and a crystalline precipitate.
 (b) specific and selective precipitating reagents.
 *(c) precipitation and coprecipitation.
 (d) peptization and coagulation.
 *(e) occlusion and mixed-crystal formation.
 (f) nucleation and particle growth.

6-2. Define
 *(a) digestion.
 (b) adsorption.
 *(c) reprecipitation.
 (d) precipitation from homogeneous solution.
 *(e) electric double layer.
 (f) specific surface area.
 *(g) relative supersaturation.

*6-3. What are the structural characteristics of a chelating agent?

6-4. How can the relative supersaturation be varied during precipitate formation?

*6-5. An aqueous solution contains $NaNO_3$ and KSCN. The thiocyanate ion is precipitated as AgSCN by addition of $AgNO_3$. After an excess of the precipitating reagent has been added,
 (a) what is the charge on the surface of the coagulated colloidal particles?
 (b) what is the source of the charge?
 (c) what ions predominate in the counter-ion layer?

6-6. Briefly explain why the AgCl produced in a gravimetric silver determination is inherently purer than the AgCl produced in a gravimetric chloride determination.

*6-7. What is peptization and how is it avoided?

6-8. Suggest a precipitation method for the separation of K^+ from Na^+ and Li^+.

---

[7] ® Thomas Scientific, Swedesboro, NJ.
* Answers to the asterisked problems are given in the answer section at the back of the book.

6-9. Propose a method for the homogeneous precipitation of
*(a) Al^{3+} from a solution that also contains Na^+.
(b) Ni^{2+} from a solution that also contains Cu^{2+}.
*(c) Pb^{2+} from a solution that also contains Ba^{2+}.
(d) Cu^{2+} from a solution that also contains Ca^{2+}.

6-10. Use chemical formulas to generate the stoichiometric factor needed to express the results of a gravimetric analysis in terms of the substance on the left if the weighing form is the substance on the right:

Sought	Weighed	Sought	Weighed
*(a) CO_2	$BaCO_3$	*(f) UO_2	U_3O_8
(b) Mg	$Mg_2P_2O_7$	*(g) $C_8H_6O_3Cl_2$	$AgCl$
*(c) K_2O	$(C_6H_5)_4BK$	*(h) $CoSiF_6 \cdot 6H_2O$	Co_3O_4
(d) Bi	Bi_2O_3	*(i) $CoSiF_6 \cdot 6H_2O$	H_2O
*(e) H_2S	$CdSO_4$	*(j) $CoSiF_6 \cdot 6H_2O$	$PbClF$

6-11. Briefly explain why the numerically smallest stoichiometric factor is associated with the precipitate that yields the greatest mass for a given mass of analyte.

*6-12. What mass of Ag_2CrO_4 can be produced from 1.200 g of
(a) $AgNO_3$?   (b) K_2CrO_4?

6-13. What mass of $Mg(OH)_2$ can be produced from 0.750 g of
(a) $Ba(OH)_2$?   (b) $MgCl_2$?

6-14. What mass of $AgCl$ can be produced from a 0.400-g sample that assays 41.3%
*(a) KCl?   (b) $MgCl_2$?   *(c) $FeCl_3$?

*6-15. Solubility products for $AgIO_3$ and $Ce(IO_3)_3$ are 3.0×10^{-8} and 3.2×10^{-10} respectively. Which compound is the more soluble in water and by how much?

6-16. Solubility products for PbI_2 and $PbSO_4$ are 7.9×10^{-9} and 1.6×10^{-8} respectively. Which compound is the more soluble and by how much?

6-17. What mass of CO_2 is evolved from a 1.204-g sample that is 36.0% $MgCO_3$ and 44.0% K_2CO_3 by mass?

*6-18. A 50.0-mL portion of a solution containing 0.200 g of $BaCl_2 \cdot 2H_2O$ is mixed with 50.0 mL of a solution containing 0.300 g of $NaIO_3$. Assume that the solubility of $Ba(IO_3)_2$ in water is negligibly small and calculate
(a) the mass of the precipitated $Ba(IO_3)_2$.
(b) the mass of the unreacted compound that remains in solution.

6-19. When a 100.0-mL portion of a solution containing 0.500 g of $AgNO_3$ is mixed with 100.0 mL of a solution containing 0.300 g of K_2CrO_4, a bright red precipitate of Ag_2CrO_4 forms.
(a) Assuming the solubility of Ag_2CrO_4 is negligible, calculate the mass of the precipitate.
(b) Calculate the mass of the unreacted component that remains in solution.

*6-20. Treatment of a 0.4000-g sample of impure potassium chloride with an excess of $AgNO_3$ resulted in the formation of 0.7332 g of $AgCl$. Calculate the percentage of KCl in the sample.

6-21. The aluminum in a 1.200-g sample of impure ammonium aluminum sulfate was precipitated with aqueous ammonia as the hydrous $Al_2O_3 \cdot xH_2O$. The precipitate was filtered and ignited at 1000°C to give anhydrous Al_2O_3, which weighed 0.1798 g. Express the result of this analysis in terms of
(a) % $NH_4Al(SO_4)_2$.   (b) % Al_2O_3.   (c) % Al.

*6-22. A 0.1799-g sample of an organic compound was burned in a stream of oxygen, and the CO_2 produced was collected in a solution of barium hydroxide. Calculate the percentage of carbon in the sample if 0.5613 g of $BaCO_3$ was formed.

6-23. A 0.7406-g sample of impure magnesite, $MgCO_3$, was decomposed with HCl; the liberated CO_2 was collected on calcium oxide and found to weigh 0.1881 g. Calculate the percentage of magnesium in the sample.

6-24. The hydrogen sulfide in a 50.0-g sample of crude petroleum was removed by distillation and collected in a solution of $CdCl_2$. The precipitated CdS was then filtered, washed, and ignited to $CdSO_4$. Calculate the percentage of H_2S in the sample if 0.108 g of $CdSO_4$ was recovered.

*6-25. Ammoniacal nitrogen can be determined by treatment of the sample with chloroplatinic acid; the product is slightly soluble ammonium chloroplatinate:

$$H_2PtCl_6 + 2NH_4^+ \rightarrow (NH_4)_2PtCl_6 + 2H^+$$

The precipitate decomposes on ignition, yielding metallic platinum and gaseous products:

$$(NH_4)_2PtCl_6 \rightarrow Pt(s) + 2Cl_2(g) + 2NH_3(g) + 2 + 2HCl(g)$$

Calculate the percentage of ammonia if a 0.2213-g sample gave rise to 0.5881 g of platinum.

*6-26. The sulfur in an 8-tablet sample of the hypnotic drug captodiamine, $C_{21}H_{29}NS_2$ (359.6 g/mol) was converted to sulfate and determined gravimetrically. Calculate the average mass of captodiamine per tablet if 0.3343 g of $BaSO_4$ was recovered.

6-27. The mercury in a 0.7152-g sample was precipitated with an excess of paraperiodic acid, H_5IO_6:

$$5Hg^{2+} + 2H_5IO_6 \rightarrow Hg_5(IO_6)_2 + 10H^+$$

The precipitate was filtered, washed free of precipitating agent, dried, and weighed, 0.3408 g being recovered. Calculate the percentage of Hg_2Cl_2 in the sample.

*6-28. The phosphorus in a 0.2374-g sample was precipitated as the slightly soluble $(NH_4)_3PO_4 \cdot 12MoO_3$. This precipitate was filtered, washed, and then redissolved in acid. Treatment of the resulting solution with an excess of Pb^{2+} resulted in the formation of 0.2752 g of $PbMoO_4$. Express the results of this analysis in terms of percent P_2O_5.

6-29. A 0.6447-g portion of manganese dioxide was added to an acidic solution in which 1.1402 g of a chloride-containing sample was dissolved. Evolution of chlorine took place as a consequence of the following reaction:

$$MnO_2(s) + 2Cl^- + 4H^+ \rightarrow Mn^{2+} + Cl_2(g) + 2H_2O$$

After the reaction was complete, the excess MnO_2 was collected by filtration, washed, and weighed, 0.3521 g being recovered. Express the results of this analysis in terms of percent aluminum chloride.

*6-30. Nitrobenzene, $C_6H_5NO_2$ (123.11 g/mol), is quantitatively reduced to aniline $C_6H_5NH_2$ (93.12 g/mol) with metallic tin:

$$3Sn(s) + 2C_6H_5NO_2 + 12H^+ \rightarrow$$
$$2C_6H_5NH_2 + 4H_2O + 3Sn^{4+}$$

A 0.5078-g sample of impure nitrobenzene was treated with 1.044 g of tin. When reaction was complete, the residual tin was found to weigh 0.338 g. Calculate the percentage of nitrobenzene in the sample.

6-31. A series of sulfate samples is to be analyzed by precipitation as $BaSO_4$. If it is known that the sulfate content in these samples ranges between 20% and 55%, what minimum sample mass should be taken to ensure that a precipitate mass no smaller than 0.300 g is produced? What is the maximum precipitate mass to be expected if this quantity of sample is taken?

*6-32. The success of a particular catalyst is highly dependent on its zirconium content. The starting material for this preparation is received in batches that assay between 68% and 84% $ZrCl_4$. Routine analysis based on precipitation of AgCl is feasible, it having been established that there are no sources of chloride ion other than the $ZrCl_4$ in the sample.
(a) What sample mass should be taken to ensure a AgCl precipitate that weighs at least 0.400 g?
(b) If this sample mass is used, what is the maximum mass of AgCl that can be expected in this analysis?
(c) To simplify calculations, what sample mass should be taken in order to have the percentage of $ZrCl_4$ exceed the mass of AgCl produced by a factor of 100?

6-33. The addition of dimethylglyoxime, $H_2C_4H_6O_2N_2$, to a solution containing nickel(II) ion gives rise to a precipitate:

$$Ni^{2+} + 2H_2C_4H_6O_2N_2 \rightarrow 2H^+ + Ni(HC_4H_6O_2N_2)_2$$

Nickel dimethylglyoxime is a bulky precipitate that is inconvenient to manipulate in amounts greater than 175 mg. The amount of nickel in a type of permanent-magnet alloy ranges between 24% and 35%. Calculate the sample mass that should not be exceeded when analyzing these alloys for their nickel content.

*6-34. A 0.6407-g sample containing chloride and iodide ions gave a silver halide precipitate weighing 0.4430 g. This precipitate was then strongly heated in a stream of Cl_2 gas to convert the AgI to AgCl; on completion of this treatment, the precipitate weighed 0.3181 g. Calculate the percentage of chloride and iodide in the sample.

6-35. A 6.881-g sample containing magnesium chloride and sodium chloride was dissolved in sufficient water to give 500 mL of solution. Analysis for the chloride content of a 50.0-mL aliquot resulted in the formation of 0.5923 g of AgCl. The magnesium in a second 50.0-mL aliquot was precipitated as $MgNH_4PO_4$; on ignition 0.1796 g of $Mg_2P_2O_7$ was found. Calculate the percent of $MgCl_2 \cdot 6H_2O$ and of NaCl in the sample.

*6-36. The iodide in a sample that also contained chloride was converted to iodate by treatment with an excess of bromine:

$$3H_2O + 3Br_2 + I^- \rightarrow 6Br^- + IO_3^- + 6H^+$$

The unused bromine was removed by boiling; an excess of barium ion was then added to precipitate the iodate:

$$Ba^{2+} + 2IO_3^- \rightarrow Ba(IO_3)_2$$

In the analysis of a 2.72-g sample, 0.0720 g of barium iodate was recovered. Express the results of this analysis as percent potassium iodide.

6-37. Several alloys that contained only Ag and Cu were analyzed by dissolving weighed quantities in HNO_3, introducing an excess of IO_3^-, and bringing the filtered mixture of $AgIO_3$ and $Cu(IO_3)_2$ to constant mass. Use the accompanying data to calculate the percentage composition of the alloys.

	Mass Sample, g	Mass Precipitate, g
*(a)	0.2175	0.7391
(b)	0.1948	0.7225
*(c)	0.2473	0.7443
(d)	0.2386	0.9962
*(e)	0.1864	0.8506

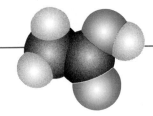

CHAPTER 7

APPLICATION OF EQUILIBRIUM CALCULATIONS TO COMPLEX SYSTEMS

Aqueous solutions encountered in the laboratory often contain several species that interact with one another and water to yield two or more equilibria that function simultaneously. For example, when water is saturated with the sparingly soluble barium sulfate, three equilibria develop:

$$BaSO_4(s) \rightleftharpoons Ba^{2+} + SO_4^{2-} \tag{7-1}$$

$$SO_4^{2-} + H_3O^+ \rightleftharpoons HSO_4^- + H_2O \tag{7-2}$$

$$2H_2O \rightleftharpoons H_3O^+ + OH^- \tag{7-3}$$

If hydronium ions are added to this system, the second equilibrium is shifted to the right by the common-ion effect. The resulting decrease in sulfate concentration causes the first equilibrium to shift to the right as well, thus increasing the solubility of the barium sulfate.

The solubility of barium sulfate is also increased when acetate ions are added to an aqueous suspension of barium sulfate because acetate ions tend to form a soluble complex with barium ions as shown by the reaction

$$Ba^{2+} + OAc^- \rightleftharpoons BaOAc^+ \tag{7-4}$$

Again, the addition of acetate ion causes both this equilibrium and the solubility equilibrium to shift to the right, thus causing a solubility increase.

If we wish to calculate the solubility of barium sulfate in a system containing hydronium and acetate ions, we must take into account not only the solubility equilibrium, but the other three equilibria as well. We find, however, that employing four equilibrium-constant expressions to calculate a solubility is much more difficult and complex than the rather simple procedure that we illustrated in Examples 3-3, 3-4, and 3-5. In order to solve this type of problem, we will find it helpful to use a systematic approach that is described in the section that follows. We then use this approach to illustrate the effect of pH and complex formation on the solubility of typical analytical precipitates. In subsequent chapters, we will use this same systematic method for solution of problems involving multiple equilibria of several types.

7A   A SYSTEMATIC METHOD FOR SOLVING MULTIPLE-EQUILIBRIUM PROBLEMS

Solution of a multiple-equilibrium problem requires us to develop as many independent equations as there are participants in the system being studied. For example, if we wish to compute the solubility of barium sulfate in a solution of acid we need to be able to calculate the concentration of all of the species present in the solution. Their number is five: Ba^{2+}, SO_4^{2-}, HSO_4^-, H_3O^+, and OH^-. In order to calculate the solubility of barium sulfate in this solution rigorously, it would then be necessary to develop five independent algebraic equations containing these five unknowns and to solve them simultaneously.

Three types of algebraic equations are used in solving multiple-equilibrium problems: (1) equilibrium-constant expressions, (2) *mass-balance* equations, and (3) a single *charge-balance* equation. We have already shown in Section 3B how equilibrium-constant expressions are written; we must now turn our attention to the development of the other two types of equations.

> The term "mass-balance equation," although widely used, is misleading because such equations are really based on balancing *concentrations* rather than *masses*.

7A-1   Mass-Balance Equations

Mass-balance equations relate the equilibrium concentrations of various species in a solution to one another and to the analytical concentrations of the various solutes. They are derived from information about how the solution was prepared and from a knowledge of the equilibria established in the solution.

Example 7-1

Write mass-balance expressions for a 0.0100 M solution of HCl that is in equilibrium with an excess of solid $BaSO_4$.

From our general knowledge of the behavior of aqueous solutions, we can write equations for three equilibria that must be present in this solution. These are given as Equations 7-1, 7-2, and 7-3.

Because the only source for the two sulfate species is the dissolved $BaSO_4$, we may write

$$[Ba^{2+}] = [SO_4^{2-}] + [HSO_4^-]$$

Because hydronium ion is produced from both the HCl and the dissociation of

If $[H_3O^+]_{HCl}$ is the hydronium ion concentration resulting from the dissociation of HCl, and $[H_3O^+]_{H_2O}$ is that resulting from the dissociation of water, then

$$[H_3O^+] = [H_3O^+]_{total}$$
$$= [H_3O^+]_{HCl} + [H_3O^+]_{H_2O}$$

But

$$[H_3O^+]_{HCl} = c_{HCl}$$

and

$$[H_3O^+]_{H_2O} = [OH^-]$$

So

$$[H_3O^+] = c_{HCl} + [OH^-]$$

water we may write a second mass-balance expression. Thus,

$$[H_3O^+] = c_{HCl} + [OH^-] = 0.0100 + [OH^-]$$

Here, the second term on the right side of the equation accounts for the concentration of hydronium ions that result from the dissociation of water.

Example 7-2

Write mass-balance expressions for the system formed when a 0.010 M NH_3 solution is saturated with AgBr.

Here, equations for the pertinent equilibria in the solution are

$$AgBr(s) \rightleftharpoons Ag^+ + Br^-$$
$$Ag^+ + 2NH_3 \rightleftharpoons Ag(NH_3)_2^+$$
$$NH_3 + H_2O \rightleftharpoons NH_4^+ + OH^-$$
$$2H_2O \rightleftharpoons H_3O^+ + OH^-$$

Because the only source of Br^-, Ag^+, and $Ag(NH_3)_2^+$ is AgBr and because silver and bromide ions are present in a 1:1 ratio in the starting material, it follows that one mass-balance equation is

$$[Ag^+] + [Ag(NH_3)_2^+] = [Br^-]$$

where the bracketed terms are molar species concentrations. Also, we know that the only source of ammonia-containing species is the 0.010 M NH_3. Therefore,

$$c_{NH_3} = [NH_3] + [NH_4^+] + 2[Ag(NH_3)_2^+] = 0.010$$

From the last two equilibria, we see one hydroxide ion is formed for each NH_4^+ and each hydronium ion. Therefore,

$$[OH^-] = [NH_4^+] + [H_3O^+]$$

7A-2 Charge-Balance Equation

We know that electrolyte solutions are electrically neutral even though they may contain many millions of charged ions. Solutions are neutral because the *molar concentration of positive charge* in an electrolyte solution always equals *the molar concentration of negative charge*. That is, for any solution containing electrolytes, we may write

no. mol/L positive charge = no. mol/L negative charge

This equation represents the charge-balance condition and is called the charge-balance equation. To be useful for equilibrium calculations, the equality must be expressed in terms of the molar concentrations of the charged species in the solution.

How much charge is contributed to a solution by 1 mol of Na^+? Or, how about 1 mol of Mg^{2+} or 1 mol of PO_4^{3-}? The concentration of charge contributed to a solution by an ion is equal to the molar concentration of that ion multiplied by its charge. Thus the molar concentration of positive charge in a solution due to the presence of sodium ions is

$$\frac{\text{mol positive charge}}{\text{L}} = \frac{1 \text{ mol positive charge}}{\text{mol } Na^{\pm}} \times \frac{\text{mol } Na^{\pm}}{\text{L}}$$

$$= 1 \times [Na^+]$$

The concentration of positive charge due to magnesium ions is

$$\frac{\text{mol positive charge}}{\text{L}} = \frac{2 \text{ mol positive charge}}{\text{mol } Mg^{2\pm}} \times \frac{\text{mol } Mg^{2\pm}}{\text{L}}$$

$$= 2 \times [Mg^{2+}]$$

since each mole of magnesium ion contributes 2 mol of positive charge to the solution. We may also write for phosphate ion

$$\frac{\text{mol negative charge}}{\text{L}} = \frac{3 \text{ mol negative charge}}{\text{mol } PO_4^{3-}} \times \frac{\text{mol } PO_4^{3-}}{\text{L}}$$

$$= 3 \times [PO_4^{3-}]$$

Now, consider how we would write a charge-balance equation for a 0.100 M solution of sodium chloride. Positive charges in this solution are supplied by Na^+ from the solute and H_3O^+ from dissociation of water. Negative charges come from Cl^- and OH^-. The molarity of positive and negative charges are

$$\text{mol/L positive charge} = [Na^+] + [H_3O^+] = 0.100 + 1 \times 10^{-7}$$

$$\text{mol/L negative charge} = [Cl^-] + [OH^-] = 0.100 + 1 \times 10^{-7}$$

We write the charge-balance equation by equating the concentrations of positive and negative charges. That is,

$$[Na^+] + [H_3O^+] = [Cl^-] + [OH^-] = 0.100 + 1 \times 10^{-7}$$

Let us now consider a solution that has an analytical concentration of magnesium chloride of 0.100 M. Here, the molarities of positive and negative charge are given by

$$\text{mol/L positive charge} = 2[Mg^{2+}] + [H_3O^+] = 2 \times 0.100 + 1 \times 10^{-7}$$

$$\text{mol/L negative charge} = \quad [Cl^-] + [OH^-] \quad = 2 \times 0.100 + 1 \times 10^{-7}$$

In the first equation, the molar concentration of magnesium ions is multiplied by two (2×0.100) because 1 mol of that ion contributes 2 mol of positive charge to the solution. In the second equation the molar chloride ion concentration is twice that of the magnesium chloride concentration or 2×0.100. To obtain the charge-balance equation, we equate the concentration of positive charge with

Always remember that a charge-balance equation is based on the equality in *molar charge concentrations* and that to obtain the charge concentration of an ion, you must multiply the molar concentration of the ion by its charge.

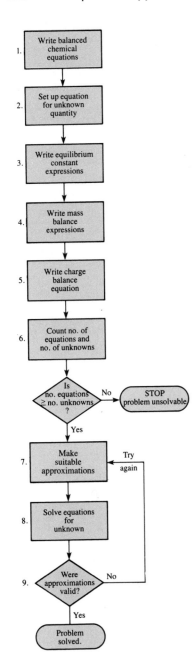

Figure 7-1

A systematic method for solving multiple-equation problems.

In some systems, a charge-balance equation cannot be written, is of no use, or is identical to one of the mass-balance equations.

the concentration of negative charge to obtain

$$2[Mg^{2+}] + [H_3O^+] = [Cl^-] + [OH^-] = 0.200 + 1 \times 10^{-7}$$

For a neutral solution, $[H_3O^+]$ and $[OH^-]$ are very small and equal so that we can ordinarily simplify the charge-balance equation to

$$2[Mg^{2+}] = [Cl^-] = 0.200$$

Example 7-3

Write a charge-balance equation for the system in Example 7-2.

$$[Ag^+] + [Ag(NH_3)_2^+] + [H_3O^+] + [NH_4^+] = [OH^-] + [Br^-]$$

Example 7-4

Write a charge-balance equation for a solution that contains NaCl, Ba(ClO$_4$)$_2$, and Al$_2$(SO$_4$)$_3$.

$$[Na^+] + 2[Ba^+] + 3[Al^{3+}] + [H_3O^+] = [ClO_4^-] + [NO_3^-] + 2[SO_4^{2-}] + [OH^-]$$

7A-3 Steps for Solving Problems Involving Several Equilibria

1. Write a set of balanced chemical equations for all pertinent equilibria.
2. State in terms of equilibrium concentrations the quantity that is being sought.
3. Write equilibrium-constant expressions for all equilibria written in step 1, and find numerical values for the constants in tables of equilibrium constants.
4. Write mass-balance expressions for the system.
5. If possible, write a charge-balance expression for the system.
6. Count the number of unknown concentrations in the equations in steps 3, 4, and 5, and compare this number with the number of independent equations. If the number of equations is equal to the number of unknowns, proceed to step 7. If the numbers are not equal, look for additional equations. If you cannot write enough equations, try to eliminate unknowns by suitable approximations regarding the concentration of one or more of the unknowns. If you cannot make approximations, the problem cannot be solved.
7. Make suitable approximations to simplify the algebra.
8. Solve the algebraic equations for the equilibrium concentrations needed to give a provisional answer as defined in step 2.
9. Check the validity of the approximations made in step 7 using the provisional concentrations computed in step 8.

Step 6 is critical because it shows whether or not an exact solution to the problem is possible. If the number of unknowns is identical to the number of equations, the problem has been reduced to one of *algebra* alone. That is, you can find answers with sufficient perseverance. On the other hand, if there are not enough equations even after approximations are made, you should abandon the problem.

7A-4 The Use of Approximations in Solving Equilibrium Calculations

When step 6 of the systematic approach is complete, we have a *mathematical* problem of solving several nonlinear simultaneous equations. This job is often formidable, tedious, and time-consuming unless a suitable computer program is available, or unless approximations can be found that decrease the number of unknowns and equations. In this section, we consider in general terms how equations describing equilibrium relationships can be simplified by suitable approximations.

Bear in mind that *only* the mass-balance and charge-balance equations can be simplified because only in these equations do the concentration terms appear as sums or differences rather than as products or quotients. It is always possible to assume that one or more of the terms in a sum or difference is so much smaller than the others that it can be ignored without significantly affecting the equality. The assumption that a concentration term in an equilibrium-constant expression is zero makes the expression meaningless.

Many students find step 7 to be the most troublesome because they fear that making invalid approximations will lead to serious errors in their computed results. Such fears are groundless. Experienced scientists are often as puzzled as beginners when making an approximation that simplifies an equilibrium calculation. Nonetheless, they make such approximations fearlessly because they know that the effects of an invalid assumption will become obvious by the time a computation is completed. Generally, you should try simplifying assumptions at the outset and compute provisional answers even though you may have doubts as to the validity of the assumptions. If an assumption leads to an intolerable error (which is easily recognized), a recalculation without the faulty approximation is then performed. Usually, it is more efficient to try a questionable assumption at the outset than to make a more time-consuming and tedious calculation without the assumption.

7B THE CALCULATION OF SOLUBILITY BY THE SYSTEMATIC METHOD

The use of the systematic method is illustrated in this section with examples involving the solubility of precipitates under various conditions. In later chapters we apply this method to other types of equilibria.

7B-1 Metal Hydroxides

Examples 7-5 and 7-6 involve calculating the solubilities of two metal hydroxides. The first is relatively soluble compared with the second. These examples illustrate the importance of making approximations and provisional calculations in the systematic procedure for solving mass-law problems.

Do not waste time starting the algebra in an equilibrium calculation until you are absolutely sure that you have enough independent equations to make the solution feasible.

Several software packages are now available for solving multiple nonlinear simultaneous equations rigorously. Three such programs are Mathcad™, Mathematica™, and TK Solver Plus™. When software such as these is available, steps 7, 8, and 9 can be dispensed with, and the results can be obtained directly.

Approximations can only be made in charge-balance and mass-balance equations—never in equilibrium-constant expressions.

Never be afraid to make an assumption in attempting to solve an equilibrium problem. If the assumption is not valid, you will know it as soon as you have an approximate answer.

Example 7-5

Calculate the molar solubility of $Mg(OH)_2$ in water.

Step 1: Pertinent Equilibria
Two equilibria that need to be considered are

$$Mg(OH)_2(s) \rightleftharpoons Mg^{2+} + 2OH^-$$
$$2H_2O \rightleftharpoons H_3O^+ + OH^-$$

Step 2: Definition of Unknown
Since 1 mol of Mg^{2+} is formed for each mole of $Mg(OH)_2$ dissolved,

$$\text{solubility } Mg(OH)_2 = [Mg^{2+}]$$

Step 3: Equilibrium-Constant Expressions

$$K_{sp} = [Mg^{2+}][OH^-]^2 = 7.1 \times 10^{-12} \qquad \textbf{(7-5)}$$
$$K_w = [H_3O^+][OH^-] = 1.00 \times 10^{-14} \qquad \textbf{(7-6)}$$

Step 4: Mass-Balance Expression

We arrived at Equation 7-7 with the following reasoning. If $[OH^-]_{H_2O}$ and $[OH^-]_{Mg(OH)_2}$ are the concentrations of OH^- produced from $Mg(OH)_2$ and H_2O respectively, then

$$[OH^-]_{H_2O} = [H_3O^+]$$
$$[OH^-]_{Mg(OH)_2} = 2[Mg^{2+}]$$
$$[OH^-]_{total} = [OH^-]_{H_2O} + [OH^-]_{Mg(OH)_2}$$
$$= [H_3O^+] + 2[Mg^{2+}]$$

$$[OH^-] = 2[Mg^{2+}] + [H_3O^+] \qquad \textbf{(7-7)}$$

The first term on the right-hand side of Equation 7-7 represents the hydroxide ion concentration resulting from dissolved $Mg(OH)_2$ in the solution, and the second term is the hydroxide ion concentration resulting from the dissociation of water.

Step 5: Charge-Balance Expression

$$[OH^-] = 2[Mg^{2+}] + [H_3O^+]$$

Note that this equation is identical to Equation 7-7. Often a mass-balance and a charge-balance equation are the same.

Step 6: Number of Independent Equations and Unknowns
We have developed three independent algebraic equations (Equations 7-5, 7-6, and 7-7) and have three unknowns ($[Mg^{2+}]$, $[OH^-]$, and $[H_3O^+]$). Therefore, the problem can be solved rigorously.

Step 7: Approximations
We can make approximations only in Equation 7-7. Since the solubility-product constant for $Mg(OH)_2$ is relatively large, the solution will be somewhat basic. Therefore, it is reasonable to assume that $[H_3O^+] \ll [Mg^{2+}]$. Equation 7-7 then simplifies to

$$2[Mg^{2+}] \cong [OH^-] \qquad \textbf{(7-8)}$$

Step 8: Solution to Equations
Substitution of Equation 7-8 into Equation 7-5 gives

$$[Mg^{2+}](2[Mg^{2+}])^2 = 7.1 \times 10^{-12}$$

$$[Mg^{2+}]^3 = \frac{7.1 \times 10^{-12}}{4} = 1.78 \times 10^{-12}$$

$$[Mg^{2+}] = \text{solubility} = 1.21 \times 10^{-4} = 1.2 \times 10^{-4}\,\text{mol/L}$$

Step 9: Check of the Assumptions
Substitution into Equation 7-8 yields

$$[OH^-] = 2 \times 1.21 \times 10^{-4} = 2.42 \times 10^{-4}$$

$$[H_3O^+] = \frac{1.00 \times 10^{-14}}{2.42 \times 10^{-4}} = 4.1 \times 10^{-11}$$

Thus our assumption that $4.1 \times 10^{-11} \ll 1.2 \times 10^{-4}$ is valid.

Example 7-6

Calculate the solubility of $Fe(OH)_3$ in water. Proceeding by the systematic approach used in Example 7-5 gives

Step 1: Pertinent Equilibrium

$$Fe(OH)_3(s) \rightleftharpoons Fe^{3+} + 3OH^-$$

$$2H_2O \rightleftharpoons H_3O^+ + OH^-$$

Step 2: Definition of Unknown

$$\text{solubility} = [Fe^{3+}]$$

Step 3: Equilibrium-Constant Expressions

$$K_{sp} = [Fe^{3+}][OH^-]^3 = 2 \times 10^{-39}$$

$$K_w = [H_3O^+][OH^-] = 1.00 \times 10^{-14}$$

Steps 4 and 5
As in Example 7-5, the mass-balance equation and the charge-balance equations are identical. That is,

$$[OH^-] = 3[Fe^{3+}] + [H_3O^+]$$

Step 6
We see that we have enough equations to calculate values for the three unknowns.

Step 7: Approximations
As in Example 7-5, let us assume $[H_3O^+] \ll 3[Fe^{3+}]$, so that

$$3[Fe^{3+}] \approx [OH^-]$$

Step 8: Solution to Equations

Substituting this equation into the solubility-product expression gives

$$[Fe^{3+}](3[Fe^{3+}])^3 = 2 \times 10^{-39}$$

$$[Fe^{3+}] = \left(\frac{2 \times 10^{-39}}{27}\right)^{1/4} = 9 \times 10^{-11}$$

$$\text{solubility} = [Fe^{3+}] = 9 \times 10^{-11} \text{ mol/L}$$

Step 9: Check of the Assumptions

From the assumption made in step 7, we can calculate a provisional value of $[OH^-]$. That is,

$$[OH^-] \approx 3[Fe^{3+}] = 3 \times 9 \times 10^{-11} = 3 \times 10^{-10}$$

Let us use this value of $[OH^-]$ to compute a *provisional* value for $[H_3O^+]$:

$$[H_3O^+] = \frac{1.00 \times 10^{-14}}{3 \times 10^{-10}} = 3 \times 10^{-5}$$

But 3×10^{-5} is not much smaller than three times our provisional value of $[Fe^{3+}]$. This discrepancy means that our assumption was invalid and the provisional values for $[Fe^{3+}]$, $[OH^-]$, and $[H_3O^+]$ are all significantly in error. Therefore, let us go back to step 7 and assume that

$$3[Fe^{3+}] \ll [H_3O^+]$$

Now the mass-balance expression becomes

$$[H_3O^+] = [OH^-]$$

Substituting this equality into the expression for K_w gives

$$[H_3O^+] = [OH^-] = \sqrt{K_w} = 1.00 \times 10^{-7}$$

Substituting this number into the solubility-product expression developed in step 3 gives

$$[Fe^{3+}] = \frac{2 \times 10^{-39}}{(1.00 \times 10^{-7})^3} = 2 \times 10^{-18} \text{ mol/L}$$

In this case we have assumed that $3[Fe^{3+}] \ll [OH^-]$ or $3 \times 2 \times 10^{-18} \ll 10^{-7}$. Thus our assumption is valid and we may write

$$\text{solubility} = 2 \times 10^{-18} \text{ mol/L}$$

Note the very large—and easily identified—error introduced by the invalid assumption.

Example 7-6 illustrates how easily the effects of an invalid assumption are detected. You may be bothered that we used a faulty value for $[OH^-]$ to obtain $[H_3O^+]$, which we then compared with $3[Fe^{3+}]$. But the point is that the faulty value of $[OH^-]$ also resulted from invalid assumption. Had the assumption been valid we would have had an internally consistent set of calculated concentrations. The presence of even one internal inconsistency is a clear indication of an invalid assumption. A valid assumption leads to concentrations that satisfy all of the algebraic equations written in steps 3, 4, and 5.

7B-2 The Effect of pH on Solubility

The solubility of precipitates containing an anion with basic properties, a cation with acidic properties, or both will be dependent on pH. The example that follows illustrates how the effect of pH on solubility can be treated in quantitative terms.

All precipitates that contain an anion that is the conjugate base of a weak acid are more soluble at low pH than at high pH.

Solubility Calculations When the pH Is Fixed and Known

Analytical precipitations are frequently performed in buffered solutions in which the pH is fixed at some predetermined and known value. The calculation of solubility under this circumstance is illustrated by Example 7-7.

A *buffer* keeps the pH of a solution constant (see Section 10C-2).

Example 7-7

Calculate the molar solubility of calcium oxalate in a solution that has been buffered so that its pH is constant and equal to 4.00.

Step 1: Pertinent Equilibria

$$CaC_2O_4 \rightleftharpoons Ca^{2+} + C_2O_4^{2-} \tag{7-9}$$

Oxalate ions react with water to form $HC_2O_4^-$ and $H_2C_2O_4$. Thus, two other equilibria in this solution are

$$H_2C_2O_4 + H_2O \rightleftharpoons H_3O^+ + HC_2O_4^- \tag{7-10}$$

$$HC_2O_4^- + H_2O \rightleftharpoons H_3O^+ + C_2O_4^{2-} \tag{7-11}$$

Step 2: Definition of the Unknown

Calcium oxalate is a strong electrolyte so that its molar analytical concentration is equal to the equilibrium calcium ion concentration. That is,

$$\text{solubility} = [Ca^{2+}] \tag{7-12}$$

Step 3: Equilibrium-Constant Expressions

$$K_{sp} = [Ca^{2+}][C_2O_4^{2-}] = 1.7 \times 10^{-9} \tag{7-13}$$

$$K_1 = \frac{[H_3O^+][HC_2O_4^-]}{[H_2C_2O_4]} = 5.60 \times 10^{-2} \tag{7-14}$$

$$K_2 = \frac{[H_3O^+][C_2O_4^{2-}]}{[HC_2O_4^-]} = 5.42 \times 10^{-5} \tag{7-15}$$

Step 4: Mass-Balance Expressions
Because CaC_2O_4 is the only source of Ca^{2+} and the three oxalate species,

$$[Ca^{2+}] = [C_2O_4^{2-}] + [HC_2O_4^-] + [H_2C_2O_4] \qquad (7\text{-}16)$$

Moreover, the problem states that the pH is 4.00. Thus,

$$[H_3O^+] = 1.00 \times 10^{-4}$$

Step 5: Charge-Balance Expression
A buffer is required to maintain the pH at 4.00. The buffer usually consists of some weak acid HA and its conjugate base, A^- (Section 10C-1). The nature of the three species and their concentrations have not been specified, however, so we do not have enough information to write a charge-balance equation.

Step 6: Number of Independent Equations and Unknowns
We have four unknowns ($[Ca^{2+}]$, $[C_2O_4^{2-}]$, $[HC_2O_4^-]$, and $[H_2C_2O_4]$) as well as four independent algebraic relationships (Equations 7-13, 7-14, 7-15, and 7-16). Therefore, an exact solution can be obtained, and the problem becomes one of algebra.

Step 7: Approximations
We can easily obtain an exact solution in this case, so we will not bother with approximations.

Step 8: Solution of the Equations
Let us substitute Equations 7-14 and 7-15 into 7-16 to develop a relationship between $[Ca^{2+}]$, $[C_2O_4^{2-}]$, and $[H_3O^+]$. Thus, we rearrange Equation 7-15 to give

$$[HC_2O_4^-] = \frac{[H_3O^+][C_2O_4^{2-}]}{K_2}$$

Substituting numerical values for $[H_3O^+]$ and K_2 gives

$$[HC_2O_4^-] = \frac{1.00 \times 10^{-4}\,[C_2O_4^{2-}]}{5.42 \times 10^{-5}} = 1.85\,[C_2O_4^{2-}]$$

Substituting this relationship into Equation 7-14 and rearranging gives

$$[H_2C_2O_4] = \frac{[H_3O^+][C_2O_4^{2-}] \times 1.85}{K_1}$$

Substituting numerical values for $[H_3O^+]$ and K_1 yield

$$[H_2C_2O_4] = \frac{1.85 \times 10^{-4}\,[C_2O_4^{2-}]}{5.60 \times 10^{-2}} = 3.30 \times 10^{-3}\,[C_2O_4^{2-}]$$

Substituting these expressions for $[HC_2O_4^-]$ and $[H_2C_2O_4]$ into Equation 7-16 gives

$$[Ca^{2+}] = [C_2O_4^{2-}] + 1.85\,[C_2O_4^{2-}] + 3.30 \times 10^{-3}\,[C_2O_4^{2-}]$$
$$= 2.85\,[C_2O_4^{2-}]$$

or

$$[C_2O_4^{2-}] = [Ca^{2+}]/2.85$$

Substituting into Equation 7-13 gives

$$\frac{[Ca^{2+}][Ca^{2+}]}{2.85} = 1.7 \times 10^{-9}$$

$$[Ca^{2+}] = \text{solubility} = \sqrt{2.85 \times 1.7 \times 10^{-9}} = 7.0 \times 10^{-5} \, \text{mol/L}$$

Solubility Calculations When the pH Is Variable

Saturating an unbuffered solution with a sparingly soluble salt containing a basic anion or an acidic cation causes the pH of the solution to change. For example, pure water saturated with barium carbonate is basic as a consequence of the reactions

$$BaCO_3(s) \rightleftharpoons Ba^{2+} + CO_3^{2-}$$

$$CO_3^{2-} + H_2O \rightleftharpoons HCO_3^- + OH^-$$

$$HCO_3^- + H_2O \rightleftharpoons H_2CO_3 + OH^-$$

In contrast to Example 7-6, the hydroxide ion concentration now becomes an unknown, and an additional algebraic equation must therefore be developed if the solubility of barium carbonate is to be calculated.

It is not difficult to write the algebraic relationships we need to calculate the solubility of such a precipitate (see Feature 7-1). Solving the equations exactly, however, is tedious unless appropriate software is available.

7B-3   The Solubility of Precipitates in the Presence of Complexing Agents

The solubility of a precipitate may increase dramatically in the presence of reagents that form complexes with the anion or the cation of the precipitate. For example, fluoride ions prevent the quantitative precipitation of aluminum hydroxide even though the solubility product of this precipitate is remarkably small (3×10^{-34}). The cause of the increase in solubility is shown by the equations

The solubility of a precipitate always increases in the presence of a complexing agent that reacts with the cation of the precipitate.

$$Al(OH)_3(s) \rightleftharpoons Al^{3+} + 3\ OH^-$$
$$+$$
$$6F^-$$
$$\updownarrow$$
$$AlF_6^{3-}$$

The fluoride complex is sufficiently stable to permit fluoride ions to compete successfully with hydroxide ions for aluminum ions.

Complex Formation with an Ion That Is Common to the Precipitate

Many precipitates react with the precipitating reagent to form soluble complexes. In a gravimetric analysis, this tendency may have the unfortunate effect of

Feature 7-1
WRITE ENOUGH INDEPENDENT EQUATIONS TO CALCULATE THE SOLUBILITY OF $BaCO_3$ IN WATER

$$\text{solubility} = [Ba^{2+}] = [CO_3^{2-}] + [HCO_3^-] + [H_2CO_3]$$

Equilibrium-constant expressions

$$K_{sp} = [Ba^{2+}][CO_3^{2-}] = 5.0 \times 10^{-19} \qquad (7\text{-}17)$$

$$K_2 = \frac{[H_3O^+][CO_3^{2-}]}{[HCO_3^-]} = 4.69 \times 10^{-11} \qquad (7\text{-}18)$$

$$K_1 = \frac{[H_3O^+][HCO_3^-]}{[H_2CO_3]} = 4.45 \times 10^{-7} \qquad (7\text{-}19)$$

$$K_w = [H_3O^+][OH^-] = 1.00 \times 10^{-14} \qquad (7\text{-}20)$$

Mass-balance equation

$$[Ba^{2+}] = [CO_3^{2-}] + [HCO_3^-] + [H_2CO_3] \qquad (7\text{-}21)$$

Charge-balance equation

$$2[Ba^{2+}] + [H_3O^+] = 2[CO_3^{2-}] + [HCO_3^-] + [OH^-] \qquad (7\text{-}22)$$

We now have 6 unknowns ($[Ba^{2+}]$, $[CO_3^{2-}]$, $[HCO_3^-]$, $[H_2CO_3]$, $[H_3O^+]$, and $[OH^-]$) and 6 equations (7-17 through 7-22). Thus, the problem can be solved exactly.

reducing the recovery of analytes if too large an excess of reagent is used. For example, in solutions containing high concentrations of chloride, silver chloride forms chloro complexes such as $AgCl_2^-$ and $AgCl_3^{2-}$. The effect of these complexes is illustrated in Figure 7-2, in which the experimentally determined solubility of silver chloride is plotted against the logarithm of the potassium chloride concentration. For low anion concentrations, the experimental solubilities do not differ greatly from those calculated with the solubility-product constant for silver chloride; beyond a chloride ion concentration of about 10^{-3} M, however, the calculated solubilities approach zero while the measured values rise steeply. Note that the solubility of silver chloride is about the same in 0.3 M KCl as in pure water, and is about eight times that figure in a 1 M solution. We can describe these effects quantitatively if the compositions of the complexes and their formation constants are known.

Increases in solubility caused by large excesses of a common ion are not unusual. Of particular interest are amphoteric hydroxides, which are sparingly soluble in dilute base but are redissolved by excess hydroxide ion. The hydroxides of zinc and aluminum, for example, are converted to the soluble zincate and aluminate ions upon treatment with excess base. For zinc, the equilibria can be

In gravimetric procedures, a small excess of precipitating agent minimizes solubility losses but a large excess often causes increased losses due to complex formation.

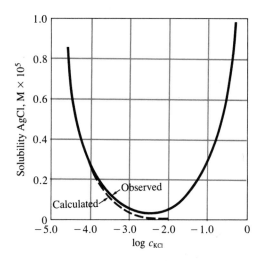

Figure 7-2
Solubility of silver chloride in potassium chloride solutions. The dashed curve is calculated from K_{sp}; the solid curve is plotted from experimental data of A. Pinkus and A. M. Timmermans, *Bull. Soc. Belges*, **1937**, *46*, 46–73.

represented as

$$Zn^{2+} + 2OH^- \rightleftharpoons Zn(OH)_2(s)$$

$$Zn(OH)_2(s) + 2OH^- \rightleftharpoons Zn(OH)_4^{2-}$$

As with silver chloride, the solubilities of amphoteric hydroxides pass through minima and then increase rapidly with increasing concentrations of base. The hydroxide ion concentration at which the solubility is a minimum can be calculated, provided equilibrium constants for the reactions are available.

Quantitative Treatment of the Effect of Complex Formation on Solubility

Solubility calculations for a precipitate in the presence of a complexing reagent are similar in principle to those discussed in the previous section. Formation constants for the complexes involved must be available.[1]

7C   SEPARATION OF IONS BASED ON SOLUBILITY DIFFERENCES; SULFIDE SEPARATIONS

Several precipitating agents permit separation of ions based on solubility differences. Such separations require close control of the active reagent concentration at a suitable and predetermined level. Most often, such control is achieved by controlling the pH of the solution with suitable buffers. This technique is applicable to anionic reagents in which the anion is the conjugate base of a weak acid. Examples include sulfide ion (the conjugate base of hydrogen sulfide), hydroxide ion (the conjugate base of water), and the anions of several organic weak acids.

Sulfide ion forms precipitates with heavy-metal cations that have solubility products ranging from 10^{-10} to 10^{-50} or smaller. In addition, the concentration of S^{2-} can be varied over a range of about 0.1 M to 10^{-22} M by controlling the pH of a saturated solution of hydrogen sulfide. These two properties make possible

---

[1] For an example of such a calculation, see P. Hadjiioannou, G. D. Christian, C. E. Efstathiou, and D. P. Nikolelis, *Problem Solving in Analytical Chemistry*, pp. 145–149. New York: Pergamon, 1988.

Feature 7-2

IMMUNOASSAY: EQUILIBRIA IN THE SPECIFIC DETERMINATION OF DRUGS

The determination of drugs in the human body is a matter of great importance in drug therapy and in the detection and prevention of drug abuse. The diversity of drugs and their typical low levels of concentration in body

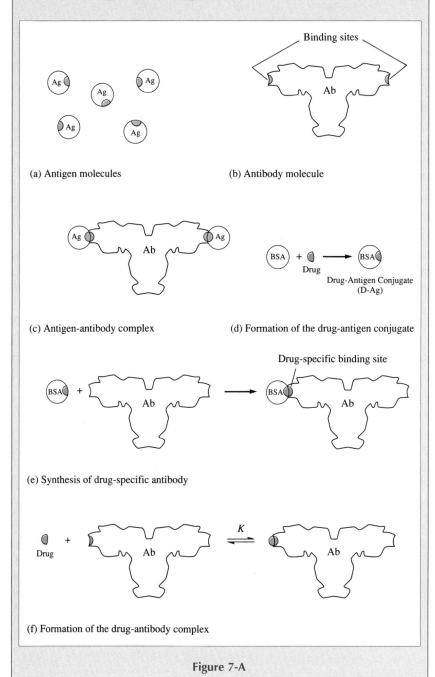

(a) Antigen molecules

(b) Antibody molecule

(c) Antigen-antibody complex

(d) Formation of the drug-antigen conjugate

(e) Synthesis of drug-specific antibody

(f) Formation of the drug-antibody complex

Figure 7-A

fluids make it difficult to identify them and measure their concentrations. Fortunately, it is possible to harness one of nature's own mechanisms, the immune response, to determine quantitatively a variety of therapeutic and illicit drugs.

When a foreign substance, or antigen (Ag), shown schematically in Figure 7-A(a) is introduced into the body of a mammal, the immune system synthesizes protein-based molecules (Figure 7-A(b)) called antibodies (Ab), which specifically bind to the antigen molecules via electrostatic interactions, hydrogen bonding, and other non-covalent short-range forces. These massive molecules (molar mass $\cong$ 150,000) form a complex with antigens as shown in the reaction below and in Figure 7-A(c).

$$Ag + Ab \rightleftharpoons AgAb \qquad K = \frac{[AgAb]}{[Ag][Ab]}$$

The immune system does not recognize relatively small molecules, so we must use a trick to prepare antibodies with binding sites that are specific for a particular drug. As shown in Figure 7-A(d), we attach the drug covalently to an antigenic carrier molecule such as bovine serum albumin (BSA), a protein that is obtained from the blood of cattle.

$$D + Ag \rightarrow D\text{-}Ag$$

When the resulting drug-antigen conjugate (D-Ag) is injected into the bloodstream of a rabbit, the immune system of the rabbit synthesizes antibodies with binding sites that are specific for the drug, as illustrated in Figure 7-A(e). Approximately three weeks following injection of the antigen, blood is drawn from the rabbit, the serum is separated from the blood, and the antibodies of interest are separated from the serum and other antibodies, usually by chromatographic methods (see Chapter 27). It is important to note that once the drug-specific antibody has been synthesized by the immune system of the rabbit, the drug can bind directly to the antibody without the aid of the carrier molecule, as shown in Figure 7-A(f). This direct drug-antibody binding forms the basis for the specific determination of the drug.

The measurement step of the immunoassay is accomplished by mixing the sample containing the drug with a measured amount of drug-specific antibody. At this point, the quantity of Ab-D must be determined by adding a standard sample of the drug that has been chemically altered to contain a detectable *label*. Typical labels are enzymes, fluorescent or chemiluminescent molecules, or radioactive atoms. For our example, we will assume that a fluorescent molecule has been attached to the drug to produce the labeled drug D*.[1] If the amount of the antibody is somewhat less than the sum of the amounts of D and D*, then D and D* compete for the antibody as shown in the following equilibria.

$$D^* + Ab \rightleftharpoons Ab\text{-}D^* \qquad K^* = \frac{[Ab\text{-}D^*]}{[D^*][Ab]}$$

$$D + Ab \rightleftharpoons Ab\text{-}D \qquad K = \frac{[Ab\text{-}D]}{[D][Ab]}$$

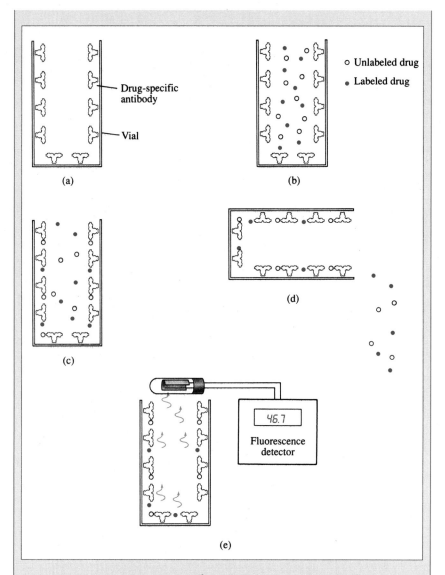

Figure 7-B

It is important to select a label that does not substantially alter the affinity of the drug for the antibody so that the labeled and unlabeled drugs bind with the antibody equally well. If this is true, then $K = K^*$. Typical values for equilibrium constants of this type, called *binding constants*, range from 10^7 to 10^{12}. The larger the concentration of the unknown, unlabeled drug, the smaller the concentration of Ab-D*, and vice-versa. This inverse relationship between D and Ab-D* forms the basis for the quantitative determination of the drug. We can find the amount of D if we measure *either* Ab-D* or D*. To differentiate between bound drug and unbound labeled drug, it is necessary to separate them prior to measurement. The amount of Ab-D* can then be found by using a fluorescence detector to measure the intensity of the

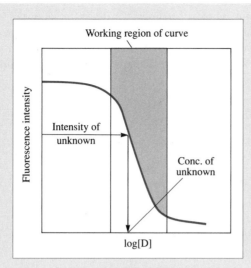

Figure 7-C

fluorescence resulting from the Ab-D*. A determination of this type using a fluorescent drug and radiation detection is called a *fluorescence immunoassay*. Determinations of this type are very sensitive and selective.

One convenient way to separate D* and Ag-D* is to prepare polystyrene vials that are coated on the inside with antibody molecules, as illustrated in Figure 7-B(a). A sample of blood serum, urine, or other body fluid containing an unknown concentration of D along with a volume of solution containing labeled drug D* are added to the vial as depicted in Figure 7-B(b). After equilibrium is achieved in the vial (Figure 7-B(c)), the solution containing residual D and D* is decanted and the vial is rinsed, leaving an amount of D* bound to the antibody that is inversely proportional to the concentration of D in the sample (Figure 7-B(d)). Finally, the fluorescence intensity of the bound D* is determined using a fluorometer as shown in Figure 7-B(e). This procedure is repeated for several standard solutions of D to produce a noniinear working curve called a *dose-response curve* similar to the curve of Figure 7-C. The fluorescence intensity for an unknown solution of D is located on the calibration curve and the concentration is read from the concentration axis.

Immunoassay is a powerful tool in the clinical laboratory and is one of the most widely used of all analytical techniques. Reagent kits for many different immunoassays are available commercially as are automated instruments for carrying out fluorescent immunoassays and immunoassays of other types. In addition to drugs, vitamins, proteins, growth hormones, pregnancy hormones, cancer and other disease indicators, and pesticide residues in natural waters and food are determined by immunoassay.

---

[1] For a discussion of molecular fluorescence, see Chapter 23.

a number of useful cation separations. To illustrate the use of hydrogen sulfide to separate cations based on pH control, let us consider the precipitation of the divalent cation M^{2+} from a solution that is kept saturated with hydrogen sulfide by bubbling the gas continuously through the solution. The important equilibria in this solution are:

$$MS(s) \rightleftharpoons M^{2+} + S^{2-} \qquad K_{sp} = [M^{2+}][S^{2-}]$$

$$H_2S + H_2O \rightleftharpoons H_3O^+ + HS^- \qquad K_1 = \frac{[H_3O^+][HS^-]}{[H_2S]} = 9.6 \times 10^{-8}$$

$$HS^- + H_2O \rightleftharpoons H_3O^+ + S^{2-} \qquad K_2 = \frac{[H_3O^+][S^{2-}]}{[HS^-]} = 1.3 \times 10^{-14}$$

We may also write

$$\text{solubility} = [M^{2+}]$$

The concentration of hydrogen sulfide in a saturated solution of the gas is approximately 0.1 M. Thus, we may write as a mass-balance expression

$$[S^{2-}] + [HS^-] + [H_2S] = 0.1$$

Because we know the hydronium ion concentration, we have four unknowns, the concentration of the metal ion and the three sulfide species.

We can simplify the calculation greatly by assuming that $([S^{2-}] + [HS^-]) \ll [H_2S]$ so that

$$[H_2S] \cong 0.10 \text{ mol/L}$$

The two dissociation constant expressions for hydrogen sulfide may be multiplied together to give an expression for the overall dissociation of hydrogen sulfide to sulfide ion:

$$H_2S + 2H_2O \rightleftharpoons 2H_3O^+ + S^{2-} \qquad K_1K_2 = \frac{[H_3O^+]^2[S^{2-}]}{[H_2S]} = 1.2 \times 10^{-21}$$

The constant for this overall reaction is simply the product of K_1 and K_2.

Substituting the numerical value for $[H_2S]$ into this equation gives

$$\frac{[H_3O^+]^2[S^{2-}]}{0.10} = 1.2 \times 10^{-21}$$

On rearranging this equation, we obtain

$$[S^{2-}] = \frac{1.2 \times 10^{-22}}{[H_3O^+]^2} \qquad \qquad \textbf{(7-23)}$$

Thus, we see that the sulfide ion concentration of a saturated hydrogen sulfide solution varies inversely as the square of the hydrogen ion concentration. Figure

7-3, which was obtained with this equation, reveals that the sulfide ion concentration of an aqueous solution can be varied by over 20 orders of magnitude within the pH range from 1 to 11.

Substituting Equation 7-23 into the solubility-product expression gives

$$K_{sp} = \frac{[M^{2+}] \times 1.2 \times 10^{-22}}{[H_3O^+]^2}$$

$$[M^{2+}] = \text{solubility} = \frac{[H_3O^+]^2 \, K_{sp}}{1.2 \times 10^{-22}}$$

Thus, the solubility of a divalent metal sulfide increases as the square of the hydronium ion concentration.

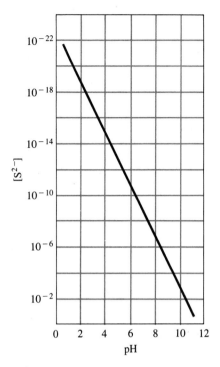

Figure 7-3

Sulfide ion concentration as a function of pH in a saturated H_2S solution.

Example 7-8

Cadmium sulfide is less soluble than thallium(I) sulfide. Find the conditions under which Cd^{2+} and Tl^+ can, in theory, be separated quantitatively with H_2S from a solution that is 0.1 M in each cation.

The constants for the two solubility equilibria are:

$$CdS(s) \rightleftharpoons Cd^{2+} + S^{2-} \qquad K_{sp} = [Cd^{2+}][S^{2-}] = 1 \times 10^{-27}$$

$$Tl_2S(s) \rightleftharpoons 2Tl^+ + S^{2-} \qquad K_{sp} = [Tl^+]^2[S^{2-}] = 6 \times 10^{-22}$$

Since CdS precipitates at a lower $[S^{2-}]$ than does Tl_2S, we first compute the sulfide ion concentration necessary for quantitative removal of Cd^{2+} from solution. In order to make such a calculation, we must first specify what constitutes a quantitative removal. The decision here is arbitrary and depends on the purpose of the separation. In this example, we shall consider a separation to be quantitative when all but 1 part in 1000 of the Cd^{2+} has been removed; that is, the concentration of the cation has been lowered to 1.00×10^{-4} M. Substituting this value into the solubility-product expression gives

$$10^{-4} [S^{2-}] = 1 \times 10^{-27}$$

$$[S^{2-}] = 2 \times 10^{-23}$$

Thus, if we maintain the sulfide concentration at this level or greater, we may assume that quantitative removal of the cadmium will take place. Next, we compute the $[S^{2-}]$ needed to initiate precipitation of Tl_2S from a 0.1 M solution. Precipitation will begin when the solubility product is just exceeded. Since the solution is 0.1 in Tl^+,

$$(0.1)^2 [S^{2-}] = 6 \times 10^{-22}$$

$$[S^{2-}] = 6 \times 10^{-20}$$

These two calculations show that quantitative precipitation of Cd^{2+} takes place if $[S^{2-}]$ is made greater than 1×10^{-23}. No precipitation of Tl^+ occurs, however, until $[S^{2-}]$ becomes greater than 6×10^{-20} M.

Substituting these two values for $[S^{2-}]$ into Equation 7-23 permits calculation of the $[H_3O^+]$ range required for the separation.

$$[H_3O^+]^2 = \frac{1.2 \times 10^{-22}}{1 \times 10^{-23}} = 12$$

$$[H_3O^+] = 3.5$$

and

$$[H_3O^+]^2 = \frac{1.2 \times 10^{-22}}{6 \times 10^{-20}} = 2.0 \times 10^{-3}$$

$$[H_3O^+] = 0.045$$

By maintaining $[H_3O^+]$ between approximately 0.045 and 3.5 M, we can in theory separate CdS quantitatively from Tl_2S.

7D QUESTIONS AND PROBLEMS

*7-1. The molar equilibrium solubilities for $BaSO_4$ (233 g/mol) and $Cd(OH)_2$ (146 g/mol) are essentially identical (1.14×10^{-5} M). What can be said about numerical values for (a) their solubility-product constants; and (b) their solubilities in g/L?

7-2. Briefly account for the difference—if any—between the molar equilibrium solubilities of $Be(OH)_2$ (43.0 g/mol) and $Eu(OH)_3$ (203 g/mol), which have nearly identical solubility-product constants (7.0×10^{-22} and 8.9×10^{-22}, respectively).

*7-3. Briefly explain why calculation of the molar solubility for $Mg(OH)_2$ ($K_{sp} = 7.0 \times 10^{-12}$) differs from the same calculation for $Pt(OH)_2$ (1×10^{-35}).

7-4. Briefly show how the molar sulfide ion concentration in a saturated solution of H_2S is related to the hydronium ion concentration.

*7-5. Why are simplifying approximations to equilibrium calculations limited to sums and differences?

7-6. Why do the molar concentrations of some species appear as multiples in charge-balance equations?

7-7. Write the mass-balance expression for a solution that is
   *(a) 0.10 M in H_3PO_4.
   (b) 0.10 M in Na_2HPO_4.
   *(c) 0.100 M in HNO_2 and 0.0500 M in $NaNO_2$.
   (d) 0.025 M in NaF and saturated with CaF_2.
   *(e) 0.100 M in NaOH and saturated with $Zn(OH)_2$, which undergoes the reaction:

   $$Zn(OH)_2(s) + 2OH^- \rightleftharpoons Zn(OH)_4^{2-}$$

   (f) saturated with $MgCO_3$.
   *(g) saturated with CaF_2.

7-8. Write the charge-balance equations for the solutions in Problem 7-7.

7-9. Calculate the molar solubility of $BaSO_4$ in a solution having an $[H_3O^+]$ of
   *(a) 2.0 M.   *(c) 0.50 M.
   (b) 1.0 M.   (d) 0.10 M.

7-10. Calculate the molar solubility of AgCN in a solution that has a fixed pH of
   (a) 3.60.   (c) 8.00.
   (b) 5.85.   (d) 9.12.

*7-11. Calculate the molar solubility of CuS in a solution in which $[H_3O^+]$ is held constant at (a) 1.0×10^{-1} M; and (b) 1.0×10^{-4} M.

7-12. Calculate the concentration of CdS in a solution in which the $[H_3O^+]$ is held constant at (a) 1.0×10^{-1} M; and (b) 1.0×10^{-4} M.

*7-13. What is the equilibrium solubility of MnS in a solution in which the $[H_3O^+]$ is held constant at (a) 1.0×10^{-5} M; and (b) 1.0×10^{-8} M? (Use $K_{sp} = 3 \times 10^{-14}$.)

7-14. The solubility-product constant for $CdCO_3$ is 1.8×10^{-14}. Calculate the equilibrium solubility of $CdCO_3$ in a solution with a constant hydronium ion concentration of
   *(a) 1.00×10^{-4} M.   *(c) 1.00×10^{-9} M.
   (b) 1.00×10^{-7} M.   (d) 1.00×10^{-11} M.

7-15. The solubility-product constant for $PbCO_3$ is 7.4×10^{-14}. Calculate the equilibrium solubility of $PbCO_3$ in a solution that has a constant hydronium ion concentration of
   *(a) 1.00×10^{-3} M.   *(c) 1.00×10^{-8} M.
   (b) 1.00×10^{-6} M.   (d) 1.00×10^{-10} M.

7-16. Calculate the molar solubility of Ag_2CO_3 in a solution that has a H_3O^+ concentration of

*(a) 1.0×10^{-6} M.   *(c) 1.0×10^{-9} M.

(b) 1.0×10^{-7} M.   (d) 1.0×10^{-11} M.

7-17. Calculate the equilibrium solubility of Ag_3AsO_4 in a solution buffered to a pH of

*(a) 3.00.   *(c) 9.00.

(b) 6.00.   (d) 12.00.

7-18. The solubility product for $Pd(OH)_2$ is 1.5×10^{-31}. Calculate the solubility of this compound in pure water.

*7-19. Calculate the solubility of CuS in pure water. Because of the low solubility, it is permissible to assume that the presence of CuS does not alter the pH of the solution, and therefore,

$$[H_3O^+] = [OH^-] = 1.0 \times 10^{-7} \text{ M}$$

7-20. The equilibrium constant for formation of $CuCl_2^-$ is given by

$$Cu^+ + 2Cl^- \rightleftharpoons CuCl_2^- \qquad K_f = \frac{[CuCl_2^-]}{[Cu^+][Cl^-]^2} = 3.2 \times 10^5$$

What is the solubility of CuCl in solutions having the following analytical NaCl concentrations:

*(a) 1.0 M? $\times 10^{-1}$ (d) 1.0×10^{-3} M?

(b) 1.0×10^{-1} M?   (e) 1.0×10^{-4} M?

*(c) 1.0×10^{-2} M?

7-21. Dilute NaOH is introduced into a solution that is 0.050 M in Cu^{2+} and 0.040 M in Mn^{2+}.

(a) Which hydroxide precipitates first?

(b) What OH^- concentration is needed to initiate precipitation of the first hydroxide?

(c) What is the concentration of the cation forming the less soluble hydroxide when the more soluble hydroxide begins to form?

*7-22. Using 1.0×10^{-6} M as the criterion for quantitative removal, determine whether it is feasible to use

(a) SO_4^{2-} to separate Ba^{2+} from Sr^{2+} in a solution that is initially 0.10 M in Sr^{2+} and 0.25 M in Ba^{2+}.

(b) SO_4^{2-} to separate Ba^{2+} and Ag^+ in a solution that is initially 0.040 M in each cation. For Ag_2SO_4, $K_{sp} = 1.6 \times 10^{-5}$.

7-23. A solution is 0.040 M in Na_2SO_4 and 0.050 M in NaOH. To this is added a dilute solution containing Pb^{2+}.

(a) Which compound precipitates first, $PbSO_4$ or $Pb(OH)_2$ [use 8×10^{-16} for K_{sp} for $Pb(OH)_2$]?

(b) What is the Pb^{2+} concentration as the first precipitate appears?

(c) What Pb^{2+} concentration is required to initiate precipitation of the more soluble substance?

(d) What is the concentration of the first anion when the more soluble precipitate begins to form?

*7-24. Silver ion is being considered as a reagent for separating I^- from SCN^- in a solution that is 0.060 M in KI and 0.070 M in NaSCN.

(a) What Ag^+ concentration is needed to lower the I^- concentration to 1.0×10^{-6} M?

(b) What is the Ag^+ concentration of the solution when AgSCN begins to precipitate?

(c) What is the ratio of SCN^- to I^- ion when AgSCN begins to precipitate?

(d) What is the ratio of SCN^- to I^- when the Ag^+ concentration is 1.0×10^{-3} M?

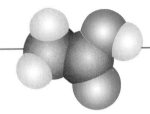

EFFECT OF ELECTROLYTES ON CHEMICAL EQUILIBRIA

The effect of electrolytes on chemical equilibria is sometimes termed the *salt effect*.

The solubility of an ionic precipitate generally increases when electrolytes are added to water in which the precipitate is suspended. This phenomenon is illustrated by the curves in Figure 8-1, which show the solubility of three precipitates as a function of the potassium nitrate concentration of the aqueous solvent. A similar effect is observed with other salts such as sodium perchlorate or lithium bromide that do not contain either of the ions making up the precipitates. In each case, the effect has as its origin the electrostatic attraction between the ions of the electrolyte and the ions of opposite charge from the precipitates. Since the electrostatic forces associated with all singly charged ions are approximately the same, the three salts exhibit essentially identical effects on equilibria.

In this chapter, we describe the conditions under which this *salt effect* is likely to be significant in magnitude and how quantitative corrections can be made for the effect.

8A VARIABLES THAT INFLUENCE THE MAGNITUDE OF THE SALT EFFECT

The position of any equilibrium that involves ionic participants will be influenced by the charge on the reactant and product ions and by a property of the solution called the *ionic strength*.

8A-1 The Effect of the Charge of Reactants and Products

Extensive studies have revealed that the magnitude of the electrolyte effect is highly dependent on the charges of the participants in an equilibrium. When only

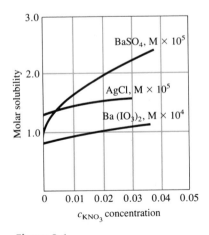

Figure 8-1

Effect of electrolyte concentration on the solubility of some salts.

neutral species are involved, the position of equilibrium is essentially independent of electrolyte concentration. With ionic participants, the magnitude of the electrolyte effect increases with charge. This generality is demonstrated by the three solubility curves in Figure 8-1. Note, for example, that in a 0.02 M solution of potassium nitrate, the solubility of barium sulfate, with its pair of doubly charged ions, is larger by a factor of 2 than it is in pure water. This same change in electrolyte concentration increases the solubility of barium iodate by a factor of only 1.25 and that of silver chloride by 1.2.

8A-2 The Effect of Ionic Strength

Systematic studies have shown that the effect of added electrolyte on equilibria is *independent* of the chemical nature of the electrolyte but depends on a property of the solution called the *ionic strength*. This quantity is defined as

$$\text{ionic strength} = \mu = \tfrac{1}{2}([A]\,Z_A^2 + [B]\,Z_B^2 + [C]\,Z_C^2 + \cdots) \qquad \textbf{(8-1)}$$

where [A], [B], [C], . . . represent the molar concentrations of ions A, B, C, . . . and Z_A, Z_B, Z_C, . . . are their charges.

Example 8-1

Calculate the ionic strength of (a) a 0.1 M solution of KNO_3 and (b) a 0.1 M solution of Na_2SO_4.
(a) For the KNO_3 solution, $[K^+]$ and $[NO_3^-]$ are 0.1 M and

$$\mu = \tfrac{1}{2}(0.1 \times 1^2 + 0.1 \times 1^2) = 0.1$$

(b) For the Na_2SO_4 solution, $[Na^+] = 0.2$ and $[SO_4^{2-}] = 0.1$. Therefore,

$$\mu = \tfrac{1}{2}(0.2 \times 1^2 + 0.1 \times 2^2) = 0.3$$

Example 8-2

What is the ionic strength of a solution that is 0.05 M in KNO_3 and 0.1 M in Na_2SO_4?

$$\mu = \tfrac{1}{2}(0.05 \times 1^2 + 0.05 \times 1^2 + 0.2 \times 1^2 + 0.1 \times 2^2) = 0.35 \text{ M}$$

It is apparent from these examples that the ionic strength of a solution of a strong electrolyte consisting solely of singly charged ions is identical with its total molar salt concentration. However, the ionic strength is greater than the molar concentration if the solution contains ions with multiple charges.

For solutions with ionic strengths of 0.1 M or less, the electrolyte effect is independent of the *kind* of ions and dependent *only on the ionic strength*. Thus, the solubility of barium sulfate is the same in aqueous sodium iodide, potassium nitrate, or aluminum chloride, provided the concentrations of these species are

Effect of Charge on Ionic Strength

Type Electrolyte	Example	Ionic Strength*
1:1	NaCl	c
1:2	$Ba(NO_3)_2$, Na_2SO_4	$3c$
1:3	$Al(NO_3)_3$, Na_3PO_4	$6c$
2:2	$MgSO_4$	$4c$

* c = molarity of the salt

such that the ionic strengths are identical. Note that this independence with respect to electrolyte species disappears at high ionic strengths.

8A-3 Source of the Salt Effect

The electrolyte effect, which we have just described, results from the electrostatic attractive and repulsive forces between the ions of an electrolyte and the ions involved in an equilibrium. These forces cause each ion from the dissociated reactant to be surrounded by a sheath of solution that contains a slight excess of electrolyte ions of opposite charge. For example, when a barium sulfate precipitate is equilibrated with a sodium chloride, each dissolved barium ion is surrounded by an ionic atmosphere that, because of electrostatic attraction and repulsion, carries a small net negative charge on the average due to repulsion of sodium ions and an attraction of chloride ions. Similarly, each sulfate ion is surrounded by an ionic atmosphere that tends to be slightly positive. These charged layers make the barium ions somewhat less positive and the sulfate ions somewhat less negative than they would be in the absence of electrolyte. The consequence of this effect is a decrease in overall attraction between barium and sulfate ions and an increase in solubility, which becomes greater as the number of electrolyte ions in the solution becomes larger. That is, the *effective concentration* of barium ions and of sulfate ions becomes less as the ionic strength of the medium becomes greater.

8B ACTIVITY COEFFICIENTS

In order to describe quantitatively the effective concentration of participants in an equilibrium at any given ionic strength, chemists use a term called *activity*, a, which for the species X is defined as

$$a_X = \gamma_X[X] \qquad (8\text{-}2)$$

The activity of a species is a measure of its effective concentration as determined by the lowering of the freezing point of water, by electrical conductivity, and by the mass action effect.

where a_X is the activity of the species X, [X] is its molar concentration, and γ_X is a dimensionless quantity called the *activity coefficient*. The activity coefficient and thus the activity of X vary with ionic strength such that substitution of a_X for [X] in any equilibrium-constant expression frees the numerical value of the constant from dependence on the ionic strength. To illustrate, if X_mY_n is a precipitate, the thermodynamic solubility-product expression is defined by the equation

$$K_{sp} = a_X^m \cdot a_Y^n \qquad (8\text{-}3)$$

Applying Equation 8-2 gives

$$K_{sp} = \gamma_X^m \cdot \gamma_Y^n \cdot [X]^m[Y]^n = \gamma_X^m \cdot \gamma_Y^n \cdot K_{sp}' \qquad (8\text{-}4)$$

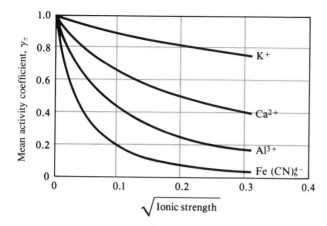

Figure 8-2
Effect of ionic strength on activity
coefficients.

Here K'_{sp} is the *concentration solubility-product constant* and K_{sp} is the thermodynamic equilibrum constant.[1] The activity coefficients γ_X and γ_Y vary with ionic strength in such a way as to keep K_{sp} numerically constant and independent of ionic strength (in contrast to the concentration constant K'_{sp}).

8B-1 Properties of Activity Coefficients

Activity coefficients have the following properties:

1. The activity coefficient of a species is a measure of the effectiveness with which that species influences an equilibrium in which it is a participant. In very dilute solutions, where the ionic strength is minimal, this effectiveness becomes constant, and the activity coefficient is unity. Under such circumstances, the activity and the molar concentration are identical (as are thermodynamic and concentration equilibrium constants). As the ionic strength increases, however, an ion loses some of its effectiveness, and its activity coefficient decreases. We may summarize this behavior in terms of Equations 8-2 and 8-3. At moderate ionic strengths, $\gamma_X < 1$; as the solution approaches infinite dilution, however, $\gamma_X \rightarrow 1$ and thus $a_X \rightarrow [X]$ and $K'_{sp} \rightarrow K_{sp}$. At high ionic strengths ($\mu > 0.1$ M), activity coefficients often increase and may even become greater than unity. Because interpretation of the behavior of solutions in this region is difficult, we shall confine our discussion to regions of low or moderate ionic strength (that is, where $\mu \leq 0.1$ M).

   The variation of typical activity coefficients as a function of ionic strength is shown in Figure 8-2.

2. In solutions that are not too concentrated, the activity coefficient for a given species is independent of the nature of the electrolyte and dependent only on the ionic strength.

3. For a given ionic strength, the activity coefficient of an ion departs farther from unity as the charge carried by the species increases. This effect is

Concentration-based equilibrium constants are commonly indicated by a prime mark regardless of the kind of equilibrium involved. For example, for the dissociation of water

$$2H_2O \rightleftharpoons H_3O^+ + OH^-$$

We may write the thermodynamic ion product constant as

$$K_w = a_{H_3O^+} \cdot a_{OH^-}$$
$$= \gamma_{H_3O^+} \cdot \gamma_{OH^-} \cdot [H_3O^+][OH^-]$$
$$= \gamma_{H_3O^+} \cdot \gamma_{OH^-} \cdot K'_w$$

where

$$K'_w = [H_3O^+][OH^-]$$

Concentration-based acid/base dissociation constants K'_a and K'_b are defined in an analogous way.

As $\mu \rightarrow 0$, $\gamma_X \rightarrow 1$, $a_X \rightarrow [X]$, and $K'_{sp} \rightarrow K_{sp}$.

---

[1] In the chapters that follow we use the prime notation only when it is necessary to distinguish between thermodynamic and concentration equilibrium constants.

shown in Figure 8-2. The activity coefficient of an uncharged molecule is approximately unity, regardless of ionic strength.

4. At any given ionic strength, the activity coefficients of ions of the same charge are approximately equal. The small variations that are observed can be correlated with the effective diameter of the hydrated ions.

5. The activity coefficient of a given ion describes its effective behavior in all equilibria in which it participates. For example, at a given ionic strength, a single activity coefficient for cyanide ion describes the influence of that species on any of the following equilibria:

$$HCN + H_2O \rightleftharpoons H_3O^+ + CN^-$$

$$Ag^+ + CN^- \rightleftharpoons AgCN(s)$$

$$Ni^{2+} + 4CN^- \rightleftharpoons Ni(CN)_4^{2-}$$

8B-2 The Debye-Hückel Equation

In 1923, P. Debye and E. Hückel used the ionic atmosphere model, described in Section 8A-3, to derive a theoretical expression that permits the calculation of activity coefficients of ions from their charge and average size.[2] This equation, which has become known as the *Debye-Hückel equation*, takes the form

When μ is less than 0.01, $1 + \sqrt{\mu} \cong 1$, and Equation 8-5 becomes

$$-\log \gamma_X = 0.51 Z_X^2 \sqrt{\mu}$$

This equation is referred to as the Debye–Hückel Limiting Law (DHLL). Thus, in solutions of very low ionic strength, the DHLL can be used to calculate approximate activity coefficients.

$$-\log \gamma_X = \frac{0.51 Z_X^2 \sqrt{\mu}}{1 + 3.3 \, \alpha_X \sqrt{\mu}} \qquad (8\text{-}5)$$

where

γ_X = activity coefficient of the species X

Z_X = charge on the species X

μ = ionic strength of the solution

α_X = effective diameter of the hydrated ion X in nanometers (10^{-9} m)

The constants 0.51 and 3.3 are applicable to aqueous solutions at 25°C; other values must be used at other temperatures.

Unfortunately, there is considerable uncertainty regarding the magnitude of α_X in Equation 8-5. Its value appears to be approximately 0.3 nm for most singly charged ions; for these species, then, the denominator of the Debye-Hückel equation simplifies to approximately $1 + \sqrt{\mu}$. For ions with higher charge, α_X may be as large as 1.0 nm. This increase in size with increase in charge makes good chemical sense. The larger the charge on an ion, the larger the number of polar water molecules that will be held in the solvation shell about the ion. It should be noted that the second term of the denominator is small with respect

Peter Debye (1884–1966) was born and educated in Europe but became Professor of Chemistry at Cornell University in 1940. He was noted for his work in several different areas of chemistry including electrolyte solutions, X-ray diffraction, and the properties of polar molecules. He received the 1936 Nobel Prize in Chemistry.

[2] P. Debye and E. Hückel, *Physik. Z.*, **1923**, *24*, 185.

Table 8-1
ACTIVITY COEFFICIENTS FOR IONS AT 25°C*

Ion	α_X, nm	Activity Coefficient at Indicated Ionic Strength				
		0.001	0.005	0.01	0.05	0.1
H_3O^+	0.9	0.967	0.933	0.914	0.86	0.83
Li^+, $C_6H_5COO^-$	0.6	0.965	0.929	0.907	0.84	0.80
Na^+, IO_3^-, HSO_3^-, HCO_3^-, $H_2PO_4^-$, $H_2AsO_4^-$, OAc^-	0.4–0.45	0.964	0.928	0.902	0.82	0.78
OH^-, F^-, SCN^-, HS^-, ClO_3^-, ClO_4^-, BrO_3^-, IO_4^-, MnO_4^-	0.35	0.964	0.926	0.900	0.81	0.76
K^+, Cl^-, Br^-, I^-, CN^-, NO_2^-, NO_3^-, $HCOO^-$	0.3	0.964	0.925	0.899	0.80	0.76
Rb^+, Cs^+, Tl^+, Ag^+, NH_4^+,	0.25	0.964	0.924	0.898	0.80	0.75
Mg^{2+}, Be^{2+}	0.8	0.872	0.755	0.69	0.52	0.45
Ca^{2+}, Cu^{2+}, Zn^{2+}, Sn^{2+}, Mn^{2+}, Fe^{2+}, Ni^{2+}, Co^{2+}, Phthalate^{2-}	0.6	0.870	0.749	0.675	0.48	0.40
Sr^{2+}, Ba^{2+}, Cd^{2+}, Hg^{2+}, S^{2-}	0.5	0.868	0.744	0.67	0.46	0.38
Pb^{2+}, CO_3^{2-}, SO_3^{2-}, $C_2O_4^{2-}$,	0.45	0.868	0.742	0.665	0.46	0.37
Hg_2^{2+}, SO_4^{2-}, $S_2O_3^{2-}$, CrO_4^{2-}, HPO_4^{2-}	0.40	0.867	0.740	0.660	0.44	0.36
Al^{3+}, Fe^{3+}, Cr^{3+}, La^{3+}, Ce^{3+}	0.9	0.738	0.54	0.44	0.24	0.18
PO_4^{3-}, $Fe(CN)_6^{3-}$,	0.4	0.725	0.50	0.40	0.16	0.095
Th^{4+}, Zr^{4+}, Ce^{4+}, Sn^{4+}	1.1	0.588	0.35	0.255	0.10	0.065
$Fe(CN)_6^{4-}$	0.5	0.57	0.31	0.20	0.048	0.021

* From J. Kielland, *J. Am. Chem. Soc.*, **1937**, *59*, 1675. By courtesy of the American Chemical Society.

to the first when the ionic strength is less than 0.01, so that at these ionic strengths, uncertainties in α_X are of little significance in calculating activity coefficients.

Kielland[3] has derived values of α_X for numerous ions from a variety of experimental data. His best values for effective diameters are given in Table 8-1. Also presented in the table are activity coefficients calculated from Equation 8-5 using these values for the size parameter.

Experimental determination of single-ion activity coefficients such as those shown in Table 8-1 is unfortunately impossible because all experimental methods give only a mean activity coefficient for the positively and negatively charged ions in a solution. In other words, it is impossible to measure the properties of individual ions in the presence of counter ions of opposite charge and solvent molecules. It should be pointed out, however, that mean activity coefficients calculated from the data in Table 8-1 agree satisfactorily with the experimental values.

Example 8-3

Use Equation 8-5 to calculate the activity coefficient for Hg^{2+} in a solution that has an ionic strength of 0.085. Use 0.5 nm for the effective diameter of the ion.

---

[3] J. Kielland, *J. Amer. Chem. Soc.*, **1937**, *59*, 1675.

Compare the calculated value with $\gamma_{Hg^{2+}}$ obtained by interpolation of data from Table 8-1.

$$-\log \gamma_{Hg^{2+}} = \frac{(0.51)(2)^2\sqrt{0.085}}{1 + (3.3)(0.5)\sqrt{0.085}} = 0.4016$$

$$\frac{1}{\gamma_{Hg^{2+}}} = 2.52$$

$$\gamma_{Hg^{2+}} = 0.397 = 0.40$$

Table 8-1 indicates that $\gamma_{Hg^{2+}} = 0.46$ when $\mu = 0.100$ and 0.38 when $\mu = 0.050$. Thus, for $\mu = 0.85$

$$\gamma_{Hg^{2+}} = 0.38 + \frac{(0.100 - 0.085)}{(0.100 - 0.050)}(0.46 - 0.38) = 0.404 = 0.40$$

The Debye-Hückel relationship and the data in Table 8-1 give satisfactory activity coefficients for ionic strengths up to about 0.1. Beyond this value, the equation fails, and experimentally determined mean-activity coefficients must be used.

8B-3 Equilibrium Calculations Using Activity Coefficients

Equilibrium calculations with activities yield values that are in better agreement with experimental results than those obtained with molar concentrations. Unless otherwise specified, equilibrium constants found in tables are based on activities and are thus thermodynamic equilibrium constants. The examples that follow illustrate how activity coefficients from Table 8-1 are applied to such data.

Example 8-4

Find the relative error introduced by neglecting activities in calculating the solubility of $Ba(IO_3)_2$ in a 0.033 M solution of $Mg(IO_3)_2$. The thermodynamic solubility product for $Ba(IO_3)_2$ is 1.57×10^{-9} (Appendix 1).

Let us first write the solubility-product expression in terms of activities:

$$K_{sp} = a_{Ba^{2+}} \times a_{IO_3^-}^2 = 1.57 \times 10^{-9}$$

where $a_{Ba^{2+}}$ and $a_{IO_3^-}^2$ are the activities of barium and iodate ions. Replacing activities in this equation by activity coefficients and concentrations from Equation 8-2 yields

$$K_{sp} = [Ba^{2+}]\gamma_{Ba^{2+}} \times [IO_3^-]^2\gamma_{IO_3^-}^2$$

where $\gamma_{Ba^{2+}}$ and $\gamma_{IO_3^-}$ are the activity coefficients for the two ions. Rearranging this expression gives

$$K'_{sp} = [Ba^{2+}][IO_3^-]^2 = \frac{K_{sp}}{\gamma_{Ba^{2+}} \times \gamma_{IO_3^-}^2} \qquad \textbf{(8-6)}$$

where K'_{sp} is the *concentration-based solubility product.*

The ionic strength of the solution is obtained by substituting into Equation 8-1:

$$\mu = \tfrac{1}{2}([Mg^{2+}] \times 2^2 + [IO_3^-] \times 1^2)$$
$$= \tfrac{1}{2}(0.033 \times 4 + 0.066 \times 1) = 0.099 \cong 0.1$$

In calculating μ, we have assumed that the Ba^{2+} and IO_3^- ions from the precipitate do not significantly affect the ionic strength of the solution. This simplification seems justified, considering the low solubility of barium iodate and the relatively high concentration of $Mg(IO_3)_2$. In situations where it is not possible to make such an assumption, the concentrations of the two ions can be approximated by solubility calculation in which activities and concentrations are assumed to be identical (as in Examples 3-3 and 3-4). These concentrations can then be introduced to give a better value for μ.

Turning now to Table 8-1, we find that at an ionic strength of 0.1,

$$\gamma_{Ba^{2+}} = 0.38 \qquad \gamma_{IO_3^-} = 0.78$$

If the calculated ionic strength did not match that of one of the columns in the table, $\gamma_{Ba^{2+}}$ and $\gamma_{IO_3^-}$ could be calculated from Equation 8-5 or by interpolation between the values in the table.

Substituting into the thermodynamic solubility-product expression gives

$$K_{sp}' = \frac{1.57 \times 10^{-9}}{(0.38)(0.78)^2} = 6.8 \times 10^{-9}$$
$$[Ba^{2+}][IO_3^-]^2 = 6.8 \times 10^{-9}$$

Proceeding now as in earlier solubility calculations,

$$\text{solubility} = [Ba^{2+}]$$
$$[IO_3^-] = 0.066$$
$$[Ba^{2+}](0.066)^2 = 6.9 \times 10^{-8}$$
$$[Ba^{2+}] = \text{solubility} = 1.56 \times 10^{-6}\,M$$

We compute solubility neglecting activities as shown earlier. That is,

$$[Ba^{2+}](0.066)^2 = 1.57 \times 10^{-9}$$
$$[Ba^{2+}] = \text{solubility} = 3.60 \times 10^{-7}\,M$$
$$\text{rel error} = \frac{3.60 \times 10^{-7} - 1.56 \times 10^{-6}}{1.56 \times 10^{-6}} \times 100\% = -77\%$$

8B-4 Effect of Ionic Strength on Other Types of Equilibria

Figure 8-3 illustrates the effect of ionic strength on three common types of chemical equilibria. Curve *A* is a plot of the product of the molar hydronium and hydroxide ion *concentrations* ($\times 10^{14}$) as a function of ionic strength. In each case, the ionic strength was varied with sodium chloride. This *concentration-*

Concentration-based equilibrium constants are commonly indicated by adding a prime mark. For example: K_w', K_{sp}', K_a'.

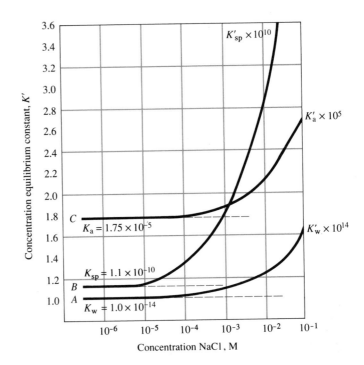

Figure 8-3
Effect of electrolyte concentration on concentration-based equilibrium constants.

As the concentration of electrolyte becomes very small, concentration-based equilibrium constants approach their thermodynamic values: K_w, K_{sp}, K_a.

based ion product is designated K'_w. At low ionic strengths, K'_w becomes independent of the electrolyte concentration and is equal to 1.00×10^{-14}, which is the *thermodynamic* ion-product constant for water, K_w. A relationship whose numerical value reaches a constant value as some concentration parameter approaches zero, such as the ion-product constant expression, is often called a *limiting law*. The numerical constant obtained at the limit is referred to as a *limiting value*.

The vertical axis for curve B in Figure 8-3 is the product of the molar concentrations of barium and sulfate ions ($\times 10^{10}$) in saturated solutions of barium sulfate. This concentration-based solubility product is designated as K'_{sp}. At low electrolyte concentrations, K'_{sp} has a limiting value of 1.1×10^{-10}, which is the accepted thermodynamic value of the solubility-product constant, K_{sp}, for barium sulfate.

Curve C is a plot of K'_a ($\times 10^5$), the concentration quotient for the equilibrium involving the dissociation of acetic acid, as a function of electrolyte concentration. Here again, the ordinate function approaches a limiting value K_a, which is the thermodynamic acid dissociation constant for acetic acid.

The dashed lines in Figure 8-3 represent ideal behavior of the solutes. Note that departures from ideality can be significant. For example, the product of the molar concentrations of hydrogen and hydroxide ion increase from 1.0×10^{-14} in pure water to about 1.7×10^{-14} in a solution that is 0.1 M in sodium chloride. The effect is even more pronounced with barium sulfate. Here, K'_{sp} in 0.1 M sodium chloride is more than double that of its limiting value.

8B-5 Omission of Activity Coefficients in Equilibrium Calculations

We will ordinarily neglect activity coefficients and simply use molar concentrations in applications of the equilibrium law. This approach simplifies the calculations and greatly decreases the amount of data needed. For most purposes, the error introduced by the assumption of unity for the activity coefficient is not large enough to lead to false conclusions. We see from the preceding examples, however, that disregard of activity coefficients may introduce a significant numerical error in calculations of this kind. Note, for example, that neglect of activities in Example 8-4 resulted in an error of about -77% relative. Be alert to the conditions under which the substitution of concentration for activity is likely to lead to the largest error. Significant discrepancies occur when the ionic strength is large (0.01 or larger) or when the ions involved have multiple charges (Table 8-1). With dilute solutions (ionic strength <0.01) of nonelectrolytes or of singly charged ions, the use of concentrations in a mass-law calculation often provides reasonably accurate results.

8C   QUESTIONS AND PROBLEMS

*8-1. Make a distinction between
  (a) activity and activity coefficient.
  (b) thermodynamic and concentration equilibrium constants.

8-2. List general properties of activity coefficients.

*8-3. Neglecting any effects caused by volume changes, would you expect the ionic strength to (1) increase, (2) decrease, or (3) remain essentially unchanged by the addition of NaOH to a dilute solution of
  (a) magnesium chloride [$Mg(OH)_2$ (s) forms]?
  (b) hydrochloric acid?
  (c) acetic acid?

8-4. Neglecting any effects caused by volume changes, would you expect the ionic strength to (1) increase, (2) decrease, or (3) remain essentially unchanged by the addition of iron(III) chloride to
  (a) HCl?
  (b) NaOH?
  (c) $AgNO_3$?

*8-5. Why is the slope of curve B in Figure 8-3 greater than that of curve A?

8-6. What is the numerical value of the activity coefficient of aqueous ammonia (NH_3) at an ionic strength of 0.1?

8-7. Calculate the ionic strength of a solution that is
  *(a) 0.040 M in $FeSO_4$.
  (b) 0.20 M in $(NH_4)_2CrO_4$.
  *(c) 0.10 M in $FeCl_3$ and 0.20 M in $FeCl_2$.
  (d) 0.060 M in $La(NO_3)_3$ and 0.030 M in $Fe(NO_3)_2$.

8-8. Use Equation 8-5 to calculate the activity coefficient of
  *(a) Fe^{3+} at $\mu = 0.075$.   *(c) Ce^{4+} at $\mu = 0.080$.
  (b) Pb^{2+} at $\mu = 0.012$.   *(d) Sn^{4+} at $\mu = 0.060$.

8-9. Calculate activity coefficients for the species in Problem 8-8 by linear interpolation of the data in Table 8-1.

8-10. For a solution in which μ is 5.0×10^{-2}, calculate K'_{sp} for
  *(a) AgSCN.   *(c) $La(IO_3)_3$.
  (b) PbI_2.   *(d) $MgNH_4PO_4$.

*8-11. Use activities to calculate the molar solubility of $Zn(OH)_2$ in
  (a) 0.0100 M KCl.
  (b) 0.0167 M K_2SO_4.
  (c) the solution that results when you mix 20.0 mL of 0.250 M KOH with 80.0 mL of 0.0250 M $ZnCl_2$.
  (d) the solution that results when you mix 20.0 mL of 0.100 M KOH with 80.0 mL of 0.0250 M $ZnCl_2$.

*8-12. Calculate the solubilities of the following compounds in a 0.0333 M solution of $Mg(ClO_4)_2$ employing (1) activities and (2) molar concentrations:
  (a) AgSCN.
  (b) PbI_2.
  (c) $BaSO_4$.
  (d) $Cd_2Fe(CN)_6$.

$$[Cd_2Fe(CN)_6(s) \rightleftharpoons 2Cd^{2+} + Fe(CN)_6^{4-}$$
$$K_{sp} = 3.2 \times 10^{-17}]$$

8-13. Calculate the solubilities of the following compounds in a 0.0167 M solution of $Ba(NO_3)_2$ using (1) activities and (2) molar concentrations:
  (a) $AgIO_3$.   (c) $BaSO_4$.
  (b) $Mg(OH)_2$.   (d) $La(IO_3)_2$.

---

* Answers to the asterisked problems are given in the answer section at the back of the book.

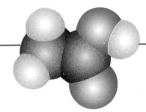

TITRIMETRIC METHODS OF ANALYSIS

The three types of quantitative titrimetry include: volumetric, gravimetric, and coulometric. Volumetric titrimetry is by far the most widely used.

Titrimetric methods include a large and powerful group of quantitative procedures that are based on measuring the amount of a reagent of known concentration that is consumed by the analyte. *Volumetric titrimetry* involves measuring the volume of a solution of known concentration that is needed to react as completely as possible with the analyte. *Gravimetric titrimetry* differs only in that the mass of the reagent is measured instead of its volume. In *coulometric titrimetry*, the "reagent" is a constant direct electrical current of known magnitude that reacts with the analyte. In this technique, the time required to complete the electrochemical reaction is measured.

Titrimetric methods are widely used for routine analyses because they are rapid, convenient, accurate, and readily automated. Volumetric titrimetry is introduced in this chapter; additional information concerning theory and applications is provided in Chapters 10 through 18. Coulometric titrimetry is considered in Section 19D-5.

9A SOME GENERAL ASPECTS OF VOLUMETRIC TITRIMETRY[1]

9A-1 Definition of Terms

A *standard solution* (or a *standard titrant*) is a reagent of known concentration that is used to carry out a titrimetric analysis. A *titration* is performed by slowly adding a standard solution from a buret or other liquid-dispersing device to a solution of the analyte until the reaction between the two is judged complete. The volume of reagent needed to complete the titration is determined from the difference between the initial and final buret readings.

The *equivalence point* in a titration is reached when the amount of added titrant is chemically equivalent to the amount of analyte in the sample. For example, the equivalence point in the titration of sodium chloride with silver nitrate occurs after exactly 1 mol of silver ion has been added for each mole of chloride ion in the sample. The equivalence point in the titration of sulfuric acid with sodium hydroxide is reached after 2 mol of base have been introduced for each mole of acid.

It is sometimes necessary to add an excess of the standard titrant and then determine the excess amount by *back-titration* with a second standard titrant. Here, the equivalence point corresponds to the point at which the amount of initial titrant is chemically equivalent to the amount of analyte plus the amount of back-titrant.

Back-titrations are often required when the rate of reaction between the analyte and reagent is slow or when the standard solution lacks stability.

9A-2 Equivalence Points and End Points

The equivalence point of a titration is a theoretical point that cannot be determined experimentally. Instead, we can only estimate its position by observing some physical change associated with the condition of equivalence. This change is called the *end point* for the titration. Every effort is made to ensure that any volume or mass difference between the equivalence point and the end point is small. Such differences do exist, however, as a result of inadequacies in the physical changes and in our ability to observe them. The difference in volume or mass between the equivalence point and the end point is the *titration error*.

Indicators are often added to the analyte solution in order to give an observable physical change (the end point) at or near the equivalence point. We shall see that large changes in the relative concentration of analyte or titrant occur in the equivalence-point region. These concentration changes cause the indicator to change in appearance. Typical indicator changes include the appearance or disappearance of a color, a change in color, or the appearance or disappearance of turbidity.

We often use instruments to detect end points. These respond to certain properties of the solution that change in a characteristic way during the titration. Among such instruments are voltmeters, ammeters, and ohmmeters; colorimeters; pH meters; temperature recorders; and refractometers.

In volumetric methods, the titration error E_t is given by

$$E_t = V_{ep} - V_{eq}$$

where V_{eq} is the theoretical volume of reagent required to reach the equivalence point and V_{ep} is the actual volume used to arrive at the end point.

9A-3 Primary Standards

A *primary standard* is a highly purified compound that serves as a reference material in all volumetric and mass titrimetric methods. The accuracy of the method is critically dependent on the properties of this compound. Important requirements for a primary standard are:

1. High purity. Established methods for confirming purity should be available.
2. Stability toward air.
3. Absence of hydrate water so that the composition of the solid does not change with variations in relative humidity.
4. Availability at modest cost.
5. Reasonable solubility in the titration medium.
6. Reasonably large molar mass so that the relative error associated with weighing the standard is minimized.

Compounds that meet or even approach these criteria are very few, and only a limited number of primary standard substances are available to the chemist. As a consequence, less pure compounds must sometimes be used instead of a primary standard. The purity of such a *secondary standard* must be established by careful analysis.

9B STANDARD SOLUTIONS

Standard solutions play a central role in all titrimetric methods of analysis. Therefore, we need to consider the desirable properties for such solutions, how they are prepared, and how their concentrations are expressed.

9B-1 Desirable Properties of Standard Solutions

The ideal standard solution for a titrimetric method will

1. be sufficiently stable so that it is only necessary to determine its concentration once;
2. react rapidly with the analyte so that the time required between additions of reagent is minimized;
3. react more or less completely with the analyte so that satisfactory end points are realized;
4. undergo a selective reaction with the analyte that can be described by a simple balanced equation.

Few reagents meet all these ideals perfectly.

9B-2 Methods for Establishing the Concentrations of Standard Solutions

In a standardization, the concentration of a solution is determined by using it to titrate a carefully measured quantity of a primary standard.

The accuracy of a titrimetric method can be no better than the accuracy of the concentration of the standard solution used in the titration. Two basic methods are used to establish the concentrations of such solutions. The first is the *direct method*, in which a carefully weighed quantity of a primary standard is dissolved in a suitable solvent and diluted to an exactly known volume in a volumetric flask. The second is by *standardization*, in which the titrant to be standardized is used to titrate (1) a weighed quantity of a primary standard, (2) a weighed quantity of a secondary standard, or (3) a measured volume of another standard solution. A titrant that is standardized against a secondary standard or against another standard solution is sometimes referred to as a *secondary standard solution*. The concentration of a secondary standard solution is subject to a larger uncertainty than that for a primary standard solution. If there is a choice, solutions are best prepared by the direct method. On the other hand, many reagents lack the properties required for a primary standard and therefore require standardization.

9B-3 Methods for Expressing the Concentrations of Volumetric Standard Solutions

The concentrations of volumetric standard solutions are generally expressed in units of either *molarity* c or *normality* c_N. Molarity gives the number of moles of reagent contained in 1 L of solution; normality gives the number of *equivalents* of reagent in the same volume.

Throughout this book we shall base volumetric calculations exclusively on molarity and molar masses. We have included in Appendix 6, however, a discussion of how volumetric calculations are carried out based on normality and equivalent weights because the reader may encounter these terms and their uses in the industrial and health science literature.

9C VOLUMETRIC CALCULATIONS

9C-1 Some Useful Algebraic Relationships

Most volumetric calculations are based on two pairs of simple equations that are derived from definitions of millimole, mole, and molar concentration. For the chemical species A, we may write

$$\text{amount A(mmol)} = \frac{\text{mass A (g)}}{\text{millimolar mass A (g/mmol)}} \tag{9-1}$$

$$\text{amount A(mol)} = \frac{\text{mass A (g)}}{\text{molar mass A (g/mol)}} \tag{9-2}$$

The second pair is derived from the definition of molar concentration. That is,

$$\text{amount A(mmol)} = V(\text{mL}) \times c_A(\text{mmol A/mL}) \tag{9-3}$$

$$\text{amount A(mol)} = V(\text{L}) \times c_A(\text{mol A/L}) \tag{9-4}$$

where V is the volume of the solution.

You should use Equations 9-1 and 9-3 when volumes are measured in milliliters and Equations 9-2 and 9-4 when the units are liters.

9C-2 Calculation of the Molarity of Standard Solutions

The following four examples illustrate how volumetric reagents are prepared.

Example 9-1

Describe the preparation of 5.000 L of 0.1000 M Na_2CO_3 (105.99 g/mol) from the primary standard solid.

Since the volume is in liters, we base our calculations on the mole rather than on the millimole. Thus to obtain the amount of Na_2CO_3 needed, we write

$$\text{amount } Na_2CO_3 = V_{\text{soln}}(\text{L}) \times c_{Na_2CO_3}(\text{mol/L})$$

$$= 5.000 \text{ L} \times \frac{0.1000 \text{ mol } Na_2CO_3}{\text{L}} = 0.5000 \text{ mol } Na_2CO_3$$

To obtain the mass of Na_2CO_3, we rearrange Equation 9-2 to give

$$\text{mass Na}_2\text{CO}_3 = 0.5000 \text{ mol Na}_2\text{CO}_3 \times \frac{105.99 \text{ g Na}_2\text{CO}_3}{\text{mol Na}_2\text{CO}_3} = 53.00 \text{ g Na}_2\text{CO}_3$$

Therefore the solution is prepared by dissolving 53.00 g of Na_2CO_3 in water and diluting to exactly 5.000 L.

Feature 9-1
UNITS IN USING EQUATIONS 9-1 THROUGH 9-4

It is useful to know that any combination of grams, moles, and liters can be replaced with any analogous combination expressed in milligrams, millimoles, and milliliters. For example, a 0.1 M solution contains 0.1 mol of a species per L or 0.1 mmol per mL. Similarly, the number of moles of a compound is equal to the mass in grams of that compound divided by its molar mass in grams or the mass in milligrams divided by its millimolar mass in milligrams.

Example 9-2

A standard 0.0100 M solution of Na^+ is required for calibrating a flame photometric method for determining the element. Describe how 500 mL of this solution can be prepared from primary standard Na_2CO_3.

We wish to compute the mass of reagent required to give a species molarity of 0.0100. Here, we will use millimoles since the volume is in milliliters. Because Na_2CO_3 dissociates to give two Na^+ ions, we can write that the number of millimoles of Na_2CO_3 needed is

$$\text{amount Na}_2\text{CO}_3 = 500 \text{ mL} \times \frac{0.0100 \text{ mmol Na}^+}{\text{mL}} \times \frac{1 \text{ mmol Na}_2\text{CO}_3}{2 \text{ mmol Na}^+}$$

$$= 2.50 \text{ mmol}$$

From the definition of millimole, we write

$$\text{mass Na}_2\text{CO}_3 = 2.50 \text{ mmol Na}_2\text{CO}_3 \times 0.10599 \frac{\text{g Na}_2\text{CO}_3}{\text{mmol Na}_2\text{CO}_3} = 0.265 \text{ g}$$

The solution is therefore prepared by dissolving 0.265 g of Na_2CO_3 in water and diluting to 500 mL.

Example 9-3

How would you prepare 50.0 mL portions of standard solutions that are 0.00500 M, 0.00200 M, and 0.00100 M in Na^+ from the solution in Example 9-2?

The number of millimoles of Na^+ taken from the concentrated solution must equal the number in the diluted solutions. Thus,

$$\text{amount } Na^+ \text{ from concd soln} = \text{amount } Na^+ \text{ in dil soln}$$

Recall that the number of millimoles is equal to number of millimoles per milliliter times the number of milliliters. That is,

$$V_{concd} \times c_{concd} = V_{dil} \times c_{dil}$$

A useful relationship is $V_{concd} \times c_{concd} = V_{dil} \times c_{dil}$.

where V_{concd} and V_{dil} are the volumes in milliliters of the concentrated and diluted solutions respectively and c_{concd} and c_{dil} are their Na^+ molar concentrations. This equation rearranges to

$$V_{concd} = \frac{V_{dil} \times c_{dil}}{c_{concd}} = \frac{50.0 \text{ mL} \times 0.00500 \text{ mmol } Na^+/mL}{0.0100 \text{ mmol } Na^+/mL} = 25.0 \text{ mL}$$

Thus, to produce 50.0 mL of 0.00500 M Na^+, 25.0 mL of the concentrated solution should be diluted to exactly 50.0 mL.

Repeat the calculation for the other two molarities to confirm that diluting 10.0 and 5.00 mL of the concentrated solution to 50.0 mL produces the desired solutions.

Example 9-4

Describe how you would prepare 2.0 L of approximate 0.25 M $HClO_4$ from the concentrated reagent, which has a specific gravity of 1.07 g/mL and contains 71% (w/w) $HClO_4$.

Here we proceed as in Example 2-11 and calculate the molarity of the concentrated reagent and then calculate the number of moles of $HClO_4$ that we must take. We then divide the second figure by the first to obtain the volume of the concentrated acid to take. Thus, to determine the molarity of the concentrated reagent, we write

$$c_{HClO_4} = \frac{1.67 \text{ g reagent}}{\text{mL reagent}} \times \frac{71 \text{ g } HClO_4}{100 \text{ g reagent}} \times \frac{1 \text{ mmol } HClO_4}{0.10046 \text{ g } HClO_4} = 11.8 \text{ M}$$

$$\text{no. mmol } HClO_4 \text{ required} = 2000 \text{ mL} \times 0.25 \frac{\text{mmol } HClO_4}{\text{mL } HClO_4}$$

$$= 500 \text{ mmol } HClO_4$$

$$\text{vol concd reagent} = \frac{500 \text{ mmol } HClO_4}{11.8 \text{ mmol } HClO_4/mL \text{ concd reagent}}$$

$$= 42.4 \text{ mL concd reagent}$$

Dilute about 42 mL of the concentrated reagent to 2.0 L.

9C-3 Treatment of Titration Data

In this section, we describe two types of volumetric calculations. The first involves computing the molarity of solutions that have been standardized against either a primary standard or another standard solution. The second involves calculating the amount of analyte in a sample from titration data. Both types are based on three algebraic relationships. Two of these are Equations 9-1 and 9-3, both of which are based on millimoles and milliliters. The third relationship is the stoichiometric ratio of the number of millimoles of the analyte and the number of millimoles of titrant.

Calculation of Molarities from Standardization Data

Examples 9-5 and 9-6 illustrate how standardization data are treated.

Example 9-5

Exactly 50.00 mL of an HCl solution required 29.71 mL of 0.01963 M $Ba(OH)_2$ to reach an end point with bromocresol green indicator. Calculate the molarity of the HCl.

In the titration, 1 mmol of $Ba(OH)_2$ reacts with 2 mmol of HCl, and thus the stoichiometric ratio is

$$\text{stoichiometric ratio} = \frac{2 \text{ mmol HCl}}{1 \text{ mmol Ba(OH)}_2}$$

The number of millimoles of the standard is obtained by substituting into Equation 9-3:

$$\text{amount Ba(OH)}_2 = 29.71 \ \cancel{\text{mL Ba(OH)}_2} \times 0.01963 \ \frac{\text{mmol Ba(OH)}_2}{\cancel{\text{mL Ba(OH)}_2}}$$

To obtain the number of millimoles of HCl, we multiply this result by the stoichiometric ratio derived initially:

$$\text{amount HCl} = (29.71 \times 0.01963) \ \cancel{\text{mmol Ba(OH)}_2} \times \frac{2 \text{ mmol HCl}}{1 \ \cancel{\text{mmol Ba(OH)}_2}}$$

To obtain the number of millimoles of HCl per mL, we divide by the volume of the acid. Thus,

$$c_{\text{HCl}} = \frac{(29.71 \times 0.01963 \times 2) \text{mmol HCl}}{50.0 \text{ mL HCl}}$$

$$= 0.023328 \ \frac{\text{mmol HCl}}{\text{mL HCl}} = 0.02333 \text{ M}$$

Example 9-6

Titration of 0.2121 g of pure $Na_2C_2O_4$ (134.00 g/mol) required 43.31 mL of $KMnO_4$. What is the molarity of the $KMnO_4$ solution? The chemical reaction is

$$2MnO_4^- + 5C_2O_4^{2-} + 16H^+ \rightarrow 2Mn^{2+} + 10CO_2 + 8H_2O$$

From this equation, we see that the stoichiometric ratio is

$$\text{stoichiometric ratio} = \frac{2 \text{ mmol } KMnO_4}{5 \text{ mmol } Na_2C_2O_4}$$

The amount of primary standard $Na_2C_2O_4$ is given by Equation 9-1:

$$\text{amount } Na_2C_2O_4 = 0.2121 \text{ g } \cancel{Na_2C_2O_4} \times \frac{1 \text{ mmol } Na_2C_2O_4}{0.13400 \text{ g } \cancel{Na_2C_2O_4}}$$

To obtain the number of millimoles of $KMnO_4$, we multiply this result by the stoichiometric factor:

$$\text{amount } KMnO_4 = \frac{0.2121}{0.1340} \text{ mmol } \cancel{Na_2C_2O_4} \times \frac{2 \text{ mmol } KMnO_4}{5 \text{ mmol } \cancel{Na_2C_2O_4}}$$

The molarity is then obtained by dividing by the volume of $KMnO_4$ consumed. Thus,

$$c_{KMnO_4} = \frac{\left(\dfrac{0.2121}{0.13400} \times \dfrac{2}{5}\right) \text{ mmol } KMnO_4}{43.31 \text{ mL } KMnO_4} = 0.01462 \text{ M}$$

Note that units are carried through all calculations as a check on the correctness of the relationships used in Examples 9-5 and 9-6.

Calculation of Quantity of Analyte from Titration Data

As shown by the examples that follow, the same systematic approach just described is also used to compute analyte concentrations from titration data.

Example 9-7

A 0.8040-g sample of an iron ore is dissolved in acid. The iron is then reduced to Fe^{2+} and titrated with 47.22 mL of 0.02242 M $KMnO_4$ solution. Calculate the results of this analysis in terms of (a) % Fe (55.847 g/mol); and (b) % Fe_3O_4 (231.54 g/mol). The reaction of the analyte with the reagent is described by the equation

$$MnO_4^- + 5Fe^{2+} + 8H^+ \rightarrow Mn^{2+} + 5Fe^{3+} + 4H_2O$$

(a)

$$\text{stoichiometric ratio} = \frac{5 \text{ mmol Fe}^{2+}}{1 \text{ mmol KMnO}_4}$$

$$\text{amount KMnO}_4 = 47.22 \text{ mL KMnO}_4 \times \frac{0.02242 \text{ mmol KMnO}_4}{\text{mL KMnO}_4}$$

$$\text{amount Fe}^{2+} = (47.22 \times 0.02242) \text{ mmol KMnO}_4 \times \frac{5 \text{ mmol Fe}^{2+}}{\text{mmol KMnO}_4}$$

The mass of Fe^{2+} is then given by

$$\text{mass Fe}^{2+} = (47.22 \times 0.02242 \times 5) \text{ mmol Fe}^{2+} \times 0.055847 \frac{\text{g Fe}^{2+}}{\text{mmol Fe}^{2+}}$$

$$\text{percent Fe}^{2+} = \frac{(47.22 \times 0.02242 \times 5 \times 0.055847) \text{ g Fe}^{2+}}{0.8040 \text{ g sample}} \times 100\%$$

$$= 36.77\%$$

The symbol ≡ means equivalent to.

(b) In order to derive a stoichiometric ratio, we note that

$$5 \text{ Fe}^{2+} \equiv 1 \text{ MnO}_4^-$$

Therefore,

$$5\text{Fe}_3\text{O}_4 \equiv 15\text{Fe}^{2+} \equiv 3\text{MnO}_4^-$$

and

$$\text{stoichiometric ratio} = \frac{5 \text{ mmol Fe}_3\text{O}_4}{3 \text{ mmol KMnO}_4}$$

As in part (a),

$$\text{amount KMnO}_4 = 47.22 \text{ mL KMnO}_4 \times 0.02242 \text{ mmol KMnO}_4 / \text{ mL KMnO}_4$$

$$\text{amount Fe}_3\text{O}_4 = (47.22 \times 0.02242) \text{ mmol KMnO}_4 \times \frac{5 \text{ mmol Fe}_3\text{O}_4}{3 \text{ mmol KMnO}_4}$$

$$\text{mass Fe}_3\text{O}_4 = (47.22 \times 0.02242 \times \tfrac{5}{3}) \text{ mmol Fe}_3\text{O}_4$$

$$\times 0.23154 \frac{\text{g Fe}_3\text{O}_4}{\text{mmol Fe}_3\text{O}_4}$$

$$\text{percent Fe}_3\text{O}_4 = \frac{(47.22 \times 0.02242 \times \tfrac{5}{3}) \times 0.23154 \text{ g Fe}_3\text{O}_4}{0.8040 \text{ g sample}} \times 100\%$$

$$= 50.81\%$$

Example 9-8

The organic matter in a 3.776-g sample of a mercuric ointment is decomposed with HNO_3. After dilution, the Hg^{2+} is titrated with 21.30 mL of a 0.1144 M solution of NH_4SCN. Calculate the percent Hg (200.59 g/mol) in the ointment.

This titration involves the formation of a stable neutral complex, $Hg(SCN)_2$:

$$Hg^{2+} + 2SCN^- \rightarrow Hg(SCN)_2(aq)$$

At the equivalence point,

$$\text{stoichiometric ratio} = \frac{1 \text{ mmol Hg}^{2+}}{2 \text{ mmol NH}_4\text{SCN}}$$

$$\text{amount NH}_4\text{SCN} = 21.30 \text{ mL NH}_4\text{SCN} \times 0.1144 \frac{\text{mmol NH}_4\text{SCN}}{\text{mL NH}_4\text{SCN}}$$

$$\text{amount Hg}^{2+} = (21.30 \times 0.1144) \text{ mmol NH}_4\text{SCN} \times \frac{1 \text{ mmol Hg}^{2+}}{2 \text{ mmol NH}_4\text{SCN}}$$

$$\text{mass Hg}^{2+} = (21.30 \times 0.1144 \times \tfrac{1}{2}) \text{ mmol Hg}^{2+} \times \frac{0.20059 \text{ g Hg}^{2+}}{\text{mmol Hg}^{2+}}$$

$$\text{percent Hg} = \frac{(21.30 \times 0.1144 \times \tfrac{1}{2}) \times 0.20059 \text{ g Hg}^{2+}}{3.776 \text{ g sample}} \times 100\%$$

$$= 6.472\% = 6.47\%$$

Feature 9-2
ANOTHER APPROACH TO EXAMPLE 9-7(a)

Some people find it easier to write out the solution to a problem so that the units in the denominator of each succeeding term eliminate the units in the numerator of the preceding one until the units of the answer are obtained. For example, the solution to part (a) of Example 9-7 can be written

$$47.22 \text{ mL KMnO}_4^- \times \frac{0.02242 \text{ mmol KMnO}_4}{\text{mL KMnO}_4^-} \times \frac{5 \text{ mmol Fe}}{1 \text{ mmol KMnO}_4^-}$$

$$\times \frac{0.05585 \text{ g Fe}}{\text{mmol Fe}} \times \frac{1}{0.8040 \text{ g sample}} \times 100\%$$

$$= 36.77\% \text{ Fe}$$

Example 9-9

A 0.4755-g sample containing $(NH_4)_2C_2O_4$ and inert materials was dissolved in H_2O and made strongly alkaline with KOH, which converted NH_4^+ to NH_3. The liberated NH_3 was distilled into exactly 50.00 mL of 0.05035 M H_2SO_4. The excess H_2SO_4 was back-titrated with 11.13 mL of 0.1214 M NaOH. Calculate (a) the % N (14.007 g/mol) and (b) the % $(NH_4)_2C_2O_4$ (124.10 g/mol) in the sample.

(a) The H_2SO_4 reacts with both NH_3 and NaOH, and two stoichiometric ratios

Feature 9-3
ROUNDING THE ANSWER TO EXAMPLE 9-8

You should note that the input data for Example 9-8 all contained four or more significant figures, but the answer was rounded to three. Why is this?

Let us proceed as we did on page 75 and make the rounding decision by doing a couple of rough calculations in our heads. We will assume that the input data are uncertain to 1 part in the last significant figure. The largest *relative* error will then be associated with the molarity of the reagent. Here, the relative uncertainty is 0.0001/0.1144. But we really do not need to know the error this accurately, and we can simply figure that the uncertainty is about 1 part in 1000 (compared with about 1 part in 2000 for the volume and 1 part in 3700 for the mass). We then assume the calculated result is uncertain to about the same amount as the least accurate measurement or 1 part in 1000. The absolute uncertainty of the final result is then $6.472\% \times 1/1000 = 0.0065 = 0.01\%$, and we round to the second figure to the right of the decimal point. Thus, we report 6.47%.

You should practice making this rough type of rounding decision whenever you make a computation.

can be derived. These are

$$\frac{2 \text{ mmol NH}_3}{1 \text{ mmol H}_2\text{SO}_4} \quad \text{and} \quad \frac{1 \text{ mmol H}_2\text{SO}_4}{2 \text{ mmol NaOH}}$$

$$\text{total amount H}_2\text{SO}_4 = 50.00 \text{ mL H}_2\text{SO}_4 \times 0.05035 \frac{\text{mmol H}_2\text{SO}_4}{\text{mL H}_2\text{SO}_4}$$

$$= 2.5175 \text{ mmol H}_2\text{SO}_4$$

The amount of H_2SO_4 consumed by the NaOH in the back-titration is

$$\text{amount H}_2\text{SO}_4 = (11.13 \times 0.1214) \text{ mmol NaOH} \times \frac{1 \text{ mmol H}_2\text{SO}_4}{2 \text{ mmol NaOH}}$$

$$= 0.6756 \text{ mmol H}_2\text{SO}_4$$

The amount of H_2SO_4 that reacted with NH_3 is then

$$\text{amount H}_2\text{SO}_4 = (2.5175 - 0.6756) \text{ mmol H}_2\text{SO}_4 = 1.8419 \text{ mmol H}_2\text{SO}_4$$

The amount of N, which is equal to the number of millimoles of NH_3, is

$$\text{amount N} = \text{no. mmol NH}_3 = 1.8419 \text{ mmol H}_2\text{SO}_4 \times \frac{2 \text{ mmol N}}{1 \text{ mmol H}_2\text{SO}_4}$$

$$= 3.6838 \text{ mmol N}$$

$$\text{percent N} = \frac{3.6838 \text{ mmol N} \times 0.014007 \text{ g N/mmol N}}{0.4755 \text{ g sample}} \times 100\%$$

$$= 10.85\%$$

(b) Since each millimole of $(NH_4)_2C_2O_4$ produces 2 mmol of NH_3, which reacts with 1 mmol of H_2SO_4,

$$\text{stoichiometric ratio} = \frac{1 \text{ mmol } (NH_4)_2C_2O_4}{1 \text{ mmol } H_2SO_4}$$

$$\text{amount } (NH_4)_2C_2O_4 = 1.8419 \text{ mmol } H_2SO_4 \times \frac{1 \text{ mmol } (NH_4)_2C_2O_4}{1 \text{ mmol } H_2SO_4}$$

$$\text{mass } (NH_4)_2C_2O_4 = 1.81419 \text{ mmol } (NH_4)_2C_2O_4 \times \frac{0.12410 \text{ g } (NH_4)_2C_2O_4}{\text{mmol } (NH_4)_2C_2O_4}$$

$$= 0.22858 \text{ g}$$

$$\text{percent } (NH_4)_2C_2O_4 = \frac{0.22858 \text{ g } (NH_4)_2C_2O_4}{0.4755 \text{ g sample}} \times 100\% = 48.07\%$$

Example 9-10

The CO in a 20.3-L sample of gas was converted to CO_2 by passage over iodine pentoxide heated to 150°C:

$$I_2O_5(s) + 5CO(g) \rightarrow 5CO_2(g) + I_2(g)$$

The iodine distilled at this temperature and was collected in an absorber containing 8.25 mL of 0.01101 M $Na_2S_2O_3$:

$$I_2(aq) + 2S_2O_3^{2-}(aq) \rightarrow 2I^-(aq) + S_4O_6^{2-}(aq)$$

The excess $Na_2S_2O_3$ was back-titrated with 2.16 mL of 0.00947 M I_2 solution. Calculate the number of milligrams of CO (28.01 g/mol) per liter of sample.

Based on the two reactions, the stoichiometric ratios are

$$\frac{5 \text{ mmol CO}}{1 \text{ mmol } I_2} \quad \text{and} \quad \frac{2 \text{ mmol } Na_2S_2O_3}{1 \text{ mmol } I_2}$$

We divide the first ratio by the second to get a third useful ratio:

$$\frac{5 \text{ mmol CO}}{2 \text{ mmol } Na_2S_2O_3}$$

This relationship reveals that 5 mmol of CO are responsible for the consumption of 2 mmol $Na_2S_2O_3$. The total amount of $Na_2S_2O_3$ is

$$\text{amount } Na_2S_2O_3 = 8.25 \text{ mL } Na_2S_2O_3 \times 0.01101 \frac{\text{mmol } Na_2S_2O_3}{\text{mL } Na_2S_2O_3}$$

$$= 0.09083 \text{ mmol } Na_2S_2O_3$$

The amount of $Na_2S_2O_3$ consumed in the back-titration is

$$\text{amount } Na_2S_2O_3 = 2.16 \text{ mL } I_2 \times 0.00947 \frac{\text{mmol } I_2}{\text{mL } I_2} \times \frac{2 \text{ mmol } Na_2S_2O_3}{\text{mmol } I_2}$$

$$= 0.04091 \text{ mmol } Na_2S_2O_3$$

The number of millimoles of CO can then be obtained by employing the third stoichiometric ratio:

$$\text{amount CO} = (0.09083 - 0.04091) \text{ mmol } \cancel{Na_2S_2O_3} \times \frac{5 \text{ mmol CO}}{2 \text{ mmol } \cancel{Na_2S_2O_3}}$$

$$= 0.1248 \text{ mmol CO}$$

$$\text{mass CO} = 0.1248 \text{ mmol } \cancel{CO} \times \frac{28.01 \text{ mg CO}}{\text{mmol } \cancel{CO}} = 3.4956 \text{ mg}$$

$$\frac{\text{mass CO}}{\text{vol sample}} = \frac{3.4956 \text{ mg CO}}{20.3 \text{ L sample}} = 0.172 \frac{\text{mg CO}}{\text{L}}$$

9D   TITRATION CURVES IN TITRIMETRIC METHODS

> Titration curves are plots of a concentration-related variable as a function of reagent volume.

As noted in Section 9A-2, an end point is an observable physical change that occurs near the equivalence point of a titration. The two most widely used end points involve (1) changes in color due to the reagent, the analyte, or an indicator and (2) a change in potential of an electrode that responds to the concentration of the reagent or the analyte.

> The vertical axis in a sigmoidal titration curve is either the p-function of the analyte or reagent or else the potential of an analyte- or reagent-sensitive electrode.

To help us understand the theoretical basis of end points and the sources of titration errors we will develop a *titration curve* for the system under consideration. Titration curves consist of a plot of reagent volume as the horizontal axis and some function of the analyte or reagent concentration as the vertical.

> The vertical axis for a linear-segment titration curve is an instrumental signal that is proportional to the concentration of the analyte or reagent.

9D-1 Types of Titration Curves

Two general types of titration curves (and thus two general types of end points) are encountered in titrimetric methods. In the first type, called a *sigmoidal curve*, important observations are confined to a small region (typically ± 0.1 to ± 0.5 mL) surrounding the equivalence point. A sigmoidal curve, in which the p-function of analyte (or sometimes the reagent) is plotted as a function of reagent volume, is shown in Figure 9-1a. In the second type, called a *linear-segment curve*, measurements are made on both sides of, but well away from, the equivalence point. Measurements near equivalence are avoided. In this type of curve, the vertical axis is an instrument reading that is directly proportional to the concentration of the analyte or the reagent. A typical linear segment curve is found in Figure 9-1b. The sigmoidal type offers the advantages of speed and convenience. The linear segment type is advantageous for reactions that are complete only in the presence of a significant excess of the reagent or analyte.

In this, and the several chapters that follow, we will be dealing exclusively with sigmoidal titration curves. Linear-segment curves are considered in Section 22B-3.

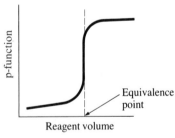

(a) Sigmoidal curve

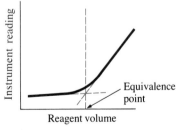

(b) Linear-segment curve

Figure 9-1
Two types of titration curves.

9D-2 Concentration Changes During Titrations

The equivalence point in a titration is characterized by major changes in the *relative* concentrations of reagent and analyte. Table 9-1 illustrates this phenome-

Table 9-1

CONCENTRATION CHANGES DURING A TITRATION
OF 50.00 mL OF 0.1000 M HCl

Volume of 0.100 M NaOH, mL	$[H_3O^+]$ mol/L	Volume of NaOH to Cause a Tenfold Decrease in $[H_3O^+]$, mL	pH	pOH
0.00	1.000×10^{-1}		1.00	13.00
40.91	1.000×10^{-2}	40.91	2.00	12.00
49.01	1.000×10^{-3}	8.11	3.00	11.00
49.90	1.000×10^{-4}	0.89	4.00	10.00
49.99	1.000×10^{-5}	0.09	5.00	9.00
49.999	1.000×10^{-6}	0.009	6.00	8.00
50.00	1.000×10^{-7}	0.001	7.00	7.00
50.001	1.000×10^{-8}	0.001	8.00	6.00
50.01	1.000×10^{-9}	0.009	9.00	5.00
50.10	1.000×10^{-10}	0.09	10.00	4.00
51.01	1.000×10^{-11}	0.91	11.00	3.00
61.11	1.000×10^{-12}	10.10	12.00	2.00

At the beginning of the titration described in Table 9-1, about 41 mL of reagent brings about a tenfold decrease in the concentration of A; only 0.001 mL is required to cause this same change at the equivalence point.

non. The data in the second column of the table show the changes in concentration of hydronium ion as a 50.00-mL aliquot of a 0.1000 M solution of hydrochloric acid is titrated with a 0.1000 M solution of sodium hydroxide. The neutralization reaction is described by the equation

$$H_3O^+ + OH^- \rightleftharpoons 2H_2O \tag{9-5}$$

In order to emphasize the changes in *relative* concentration that occur in the equivalence region, the volume increments computed are those required to cause

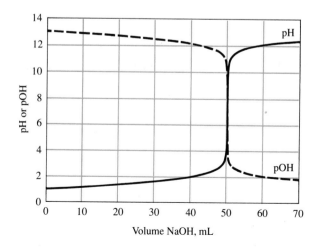

Figure 9-2

Titration curve from data in Table 9-1.

Feature 9-4

How did we calculate the volumes of NaOH shown in the first column of Table 9-1?

Up until the equivalence point, $[H_3O^+]$ will equal the molar concentration of unreacted HCl (c_{HCl}). The concentration of HCl is equal to the original number of millimoles of HCl (50.00×0.1000) minus the number of millimoles of NaOH added ($V_{NaOH} \times 0.1000$), divided by the total volume of the solution. That is,

$$c_{HCl} = [H_3O^+] = \frac{(50.00 \times 0.1000 - V_{NaOH} \times 0.1000) \text{ mmol HCl}}{(50.00 + V_{NaOH}) \text{ mL soln}}$$

where V_{NaOH} is the volume of 0.1000 M NaOH added. This equation reduces to

$$50.00 [H_3O^+] + V_{NaOH} [H_3O^+] = 5.000 - 0.1000 V_{NaOH}$$

Collecting the terms containing V_{NaOH} gives

$$V_{NaOH}(0.1000 + [H_3O^+]) = 5.000 - 50.00 [H_3O^+]$$

or

$$V_{NaOH} = \frac{5.000 - 50.00 [H_3O^+]}{0.1000 + [H_3O^+]}$$

Thus to obtain $[H_3O^+] = 1.000 \times 10^{-2}$, we find

$$V_{NaOH} = \frac{5.000 - 50.00 \times 1.000 \times 10^{-2}}{0.1000 + 1.000 \times 10^{-2}} = 40.91 \text{ mL}$$

Challenge: Use the same reasoning to show that beyond the equivalence point

$$V_{NaOH} = \frac{50.00 [OH^-] + 5.000}{0.1000 - [OH^-]}$$

tenfold decreases in the concentration of H_3O^+ (or tenfold increases in hydroxide ion concentration). Thus, we see in the third column that an addition of 40.91 mL of base is needed to decrease the concentration of the hydronium ion by one order of magnitude from 0.100 M to 0.0100 M. An addition of only 8.11 mL is required to lower the concentration by another factor of 10 to 0.00100 M; 0.89 mL causes yet another tenfold decrease. Corresponding increases in hydroxide concentration occur at the same time. End-point detection, then, is based on this large change in *relative* concentration of the analyte (or the reagent) that occurs at the equivalence point for every type of titration.

The large relative concentration changes that occur in the region of chemical equivalence are shown by plotting the negative logarithm of the analyte or the reagent concentration (the p-function) against reagent volume as has been done in Figure 9-2. The data for this plot are found in the fourth column of Table 9-1. Titration curves for reactions involving complex formation, precipitation, and oxidation/reduction all exhibit the same sharp change in p-function as those shown in Figure 9-2 in the equivalence-point region. Titration curves define the properties required of an indicator and allow us to estimate the error associated with titration methods.

9E   QUESTIONS AND PROBLEMS

9-1. Write two equations that—along with the stoichiometric factor—form the basis for the calculations of volumetric titrimetry.

9-2. Define
*(a) millimole.
(b) titration.
*(c) stoichiometric factor.
(d) titration error.

9-3. Distinguish between
*(a) the equivalence point and the end point of a titration.
(b) the density and the specific gravity of a solution.
*(c) a primary standard and a secondary standard.

***9-4.** Briefly explain why milligrams of solute per liter and parts per million can be used interchangeably to describe the concentration of a dilute aqueous solution.

9-5. Calculations of volumetric analysis ordinarily consist of transforming the quantity of titrant used to a chemically equivalent quantity of analyte through use of a stoichiometric factor. Use chemical formulas (no calculations required) to express this factor to calculate the percentage of
*(a) hydrazine in rocket fuel determined by titration with standard iodine. Reaction:

$$H_2NNH_2 + 2I_2 \rightarrow N_2(g) + 4I^- + 4H^+$$

(b) hydrogen peroxide in a cosmetic preparation determined by titration with standard permanganate. Reaction:

$$5H_2O_2 + 2MnO_4^- + 6H^+ \rightarrow 2Mn^{2+} + 5O_2(g) + 8H_2O$$

*(c) boron in a sample of borax, $Na_2B_4O_7 \cdot 10H_2O$ determined by titration with standard acid. Reaction:

$$B_4O_7^{2-} + 2H^+ + 5H_2O \rightarrow 4H_3BO_3$$

(d) sulfur in an agricultural spray that was converted to thiocyanate with an unmeasured excess of cyanide:

$$S(s) + CN^- \rightarrow SCN^-$$

After removal of the excess cyanide, the thiocyanate was titrated with a standard potassium iodate solution in strong HCl. Reaction:

$$2SCN^- + 3IO_3^- + 2H^+ + 6Cl^- \rightarrow$$
$$2SO_4^{2-} + 2CN^- + 3ICl_2^- + H_2O$$

***9-6.** Calculate the molar concentration of a solution that is 50.0% in NaOH (w/w) and has a specific gravity of 1.52.

9-7. Calculate the molar concentration of a 20.0% solution (w/w) of KCl that has a specific gravity of 1.13.

9-8. Calculate the molar analytical concentration of solute in an aqueous solution that is
*(a) 11.00% (w/w) NH_3 and has a density of 0.9538.
(b) 18.00% (w/w) KBr and has a density of 1.149.
*(c) 28.00% (w/w) ethylene glycol (62.07 g/mol) and has a density of 1.0350.
(d) 15.00% (w/w) sucrose (342.5 g/mol) and has a density of 1.0592.

***9-9.** How would you prepare
(a) 500 mL of 16.0% (w/v) aqueous ethanol (46.1 g/mol)?
(b) 500 mL of 16.0% (v/v) aqueous ethanol?
(c) 500 g of 16.0% (w/w) aqueous ethanol?

9-10. Describe the preparation of
(a) 250 mL of 20.0% (w/v) aqueous acetone (58.05 g/mol).
(b) 250 mL of 20.0% (v/v) aqueous acetone.
(c) 250 mL of 20.0% (w/w) aqueous acetone.

***9-11.** A solution of $HClO_4$ was standardized by dissoving 0.3745 g of primary-standard-grade HgO in a solution of KBr:

$$HgO(s) + 4Br^- + H_2O \rightarrow HgBr_4^{2-} + 2OH^-$$

The liberated OH^- was neutralized with 37.79 mL of the acid. Calculate the molarity of the $HClO_4$.

9-12. A 0.3367-g sample of primary-standard-grade Na_2CO_3 required 28.66 mL of a H_2SO_4 solution to reach the end point in the reaction

$$CO_3^{2-} + 2H^+ \rightarrow H_2O + CO_2(g)$$

What is the molarity of the H_2SO_4?

*9-13. A 0.3396-g sample that assayed 96.4% Na_2SO_4 consumed 37.70 mL of a barium chloride solution. Reaction:

$$Ba^{2+} + SO_4^{2-} \rightarrow BaSO_4$$

Calculate the analytical molarity of $BaCl_2$ in the solution.

*9-14. A 0.4793-g sample of primary-standard Na_2CO_3 was treated with 40.00 mL of dilute perchloric acid. The solution was boiled to remove CO_2, following which the excess $HClO_4$ was back-titrated with 8.70 mL of dilute NaOH. In a separate experiment it was established that 27.43 mL of the $HClO_4$ neutralized a 25.00-mL portion of the NaOH. Calculate the molarities of the $HClO_4$ and NaOH.

9-15. Titration of 50.00 mL of 0.05251 M $Na_2C_2O_4$ required 38.71 mL of a potassium permanganate solution:

$$2MnO_4^- + 5H_2C_2O_4 + 6H^+ \rightarrow 2Mn^{2+} + 10CO_2(g) + 8H_2O$$

Calculate the molarity of the $KMnO_4$ solution.

*9-16. Titration of the I_2 produced from 0.1238 g of primary-standard KIO_3 required 41.27 mL of sodium thiosulfate:

$$IO_3^- + 5I^- + 6H^+ \rightarrow 3I_2 + 3H_2O$$
$$I_2 + 2S_2O_3^{2-} \rightarrow 2I^- + S_4O_6^{2-}$$

Calculate the concentration of the $Na_2S_2O_3$.

*9-17. A 4.476-g sample of a petroleum product was burned in a tube furnace, and the SO_2 produced was collected in 3% H_2O_2. Reaction:

$$SO_2(g) + H_2O_2 \rightarrow H_2SO_4$$

A 25.00-mL portion of 0.00923 M NaOH was introduced into the solution of H_2SO_4, and then the excess base was back-titrated with 13.33 mL of 0.01007 M HCl. Calculate the parts per million of sulfur in the sample.

9-18. A 100.0-mL sample of spring water was treated to convert any iron present to Fe^{2+}. Addition of 25.00 mL of 0.002107 M $K_2Cr_2O_7$ resulted in the reaction

$$6Fe^{2+} + Cr_2O_7^{2-} + 14H^+ \rightarrow 6Fe^{3+} + 2Cr^{3+} + 7H_2O$$

The excess $K_2Cr_2O_7$ was back-titrated with 7.47 mL of a 0.00979 M Fe^{2+} solution. Calculate the parts per million of iron in the sample.

*9-19. The arsenic in a 1.223-g sample of a pesticide was converted to H_3AsO_4 by suitable treatment. The acid was then neutralized, and exactly 40.00 mL of 0.07891 M $AgNO_3$ were added to precipitate the arsenic quantitatively as Ag_3AsO_4. The excess Ag^+ in the filtrate and washings from the precipitate was titrated with 11.27 mL of 0.1000 M KSCN; the reaction was

$$Ag^+ + SCN^- \rightarrow AgSCN(s)$$

Calculate the percent As_2O_3 in the sample.

*9-20. The thiourea in a 1.455-g sample of organic material was extracted into a dilute H_2SO_4 solution and titrated with 37.31 mL of 0.009372 M Hg^{2+} via the reaction

$$4(NH_2)_2CS + Hg^{2+} \rightarrow [(NH_2)_2CS]_4Hg^{2+}$$

Calculate the percent $(NH_2)_2CS$ (76.12 g/mol) in the sample.

9-21. The ethyl acetate concentration in an alcoholic solution was determined by diluting a 10.00-mL sample to exactly 100 mL. A 20.00-mL portion of the diluted solution was refluxed with 40.00 mL of 0.04672 M KOH:

$$CH_3COOC_2H_5 + OH^- \rightarrow CH_3COO^- + C_2H_5OH$$

After cooling, the excess OH^- was back-titrated with 3.41 mL of 0.05042 M H_2SO_4. Calculate the number of grams of ethyl acetate (88.11 g/mol) per 100 mL of the original sample.

*9-22. A solution of $Ba(OH)_2$ was standardized against 0.1016 g of primary-standard-grade benzoic acid C_6H_5COOH (122.12 g/mol). An end point was observed after addition of 44.42 mL of base.
(a) Calculate the molarity of the base.
(b) Calculate the standard deviation of the molarity if the standard deviation for weighing was ± 0.2 mg and that for the volume measurement was ± 0.03 mL.
(c) Assuming an error of -0.3 mg in the weighing, calculate the absolute and relative systematic error in the molarity.

9-23. A 0.1475 M solution of $Ba(OH)_2$ was used to titrate the acetic acid (60.05 g/mol) in a dilute aqueous solution. The following results were obtained.

Sample	Sample Volume, mL	$Ba(OH)_2$ Volume, mL
1	50.00	43.17
2	49.50	42.68
3	25.00	21.47
4	50.00	43.33

(a) Calculate the mean w/v percentage of acetic acid in the sample.
(b) Calculate the standard deviation for the results.
(c) Calculate the 90% confidence interval for the mean.
(d) At the 90% confidence level, could any of the results be discarded?
(e) Assume that the buret used to measure out the acetic acid had a systematic error of -0.05 mL at all volumes delivered. Calculate the systematic error in the mean result.

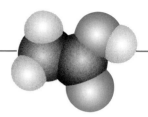

CHAPTER 10

THEORY OF NEUTRALIZATION TITRATIONS

Standard solutions of strong acids and strong bases are used extensively for determining analytes that are themselves acids or bases or analytes that can be converted to such species by chemical treatment. This chapter deals with theoretical aspects of titrations with such reagents.

10A SOLUTIONS AND INDICATORS FOR ACID/BASE TITRATIONS

Before showing how titration curves for acid/base reactions are derived and used, we will describe the types of standard solutions and indicators that are commonly used for neutralization titrations.

10A-1 Standard Solutions

The standard solutions employed in neutralization titrations are strong acids or strong bases because these substances react more completely with an analyte than do their weaker counterparts and thus yield sharper end points. Standard solutions of acids are prepared by diluting concentrated hydrochloric, perchloric, or sulfuric acids. Nitric acid is seldom used because its oxidizing properties offer the potential for undesirable side reactions. It should be pointed out that hot concentrated perchloric and sulfuric acids are potent oxidizing agents and thus are hazardous. Fortunately, however, cold dilute solutions of these reagents are relatively harmless and can be used in the analytical laboratory without any special precautions other than eye protection.

Standard basic solutions are ordinarily prepared from solid sodium, potassium, and occasionally barium hydroxides. Remember that eye protection should *always* be worn when handling dilute solutions of these reagents.

10A-2 Acid/Base Indicators

You will find a list of common acid/base indicators and their colors inside the front cover of this book.

Many substances, both naturally occurring and synthetic, display colors that depend on the pH of the solutions in which they are dissolved. Some of these substances, which have been used for centuries to indicate the acidity or alkalinity of water, still find current application as acid/base indicators.

The coloring matter in red cabbage makes an excellent acid/base indicator.

An acid/base indicator is a weak organic acid or a weak organic base whose undissociated form differs in color from its conjugate base or its conjugate acid form. For example, the behavior of an acid-type indicator, HIn, is described by the equilibrium

$$\underset{\substack{\text{acid} \\ \text{color}}}{HIn} + H_2O \rightleftharpoons \underset{\substack{\text{base} \\ \text{color}}}{In^-} + H_3O^+$$

Here, internal structural changes accompany dissociation and cause the color change (Figure 10-1). The equilibrium for a base-type indicator, In, is

$$\underset{\substack{\text{base} \\ \text{color}}}{In} + H_2O \rightleftharpoons \underset{\substack{\text{acid} \\ \text{color}}}{InH^+} + OH^-$$

In the paragraphs that follow, we focus on the behavior of acid-type indicators. However, the discussion can be readily applied to base-type indicators as well.

The equilibrium-constant expression for the dissociation of an acid-type indicator takes the form

$$K_a = \frac{[H_3O^+][In^-]}{[HIn]} \tag{10-1}$$

Rearranging leads to

$$[H_3O^+] = K_a \frac{[HIn]}{[In^-]} \tag{10-2}$$

We see then that the hydronium ion concentration determines the ratio of acid and conjugate base form of the indicator.

The human eye is not very sensitive to color differences in a solution containing a mixture of In$^-$ and HIn, particularly when the ratio [In$^-$]/[HIn] is greater than about 10 or smaller than about 0.1. Consequently, the color imparted to a solution by a typical indicator appears to the average observer to change rapidly only within the limited concentration ratio of approximately 10 to 0.1. At greater or smaller ratios, the color becomes essentially constant to the human eye and independent of the ratio. Therefore, we can write that the average indicator HIn exhibits its pure acid color when

$$\frac{[HIn]}{[In^-]} \geq \frac{10}{1}$$

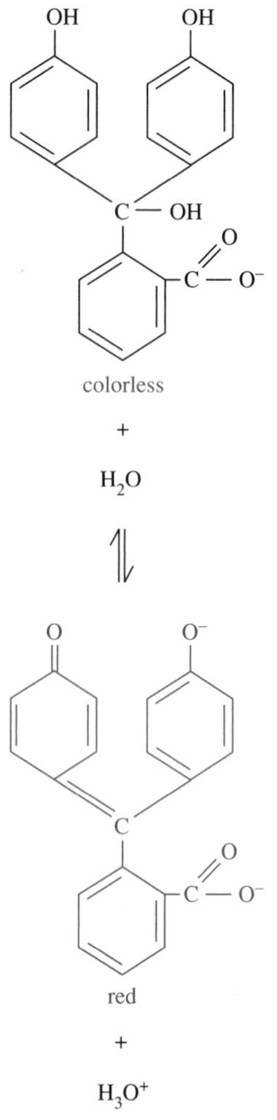

Figure 10-1
Color change for phenolphthalein.

and its base color when

$$\frac{[\text{HIn}]}{[\text{In}^-]} \leq \frac{1}{10}$$

The color appears to be intermediate for ratios between these two values. These ratios vary considerably from indicator to indicator. Furthermore, people differ significantly in their ability to distinguish between colors, with a color-blind person representing one extreme.

   If the two concentration ratios are substituted into Equation 10-2, the range of hydronium ion concentrations needed to cause the complete indicator color change can be evaluated. Thus, for the full acid color,

$$[\text{H}_3\text{O}^+] \geq K_\text{a} \frac{10}{1}$$

and similarly for the full base color,

$$[\text{H}_3\text{O}^+] \leq K_\text{a} \frac{1}{10}$$

To obtain the indicator range, we take the negative logarithms of the two expressions:

$$\text{indicator pH range} = -\log 10K_\text{a} \quad \text{to} \quad -\log \frac{K_\text{a}}{10}$$
$$= -1 + \text{p}K_\text{a} \quad \text{to} \quad -(-1) + \text{p}K_\text{a}$$

$$\text{indicator pH range} = \text{p}K_\text{a} \pm 1 \tag{10-3}$$

Thus, an indicator with an acid dissociation constant of 1×10^{-5} ($\text{p}K_\text{a} = 5$) typically shows a complete color change when the pH of the solution in which it is dissolved changes from 4 to 6 (see Figure 10-2). A similar relationship is easily derived for a basic-type indicator.

> The approximate pH range of most indicators is roughly $\text{p}K_\text{a} \pm 1$.

Variables That Influence the Behavior of Indicators

The pH over which an indicator changes color depends on temperature, ionic strength, and the presence of organic solvents and colloidal particles. Some of these effects, particularly the latter two, can cause the transition range to shift by one or more pH units.[1]

Some Common Acid/Base Indicators

The list of acid/base indicators is large and includes a number of organic structures. Indicators are available for any desired pH range. A few common indicators and

---

[1] For a discussion of these effects, see H. A. Laitinen and W. E. Harris, *Chemical Analysis*, 2nd ed., pp. 48–51. New York: McGraw-Hill, 1975.

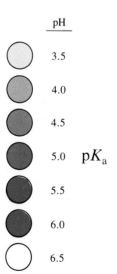

Figure 10-2
Indicator color as a function of pH ($\text{p}K_\text{a} = 5.0$).

their properties are listed in Table 10-1. Note that the transition ranges vary from 1.1 to 2.2 with the average being about 1.6 units.

10B TITRATION CURVES FOR STRONG ACIDS AND STRONG BASES

The hydronium ions in an aqueous solution of a strong acid have two sources: (1) the reaction of the acid with water and (2) the dissociation of water. In all but the most dilute solutions, however, the contribution from the strong acid far exceeds that from the solvent. Thus, for a solution of HCl with a concentration greater than about 1×10^{-6} M, we can write

> In solutions of a strong acid that are more concentrated than about 1×10^{-6} M, we may assume that the equilibrium concentration of $[H_3O^+]$ is equal to the analytical concentration of the acid. The same is true for the hydroxide concentration in solutions of a strong base.

$$[H_3O^+] = c_{HCl} + [OH^-] \cong c_{HCl}$$

where $[OH^-]$ represents the contribution of hydronium ions from the dissociation of water. An analogous relationship applies for a solution of a strong base, such as sodium hydroxide. That is,

$$[OH^-] = c_{NaOH} + [H_3O^+] \cong c_{NaOH}$$

10B-1 The Titration of a Strong Acid with a Strong Base

> Before the equivalence point we calculate the pH from the molar concentration of unreacted acid.

> At the equivalence point, the solution is neutral and pH = 7.00.

To derive a titration curve for a solution of a strong acid with a strong base, three types of calculations are required, each corresponding to a distinct stage in the titration: (1) preequivalence, (2) equivalence, and (3) postequivalence. In

Table 10-1
SOME IMPORTANT ACID/BASE INDICATORS

Common Name	Transition Range, pH	pK_a*	Color Change†	Indicator Type‡
Thymol blue	1.2–2.8	1.65§	R–Y	1
	8.0–9.6	8.90§	Y–B	
Methyl yellow	2.9–4.0		R–Y	2
Methyl orange	3.1–4.4	3.46§	R–O	2
Bromocresol green	3.8–5.4	4.66§	Y–B	1
Methyl red	4.2–6.3	5.00§	R–Y	2
Bromocresol purple	5.2–6.8	6.12§	Y–P	1
Bromothymol blue	6.2–7.6	7.10§	Y–B	1
Phenol red	6.8–8.4	7.81§	Y–R	1
Cresol purple	7.6–9.2		Y–P	1
Phenolphthalein	8.3–10.0		C–R	1
Thymolphthalein	9.3–10.5		C–B	1
Alizarin yellow GG	10–12		C–Y	2

* At ionic strength of 0.1.
† B = blue; C = colorless; O = orange; P = purple; R = red; Y = yellow.
‡ (1) Acid type: $HIn + H_2O \rightleftharpoons H_3O^+ + In^-$
‡ (2) Base type: $In + H_2O \rightleftharpoons InH^+ + OH^-$
§ For the reaction $InH^+ + H_2O \rightleftharpoons H_3O^+ + In$

the preequivalence stage, we compute the concentration of the analyte from its starting concentration of the acid and the volumetric data. At the equivalence point, the hydronium and hydroxide ions are present in equal concentrations, and the hydronium ion concentration is derived directly from the ion-product constant for water. In the postequivalence stage, the analytical concentration of the excess base is computed, and the hydroxide ion concentration is assumed to be identical to this concentration. A convenient way of converting hydroxide concentration to pH can be developed by taking the negative logarithm of each side of the ion-product constant for water. Thus,

Beyond the equivalence point, we first calculate pOH and then pH.

$$Remember\ pH = pK_w - pOH$$
$$= 14.00 - pOH$$

$$K_w = [H_3O^+][OH^-]$$

$$-\log K_w = -\log [H_3O^+][OH^-] = -\log [H_3O^+] - \log [OH^-]$$

$$pK_w = pH + pOH$$

$$-\log 10^{-14} = 14.00 = pH + pOH$$

Example 10-1

Derive a titration curve for the titration of 50.00 mL of 0.0500 M HCl with 0.1000 M NaOH.

Initial point
At the outset, the solution is 0.0500 M in H_3O^+, and

$$pH = -\log [H_3O^+] = -\log 0.0500 = 1.30$$

After addition of 10.00 mL of reagent
The hydronium ion concentration is decreased as a result of both reaction with the base and dilution. Thus, the analytical concentration of HCl is

$$c_{HCl} = \frac{\text{no. mmol HCl remaining after addition of NaOH}}{\text{total volume soln}}$$

$$= \frac{\text{original no. mmol HCl} - \text{no. mmol NaOH added}}{\text{total volume soln}}$$

$$= \frac{(50.00\ mL \times 0.0500\ M) - (10.00\ mL \times 0.1000\ M)}{50.00\ mL + 10.00\ mL}$$

$$= \frac{(2.500\ mmol - 1.000\ mmol)}{60.00\ mL} = 2.500 \times 10^{-2}\ M$$

$$[H_3O^+] = 2.50 \times 10^{-2}$$

and $pH = -\log [H_3O^+] = -\log 2.500 \times 10^{-2} = 1.602$

Additional points defining the curve in the region before the equivalence point are obtained in the same way. The results of such calculations are shown in the second column of Table 10-2.

Table 10-2
CHANGES IN pH DURING THE TITRATION OF A STRONG ACID WITH A STRONG BASE

Volume of NaOH, mL	pH	
	50.00 mL of 0.0500 M HCl with 0.1000 M NaOH	50.00 mL of 0.000500 M HCl with 0.001000 M NaOH
0.00	1.30	3.30
10.00	1.60	3.60
20.00	2.15	4.15
24.00	2.87	4.87
24.90	3.87	5.87
25.00	7.00	7.00
25.10	10.12	8.12
26.00	11.12	9.12
30.00	11.80	9.80

Equivalence point

At the equivalence point, neither HCl nor NaOH is in excess, and so the concentrations of hydronium and hydroxide ions must be equal. Substituting this equality into the ion-product constant for water yields

$$[H_3O^+] = \sqrt{K_w} = \sqrt{1.00 \times 10^{-14}} = 1.00 \times 10^{-7}$$
$$pH = -\log 1.00 \times 10^{-7} = 7.00$$

After addition of 25.10 mL of reagent

The solution now contains an excess of NaOH, and we can write

$$c_{NaOH} = \frac{25.10 \times 0.1000 - 50.00 \times 0.0500}{75.10} = 1.33 \times 10^{-4}\,M$$

and the equilibrium concentration of hydroxide ion is

$$[OH^-] = c_{NaOH} = 1.33 \times 10^{-4}\,M$$
$$pOH = -\log 1.33 \times 10^{-4} = 3.88$$

and

$$pH = 14.00 - 3.88 = 10.12$$

Additional data defining the curve beyond the equivalence point are computed in the same way. The results of such computations are shown in Table 10-2.

The Effect of Concentration

The effects of reagent and analyte concentration on the neutralization titration curves for strong acids are shown by the two sets of data in Table 10-2 and the plots in Figure 10-3. With 0.1 M NaOH as the titrant (curve *A*), the change in pH in the equivalence-point region is large. With 0.001 M NaOH, the change is markedly less but still pronounced.

Feature 10-1
DERIVING TITRATION CURVES FROM
THE CHARGE-BALANCE EQUATION

In Example 10-1 we generated an acid/base titration curve from the reaction stoichiometry. We can show that all points on the curve can be derived from the charge-balance equation also.

For the system treated in Example 10-1, the charge-balance equation is given by

$$[H_3O^+] + [Na^+] = [OH^-] + [Cl^-]$$

where the sodium and chloride ion concentrations are given by

$$[Na^+] = \frac{V_{NaOH}c_{NaOH}}{V_{NaOH} + V_{HCl}}$$

$$[Cl^-] = \frac{V_{HCl}c_{HCl}}{V_{NaOH} + V_{HCl}}$$

For volumes of NaOH short of the equivalence point $[OH^-] \ll [Cl^-]$ and we can rewrite the charge-balance equation in the form

$$[H_3O^+] \cong [Cl^-] - [Na^+]$$

and

$$[H_3O^+] = \frac{V_{HCl}c_{HCl}}{V_{HCl} + V_{NaOH}} - \frac{V_{NaOH}c_{NaOH}}{V_{HCl} + V_{NaOH}} = \frac{V_{HCl}c_{HCl} - V_{NaOH}c_{NaOH}}{V_{HCl} + V_{NaOH}}$$

At the equivalence point, $[Na^+] = [Cl^-]$ and

$$[H_3O^+] = [OH^-]$$
$$[H_3O^+] = \sqrt{K_w}$$

Beyond the equivalence point, $[H_3O^+] \ll [Na^+]$, and the charge-balance equation rearranges to

$$[OH^-] \cong [Na^+] - [Cl^-]$$

$$= \frac{V_{NaOH}c_{NaOH}}{V_{NaOH} + V_{HCl}} - \frac{V_{HCl}c_{HCl}}{V_{NaOH} + V_{HCl}} = \frac{V_{NaOH}c_{NaOH} - V_{HCl}c_{HCl}}{V_{NaOH} + V_{HCl}}$$

Indicator Choice

Figure 10-3 shows that the selection of an indicator is not critical when the reagent concentration is approximately 0.1 M. Here, the volume differences in titrations with the three indicators shown are of the same magnitude as the uncertainties associated with reading the buret and therefore are negligible. Note, however, that bromocresol green is clearly unsuited for a titration involving the

Feature 10-2
SIGNIFICANT FIGURES IN TITRATION CURVE CALCULATIONS

Concentrations calculated in the equivalence-point region of titration curves are generally of low precision because they are based on small differences between large numbers. For example, in the calculation of c_{NaOH} after introduction of 25.10 mL of NaOH in Example 10-1, the numerator (2.510 − 2.500 = 0.010) is known to only two significant figures. To minimize rounding error, however, three digits were retained in c_{NaOH} (1.33 × 10⁻⁴), and rounding was postponed until pOH and pH were computed.

In rounding the calculated values for p-functions, you should remember (Section 4E-2) that it is the *mantissa of a logarithm (that is, the number to the right of the decimal point) that should be rounded to include only significant figures* because the characteristic (the number to the left of the decimal point) serves merely to locate the decimal point. Fortunately, the large changes in p-functions characteristic of most equivalence points are not obscured by the limited precision of the calculated data. Generally, in computing data for titration curves, we will round p-functions to two places to the right of the decimal point regardless of whether such rounding is called for.

0.001 M reagent because the color change occurs over a 5-mL range well before the equivalence point. The use of phenolphthalein is subject to similar objections. Of the three indicators, then, only bromothymol blue would provide a satisfactory end point with a minimal systematic titration error when the more dilute solution is titrated.

10B-2 The Titration of a Strong Base with a Strong Acid

Titration curves for strong bases are derived in an analogous way to that for strong acids. Short of the equivalence point, the solution is highly basic, the hydroxide ion concentration being numerically equal to the analytical molarity of the base. The solution is neutral at the equivalence point and becomes acidic in the region beyond the equivalence point. The hydronium ion concentration is

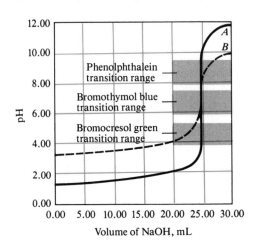

Figure 10-3

Titration curves for HCl with NaOH. *A*: 50.00 mL of 0.0500 M HCl with 0.1000 M NaOH. *B*: 50.00 mL of 0.000500 M HCl with 0.001000 M NaOH.

equal to the analytical concentration of the excess strong acid. A curve for the titration of a strong base with 0.1 M hydrochloric acid is shown later in the chapter in Figure 10-8. Indicator selection is based on the considerations described for the titration of a strong acid with a strong base.

10C   BUFFER SOLUTIONS

Whenever a weak acid is titrated with a strong base or a weak base with a strong acid, a *buffer solution* consisting of a conjugate acid/base pair is formed. Thus, before we can show how titration curves for weak acids and weak bases are derived we must investigate in some detail the properties and behavior of buffer solutions. By definition, a *buffer solution* is a solution of a conjugate acid/base pair that resists changes in pH. Chemists employ buffers whenever they need to maintain the pH of a solution at a constant and predetermined level. You will find many references to the use of buffers throughout this book.

Buffers are used in all types of chemistry whenever it is desirable to maintain the pH of a solution at a relatively constant and predetermined level.

Buffered aspirin contains buffers to help prevent stomach irritation from the acidity of the carboxylic acid group in aspirin.

aspirin

10C-1   Calculation of the pH of Buffer Solutions

Weak Acid/Conjugate Base Buffers

A solution containing a weak acid, HA, and its conjugate base, A^-, may be acidic, neutral, or basic, depending on the position of two competitive equilibria:

$$HA + H_2O \rightleftharpoons H_3O^+ + A^- \qquad K_a = \frac{[H_3O^+][A^-]}{[HA]} \qquad (10\text{-}4)$$

$$A^- + H_2O \rightleftharpoons OH^- + HA \qquad K_b = \frac{[OH^-][HA]}{[A^-]} = \frac{K_w}{K_a} \qquad (10\text{-}5)$$

If the first equilibrium lies farther to the right than the second, the solution is acidic. If the second equilibrium is more favorable, the solution is basic. These two equilibrium-constant expressions show that the relative concentrations of the hydronium and hydroxide ions depend not only on the magnitudes of K_a and K_b but also on the ratio of the concentrations of the acid and its conjugate base.

In order to compute the pH of a solution containing both an acid, HA, and its salt, NaA, we need to express the equilibrium concentrations of HA and NaA in terms of their analytical concentrations, c_{HA} and c_{NaA}. An examination of the two equilibria reveals that the first reaction decreases the concentration of HA by an amount equal to $[H_3O^+]$, whereas the second increases the HA concentration by an amount equal to $[OH^-]$. Thus, the species concentration of HA is related to its analytical concentration by the mass-balance equation

$$[HA] = c_{HA} - [H_3O^+] + [OH^-] \qquad (10\text{-}6)$$

Similarly, the first equilibrium will increase the concentration of A^- by an amount equal to $[H_3O^+]$, and the second will decrease this concentration by the amount $[OH^-]$. Thus the equilibrium concentration is given by a second mass-balance equation

$$[A^-] = c_{NaA} + [H_3O^+] - [OH^-] \qquad (10\text{-}7)$$

Feature 10-3
APPLICATION OF THE SYSTEMATIC METHOD TO
BUFFER CALCULATIONS

Equations 10-6 and 10-7 can also be derived from mass- and charge-balance expressions. Thus, mass-balance considerations require that

$$c_{HA} + c_{NaA} = [HA] + [A^-]$$

Electrical neutrality considerations require that

$$[Na^+] + [H_3O^+] = [A^-] + [OH^-]$$

but

$$[Na^+] = c_{NaA}$$

Therefore, the charge-balance equation is

$$c_{NaA} + [H_3O^+] = [A^-] + [OH^-]$$

which rearranges to Equation 10-7:

$$[A^-] = c_{NaA} + [H_3O^+] - [OH^-]$$

Subtracting the first equation from the fourth and rearranging gives

$$[HA] = c_{HA} - [H_3O^+] + [OH^-]$$

which is identical to Equation 10-6.

Because of the inverse relationship between $[H_3O^+]$ and $[OH^-]$, it is *always* possible to eliminate one or the other from Equations 10-6 and 10-7. Moreover, the *difference* in concentration between these two species is often so small relative to the molar concentrations of acid and conjugate base that Equations 10-6 and 10-7 simplify to

$$[HA] \cong c_{HA} \tag{10-8}$$

$$[A^-] \cong c_{NaA} \tag{10-9}$$

Substituting Equations 10-8 and 10-9 into the dissociation-constant expression and rearranging yields

$$[H_3O^+] = K_a \frac{c_{HA}}{c_{NaA}} \tag{10-10}$$

The assumption leading to Equations 10-8 and 10-9 sometimes breaks down with acids or bases that have dissociation constants greater than about 10^{-3} or when the molar concentration of either the acid or its conjugate base (or both) is very small. In these circumstances, either $[OH^-]$ or $[H_3O^+]$ must be retained in Equations 10-6 and 10-7, depending on whether the solution is acidic or basic. In any case, Equations 10-8 and 10-9 should always be used initially. The provisional values for $[H_3O^+]$ and $[OH^-]$ can then be employed to test the assumptions.

Within the limits imposed by the assumptions made in its derivation, Equation 10-10 says that the hydronium ion concentration of a solution containing a weak acid and its conjugate base is dependent only on the *ratio* of the molar concentrations of these two solutes. Furthermore, this ratio is *independent of dilution* because the concentration of each component changes proportionately when the volume changes.

Example 10-2

What is the pH of a solution that is 0.400 M in formic acid and 1.00 M in sodium formate?

The equilibrium governing the hydronium ion concentration in this solution is

$$H_2O + HCOOH \rightleftharpoons H_3O^+ + HCOO^-$$

for which (Appendix 2)

$$K_a = \frac{[H_3O^+][HCOO^-]}{[HCOOH]} = 1.80 \times 10^{-4}$$

$$[HCOO^-] \cong c_{HCOO^-} = 1.00$$

$$[HCOOH] \cong c_{HCOOH} = 0.400$$

Feature 10-4

The Henderson-Hasselbalch equation is an alternative form of Equation 10-10 that is frequently encountered in the biological literature and biochemical texts. It is obtained by expressing each term in Equation 10-10 in the form of its negative logarithm and inverting the concentration ratio to keep all signs positive:

$$-\log[H_3O^+] = -\log K_a - \log \frac{c_{HA}}{c_{NaA}}$$

Therefore,

$$pH = pK_a + \log \frac{c_{NaA}}{c_{HA}} \qquad \textbf{(10-11)}$$

Substituting into Equation 10-10 gives

$$[H_3O^+] = 1.80 \times 10^{-4} \times \frac{0.400}{1.00} = 7.20 \times 10^{-5}$$

Note that the assumptions that $[H_3O^+] \ll c_{HCOOH}$ and $[H_3O^+] \ll c_{HCOO^-}$ are valid. Thus,

$$pH = -\log 7.20 \times 10^{-5} = 4.14$$

Weak Base/Conjugate Acid Buffers

As shown in Example 10-3, Equations 10-6 and 10-7 also apply to buffer systems consisting of a weak base and its conjugate acid. Furthermore, in most cases it is possible to simplify these equations so that Equation 10-10 can be used.

Example 10-3

Calculate the pH of a solution that is 0.200 M in NH_3 and 0.300 M in NH_4Cl. In Appendix 2, we find that the acid dissociation constant K_a for NH_4^+ is 5.70×10^{-10}.

The equilibria we must consider are

$$NH_4^+ + H_2O \rightleftharpoons NH_3 + H_3O^+ \qquad K_a = 5.70 \times 10^{-10}$$

$$NH_3 + H_2O \rightleftharpoons NH_4^+ + OH^- \qquad K_b = \frac{K_w}{K_a} = \frac{1.00 \times 10^{-14}}{5.70 \times 10^{-10}} = 1.75 \times 10^{-5}$$

Using the arguments that led to Equations 10-6 and 10-7, we obtain

$$[NH_4^+] = c_{NH_4Cl} + [OH^-] - [H_3O^+] \cong c_{NH_4Cl} + [OH^-]$$

$$[NH_3] = c_{NH_3} + [H_3O^+] - [OH^-] \cong c_{NH_3} - [OH^-]$$

Here, we have assumed on the basis of the relative sizes of the two equilibrium constants that $[H_3O^+]$ is negligibly small and therefore can be neglected.

Let us also assume that $[OH^-]$ is much smaller than c_{NH_4Cl} and c_{NH_3} so that

$$[NH_4^+] = c_{NH_4Cl} = 0.300$$

$$[NH_3] = c_{NH_3} = 0.200$$

Substituting into the acid-dissociation constant for NH_4^+, we obtain a relationship similar to Equation 10-10. That is,

$$[H_3O^+] = \frac{K_a \times [NH_4^+]}{[NH_3]} = \frac{5.70 \times 10^{-10} \times c_{NH_4^+}}{c_{NH_3}}$$

$$= \frac{5.70 \times 10^{-10} \times 0.300}{0.200} = 8.55 \times 10^{-10}$$

To check the validity of our approximations, we calculate $[OH^-]$. Thus,

$$[OH^-] = 1.00 \times 10^{-14}/(8.55 \times 10^{-10}) = 1.17 \times 10^{-5}$$

which is certainly much smaller than either $c_{NH_4^+}$ or c_{NH_3}. Thus we may write

$$pH = -\log 8.55 \times 10^{-10} = 9.07$$

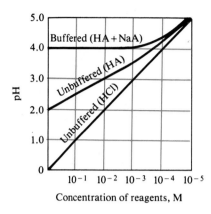

Figure 10-4
The effect of dilution on the pH of buffered and unbuffered solutions. The dissociation constant for HA is 1.00×10^{-4}. Initial solute concentrations are 1.00 M.

Buffers do not maintain pH at an absolutely constant value, but changes in pH are relatively small when small amounts of acid or base are added.

10C-2 Properties of Buffer Solutions

In this section we illustrate the resistance of buffers to changes of pH brought about by dilution or addition of strong acids or bases.

The Effect of Dilution

The pH of a buffer solution remains essentially independent of dilution until the concentrations of the species it contains are decreased to the point where the approximations used to develop Equations 10-8 and 10-9 become invalid. Figure 10-4 contrasts the behavior of buffered and unbuffered solutions with dilution. For each, the initial solute concentrations are 1.00 M. The buffered solution resists changes in pH during dilution; the unbuffered solution does not.

The Effect of Added Acids and Bases

Example 10-4 illustrates a second property of buffer solutions, their resistance to pH change after addition of small amounts of strong acids or bases.

Example 10-4

Calculate the pH change that takes place when a 100-mL portion of (a) 0.0500 M NaOH and (b) 0.0500 M HCl is added to 400 mL of the buffer solution that was described in Example 10-3.

(a) Addition of NaOH converts part of the NH_4^+ in the buffer to NH_3:

$$NH_4^+ + OH^- \rightleftharpoons NH_3 + H_2O$$

The analytical concentrations of NH_3 and NH_4Cl then become

$$c_{NH_3} = \frac{400 \times 0.200 + 100 \times 0.0500}{500} = \frac{85.0}{500} = 0.170 \text{ M}$$

$$c_{NH_4Cl} = \frac{400 \times 0.300 - 100 \times 0.0500}{500} = \frac{115}{500} = 0.230 \text{ M}$$

When substituted into the acid-dissociation constant expression for NH_4^+, these values yield

$$[H_3O^+] = 5.70 \times 10^{-10} \times \frac{0.230}{0.170} = 7.71 \times 10^{-10}$$

$$pH = -\log 7.71 \times 10^{-10} = 9.11$$

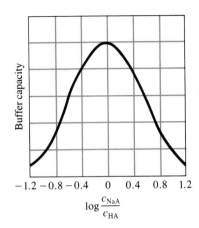

Figure 10-5
Buffer capacity as a function of the ratio c_{NaA}/c_{HA}.

and the change in pH is

$$\Delta pH = 9.11 - 9.07 = 0.04$$

(b) Addition of HCl converts part of the NH_3 to NH_4^+; thus,

$$NH_3 + H_3O^+ \rightleftharpoons NH_4^+ + H_2O$$

$$c_{NH_3} = \frac{400 \times 0.200 - 100 \times 0.0500}{500} = \frac{75}{500} = 0.150 \text{ M}$$

$$c_{NH_4^+} = \frac{400 \times 0.300 + 100 \times 0.0500}{500} = \frac{125}{500} = 0.250 \text{ M}$$

$$[H_3O^+] = 5.70 \times 10^{-10} \times \frac{0.250}{0.150} = 9.50 \times 10^{-10}$$

$$pH = -\log 9.50 \times 10^{-10} = 9.02$$

$$\Delta pH = 9.02 - 9.07 = -0.05$$

It is interesting to contrast the behavior of an unbuffered solution with a pH of 9.07 to that of the buffer in Example 10-4. It can be shown that adding the same quantity of base to the unbuffered solution would increase the pH to 12.00—a pH change of 2.93 units. Adding the acid would decrease the pH by slightly more than 7 units.

Buffer Capacity

Figure 10-4 and Example 10-4 demonstrate that a solution containing a conjugate acid/base pair possesses remarkable resistance to changes in pH. The ability of a buffer to prevent a significant change in pH is directly related to the total concentration of the buffering species as well as to their concentration ratio. For example, the pH of a 400-mL portion of a buffer formed by diluting the solution described in Example 10-3 by 10 would change by about 0.4 to 0.5 unit when treated with 100 mL of 0.0500 M sodium hydroxide or 0.0500 M hydrochloric acid. We have shown in Example 10-4 that the change is only about 0.04 to 0.05 unit for the more concentrated buffer.

> Buffer capacity is the number of moles of strong acid or strong base that 1 L of the buffer can absorb per unit change in pH.

The *buffer capacity* of a solution is defined as the number of moles of a strong acid or a strong base that causes 1.00 L of the buffer to undergo a 1.00-unit change in pH. The capacity of a buffer depends not only on the total concentration of the two buffer components but also on their concentration ratio. Buffer capacity falls off at a moderately rapid rate as the concentration ratio of acid to conjugate base becomes larger or smaller than unity (see Figure 10-5). For this reason, the pK_a of the acid chosen for a given application should lie within ± 1 unit of the desired pH in order for the buffer to have a reasonable capacity.

Preparation of Buffers

In principle, a buffer solution of any desired pH can be prepared by combining calculated quantities of a suitable conjugate acid/base pair. In practice, however, the pH values of buffers prepared from theoretical generated formulas differ from the predicted values because of uncertainties in the numerical values of

Feature 10-5

ACID RAIN AND THE BUFFER CAPACITY OF LAKES

Acid rain has been the subject of considerable controversy over the past two decades. Acid rain forms when the gaseous oxides of nitrogen and sulfur dissolve in water droplets in the air. These gases form at high temperatures in power plants, automobiles, and other combustion sources. The combustion products pass into the atmosphere where they react with water to form nitric acid and sulfuric acid as shown below.

$$4NO_2(g) + 2H_2O(l) + O_2(g) \rightarrow 4HNO_3(aq)$$

$$SO_3(g) + H_2O(l) \rightarrow H_2SO_4(aq)$$

Eventually the droplets coalesce with other droplets of acid and droplets of water to form acid rain. The profound effects of acid rain have been highly publicized. Stone buildings and monuments literally dissolve away as acid rain flows over their surfaces. Forests are slowly being killed off in some locations. To illustrate the effects on aquatic life, let us consider the changes in pH that have occurred in the lakes of the Adirondack Mountains area of New York illustrated in the bar graphs of Figure 10-A.

Figure 10-5

Buffer capacity as a function of the ratio c_{NaA}/c_{HA}.

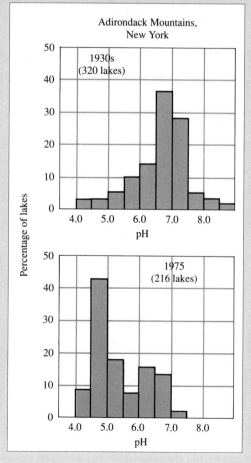

Adirondack Mountains, New York

Figure 10-A

The graphs show the distribution of pH in these lakes, which were studied first in the 1930s and then again in 1975.[1] The shift in pH of the lakes over a 40-year period is dramatic. The average pH of the lakes changed from 6.4 to about 5.1, which represents a 20-fold change in the hydronium ion concentration. Such changes in pH have a profound effect upon aquatic life as shown by a study of the fish population in lakes in the same area.[2] In the graph of Figure 10-B, the number of lakes is plotted as a function of pH. The gray bars represent lakes containing fish, and lakes having no fish are in color. There is a distinct correlation between pH changes in the lakes and diminished fish population.

Many factors contribute to pH changes in groundwater and lakes in a given geographical area. These include the prevailing wind patterns and weather, types of soils, water sources, nature of the terrain, characteristics of plant life, human activity, and the geological characteristics of the area. The susceptibility of natural water to acidification is largely determined by its buffer capacity, and the principal buffer of natural water is bicarbonate ion ($K_1 = 4.45 \times 10^{-7}$, $K_2 = 4.69 \times 10^{-11}$). Recall that the buffer capacity of a solution is proportional to the concentration of the buffering agent. So, the higher the concentration of dissolved bicarbonate, the greater is the capacity of the water to neutralize acid from acid rain. The most important source of bicarbonate ion in natural water is limestone, or calcium carbonate, which reacts with hydronium ion as shown in the following equation.

$$CaCO_3(s) + H_3O^+(aq) \rightleftharpoons HCO_3^-(aq) + Ca^{2+}(aq)$$

Limestone-rich areas have lakes with relatively high concentrations of dissolved bicarbonate, and thus low susceptibility to acidification. Granite, sandstone, shale, and other rock containing little or no calcium carbonate are associated with lakes having high susceptibility to acidification. The map of the United States shown in Figure 10-C vividly illustrates the correlation between the absence of limestone-bearing rocks and the acidification of groundwaters.[3]

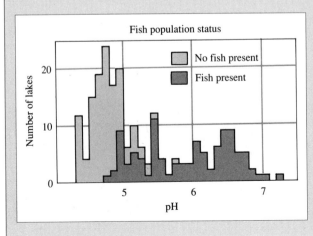

Figure 10-B

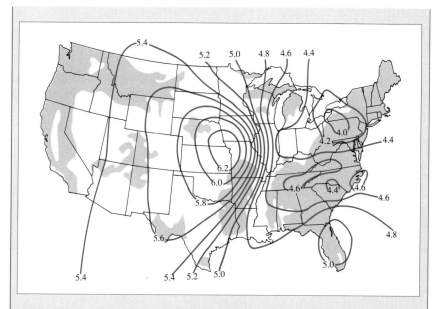

Figure 10-C

Areas containing little limestone are shaded in color; areas rich in limestone are white. Contour lines of equal pH for groundwater during the period 1978–1979 are superimposed on the map. The Adirondack Mountains area, located in northeastern New York, contains little limestone and exhibits pH in the range of 4.2 to 4.4. The low buffer capacity of the lakes in this region combined with the low pH of precipitation appears to have caused the decline of fish populations. Similar correlations among acid rain, buffer capacity of lakes, and wildlife decline occur throughout the industrialized world.

Although natural sources such as volcanos produce sulfur trioxide, and lightning discharges in the atmosphere generate nitrogen dioxide, large quantities of these compounds come from the burning of high-sulfur coal and from automobile emissions. To minimize emissions of these pollutants, some states have enacted legislation imposing strict standards on automobiles sold and operated within their borders. Some states have required the installation of scrubbers to remove oxides of sulfur from the emissions of coal-fired power plants. To minimize the effects of acid rain on lakes, powdered limestone is dumped on their surfaces to increase the buffer capacity of the water. Solutions to these problems require the expenditure of much time, energy, and money. We must sometimes make difficult economic decisions in order to preserve the quality of our environment and to reverse trends that have operated for many decades. For a comprehensive discussion of the problem of acid rain, refer to the book cited below.[4]

[1] R. F. Wright and E. T. Gjessing, *Ambio*, **1976**, *5*, 219.
[2] C. L. Schofield, *Ambio*, **1976**, *5*, 228.
[3] J. Root, et al., cited in *The Effects of Air Pollution and Acid Rain on Fish, Wildlife, and Their Habitats—Introduction*. U.S. Fish and Wildlife Service, Biological Services Program, Eastern Energy and Land Use Team, M. A. Peterson, Ed., p. 63. U.S. Government Publication FWS/OBS-80/40.3.
[4] C. C. Park, *Acid Rain*. New York: Methuen, 1987.

many dissociation constants and from the simplifications used in calculations. Of more importance is the fact that the ionic strength of a buffer is usually so high that good values for the activity coefficients of the ions in the solution cannot be obtained from the Debye-Hückel relationship. Because of these uncertainties, we prepare buffers by making up a solution of approximately the desired pH and then adjust by adding strong acid or strong base until the required pH is indicated by a pH meter.

An alternative is empirically derived recipes for preparing buffer solutions of known pH, available in chemical handbooks and reference works.[2]

Buffers are also used to standardize pH meters.

Buffers are of tremendous importance in biological and biochemical studies where a low but constant concentration of hydronium ions (10^{-6} to 10^{-10} M) must be maintained throughout experiments. Several biological supply houses offer a variety of such buffers.

10D TITRATION CURVES FOR WEAK ACIDS

Four distinctly different types of calculations are needed to derive a titration curve for a weak acid (or a weak base):

1. At the beginning, the solution contains only a weak acid or a weak base, and the pH is calculated from the concentration of the solute and its dissociation constant.
2. After various increments of titrant have been added (in quantities up to, but not including, an equivalent amount), the solution consists of a series of buffers. The pH of each buffer can be calculated from the analytical concentrations of the conjugate base or acid and the residual concentrations of the weak acid or base.
3. At the equivalence point, the solution contains only the conjugate of the weak acid or base being titrated (that is, a salt), and the pH is calculated from the concentration of this product.
4. Beyond the equivalence point, the excess of strong base or acid titrant represses the basic or acidic character of the reaction product to such an extent that the pH is governed largely by the concentration of the excess titrant.

Example 10-5

Derive a curve for the titration of 50.00 mL of 0.1000 M acetic acid ($K_a = 1.75 \times 10^{-5}$) with 0.1000 M sodium hydroxide.

Initial pH
Initially, we must calculate the pH of a 0.1000 M solution of HOAc using Equation 3-20:

$$[H_3O^+] = \sqrt{K_a c_{HOAc}} = \sqrt{1.75 \times 10^{-5} \times 0.1000} = 1.32 \times 10^{-3}$$
$$pH = -\log 1.32 \times 10^{-3} = 2.88$$

---

[2] See, for example, L. Meites, Ed., *Handbook of Analytical Chemistry*, pp. **5**–112 and **11**–3 to **11**–8. New York: McGraw-Hill, 1963.

pH After Addition of 10.00 mL of Reagent

A buffer solution consisting of NaOAc and HOAc has now been produced. The analytical concentrations of the two constituents are

$$c_{HOAc} = \frac{50.00 \text{ mL} \times 0.1000 \text{ M} - 10.00 \text{ mL} \times 0.1000 \text{ M}}{60.00 \text{ mL}} = \frac{4.000}{60.00} \text{ M}$$

$$c_{NaOAc} = \frac{10.00 \text{ mL} \times 0.1000 \text{ M}}{60.00 \text{ mL}} = \frac{1.000}{60.00} \text{ M}$$

We substitute these concentrations into the dissociation-constant expression for acetic acid and obtain

$$\frac{[H_3O^+](1.000/60.00)}{4.000/60.00} = K_a = 1.75 \times 10^{-5}$$

$$[H_3O^+] = 7.00 \times 10^{-5}$$

$$pH = 4.16$$

Calculations similar to this provide points on the curve throughout the buffer region. Data from such calculations are given in column 2 of Table 10-3.

Equivalence Point pH

At the equivalence point, all the acetic acid has been converted to sodium acetate. The solution is therefore similar to one formed by dissolving that base in water, and the pH calculation is identical to that shown in Example 3-10 for a weak base. In the present example, the NaOAc concentration is 0.0500 M. Thus,

$$OAc^- + H_2O \rightleftharpoons HOAc + OH^-$$

$$[OH^-] = [HOAc]$$

$$[OAc^-] = 0.0500 - [OH^-] \cong 0.0500$$

Table 10-3
CHANGES IN pH DURING THE TITRATION OF A WEAK ACID WITH A STRONG BASE

Volume of NaOH, mL	pH	
	50.00 mL of 0.1000 M HOAc Titrated with 0.1000 M NaOH	50.00 mL of 0.001000 M HOAc Titrated with 0.001000 M NaOH
0.00	2.88	3.91
10.00	4.16	4.30
25.00	4.76	4.80
40.00	5.36	5.38
49.00	6.45	6.46
49.90	7.46	7.47
50.00	8.73	7.73
50.10	10.00	8.09
51.00	11.00	9.00
60.00	11.96	9.96
75.00	12.30	10.30

Substituting in the base dissociation-constant expression for OAc$^-$ gives

$$\frac{[OH^-]^2}{0.0500} = \frac{K_w}{K_a} = \frac{1.00 \times 10^{-14}}{1.75 \times 10^{-5}} = 5.71 \times 10^{-10}$$

$$[OH^-] = \sqrt{0.0500 \times 5.71 \times 10^{-10}} = 5.34 \times 10^{-6}$$

$$pH = 14.00 - (-\log 5.34 \times 10^{-6}) = 8.73$$

pH After Addition of 50.10 mL of Base

At the half-neutralization point in the titration of a weak acid $[H_3O^+] = K_a$ or $pH = pK_a$.

After the addition of 50.10 mL of NaOH, both the excess base and the acetate ion are sources of the hydroxide ion. The contribution of the latter is small, however, because the excess of strong base suppresses the formation of hydroxide from the reaction of acetate ion with water. This fact becomes evident when we consider that the hydroxide ion concentration is only 5.34×10^{-6} M at the equivalence point; once an excess of strong base is added, the contribution from the reaction of the acetate is even smaller. Thus,

At the half-neutralization point in the titration of a weak base $[OH^-] = K_b$ or $pOH = pK_b$.

$$[OH^-] \cong c_{NaOH} = \frac{50.10 \text{ mL} \times 0.1000 \text{ M} - 50.00 \text{ mL} \times 0.1000 \text{ M}}{100.0 \text{ mL}}$$

$$= 1.00 \times 10^{-4} \text{ M}$$

$$pH = 14.00 - (-\log 1.00 \times 10^{-4}) = 10.00$$

Titration curves for strong and weak acids with strong base are identical just slightly beyond the equivalence point. The same is true for the titration of strong and weak bases with strong acid.

Note that the titration curve for a weak acid with a strong base is identical with that for a strong acid with a strong base in the region beginning slightly beyond the equivalence point.

Note that the analytical concentrations of acid and conjugate base are identical when an acid has been half neutralized (in Example 10-5, after the addition of exactly 25.00 mL of base). Thus, these terms cancel in the equilibrium-constant expression, and the hydronium ion concentration is numerically equal to the dissociation constant. Likewise, in the titration of a weak base, the hydroxide ion concentration is numerically equal to the dissociation constant of the base at the midpoint in the titration curve. In addition, the buffer capacities of each of the solutions are at a maximum at this point.

Feature 10-6
DETERMINATION OF DISSOCIATION CONSTANTS FOR WEAK ACIDS AND BASES

The dissociation constants of weak acids or weak bases are often determined by monitoring the pH of the solution while the acid or base is being titrated. A pH meter and a glass electrode are used for the measurements. For an acid, the measured pH when the acid is exactly half neutralized is numerically equal to pK_a. For a weak base, the pH at half neutralization must be converted to pOH, which is then equal to pK_b.

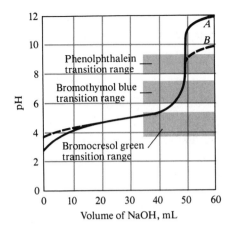

Figure 10-6
Curve for the titration of acetic acid with sodium hydroxide. *A*: 0.1000 M acid with 0.1000 M base. *B*: 0.001000 M acid with 0.001000 M base.

10D-1 The Effect of Concentration

The second and third columns of Table 10-3 contain pH data for the titration of 0.1000 M and of 0.001000 M acetic acid with sodium hydroxide solutions of the same two concentrations. In deriving the data for the more dilute acid, none of the approximations shown in Example 10-5 was valid, and solution of a quadratic equation was necessary throughout.

Figure 10-6 is a plot of the data in Table 10-3. Note that the initial pH values are higher and the equivalence-point pH is lower for the more dilute solution. At intermediate titrant volumes, however, the pH values differ only slightly because of the buffering action of the acetic acid/sodium acetate system that predominates in this region. Figure 10-6 confirms graphically the fact that the pH of buffers is largely independent of dilution.

10D-2 The Effect of Reaction Completeness

Titration curves for 0.1000 M solutions of acids with different dissociation constants are shown in Figure 10-7. Note that the pH change in the equivalence-point region becomes smaller as the acid becomes weaker—that is, as the reaction

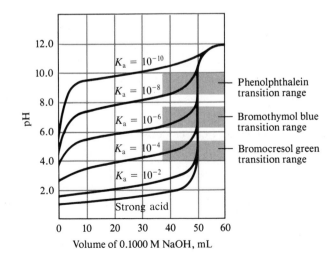

Figure 10-7
The effect of acid strength on titration curves. Each curve represents the titration of 50.0 mL of 0.1000 M acid with 0.1000 M NaOH.

between the acid and the base becomes less complete. The effects of the reactant concentrations and of the completeness of a reaction illustrated by Figures 10-6 and 10-7 are analogous to these effects on strong acid titrations shown in Figure 10-3.

10D-3 Indicator Choice; The Feasibility of Titration

Figures 10-6 and 10-7 show clearly that the choice of indicator for the titration of a weak acid is more limited than that for a strong acid. For example, from Figure 10-6 it is obvious that bromocresol green is totally unsuited for titration of 0.1000 M acetic acid. Bromothymol blue is also unsatisfactory because its full color change occurs over a range from about 47 mL to 50 mL of 0.1000 M base. On the other hand, an indicator exhibiting a color change in the basic region, such as phenolphthalein, should provide a sharp end point with a minimal titration error.

The end-point pH change associated with the titration of 0.001000 M acetic acid (curve *B*, Figure 10-6) is so small that a significant titration error is likely to be introduced regardless of indicator. However, use of an indicator with a transition range between that of phenolphthalein and that of bromothymol blue in conjunction with a suitable color comparison standard makes it possible to establish the end point in this titration with a reproducibility of a few percent relative.

Figure 10-7 illustrates that similar problems occur as the strength of the acid being titrated decreases. Precision on the order of ± 2 ppt can be achieved in the titration of a 0.1000 M acid solution with a dissociation constant of 10^{-8} provided a suitable color comparison standard is available. With more concentrated solutions, somewhat weaker acids can be titrated with reasonable precision.

10E   TITRATION CURVES FOR WEAK BASES

The derivation of a curve for the titration of a weak base is analogous to that of a weak acid.

Example 10-6

A 50.00-mL aliquot of 0.0500 M NaCN is titrated with 0.1000 M HCl. The reaction is

$$CN^- + H_3O^+ \rightleftharpoons HCN + H_2O$$

Calculate the pH after the addition of (a) 0.00; (b) 10.00; (c) 25.00; and (d) 26.00 mL of acid.

(a) Initial pH

The pH of a solution of NaCN can be derived by the method shown in Example 3-10:

$$CN^- + H_2O \rightleftharpoons HCN + OH^-$$

$$K_b = \frac{[OH^-][HCN]}{[CN^-]} = \frac{K_w}{K_a} = \frac{1.00 \times 10^{-14}}{6.2 \times 10^{-10}} = 1.61 \times 10^{-5}$$

$$[OH^-] = [HCN]$$

$$[CN^-] = c_{NaCN} - [OH^-] \cong c_{NaCN} = 0.0500$$

Substituting into the dissociation-constant expression and rearranging gives

$$[OH^-] = \sqrt{K_b c_{NaCN}} = \sqrt{1.61 \times 10^{-5} \times 0.0500} = 8.97 \times 10^{-4}$$
$$pH = 14.00 - (-\log 8.97 \times 10^{-4}) = 10.95$$

(b) pH After Addition of 10.00 mL of Reagent
Addition of acid produces a buffer with a composition given by

$$c_{NaCN} = \frac{50.00 \times 0.0500 - 10.00 \times 0.1000}{60.00} = \frac{1.500}{60.00} M$$

$$c_{HCN} = \frac{10.00 \times 0.1000}{60.00} = \frac{1.000}{60.00} M$$

These values are then substituted into the expression for the acid dissociation constant of HCN to give $[H_3O^+]$ directly:

$$[H_3O^+] = \frac{6.2 \times 10^{-10} \times (1.000/60.00)}{1.500/60.00} = 4.13 \times 10^{-10}$$
$$pH = -\log 4.13 \times 10^{-10} = 9.38$$

(c) pH After Addition of 25.00 mL of Reagent
This volume corresponds to the equivalence point, where the principal solute species is the weak acid HCN. Thus,

$$c_{HCN} = \frac{25.00 \times 0.1000}{75.00} = 0.03333 M$$

Applying Equation 3-20 gives

$$[H_3O^+] = \sqrt{K_a c_{HCN}} = \sqrt{6.2 \times 10^{-10} \times 0.03333} = 4.45 \times 10^{-6}$$
$$pH = -\log 4.55 \times 10^{-6} = 5.34$$

(d) pH After Addition of 26.00 mL of Reagent
The excess of strong acid now present represses the dissociation of the HCN to the point where its contribution to the pH is negligible. Thus,

$$[H_3O^+] = c_{HCl} = \frac{26.00 \times 0.1000 - 50.00 \times 0.0500}{76.00} = 1.32 \times 10^{-3}$$
$$pH = -\log 1.32 \times 10^{-3} = 2.88$$

Figure 10-8 shows theoretical curves for a series of weak bases of different strengths. Clearly, indicators with *acidic* transition ranges must be used for weak bases.

When titrating a weak base, an indicator with an acidic transition range is used. When titrating a weak acid, the indicator should have a basic transition range.

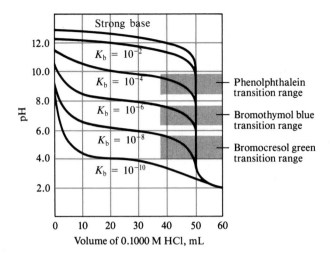

Figure 10-8
The effect of base strength on titration curves. Each curve represents the titration of 50.0 mL of 0.1000 M base with 0.1000 M HCl.

10F   THE COMPOSITION OF BUFFER SOLUTIONS AS A FUNCTION OF pH

The changes in composition that occur while a solution of a weak acid or a weak base is being titrated are sometimes of interest and can be visualized by plotting the *relative* concentration of the weak acid as well as the relative concentration of its conjugate base as a function of the pH of the solution. These relative concentrations are called *alpha values*. For example, if we let c_T be the sum of the analytical concentrations of acetic acid and sodium acetate at any point in the titration curve derived in Example 10-5, we may write

$$c_T = c_{HOAc} + c_{NaOAc} \tag{10-12}$$

We then define α_0 as

$$\alpha_0 = \frac{[HOAc]}{c_T} \tag{10-13}$$

and α_1 as

$$\alpha_1 = \frac{[OAc^-]}{c_T} \tag{10-14}$$

Alpha values are unitless ratios. The sum of all alpha values for a given compound must equal unity. That is,

$$\alpha_0 + \alpha_1 = 1$$

Alpha values are determined by $[H_3O^+]$ and K_a alone and are independent of c_T. To obtain expressions for α_0, we rearrange the dissociation-constant expression to

$$[OAc^-] = \frac{K_a[HOAc]}{[H_3O^+]}$$

From mass-balance consideration, we write

$$c_T = [HOAc] + [OAc^-] \qquad \textbf{(10-15)}$$

Substituting the previous equation into Equation 10-16 gives

$$c_T = [HOAc] + \frac{K_a[HOAc]}{[H_3O^+]} = [HOAc]\left(\frac{[H_3O^+] + K_a}{[H_3O^+]}\right)$$

Upon rearrangement we obtain

$$\frac{[HOAc]}{c_T} = \frac{[H_3O^+]}{[H_3O^+] + K_a}$$

But by definition $[HOAc]/c_T = \alpha_0$ (Equation 10-14) or

$$\alpha_0 = \frac{[HOAc]}{c_T} = \frac{[H_3O^+]}{[H_3O^+] + K_a} \qquad \textbf{(10-16)}$$

To obtain an expression for α_1, we rearrange the dissociation-constant expression to

$$[HOAc] = \frac{[H_3O^+][OAc^-]}{K_a}$$

and substitute into Equation 10-16

$$c_T = \frac{[H_3O^+][OAc^-]}{K_a} + [OAc^-] = [OAc^-]\left(\frac{[H_3O^+] + K_a}{K_a}\right)$$

Rearranging this gives α_1 as defined by Equation 10-15

$$\alpha_1 = \frac{[OAc^-]}{c_T} = \frac{K_a}{[H_3O^+] + K_a} \qquad \textbf{(10-17)}$$

Alternatively,

$$\alpha_1 = 1 - \alpha_0$$
$$= 1 - \frac{[H_3O^+]}{[H_3O^+] + K_a}$$
$$= \frac{[\cancel{H_3O^+}] + K_a - [\cancel{H_3O^+}]}{[H_3O^+] + K_a}$$
$$= \frac{K_a}{[H_3O^+] + K_a}$$

Note that the denominator is the same in Equations 10-17 and 10-18.

The solid straight lines labeled α_0 and α_1 in Figure 10-9 were computed with Equations 10-17 and 10-18 using values for $[H_3O^+]$ shown in column 2 of Table 10-3. The actual titration curve is shown as the curved line in Figure 10-9. Note that at the outset of the titration α_0 is nearly 1 (0.987) meaning that 98.7% of the acetate-containing species is present as HOAc and only 1.3% is present as OAc^-. At the equivalence point, α_0 has decreased to 1.1×10^{-4} and α_1 approaches 1. Thus, only about 0.011% of the acetate-containing species is HOAc. Note that when the acid is half neutralized (25.00 mL), α_0 and α_1 are each 0.5.

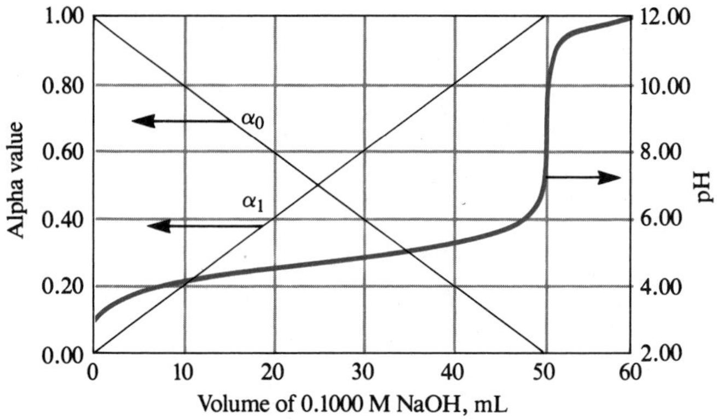

Figure 10-9
The straight lines show the change in relative amounts of HOAc (α_0) and OAc$^-$ (α_1) during the titration of 50.00 mL of 0.1000 M acetic acid. The curved line is the titration curve for the system.

10G   QUESTIONS AND PROBLEMS†

*10-1. Consider curves for the titration of 0.10 M NaOH and 0.010 M NH$_3$ with 0.10 M HCl.
  (a) Briefly account for the differences between curves for the two titrations.
  (b) In what respect will the two curves be indistinguishable?

10-2. What factors affect end-point sharpness in an acid/base titration?

*10-3. Why does the typical acid/base indicator exhibit its color change over a range of about 2 pH units?

10-4. What variables can cause the pH range of an indicator to shift?

*10-5. Why are the standard reagents used in neutralization titrations generally strong acids and bases rather than weak acids and bases?

10-6. What is a buffer solution and what are its properties?

*10-7. Define buffer capacity.

10-8. Which has the greater buffer capacity: (a) a mixture containing 0.100 mol of NH$_3$ and 0.200 mol of NH$_4$Cl; or (b) a mixture containing 0.0500 mol of NH$_3$ and 0.100 mol of NH$_4$Cl?

*10-9. Consider solutions prepared by
  (a) dissolving 8.00 mmol of NaOAc in 200 mL of 0.100 M HOAc.
  (b) adding 100 mL of 0.0500 M NaOH to 100 mL of 0.175 M HOAc.
  (c) adding 40.0 mL of 0.1200 M HCl to 160.0 mL of 0.0420 M NaOAc.
  In what respects do these solutions resemble one another? How do they differ?

10-10. Consult Appendix 2 and pick out a suitable acid/base pair to prepare a buffer with a pH
  *(a) 3.5.   (b) 7.6.   *(c) 9.3.   (d) 5.1.

10-11. Which solute would provide the sharper end point in a titration with 0.10 M HCl?
  *(a) 0.10 M NaOCl or 0.10 M hydroxylamine
  (b) 0.10 M NH$_3$ or 0.10 M sodium phenolate
  *(c) 0.10 M methylamine or 0.10 M hydroxylamine
  (d) 0.10 M hydrazine or 0.10 M NaCN

10-12. Which solute would provide the sharper end point in a titration with 0.10 M NaOH?
  *(a) 0.10 M nitrous acid or 0.10 M iodic acid
  (b) 0.10 M anilinium hydrochloride (C$_6$H$_5$NH$_3$Cl) or 0.10 M benzoic acid
  *(c) 0.10 M hypochlorous acid or 0.10 M pyruvic acid
  (d) 0.10 M salicylic acid or 0.10 M acetic acid

10-13. Before glass electrodes and pH meters became so widely used, pH was often determined by measuring the concentration of the acid and base forms of the indicator colorimetrically. If bromothymol blue is introduced into a solution and the concentration ratio of acid to base form is found to be 1.43, what is the pH of the solution?

*10-14. The procedure described in 10-13 was used to determine pH with bromocresol green as the indicator. The concentration ratio of the acid to base form of the indicator was 1.64. Calculate the pH of the solution.

10-15. Values for K_w at 0°C, 50°C, and 100°C are 1.14×10^{-15}, 5.47×10^{-14}, and 4.9×10^{-13}, respectively. Calculate the pH for a neutral solution at each of these temperatures.

10-16. Calculate pK_w at
  *(a) 0°C.   (b) 50°C.   (c) 100°C.

10-17. Calculate the pH of a 1.00×10^{-2} M NaOH solution at
  *(a) 0°C.   (b) 50°C.   (c) 100°C.

---

† Unless otherwise instructed, round all calculated values for pH and pOH to two figures to the right of the decimal point.
* Answers to the asterisked problems are given in the answer section at the back of the book.

*10-18. What is the pH of an aqueous solution that is 14.0% HCl by weight and has a density of 1.054 g/mL?

10-19. Calculate the pH of a solution that contains 9.00% (w/w) NaOH and has a density of 1.098 g/mL.

*10-20. What is the pH of a solution that is 2.0×10^{-8} M in NaOH? (*Hint*: In such a dilute solution you must take into account the contribution of H_2O to the hydroxide ion concentration.)

10-21. What is the pH of a 2.0×10^{-8} M HCl solution?

*10-22. Calculate the pH of the solution that results after mixing 20.0 mL of 0.2000 M HCl with 25.0 mL of
  (a) distilled water.
  (b) 0.132 M $AgNO_3$.
  (c) 0.132 M NaOH.
  (d) 0.132 M NH_3.
  (e) 0.232 M NaOH.

10-23. What is the pH of the solution that results when 0.102 g of $Mg(OH)_2$ is mixed with
  (a) 75.0 mL of 0.0600 M HCl?
  (b) 15.0 mL of 0.0600 M HCl?
  (c) 30.0 mL of 0.0600 M HCl?
  (d) 30.0 mL of 0.0600 M $MgCl_2$?

*10-24. Calculate the hydronium ion concentration and pH of a solution that is 0.0500 M in HCl
  (a) neglecting activities.
  (b) using activities.

10-25. Calculate the hydroxide ion concentration and the pH of a 0.0167 M $Ba(OH)_2$ solution
  (a) neglecting activities.
  (b) using activities.

*10-26. Calculate the pH of a HOCl solution that is (a) 1.00×10^{-1} M, (b) 1.00×10^{-2} M, (c) 1.00×10^{-4} M.

10-27. Calculate the pH of a NaOCl solution that is (a) 1.00×10^{-1} M, (b) 1.00×10^{-2} M, (c) 1.00×10^{-4} M.

*10-28. Calculate the pH of an ammonia solution that is (a) 1.00×10^{-1} M, (b) 1.00×10^{-2} M, (c) 1.00×10^{-4} M.

10-29. Calculate the pH of an NH_4Cl solution that is (a) 1.00×10^{-1} M, (b) 1.00×10^{-2} M, (c) 1.00×10^{-4} M.

*10-30. Calculate the pH of a solution in which the concentration of piperidine is (a) 1.00×10^{-1} M, (b) 1.00×10^{-2} M, (c) 1.00×10^{-4} M.

10-31. Calculate the pH of an iodic acid solution that is (a) 1.00×10^{-1} M, (b) 1.00×10^{-2} M, (c) 1.00×10^{-4} M.

*10-32. Calculate the pH of a solution prepared by
  (a) dissolving 43.0 g of lactic acid in water and diluting to 500 mL.
  (b) diluting 25.0 mL of the solution in (a) to 250 mL.
  (c) diluting 10.0 mL of the solution in (b) to 1.00 L.

10-33. Calculate the pH of a solution prepared by
  (a) dissolving 1.05 g of picric acid, $(NO_2)_3C_6H_2OH$ (229.11 g/mol), in 100 mL of water.
  (b) diluting 10.0 mL of the solution in (a) to 100 mL.
  (c) diluting 10.0 mL of the solution in (b) to 1.00 L.

*10-34. Calculate the pH of the solution that results when 20.0 mL of 0.200 M formic acid are

  (a) diluted to 45.0 mL with distilled water.
  (b) mixed with 25.0 mL of 0.160 M NaOH solution.
  (c) mixed with 25.0 mL of 0.200 M NaOH solution.
  (d) mixed with 25.0 mL of 0.200 sodium formate solution.

10-35. Calculate the pH of the solution that results when 40.0 mL of 0.100 M NH_3 are
  (a) diluted to 60.0 mL with distilled water.
  (b) mixed with 20.0 mL of 0.200 M HCl solution.
  (c) mixed with 20.0 mL of 0.250 M HCl solution.
  (d) mixed with 20.0 mL of 0.200 M NH_4Cl solution.
  (e) mixed with 20.0 mL of 0.100 M HCl solution.

10-36. A solution is 0.0500 M in NH_4Cl and 0.0300 M in NH_3. Calculate its OH^- concentration and its pH
  (a) neglecting activities.
  (b) taking activities into account.

*10-37. What is the pH of a solution that is
  (a) prepared by dissolving 9.20 g of lactic acid (90.08 g/mol) and 11.15 g of sodium lactate (112.06 g/mol) in water and diluting to 1.00 L?
  (b) 0.0550 M in acetic acid and 0.0110 M in sodium acetate?
  (c) prepared by dissolving 3.00 g of salicylic acid, $C_6H_4(OH)COOH$ (138.12 g/mol), in 50.0 mL of 0.1130 M NaOH and diluting to 500.0 mL?
  (d) 0.0100 M in picric acid and 0.100 M in sodium picrate?

10-38. What is the pH of a solution that is
  (a) prepared by dissolving 3.30 g of $(NH_4)_2SO_4$ in water, adding 125.0 mL of 0.1011 M NaOH, and diluting to 500.0 mL?
  (b) 0.120 M in piperidine and 0.080 M in its chloride salt?
  (c) 0.050 M in ethylamine and 0.167 M in its chloride salt?
  (d) prepared by dissolving 2.32 g of aniline (93.13 g/mol) in 100 mL of 0.0200 M HCl and diluting to 250.0 mL?

10-39. Calculate the change in pH that occurs in each of the solutions listed below as a result of a tenfold dilution with water. Round calculated values for pH to three figures to the right of the decimal point.
  *(a) H_2O
  (b) 0.0500 M HCl
  *(c) 0.0500 M NaOH
  (d) 0.0500 M CH_3COOH
  *(e) 0.0500 M CH_3COONa
  (f) 0.0500 M CH_3COOH + 0.0500 M CH_3COONa
  *(g) 0.500 M CH_3COOH + 0.500 M CH_3COONa

*10-40. Calculate the change in pH that occurs when 1.00 mmol of a strong acid is added to 100 mL of the solutions listed in 10-39.

*10-41. Calculate the change in pH that occurs when 1.00 mmol of a strong base is added to 100 mL of the solutions listed in 10-39. Calculate values to three decimal places.

10-42. Calculate the change in pH to three decimal places that occurs when 0.500 mmol of a strong acid is added to 100

mL of
(a) 0.0200 M lactic acid + 0.0800 M sodium lactate.
*(b) 0.0800 M lactic acid + 0.0200 M sodium lactate.
(c) 0.0500 M lactic acid + 0.0500 M sodium lactate.

*10-43. What weight of sodium formate must be added to 400 mL of 1.00 M formic acid to produce a buffer solution that has a pH of 3.50?

10-44. What weight of sodium glycolate should be added to 300 mL of 1.00 M glycolic acid to produce a buffer solution with a pH of 4.00?

*10-45. What volume of 0.200 M HCl must be added to 250 mL of 0.300 M sodium mandelate to produce a buffer solution with a pH of 3.37?

10-46. What volume of 2.00 M NaOH must be added to 300 mL of 1.00 M glycolic acid to produce a buffer solution having a pH of 4.00?

*10-47. A 50.00-mL aliquot of 0.1000 M NaOH is titrated with 0.1000 M HCl. Calculate the pH of the solution after the addition of 0.00, 10.00, 25.00, 40.00, 45.00, 49.00, 50.00, 51.00, 55.00, and 60.00 mL of acid and prepare a titration curve from the data.

*10-48. In a titration of 50.00 mL of 0.05000 M formic acid with 0.1000 M KOH, the titration error must be smaller than ±0.05 mL. What indicator can be chosen to realize this goal?

10-49. In a titration of 50.00 mL of 0.1000 M ethylamine with 0.1000 M $HClO_4$, the titration error must be no more than ±0.05 mL. What indicator can be chosen to realize this goal?

10-50. Calculate the pH after additon of 0.00, 5.00, 15.00, 25.00, 40.00, 45.00, 49.00, 50.00, 51.00, 55.00, and 60.00 mL of 0.1000 M NaOH in the titration of 50.00 mL of

*(a) 0.1000 M HNO_2.
(b) 0.1000 M lactic acid.
*(c) 0.1000 M pyridinium chloride.

10-51. Calculate the pH after addition of 0.00, 5.00, 15.00, 25.00, 40.00, 45.00, 49.00, 50.00, 51.00, 55.00, and 60.00 mL of 0.1000 M HCl in the titration of 50.00 mL of
*(a) 0.1000 M ammonia.
(b) 0.1000 M hydrazine.
(c) 0.1000 M sodium cyanide.

10-52. Calculate the pH after addition of 0.00, 5.00, 15.00, 25.00, 40.00, 49.00, 50.00, 51.00, 55.00, and 60.00 mL of reagent in the titration of 50.0 mL of
*(a) 0.1000 M anilinium chloride with 0.1000 M NaOH.
(b) 0.01000 M chloroacetic acid with 0.01000 M NaOH.
*(c) 0.1000 M hypochlorous acid with 0.1000 M NaOH.
(d) 0.1000 M hydroxylamine with 0.1000 M HCl.
Construct titration curves from the data.

10-53. Calculate α_0 and α_1 for
*(a) acetic acid species in a solution with a pH of 5.320.
(b) picric acid species in a solution with a pH of 1.250.
*(c) hypochlorous acid species in a solution with a pH of 7.000.
(d) hydroxylamine acid species in a solution with a pH of 5.120.
*(e) piperidine species in a solution with a pH of 10.080.

*10-54. Calculate the equilibrium concentration of undissociated HCOOH in a formic acid solution with an analytical formic acid concentration of 0.0850 and a pH of 3.200.

10-55. Calculate the equilibrium concentration of methylamine in a solution that has a molar analytical CH_3NH_2 concentration of 0.120 and a pH of 11.471.

10-56. Supply the missing data for the table below.

Acid	Molar Analytical Concentration, c_T ($c_T = c_{HA} + c_{A^-}$)	pH	[HA]	[A⁻]	α_0	α_1
Lactic	0.120				0.640	
Iodic	0.200					0.765
Butanoic		5.00	0.0644			
Hypochlorous	0.280	7.00				
Nitrous				0.105	0.413	0.587
Hydrogen cyanide			0.145	0.221		
Sulfamic	0.250	1.20				

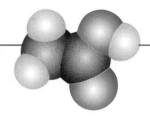

CHAPTER 11

TITRATION CURVES FOR POLYFUNCTIONAL ACIDS AND POLYFUNCTIONAL BASES

In this chapter, we describe methods for deriving titration curves for acids or bases that have two or more acidic or basic functional groups.

11A POLYFUNCTIONAL ACIDS AND BASES

Phosphoric acid is a typical polyfunctional acid. In aqueous solution it undergoes the following three dissociation reactions:

$$H_3PO_4 + H_2O \rightleftharpoons H_2PO_4^- + H_3O^+ \qquad K_{a1} = \frac{[H_3O^+][H_2PO_4^-]}{[H_3PO_4]}$$

$$K_{a1} = 7.11 \times 10^{-3}$$

$$H_2PO_4^- + H_2O \rightleftharpoons HPO_4^{2-} + H_3O^+ \qquad K_{a2} = \frac{[H_3O^+][HPO_4^{2-}]}{[H_2PO_4^-]}$$

$$K_{a1} = 6.32 \times 10^{-8}$$

$$HPO_4^{2-} + H_2O \rightleftharpoons PO_4^{3-} + H_3O^+ \qquad K_{a3} = \frac{[H_3O^+][PO_4^{3-}]}{[HPO_4^{2-}]}$$

$$K_{a1} = 4.5 \times 10^{-13}$$

With this acid, as with other polyprotic acids, $K_{a1} > K_{a2} > K_{a3}$.

Polyfunctional bases are also common, an example being sodium carbonate. Carbonate ion, the conjugate base of the hydrogen carbonate ion, is involved in

Throughout the remainder of this chapter we will find it useful to use K_{a1}, K_{a2}, and so forth to represent the first and second dissociation constants of acids and K_{b1}, K_{b2}, and so forth to represent the stepwise constants for bases.

Generally, $K_{a1} > K_{a2}$, often by a factor of 10^4 to 10^5 because of electrostatic forces. That is, the first dissociation involves separating a single positively charged hydronium ion from a singly charged anion. In the second step, a hydronium ion is separated from a doubly charged anion, a process that requires considerably more energy.

195

A second reason that $K_{a1} > K_{a2}$ is a statistical one. In the first step, a proton can be removed from two locations, whereas in the second step, only from one. Thus, the first dissociation is twice as probable as the second.

Feature 11-1
COMBINING EQUILIBRIUM-CONSTANT EXPRESSIONS

When two adjacent stepwise equilibria are *added*, the equilibrium constant for the resulting overall reaction is the *product* of the two constants. Thus for the first two dissociation equilibria for H_3PO_4, we may write

$$H_3PO_4 + H_2O \rightleftharpoons H_2PO_4^- + H_3O^+$$
$$\underline{H_2PO_4^- + H_2O \rightleftharpoons HPO_4^{2-} + H_3O^+}$$
$$H_3PO_4 + 2H_2O \rightleftharpoons HPO_4^{2-} + 2H_3O^+$$

and

$$K_{a1}K_{a2} = \frac{[H_3O^+]^2[HPO_4^{2-}]}{[H_3PO_4]}$$
$$= 7.11 \times 10^{-3} \times 6.32 \times 10^{-8} = 4.49 \times 10^{-10}$$

Similarly, for the reaction

$$H_3PO_4 + 3H_2O \rightleftharpoons 3H_3O^+ + PO_4^{3-}$$

we may write

$$K_{a1}K_{a2}K_{a3} = \frac{[H_3O^+]^3[PO_4^{3-}]}{H_3PO_4}$$
$$= 7.11 \times 10^{-3} \times 6.32 \times 10^{-8} \times 4.5 \times 10^{-13} = 2.0 \times 10^{-22}$$

the stepwise equilibria:

$$CO_3^{2-} + H_2O \rightleftharpoons HCO_3^- + OH^- \qquad K_{b1} = \frac{[HCO_3^-][OH^-]}{[CO_3^{2-}]} = \frac{K_w}{K_{a2}}$$
$$= \frac{1.00 \times 10^{-14}}{4.69 \times 10^{-11}} = 2.13 \times 10^{-4}$$

$$HCO_3^- + H_2O \rightleftharpoons H_2CO_3 + OH^- \qquad K_{b2} = \frac{[H_2CO_3][OH^-]}{[HCO_3^-]} = \frac{K_w}{K_{a1}}$$
$$= \frac{1.00 \times 10^{-14}}{4.45 \times 10^{-7}} = 2.25 \times 10^{-8}$$

where K_{a1} and K_{a2} are the first and second dissociation constants for carbonic acid and K_{b1} and K_{b2} are the first and second dissociation constants of the base CO_3^{2-}.

The overall basic dissociation reaction of sodium carbonate is described by the equations

$$CO_3^{2-} + 2H_2O \rightleftharpoons H_2CO_3 + 2OH^- \qquad K_{b1}K_{b2} = \frac{[H_2CO_3][OH^-]^2}{[CO_3^{2-}]}$$

$$= 2.13 \times 10^{-4} \times 2.25 \times 10^{-8}$$

$$= 4.79 \times 10^{-12}$$

The pH of polyfunctional systems, such as phosphoric acid or sodium carbonate, can be computed rigorously through use of the systematic approach to multiequilibrium problems described in Chapter 7. Solution of the several simultaneous equations that are involved is difficult and time consuming, however. Fortunately, simplifying assumptions can be invoked when the successive equilibrium constants for the acid (or base) differ by a factor of about 10^3 (or more). With one exception, these assumptions make it possible to derive pH data by the techniques we have discussed in earlier chapters. The exception, which we now consider, is the pH of solutions of amphiprotic salts of the type NaHA.

Challenge: Write a sufficient number of equations to make possible the calculation of all of the species present in a solution containing known molar analytical concentrations of Na_2CO_3 and $NaHCO_3$.

11B CALCULATION OF THE pH OF SOLUTIONS OF NaHA

Thus far, we have not considered how to calculate the pH of solutions of salts that have both acidic and basic properties—that is, salts that are *amphiprotic*. Such salts are formed during neutralization titration of polyfunctional acids and bases. For example, when 1 mol of NaOH is added to a solution containing 1 mol of the acid H_2A, 1 mol of NaHA is formed. The pH of this solution is determined by two equilibria established between HA^- and water:

An amphiprotic salt is a species that can act as an acid and as a base when dissolved in a suitable solvent.

$$HA^- + H_2O \rightleftharpoons A^{2-} + H_3O^+$$

and

$$HA^- + H_2O \rightleftharpoons H_2A + OH^-$$

One of these reactions produces hydronium ions and the other hydroxide ions. A solution of NaHA will be acidic or basic, depending on the relative magnitude of the equilibrium constants for these processes:

$$K_{a2} = \frac{[H_3O^+][A^{2-}]}{[HA^-]} \tag{11-1}$$

$$K_{b2} = \frac{K_w}{K_{a1}} = \frac{[H_2A][OH^-]}{[HA^-]} \tag{11-2}$$

where K_{a1} and K_{a2} are the acid dissociation constants for H_2A. If K_{b2} is greater than K_{a2}, the solution is basic; otherwise, it is acidic.

A solution of NaHA contains five species ($[H_3O^+]$, $[OH^-]$, $[H_2A]$, $[HA^-]$, and $[A^{2-}]$), and therefore five independent equations are needed for a rigorous computation of the hydronium ion concentration. In Feature 11-2, we show that an approximate value for $[H_3O^+]$ is given by the equation

$$[H_3O^+] = \sqrt{\frac{K_{a2}c_{NaHA} + K_w}{1 + c_{NaHA}/K_{a1}}}$$ (11-3)

where c_{NaHA} is the molar concentration of the salt NaHA. This equation should be used when NaHA is the *only* solute species that contributes significantly to the pH of the solution. It is not valid for very dilute solutions of NaHA or for systems in which K_{a2} or K_w/K_{a1} is relatively large.

Feature 11-2
HOW WAS EQUATION 11-3 GENERATED?

A solution of NaHA can be described in terms of mass balance:

$$c_{NaHA} = [H_2A] + [HA^-] + [A^{2-}]$$ (11-4)

and charge balance:

$$[Na^+] + [H_3O^+] = [HA^-] + 2[A^{2-}] + [OH^-]$$

Since the sodium ion concentration is equal to the molar analytical concentration of the salt, the last equation can be rewritten as

$$c_{NaHA} + [H_3O^+] = [HA^-] + 2[A^{2-}] + [OH^-]$$ (11-5)

We now have four algebraic equations (Equations 11-4 and 11-5 and the two dissociation-constant expressions for H_2A) and need one additional to solve for the five unknowns. The ion-product constant for water serves this purpose:

$$K_w = [H_3O^+][OH^-]$$

The rigorous computation of the hydronium ion concentration from these five equations is difficult. However, a reasonable approximation, applicable to solutions of most acid salts, can be obtained as follows.

We first subtract the mass-balance equation from the charge-balance equation.

$$\begin{aligned} c_{NaHA} + [H_3O^+] &= [HA^-] + 2[A^{2-}] + [OH^-] \quad \text{charge balance} \\ c_{NaHA} &= [H_2A] + [HA^-] + [A^{2-}] \quad \text{mass balance} \\ \hline [H_3O^+] &= [A^{2-}] + [OH^-] - [H_2A] \end{aligned}$$ (11-6)

We then rearrange the acid-dissociation constant expressions for H_2A to obtain

$$[H_2A] = \frac{[H_3O^+][HA^-]}{K_{a1}}$$

and for HA^- to give

$$[A^{2-}] = \frac{K_{a2}[HA^-]}{[H_3O^+]}$$

Substituting these expressions and the expression for K_w into Equation 11-6 yields

$$[H_3O^+] = \frac{K_{a2}[HA^-]}{[H_3O^+]} + \frac{K_w}{[H_3O^+]} - \frac{[H_3O^+][HA^-]}{K_{a1}}$$

Multiplication through by $[H_3O^+]$ gives

$$[H_3O^+]^2 = K_{a2}[HA^-] + K_w - \frac{[H_3O^+[^2[HA^-]}{K_{a1}}$$

We collect terms to obtain

$$[H_3O^+]^2\left(\frac{[HA^-]}{K_{a1}} + 1\right) = K_{a2}[HA^-] + K_w$$

Finally, this equation rearranges to

$$[H_3O^+] = \sqrt{\frac{K_{a2}[HA^-] + K_w}{1 + [HA^-]/K_{a1}}} \qquad (11\text{-}7)$$

Under most circumstances, we can assume that

$$[HA^-] \cong c_{NaHA} \qquad (11\text{-}8)$$

Introduction of this relationship into Equation 11-7 gives

$$[H_3O^+] = \sqrt{\frac{K_{a2}c_{NaHA} + K_w}{1 + c_{NaHA}/K_{a1}}} \qquad (11\text{-}9)$$

It is important to understand that the approximation shown as Equation 11-8 requires that $[HA^-]$ be much larger than any of the other equilibrium concentrations in Equations 11-4 and 11-5. This assumption is not valid for very dilute solutions of NaHA or when K_{a2} or K_w/K_{a1} is relatively large.

Frequently, the ratio c_{NaHA}/K_{a1} is much larger than unity and $K_{a2}c_{NaHA}$ is considerably greater than K_w in Equation 11-3. With these assumptions, the equation simplifies to

Make sure that you always check the assumptions that are inherent in Equation 11-10.

$$[H_3O^+] \cong \sqrt{K_{a1}K_{a2}} \qquad (11\text{-}10)$$

Note that Equation 11-10 does not contain c_{NaHA}, which implies that the pH of solutions of this type remains constant over a considerable range of NaHA concentration.

Example 11-1

Calculate the hydronium ion concentration of a 0.100 M $NaHCO_3$ solution.

We first examine the assumptions leading to Equation 11-10. The dissociation constants for H_2CO_3 are $K_{a1} = 4.45 \times 10^{-7}$ and $K_{a2} = 4.69 \times 10^{-11}$. Clearly, c_{NaHA}/K_{a1} in the denominator is much larger than unity; in addition, $K_{a2}c_{NaHA}$ has a value of 4.69×10^{-12}, which is substantially greater than K_w. Thus Equation 11-10 applies and

$$[H_3O^+] = \sqrt{4.45 \times 10^{-7} \times 4.69 \times 10^{-11}} = 4.6 \times 10^{-9}$$

Example 11-2

Calculate the hydronium ion concentration of a 1.00×10^{-3} M Na_2HPO_4 solution.

The pertinent dissociation constants are K_{a2} and K_{a3}, which both contain $[HPO_4^{2-}]$. Their values are $K_{a2} = 6.32 \times 10^{-8}$ and $K_{a3} = 4.5 \times 10^{-13}$. Considering again the assumptions that led to Equation 11-10, we find that $(1.0 \times 10^{-3})/(6.32 \times 10^{-8})$ is much larger than 1 so that the denominator can be simplified. However, the product $K_{a3}c_{Na_2HPO_4}$ is by no means much larger than K_w. We therefore use a partially simplified version of Equation 11-3:

$$[H_3O^+] = \sqrt{\frac{4.5 \times 10^{-13} \times 1.00 \times 10^{-3} + 1.00 \times 10^{-14}}{(1.00 \times 10^{-3})/(6.32 \times 10^{-8})}} = 8.1 \times 10^{-10}$$

Use of Equation 11-10 yields a value of 1.7×10^{-10} M.

Example 11-3

Find the hydronium ion concentration of a 0.0100 M NaH_2PO_4 solution.

The two dissociation constants of importance (those containing $[H_2PO_4^-]$) are $K_{a1} = 7.11 \times 10^{-3}$ and $K_{a2} = 6.32 \times 10^{-8}$. We see that the denominator of Equation 11-3 cannot be simplified, but the numerator reduces to $K_{a2}c_{NaH_2PO_4}$. Thus, Equation 11-3 becomes

$$[H_3O^+] = \sqrt{\frac{6.32 \times 10^{-8} \times 1.00 \times 10^{-2}}{1.00 + (1.00 \times 10^{-2})/(7.11 \times 10^{-3})}} = 1.62 \times 10^{-5}$$

Use of Equation 11-10 yields a value of 2.12×10^{-5}.

11C   TITRATION CURVES FOR POLYFUNCTIONAL ACIDS

Compounds with two or more acid functional groups can yield multiple end points in a titration, provided the functional groups differ sufficiently in strengths as acids. The computational techniques described in Chapter 10 permit derivation of reasonably accurate theoretical titration curves for polyprotic acids if the ratio K_{a1}/K_{a2} is somewhat greater than 10^3. If this ratio is smaller, the error, particularly in the region of the first equivalence point, becomes excessive, and a more rigorous treatment of the equilibrium relationships is required.

Figure 11-1 shows the titration curve for a diprotic acid H_2A with dissociation constants of $K_{a1} = 1.00 \times 10^{-3}$ and $K_{a2} = 1.00 \times 10^{-7}$. Because the K_{a1}/K_{a2} is significantly greater than 10^3, we can derive this curve (except for the first equivalence point) using the techniques developed in Chapter 10 for simple monoprotic weak acids. Thus, to obtain the initial pH (point A), we treat the system as if it contained a single monoprotic acid with a dissociation constant of $K_{a1} = 1.00 \times 10^{-3}$. In region B we have the equivalent of a simple buffer solution consisting of the weak acid H_2A and its conjugate base NaHA. That is, we assume that the concentration of A^{2-} is negligible with respect to the other two A-containing species and employ Equation 9-10 to obtain $[H_3O^+]$. At the first equivalence point (point C), we have a solution of an acid salt and use Equation 11-3 or one of its simplifications to compute the hydronium ion concentration. In the region labeled D, we have a second buffer consisting of a weak acid HA^- and its conjugate base Na_2A and we calculate the pH employing the second dissociation constant, $K_{a2} = 1.00 \times 10^{-7}$. At point E, the solution contains the conjugate base of a weak acid with a dissociation constant of 1.00×10^{-7}. That is, we assume that the hydroxide concentration of the solution is determined

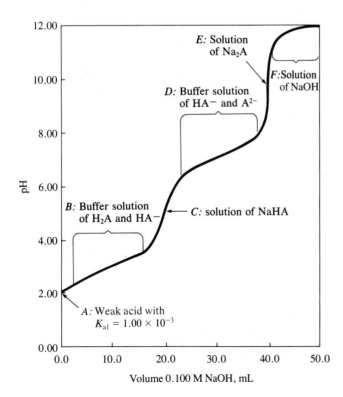

Figure 11-1

Titration of 20.0 mL of 0.100 M H_2A with 0.100 M NaOH. For H_2A, $K_{a1} = 1.00 \times 10^{-3}$ and $K_{a2} = 1.00 \times 10^{-7}$. Method of pH calculation is shown for several points and regions on the titration curve.

solely by the reaction of A^{2-} with water to form HA^- and OH^-. Finally in the region labeled F, we compute the hydroxide concentration from the molarity of the NaOH and derive the pH from this quantity.

Example 11-4

Derive a curve for the titration of 25.00 mL of 0.1000 M maleic acid, $HOOC{-}CH{=}CH{-}COOH$, with 0.1000 M NaOH.

Symbolizing the acid as H_2M, we can write the two dissociation equilibria as

$$H_2M + H_2O \rightleftharpoons H_3O^+ + HM^- \qquad K_{a1} = 1.3 \times 10^{-2}$$
$$HM^- + H_2O \rightleftharpoons H_3O^+ + M^{2-} \qquad K_{a2} = 5.9 \times 10^{-7}$$

Because the ratio K_{a1}/K_{a2} is large (2×10^4), we proceed as just described.

Initial pH
Only the first dissociation makes an appreciable contribution to $[H_3O^+]$. Thus,

$$[H_3O^+] \cong [HM^-]$$

Mass balance requires that

$$[H_2M] + [HM^-] \cong 0.1000$$

or

$$[H_2M] = 0.1000 - [HM^-] = 0.1000 - [H_3O^+]$$

Substituting these relationships into the expression for K_{a1} gives

$$K_{a1} = 1.3 \times 10^{-2} = \frac{[H_3O^+]^2}{0.1000 - [H_3O^+]}$$

Rearranging yields

$$[H_3O^+]^2 + 1.3 \times 10^{-2}\,[H_3O^+] - 1.3 \times 10^{-3} = 0$$

Because K_{a1} for maleic acid is large, we must solve the quadratic equation exactly or by successive approximations. When we do so, we obtain

$$[H_3O^+] = 3.01 \times 10^{-2}$$
$$pH = 2 - \log 3.01 = 1.52$$

First Buffer Region
The addition of 5.00 mL of base results in the formation of a buffer consisting of the weak acid H_2M and its conjugate base HM^-. To the extent that dissociation of HM^- to give M^{2-} is negligible, the solution can be treated as a simple buffer

system. Thus applying Equations 10-8 and 10-9 gives

$$c_{NaHM} \cong [HM^-] = \frac{5.00 \times 0.1000}{30.00} = 1.67 \times 10^{-2} \text{ M}$$

$$c_{H_2M} \cong [H_2M] = \frac{25.00 \times 0.1000 - 5.00 \times 0.1000}{30.00} = 6.67 \times 10^{-2} \text{ M}$$

Substitution of these values into the equilibrium-constant expression for K_{a1} yields a tentative value of 5.2×10^{-2} M for $[H_3O^+]$. It is clear, however, that the approximation $[H_3O^+] \ll c_{H_2M}$ or c_{HM^-} is not valid; therefore Equations 10-6 and 10-7 must be used, and

$$[HM^-] = 1.67 \times 10^{-2} + [H_3O^+] - [OH^-]$$
$$[H_2M] = 6.67 \times 10^{-2} - [H_3O^+] + [OH^-]$$

Because the solution is quite acidic, the approximation that $[OH^-]$ is very small is surely justified. Substitution of these expressions into the dissociation-constant relationship gives

$$\frac{[H_3O^+](1.67 \times 10^{-2} + [H_3O^+])}{6.67 \times 10^{-2} - [H_3O^+]} = 1.3 \times 10^{-2} = K_{a1}$$

$$[H_3O^+]^2 + 2.97 \times 10^{-2}[H_3O^+] - 8.67 \times 10^{-4} = 0$$

$$[H_3O^+] = 1.81 \times 10^{-2}$$

$$pH = -\log 1.81 \times 10^{-2} = 1.74$$

Additional points in the first buffer region can be computed in a similar way.

First Equivalence Point

At the first equivalence point,

$$[HM^-] \cong c_{NaHM} = \frac{25.00 \times 0.1000}{50.00} = 5.00 \times 10^{-2}$$

Simplification of the numerator in Equation 11-3 is clearly justified. On the other hand, the second term in the denominator is not $\ll 1$. Hence,

$$[H_3O^+] \cong \sqrt{\frac{K_{a2}c_{NaHM}}{1 + c_{NaHM}/K_{a1}}} = \sqrt{\frac{5.9 \times 10^{-7} \times 5.00 \times 10^{-2}}{1 + (5.00 \times 10^{-2})/(1.3 \times 10^{-2})}}$$

$$= 7.80 \times 10^{-5}$$

$$pH = -\log 7.80 \times 10^{-5} = 4.11$$

Second Buffer Region

Further additions of base to the solution create a new buffer system consisting of HM^- and M^{2-}. When enough base has been added so that the reaction of HM^- with water to give OH^- can be neglected (a few tenths of a milliliter beyond the first equivalence point), the pH of the mixture is readily obtained from K_{a2}.

With the introduction of 25.50 mL of NaOH, for example,

$$[M^{2-}] \cong c_{Na_2M} \cong \frac{(25.50 - 25.00)(0.1000)}{50.50} = \frac{0.050}{50.50} \, M$$

and the molar concentration of NaHM is

$$[HM^-] \cong c_{NaHM} \cong \frac{(25.00 \times 0.1000) - (25.50 - 25.00)(0.1000)}{50.50} = \frac{2.45}{50.50} \, M$$

Substituting these values into the expression for K_{a2} gives

$$\frac{[H_3O^+](0.050/50.50)}{2.45/50.50} = 5.9 \times 10^{-7}$$

$$[H_3O^+] = 2.89 \times 10^{-5}$$

The assumption that $[H_3O^+]$ is small relative to c_{HM^-} and $c_{M^{2-}}$ is valid and pH = 4.54.

Second Equivalence Point

After the addition of 50.00 mL of 0.1000 M sodium hydroxide, the solution is 0.0333 M in Na$_2$M. Reaction of the base M^{2-} with water is the predominant equilibrium in the system and the only one that we need to take into account. Thus,

$$M^{2-} + H_2O \rightleftharpoons OH^- + HM^-$$

$$\frac{[OH^-][HM^-]}{[M^{2-}]} = \frac{K_w}{K_{a2}} = \frac{1.00 \times 10^{-14}}{5.9 \times 10^{-7}} = 1.69 \times 10^{-8} = K_{b1}$$

$$[OH^-] \cong [HM^-]$$

$$[M^{2-}] = 0.0333 - [OH^-] \cong 0.0333$$

$$\frac{[OH^-]^2}{0.0333} = \frac{1.00 \times 10^{-14}}{5.9 \times 10^{-7}}$$

$$[OH^-] = 2.38 \times 10^{-5}$$

$$pH = 14.00 - (-\log 2.38 \times 10^{-5}) = 9.38$$

pH Beyond the Second Equivalence Point

Further additions of sodium hydroxide repress the basic dissociation of M^{2-}. The pH is calculated from the concentration of NaOH added in excess of that required for the complete neutralization of H$_2$M. Thus when 51.00 mL of NaOH have been added, we have a 1.00-mL excess of 0.1000 M NaOH and

$$[OH^-] = \frac{1.00 \times 0.1000}{76.00} = 1.32 \times 10^{-3}$$

$$pH = 14.00 - (-\log 1.32 \times 10^{-3}) = 11.12$$

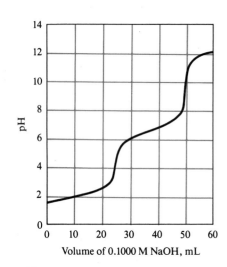

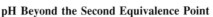

Figure 11-2

Titration curve for 25.00 mL of 0.1000 M maleic acid, H$_2$M, with 0.1000 M NaOH.

Figure 11-2 is the titration curve for 0.1000 M maleic acid derived as shown in Example 11-4. Two end points are apparent, either of which in principle could be used as a measure of the concentration of the acid. The second end point is clearly more satisfactory, however, inasmuch as the pH change is more pronounced.

Figure 11-3 shows titration curves for three other polyprotic acids. These curves illustrate that a well-defined end point corresponding to the first equivalence point is observed only when the degree of dissociation of the two acids is sufficiently different. The ratio of K_{a1} to K_{a2} for oxalic acid (curve B) is approximately 1000. The curve for this titration shows an inflection corresponding to the first equivalence point. However, the magnitude of the pH change is too small to permit precise location of equivalence with an indicator. The second end point, however, provides a means for the accurate determination of oxalic acid.

Curve A is the theoretical titration curve for triprotic phosphoric acid. Here, the ratio K_{a1}/K_{a2} is approximately 10^5, as is K_{a2}/K_{a3}. Because these ratios are large, two well-defined end points, either of which is satisfactory for analytical purposes, are observed. An acid-range indicator will provide a color change when 1 mol of base has been introduced for each mole of acid. A base-range indicator will require 2 mol of base per mole of acid. The third hydrogen of phosphoric acid is so slightly dissociated ($K_{a3} = 4.5 \times 10^{-13}$) that no practical end point is associated with its neutralization. The buffering effect of the third dissociation is noticeable however, and causes the pH for curve A to be lower than the pH for the other two curves in the region beyond the second equivalence point.

Curve C is the titration curve for sulfuric acid, a substance that has one fully dissociated proton and one that is dissociated to a relatively large extent ($K_{a2} = 1.02 \times 10^{-2}$). Because of the similarity in strengths of the two acids, only a single end point, corresponding to the titration of both protons, is observed.

In general, the titration of acids or bases that have two reactive groups yields individual end points that are of practical value only when the ratio between the two dissociation constants is at least 10^4. If the ratio is much smaller than this, the pH change at the first equivalence point will prove less satisfactory for an analysis.

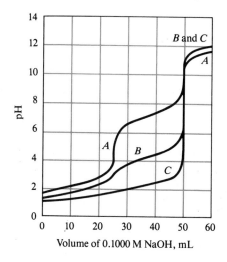

Figure 11-3

Curves for the titration of polyprotic acids. A 0.1000 M NaOH solution is used to titrate 25.00 mL of 0.1000 M H_3PO_4 (A), 0.1000 M oxalic acid (B), and 0.1000 M H_2SO_4 (C).

Feature 11-3

Sulfuric acid is unusual in that one of its protons behaves as a strong acid and the other as a weak acid ($K_{a2} = 1.02 \times 10^{-2}$). Let us consider how the hydronium ion concentration of sulfuric acid solutions is computed using a 0.0400 M solution as an example.

We will first assume that the dissociation of HSO_4^- is negligible because of the large excess of H_3O^+ resulting from the complete dissociation of H_2SO_4. Therefore,

$$[H_3O^+] \cong [HSO_4^-] \cong 0.0400$$

However, an estimate of $[SO_4^{2-}]$ based on this approximation and the expression for K_{a2} reveals that

$$\frac{0.0400\ [SO_4^{2-}]}{0.0400} = 1.02 \times 10^{-2}$$

Clearly, $[SO_4^{2-}]$ is *not* small relative to $[HSO_4^-]$, and a more rigorous solution is required.

From stoichiometric considerations, it is necessary that

$$[H_3O^+] = 0.0400 + [SO_4^{2-}]$$

The first term on the right is the concentration of H_3O^+ from dissociation of the H_2SO_4 to HSO_4^-. The second term is the contribution of the dissociation of HSO_4^-. Rearrangement yields

$$[SO_4^{2-}] = [H_3O^+] - 0.0400$$

Mass-balance considerations require that

$$c_{H_2SO_4} = 0.0400 = [HSO_4^-] + [SO_4^{2-}]$$

Combining the last two equations and rearranging yields

$$[HSO_4^-] = 0.0800 - [H_3O^+]$$

Introduction of these equations for $[SO_4^{2-}]$ and $[HSO_4^-]$ into the expression for K_{a2} yields

$$\frac{[H_3O^+]([H_3O^+] - 0.0400)}{0.0800 - [H_3O^+]} = 1.02 \times 10^{-2}$$

$$[H_3O^+]^2 - (0.0298)[H_3O^+] - 8.16 \times 10^{-4} = 0$$

$$[H_3O^+] = 0.0471$$

11D TITRATION CURVES FOR POLYFUNCTIONAL BASES

The derivation of a titration curve for a polyfunctional base involves no new principles. To illustrate, consider the titration of a sodium carbonate solution with standard hydrochloric acid. The important equilibrium constants are

$$CO_3^{2-} + H_2O \rightleftharpoons OH^- + HCO_3^- \qquad K_{b1} = \frac{K_w}{K_{a2}} = \frac{1.00 \times 10^{-14}}{4.69 \times 10^{-11}} = 2.13 \times 10^{-4}$$

$$HCO_3^- + H_2O \rightleftharpoons OH^- + H_2CO_3 \qquad K_{b2} = \frac{K_w}{K_{a1}} = \frac{1.00 \times 10^{-14}}{4.45 \times 10^{-7}} = 2.25 \times 10^{-8}$$

The reaction of carbonate ion with water governs the initial pH of the solution, which can be computed by the method shown for the second equivalence point in Example 11-4. With the first additions of acid, a carbonate/hydrogen carbonate buffer is established. In this region, the pH can be computed from *either* the hydroxide ion concentration calculated from K_{b1} *or* the hydronium ion concentration calculated from K_{a2}. Because we are usually interested in calculating $[H_3O^+]$ and pH, the expression for K_{a2} is easier to use.

Sodium hydrogen carbonate is the principal solute species at the first equivalence point, and Equation 11-10 is used to compute the hydronium ion concentration (see Example 11-1). With the addition of more acid, a new buffer consisting of sodium hydrogen carbonate and carbonic acid is formed. The pH of this buffer is readily obtained from either K_{b2} or K_{a1}.

At the second equivalence point, the solution consists of carbonic acid and sodium chloride. The carbonic acid can be treated as a simple weak acid having a dissociation constant K_{a1}. Finally, after excess hydrochloric acid has been introduced, the dissociation of the weak acid is repressed to a point where the hydronium ion concentration is essentially that of the molar concentration of the strong acid.

Figure 11-4 illustrates that two end points are observed in the titration of sodium carbonate, the second being appreciably sharper than the first. It is apparent that the individual components in mixtures of sodium carbonate and sodium hydrogen carbonate can be determined by neutralization methods.

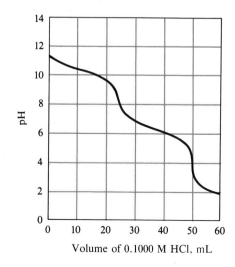

Figure 11-4

Curve for the titration of 25.00 mL of 0.1000 M Na_2CO_3 with 0.1000 M HCl.

11E THE COMPOSITION OF SOLUTIONS OF A POLYPROTIC ACID AS A FUNCTION OF pH

In Section 10F, we showed how alpha values are useful in visualizing the changes in the concentration of various species that occur in a titration of a simple weak acid. Alpha values can also be derived for polyfunctional acids and bases. For example, if we let c_T be the sum of the molar concentrations of the maleate-containing species in the solution throughout the titration described in Example 11-4, the alpha value for the free acid is

$$\alpha_0 = \frac{[H_2M]}{c_T}$$

where

$$c_T = [H_2M] + [HM^-] + [M^{2-}] \tag{11-11}$$

The alpha values for HM^- and M^{2-} are given by similar equations:

$$\alpha_1 = \frac{[HM^-]}{c_T}$$

$$\alpha_2 = \frac{[M^{2-}]}{c_T}$$

As noted earlier, the sum of the alpha values for a system must equal unity:

$$\alpha_0 + \alpha_1 + \alpha_2 = 1$$

The alpha values for the maleic acid system are readily expressed in terms of $[H_3O^+]$, K_{a1}, and K_{a2}. To obtain such expressions, we follow the method used to

---

Feature 11-4
A GENERAL EXPRESSION FOR ALPHA VALUES

For the weak acid H_nA, the denominator in all alpha-value expressions takes the form:

$$[H_3O^+]^n + K_{a1}[H_3O^+]^{(n-1)} + K_{a1}K_{a2}[H_3O^+]^{(n-2)} + \cdots K_{a1}K_{a2} \cdots K_{an}$$

The numerator for α_0 is the first term in the denominator; for α_1, it is the second term; and so forth. Thus, if we let D be the denominator, $\alpha_0 = [H_3O^+]^n/D$ and $\alpha_1 = K_{a1}[H_3O^+]^{(n-1)}/D$.

Alpha values for polyfunctional bases are generated in an analogous way, with the equations written in terms of base dissociation constants and $[OH^-]$.

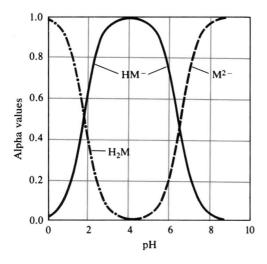

Figure 11-5

Composition of H_2M solutions as a function of pH.

derive Equations 10-16 and 10-17 in Section 10F and obtain

Challenge: Derive Equations 11-12, 11-13, and 11-14.

$$\alpha_0 = \frac{[H_3O^+]^2}{[H_3O^+]^2 + K_{a1}[H_3O^+] + K_{a1}K_{a2}} \qquad \textbf{(11-12)}$$

$$\alpha_1 = \frac{K_{a1}[H_3O^+]}{[H_3O^+]^2 + K_{a1}[H_3O^+] + K_{a1}K_{a2}} \qquad \textbf{(11-13)}$$

$$\alpha_2 = \frac{K_{a1}K_{a2}}{[H_3O^+]^2 + K_{a1}[H_3O^+] + K_{a1}K_{a2}} \qquad \textbf{(11-14)}$$

Note that the denominator is the same for each expression. Note also that the fractional amount of each species is fixed at any pH and is *independent* of the total concentration, c_T.

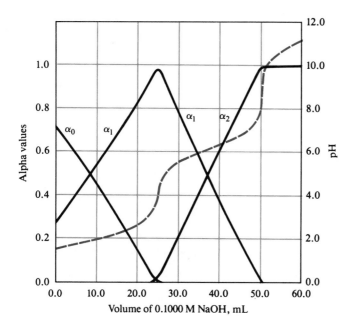

Figure 11-6

Titration of 25.00 mL of 0.1000 M maleic acid with 0.1000 M NaOH. The solid curves are plots of alpha values as a function of volume. The broken curve is a plot of pH as a function of volume.

The three curves plotted in Figure 11-5 show the alpha value for each maleate-containing species as a function of pH. The solid curves in Figure 11-6 depict the same alpha values but now plotted as a function of volume of sodium hydroxide as the acid is titrated. The titration curve is also shown by the dashed line in Figure 11-6. Consideration of these curves gives a clear picture of all concentration changes that occur during the titration. For example, Figure 11-6 reveals that before the addition of any base, α_0 for H_2M is roughly 0.7 and α_1 for HM^- is approximately 0.3. For all practical purposes, α_2 is zero. Thus, approximately 70% of the maleic acid exists as H_2M and 30% as HM^-. With addition of base, the pH rises, as does the fraction of HM^-. At the first equivalence point (pH = 4.11), essentially all of the maleate is present as HM^- ($\alpha_1 \rightarrow 1$). Beyond the first equivalence point, α_1 decreases and α_2 increases. At the second equivalence point (pH = 9.38) and beyond, essentially all of the maleate exists as M^{2-}.

11F QUESTIONS AND PROBLEMS

*11-1. As its name implies, NaHA is an "acid salt" because it has a proton available to donate to a base. Briefly explain why a pH calculation for a solution of NaHA differs from that for a weak acid of the type HA.

11-2. Briefly explain why the use of Equation 11-5 is limited to calculation of the hydronium ion concentration of solutions in which NaHA is the only solute that determines the pH.

*11-3. Why is it impossible to titrate all three protons of phosphoric acid in aqueous solution?

11-4. Indicate whether an aqueous solution of the following compounds is acidic, neutral, or basic:
*(a) NH_4OAc.   *(e) $Na_2C_2O_4$.
(b) $NaNO_2$.   *(f) Na_2HPO_4.
*(c) $NaNO_3$.   *(g) NaH_2PO_4.
(d) $NaHC_2O_4$.   *(h) Na_3PO_4.

11-5. Suggest an indicator that could be used to provide an end point for the titration of the first proton in H_3AsO_4.

*11-6. Suggest an indicator that would give an end point for the titration of the first two protons in H_3AsO_4.

11-7. Suggest a method for the determination of the amounts of H_3PO_4 and NaH_2PO_4 in an aqueous solution.

11-8. Suggest a suitable indicator for a titration based on the following reactions. (Use 0.05 M if an equivalence-point concentration is needed.)
*(a) $H_2CO_3 + NaOH \rightarrow NaHCO_3 + H_2O$
(b) $H_2P + 2NaOH \rightarrow Na_2P + 2H_2O$
    (H_2P = o-phthalic acid)
*(c) $H_2T + 2NaOH \rightarrow Na_2T + 2H_2O$
    (H_2T = tartaric acid)
(d) $NH_2C_2H_4NH_2 + HCl \rightarrow NH_2C_2H_4NH_3Cl$
*(e) $NH_2C_2H_4NH_2 + 2HCl \rightarrow ClNH_3C_2H_4NH_3Cl$
(f) $H_2SO_3 + NaOH \rightarrow NaHSO_3 + H_2O$
*(g) $H_2SO_3 + 2NaOH \rightarrow Na_2SO_3 + 2H_2O$

11-9. Calculate the pH of a solution that is 0.0400 M in
*(a) H_3PO_4.   *(d) H_2SO_3.
(b) $H_2C_2O_4$.   *(e) H_2S.
*(c) H_3PO_3.   *(f) $H_2NC_2H_4NH_2$.

11-10. Calculate the pH of a solution that is 0.0400 M in
*(a) NaH_2PO_4.   *(d) $NaHSO_3$.
(b) $NaHC_2O_4$.   *(e) $NaHS$.
*(c) NaH_2PO_3.   *(f) $H_2NC_2H_4NH_3^+Cl^-$.

11-11. Calculate the pH of a solution that is 0.0400 M in
*(a) Na_3PO_4.   *(d) Na_2SO_3.
(b) $Na_2C_2O_4$.   *(e) Na_2S.
*(c) Na_2HPO_3.   *(f) $C_2H_4(NH_3^+Cl^-)_2$.

*11-12. Calculate the pH of a solution that is made up to contain the following analytical concentrations:
(a) 0.0500 M in H_3AsO_4 and 0.0200 M in NaH_2AsO_4.
(b) 0.0300 M in NaH_2AsO_4 and 0.0500 M in Na_2HAsO_4.
(c) 0.0600 M in Na_2CO_3 and 0.0300 M in $NaHCO_3$.
(d) 0.0400 M in H_3PO_4 and 0.0200 M in Na_2HPO_4.
(e) 0.0500 M in $NaHSO_4$ and 0.0400 M in Na_2SO_4.

11-13. Calculate the pH of a solution made up to contain the following analytical concentrations:
(a) 0.240 M in H_3PO_4 and 0.480 M in NaH_2PO_3.
(b) 0.0670 M in Na_2SO_3 and 0.0315 M in $NaHSO_3$.
(c) 0.640 M in $HOC_2H_4NH_2$ and 0.750 M in $HOC_2H_4NH_3Cl$.
(d) 0.0240 in $H_2C_2O_4$ (oxalic acid) and 0.0360 M in $Na_2C_2O_4$.
(e) 0.0100 M in $Na_2C_2O_4$ and 0.0400 M in $NaHC_2O_4$.

*11-14. Calculate the pH of a solution that is
(a) 0.0100 M in HCl and 0.0200 M in picric acid.
(b) 0.0100 M in HCl and 0.0200 M in benzoic acid.
(c) 0.0100 M in NaOH and 0.100 M in Na_2CO_3.
(d) 0.0100 M in NaOH and 0.100 M in NH_3.

11-15. Calculate the pH of a solution that is
  (a) 0.0100 M in $HClO_4$ and 0.0300 M in monochloro-acetic acid.
  (b) 0.0100 M in HCl and 0.0150 M in H_2SO_4.
  (c) 0.0100 M in NaOH and 0.0300 M in Na_2S.
  (d) 0.0100 M in NaOH and 0.0300 M in sodium acetate.

*11-16. Identify the principal conjugate acid/base pair and calculate the ratio between them in a solution that is buffered to pH 6.00 and contains
  (a) H_2SO_3.   **(c)** malonic acid.
  (b) citric acid.   **(d)** tartaric acid.

11-17. Identify the principal conjugate acid/base pair and calculate the ratio between them in a solution that is buffered to pH 9.00 and contains
  (a) H_2S.   **(c)** H_3AsO_4.
  (b) ethylenediamine dihydrochloride.   **(d)** H_2CO_3.

*11-18. How many grams of $Na_2HPO_4 \cdot 2H_2O$ must be added to 400 mL of 0.200 M H_3PO_4 to give a buffer of pH 7.30?

11-19. How many grams of dipotassium phthalate must be added to 750 mL of 0.0500 M phthalic acid to give a buffer of pH 5.75?

*11-20. What is the pH of the buffer formed by mixing 50.0 mL of 0.200 M NaH_2PO_4 with
  (a) 50.0 mL of 0.120 M HCl?
  (b) 50.0 mL of 0.120 M NaOH?

11-21. What is the pH of the buffer formed by adding 100 mL of 0.150 M potassium hydrogen phthalate to
  (a) 100 mL of 0.0800 M NaOH?
  (b) 100 mL of 0.0800 M HCl?

*11-22. How would you prepare 1.00 L of a buffer with a pH of 9.60 from 0.300 M Na_2CO_3 and 0.200 M HCl?

11-23. How would you prepare 1.00 L of a buffer with a pH of 7.00 from 0.200 M H_3PO_4 and 0.160 M NaOH?

*11-24. How would you prepare 1.00 L of a buffer with a pH of 6.00 from 0.500 M Na_3AsO_4 and 0.400 M HCl?

11-25. Identify by letter the curve in the following figures you would expect in the titration of a solution containing
  (a) disodium maleate, Na_2M, with standard acid.
  (b) pyruvic acid, HP, with standard base.
  (c) sodium carbonate, Na_2CO_3, with standard acid.

*11-26. Refer to 11-25. Describe the composition of a solution that would be expected to yield a curve resembling
  (a) curve B.   **(b)** curve A.   **(c)** curve E.

11-27. Refer to 11-25. Briefly explain why curve B *cannot* describe the titration of a mixture consisting of H_3PO_4 and NaH_2PO_4.

11-28. Derive a curve for the titration of 50.00 mL of a 0.1000 M solution of compound A with a 0.2000 M solution of compound B in the following list. For each titration, calculate the pH after the addition of 0.00, 12.50, 20.00, 24.00, 25.00, 26.00, 37.50, 45.00, 49.00, 50.00, 51.00, and 60.00 mL of compound B:

	A	B
*(a)	Na_2CO_3	HCl
(b)	ethylenediamine	HCl
*(c)	H_2SO_4	NaOH
(d)	H_2SO_3	NaOH

*11-29. Generate a curve for the titration of 50.00 mL of a solution in which the analytical concentration of NaOH is 0.1000 M and that for hydrazine is 0.0800 M. Calculate the pH after addition of 0.00, 10.00, 20.00, 24.00, 25.00, 26.00, 35.00, 44.00, 45.00, 46.00, and 50.00 mL of 0.2000 M $HClO_4$.

11-30. Generate a curve for the titration of 50.00 mL of a solution in which the analytical concentration of $HClO_4$ is 0.1000 M and that for formic acid is 0.0800 M. Calculate the pH after addition of 0.00, 10.00, 20.00, 24.00, 25.00, 26.00, 35.00, 44.00, 45.00, 46.00, and 50.00 mL of 0.2000 M KOH.

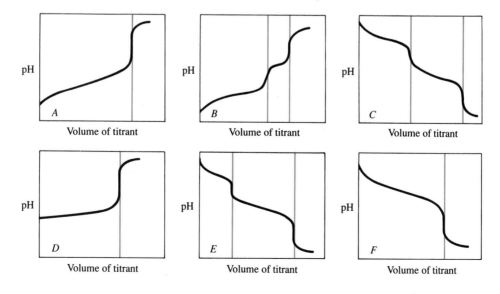

11-31. Formulate equilibrium constants for the following equilibria, giving numerical values for the constants:

*(a) $H_2AsO_4^- + H_2AsO_4^- \rightleftharpoons H_3AsO_4 + HAsO_4^{2-}$

(b) $HAsO_4^{2-} + HAsO_4^{2-} \rightleftharpoons AsO_4^{3-} + H_2AsO_4^-$

*11-32. Derive a numerical value for the equilibrium constant for the reaction:

$$NH_4^+ + OAc^- \rightleftharpoons NH_3 + HOAc$$

11-33. For pH values of 2.00, 6.00, and 10.00, calculate the alpha value for each species in an aqueous solution of

*(a) phthalic acid.   *(d) arsenic acid.

(b) phosphoric acid.   *(e) phosphorous acid.

*(c) citric acid.   *(f) oxalic acid.

11-34. Derive equations that define α_0, α_1, α_2, and α_3 for the acid H_3AsO_4.

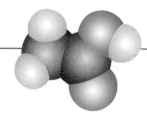

CHAPTER **12**

APPLICATIONS OF NEUTRALIZATION TITRATIONS

Neutralization titrations are widely used to determine the concentration of analytes that are themselves acids or bases or are convertible to such species by suitable treatment.[1] Water is the usual solvent for neutralization titrations because it is readily available, inexpensive, and nontoxic. Its low temperature coefficient of expansion is an added virtue. Some analytes, however, are not titratable in aqueous media because their solubilities are too low or because their strengths as acids or bases are not sufficiently great to provide satisfactory end points. Such substances can often be titrated in a solvent other than water.[2] We shall restrict our discussions to aqueous systems.

Nonaqueous solvents such as methyl alcohol and ethyl alcohol, glacial acetic acid, and methyl isobutyl ketone often make it possible to titrate acids or bases that are too weak to titrate in aqueous solution.

12A REAGENTS FOR NEUTRALIZATION REACTIONS

In Chapter 10, we noted that strong acids and strong bases cause the most pronounced change in pH at the equivalence point. For this reason, standard solutions for neutralization titrations are always prepared from these reagents.

12A-1 Preparation of Standard Acid Solutions

Hydrochloric acid is widely used for titration of bases. Dilute solutions of the reagent are stable indefinitely and do not cause troublesome precipitation reactions

Solutions of HCl, $HClO_4$, and H_2SO_4 are stable indefinitely. Restandardization is never required.

---

[1] For a review of applications of neutralization titrations, see D. Rosenthal and P. Zuman, in *Treatise on Analytical Chemistry*, 2nd ed., I. M. Kolthoff and P. J. Elving, Eds., Part I, Vol. 2, Chapter 18. New York: Wiley, 1979.
[2] For a review of nonaqueous acid/base titrimetry, see *Treatise on Analytical Chemistry*, 2nd ed., I. M. Kolthoff and P. J. Elving, Eds., Part I, Vol. 2, Chapters 19A–19E. New York: Wiley, 1979.

213

with most cations. It is reported that 0.1 M solutions of HCl can be boiled for as long as one hour without loss of acid, provided that the water lost by evaporation is periodically replaced; 0.5 M solutions can be boiled for at least ten minutes without significant loss.

Solutions of perchloric acid and sulfuric acid are also stable and are useful for titrations where chloride ion interferes by forming precipitates. Standard solutions of nitric acid are seldom encountered because of their oxidizing properties. Standard acid solutions are ordinarily prepared by diluting an approximate volume of the concentrated reagent and subsequently standardizing the diluted solution against a primary-standard base.

12A-2 The Standardization of Acids

Sodium Carbonate

Acids are frequently standardized against weighed quantities of sodium carbonate. Primary-standard-grade sodium carbonate is available commercially or can be prepared by heating purified sodium hydrogen carbonate between 270°C and 300°C for one hour:

$$2NaHCO_3(s) \rightarrow Na_2CO_3(s) + H_2O(g) + CO_2(g)$$

As was shown in Figure 11-4, two end points are observed in the titration of sodium carbonate. The first, corresponding to the conversion of carbonate to hydrogen carbonate, occurs at about pH 8.3; the second, involving the formation of carbonic acid and carbon dioxide, is observed at about pH 3.8. The second end point is always used for standardization because the change in pH is greater than that of the first. An even sharper end point can be achieved by boiling the solution briefly to eliminate the reaction product carbon dioxide. The sample is titrated to the first appearance of the acid color of the indicator (such as bromocresol green or methyl orange). At this point, the solution contains a large amount of dissolved carbon dioxide and small amounts of carbonic acid and unreacted hydrogen carbonate. Boiling effectively destroys this buffer by eliminating the carbonic acid:

$$H_2CO_3(aq) \rightarrow CO_2(g) + H_2O(l)$$

The solution then becomes alkaline again due to the residual hydrogen carbonate ion (see Figure 12-1). The titration is completed after the solution has cooled. Now, however, a substantially larger decrease in pH occurs during the final additions of acid, thus giving a more abrupt color change.

As an alternative, the acid can be introduced in an amount slightly in excess of that needed to convert the sodium carbonate to carbonic acid. The solution is boiled as before to remove carbon dioxide and cooled; the excess acid is then back-titrated with a dilute solution of base. Any indicator suitable for a strong acid/strong base titration is satisfactory. The volume ratio of acid to base must be established by an independent titration.

Other Primary Standards for Acids

Tris-(hydroxymethyl)aminomethane, $(HOCH_2)_3CNH_2$, known also as TRIS or THAM, is available in primary-standard purity from commercial sources. It

Sodium carbonate occurs naturally in large deposits as *washing soda*, $Na_2CO_3 \cdot 10H_2O$, and as *trona*, $Na_2CO_3 \cdot NaHCO_3 \cdot 2H_2O$. These minerals find wide use in many industries; for example, in the manufacture of glass. Primary standard sodium carbonate is manufactured by extensive purification of these minerals.

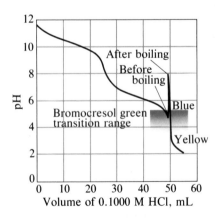

Figure 12-1
Titration of 25.00 mL of 0.1000 M Na_2CO_3 with 0.1000 M HCl. After about 49 mL of HCl have been added, the solution is boiled, causing the increase in pH shown. The change in pH on further addition of HCl is much larger.

possesses the advantage of a substantially greater mass per mole of protons consumed (121.1) than sodium carbonate (53.0).

Sodium tetraborate decahydrate and mercury(II) oxide have also been recommended as primary standards. The reaction of an acid with the tetraborate is

$$B_4O_7^{2-} + 2H_3O^+ + 3H_2O \rightarrow 4H_3BO_3$$

A high mass per proton consumed is desirable in a primary standard because a larger mass of reagent must be used, thus decreasing the relative weighing error.

TRIS

TRIS reacts in a 1:1 molar ratio with hydronium ions.

Borax, $Na_2B_4O_7 \cdot 10H_2O$, is a mineral that is mined in desert country and is widely used in cleaning preparations. A highly purified form of borax is used as a primary standard for bases.

Example 12-1

Compare the masses of (a) TRIS (121 g/mol); (b) Na_2CO_3 (106 g/mol); and (c) $Na_2B_4O_7 \cdot 10H_2O$ (381 g/mol) that should be taken to standardize an approximately 0.040 molar solution of HCl if it is desired that at least 30 mL of the acid be used for the titration.

In each case, the amount of HCl =

$$30.0 \text{ mL HCl} \times 0.040 \frac{\text{mmol HCl}}{\text{mL HCl}} = 1.2 \text{ mmol HCl}$$

(a)

$$\text{mass TRIS} = 1.2 \text{ mmol HCl} \times \frac{1 \text{ mmol TRIS}}{\text{mmol HCl}} \times \frac{0.121 \text{ g TRIS}}{\text{mmol TRIS}} = 0.15 \text{ g}$$

(b)

$$\text{mass Na}_2\text{CO}_3 = 1.2 \text{ mmol HCl} \times \frac{1 \text{ mmol Na}_2\text{CO}_3}{2 \text{ mmol HCl}} \times \frac{0.106 \text{ g Na}_2\text{CO}_3}{\text{mmol Na}_2\text{CO}_3}$$

$$\text{mass Na}_2\text{CO}_3 = 0.064 \text{ g}$$

(c)

$$\text{mass Na}_2\text{B}_4\text{O}_7 \cdot 10\text{H}_2\text{O} = 1.2 \text{ mmol HCl} \times \frac{1 \text{ mmol Na}_2\text{B}_4\text{O}_7 \cdot 10\text{H}_2\text{O}}{2 \text{ mmol HCl}}$$

$$\times \frac{0.381 \text{ g Na}_2\text{B}_4\text{O}_7 \cdot 10\text{H}_2\text{O}}{\text{mmol Na}_2\text{B}_4\text{O}_7 \cdot 10\text{H}_2\text{O}}$$

$$= 0.23 \text{ g Na}_2\text{B}_4\text{O}_7 \cdot 10\text{H}_2\text{O}$$

12A-3 Preparation of Standard Solutions of Base

Sodium hydroxide is the most common base for preparing standard solutions although potassium hydroxide and barium hydroxide are also encountered. None of these is obtainable in primary-standard purity, and so standardization is required after preparation.

The Effect of Carbon Dioxide on Standard Base Solutions

In solution as well as in the solid state, the hydroxides of sodium, potassium, and barium react rapidly with atmospheric carbon dioxide to produce the corre-

Absorption of carbon dioxide by a standardized solution of sodium or potassium hydroxide leads to a negative systematic error in analyses in which an indicator with a basic range is used. No systematic error is incurred when an indicator with an acidic range is used.

sponding carbonate:

$$CO_2(g) + 2OH^- \rightarrow CO_3^{2-} + H_2O$$

Although production of each carbonate ion uses up two hydroxide ions, the uptake of carbon dioxide by a solution of base does not necessarily alter its combining capacity for hydronium ions. Thus, at the end point of a titration that requires an acid-range indicator (such as bromocresol green), each carbonate ion produced from sodium or potassium hydroxide will have reacted with two hydronium ions of the acid (Figure 12-1):

$$CO_3^{2-} + 2H_3O^+ \rightarrow H_2CO_3 + 2H_2O$$

Because the amount of hydronium ion consumed by this reaction is identical to the amount of hydroxide lost during formation of the carbonate ion, no error is incurred.

Unfortunately, most applications of standard base require an indicator with a basic transition range (phenolphthalein, for example). Here, each carbonate ion has reacted with only one hydronium ion when the color change of the indicator is observed:

$$CO_3^{2-} + H_3O^+ \rightarrow HCO_3^- + H_2O$$

The effective concentration of the base is thus diminished by absorption of carbon dioxide, and a systematic error (called a *carbonate error*) results.

Example 12-2

A carbonate-free NaOH solution was found to be 0.05118 M immediately after preparation. Exactly 1.000 L of this solution was exposed to air for some time and absorbed 0.1962 g CO_2. Calculate the relative carbonate error that would arise in the determination of acetic acid with the contaminated solution if phenolphthalein was used as an indicator.

$$2NaOH + CO_2 \rightarrow Na_2CO_3 + H_2O$$

$$c_{Na_2CO_3} = 0.1962 \text{ g } CO_2 \times \frac{1 \text{ mol } CO_2}{44.01 \text{ g } CO_2} \times \frac{1 \text{ mol } Na_2CO_3}{\text{mol } CO_2} \times \frac{1}{1.000 \text{ L soln}}$$

$$= 4.458 \times 10^{-3} \text{ M}$$

Effective concentration c_{NaOH} of NaOH for acetic acid is

$$c_{NaOH} = 0.05118 \, \frac{\text{mol NaOH}}{L} - \frac{4.458 \times 10^{-3} \text{ mol } Na_2CO_3}{L}$$

$$\times \frac{1 \text{ mol HAc}}{\text{mol } Na_2CO_3} \times \frac{1 \text{ mol NaOH}}{\text{mol HAc}}$$

$$= 0.04672 \text{ M}$$

$$\text{rel error} = \frac{0.04672 - 0.05118}{0.05118} \times 100\% = -8.7\%$$

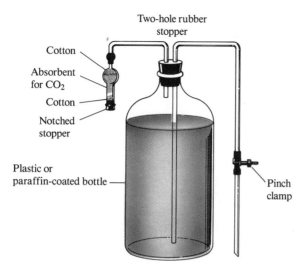

Two-hole rubber stopper

Cotton

Absorbent for CO_2

Cotton

Notched stopper

Plastic or paraffin-coated bottle

Pinch clamp

Figure 12-2

Arrangement for the storage of standard base solutions.

The solid reagents used to prepare standard solutions of base are always contaminated by significant amounts of carbonate ion. The presence of this contaminant does not cause a carbonate error provided the same indicator is used for both standardization and analysis. It does, however, lead to less sharp end points. Consequently, steps are usually taken to remove carbonate ion before a solution of a base is standardized.

The best method for preparing carbonate-free sodium hydroxide solutions takes advantage of the very low solubility of sodium carbonate in concentrated solutions of the base. An approximately 50% aqueous solution of sodium hydroxide is prepared (or purchased from commercial sources). The solid sodium carbonate is allowed to settle to give a clear liquid that is then decanted and diluted to give the desired concentration. (Alternatively, the solid is removed by vacuum filtration.)

Water that is used to prepare carbonate-free solutions of base must also be free of carbon dioxide. Distilled water, which is sometimes supersaturated with carbon dioxide, should be boiled briefly to eliminate the gas. The water is then allowed to cool to room temperature before the introduction of base because hot alkali solutions rapidly absorb carbon dioxide. Deionized water ordinarily does not contain significant amounts of carbon dioxide.

Standard solutions of base are reasonably stable as long as they are protected from contact with the atmosphere. Figure 12-2 shows an arrangement for preventing the uptake of atmospheric carbon dioxide during storage and when the reagent is dispensed. Air entering the vessel is passed over a solid absorbent for CO_2, such as soda lime or Ascarite II®.[3] The contamination that occurs as the solution is transferred from this storage bottle to the buret is ordinarily negligible.

A tightly capped polyethylene bottle usually provides sufficient short-term protection against the uptake of atmospheric carbon dioxide. Before capping, the bottle is squeezed to minimize the interior air space. Care should also be taken to keep the bottle closed except during the brief periods when the contents are

Generally, carbonate ion in standard solutions of bases is undesirable because it decreases the sharpness of end points.

WARNING: Concentrated solutions of NaOH (and KOH) are extremely corrosive to the skin. In making up standard solutions of NaOH, **a face shield, rubber gloves, and protective clothing must be worn at all times.**

Water that is in equilibrium with atmospheric constituents contains only about 1.5×10^{-5} mol CO_2/L, an amount that has a negligible effect on the strength of most standard bases. As an alternative to boiling, supersaturated solutions of CO_2 can be purged by bubbling air through the water for several hours. This process is called *sparging* and produces a solution that contains the equilibrium concentration of CO_2.

---

[3] Thomas Scientific, Swedesboro, NJ. Ascarite II consists of sodium hydroxide deposited on a nonfibrous silicate structure.

being transferred to a buret. Sodium hydroxide solutions will ultimately cause a polyethylene bottle to become brittle.

The concentration of a sodium hydroxide solution will decrease slowly (0.1% to 0.3% per week) if the base is stored in glass bottles. The loss in strength is caused by the reaction of the base with the glass to form sodium silicates. For this reason, standard solutions of base should not be stored for extended periods (longer than one or two weeks) in glass containers. In addition, bases should never be kept in glass-stoppered containers because the reaction between the base and the stopper may cause the latter to "freeze" after a brief period. Finally, to avoid the same type of freezing, burets with glass stopcocks should be promptly drained and thoroughly rinsed with water after use with standard base solutions. This problem does not occur with burets equipped with Teflon stopcocks.

12A-4   The Standardization of Bases

Several excellent primary standards are available for the standardization of bases. Most are weak organic acids that require the use of an indicator with a basic transition range.

Potassium Hydrogen Phthalate, $KHC_8H_4O_4$

Potassium hydrogen phthalate is an ideal primary standard. It is a nonhygroscopic crystalline solid with a high mass (204.2 g/mol). For most purposes, the commercial analytical-grade salt can be used without further purification. For the most exacting work, potassium hydrogen phthalate of certified purity is available from the National Institute of Standards and Technology.

Other Primary Standards for Bases

Benzoic acid can be obtained in primary-standard purity and can be used for the standardization of bases. Because its solubility in water is limited, this reagent is ordinarily dissolved in ethanol prior to dilution with water and titration. A blank should always be carried through this standardization because commercial alcohol is sometimes slightly acidic.

Potassium hydrogen iodate, $KH(IO_3)_2$, is an excellent primary standard with a high molecular mass per mole of protons. It is also a strong acid that can be titrated using virtually any indicator with a transition range between pH 4 and 10.

12B   TYPICAL APPLICATIONS OF NEUTRALIZATION TITRATIONS

Neutralization titrations are used to determine the innumerable inorganic, organic, and biological species that possess inherent acidic or basic properties. Equally important, however, are the many applications that involve conversion of an analyte to an acid or base by suitable chemical treatment followed by titration with a standard strong base or acid.

Two major types of end points find widespread use in neutralization titrations. The first is a visual end point based on indicators such as those described in Section 10A. The second is a *potentiometric* end point in which the potential of a glass/calomel electrode system is determined with a voltage-measuring device. The measured potential is directly proportional to pH. Potentiometric end points are described in Section 18G.

Solutions of bases are preferably stored in polyethylene bottles rather than glass because of the reaction between bases and glass. Such solutions should never be stored in glass-stoppered bottles; after standing for a period, removal of the stopper often becomes impossible.

Standard solutions of strong bases cannot be prepared directly by mass and must always be standardized against a primary standard acid.

Potassium hydrogen phthalate

$KH(IO_3)_2$, in contrast to all other primary standards for bases, has the advantage of being a strong acid, thus making the choice of indicator less critical.

Neutralization titrations are still among the most widely used analytical methods.

12B-1 Elemental Analysis

Several elements that are important in organic and biological systems are conveniently determined by methods that involve an acid/base titration as the final step. Generally, the elements susceptible to this type of analysis are nonmetallic and include carbon, nitrogen, chlorine, bromine, fluorine, as well as a few other less common species. Pretreatment converts the element to an inorganic acid or base that is then titrated. A few examples follow.

Nitrogen

Nitrogen is found in a wide variety of substances of interest in research, industry, and agriculture. Examples include amino acids, proteins, synthetic drugs, fertilizers, explosives, soils, potable water supplies, and dyes. Thus analytical methods for the determination of nitrogen, particularly in organic substrates, are of singular importance.

The most common method for determining organic nitrogen is the *Kjeldahl method*, which is based on a neutralization titration. The procedure is straightforward, requires no special equipment, and is readily adapted to the routine analysis of large numbers of samples. The Kjeldahl method—or one of its modifications—is the standard means for determining the protein content of grains, meats, and other biological materials. Since most proteins contain approximately the same percentage of nitrogen, multiplication of this percentage by a suitable factor (6.25 for meats, 6.38 for dairy products, and 5.70 for cereals) gives the percentage of protein in a sample.

In the Kjeldahl method, the sample is decomposed in hot, concentrated sulfuric acid to convert the bound nitrogen to ammonium ion. The resulting solution is then cooled, diluted, and made basic. The liberated ammonia is distilled, collected in an acidic solution, and determined by a neutralization titration.

Kjeldahl is pronounced *Kyell′ dahl*. Hundreds of thousands of Kjeldahl-nitrogen determinations are performed each year, primarily to provide a measure of the protein content of meats, grains, and animal feeds.

Feature 12-1
OTHER METHODS FOR DETERMINING ORGANIC NITROGEN

Two other methods in addition to the Kjeldahl method are used to determine the nitrogen content of organic materials. In the *Dumas method*, the sample is mixed with powdered copper(II) oxide and ignited in a combustion tube to give carbon dioxide, water, nitrogen, and small amounts of nitrogen oxides. A stream of carbon dioxide carries these products through a packing of hot copper, which reduces any oxides of nitrogen to the elemental state. The mixture then is passed into a gas buret filled with concentrated potassium hydroxide. The only component not absorbed by the base is nitrogen, and its volume is measured directly.

The newest method for determining organic nitrogen involves combusting the sample at 1100°C for a few minutes to convert the nitrogen to nitric oxide, NO. Ozone is then introduced into the gaseous mixture, which oxidizes the nitric oxide to nitrogen dioxide. This reaction gives off visible radiation (*chemiluminescence*), the intensity of which is proportional to the nitrogen content of the sample. An instrument for this procedure is available from commercial sources.

The Kjeldahl method was developed by a Danish chemist who first described it in 1883. J. Kjeldahl, *Z. Anal. Chem.* **1883**, *22*, 366.

The critical step in the Kjeldahl method is the decomposition with sulfuric acid, which oxidizes the carbon and hydrogen in the sample to carbon dioxide and water. The fate of the nitrogen, however, depends on its state of combination in the original sample. Amine and amide nitrogens are quantitatively converted to ammonium ion. In contrast, nitro, azo, and azoxy groups are likely to yield elemental nitrogen or various oxides of nitrogen, all of which are lost from the hot acidic medium. This loss can be avoided by pretreating the sample with a reducing agent to form products that behave as amide or amine nitrogen. In one such prereduction scheme, salicylic acid and sodium thiosulfate are added to the concentrated sulfuric acid solution containing the sample. After a brief period, the digestion is performed in the usual way.

Certain aromatic heterocyclic compounds, such as pyridine and its derivatives, are particularly resistant to complete decomposition by sulfuric acid. Such compounds yield low results as a consequence (Figure 4-3) unless special precautions are taken.

$-NO_2$ $-N{=}N-$ $-N_+{=}N-$
nitro group   azo group $\underset{O^-}{|}$
                      azoxy group

Example 12-3

A 0.7121 g sample of a wheat flour was analyzed by the Kjeldahl method. The ammonia formed by addition of concentrated base after digestion with H_2SO_4 was distilled into 25.00 mL of 0.04977 M HCl. The excess HCl was then back-titrated with 3.97 mL of 0.04012 M NaOH. Calculate the percent protein in the flour.

$$\text{no. mmol HCl} = 25.00 \text{ mL HCl} \times 0.04977 \; \frac{\text{mmol HCl}}{\text{mL HCl}} = 1.2443$$

$$\text{no. mmol NaOH} = 3.97 \text{ mL NaOH} \times 0.04012 \; \frac{\text{mmol NaOH}}{\text{mL NaOH}} = 0.1593$$

$$\text{no. mmol N} = 1.2443 - 0.1593 = 1.0850$$

$$\text{percent N} = \frac{1.0850 \text{ mmol N} \times 0.014007 \text{ g N/mmol N}}{0.7121 \text{ g sample}} \times 100\%$$

$$= 2.1341\%$$

$$\text{percent protein} = 2.1341\% \text{ N} \times \frac{5.70\% \text{ protein}}{\% \text{ N}} = 12.16\%$$

Sulfur

Sulfur dioxide in the atmosphere is often determined by drawing a sample through a hydrogen peroxide solution and then titrating the resulting sulfuric acid.

Sulfur in organic and biological materials is conveniently determined by burning the sample in a stream of oxygen. The sulfur dioxide (as well as the sulfur trioxide) formed during the oxidation is collected by distillation into a dilute solution of hydrogen peroxide:

$$SO_2(g) + H_2O_2 \rightarrow H_2SO_4$$

The sulfuric acid is then titrated with standard base.

Other Elements

Table 12-1 lists other elements that can be determined by neutralization methods.

Table 12-1
ELEMENTAL ANALYSES BASED ON NEUTRALIZATION TITRATIONS

Element	Converted to	Absorption or Precipitation Products	Titration
N	NH_3	$NH_3(g) + H_3O^+ \rightarrow NH_4^+ + H_2O$	Excess HCl with NaOH
S	SO_2	$SO_2(g) + H_2O_2 \rightarrow H_2SO_4$	NaOH
C	CO_2	$CO_2(g) + Ba(OH)_2 \rightarrow BaCO_3(s) + H_2O$	Excess $Ba(OH)_2$ with HCl
Cl (Br)	HCl	$HCl(g) + H_2O \rightarrow Cl^- + H_3O^+$	NaOH
F	SiF_4	$SiF_4(g) + H_2O \rightarrow H_2SiF_6$	NaOH
P	H_3PO_4	$12H_2MoO_4 + 3NH_4^+ + H_3PO_4 \rightarrow$ $(NH_4)_3PO_4 \cdot 12MoO_3(s) + 12H_2O + 3H^+$ $(NH_4)_3PO_4 \cdot 12MoO_3(s) + 26OH^- \rightarrow$ $HPO_4^{2-} + 12MoO_4^{2-} + 14H_2O + 3NH_3(g)$	Excess NaOH with HCl

12B-2 The Determination of Inorganic Substances

Numerous inorganic species can be determined by titration with strong acids or bases. A few examples follow.

Ammonium Salts

Ammonium salts are conveniently determined by conversion to ammonia with strong base followed by distillation. The ammonia is collected and titrated as in the Kjeldahl method.

Nitrates and Nitrites

The method just described for ammonium salts can be extended to the determination of inorganic nitrate or nitrite. These ions are first reduced to ammonium ion by Devarda's alloy (50% Cu, 45% Al, 5% Zn). Granules of the alloy are introduced into a strongly alkaline solution of the sample in a Kjeldahl flask. The ammonia is distilled after reaction is complete. Arnd's alloy (60% Cu, 40% Mg) has also been used as the reducing agent.

Carbonate and Carbonate Mixtures

The qualitative and quantitative determination of the constituents in a solution containing sodium carbonate, sodium hydrogen carbonate, and sodium hydroxide, either alone or admixed, provides interesting examples of how neutralization titrations can be employed to analyze mixtures. No more than two of these three constituents can exist in appreciable amount in any solution because reaction eliminates the third. Thus, mixing sodium hydroxide with sodium hydrogen carbonate results in the formation of sodium carbonate until one or the other (or both) of the original reactants is exhausted. If the sodium hydroxide is used up, the solution will contain sodium carbonate and sodium hydrogen carbonate; if sodium hydrogen carbonate is depleted, sodium carbonate and sodium hydroxide will remain; if equimolar amounts of sodium hydrogen carbonate and sodium hydroxide are mixed, the principal solute species will be sodium carbonate.

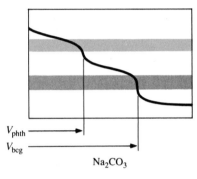

Phenolphthalein transition range

Bromocresol green transition range

V_phth
V_bcg

NaOH

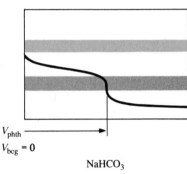

V_phth
V_bcg

Na_2CO_3

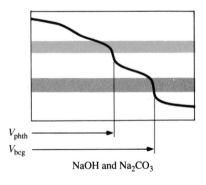

V_phth
$V_\text{bcg} = 0$

$NaHCO_3$

V_phth
V_bcg

NaOH and Na_2CO_3

Figure 12-3

Titration curves and indicator transition ranges for the analysis of mixtures containing hydroxide, carbonate, and hydrogen carbonate ions.

Table 12-2

VOLUME RELATIONSHIPS IN THE ANALYSIS OF MIXTURES CONTAINING HYDROXIDE, CARBONATE, AND HYDROGEN CARBONATE IONS

Constituent(s) in Sample	Relationship Between V_phth and V_bcg in the Titration of an Equal Volume of Sample*
NaOH	$V_\text{phth} = V_\text{bcg}$
Na_2CO_3	$V_\text{phth} = \frac{1}{2}V_\text{bcg}$
$NaHCO_3$	$V_\text{phth} = 0; V_\text{bcg} > 0$
NaOH, Na_2CO_3	$V_\text{phth} > \frac{1}{2}V_\text{bcg}$
Na_2CO_3, $NaHCO_3$	$V_\text{phth} < \frac{1}{2}V_\text{bcg}$

* V_phth = volume of acid needed for a phenolphthalein end point; V_bcg = volume of acid needed for a bromocresol green end point.

The analysis of such mixtures requires two titrations, one with an alkaline-range indicator, such as phenolphthalein, and the other with an acid-range indicator, such as bromocresol green. The composition of the solution can then be deduced from the relative volumes of acid needed to titrate equal volumes of the sample (Table 12-2 and Figure 12-3). Once the composition of the solution has been established, the volume data can be used to determine the concentration of each component in the sample.

Example 12-4

A solution contains $NaHCO_3$, Na_2CO_3, and NaOH, either alone or in permissible combination. Titration of a 50.0-mL portion to a phenolphthalein end point requires 22.1 mL of 0.100 M HCl. A second 50.0-mL aliquot requires 48.4 mL of the HCl when titrated to a bromocresol green end point. Deduce the composition and calculate the molar solute concentrations of the original solution.

If the solution contained only NaOH, the volume of acid required would be the same regardless of indicator (see Figure 12-3a). Similarly, we can rule out the presence of Na_2CO_3 alone because titration of this compound to a bromocresol green end point would consume just twice the volume of acid required to reach the phenolphthalein end point (see Figure 12-3b). In fact, however, the second titration requires 48.4 mL. Because less than half of this amount is involved in the first titration, the solution must contain some $NaHCO_3$ in addition to Na_2CO_3 (see Figure 12-3). We can now calculate the concentration of the two constituents.

When the phenolphthalein end point is reached, the CO_3^{2-} originally present is converted to HCO_3^-. Thus,

$$\text{amount } Na_2CO_3 = 22.1 \text{ mL} \times 0.100 \text{ mmol/mL} = 2.21 \text{ mmol}$$

The titration from the phenolphthalein to the bromocresol green end point (48.4 − 22.1 = 26.3 mL) involves both the hydrogen carbonate originally present and that formed by titration of the carbonate. Thus,

$$\text{amount } NaHCO_3 + \text{no. mmol } Na_2CO_3 = 26.3 \times 0.100 = 2.63 \text{ mmol}$$

Hence,

$$\text{amount NaHCO}_3 = 2.63 - 2.21 = 0.42 \text{ mmol}$$

The molar concentrations are readily calculated from these data:

$$c_{\text{Na}_2\text{CO}_3} = \frac{2.21 \text{ mmol}}{50.0 \text{ mL}} = 0.0442 \text{ M}$$

$$c_{\text{NaHCO}_3} = \frac{0.42 \text{ mmol}}{50.0 \text{ mL}} = 0.084 \text{ M}$$

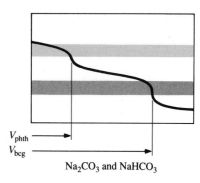

V_{phth}
V_{bcg}

Na_2CO_3 and NaHCO_3

Figure 12-3 (cont.)

The method described in Example 12-4 is not entirely satisfactory because the pH change corresponding to the hydrogen carbonate equivalence point is not sufficient to give a sharp color change with a chemical indicator. Relative errors of 1% or more must be expected as a consequence.

Compatible mixtures containing two of the following can also be analyzed in a similar way: HCl, H_3PO_4, NaH_2PO_4, Na_2HPO_4, Na_3PO_4, and $NaOH$.

12C QUESTIONS AND PROBLEMS

*12-1. The boiling points of HCl and CO_2 are nearly the same ($-85°C$ and $-78°C$). Explain why CO_2 can be removed from an aqueous solution by boiling briefly while essentially no HCl is lost even after boiling for 1 h or more.

12-2. Why is HNO_3 seldom used to prepare standard acid solutions?

*12-3. Explain how Na_2CO_3 of primary-standard grade can be prepared from primary standard $NaHCO_3$.

12-4. Why is it common practice to boil the solution near the equivalence point in the standardization of Na_2CO_3 with acid?

*12-5. Briefly explain why $KH(IO_3)_2$ would be preferred over benzoic acid as a primary standard for a 0.010 M NaOH solution.

12-6. Briefly describe the circumstance where the molarity of a sodium hydroxide solution will apparently be unaffected by the absorption of carbon dioxide.

12-7. What types of organic nitrogen-containing compounds tend to yield low results with the Kjeldahl method unless special precautions are taken?

*12-8. How would you prepare 2.00 L of
(a) 0.15 M KOH from the solid?
(b) 0.015 M $Ba(OH)_2 \cdot 8H_2O$ from the solid?
(c) 0.200 M HCl from a reagent that has a density of 1.0579 g/mL and is 11.50% HCl (w/w)?

12-9. How would you prepare 500 mL of
(a) 0.250 M H_2SO_4 from a reagent that has a density of 1.1539 g/mL and is 21.8% H_2SO_4 (w/w)?
(b) 0.30 M NaOH from the solid?
(c) 0.08000 M Na_2CO_3 from the pure solid?

*12-10. Standardization of a sodium hydroxide solution against potassium hydrogen phthalate (KHP) yielded the accompanying results.

mass KHP, g	0.7987	0.8365	0.8104	0.8039
volume NaOH, mL	38.29	39.96	38.51	38.29

Calculate
(a) the average molarity of the base.
(b) the standard deviation and the coefficient of variation for the data.

12-11. The molarity of a perchloric acid solution was established by titration against primary standard sodium carbonate (product: CO_2); the following data were obtained.

mass Na_2CO_3, g	0.2068	0.1997	0.2245	0.2137
volume $HClO_4$, mL	36.31	35.11	39.00	37.54

(a) Calculate the average molarity of the acid.
(b) Calculate the standard deviation and the coefficient of variation for the data.
(c) Is there statistical justification for disregarding the outlying result?

*12-12. If 1.000 L of 0.1500 M NaOH was unprotected from the air after standardization and absorbed 11.2 mmol of CO_2, what is its new molarity when it is standardized against a standard solution of HCl using
(a) phenolphthalein?
(b) bromocresol green?

12-13. An NaOH solution was 0.1019 M immediately after standardization. Exactly 500.0 mL of the reagent were left exposed to air for several days and absorbed 0.652 g of CO_2. Calculate the relative carbonate error in the determi-

---

* Answers to the asterisked problems are given in the answer section at the back of the book.

nation of acetic acid with this solution if the titrations were performed with phenolphthalein.

*12-14. Calculate the molar concentration of a dilute HCl solution if
(a) a 50.00-mL aliquot yielded 0.6010 g of AgCl.
(b) titration of 25.00 mL of 0.04010 M $Ba(OH)_2$ required 19.92 mL of the acid.
(c) titration of 0.2694 g of primary standard Na_2CO_3 required 38.77 mL of the acid (products: CO_2 and H_2O).

12-15. Calculate the molarity of a dilute $Ba(OH)_2$ solution if
(a) 50.00 mL yielded 0.1684 g of $BaSO_4$.
(b) titration of 0.4815 g of primary standard potassium hydrogen phthalate (KHP) required 29.41 mL of the base.
(c) addition of 50.00 mL of the base to 0.3614 g of benzoic acid required a 4.13-mL back titration with 0.05317 M HCl.

12-16. Suggest a range of sample masses for the indicated primary standard if it is desired to use between 35 and 45 mL of titrant:
*(a) 0.150 M $HClO_4$ titrated against Na_2CO_3 (CO_2 product).
(b) 0.075 M HCl titrated against $Na_2C_2O_4$:

$$Na_2C_2O_4 \rightarrow Na_2CO_3 + CO$$
$$CO_3^{2-} + 2H^+ \rightarrow H_2O + CO_2$$

*(c) 0.20 M NaOH titrated against benzoic acid.
(d) 0.030 M $Ba(OH)_2$ titrated against $KH(IO_3)_2$.
*(e) 0.040 M $HClO_4$ titrated against TRIS.
(f) 0.080 M H_2SO_4 titrated against $Na_2B_4O_7 \cdot 10H_2O$. Reaction:

$$B_4O_7^{2-} + 2H_3O^+ + 3H_2O \rightarrow 4H_3BO_3$$

*12-17. Calculate the relative standard deviation in the computed molarity of 0.0400 M HCl if this acid was standardized against the masses derived in Example 12-1 for (a) TRIS, (b) Na_2CO_3, and (c) $Na_2B_4O_7 \cdot 10H_2O$. Assume that the absolute standard deviation in the mass measurement is 0.0001 g and that this measurement limits the precision of the computed molarity.

12-18. (a) Compare the masses of potassium hydrogen phthalate (204.22 g/mol), potassium hydrogen iodate (389.91 g/mol), and benzoic acid (122.12 g/mol) needed for a 30.00-mL standardization of 0.0400 M NaOH.
(b) What would be the relative standard deviation in the molarity of the base if the standard deviation in the measurement of mass in (a) is 0.002 g and this uncertainty limits the precision of the calculation?

*12-19. A 50.00-mL sample of a white dinner wine required 21.48 mL of 0.03776 M NaOH to achieve a phenolphthalein end point. Express the acidity of the wine in terms of grams of tartaric acid ($H_2C_4H_4O_6$; 150.09 g/mol) per 100 mL. (Assume that two hydrogens of the acid are titrated.)

12-20. A 25.0-mL aliquot of vinegar was diluted to 250 mL in a volumetric flask. Titration of 50.0-mL aliquots of the diluted solution required an average of 34.88 mL of 0.09600 M NaOH. Express the acidity of the vinegar in terms of the percentage (w/v) of acetic acid.

*12-21. Titration of 0.7439-g sample of impure $Na_2B_4O_7$ required 31.64 mL of 0.1081 M HCl (see 12-16f for reaction). Express the results of this analysis in terms of percent:
(a) $Na_2B_4O_7$.
(b) $Na_2B_4O_7 \cdot 10H_2O$.
(c) B_2O_3.
(d) B.

12-22. A 0.6334-g sample of impure mercury(II) oxide was dissolved in an unmeasured excess of potassium iodide. Reaction:

$$HgO(s) + 4I^- + H_2O \rightarrow HgI_4^{2-} + 2OH^-$$

Calculate the percentage of HgO in the sample if titration of the liberated hydroxide required 42.59 mL of 0.1178 M HCl.

*12-23. The formaldehyde content of a pesticide preparation was determined by weighing 0.3124 g of the liquid sample into a flask containing 50.00 mL of 0.0996 M NaOH and 50 mL of 3% H_2O_2. Upon heating, the following reaction took place:

$$OH^- + HCHO + H_2O_2 \rightarrow HCOO^- + 2H_2O$$

After cooling, the excess base was titrated with 23.3 mL of 0.05250 M H_2SO_4. Calculate the percentage of HCHO (30.026 g/mol) in the sample.

12-24. The benzoic acid extracted from 106.3 g of catsup required a 14.76-mL titration with 0.0514 M NaOH. Express the results of this analysis in terms of percent sodium benzoate (144.10 g/mol).

*12-25. The active ingredient in Antabuse, a drug used for the treatment of chronic alcoholism, is tetraethylthiuram disulfide,

$$(C_2H_5)_2NCSSCN(C_2H_5)_2$$

(296.54 g/mol). The sulfur in a 0.4329-g sample of an Antabuse preparation was oxidized to SO_2, which was absorbed in H_2O_2 to give H_2SO_4. The acid was titrated with 22.13 mL of 0.03736 M base. Calculate the percentage of active ingredient in the preparation.

12-26. A 25.00-mL sample of a household cleaning solution was diluted to 250.0 mL in a volumetric flask. A 50.00-mL aliquot of this solution required 40.38 mL of 0.2506 M HCl to reach a bromocresol green end point. Calculate the w/v percentage of NH_3 in the sample. (Assume that all the alkalinity results from the ammonia.)

*12-27. A 0.1401-g sample of a purified carbonate was dissolved in 50.00 mL of 0.1140 M HCl and boiled to eliminate CO_2. Back-titration of the excess HCl required 24.21 mL of 0.09802 M NaOH. Identify the carbonate.

12-28. A dilute solution of an unknown weak acid required a 28.62-mL titration with 0.1084 M NaOH to reach a phenolphthalein end point. The titrated solution was evaporated to dryness. Calculate the molar mass of the acid if the sodium salt was found to weigh 0.2110 g. Assume the acid contained a single titratable proton.

*12-29. A 3.00-L sample of urban air was bubbled through a solution containing 50.0 mL of 0.0116 M $Ba(OH)_2$, which caused the CO_2 in the sample to precipitate as $BaCO_3$. The excess base was back-titrated to a phenolphthalein end point with 23.6 mL of 0.0108 M HCl. Calculate the parts per million of CO_2 in the air (that is, mL $CO_2/10^6$ mL air); use 1.98 g/L for the density of CO_2.

12-30. Air was bubbled at a rate of 30.0 L/min through a trap containing 75 mL of 1% H_2O_2 ($H_2O_2 + SO_2 \rightarrow H_2SO_4$). After 10.0 min, the H_2SO_4 was titrated with 11.1 mL of 0.00204 M NaOH. Calculate the parts per million of SO_2 (that is, mL $SO_2/10^6$ mL air) if the density of SO_2 is 0.00285 g/mL.

*12-31. The digestion of a 0.1417-g sample of a phosphorus-containing compound in a mixture of HNO_3 and H_2SO_4 resulted in the formation of CO_2, H_2O, and H_3PO_4. Addition of ammonium molybdate yielded a solid having the composition $(NH_4)_3PO_4 \cdot 12MoO_3$ (1876.3 g/mol). This precipitate was filtered, washed, and dissolved in 50.00 mL of 0.2000 M NaOH:

$$(NH_4)_3 \cdot 12MoO_3(s) + 26OH^- \rightarrow$$
$$HPO_4^{2-} + 12MoO_4^{2-} + 14H_2O + 3NH_3(g)$$

After the solution was boiled to remove the NH_3, the excess NaOH was titrated with 14.17 mL of 0.1741 M HCl to a phenolphthalein end point. Calculate the percentage of phosphorus in the sample.

*12-32. A 0.8160-g sample containing dimethylphthalate, $C_6H_4(COOCH_3)_2$ (194.19 g/mol), and unreactive species was refluxed with 50.00 mL of 0.1031 M NaOH to hydrolyze the ester groups (this process is called *saponification*).

$$C_6H_4(COOCH_3)_2 + 2OH^- \rightarrow C_6H_4(COO)_2^{2-} + 2CH_3OH$$

After the reaction was complete, the excess NaOH was back-titrated with 24.27 mL of 0.1644 M HCl. Calculate the percentage of dimethylphthalate in the sample.

*12-33. Neohetramine, $C_{16}H_{21}ON_4$ (285.37 g/mol), is a common antihistamine. A 0.1247-g sample containing this compound was analyzed by the Kjeldahl method. The ammonia produced was collected in H_3BO_3; the resulting $H_2BO_3^-$ was titrated with 26.13 mL of 0.01477 M HCl. Calculate the percentage of neohetramine in the sample.

12-34. The *Merck Index* indicates that 10 mg of guanidine, CH_5N_3, may be administered for each kilogram of body weight in the treatment of myasthenia gravis. The nitrogen in a 4-tablet sample that weighed a total of 7.50 g was converted to ammonia by a Kjeldahl digestion, followed by distillation into 100.0 mL of 0.1750 M HCl. The analysis was completed by titrating the excess acid with 11.37 mL of 0.1080 M NaOH. How many of these tablets represent a proper dose for a 48-kg patient?

*12-35. A 1.047-g sample of canned tuna was analyzed by the Kjeldahl method; 24.61 mL of 0.1180 M HCl were required to titrate the liberated ammonia. Calculate the percentage of nitrogen in the sample.

12-36. Calculate the grams of protein in a 6.50-oz can of tuna in 12-35 (1 oz = 28.3 g).

*12-37. A 0.5843-g sample of a plant food preparation was analyzed for its N content by the Kjeldahl method, the liberated NH_3 being collected in 50.00 mL of 0.1062 M HCl. The excess acid required an 11.89-mL back-titration with 0.0925 M NaOH. Express the results of this analysis in terms of percent
*(a) N.   *(c) $(NH_4)_2SO_4$.
(b) urea, H_2NCONH_2.   (d) $(NH_4)_3PO_4$.

12-38. A 0.9092-g sample of a wheat flour was analyzed by the Kjeldahl procedure. The ammonia formed was distilled into 50.00 mL of 0.05063 M HCl; a 7.46-mL back-titration with 0.04917 M NaOH was required. Calculate the percentage of protein in the flour.

*12-39. A 1.219-g sample containing $(NH_4)_2SO_4$, NH_4NO_3, and nonreactive substances was diluted to 200 mL in a volumetric flask. A 50.00-mL aliquot was made basic with strong alkali, and the liberated NH_3 was distilled into 30.00 mL of 0.08421 M HCl. The excess HCl required 10.17 mL of 0.08802 M NaOH. A 25.00-mL aliquot of the sample was made alkaline after the addition of Devarda's alloy, and the NO_3^- was reduced to NH_3. The NH_3 from both NH_4^+ and NO_3^- was then distilled into 30.00 mL of the standard acid and back-titrated with 14.16 mL of the base. Calculate the percentage of $(NH_4)_2SO_4$ and NH_4NO_3 in the sample.

*12-40. A 1.217-g sample of commercial KOH contaminated by K_2CO_3 was dissolved in water, and the resulting solution was diluted to 500.00 mL. A 50.00-mL aliquot of this solution was treated with 40.00 mL of 0.05304 M HCl and boiled to remove CO_2. The excess acid consumed 4.74 mL of 0.04983 M NaOH (phenolphthalein indicator). An excess of neutral $BaCl_2$ was added to another 50.00-mL aliquot to precipitate the carbonate as $BaCO_3$. The solution was then titrated with 28.56 mL of the acid to a phenolphthalein end point. Calculate the percentage KOH, K_2CO_3, and H_2O in the sample, assuming that these are the only compounds present.

12-41. A 0.5000-g sample containing $NaHCO_3$, Na_2CO_3, and H_2O was dissolved and diluted to 250.0 mL. A 25.00-mL aliquot was then boiled with 50.00 mL of 0.01255 M HCl. After cooling, the excess acid in the solution required 2.34 mL of 0.01063 M NaOH when titrated to a phenolphthalein end point. A second 25.00-mL aliquot was then treated with an excess of $BaCl_2$ and 25.00 mL of the base; precipitation of all the carbonate resulted, and 7.63 mL of the HCl were required to titrate the excess base. Calculate the percent composition of the mixture.

*12-42. Calculate the volume of 0.06122 M HCl needed to titrate
(a) 20.00 mL of 0.05555 M Na_3PO_4 to a thymolphthalein end point.
(b) 25.00 mL of 0.05555 M Na_3PO_4 to a bromocresol green end point.

(c) 40.00 mL of a solution that is 0.02102 M in Na_3PO_4 and 0.01655 M in Na_2HPO_4 to a bromocresol green end point.

(d) 20.00 mL of a solution that is 0.02102 M in Na_3PO_4 and 0.01655 M in NaOH to a thymolphthalein end point.

12-43. Calculate the volume of 0.07731 M NaOH needed to titrate

(a) 25.00 mL of a solution that is 0.03000 M in HCl and 0.01000 M in H_3PO_4 to a bromocresol green end point.

(b) the solution in (a) to a thymolphthalein end point.

(c) 30.00 mL of 0.06407 M NaH_2PO_4 to a thymolphthalein end point.

(d) 25.00 mL of a solution that is 0.02000 M in H_3PO_4 and 0.03000 M in NaH_2PO_4 to a thymolphthalein end point.

*12-44. A series of solutions containing NaOH, Na_2CO_3, and $NaHCO_3$, alone or in compatible combination, was titrated with 0.1202 M HCl. Tabulated below are the volumes of acid needed to titrate 25.00-mL portions of each solution to a (1) phenolphthalein and (2) bromocresol green end point. Use this information to deduce the composition of the solutions. In addition, calculate the number of milligrams of each solute per milliliter of solution.

	(1)	(2)
(a)	22.42	22.44
(b)	15.67	42.13
(c)	29.64	36.42
(d)	16.12	32.23
(e)	0.00	33.33

12-45. A series of solutions containing NaOH, Na_3AsO_4, and Na_2HAsO_4, alone or in compatible combination, was titrated with 0.08601 M HCl. Tabulated below are the volumes of acid needed to titrate 25.00-mL portions of each solution to a (1) phenolphthalein and (2) bromocresol green end point. Use this information to deduce the composition of the solutions. In addition, calculate the number of milligrams of each solute per milliliter of solution.

	(1)	(2)
(a)	0.00	18.15
(b)	21.00	28.15
(c)	19.80	39.61
(d)	18.04	18.03
(e)	16.00	37.37

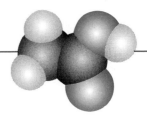

CHAPTER 13

PRECIPITATION TITRIMETRY

Precipitation titrimetry, which is based on reactions that yield ionic compounds of limited solubility, is one of the oldest analytical techniques, dating back to the mid-1800s. The slow rate of formation of most precipitates, however, limits to a handful the number of precipitating agents that can be used in titrimetry. By far the most widely used and important precipitating reagent is silver nitrate, which is used for the determination of the halogens, the halogen-like anions (SCN^-, CN^-, CNO^-), mercaptans, fatty acids, and several divalent inorganic anions. Titrimetric methods based on silver nitrate are sometimes termed *argentometric methods*. In this book, we will limit our discussion of precipitation titrimetry to argentometric methods.

> The term *argentometric* is derived from the Latin noun *argentum*, which means silver.

13A TITRATION CURVES

Titration curves for precipitation reactions are derived in a completely analogous way to the methods described in Section 10B-1 for titrations involving strong acids and strong bases. The only difference is that the solubility product of the precipitate is substituted for the ion-product constant for water. The example that follows illustrates how p-functions are derived for the pre-equivalence-point region, the equivalence point, and the post-equivalence-point region for a typical precipitation titration. Compare these calculations with those for a strong acid/base system shown in Example 10-1.

Example 13-1

Calculate the pAg of the solution during the titration of 50.00 mL of 0.0500 M NaCl with 0.1000 M $AgNO_3$ after the addition of the following volumes of reagent: (a) 0.00 mL; (b) 24.50 mL; (c) 25.00 mL; (d) 25.50 mL.

(a) Because no $AgNO_3$ has been added, $[Ag^+] = 0$ and pAg is indeterminate.

(b) At 24.50 mL added, $[Ag^+]$ is very small and cannot be computed from stoichiometric considerations, but $[Cl^-]$ can be obtained readily.

$$[Cl^-] \cong c_{NaCl} = \frac{\text{original no. mmol Cl}^- - \text{no. mol AgNO}_3}{\text{total volume of solution}}$$

$$= \frac{(50.00 \times 0.0500 - 24.50 \times 0.1000)}{50.00 + 24.50} = 6.71 \times 10^{-4}$$

$$[Ag^+] = K_{sp}/(6.71 \times 10^{-4}) = 1.82 \times 10^{-10}/(6.71 \times 10^{-4})$$

$$= 2.71 \times 10^{-7}$$

$$pAg = -\log(2.71 \times 10^{-7}) = 6.57$$

A useful relationship can be derived by taking the negative logarithm of both sides of a solubility-product expression. Thus, for silver chloride

$$-\log K_{sp} = -\log([Ag^+][Cl^-])$$
$$= -\log[Ag^+] - \log[Cl^-]$$
$$pK_{sp} = pAg + pCl$$

(c) This is the equivalence point where $[Ag^+] = [Cl^-]$ and

$$[Ag^+] = \sqrt{K_{sp}} = \sqrt{1.82 \times 10^{-10}} = 1.35 \times 10^{-5}$$

$$pAg = -\log(1.35 \times 10^{-5}) = 4.87$$

Compare this expression with the one in Example 10-1 on page 172.

This titration curve can also be derived from the charge-balance equation as shown in Feature 10-1.

(d) $[Ag^+] = c_{AgNO_3} = \dfrac{(25.50 \times 0.1000 - 50.00 \times 0.0500)}{75.50} = 6.62 \times 10^{-4}$

$$pAg = -\log(6.62 \times 10^{-4}) = 3.18$$

Figure 13-1 shows titration curves for chloride ion that have been derived employing the techniques shown in Example 13-1. Note the effect of analyte and reagent concentrations on the magnitude of the change in pAg in the equivalence-point region. The effect here is analogous to that illustrated for acid/base titrations in Figure 10-3. As shown by the shaded region in the figure, an indicator with a pAg range of 4 to 6 should provide a sharp end point in the titration of the 0.05 M chloride solution. In contrast, for the more dilute solution the end point would be drawn out over a large enough volume of reagent to make accurate establishment of the end point impossible. That is, the color change would begin at about 24 mL and be complete at about 26 mL.

Figure 13-1

Effect of titrant concentration on titration curves: *A*, 50.00 mL of 0.0500 M NaCl with 0.1000 M AgNO₃; *B*, 50.00 mL of 0.00500 M NaCl with 0.1000 M AgNO₃.

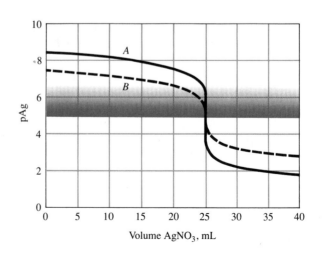

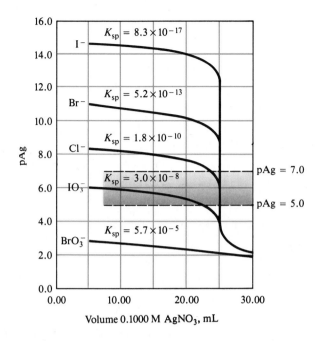

Figure 13-2

Effect of reaction completeness on titration curves. For each curve, 50.00 mL of a 0.0500 M solution of the anion was titrated with 0.1000 M $AgNO_3$.

Figure 13-2 illustrates the effect of product solubility on the sharpness of the end point in titrations with 0.1 M silver nitrate. Clearly the change in pAg at the equivalence point becomes greater as the solubility products become smaller—that is, as the reaction between the analyte and silver nitrate becomes more complete. By careful choice of indicator—one that changes color in the region of pAg from 4 to 6—titration of chloride ions is possible with a minimal titration error. Note that ions forming precipitates with solubility products much larger than about 10^{-10} do not yield satisfactory end points.

13B END POINTS FOR ARGENTOMETRIC TITRATIONS

Three types of end points are used for titrations with silver nitrate: (1) chemical indicators, (2) potentiometric, and (3) amperometric. Some chemical indicators are described in Section 13B-1. Potentiometric end points are based on measurements of the potential of a silver electrode immersed in the analyte solution; amperometric end points involve determining the current generated between a pair of silver microelectrodes in the solution of the analyte. Potentiometric indicators are discussed in Section 18G.

13B-1 Chemical Indicators for Precipitation Titrations

The end point produced by a chemical indicator usually consists of a color change or, occasionally, the appearance or disappearance of turbidity in the solution being titrated. The requirements for an indicator for a precipitation titration are analogous to those for an indicator for a neutralization titration: (1) the color change should occur over a limited range in p-function of the reagent or the analyte, and (2) the color change should take place within the steep portion of the titration curve for the analyte. For example, we see in Figure 13-2 that the indicator shown should provide a satisfactory end point for the titration of iodide and bromide ions but not chloride. An indicator with a pAg range of 6.0 to 4.0, however, should be satisfactory for chloride ions (Figure 13-1). Three indicators that have found extensive use for argentometric titrations are discussed in the sections that follow.

Chromate Ion; The Mohr Method

Sodium chromate can serve as an indicator for the argentometric determination of chloride, bromide, and cyanide ions by reacting with silver ion to form a brick-red silver chromate (Ag_2CrO_4) precipitate in the equivalence-point region. The silver ion concentration at chemical equivalence in the titration of chloride with silver ions is given by

$$[Ag^+] = \sqrt{K_{sp}} = \sqrt{1.82 \times 10^{-10}} = 1.35 \times 10^{-5}\,M$$

The chromate ion concentration required to initiate formation of silver chromate under this condition can be computed from the solubility product for silver chromate,

$$[CrO_4^{2-}] = \frac{1.2 \times 10^{-12}}{[Ag^+]^2} = \frac{1.2 \times 10^{-12}}{(1.35 \times 10^{-5})^2} = 6.6 \times 10^{-3}$$

In principle, then, an amount of chromate ion to give this concentration should be added for appearance of the red-color precipitate just after the equivalence point. In fact, however, a chromate ion concentration of 6.6×10^{-3} M imparts such an intense yellow color to the solution that formation of the red silver chromate is not readily detected and lower concentrations of chromate ion are generally used for this reason. As a consequence, excess silver nitrate is required before precipitation begins. An additional excess of the reagent must also be added to produce enough silver chromate to be seen. These two factors create a positive systematic error in the Mohr method that becomes significant in magnitude at reagent concentrations lower than about 0.1 M. A correction for this error can readily be made by a blank titration of a chloride-free suspension of calcium carbonate. Alternatively, the silver nitrate solution can be standardized against primary-standard-grade sodium chloride using the same conditions as are to be used in the analysis. This technique not only compensates for the overconsumption of reagent but also for the acuity of the analyst in detecting the appearance of the color.

The Mohr titration must be carried out at a pH of 7 to 10 because chromate ion is the conjugate base of the weak chromic acid. Consequently, in more acidic solutions the chromate ion concentration is too low to produce the precipitate at the equivalence point. Normally, a suitable pH is achieved by saturating the analyte solution with sodium hydrogen carbonate.

Adsorption Indicators; The Fajans Method

An *adsorption indicator* is an organic compound that tends to be adsorbed onto the surface of the solid in a precipitation titration. Ideally, the adsorption (or desorption) occurs near the equivalence point and results not only in a color change but also in a transfer of color from the solution to the solid (or the reverse).

The Mohr method was first described in 1865 by K. F. Mohr, a German pharmaceutical chemist who did much pioneering work in the development of titrimetry.

Mohr method for chloride.

$$Ag^+ + Cl^- \rightleftharpoons AgCl(s)$$
white

$$2Ag^+ + CrO_4^{2-} \rightleftharpoons Ag_2CrO_4(s)$$
red

Adsorption indicators were first described by Kasimir Fajans (pronounced *Fay' yahns*), a Polish chemist, in 1926.

Fluorescein is a typical adsorption indicator that is useful for the titration of chloride ion with silver nitrate. In aqueous solution, fluorescein partially dissociates into hydronium ions and negatively charged fluoresceinate ions that are yellow-green. The fluoresceinate ion forms a bright red silver salt. Whenever this dye is used as an indicator, however, *its concentration is never large enough to precipitate as silver fluoresceinate.*

In the early stages of the titration of chloride ion with silver nitrate, the colloidal silver chloride particles are negatively charged because of adsorption of excess chloride ions (Section 6A-2). The dye anions are repelled from this surface by electrostatic repulsion and impart a yellow-green color to the solution. Beyond the equivalence point, however, the silver chloride particles strongly adsorb silver ions and thereby acquire a positive charge. Fluoresceinate anions are then attracted *into the counter-ion layer* that surrounds each colloidal silver chloride particle. The net result is the appearance of the red color of silver fluoresceinate *in the surface layer of the solution surrounding the solid.* It is important to emphasize that the color change is an *adsorption* (not a precipitation) process because the solubility product of the silver fluoresceinate is never exceeded. The adsorption is reversible, the dye being desorbed on back-titration with chloride ion.

Titrations involving adsorption indicators are rapid, accurate, and reliable, but their application is limited to the relatively few precipitation reactions in which a colloidal precipitate is formed rapidly.

fluorescein

Iron(III) Ion; The Volhard Method

In the Volhard method silver ions are titrated with a standard solution of thiocyanate ion:

$$Ag^+ + SCN^- \rightleftharpoons AgSCN(s)$$

Iron(III) serves as the indicator. The solution turns red with the first slight excess of thiocyanate ion:

$$Fe^{3+} + SCN^- \rightleftharpoons \underset{red}{FeSCN^{2+}} \qquad K_f = \frac{[Fe(SCN)^{2+}]}{[Fe^{3+}][SCN^-]} = 1.05 \times 10^3$$

The titration must be carried out in acidic solution to prevent precipitation of iron(III) as the hydrated oxide.

The indicator concentration is not critical in the Volhard titration. In fact, calculations similar to those shown in Feature 13-1 demonstrate that a titration error of one part in a thousand or less is possible if the iron(III) concentration is held between 0.002 and 1.6 M. In practice, it is found that an indicator concentration greater than 0.2 M imparts sufficient color to the solution to make detection of the thiocyanate complex difficult because of the yellow color of Fe^{3+}. Therefore, lower concentrations (usually about 0.01 M) of iron(III) ion are preferred.

The most important application of the Volhard method is for the indirect determination of halide ions. A measured excess of standard silver nitrate solution is added to the sample, and the excess silver ion is determined by back-titration with a standard thiocyanate solution. The strong acid environment required for the Volhard procedure represents a distinct advantage over other methods of halide analysis because such ions as carbonate, oxalate, and arsenate (which form

> The Volhard method was first described in 1874 by Jacob Volhard, a German chemist.

Volhard method for chloride.

$$\underset{excess}{Ag^+} + Cl^- \rightleftharpoons \underset{white}{AgCl(s)}$$

$$SCN^- + Ag^+ \rightleftharpoons \underset{white}{AgSCN(s)}$$

$$Fe^{3+} + SCN^- \rightleftharpoons \underset{red}{Fe(SCN)^{2+}}$$

> The Volhard procedure requires that the analyte solution be distinctly acidic.

Feature 13-1

From experiment it has been found that the average observer can just detect the red color of $Fe(SCN)^{2+}$ when its concentration is 6.4×10^{-6} M. In the titration of 50.0 mL of 0.050 M Ag^+ with 0.100 M KSCN, what concentration of Fe^{3+} should be used in theory to yield a titration error of zero?

For a zero titration error, the $FeSCN^{2+}$ color should appear when the concentration of Ag^+ remaining in the solution is identical to the sum of the two thiocyanate species. That is, at the equivalence point

$$[Ag^+] = [SCN^-] + [Fe(SCN)^{2+}]$$
$$= [SCN^-] + 6.4 \times 10^{-6}$$

or

$$\frac{K_{sp}}{[SCN^-]} = \frac{1.1 \times 10^{-12}}{[SCN^-]} = [SCN^-] + 6.4 \times 10^{-6}$$

which rearranges to

$$[SCN^-]^2 + 6.4 \times 10^{-6}[SCN^-] - 1.1 \times 10^{-12} = 0$$
$$[SCN^-] = 1.7 \times 10^{-7}$$

The formation constant for $FeSCN^{2+}$ is

$$K_f = \frac{[Fe(SCN)^{2+}]}{[Fe^{3+}][SCN^-]} = 1.05 \times 10^3$$

If we now substitute the $[SCN^-]$ necessary to give a detectable concentration of $FeSCN^{2+}$ at the equivalence point, we obtain

$$1.05 \times 10^3 = \frac{6.4 \times 10^{-6}}{[Fe^{3+}]1.7 \times 10^{-7}}$$

$$[Fe^{3+}] = 0.036$$

slightly soluble silver salts in neutral media but not in acidic media) do not interfere.

Silver chloride is more soluble than silver thiocyanate. As a consequence, in chloride determinations by the Volhard method, the reaction

$$AgCl(s) + SCN^- \rightleftharpoons AgSCN(s) + Cl^-$$

occurs to a significant extent near the end of the back-titration of the excess silver ion. This reaction causes the end point to fade and results in an overconsumption of thiocyanate ion, which in turn leads to low values for the chloride analysis. This error can be circumvented by filtering the silver chloride before undertaking the back-titration. Filtration is not required in the determination of other halides because they all form silver salts that are less soluble than silver thiocyanate.

Example 13-2

The As in a 9.13-g sample of pesticide was converted to AsO_4^{3-} and precipitated as Ag_3AsO_4 with 50.00 mL of 0.02015 M $AgNO_3$. The excess Ag^+ was then titrated with 4.75 mL of 0.04321 M KSCN. Calculate the percentage of As_2O_3 in the sample.

$$\text{no. mmol } AgNO_3 = 50.00 \text{ mL } \cancel{AgNO_3} \times 0.02015 \frac{\text{mmol } AgNO_3}{\text{mL } \cancel{AgNO_3}} = 1.0075$$

$$\text{no. mmol KSCN} = 4.75 \text{ mL } \cancel{KSCN} \times 0.04321 \frac{\text{mmol KSCN}}{\text{mL } \cancel{KSCN}} = 0.2052$$

$$\text{no. mmol } AgNO_3 \text{ consumed by } AsO_4^{3-} = 0.8023$$

$$As_2O_3 \equiv 2AsO_4^{3-} \equiv 6AgNO_3$$

$$\text{percent } As_2O_3 = \frac{0.8023 \text{ mmol } \cancel{AgNO_3} \times \frac{1 \text{ mmol } As_2O_3}{6 \text{ mmol } AgNO_3} \times 0.1978 \frac{\text{g } As_2O_3}{\text{mmol } As_2O_3}}{9.13 \text{ g sample}} \times 100\%$$

$$= 0.2897\%$$

Table 13-1
TYPICAL ARGENTOMETRIC PRECIPITATION METHODS

Substance Being Determined	End Point	Remarks
AsO_4^{3-}, Br^-, I^-, CNO^-, SCN^-	Volhard	Removal of silver salt not required
CO_3^{2-}, CrO_4^{2-}, CN^-, Cl^-, $C_2O_4^{2-}$, PO_4^{3-}, S^{2-}, NCN^{2-}	Volhard	Removal of silver salt required before back-titration of excess Ag^+
BH_4^-	Modified Volhard	Titration of excess Ag^+ following $BH_4^- + 8Ag^+ + 8OH^- \rightarrow 8Ag(s) + H_2BO_3^- + 5H_2O$
Epoxide	Volhard	Titration of excess Cl^- following hydrohalogenation
K^+	Modified Volhard	Precipitation of K^+ with known excess of $B(C_6H_5)_4^-$, addition of excess Ag^+ giving $AgB(C_6H_5)_4(s)$, and back-titration of the excess
Br^-, Cl^-	$2 Ag^+ + CrO_4^{2-} \rightarrow Ag_2CrO_4(s)$ red	In neutral solution
Br^-, Cl^-, I^-, SeO_3^{2-}	Adsorption indicator	
$V(OH)_4^+$, fatty acids, mercaptans	Electroanalytical	Direct titration with Ag^+
Zn^{2+}	Modified Volhard	Precipitation as $ZnHg(SCN)_4$, filtration, dissolution in acid, addition of excess Ag^+, back-titration of excess Ag^+
F^-	Modified Volhard	Precipitation as $PbClF$, filtration, dissolution in acid, addition of excess Ag^+, back-titration of excess Ag^+

13C APPLICATIONS OF STANDARD SILVER NITRATE SOLUTIONS

Table 13-1 lists some typical applications of precipitation titrations in which silver nitrate is the standard solution. In most of these methods, the analyte is precipitated with a measured excess of silver nitrate and the excess determined by a Volhard titration with standard potassium thiocyanate.

Both silver nitrate and potassium thiocyanate are obtainable in primary-standard quality. However, the latter is somewhat hygroscopic, and thiocyanate solutions are ordinarily standardized against silver nitrate. Both silver nitrate and potassium thiocyanate solutions are stable indefinitely.

13D QUESTIONS AND PROBLEMS

*13-1. In what respect is the Fajans method superior to the Volhard method for the titration of chloride ion?

13-2. Briefly explain why the sparingly soluble product must be removed by filtration before you back-titrate the excess silver ion in the Volhard determination of
(a) chloride ion.
(b) cyanide ion.
(c) carbonate ion.

*13-3. Why does a Volhard determination of iodide ion require fewer steps than a Volhard determination of
(a) carbonate ion?
(b) cyanide ion?

13-4. Why does the charge on the surface of precipitate particles change sign at the equivalence point in a titration?

*13-5. Outline a method for the determination of K^+ based on argentometry. Write balanced equations for the reactions.

13-6. Suggest an argentometric method for the determination of F^-. Write balanced equations for the reactions.

13-7. A silver nitrate solution contains 14.77 g of primary-standard $AgNO_3$ in 1.00 L. What volume of this solution will be needed to react with
*(a) 0.2631 g of NaCl?
(b) 0.1799 g of Na_2CrO_4?
*(c) 64.13 mg of Na_3AsO_4?
(d) 381.1 mg of $BaCl_2 \cdot 2H_2O$?
*(e) 25.00 mL of 0.05361 M Na_3PO_4?
(f) 50.00 mL of 0.01808 M H_2S?

13-8. What is the molar analytical concentration of a silver nitrate solution if a 25.00-mL aliquot reacts with the amount of solutes listed in 13-7?

13-9. What minimum volume of 0.09621 M $AgNO_3$ will be needed to assure an excess of silver ion in the titration of
*(a) an impure NaCl sample that weighs 0.2513 g?
(b) a 0.3462-g sample that is 74.52% (w/w) $ZnCl_2$?
*(c) 25.00 mL of 0.01907 M $AlCl_3$?

13-10. A Fajans titration of a 0.7908-g sample required 45.32 mL of 0.1046 M $AgNO_3$. Express the results of this analysis in

terms of the percentage of
(a) Cl^-.
(b) $BaCl_2 \cdot 2H_2O$.
(c) $ZnCl_2 \cdot 2NH_4Cl$ (243.28 g/mol).

*13-11. Titration of a 0.485-g sample by the Mohr method required 36.8 mL of standard 0.1060 M $AgNO_3$ solution. Calculate the percentage of chloride in the sample.

13-12. A 0.1064-g sample of a pesticide was decomposed by the action of sodium biphenyl in toluene. The liberated Cl^- was extracted with water and titrated with 23.28 mL of 0.03337 M $AgNO_3$ using an adsorption indicator. Express the results of this analysis in terms of percent aldrin, $C_{12}H_8Cl_6$ (364.92 g/mol).

*13-13. A 100-mL sample of brackish water was made ammoniacal and the sulfide it contained titrated with 8.47 mL of 0.01310 M $AgNO_3$. The net reaction is

$$2Ag^+ + S^{2-} \rightarrow Ag_2S(s)$$

Calculate the parts per million of H_2S in the water.

13-14. A 2.000-L water sample was evaporated to a small volume and treated with an excess of sodium tetraphenylboron, $NaB(C_6H_5)_4$. The precipitated $KB(C_6H_5)_4$ was filtered and then redissolved in acetone. The analysis was completed by a Mohr titration, with 37.90 mL of 0.03981 M $AgNO_3$ being used. The net reaction is

$$KB(C_6H_5)_4 + Ag^+ \rightarrow AgB(C_6H_5)_4(s) + K^+$$

Express the results of this analysis in terms of parts per million of K (that is, mg K/L).

*13-15. The phosphate in a 4.258-g sample of plant food was precipitated as Ag_3PO_4 through the addition of 50.00 mL of 0.0820 M $AgNO_3$:

$$3Ag^+ + HPO_4^{2-} \rightarrow Ag_3PO_4(s) + H^+$$

The solid was filtered and washed, following which the filtrate and washings were diluted to exactly 250.0 mL.

Titration of a 50.00-mL aliquot of this solution required a 4.64-mL back-titration with 0.0625 M KSCN. Express the results of this analysis in terms of the percentage of P_2O_5.

*13-16. The monochloroacetic acid ($ClCH_2COOH$) preservative in 100.0 mL of a carbonated beverage was extracted into diethyl ether and then returned to aqueous solution as $ClCH_3COO^-$ by extraction with 1 M NaOH. This aqueous extract was acidified and treated with 50.00 mL of 0.04521 M $AgNO_3$. The reaction is

$$ClCH_2COOH + Ag^+ + H_2O \rightarrow$$
$$HOCH_2COOH + H^+ + AgCl(s)$$

After filtration of the AgCl, titration of the filtrate and washings required 10.43 mL of an NH_4SCN solution. Titration of a blank taken through the entire process used 22.98 mL of the NH_4SCN. Calculate the weight (in milligrams) of $ClCH_2COOH$ in the sample.

13-17. An analysis for borohydride ion is based on its reaction with Ag^+:

$$BH_4^- + 8Ag^+ + 8OH^- \rightarrow H_2BO_3^- + 8Ag(s) + 5H_2O$$

The purity of a quantity of KBH_4 for use in an organic synthesis was established by diluting 3.213 g of the material to exactly 500.0 mL, treating a 100.0-mL aliquot with 50.00 mL of 0.2221 M $AgNO_3$, and titrating the excess silver ion with 3.36 mL of 0.0397 M KSCN. Calculate the percent purity of the KBH_4 (53.941 g/mol).

13-18. What volume of 0.04642 M KSCN would be needed if the analysis in 13-17 were completed by filtering off the metallic Ag, dissolving it in acid, diluting to 250.0 mL, and titrating a 50.00-mL aliquot?

*13-19. An adsorption indicator is to be used in the routine determination of chloride. It is desired that the volume of standard $AgNO_3$ used in the titrations be numerically equal to the percent Cl^- when 0.2500-g samples are used. What should be the molarity of the $AgNO_3$ solution?

*13-20. The Association of Official Analytical Chemists recommends a Volhard titration for analysis of the insecticide heptachlor, $C_{10}H_5Cl_7$. The percentage of heptachlor is given by

$$\% \text{ heptachlor} = \frac{(V_{Ag} \times c_{Ag} - V_{SCN} \times c_{SCN}) \times 37.33}{\text{mass sample}}$$

What does this calculation reveal concerning the stoichiometry of this titration?

13-21. A carbonate fusion was needed to free the Bi from a 0.6423-g sample containing the mineral eulytite ($2Bi_2O_3 \cdot 3SiO_2$). The fused mass was dissolved in dilute acid, following which the Bi^{3+} was titrated with 27.36 mL of 0.03369 M NaH_2PO_4. The reaction is

$$Bi^{3+} + H_2PO_4^- \rightarrow BiPO_4(s) + 2H^+$$

Calculate the percentage purity of eulytite (1112 g/mol) in the sample.

13-22. The theobromine ($C_7H_8N_4O_2$) in a 2.95-g sample of ground cocoa beans was converted to the sparingly soluble silver salt $C_7H_7N_4O_2Ag$ by warming in an ammoniacal solution containing 25.0 mL of 0.0100 M $AgNO_3$. After reaction was complete, all solids were removed by filtration. Calculate the percentage of theobromine (180.1 g/mol) in the sample if the combined filtrate and washings required a back-titration with 7.69 mL of 0.0108 M KSCN.

*13-23. A 20-tablet sample of soluble saccharin was treated with 20.00 mL of 0.08181 M $AgNO_3$. The reaction is

After removal of the solid, titration of the filtrate and washings required 2.81 mL of 0.04124 M KSCN. Calculate the average number of milligrams of saccharin (205.17 g/mol) in each tablet.

*13-24. The formaldehyde in a 5.00-g sample of a seed disinfectant was steam-distilled and the aqueous distillate was collected in a 500-mL volumetric flask. After dilution to volume, a 25.0-mL aliquot was treated with 30.0 mL of 0.121 M KCN solution to convert the formaldehyde to potassium cyanohydrin.

$$K^+ + CH_2O + CN^- \rightarrow KOCH_2CN$$

The excess KCN was then removed by addition of 40.0 mL of 0.100 M $AgNO_3$.

$$2CN^- + 2Ag^+ \rightarrow Ag_2(CN)_2(s)$$

The excess Ag^+ in the filtrate and washings required a 16.1-mL titration with 0.134 M NH_4SCN. Calculate the percent CH_2O in the sample.

*13-25. The action of an alkaline I_2 solution on the rodenticide warfarin, $C_{19}H_{16}O_4$ (308.34 g/mol), results in the formation of 1 mol of iodoform, CHI_3 (393.73 g/mol), for each mole of the parent compound reacted. Analysis for warfarin can then be based on the reaction between CHI_3 and Ag^+:

$$CHI_3 + 3Ag^+ + H_2O \rightarrow 3AgI(s) + 3H^+ + CO(g)$$

The CHI_3 produced from a 13.96-g sample was treated with 25.00 mL of 0.02979 M $AgNO_3$, and the excess Ag^+ was then titrated with 2.85 mL of 0.05411 M KSCN. Calculate the percentage of warfarin in the sample.

13-26. A 5.00-mL aqueous suspension of elemental selenium was treated with 25.00 mL of ammoniacal 0.0360 M $AgNO_3$. Reaction:

$$6Ag(NH_3)_2^+ + 3Se(s) + 3H_2O \rightarrow$$
$$2Ag_2Se(s) + Ag_2SeO_3(s) + 6NH_4^+$$

After this reaction was complete, nitric acid was added to dissolve the Ag_2SeO_3 but not the Ag_2Se. The Ag^+ from the dissolved Ag_2SeO_3 and the excess reagent required 16.74 mL of 0.01370 M KSCN in a Volhard titration. How many milligrams of Se were contained in each milliliter of sample?

13-27. A 2.4414-g sample containing KCl, K_2SO_4, and inert materials was dissolved in sufficient water to give 250.0 mL of solution. A Mohr titration of a 50.00-mL aliquot required 41.36 mL of 0.05818 M $AgNO_3$. A second 50.00-mL aliquot was treated with 40.00 mL of 0.1083 M $NaB(C_6H_5)_4$. The reaction is

$$NaB(C_6H_5)_4 + K^+ \rightarrow KB(C_6H_5)_4(s) + Na^+$$

The solid was filtered, redissolved in acetone, and titrated with 49.98 mL of the $AgNO_3$ solution (see 13-14). Calculate the percentages of KCl and K_2SO_4 in the sample.

*13-28. A 1.998-g sample containing Cl^- and ClO_4^- was dissolved in sufficient water to give 250.0 mL of solution. A 50.00-mL aliquot required 13.97 mL of 0.08551 M $AgNO_3$ to titrate the Cl^-. A second 50.00-mL aliquot was treated with $V_2(SO_4)_3$ to reduce the ClO_4^- to Cl^-:

$$ClO_4^- + 4V_2(SO_4)_3 + 4H_2O \rightarrow$$
$$Cl^- + 12SO_4^{2-} + 8VO^{2+} + 8H^+$$

Titration of the reduced sample required 40.12 mL of the $AgNO_3$ solution. Calculate the percentages of Cl^- and ClO_4^- in the sample.

13-29. For each of the following precipitation titrations, calculate the cation and anion concentrations at equivalence as well as at reagent volumes corresponding to ±20.00 mL, ±10.00 mL, and ±1.00 mL of equivalence. Construct a titration curve from the data, plotting the p-function of the cation versus reagent volume.

*(a) 25.00 mL of 0.05000 M $AgNO_3$ with 0.02500 M NH_4SCN

(b) 20.00 mL of 0.06000 M $AgNO_3$ with 0.03000 M KI

*(c) 30.00 mL of 0.07500 M $AgNO_3$ with 0.07500 M NaCl

(d) 35.00 mL of 0.4000 M Na_2SO_4 with 0.2000 M $Pb(NO_3)_2$

*(e) 40.00 mL of 0.02500 M $BaCl_2$ with 0.05000 M Na_2SO_4

(f) 50.00 mL of 0.2000 M NaI with 0.4000 M $TlNO_3$ (K_{sp} for TlI = 6.5×10^{-8})

13-30. Calculate the silver ion concentration after the addition of 5.00, 15.00, 25.00, 30.00, 35.00, 39.00, 40.00, 41.00, 45.00, and 50.00 mL of 0.05000 M $AgNO_3$ to 50.0 mL of 0.0400 M KBr. Construct a titration curve from these data plotting pAg as a function of titrant volume.

CHAPTER 14

COMPLEX-FORMATION TITRATIONS

Complex-formation reagents are widely used for titrating cations. The most useful of these reagents are organic compounds that contain several electron-donor groups that are capable of forming multiple covalent bonds with metal ions.

14A COMPLEX-FORMATION REACTIONS

Most metal ions react with electron-pair donors to form coordination compounds or complexes. The donor species, or *ligand*, must have at least one pair of unshared electrons available for bond formation. Water, ammonia, and halide ions are common inorganic ligands.

A ligand is an ion or a molecule that forms a covalent bond with a cation or a neutral metal atom by donating a pair of electrons, which are then shared by the two.

The number of covalent bonds that a cation tends to form with electron donors is its *coordination number*. Typical values for coordination numbers are two, four, and six. The species formed as a result of coordination can be electrically positive, neutral, or negative. For example, copper(II), which has a coordination number of four, forms a cationic ammine complex, $Cu(NH_3)_4^{2+}$; a neutral complex with glycine, $Cu(NH_2CH_2COO)_2$; and an anionic complex with chloride ion, $CuCl_4^{2-}$.

Titrimetric methods based on complex formation (sometimes called *complexometric methods*) have been used for more than a century. The truly remarkable growth in their analytical application began in the 1940s, however, and is based on a particular class of coordination compounds called *chelates*. A chelate is produced when a metal ion coordinates with two or more donor groups of a single ligand to form a five- or six-membered heterocyclic ring. The copper complex of glycine, mentioned in the previous paragraph, is an example. Here, copper bonds to both the oxygen of the carboxyl group and the nitrogen of the

Chelate is pronounced *kee'late* and is derived from the Greek word for claw.

amine group:

$$Cu^{2+} + 2H-\overset{\overset{\displaystyle NH_2}{|}}{\underset{\underset{\displaystyle H}{|}}{C}}-\overset{\overset{\displaystyle O}{\|}}{C}-OH \longrightarrow \quad \text{[Cu complex]} \quad + 2H^+$$

Dentate (Latin) means having tooth-like projections.

A ligand that has a single donor group, such as ammonia, is called *unidentate* (single-toothed), whereas one such as glycine, which has two groups available for covalent bonding, is called *bidentate*. *Tridentate, tetradentate, pentadentate*, and *hexadentate* chelating agents are also known.

As titrants, multidentate ligands, particularly those having four or six donor groups, have two advantages over their unidentate counterparts. First, they generally react more completely with cations and thus provide sharper end points. Second, they ordinarily react with metal ions in a single-step process, whereas complex formation with unidentate ligands usually involves two or more intermediate species.

The advantage of a single-step reaction is illustrated by the titration curves shown in Figure 14-1. Each of the titrations involves a reaction that has an overall equilibrium constant of 10^{20}. Curve *A* is derived for a reaction in which a metal ion M having a coordination number of four reacts with a tetradentate ligand D to form the complex of MD (here we have omitted the charges on the two reactants for convenience). Curve *B* is for the reaction of M with a hypothetical bidentate ligand B to give MB_2 in two steps. The formation constant for the first step is 10^{12} and for the second 10^8. Curve *C* involves a unidentate ligand, A, that forms MA_4 in four steps with successive formation constants of 10^8, 10^6, 10^4, and 10^2. These curves demonstrate that a much sharper end point is obtained with a reaction that takes place in a single step. For this reason, multidentate ligands are ordinarily preferred for complexometric titrations.

Tetradentate or hexadentate ligands are more satisfactory as titrants than ligands with a lesser number of donor groups because their reactions with cations are more complete and because they tend to form 1 : 1 complexes.

14B TITRATIONS WITH AMINOPOLYCARBOXYLIC ACIDS

Tertiary amines that also contain carboxylic acid groups form remarkably stable chelates with many metal ions.[1] Schwarzenbach first recognized their potential as analytical reagents in 1945. Since this original work, investigators throughout the world have described applications of these compounds to the volumetric determination of most of the metals in the periodic table.

14B-1 Ethylenediaminetetraacetic Acid (EDTA)

Ethylenediaminetetraacetic acid [also called (ethylenedinitrilo)tetraacetic acid], which is commonly shortened to EDTA, is the most widely used complexometric

---

[1] These reagents are the subject of several excellent monographs. See, for example, R. Pribil, *Applied Complexometry*. New York: Pergamon, 1982; A. Ringbom and E. Wanninen, in *Treatise on Analytical Chemistry*, 2nd ed., I. M. Kolthoff and P. J. Elving, Eds., Part I, Vol. 2, Chapter 11. New York: Wiley, 1979.

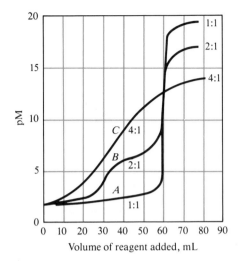

Figure 14-1
Curves for complex-formation titrations. Titration of 60.0 mL of a solution that is 0.020 M in M with (*A*) a 0.020 M solution of the tetraedentate ligand D to give MD as product; (*B*) a 0.040 M solution of the bidentate ligand B to give MB_2; and (*C*) a 0.080 M solution of the unidentate ligand A to give MA_4. The overall formation constant for each product is 1.0×10^{20}.

titrant. EDTA has the structure (see also figure in the margin)

$$HOOC-CH_2 \diagdown \qquad \qquad \diagup CH_2-COOH$$
$$:N-CH_2-CH_2-N:$$
$$HOOC-CH_2 \diagup \qquad \qquad \diagdown CH_2-COOH$$

The EDTA molecule is a hexadentate ligand that has six potential sites for bonding a metal ion: the four carboxyl groups and the two amino groups, each of the latter with an unshared pair of electrons.

EDTA, a hexadentate ligand, is among the most important and widely used reagents in titrimetry.

Acidic Properties of EDTA

The dissociation constants for the acidic groups in EDTA are $K_1 = 1.02 \times 10^{-2}$, $K_2 = 2.14 \times 10^{-3}$, $K_3 = 6.92 \times 10^{-7}$, and $K_4 = 5.50 \times 10^{-11}$. It is of interest that the first two constants are of the same order of magnitude, which suggests that the two protons involved dissociate from opposite ends of the rather long molecule. As a consequence of their physical separation, the negative charge created by the first dissociation does not greatly affect the removal of the second proton. The same cannot be said for the dissociation of the other two protons, however, which are much closer to the negatively charged carboxylate ions created by the initial dissociations.

The various EDTA species are often abbreviated H_4Y, H_3Y^-, H_2Y^{2-}, HY^{3-}, and Y^{4-}. Figure 14-2 illustrates how the relative amounts of these five species vary as a function of pH. It is apparent that H_2Y^{2-} predominates in moderately acidic medium (pH 3 to 6). Only at pH values greater than 10 does Y^{4-} become a major component of solutions.

Reagents

The free acid, H_4Y, and the dihydrate of the sodium salt, $Na_2H_2Y \cdot 2H_2O$, are commercially available in reagent quality. The former can serve as a primary

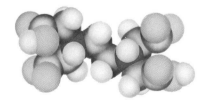

Space-filling model of H_4Y zwitterion.

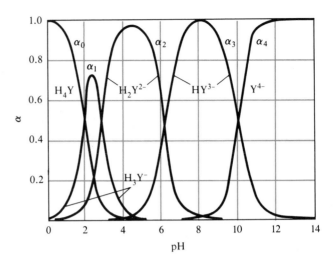

Figure 14-2
Composition of EDTA solutions as a function of pH.

Standard EDTA solutions are ordinarily prepared by dissolving weighed quantities of $Na_2H_2Y \cdot 2H_2O$ and diluting to the mark in a volumetric flask.

Nitrilotriacetic acid (NTA) is the second most common aminopolycarboxylic acid used for titrimetry. It is a tetradentate chelating agent having the structure

$$HOOC—CH_2\diagdown\quad\diagup CH_2—COOH$$
$$N$$
$$\diagdown CH_2—COOH$$

The reaction of the EDTA anion with a metal ion M^{n+} is described by the equation

$$M^{n+} + Y^{4-} \rightleftharpoons MY^{(n-4)+}$$

standard after it has been dried for several hours at 130°C to 145°C. It is then dissolved in the minimum amount of base required for complete solution.

Under normal atmospheric conditions, the dihydrate $Na_2H_2Y \cdot 2H_2O$ contains 0.3% moisture in excess of the stoichiometric amount. For all but the most exacting work, this excess is sufficiently reproducible to permit use of a corrected weight of the salt in the direct preparation of a standard solution. If necessary, the pure dihydrate can be prepared by drying at 80°C for several days in an atmosphere of 50% relative humidity.

A number of compounds that are chemically related to EDTA have also been investigated but do not appear to offer advantages over the original species. We shall thus confine our discussion to the properties and applications of EDTA.

14B-2 Complexes of EDTA and Metal Ions

Solutions of EDTA are particularly valuable as titrants because the reagent combines with metal ions in a 1 : 1 ratio, regardless of the charge on the cation. For example, formation of the silver and aluminum complexes is described by the equations

$$Ag^+ + Y^{4-} \rightleftharpoons AgY^{3-}$$
$$Al^{3+} + Y^{4-} \rightleftharpoons AlY^-$$

EDTA is a remarkable reagent not only because it forms chelates with all cations but also because most of these chelates are sufficiently stable to form the basis for a titrimetric method. This great stability undoubtedly results from the several complexing sites within the molecule that give rise to a cage-like structure in which the cation is effectively surrounded and isolated from solvent molecules. One form of the complex is depicted in Figure 14-4. Note that six donor atoms are involved in bonding the divalent metal ion.

Table 14-1 lists formation constants K_{MY} for common EDTA complexes. Note that the constant refers to the equilibrium involving the species Y^{4-} with the metal ion:

$$M^{n+} + Y^{4-} \rightleftharpoons MY^{(n-4)+} \qquad K_{MY} = \frac{[MY^{(n-4)+}]}{[M^{n+}][Y^{4-}]} \qquad \textbf{(14-1)}$$

Feature 14-1
SPECIES PRESENT IN A SOLUTION OF EDTA

When it is dissolved in water, EDTA behaves like an amino acid, such as glycine (see Feature 3-1). With EDTA, however, a double zwitterion forms, which has the structure shown in Figure 14-3a. Note that the net charge on this species is zero and that it contains four acidic protons, two associated with two of the carboxyl groups and the other two with the two amine groups. For simplicity, we generally formulate the double zwitterion as H_4Y, where Y^{4-} is the fully deprotonated ion of Figure 14-3e. The first and second steps in the dissociation process involve successive loss of protons from the two carboxylic acid groups; the third and fourth steps involve dissociation of the protonated amine groups. The structures of H_3Y^-, H_2Y^{2-}, and HY^{3-} are shown in Figure 14-3b, c, and d.

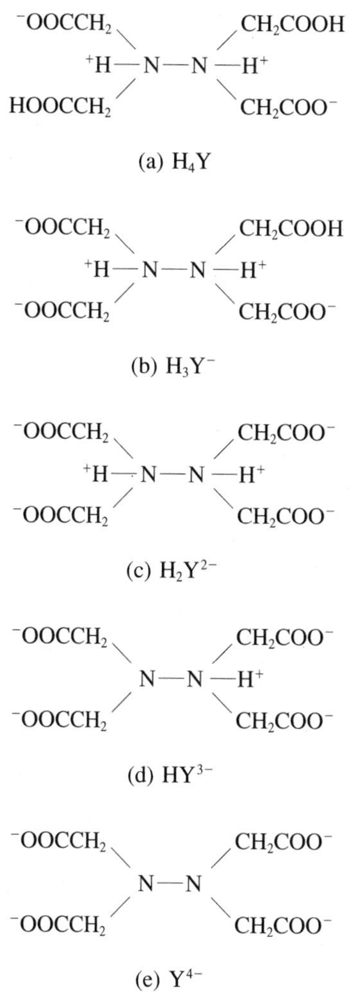

(a) H_4Y

(b) H_3Y^-

(c) H_2Y^{2-}

(d) HY^{3-}

(e) Y^{4-}

Figure 14-3
Structure of H_4Y and its dissociation products.

14B-3 Equilibrium Calculations Involving EDTA

A titration curve for the reaction of a cation M^{n+} with EDTA consists of a plot of pM versus reagent volume. Values for pM are readily computed in the early stage of a titration by assuming that the equilibrium concentration of M^{n+} is equal to its analytical concentration, which in turn is readily derived from stoichiometric data.

Calculation of M^{n+} at and beyond the equivalence point requires the use of Equation 14-1. The computation in this region is troublesome and time-consuming if the pH is unknown and variable because both $[MY^{(n-4)+}]$ and $[M^{n+}]$ are pH dependent. Fortunately EDTA titrations are always performed in solutions that are buffered to a known pH to avoid interferences by other cations or to ensure

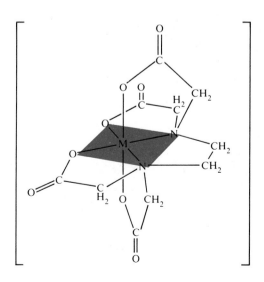

Figure 14-4
Structure of a metal/EDTA chelate.

Table 14-1
FORMATION CONSTANTS FOR EDTA COMPLEXES*

Cation	K_{MY}	$\log K_{MY}$	Cation	K_{MY}	$\log K_{MY}$
Ag^+	2.1×10^7	7.32	Cu^{2+}	6.3×10^{18}	18.80
Mg^{2+}	4.9×10^8	8.69	Zn^{2+}	3.2×10^{16}	16.50
Ca^{2+}	5.0×10^{10}	10.70	Cd^{2+}	2.9×10^{16}	16.46
Sr^{2+}	4.3×10^8	8.63	Hg^{2+}	6.3×10^{21}	21.80
Ba^{2+}	5.8×10^7	7.76	Pb^{2+}	1.1×10^{18}	18.04
Mn^{2+}	6.2×10^{13}	13.79	Al^{3+}	1.3×10^{16}	16.13
Fe^{2+}	2.1×10^{14}	14.33	Fe^{3+}	1.3×10^{25}	25.1
Co^{2+}	2.0×10^{16}	16.31	V^{3+}	7.9×10^{25}	25.9
Ni^{2+}	4.2×10^{18}	18.62	Th^{4+}	1.6×10^{23}	23.2

* Data from G. Schwarzenbach, *Complexometric Titrations*, p. 8. London: Chapman and Hall, 1957. With permission. Constants valid at 20°C and an ionic strength of 0.1.

satisfactory indicator behavior. Calculating $[M^{n+}]$ in a buffered solution containing EDTA is a relatively straightforward procedure provided the pH is known. In this computation, use is made of the alpha value for H_4Y. Recall (Section 11E) that α_4 for H_4Y would be defined as

$$\alpha_4 = \frac{[Y^{4-}]}{c_T} \tag{14-2}$$

where c_T is the total molar concentration of *uncomplexed* EDTA:

$$c_T = [Y^{4-}] + [HY^{3-}] + [H_2Y^{2-}] + [H_3Y^{3-}] + [H_4Y]$$

Conditional Formation Constants

Conditional, or *effective*, *formation constants* are pH-dependent equilibrium constants that *apply at a single pH only*. To obtain the conditional constant for the equilibrium shown in Equation 14-1, we substitute $\alpha_4 c_T$ from Equation 14-2 for $[Y^{4-}]$ in the formation-constant expression (Equation 14-1):

Conditional formation constants are pH-dependent.

$$M^{n+} + Y^{4-} \rightleftharpoons MY^{(n-4)+} \qquad K_{MY} = \frac{[MY^{(n-4)+}]}{[M^{n+}]\alpha_4 c_T} \tag{14-3}$$

Combining the two constants, α_4 and K_{MY}, yields a new constant:

$$K'_{MY} = \alpha_4 K_{MY} = \frac{[MY^{(n-4)+}]}{[M^{n+}]c_T} \tag{14-4}$$

where K'_{MY}, the conditional formation constant, describes equilibrium relationships *only at the pH for which α_4 is applicable.*

Conditional constants are readily computed and provide a simple means by which the equilibrium concentrations of the metal ion and the complex can be calculated at the equivalence point and where there is an excess of chelating reagent. Note that replacement of $[Y^{4-}]$ with c_T in the equilibrium-constant expression greatly simplifies calculations because c_T is easily determined from the reaction stoichiometry, whereas $[Y^{4-}]$ is not.

The Computation of α_4 Values for EDTA Solutions

An expression for calculating α_4 at a given hydrogen ion concentration is derived by the method demonstrated in Section 11E (see Feature 11-4). Thus, α_4 for EDTA is

$$\alpha_4 = \frac{K_1K_2K_3K_4}{[H^+]^4 + K_1[H^+]^3 + K_1K_2[H^+]^2 + K_1K_2K_3[H^+] + K_1K_2K_3K_4} \quad (14\text{-}5)$$

$$\alpha_4 = \frac{K_1K_2K_3K_4}{D} \quad (14\text{-}6)$$

where K_1, K_2, K_3, and K_4 are the four dissociation constants for H_4Y and D is the denominator of Equation 14-5.

Table 14-2 lists α_4 at selected pH values. Note that only about 4×10^{-12} percent of EDTA exists as Y^{4-} at pH 2.00. Example 14-1 illustrates how $[Y^{4-}]$ is calculated for a solution of known pH.

Example 14-1

Calculate the molar Y^{4-} concentration in a 0.0200 M EDTA solution that has been buffered to a pH of 10.00.

At pH 10.00, α_4 is 0.35 (Table 14-2). Thus,

$$[Y^{4-}] = \alpha_4 c_T = (0.35)(0.0200) = 7.0 \times 10^{-3} \text{ M}$$

Calculation of the Cation Concentration in EDTA Solutions

Example 14-2 demonstrates how the cation concentration in a solution of an EDTA complex is computed. Example 14-3 illustrates this calculation for a solution that contains an excess of EDTA.

Example 14-2

Calculate the equilibrium concentration of Ni^{2+} in a solution with an analytical NiY^{2-} concentration of 0.0150 M at pH (a) 3.0 and (b) 8.0.

From Table 14-1,

$$Ni^{2+} + Y^{4-} \rightleftharpoons NiY^{2-} \qquad K_{MY} = \frac{[NiY^{2-}]}{[Ni^{2-}][Y^{4-}]} = 4.2 \times 10^{18}$$

Table 14-2
VALUES FOR α_4 FOR EDTA AT SELECTED pH VALUES

pH	α_4	pH	α_4
2.0	3.7×10^{-14}	7.0	4.8×10^{-4}
3.0	2.5×10^{-11}	8.0	5.4×10^{-3}
4.0	3.6×10^{-9}	9.0	5.2×10^{-2}
5.0	3.5×10^{-7}	10.0	3.5×10^{-1}
6.0	2.2×10^{-5}	11.0	8.5×10^{-1}
		12.0	9.8×10^{-1}

Alpha values for the other EDTA species are obtained in a similar way and are found to be

$$\alpha_0 = [H^+]^4/DK_1 \qquad \alpha_2 = K_1K_2[H^+]^2/DK$$
$$\alpha_1 = K_1[H^+]^3/D \qquad \alpha_3 = K_1K_2K_3[H^+]/D$$

Only α_4 is needed, however, in deriving titration curves.

In this chapter and those that follow, we shall revert to the use of H^+ as a convenient shorthand notation for H_3O^+; from time to time, we shall refer to the species represented by H^+ as the hydrogen ion.

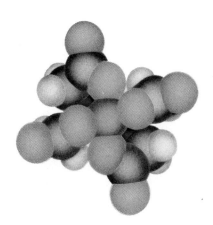

Space-filling model of NiY^{2-}.

The equilibrium concentration of NiY^{4-} is equal to the analytical concentration of the complex minus the concentration lost by dissociation. The latter is identical to the equilibrium nickel ion concentration. Thus,

$$[NiY^{4-}] = 0.0150 - [Ni^{2+}]$$

If we assume $[Ni^{2+}] \ll 0.0150$, an assumption that is almost certainly valid in light of the large formation constant of the complex, the foregoing equation simplifies to

$$[NiY^{2-}] \cong 0.0150$$

Since the complex is the only source of both Ni^{2+} and the EDTA species,

$$[Ni^{2+}] = [Y^{4-}] + [HY^{3-}] + [H_2Y^{2-}] + [H_3Y^-] + [H_4Y] = c_T$$

Substitution of this equality into Equation 14-4 gives

$$K'_{MY} = \frac{[NiY^{2-}]}{[Ni^{2+}]c_T} = \frac{[NiY^{2-}]}{[Ni^{2+}]^2} = \alpha_4 K_{MY}$$

(a) Table 14-2 indicates that α_4 is 2.5×10^{-11} at pH 3.0. Substitution of this value and the concentration of NiY^{2-} into the equation for K'_{MY} gives

$$\frac{0.0150}{[Ni^{2+}]^2} = 2.5 \times 10^{-11} \times 4.2 \times 10^{18} = 1.05 \times 10^8$$

$$[Ni^{2+}] = \sqrt{1.43 \times 10^{-10}} = 1.2 \times 10^{-5}\ M$$

Our assumption that $[Ni^{2+}] \ll [NiY^{2-}]$ is valid.

Note that $[Ni^{2+}] \ll 0.0150$, as assumed.

(b) At pH 8.0, the conditional constant is much larger. Thus,

$$K'_{MY} = 5.4 \times 10^{-3} \times 4.2 \times 10^{18} = 2.27 \times 10^{16}$$

and substitution into the equation for K'_{MY} followed by rearrangement gives

$$[Ni^{2+}] = \sqrt{0.0150/(2.27 \times 10^{16})} = 8.1 \times 10^{-10}\ M$$

Example 14-3

Calculate the concentration of Ni^{2+} in a solution that was prepared by mixing 50.0 mL of 0.0300 M Ni^{2+} with 50.0 mL of 0.0500 M EDTA. The mixture was buffered to a pH of 3.00.

Here, the solution has an excess of EDTA, and the analytical concentration of the complex is determined by the amount of Ni^{2+} originally present. Thus,

$$c_{NiY^{2-}} = 50.0 \times \frac{0.03000}{100} = 0.0150\ M$$

$$c_{EDTA} = \frac{50.0 \times 0.0500 - 50.0 \times 0.0300}{100} = 0.0100\ M$$

Again let us assume that $[Ni^{2+}] \ll [NiY^{2-}]$ so that

$$[NiY^{2-}] = 0.0150 - [Ni^{2+}] \cong 0.0150$$

At this point, the total concentration of uncomplexed EDTA is given by its molarity:

$$c_T = 0.0100 \text{ M}$$

Substitution into Equation 14-4 gives

$$K'_{MY} = \frac{0.0150}{[Ni^{2+}]0.0100} = \alpha_4 K_{MY}$$
$$= 2.5 \times 10^{-11} \times 4.2 \times 10^{18} = 1.05 \times 10^8$$

$$[Ni^{2+}] = \frac{0.0150}{0.0100 \times 1.05 \times 10^8} = 1.4 \times 10^{-8} \text{ M}$$

Again, our assumption that $[Ni^{2+}] \ll [NiY^{2-}]$ is seen to be valid.

14B-4 The Derivation of EDTA Titration Curves

Example 14-4 demonstrates how an EDTA titration curve for a metal ion is derived for a solution of fixed pH.

Example 14-4

Derive a curve (pCa as a function of volume of EDTA) for the titration of 50.0 mL of 0.00500 M Ca^{2+} with 0.0100 M EDTA in a solution buffered to a constant pH of 10.0.

Calculation of a Conditional Constant

The conditional formation constant for the calcium/EDTA complex at pH 10 is obtained from the formation constant of the complex (Table 14-1) and the α_4 value for EDTA at pH 10 (Table 14-2). Thus, substitution into Equation 14-4 gives

$$K'_{CaY} = \frac{[CaY^{2-}]}{[Ca^{2+}]c_T} = \alpha_4 K_{CaY}$$
$$= 0.35 \times 5.0 \times 10^{10} = 1.75 \times 10^{10}$$

Preequivalence-Point Values for pCa

Before the equivalence point is reached, the equilibrium concentration of Ca^{2+} is equal to the sum of the contributions from the untitrated excess of the cation and from the dissociation of the complex, the latter being numerically equal to c_T. It is ordinarily reasonable to assume that c_T is small relative to the analytical concentration of the uncomplexed calcium ion. Thus, for example, after the addition of 10.0 mL of reagent,

$$[Ca^{2+}] = \frac{50.0 \times 0.00500 - 10.0 \times 0.0100}{60.0} + c_T \cong 2.50 \times 10^{-3} \text{ M}$$

$$pCa = -\log 2.50 \times 10^{-3} = 2.60$$

Other preequivalence-point data are derived in this same way.

The Equivalence-Point pCa

Following the method shown in Example 14-2, we first compute the analytical concentration of CaY^{2-}:

$$c_{CaY^{2-}} = \frac{50.0 \times 0.00500}{50.0 + 25.0} = 3.33 \times 10^{-3}\,M$$

The only source of Ca^{2+} ions is the dissociation of this complex. It also follows that the calcium ion concentration must be identical to the sum of the concentrations of the uncomplexed EDTA ions, c_T. Thus,

$$[Ca^{2+}] = c_T$$
$$[CaY^{2-}] = 0.00333 - [Ca^{2+}] \cong 0.00333\,M$$

Substituting into the conditional formation-constant expression gives

$$\frac{[CaY^{2-}]}{[Ca^{2+}]c_T} = \frac{0.00333}{[Ca^{2+}]^2} = 1.75 \times 10^{10}$$

$$[Ca^{2+}] = \sqrt{\frac{0.00333}{1.75 \times 10^{10}}} = 4.36 \times 10^{-7}\,M$$

$$pCa = -\log 4.36 \times 10^{-7} = 6.36$$

Postequivalence-Point pCa

Beyond the equivalence point, analytical concentrations of CaY^{2-} and EDTA are obtained directly from the stoichiometric data. A calculation similar to that in Example 14-3 is then performed. Thus, after the addition of 35.0 mL of reagent,

$$c_{CaY^{2-}} = \frac{50.0 \times 0.00500}{50.0 + 35.0} = 2.94 \times 10^{-3}\,M$$

$$c_{EDTA} = \frac{35.0 \times 0.0100 - 50.0 \times 0.00500}{85.0} = 1.18 \times 10^{-3}\,M$$

As an approximation, we can write

$$[CaY^{2-}] = 2.94 \times 10^{-3} - [Ca^{2+}] \cong 2.94 \times 10^{-3}$$
$$c_T = 1.18 \times 10^{-3} + [Ca^{2+}] \cong 1.18 \times 10^{-3}\,M$$

and substitution into the conditional formation-constant expression gives

$$K'_{CaY} = \frac{2.94 \times 10^{-3}}{[Ca^{2+}] \times 1.18 \times 10^{-3}} = 1.75 \times 10^{10}$$

$$[Ca^{2+}] = \frac{2.94 \times 10^{-3}}{1.18 \times 10^{-3} \times 1.75 \times 10^{10}} = 1.42 \times 10^{-10}$$

$$pCa = -\log 1.42 \times 10^{-10} = 9.85$$

The approximation was clearly valid.

We compute other postequivalence-point data in this same way.

Curve A in Figure 14-5 is a plot of data for the titration in Example 14-4. Curve B is the titration curve for a solution of magnesium ion under identical conditions. The formation constant for the EDTA complex of magnesium is smaller than that of the calcium complex. Consequently, the reaction of calcium ion with the EDTA is more complete, and a larger change in p-function is observed in the equivalence region. The effect is analogous to what we have seen earlier for precipitation and neutralization titrations.

Figure 14-6 provides titration curves for calcium ion in solutions buffered to various pH levels. Recall that α_4, and hence K'_{CaY}, becomes smaller as the pH decreases. The less favorable equilibrium constant leads to a smaller change in pCa in the equivalence-point region. It is apparent from Figure 14-6 that an adequate end point in the titration of calcium requires a pH of about 8 or greater. As shown in Figure 14-7, however, cations with larger formation constants provide good end points even in acidic media. Figure 14-8 shows the minimum permissible pH for a satisfactory end point in the titration of various metal ions in the absence of competing complexing agents. Note that a moderately acidic environment is satisfactory for many divalent heavy-metal cations and that a strongly acidic medium can be tolerated in the titration of such ions as iron(III) and indium(III).

End points for EDTA titrations become less sharp with pH decreases because the complex-formation reaction is less complete under these circumstances.

Figure 14-8 shows that most cations having a +3 or +4 charge can be titrated in a distinctly acidic solution.

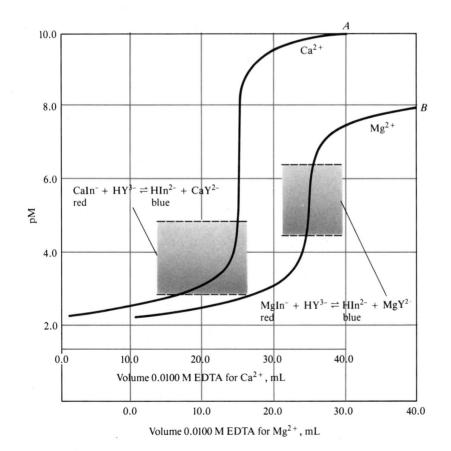

Figure 14-5

EDTA titration curves for 50.00 mL of 0.00500 M Ca^{2+} (K' for CaY^{2-} = 1.75×10^{10}) and Mg^{2+} (K' for MgY^{2-} = 1.72×10^{8}) at pH 10.0. The shaded areas show the transition range for Eriochrome Black T.

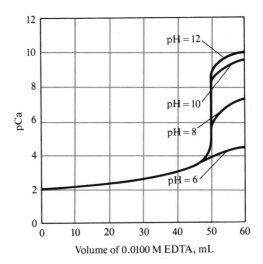

Figure 14-6

Influence of pH on the titration of 0.0100 M Ca^{2+} with 0.0100 M EDTA.

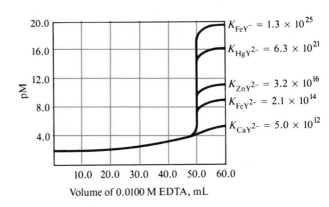

Figure 14-7

Titration curves for 50.0 mL of 0.0100 M cation solutions at pH 6.0.

14B-5 The Effect of Other Complexing Agents on EDTA Titration Curves

Many cations form hydrous oxide precipitates at the pH required for their titration with EDTA. When this problem is encountered, an auxiliary complexing agent is needed to keep the cation in solution. For example, zinc(II) is ordinarily titrated in a medium that has fairly high concentrations of ammonia and ammonium chloride. These species buffer the solution to a pH that will ensure a complete reaction between cation and titrant; in addition, ammonia forms ammine complexes with zinc(II) and prevents formation of the sparingly soluble zinc hydroxide, particularly in the early stages of the titration. A somewhat more realistic description of the reaction is then

$$Zn(NH_3)_4^{2+} + HY^{3-} \rightarrow ZnY^{2-} + 3NH_3 + NH_4^+$$

The solution also contains such other zinc/ammonia species as $Zn(NH_3)_3^{2+}$, $Zn(NH_3)_2^{2+}$, and $Zn(NH_3)^{2+}$. Calculation of pZn in a solution that contains ammonia must take these species into account.[2] Qualitatively, complexation of a cation

Often, auxiliary complexing agents must be used in EDTA titrations to prevent precipitation of the analyte as a hydrous oxide. Such reagents cause end points to be less sharp.

---

[2] See, for example, D. A. Skoog, D. M. West, and F. J. Holler, *Fundamentals of Analytical Chemistry*, 6th ed., pp. 296–301. Philadelphia: Saunders College Publishing, 1992.

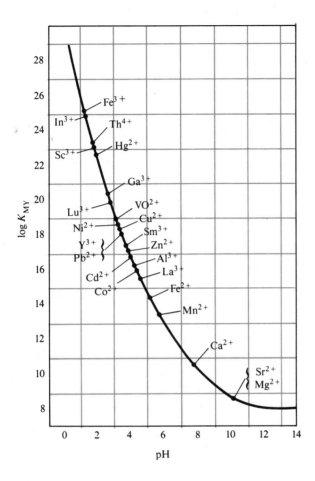

Figure 14-8
Minimum pH needed for satisfactory titration of various cations with EDTA. (From C. N. Reilley and R. W. Schmid, *Anal. Chem.*, **1958**, *30*, 947. With permission of the American Chemical Society.)

by an auxiliary complexing reagent causes preequivalence pM values to be larger than in a comparable solution with no such reagent.

Figure 14-9 shows two theoretical curves for the titration of zinc(II) with EDTA at pH 9.00. The equilibrium concentration of ammonia was 0.100 M for one titration and 0.0100 M for the other. Note that the presence of ammonia decreases the change in pZn near the equivalence point. For this reason, the concentration of auxiliary complexing reagents should always be kept to the minimum required to prevent precipitation of the analyte. Note that the auxiliary complexing agent does not affect pZn beyond the equivalence point. On the other hand, keep in mind that α_4, and thus pH, plays an important role in defining this part of the titration curve (Figure 14-6).

14B-6 Indicators for EDTA Titrations

Reilley and Barnard[3] have listed nearly 200 organic compounds that have been investigated as indicators for metal ions in EDTA titrations. In general, these indicators are organic dyes that form colored chelates with metal ions in a pM

---

[3] C. N. Reilley and A. J. Barnard Jr., in *Handbook of Analytical Chemistry*, L. Meites, Ed., p. **3**–77. New York: McGraw-Hill, 1963.

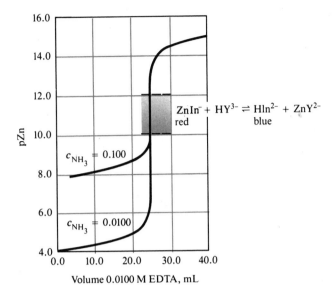

Figure 14-9

Influence of ammonia concentration on the end point for the titration of 50.0 mL of 0.00500 M Zn^{2+}. Solutions buffered to pH 9.00. The shaded area shows the transition range for Eriochrome Black T.

range that is characteristic of the particular cation and dye. The complexes are often intensely colored and are discernible to the eye at concentrations in the range of 10^{-6} to 10^{-7} M.

Eriochrome Black T is a typical metal-ion indicator that is used in the titration of several common cations. As shown in Figure 14-10, this compound contains a sulfonic acid group, which is completely dissociated in water and two phenolic groups that only partially dissociate. Its behavior as a weak acid is described by the equations

$$H_2O + \underset{red}{H_2In^-} \rightleftharpoons \underset{blue}{HIn^{2-}} + H_3O^+ \qquad K_1 = 5 \times 10^{-7}$$

$$H_2O + \underset{blue}{HIn^{2-}} \rightleftharpoons \underset{orange}{In^{3-}} + H_3O^+H \qquad K_2 = 2.8 \times 10^{-12}$$

Note that the acids and their conjugate bases have different colors. Thus, Eriochrome Black T behaves as an acid/base indicator as well as a metal-ion indicator.

The metal complexes of Eriochrome Black T are generally red, as is H_2In^-. Thus, for metal ion detection, it is necessary to adjust the pH to 7 or above so that the blue form of the species, HIn^{2-}, predominates in the absence of a metal ion. Until the equivalence point in a titration, the indicator complexes the excess metal ion so that the solution is red. With the first slight excess of EDTA, the solution turns blue as a consequence of the reaction

$$\underset{red}{MIn^-} + HY^{3-} \rightleftharpoons \underset{blue}{HIn^{2-}} + MY^{2-}$$

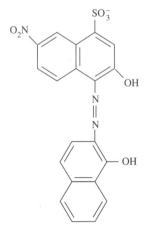

Figure 14-10

Eriochrome Black T.

Eriochrome Black T forms red complexes with more than two dozen metal ions, but the formation constants of only a few are appropriate for end-point detection. Transition ranges for magnesium and calcium are indicated on the titration curves in Figure 14-5. The indicator is clearly ideal for the titration of magnesium and zinc (Figure 14-9); it is totally unsatisfactory for calcium, however.

A limitation of Eriochrome Black T is that its solutions decompose slowly on standing; refrigeration slows this process. It is claimed that solutions of Calmagite, an indicator that for all practical purposes is identical in behavior to Eriochrome Black T, do not suffer this disadvantage. Many other metal indicators have been developed for EDTA titrations.[4] In contrast to Eriochrome Black T, some of these indicators can be employed in strongly acidic media.

14B-7 Types of EDTA Titrations

Direct Titration Procedures

Approximately 40 cations can be determined by direct titration with standard EDTA solutions using metal-ion indicators.[5] Often, cations can be determined by direct titration even when no satisfactory indicator for them is available. For example, calcium ion indicators are not nearly as satisfactory as those for magnesium ion. Calcium can be determined by direct titration, however, by introducing a small amount of magnesium chloride into the EDTA solution. The titration is then carried out with Eriochrome Black T indicator. During the early stages of the titration calcium ions displace magnesium ions from the EDTA because the calcium chelate is considerably more stable than the magnesium complex. After the calcium ions have been fully complexed, the EDTA then recombines with magnesium ions, ultimately giving an Eriochrome Black T end point. To use this approach, it is necessary to standardize the magnesium-containing EDTA against primary standard calcium carbonate.

Direct titration is also used to determine those metal ions for which specific ion electrodes are available. Electrodes of this type are described in Section 18D.

> Direct titration procedures with a metal-ion indicator are the easiest and the most convenient to use.

Back-Titration Methods

Back-titration procedures are useful for the determination of cations that form stable EDTA complexes and for which a satisfactory indicator is not available. The method is also useful for cations that react only slowly with EDTA. A measured excess of standard EDTA solution is added to the analyte solution. After the reaction is judged complete, the excess EDTA is back-titrated with a standard magnesium or zinc ion solution to an Eriochrome Black T or Calmagite end point.[6] For this procedure to be effective, the formation constant for the chelate of the analyte must be larger than that for magnesium or for zinc.

Back-titration is also useful for analyzing samples that contain anions that would otherwise form sparingly soluble precipitates with the analyte under the conditions needed for satisfactory complexation. Here, the excess EDTA prevents precipitate formation.

> Back-titration procedures are resorted to when no suitable indicator is available, when the reaction between analyte and EDTA is slow, or when the analyte forms precipitates at the pH required for its titration.
>
> Displacement titrations are used when no indicator for an analyte is available.

Displacement Methods

In displacement titrations, an unmeasured excess of a solution containing the magnesium or zinc complex of EDTA is introduced into the analyte solution. If

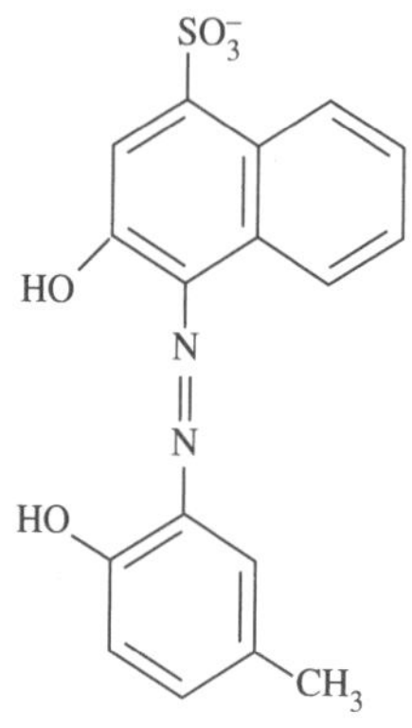

Figure 14-11
Calmagite.

---

[4] See, for example, L. Meites, *Handbook of Analytical Chemistry*, pp. **3**–101 to **3**–165. New York: McGraw-Hill, 1963.
[5] C. N. Reilley and A. J. Barnard Jr., in *Handbook of Analytical Chemistry*, L. Meites, Ed., pp. **3**–167 to **3**–200. New York: McGraw-Hill, 1963.
[6] For a recent analysis of the back-titration procedure, see C. Macca and M. Fiorana, *J. Chem. Educ.*, **1986**, *63*, 121.

the analyte forms a more stable complex than that of magnesium or zinc, the following displacement reaction occurs:

$$MgY^{2-} + M^{2+} \rightarrow MY^{2-} + Mg^{2+}$$

where M^{2+} represents the analyte cation. The liberated magnesium or zinc is then titrated with a standard EDTA solution.

Displacement methods are useful where no satisfactory indicator is available for the metal ion being determined.

14B-8 The Scope of EDTA Titrations

Complexometric titrations with EDTA have been applied to the determination of virtually every metal cation with the exception of the alkali metal ions. Because EDTA complexes most cations, the reagent might appear at first glance to be totally lacking in selectivity. In fact, however, considerable control over interferences can be realized by pH regulation. For example, trivalent cations can usually be titrated without interference from divalent species by maintaining the solution at a pH of about 1 (Figure 14-8). At this pH, the less stable divalent chelates do not form to any significant extent, but the trivalent ions are quantitatively complexed.

Similarly, ions such as cadmium and zinc, which form more stable EDTA chelates than does magnesium, can be determined in the presence of the latter ion by buffering the mixture to pH 7 before titration. Eriochrome Black T serves as an indicator for the cadmium or zinc end points without interference from magnesium because the indicator chelate with magnesium is not formed at this pH.

Finally, interference from a particular cation can sometimes be eliminated by adding a suitable *masking agent*, an auxiliary ligand that preferentially forms highly stable complexes with the potential interference.[7] For example, cyanide ion is often employed as a masking agent to permit the titration of magnesium and calcium ions in the presence of ions such as cadmium, cobalt, copper, nickel, zinc, and palladium. All of the latter form sufficiently stable cyanide complexes to prevent reaction with EDTA. Feature 14-2 illustrates how masking and demasking reagents are used to improve the selectivity of EDTA reactions.

14B-9 The Determination of Water Hardness

Historically, water "hardness" was defined in terms of the capacity of cations in the water to replace the sodium or potassium ions in soaps and form sparingly soluble products. Most multiply charged cations share this undesirable property. In natural waters, however, the concentrations of calcium and magnesium ions generally far exceed that of any other metal ion. Consequently, hardness is now expressed in terms of the concentration of calcium carbonate that is equivalent to the total concentration of all the multivalent cations in the sample.

> A *masking agent* is a complexing agent that reacts selectively with a component in a solution and in so doing prevents that component from interfering in an analysis.

> Hard water contains calcium, magnesium, and heavy-metal ions that form precipitates with soap (but not detergents).

---

[7] For further information, see D. D. Perrin, *Masking and Demasking of Chemical Reactions*. New York: Wiley-Interscience, 1970; and C. N. Reilley and A. J. Barnard Jr., in *Handbook of Analytical Chemistry*, L. Meites, Ed., pp. **3**–208 to **3**–225. New York: McGraw-Hill, 1963.

The determination of hardness is a useful analytical test that provides a measure of the quality of water for household and industrial uses. The test is particularly important to industry because when heated hard water precipitates calcium carbonate, which then clogs boilers and pipes.

Water hardness is ordinarily determined by an EDTA titration after the sample has been buffered to pH 10. Magnesium, which forms the least stable EDTA complex of all of the common multivalent cations in typical water samples, is not titrated until enough reagent has been added to complex all of the other cations in the sample. Therefore, a magnesium ion indicator, such as Calmagite or Eriochrome Black T, can serve as indicator in water-hardness titrations. Often, a small concentration of the magnesium/EDTA chelate is incorporated in the buffer or in the titrant to ensure the presence of sufficient magnesium ions for satisfactory indicator action.

Test kits for determining the hardness of household water are available commercially. They consist of a vessel calibrated to contain a known volume of water, a measuring scoop to deliver an appropriate amount of a solid buffer mixture, an indicator solution, and a bottle of standard EDTA, which is equipped with a

Feature 14-2
HOW CAN MASKING AND DEMASKING AGENTS BE USED TO ENHANCE THE SELECTIVITY OF EDTA TITRATIONS?

Lead, magnesium, and zinc can be determined on a single sample by two titrations with standard EDTA and one titration with standard Mg^{2+}. The sample is first treated with an excess of NaCN, which masks Zn^{2+} and prevents it from reacting with EDTA.

$$Zn^{2+} + 4CN^- \rightleftharpoons Zn(CN)_4^{2-}$$

The Pb^{2+} and Mg^{2+} are then titrated with standard EDTA. After the equivalence point has been reached, a solution of the complexing agent BAL (2-3-dimercapto-1-propanol, $CH_2SHCHSHCH_2OH$), which we will write as $R(SH)_2$, is added to the solution. This bidentate ligand reacts selectively to form a complex with Pb^{2+} that is much more stable than PbY^{2-}:

$$PbY^{2-} + 2R(SH)_2 \rightarrow Pb(RS)_2 + 2H^+ + Y^{4-}$$

The liberated Y^{4-} is then titrated with a standard solution of Mg^{2+}. Finally the zinc is demasked by adding formaldehyde

$$Zn(CN)_4^{2-} + 4HCHO + 4H_2O \rightarrow Zn^{2+} + 4HOCH_2CN + 4OH^-$$

The liberated Zn^{2+} is then titrated with the standard EDTA solution.

Suppose the initial titration of Mg^{2+} and Pb^{2+} required 42.22 mL of 0.02064 M EDTA. Titration of the Y^{4-} liberated by the BAL consumed 19.35 mL of 0.007657 M Mg^{2+}. Finally, after addition of formaldehyde, the liberated Zn^{2+} was titrated with 28.63 mL of the EDTA. Calculate the percent of the three elements if a 0.4085-g sample was used.

The initial titration reveals the number of millimoles of Pb^{2+} and Mg^{2+} present. That is,

$$\text{no. mmol } (Pb^{2+} + Mg^{2+}) = 42.22 \times 0.02064 = 0.87142$$

The second titration gives the number of millimoles of Pb^{2+}. Thus,

$$\text{no. mmol } Pb^{2+} = 19.35 \times 0.007657 = 0.14816$$
$$\text{no. mmol } Mg^{2+} = 0.87142 - 0.14816 = 0.72326$$

Finally from the third titration we obtain

$$\text{no. mmol } Zn^{2+} = 28.63 \times 0.02064 = 0.59092$$

To obtain the percentages, we write

$$\frac{0.14816 \text{ mmol Pb} \times 0.2072 \text{ g Pb/mmol Pb}}{0.4085 \text{ g sample}} \times 100\% = 7.515\% \text{ Pb}$$

$$\frac{0.72326 \text{ mmol Mg} \times 0.024305 \text{ g Mg/mmol Mg}}{0.4085 \text{ g sample}} \times 100\% = 4.303\% \text{ Mg}$$

$$\frac{0.59092 \text{ mmol Zn} \times 0.06539 \text{ g Zn/mmol Zn}}{0.4085 \text{ g sample}} \times 100\% = 9.459\% \text{ Zn}$$

medicine dropper. The drops of standard reagent needed to cause a color change are counted. The concentration of the EDTA solution is ordinarily such that one drop corresponds to one grain (about 0.065 g) of calcium carbonate per gallon of water.

14C QUESTIONS AND PROBLEMS

14-1. Define
*(a) chelate.
(b) tetradentate chelating agent.
*(c) ligand.
(d) coordination number.
*(e) conditional formation constant.
(f) NTA.
*(g) water hardness.
(h) EDTA displacement titration.

*14-2. Describe three general methods for performing EDTA titrations. What are the advantages of each?

14-3. Why are multidentate ligands preferable to unidentate ligands for complexometric titrations?

14-4. Write chemical equations and equilibrium-constant expressions for the stepwise formation of
*(a) $Ag(S_2O_3)_2^{3-}$.   *(c) $Cd(NH_3)_4^{2+}$.
(b) AlF_6^{3-}.   (d) $Ni(SCN)_3^-$.

*14-5. Explain how stepwise and overall formation constants are related.

14-6. Propose a complexometric method for the determination of the individual components in a solution containing In^{3+}, Zn^{2+}, and Mg^{2+}.

14-7. Why is a small amount of MgY^{2-} often added to a water specimen that is to be titrated for hardness?

*14-8. An EDTA solution was prepared by dissolving 3.853 g

---

* Answers to the asterisked problems are given in the answer section at the back of the book.

of purified and dried $Na_2H_2Y \cdot 2H_2O$ in sufficient water to give 1.000 L. Calculate the molar concentration, given that the solute contained 0.3% excess moisture (page 240).

14-9. A solution was prepared by dissolving about 3.0 g of $Na_2H_2Y \cdot 2H_2O$ in approximately 1 L of water and standardizing against 50.00-mL aliquots of 0.004517 M Mg^{2+}. An average titration of 32.22 mL was required. Calculate the molar concentration of the EDTA.

14-10. Calculate the volume of 0.0500 M EDTA needed to titrate
  *(a) 26.37 mL of 0.0741 M $Mg(NO_3)_2$.
  (b) the Ca in 0.2145 g of $CaCO_3$.
  *(c) the Ca in a 0.4397-g mineral specimen that is 81.4% brushite, $CaHPO_4 \cdot 2H_2O$ (172.09 g/mol).
  (d) the Mg in a 0.2080-g sample of the mineral hydromagnesite, $3MgCO_3 \cdot Mg(OH)_2 \cdot 3H_2O$ (365.3 g/mol).
  *(e) the Ca and Mg in a 0.1557-g sample that is 92.5% dolomite, $CaCO_3 \cdot MgCO_3$ (184.4 g/mol).

14-11. A solution contains 1.694 mg of $CoSO_4$ (155.0 g/mol) per milliliter. Calculate
  (a) the volume of 0.008640 M EDTA needed to titrate a 25.00-mL aliquot of this solution.
  (b) the volume of 0.009450 M Zn^{2+} needed to titrate the excess reagent after addition of 50.00 mL of 0.008640 M EDTA to a 25.00-mL aliquot of this solution.
  (c) the volume of 0.008640 M EDTA needed to titrate the Zn^{2+} displaced by Co^{2+} following addition of an unmeasured excess of ZnY^{2-} to a 25.00-mL aliquot of the $CoSO_4$ solution. Reaction:

$$Co^{2+} + ZnY^{2-} \rightarrow CoY^{2-} + Zn^{2+}$$

***14-12.** The Zn in a 0.7556-g sample of foot powder was titrated with 21.27 mL of 0.01645 M EDTA. Calculate the percent Zn in this sample.

14-13. The Cr plating on a surface that measured 3.00 × 4.00 cm was dissolved in HCl. The pH was suitably adjusted, following which 15.00 mL of 0.01768 M EDTA were introduced. The excess reagent required a 4.30-mL back-titration with 0.008120 M Cu^{2+}. Calculate the average weight of Cr on each square centimeter of surface.

***14-14.** The Tl in a 9.76-g sample of rodenticide was oxidized to the trivalent state and treated with an unmeasured excess of Mg/EDTA solution. Reaction:

$$Tl^{3+} + MgY^{2-} \rightarrow TlY^- + Mg^{2+}$$

Titration of the liberated Mg^{2+} required 13.34 mL of 0.03560 M EDTA. Calculate the percentage of Tl_2SO_4 (504.8 g/mol) in the sample.

14-15. An EDTA solution was prepared by dissolving approximately 4 g of the disodium salt in approximately 1 L of water. An average of 42.35 mL of this solution was required to titrate 50.00-mL aliquots of a standard that contained 0.7682 g of $MgCO_3$ per liter. Titration of a 25.00-mL sample of mineral water at pH 10 required 18.81 mL of the EDTA solution. A 50.00-mL aliquot

of the mineral water was rendered strongly alkaline to precipitate the magnesium as $Mg(OH)_2$. Titration with a calcium-specific indicator required 31.54 mL of the EDTA solution. Calculate
  (a) the molarity of the EDTA solution.
  (b) the ppm of $CaCO_3$ in the mineral water.
  (c) the ppm of $MgCO_3$ in the mineral water.

***14-16.** A 50.00-mL aliquot of a solution containing iron(II) and iron(III) required 13.73 mL of 0.01200 M EDTA when titrated at pH 2.0 and 29.62 mL when titrated at pH 6.0. Express the concentration of the solution in terms of the parts per million of each solute.

14-17. A 24-h urine specimen was diluted to 2.000 L. After the solution was buffered to pH 10, a 10.00-mL aliquot was titrated with 26.81 mL of 0.003474 M EDTA. The calcium in a second 10.00-mL aliquot was isolated as $CaC_2O_4(s)$, redissolved in acid, and titrated with 11.63 mL of the EDTA solution. Assuming that 15 to 300 mg of magnesium and 50 to 400 mg of calcium per day are normal, did this specimen fall within these ranges?

***14-18.** A 1.509-g sample of a Pb/Cd alloy was dissolved in acid and diluted to exactly 250.0 mL in a volumetric flask. A 50.00-mL aliquot of the diluted solution was brought to a pH of 10.0 with an NH_4^+/NH_3 buffer; the subsequent titration involved both cations and required 28.89 mL of 0.06950 M EDTA. A second 50.00-mL aliquot was brought to a pH of 10.0 with an HCN/NaCN buffer, which also served to mask the Cd^{2+}; 11.56 mL of the EDTA solution were needed to titrate the Pb^{2+}. Calculate the percent Pb and Cd in the sample.

14-19. A 0.6004-g sample of Ni/Cu condenser tubing was dissolved in acid and diluted to 100.0 mL in a volumetric flask. Titration of both cations in a 25.00-mL aliquot of this solution required 45.81 mL of 0.05285 M EDTA. Mercaptoacetic acid and NH_3 were then introduced; production of the Cu complex with the former resulted in the release of an equivalent amount of EDTA, which required a 22.85-mL titration with 0.07238 M Mg^{2+}. Calculate the percent Cu and Ni in the alloy.

***14-20.** Calamine, which is used for relief of skin irritations, is a mixture of zinc and iron oxides. A 1.022-g sample of dried calamine was dissolved in acid and diluted to 250.0 mL. Potassium fluoride was added to a 10.00-mL aliquot of the diluted solution to mask the iron; after suitable adjustment of the pH, Zn^{2+} consumed 38.71 mL of 0.01294 M EDTA. A second 50.00-mL aliquot was suitably buffered and titrated with 2.40 mL of 0.002727 M ZnY^{2-} solution:

$$Fe^{3+} + ZnY^{2-} \rightarrow FeY^- + Zn^{2+}$$

Calculate the percentages of ZnO and Fe_2O_3 in the sample.

14-21. A 3.650-g sample containing bromate and bromide was dissolved in sufficient water to give 250.0 mL. After acidification, silver nitrate was introduced to a 25.00-mL aliquot to precipitate AgBr, which was filtered, washed, and then redissolved in an ammoniacal solution of potassium tetracyanonickelate(II):

$$Ni(CN)_4^{2-} + 2AgBr(s) \rightarrow 2Ag(CN)_2^- + Ni^{2+} + 2Br^-$$

The liberated nickel ion required 26.73 mL of 0.02089 M EDTA. The bromate in a 10.00-mL aliquot was reduced to bromide with arsenic(III) prior to the addition of silver nitrate. The same procedure was followed, and the released nickel ion was titrated with 21.94 mL of the EDTA solution. Calculate the percentages of NaBr and NaBrO₃ in the sample.

*14-22. The potassium ion in a 250.0-mL sample of mineral water was precipitated with sodium tetraphenylboron:

$$K^+ + B(C_6H_4)_4^- \rightarrow KB(C_6H_5)_4(s)$$

The precipitate was filtered, washed, and redissolved in an organic solvent. An excess of the mercury(II)/EDTA chelate was added:

$$4HgY^{2-} + B(C_6H_4)_4^- + 4H_2O \rightarrow$$
$$H_3BO_3 + 4C_6H_5Hg^+ + 4HY^{3-} + OH^-$$

The liberated EDTA was titrated with 29.64 mL of 0.05581 M Mg²⁺. Calculate the potassium ion concentration in parts per million.

14-23. Chromel is an alloy composed of nickel, iron, and chromium. A 0.6472-g sample was dissolved and diluted to 250.0 mL. When a 50.00-mL aliquot of 0.05182 M EDTA was mixed with an equal volume of the diluted sample, all three ions were chelated, and a 5.11-mL back-titration with 0.06241 M copper(II) was required. The chromium in a second 50.00-mL aliquot was masked through the addition of hexamethylenetetramine; titration of the Fe and Ni required 36.28 mL of 0.05182 M EDTA. Iron and chromium were masked with pyrophosphate in a third 50.0-mL aliquot, and the nickel was titrated with 25.91 mL of the EDTA solution. Calculate the percentages of nickel, chromium, and iron in the alloy.

*14-24. A 0.3284-g sample of brass (containing lead, zinc, copper, and tin) was dissolved in nitric acid. The sparingly soluble SnO₂ · 4H₂O was removed by filtration, and the combined filtrate and washings were then diluted to 500.0 mL. A 10.00-mL aliquot was suitably buffered; titration of the lead, zinc, and copper in this aliquot required 37.56 mL of 0.002500 M EDTA. The copper in a 25.00-mL aliquot was masked with thiosulfate; the lead and zinc were then titrated with 27.67 mL of the EDTA solution. Cyanide ion was used to mask the copper and zinc in a 100-mL aliquot; 10.80 mL of the EDTA solution were needed to titrate the lead ion. Determine the composition of the brass sample; evaluate the percentage of tin by difference.

*14-25. Calculate conditional constants for the formation of the EDTA complex of Fe²⁺ at a pH of (a) 6.0, (b) 8.0, (c) 10.0.

14-26. Calculate conditional constants for the formation of the EDTA complex of Ba²⁺ at a pH of (a) 7.0, (b) 9.0, (c) 11.0.

*14-27. Derive a titration curve for 50.00 mL of 0.01000 M Sr²⁺ with 0.02000 M EDTA in a solution buffered to pH 11.0. Calculate pSr values after the addition of 0.00, 10.00, 24.00, 24.90, 25.00, 25.10, 26.00, and 30.00 mL of titrant.

14-28. Derive a titration curve for 50.00 mL of 0.0150 M Fe²⁺ with 0.0300 M EDTA in a solution buffered to pH 7.0. Calculate pFe values after the addition of 0.00, 10.00, 24.00, 24.90, 25.00, 25.10, 26.00, and 30.00 mL of titrant.

CHAPTER 15

AN INTRODUCTION TO ELECTROCHEMISTRY

We now turn our attention to several analytical methods that are based on oxidation/reduction reactions. These methods, which are described in Chapters 16 through 19, include oxidation/reduction titrimetry, potentiometry, coulometry, and electrogravimetry. Fundamentals of electrochemistry that are necessary for the treatment of the theory of these procedures are presented in this chapter.

15A OXIDATION/REDUCTION REACTIONS

An oxidation/reduction reaction is one in which electrons are transferred from one reactant to another. An example is the oxidation of iron(II) ions by cerium(IV) ions. The reaction is described by the equation

$$Ce^{4+} + Fe^{2+} \rightleftharpoons Ce^{3+} + Fe^{3+} \tag{15-1}$$

Here, the Ce^{4+} ion removes an electron from the Fe^{2+} ion to form Ce^{3+} and Fe^{3+} ions. A substance like Ce^{4+} that has a strong affinity for electrons, and thus tends to remove them from other species, is called an *oxidizing agent*, or an *oxidant*. A *reducing agent*, or *reductant*, is a species like Fe^{2+} that readily donates electrons to another species. With regard to Equation 15-1, we say that Fe^{2+} is *oxidized* by Ce^{4+}. In the same way, Ce^{4+} is *reduced* by Fe^{2+}.

We can split any oxidation/reduction equation into two *half-reactions* that show clearly which species gains electrons and which loses them. For example, Equation 15-1 is the sum of the two half-reactions

$$Ce^{4+} + e^- \rightleftharpoons Ce^{3+} \qquad \text{(reduction of } Ce^{4+}\text{)}$$
$$Fe^{2+} \rightleftharpoons Fe^{3+} + e^- \qquad \text{(oxidation of } Fe^{2+}\text{)}$$

The rules for balancing half-reactions are the same as those for other reaction

Oxidation/reduction reactions are sometimes called redox reactions.

A reducing agent is an electron donor. An oxidizing agent is an electron acceptor.

It is important to understand that while we can easily write an equation for a half-reaction in which electrons are consumed or generated, we cannot observe an isolated half-reaction experimentally because there must always be a second half-reaction that serves as a source of electrons or a recipient of electrons—that is, an individual half-reaction is a theoretical concept.

257

types. In other words, the number of atoms of each element as well as the net charge on each side of the equation must be the same. Thus, for the oxidation of Fe^{2+} by MnO_4^-, the half-reactions are

$$MnO_4^- + 5e^- + 8H^+ \rightleftharpoons Mn^{2+} + 4H_2O$$

$$5Fe^{2+} \rightleftharpoons 5Fe^{3+} + 5e^-$$

In the first half-reaction, the net charge on the left side is $(-1 - 5 + 8) = +2$, which is the same as that on the right. Note also that we have multiplied the second half-reaction by 5 so that the number of electrons lost by Fe^{2+} equals the number gained by MnO_4^-. A balanced net-ionic equation for the overall reaction is then obtained by adding the two half-reactions

$$MnO_4^- + 5Fe^{2+} + 8H^+ \rightleftharpoons Mn^{2+} + 5Fe^{3+} + 4H_2O$$

15A-1 Comparison of Oxidation/Reduction Reactions with Acid/Base Reactions

Recall that in the Brønsted/Lowry concept an acid/base reaction is described by the equation

$$acid_1 + base_2 \rightleftharpoons base_1 + acid_2$$

Oxidation/reduction reactions can be viewed in a way that is analogous to the Brønsted-Lowry concept of acid/base reactions (Section 3A-2). Both involve the transfer of one or more charged particles from a donor to an acceptor—the particles being electrons in oxidation/reduction and protons in neutralization. When an acid donates a proton, it becomes a conjugate base that is capable of accepting a proton. By analogy when a reducing agent donates an electron it becomes an oxidizing agent that can accept an electron. This product could be called a conjugate oxidant (but seldom, if ever, is). We may then write a generalized equation for a redox reaction as

$$A_{red} + B_{ox} \rightleftharpoons A_{ox} + B_{red}$$

Here, B_{ox}, the oxidized form of species B, accepts electrons from A_{red} to form the new reductant, B_{red}. At the same time, reductant A_{red}, having given up electrons, becomes an oxidizing agent, A_{ox}. If we know from chemical evidence that the equilibrium in this equation lies to the right, we can state that B_{ox} is a better electron acceptor (stronger oxidant) than A_{ox}. Furthermore, A_{red} is a more effective electron donor (better reductant) than B_{red}.

Example 15-1

The following reactions proceed to the right, as written

$$2H^+ + Cd(s) \rightleftharpoons H_2(g) + Cd^{2+}$$

$$2Ag^+ + H_2(g) \rightleftharpoons 2Ag(s) + 2H^+$$

$$Cd^{2+} + Zn(s) \rightleftharpoons Cd(s) + Zn^{2+}$$

What can we deduce regarding the strengths of H^+, Ag^+, Cd^{2+}, and Zn^{2+} as electron acceptors (or oxidizing agents)?

The second reaction establishes that Ag^+ is a better electron acceptor than H^+; the first reaction demonstrates that H^+ is better than Cd^{2+}. Finally the third

Feature 15-1
BALANCING REDOX EQUATIONS

To understand all the concepts covered in this chapter, we must know how to balance oxidation/reduction reactions. In this feature, we remind you of how the process works. For practice, let us complete and balance the following equation after adding H^+, OH^-, or H_2O as needed.

$$MnO_4^- + NO_2^- \rightleftharpoons Mn^{2+} + NO_3^-$$

First we write and balance the two half-reactions involved. For MnO_4^- we first write

$$MnO_4^- \rightleftharpoons Mn^{2+}$$

To take up the 4 oxygens on the left-hand side of the equation, we add $8H^+$, which gives $4H_2O$ on the right-hand side of the equation:

$$MnO_4^- + 8H^+ \rightleftharpoons Mn^{2+} + 4H_2O$$

To balance the charge we need to add 5 electrons to the left side of the equation. Thus,

$$MnO_4^- + 8H^+ + 5e^- \rightleftharpoons Mn^{2+} + 4H_2O$$

For the other half-reaction we add one H_2O to the left side of the equation to supply the needed oxygen.

$$NO_2^- + H_2O \rightleftharpoons NO_3^- + 2H^+$$

Then we add two electrons to the right-hand side to balance the charge.

$$NO_2^- + H_2O \rightleftharpoons NO_3^- + 2H^+ + 2e^-$$

Before combining the two equations, we must multiply the first by 2 and the second by 5 so that the number of electrons lost will equal the number of electrons gained. We then add the two equations to obtain

$$2MnO_4^- + 16H^+ + 10e^- + 5NO_2^- + 5H_2O \rightleftharpoons$$
$$2Mn^{2+} + 8H_2O + 5NO_3^- + 10H^+ + 10e^-$$

which then simplifies to the balanced equation

$$2MnO_4^- + 6H^+ + 5NO_2^- \rightleftharpoons 2Mn^{2+} + 5NO_3^- + 3H_2O$$

Figure 15-1
Photograph of a "silver tree."

For an interesting illustration of this reaction, immerse a piece of copper in a solution of silver nitrate. The result is the deposition of silver on the copper in the form of a "silver tree." See Figure 15-1 and color plate 13.

> Salt bridges are widely used in electrochemistry to prevent mixing of the contents of the two electrolyte solutions making up electrochemical cells. Ordinarily, the two ends of the bridge are equipped with sintered glass disks or other devices to prevent liquid from siphoning from one part of the cell to the other.

The equilibrium-constant expression for the reaction shown in Equation 15-1 is

$$K_{eq} = \frac{[Cu^{2+}]}{[Ag^+]^2} = 4.14 \times 10^{15} \quad (15\text{-}3)$$

This expression applies regardless of whether the reaction occurs directly between reactants or within an electrochemical cell.

At equilibrium, the two half-reactions in a cell continue, but their rates become equal.

equation shows that Cd^{2+} is more effective than Zn^{2+}. Thus, the order of oxidizing strength is $Ag^+ > H^+ > Cd^{2+} > Zn^{2+}$.

15A-2 Oxidation/Reduction Reactions in Electrochemical Cells

Many oxidation/reduction reactions can be carried out in either of two ways that are physically quite different. In one, the reaction is performed by direct contact between the oxidant and the reductant in a suitable container. In the second, the reaction is carried out in an electrochemical cell in which the reactants do not come in direct contact with one another. An example of the first process involves immersing a strip of copper in a solution containing silver nitrate. Here, silver ions migrate to the metal and are reduced:

$$Ag^+ + e^- \rightarrow Ag(s)$$

At the same time an equivalent quantity of copper is oxidized:

$$Cu(s) \rightleftharpoons Cu^{2+} + 2e^-$$

Multiplication of the silver half-reaction by two and addition of the two half-reactions yields a net-ionic equation for the overall process.

$$2Ag^+ + Cu(s) \rightleftharpoons 2Ag(s) + Cu^{2+} \quad (15\text{-}2)$$

A unique aspect of oxidation/reduction reactions is that the transfer of electrons—and thus an identical net reaction—can often be brought about in an *electrochemical cell* in which the oxidizing agent and the reducing agent are physically separated from one another. Figure 15-2a shows such an arrangement. Note that a *salt bridge* isolates the reactants but maintains electrical contact between the two halves of the cell. The bridge is necessary to prevent Ag^+ from reacting directly with the copper metal. An external metallic conductor connects the two metals. In this cell, metallic copper is oxidized, silver ions are reduced, and electrons flow through the external circuit to the silver electrode. The voltmeter measures the potential difference between the two metals at any instant and is a measure of the tendency of the cell reaction to proceed toward equilibrium. As the reaction goes on, this tendency, and thus the potential, decrease continuously and approach zero as the state of equilibrium for the overall reaction is approached.

When zero voltage is reached, the concentrations of Cu(II) and Ag(I) ions will have values that satisfy the equilibrium-constant expression shown in Equation 15-3. At this point, no further net flow of electrons will occur. *It is important to realize that the overall reaction and its position of equilibrium are totally independent of the way the reaction is carried out*, whether it is by direct reaction in a solution or by indirect action in an electrochemical cell.

15B ELECTROCHEMICAL CELLS

Oxidation/reduction equilibria are conveniently studied by measuring the potentials of electrochemical cells in which the two half-reactions making up the

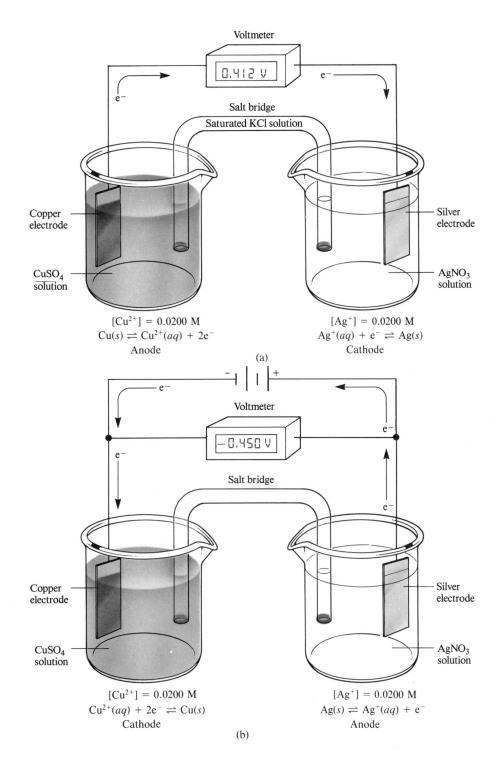

Figure 15-2
(a) A galvanic cell; (b) an electrolytic
cell.

The electrodes in some cells share a common electrolyte; these are known as *cells without liquid junction*. For an example of such a cell, see Figure 16-2 and Example 16-4.

equilibrium are participants. For this reason, we need to consider some of the characteristics of cells.

An electrochemical cell consists of two conductors called *electrodes*, each of which is immersed in an electrolyte solution. In most of the cells that will be of interest to us, the solutions surrounding the two electrodes are different and must be separated to avoid direct reaction between the reactants. The most common way of avoiding mixing is to insert a salt bridge, such as that shown in Figure 15-2, between the solutions. Conduction of electricity from one electrolyte solution to the other then occurs by migration of potassium ions in the bridge in one direction and chloride ions in the other. However, there is no direct contact between copper metal and silver ions.

15B-1 Cathodes and Anodes

A cathode is the electrode where reduction occurs. An anode is the electrode where oxidation occurs.

The *cathode* in an electrochemical cell is the electrode at which a reduction reaction occurs. The *anode* is the electrode at which an oxidation takes place.

Examples of typical cathodic reactions include:

$$Ag^+ + e^- \rightleftharpoons Ag(s)$$
$$Fe^{3+} + e^- \rightleftharpoons Fe^{2+}$$
$$NO_3^- + 10H^+ + 8e^- \rightleftharpoons NH_4^+ + 3H_2O$$

The reaction $2H^+ + 2e^- \rightleftharpoons H_2(g)$ occurs at a cathode when an aqueous solution contains no easily reduced species.

We can make these reactions occur by applying a suitable potential to an inert electrode such as platinum. Note that the third reaction reveals that anions can migrate to a cathode and be reduced.

Typical anodic reactions include:

$$Cu(s) \rightleftharpoons Cu^{2+} + 2e^-$$
$$2Cl^- \rightleftharpoons Cl_2(g) + 2e^-$$
$$Fe^{2+} \rightleftharpoons Fe^{3+} + e^-$$

The first reaction requires a copper anode, but the other two can be carried out at the surface of an inert platinum electrode.

15B-2 Types of Electrochemical Cells

Electrochemical cells are either galvanic or electrolytic. *Galvanic*, or *voltaic*, cells store electrical energy. The reactions at the two electrodes in such cells tend to proceed spontaneously and produce a flow of electrons from the anode to the cathode via an external conductor. The cell shown in Figure 15-2a is a galvanic cell that develops a potential of about 0.412 V when electrons move through the external circuit from the copper anode to the silver cathode.

An *electrolytic* cell, in contrast, requires an external source of electrical energy for operation. The cell just considered can be operated electrolytically by connecting the positive terminal of a battery having a potential of somewhat greater than 0.412 V to the silver electrode and the negative terminal to the copper electrode. Here, the direction of the current is reversed, and the reactions at the electrodes are reversed as well (see Figure 15-2b). Oxidation then occurs at the silver anode

Allessandro Volta (1745–1827), Italian physicist, was the inventor of the first battery, the so-called voltaic pile. It consisted of alternating disks of copper and zinc separated by disks of cardboard soaked with salt solution. In honor of his many contributions to electrical science, the unit of potential difference, the volt, is named for Volta.

and reduction takes place at the copper cathode. That is,

$$2Ag(s) + Cu^{2+} \rightleftharpoons 2Ag^+ + Cu(s) \qquad \text{(15-4)}$$

The cell in Figure 15-2 is an example of a *reversible cell* in which the direction of the electrochemical reaction is reversed when the direction of electron flow is changed. In an *irreversible* cell, changing the direction of current causes entirely different half-reactions to occur at one or both electrodes.

15B-3 Schematic Representation of Cells

Chemists frequently use a shorthand notation to describe electrochemical cells. The cell in Figure 15-2a, for example, is described by

$$Cu|Cu^{2+}(0.0200\ M)\|Ag^+(0.0200\ M)|Ag \qquad \text{(15-5)}$$

By convention, the anode is *always* displayed on the left in these representations. A single vertical line indicates a phase boundary, or interface, at which a potential develops. For example, the first vertical line in this schematic indicates that a potential develops at the phase boundary between the copper anode and the copper sulfate solution. The double vertical line represents two phase boundaries, one at each end of the salt bridge. A *liquid-junction potential* develops at each of these interfaces. This potential results from differences in rates at which the ions in the cell compartments and the salt bridge migrate across the interfaces. A liquid-junction potential can amount to as much as several hundredths of a volt, but the junction potentials at the two ends of a salt bridge tend to cancel each other. The net effect on the overall potential of a cell is thus a few millivolts or less. For our purposes, we will neglect the contribution of liquid-junction potentials to the total potential of the cell.

The Fe^{2+} half-reaction may seem somewhat unusual because a cation rather than an anion migrates to the electrode and gives up an electron. As we shall see, oxidation of a cation at an anode or reduction of an anion at a cathode is not uncommon.

The reaction $2H_2O \rightleftharpoons O_2(g) + 4H^+ + 4e^-$ occurs at an anode when an aqueous solution contains no more easily oxidized species.

Galvanic cells store electrical energy; electrolytic cells consume electricity.

For both galvanic and electrolytic cells, remember that *reduction always takes place at the cathode*, and *oxidation always takes place at the anode*.

In a reversible cell, reversing the current reverses the cell reaction. In an irreversible cell, reversing the current causes a different half-reaction to occur at one or both of the electrodes.

A liquid-junction potential is a potential that develops across the interface between two solutions that differ in their electrolyte composition.

Feature 15-2
THE DANIELL GRAVITY CELL

The Daniell gravity cell was one of the earliest galvanic cells that found widespread practical application. It was used in the mid-1800s as a battery to power telegraphic communication systems. As shown in Figure 15-3 the cathode was a piece of copper immersed in a saturated solution of copper sulfate. A much less dense solution of dilute zinc sulfate was layered on top of the copper sulfate, and a massive zinc electrode was located in this solution. The electrode reactions were

$$Zn(s) \rightleftharpoons Zn^{2+} + 2e^- \qquad \text{anode}$$
$$Cu^{2+} + 2e^- \rightleftharpoons Cu(s) \qquad \text{cathode}$$

This cell develops an initial voltage of 1.18 V, which gradually decreases with use.

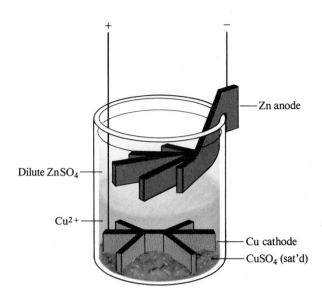

Figure 15-3
A Daniell gravity cell.

An alternative way of writing the cell shown in Figure 15-2a is

$$Cu|CuSO_4(0.0200 \text{ M})\|AgNO_3(0.0200 \text{ M})|Ag$$

Here, the compounds used to prepare the cell are indicated rather than the active participants in the cell half-reactions.

The cell in Figure 15-2b can be written as

$$Ag|Ag^+(0.0200 \text{ M})\|Cu^{2+}(0.0200 \text{ M})|Cu$$

Placing the silver half-cell on the left indicates that the cell is being operated in such a way as to make the silver couple operate as the anode.

15B-4 Currents in Electrochemical Cells

As shown in Figure 15-4, electricity is transported through an electrochemical cell by three mechanisms:

In a cell, electricity is carried by movement of anions toward the anode and cations toward the cathode.

1. Electrons carry electricity within the electrodes as well as the external conductor.
2. Anions and cations carry electricity within the cell. As shown in Figure 15-4, copper ions, silver ions, and other positively charged species move away from the copper anode and toward the silver cathode, whereas anions, such as sulfate and hydrogen sulfate ions, are attracted toward the copper anode. Within the salt bridge, chloride ions migrate toward and into the copper compartment, whereas potassium ions move in the opposite direction.
3. The ionic conduction of the solution is coupled to the electronic conduction in the electrodes by the reduction reaction at the cathode and the oxidation reaction at the anode.

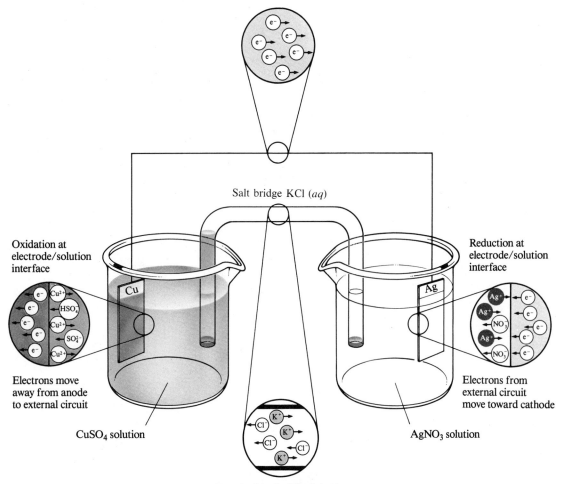

Figure 15-4
Movement of charge in a galvanic cell.

15C   ELECTRODE POTENTIALS

The potential difference that develops between the cathode and the anode of the cell in Figure 15-5a is a measure of the tendency for the reaction

$$2Ag^+ + Cu(s) \rightleftharpoons 2Ag(s) + Cu^{2+}$$

to proceed from a nonequilibrium state to the condition of equilibrium. Thus, as shown in the upper figure, when the copper and silver ion concentrations in the cell are both 0.0200 M, a potential of 0.412 V develops, which shows that this reaction is far from equilibrium. This potential becomes smaller and smaller as the reaction proceeds, until at equilibrium the meter reads 0.000 V. As shown in Figure 15-5b, the copper ion equilibrium concentration is then just slightly less than 0.0300 M and the silver ion concentration is 2.7×10^{-9} M.

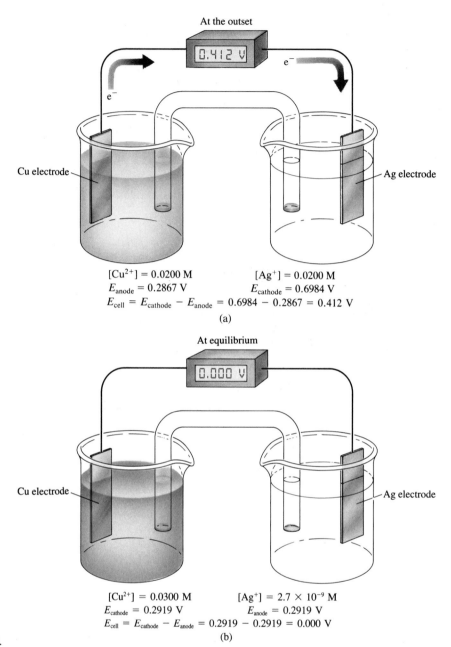

At the outset

0.412 V

e^-

e^-

Cu electrode

Ag electrode

$[Cu^{2+}] = 0.0200\ M$ $[Ag^+] = 0.0200\ M$
$E_{anode} = 0.2867\ V$ $E_{cathode} = 0.6984\ V$
$E_{cell} = E_{cathode} - E_{anode} = 0.6984 - 0.2867 = 0.412\ V$

(a)

At equilibrium

0.000 V

Cu electrode

Ag electrode

$[Cu^{2+}] = 0.0300\ M$ $[Ag^+] = 2.7 \times 10^{-9}\ M$
$E_{cathode} = 0.2919\ V$ $E_{anode} = 0.2919\ V$
$E_{cell} = E_{cathode} - E_{anode} = 0.2919 - 0.2919 = 0.000\ V$

(b)

Figure 15-5

Change in cell potential after passage
of current until equilibrium is reached.

The potential of a cell such as that shown in Figure 15-5a is the difference
between two *half-cell* or *single-electrode potentials*. One of these potentials is
associated with the half-reaction at the silver cathode ($E_{cathode}$) and the other with
the half-reaction at the copper anode (E_{anode}).

Although we cannot determine absolute potentials of electrodes such as these
(see Feature 15-3), we can easily determine *relative* electrode potentials. For
example, if we replace the copper anode in the cell in Figure 15-1 with a cadmium
electrode immersed in a cadmium sulfate solution, the voltmeter reads about 0.7

V greater than the original cell. Since the cathode compartment remains unaltered, we conclude that the half-cell potential for the oxidation of cadmium is about 0.7 V greater than that for copper (that is, cadmium is a stronger reductant than is copper). The tendency for other ions to take on electrons can also be compared by substituting other cathode half-cells while keeping the anode half-cell unchanged.

15C-1 The Standard Hydrogen Reference Electrode

For relative electrode potential data to be widely applicable and useful, we must agree on a reference half-cell against which all others are compared. Such an electrode must be easy to construct, be reversible, and be highly reproducible in its behavior. The *standard hydrogen electrode* (SHE) meets these specifications and has been used throughout the world for many years as a universal reference electrode. It is a typical *gas electrode.*

Figure 15-7 shows how a hydrogen electrode is constructed. The metal conductor is a piece of platinum that has been coated, or *platinized*, with finely divided platinum (*platinum black*) to increase its specific surface area. This electrode is immersed in an aqueous acid solution of known, constant hydrogen ion activity. The solution is kept saturated with hydrogen by bubbling the gas at constant pressure over the surface of the electrode. The platinum serves only as the site where electrons are transferred and does not take part in the electrochemical reaction. The half-reaction responsible for the potential that develops at this electrode is

$$2H^+(aq) + 2e^- \rightleftharpoons H_2(g) \qquad \textbf{(15-6)}$$

> The standard hydrogen electrode is sometimes called the normal hydrogen electrode.

> SHE is the abbreviation for standard hydrogen electrode.

> Platinum black is a layer of finely divided platinum that is formed on the surface of a smooth platinum electrode by electrolytic deposition of the metal from a solution of chloroplatinic acid H_2PtCl_6. The platinum black provides a large specific surface area of platinum at which reaction can occur.

> Platinum black catalyzes the reaction shown in Equation 15-6. Remember that catalysts do not change the position of equilibrium but simply shorten the time it takes to reach equilibrium.

Feature 15-3
WHY ABSOLUTE ELECTRODE POTENTIALS CANNOT BE MEASURED

Although it is not difficult to measure *relative* half-cell potentials, it is impossible to determine absolute half-cell potentials because all voltage-measuring devices measure only *differences* in potential. To measure the potential of an electrode, one contact of a voltmeter is connected to the electrode in question. The other contact from the meter must then be brought into electrical contact with the solution in the electrode compartment via another conductor. This second contact, however, inevitably involves a solid/solution interface that acts as a second half-cell at which chemical change *must occur* if charge is to flow and the potential is to be measured. A potential is associated with this second reaction. Thus an absolute half-cell potential is not obtained but rather is the difference between the half-cell potential of interest and a half-cell made up of the second contact and the solution.

Our inability to measure absolute half-cell potentials presents no real obstacle because relative half-cell potentials are just as useful provided they are all measured against the same half-cell. Relative half-cell potentials can be combined to give cell potentials. We can also use them to calculate equilibrium constants and generate titration curves.

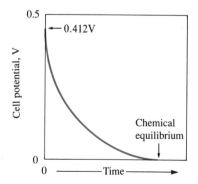

Figure 15-6
Potential of cell shown in Figure 15-5a as a function of time.

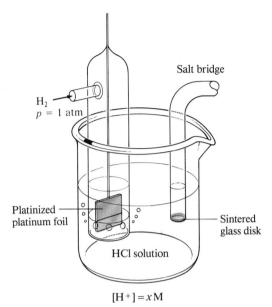

Salt bridge

H_2
$p = 1$ atm

Platinized
platinum foil

Sintered
glass disk

HCl solution

Figure 15-7
A hydrogen gas electrode.

$[H^+] = x\,M$

The reaction shown as Equation 15-6
involves two equilibria:

$$2H^+ + 2e^- \rightleftharpoons H_2(aq)$$

$$H_2(aq) \rightleftharpoons H_2(g)$$

The continuous stream of gas at constant
pressure provides the solution with a constant molecular hydrogen concentration.

The standard hydrogen electrode is reversible and can act either as an anode or as a
cathode, depending on the electrode to
which it is connected.

As an anode, the hydrogen electrode shown in Figure 15-7 can be represented
schematically as

$$Pt,H_2(p = 1.00 \text{ atm})|([H^+] = x \text{ M})\|$$

Here, hydrogen gas is specified as having a partial pressure of one atmosphere
and the concentration of hydrogen ions in the solution is x M.

The hydrogen electrode is reversible and acts either as an anode or as a cathode,
depending on the half-cell with which it is coupled. Hydrogen is oxidized to
hydrogen ion when the electrode is the anode. Hydrogen ion is reduced to
molecular hydrogen when it acts as a cathode.

The potential of a hydrogen electrode depends on temperature and the activities
of hydrogen ion and molecular hydrogen in the solution. The activity of molecular
hydrogen is proportional to the pressure of the gas that is used to keep the solution
saturated with hydrogen. For the *standard hydrogen electrode* (SHE), the activity
of hydrogen ion is specified as unity and the partial pressure of the gas is
specified as one atmosphere. *By convention, the potential of the standard hydrogen
electrode is assigned a value of 0.000 V at all temperatures.* As a consequence
of this definition, any potential developed in a galvanic cell consisting of a
standard hydrogen electrode and some other electrode is attributed entirely to
the other electrode.

Several other reference electrodes that are more convenient for routine measurements have been developed. Some of these are described in Section 18B.

At $p_{H_2} = 1.00$, $a_{H^+} = 1.00$, and at all
temperatures, the potential of the hydrogen electrode is assigned a value of
exactly 0.000 V and is called the standard hydrogen electrode.

15C-2 Definition of Electrode Potential and Standard Electrode Potential

An *electrode potential* is defined as the potential of a cell consisting of the
electrode of interest *acting as a cathode* and the standard hydrogen electrode
acting as an anode. We must emphasize that, in spite of its name, *an electrode
potential is in fact the potential of an electrochemical cell involving a carefully*

An electrode potential is the potential
of a cell that has a standard hydrogen
electrode *as the anode.*

defined reference electrode. It could be more properly called a "relative electrode potential."

The *standard electrode potential* of a half-reaction E^0 is defined as its electrode potential when the activities of the reactants and products are all unity. The cell in Figure 15-8 illustrates the definition of the standard electrode potential for the half-reaction

$$Ag^+ + e^- \rightleftharpoons Ag(s)$$

Here, the half-cell on the right consists of a strip of pure silver in contact with a solution that has a silver ion activity of 1.00; the electrode on the left is the standard hydrogen electrode.

The cell shown in Figure 15-8 can be represented schematically as

$$Pt,H_2(p = 1.00\ atm)|H^+(a_{H^+} = 1.00\ M)\|Ag^+(a_{Ag^+} = 1.00)|Ag$$

or alternatively as

$$SHE\|Ag^+(a_{Ag^+} = 1.00)|Ag$$

(Again, SHE is a shorthand notation for the standard hydrogen electrode.) This galvanic cell develops a potential of 0.799 V with the silver electrode functioning *as the cathode*; that is, the spontaneous cell reaction is

$$2Ag^+ + H_2(g) \rightleftharpoons 2Ag(s) + 2H^+$$

Because the silver electrode serves as the cathode, the measured potential is *by definition* the standard electrode potential for the silver half-reaction (or the silver *couple*). Note that the silver electrode is positive with respect to the hydrogen

A half-cell is sometimes called a *couple*.

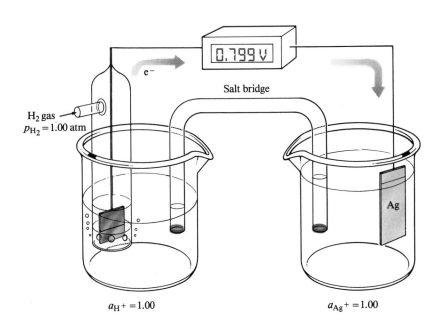

H₂ gas
$p_{H_2} = 1.00$ atm

0.799 V

e⁻

Salt bridge

Ag

$a_{H^+} = 1.00$

$a_{Ag^+} = 1.00$

Figure 15-8
Definition of the standard electrode potential for $Ag^+ + e^- \rightleftharpoons Ag(s)$.

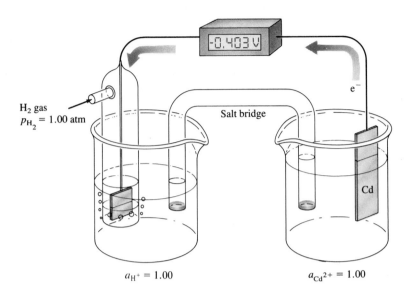

Figure 15-9

Definition of the standard electrode potential for $Cd^{2+} + 2e^- \rightleftharpoons Cd(s)$.

anode (that is, electrons flow from the negative hydrogen anode to the silver cathode). Therefore, the electrode potential is given a positive sign, and we write

$$Ag^+ + e^- \rightleftharpoons Ag(s) \qquad E^0_{Ag^+} = +0.799 \text{ V}$$

Figure 15-9 illustrates the definition of the standard electrode potential for the half-reaction

$$Cd^{2+} + 2e^- \rightleftharpoons Cd(s)$$

In contrast to the silver electrode, the cadmium electrode acts as the anode of the galvanic cell. That is, the spontaneous cell reaction is

$$Cd(s) + 2H^+ \rightleftharpoons H_2(g) + Cd^{2+}$$

Here, the cadmium electrode bears a negative charge with respect to the standard hydrogen electrode. In order to reverse this reaction so that the cadmium electrode acts as a cathode, a potential more negative than -0.403 V must be applied to the cell. Consequently, the standard electrode potential of the Cd/Cd^{2+} couple is *by convention* given a negative sign and is equal to -0.403 V.

A zinc electrode immersed in a solution with a zinc ion activity of unity develops a potential of -0.763 V when paired with a standard hydrogen electrode. Because the zinc electrode also behaves as an anode in the galvanic cell, its electrode potential is also negative.

The standard electrode potentials for the four half-cells just described can be arranged in the order

Half-Reaction	Standard Electrode Potential, V
$Ag^+ + e^- \rightleftharpoons Ag(s)$	+0.799
$2H^+ + 2e^- \rightleftharpoons H_2(g)$	+0.000
$Cd^{2+} + 2e^- \rightleftharpoons Cd(s)$	−0.403
$Zn^{2+} + 2e^- \rightleftharpoons Zn(s)$	−0.763

The magnitudes of these electrode potentials indicate the relative strength of the four ionic species as electron acceptors (oxidizing agents); that is, in decreasing strength, $Ag^+ > H^+ > Cd^{2+} > Zn^{2+}$.

15C-3 Sign Convention for Electrode Potentials

Historically, electrochemists have not always used the sign convention just described. Indeed, disagreements regarding the conventions to be used in specifying signs for half-cell processes caused much controversy and confusion in the development of electrochemistry. The International Union of Pure and Applied Chemistry (IUPAC) addressed itself to this problem at its 1953 meeting in Stockholm. The usages adopted at that meeting are collectively referred to as the *Stockholm Convention* or the *IUPAC Convention* and are now generally accepted. The sign convention described in the previous section and in the paragraphs that follow is based on the IUPAC recommendations.

Any sign convention must be based on expressing half-cell processes in a single way—that is, either as oxidations or as reductions. According to the IUPAC convention, the term *electrode potential* (or, more exactly, *relative electrode potential*) *is reserved exclusively to describe half-reactions written as reductions*. There is no objection to the use of the term *oxidation potential* to indicate a process written in the opposite sense, but it is not proper to refer to such a potential as an electrode potential.

The sign of an electrode potential is determined by the sign of its half-cell when it is coupled to a standard hydrogen electrode. When the half-cell of interest behaves spontaneously as a cathode, it is the positive electrode of the galvanic cell (Figure 15-8). Thus, its electrode potential is positive. When the half-cell of interest behaves as an anode, the electrode is negative, and so is its electrode potential (Figure 15-9).

It is important to emphasize that the electrode potential refers to a half-cell process written *as a reduction*. For the zinc and cadmium electrodes we have been considering, the spontaneous reactions are oxidations. *Note that the sign of an electrode potential indicates whether the reduction is spontaneous with respect to the standard hydrogen electrode*. The positive sign associated with the electrode potential for silver indicates that the process

$$2Ag^+ + H_2(g) \rightleftharpoons 2Ag(s) + 2H^+$$

favors the products under ordinary conditions. Similarly, the negative sign of the electrode potential for zinc means that the analogous reaction

$$Zn^{2+} + H_2(g) \rightleftharpoons Zn(s) + 2H^+$$

is not spontaneous. In other words the reaction tends to go in the other direction. That is,

$$Zn(s) + 2H^+ \rightleftharpoons Zn^{2+} + H_2(g)$$

15C-4 Effect of Concentration on Electrode Potentials: The Nernst Equation

An electrode potential is a measure of the extent to which the existing concentrations in a half-cell differ from their equilibrium values. For example, there is a

An electrode potential is, by definition, a reduction potential. An oxidation potential is the potential for the half-reaction written in the opposite way. The sign of an oxidation potential is, therefore, opposite that for a reduction potential, but the magnitude is the same.

The IUPAC sign convention is based on the actual sign of the half-cell of interest when it is connected to the standard hydrogen electrode.

greater tendency for the process

$$Ag^+ + e^- \rightleftharpoons Ag(s)$$

to occur in a concentrated solution of silver(I) than in a dilute solution of that ion. It follows that the magnitude of the electrode potential for this process must also become larger (more positive) as the silver ion concentration of a solution is increased. We now examine the quantitative relationship between concentration and electrode potential.

Consider the reversible half-reaction

$$aA + bB + \cdot \cdot \cdot + ne^- \rightleftharpoons cC + dD + \cdot \cdot \cdot \qquad \text{(15-7)}$$

where the capital letters represent formulas for the participating species (atoms, molecules, or ions), e^- represents electrons, and the lowercase italic letters indicate the number of moles of each species appearing in the half-reaction as it has been written. The electrode potential for this process is described by the equation

$$E = E^0 - \frac{RT}{nF} \ln \frac{[C]^c[D]^d \cdot \cdot \cdot}{[A]^a[B]^b \cdot \cdot \cdot} \qquad \text{(15-8)}$$

The meanings of the bracketed terms in Equations 15-8 and 15-9 are:

for a solute A,

[A] = molar concentration

for a gas B,

[B] = p_B = partial pressure in atmospheres

If one or more of the species appearing in Equation 15-8 is a pure liquid, a pure solid, or the solvent present in excess, then no bracketed term for this species appears in the quotient.

where

E^0 = the *standard electrode potential*, a constant that is characteristic for each half-reaction

R = the gas constant 8.314 J K^{-1} mol^{-1}

T = the temperature in kelvins

n = the number of moles of electrons that appear in the half-reaction for the electrode process as written

F = the faraday = 96,485 C (coulombs)

ln = the natural logarithm = 2.303 log

Substituting numerical values for the constants, converting to base 10 logarithms, and specifying 25°C for the temperature gives

$$E = E^0 - \frac{0.0592}{n} \log \frac{[C]^c[D]^d \cdot \cdot \cdot}{[A]^a[B]^b \cdot \cdot \cdot} \qquad \text{(15-9)}$$

The letters in brackets strictly represent activities, but we shall ordinarily follow our practice of substituting molar concentrations for activities in most calculations. Thus, if some participating species A is a solute, [A] is the concentration of A in moles per liter. If A is a gas, [A] in Equation 15-9 is replaced by p_A, the partial pressure of A in atmospheres. If A is a pure liquid, a pure solid, or the solvent, no term for A is included in the equation because its concentration is constant. The rationale for these assumptions is the same as that described in Section 3B, which deals with equilibrium-constant expressions.

Equation 15-8 is known as the *Nernst equation* in honor of the German chemist who was responsible for its development.

Example 15-2

Typical half-cell reactions and their corresponding Nernst expressions follow.

(1) $\qquad Zn^{2+} + 2e^- \rightleftharpoons Zn(s) \qquad E = E^0 - \dfrac{0.0592}{2} \log \dfrac{1}{[Zn^{2+}]}$

Walther Hermann Nernst (1864–1941) was a German physical chemist who made many contributions to our understanding of electrochemistry. He is probably most famous for the equation that bears his name (Equation 15-7), but he was also known for discoveries and inventions in other fields. He invented the Nernst glower, a source of infrared radiation that is shown on the stamp. Nernst received the Nobel Prize in Chemistry in 1920 for his numerous contributions to the field of chemical thermodynamics.

No term for elemental zinc is included in the logarithmic term because it is a pure second phase. Thus, the electrode potential varies linearly with the logarithm of the reciprocal of the zinc ion concentration.

(2) $\qquad Fe^{3+} + e^- \rightleftharpoons Fe^{2+} \qquad E = E^0 - \dfrac{0.0592}{1} \log \dfrac{[Fe^{2+}]}{[Fe^{3+}]}$

The potential for this couple can be measured with an inert metallic electrode immersed in a solution containing both iron species. The potential depends on the logarithm of the ratio between the molar concentrations of these ions.

(3) $\qquad 2H^+ + 2e^- \rightleftharpoons H_2(g) \qquad E = E^0 - \dfrac{0.0592}{2} \log \dfrac{p_{H_2}}{[H^+]^2}$

In this example, p_{H_2} is the partial pressure of hydrogen (in atmospheres) at the surface of the electrode. Ordinarily, its value will be the same as the atmospheric pressure.

(4)

$MnO_4^- + 5e^- + 8H^+ \rightleftharpoons Mn^{2+} + H_2O \qquad E = E^0 - \dfrac{0.0592}{5} \log \dfrac{[Mn^{2+}]}{[MnO_4^-][H^+]^8}$

Here, the potential depends not only on the concentrations of the manganese species but also on the pH of the solution.

(5) $\qquad AgCl(s) + e^- \rightleftharpoons Ag(s) + Cl^- \qquad E = E^0 - \dfrac{0.0592}{1} \log [Cl^-]$

The Nernst expression in part (5) of Example 15-2 requires an excess of solid AgCl *so that the solution is saturated* with AgCl at all times.

This half-reaction describes the behavior of a silver electrode immersed in a chloride solution that is *saturated* with AgCl. To ensure this condition, an excess of the solid must always be present. Note that this electrode reaction is the sum of the two reactions

$$AgCl(s) \rightleftharpoons Ag^+ + Cl^-$$
$$Ag^+ + e^- \rightleftharpoons Ag(s)$$

Note also that the electrode potential is independent of the amount of AgCl present as long as there is enough present to keep the solution saturated.

15C-5 The Standard Electrode Potential, E^0

Examination of Equations 15-8 and 15-9 reveals that the constant E^0 is the electrode potential whenever the activity quotient has a value of one. This constant is by definition *the standard electrode potential* for the half-reaction. Note that the quotient is always unity when the activities of the reactants and products of a half-reaction are unity.

The standard electrode potential is an important physical constant that provides quantitative information regarding the driving force for a half-cell reaction.[1] The important characteristics of these constants are:

1. The standard electrode potential is a relative quantity in the sense that it is the potential of an electrochemical cell in which the anode is the standard hydrogen electrode, whose potential has been arbitrarily set at 0 V.
2. The standard electrode potential for a half-reaction refers exclusively to a reduction reaction; that is, it is a relative reduction potential.
3. The standard electrode potential measures the relative force tending to drive the half-reaction from a state in which the reactants and products are at unit activity to a state in which the reactants and products are at their equilibrium activities relative to the standard hydrogen electrode.
4. The standard electrode is independent of the number of moles of reactant and product shown in the balanced half-reaction. Thus, the standard electrode for the half-reaction

$$Fe^{3+} + e^- \rightleftharpoons Fe^{2+} \qquad E^0 = +0.771$$

does not change if we choose to write the reaction as

$$5Fe^{3+} + 5e^- \rightleftharpoons 5Fe^{2+} \qquad E^0 = +0.771$$

Note, however, that the Nernst equation must be consistent with the half-reaction as it is written. For the first case, it will be

$$E = 0.771 - \frac{0.0592}{1} \log \frac{[Fe^{2+}]}{[Fe^{3+}]}$$

and for the second

$$E = 0.771 - \frac{0.0592}{5} \log \frac{[Fe^{2+}]^5}{[Fe^{3+}]^5}$$

5. A positive electrode potential indicates that the half-reaction in question is spontaneous with respect to the standard hydrogen electrode half-reaction. That is, the oxidant in the half-reaction is a stronger oxidant than is hydrogen ion. A negative sign indicates just the opposite.
6. The standard electrode potential for a half-reaction is temperature dependent.

The standard electrode potential for a half-reaction, E^0, is defined as the electrode potential when all reactants and products of a half-reaction have unit activity.

Note that the two log terms have identical values. That is,

$$\frac{0.0592}{1} \log \frac{[Fe^{2+}]}{[Fe^{3+}]} = \frac{0.0592}{5} \log \frac{[Fe^{2+}]^5}{[Fe^{3+}]^5}$$

---

[1] For further reading on standard electrode potentials, see R. G. Bates, in *Treatise on Analytical Chemistry*, 2nd ed., I. M. Kolthoff and P. J. Elving, Eds., Part I, Vol. 1, Chapter 13. New York: Wiley, 1978.

Table 15-1
STANDARD ELECTRODE POTENTIALS*

Reaction	E^0 at 25°C, V
$Cl_2(g) + 2e^- \rightleftharpoons 2Cl^-$	+1.359
$O_2(g) + 4H^+ + 4e^- \rightleftharpoons 2H_2O$	+1.229
$Br_2(aq) + 2e^- \rightleftharpoons 2Br^-$	+1.087
$Br_2(l) + 2e^- \rightleftharpoons 2Br^-$	+1.065
$Ag^+ + e^- \rightleftharpoons Ag(s)$	+0.799
$Fe^{3+} + e^- \rightleftharpoons Fe^{2+}$	+0.771
$I_3^- + 2e^- \rightleftharpoons 3I^-$	+0.536
$Cu^{2+} + 2e^- \rightleftharpoons Cu(s)$	+0.337
$UO_2^{2+} + 4H^+ + 2e^- \rightleftharpoons U^{4+} + 2H_2O$	+0.334
$Hg_2Cl_2(s) + 2e^- \rightleftharpoons 2Hg(l) + 2Cl^-$	+0.268
$AgCl(s) + e^- \rightleftharpoons Ag(s) + Cl^-$	+0.222
$Ag(S_2O_3)_2^{3-} + e^- \rightleftharpoons Ag(s) + 2S_2O_3^{2-}$	+0.017
$2H^+ + 2e^- \rightleftharpoons H_2(g)$	0.000
$AgI(s) + e^- \rightleftharpoons Ag(s) + I^-$	−0.151
$PbSO_4(s) + 2e^- \rightleftharpoons Pb(s) + SO_4^{2-}$	−0.350
$Cd^{2+} + 2e^- \rightleftharpoons Cd(s)$	−0.403
$Zn^{2+} + 2e^- \rightleftharpoons Zn(s)$	−0.763

* See Appendix 4 for a more extensive list.

Standard electrode potential data are available for an enormous number of half-reactions. Many have been determined directly from electrochemical measurements. Others have been computed from equilibrium studies of oxidation/reduction systems and from thermochemical data associated with such reactions. Table 15-1 contains standard electrode data from several half-reactions that we will be considering in the pages that follow. A more extensive listing is found in Appendix 4.[2]

Table 15-1 and Appendix 4 illustrate the two common ways for tabulating standard potential data. In Table 15-1, potentials are listed in decreasing numerical order. As a consequence, the species in the upper left part are the most effective electron acceptors, as indicated by their large positive E^0 values. They are, therefore, the strongest *oxidizing agents*. As we proceed down the left side of such a table, each succeeding species is less effective as an electron acceptor than the one above it. The half-cell reactions at the bottom of the table have little or no tendency to take place as they are written. On the other hand, they do tend to occur in the opposite direction. The most effective *reducing agents*, then, are those species that appear in the lower right portion of the table.

Compilations of electrode-potential data, such as that shown in Table 15-1, provide the chemist with qualitative insights into the extent and direction of

Challenge: Based on the E^0 values in Table 15-1 for Fe^{3+} and I_3^-, which species would you expect to predominate in a solution produced by mixing iron(III) and iodide ions?

[2] Comprehensive sources for standard electrode potentials include: *Standard Electrode Potentials in Aqueous Solutions*, A. J. Bard, R. Parsons, and J. Jordan, Eds. New York: Marcel Dekker, 1985; G. Milazzo and S. Caroli, *Tables of Standard Electrode Potentials*. New York: Wiley-Interscience, 1977; M. S. Antelman and F. J. Harris, *Chemical Electrode Potentials*. New York: Plenum Press, 1982. Some compilations are arranged alphabetically by element; others are tabulated according to the numerical value of E^0.

Feature 15-4

SIGN CONVENTION IN THE OLD LITERATURE

Reference works, particularly those published before 1953, often contain tabulations of electrode potentials that are not in accord with the IUPAC recommendations. For example, in a classic source of standard-potential data compiled by Latimer,* one finds

$$Zn(s) \rightleftharpoons Zn^{2+} + 2e^- \qquad E = +0.76 \text{ V}$$
$$Cu(s) \rightleftharpoons Cu^{2+} + 2e^- \qquad E = -0.34 \text{ V}$$

To convert these oxidation potentials to electrode potentials as defined by the IUPAC convention, you must mentally (1) express the half-reactions as reductions, and (2) change the signs of the potentials.

The sign convention used in a tabulation of electrode potentials may not be stated explicitly. This information can be deduced, however, by noting the direction and sign of the potential for a familiar half-reaction. If the sign agrees with the IUPAC convention, the table can be used as is; if the sign does not agree, the signs of *all* of the data must be reversed. For example, the reaction

$$O_2(g) + 4H^+ + 4e^- \rightleftharpoons 2H_2O \qquad E = +1.229 \text{ V}$$

occurs spontaneously with respect to the standard hydrogen electrode and thus carries a positive sign. If the potential for this half-reaction is negative in a table, it and all the other potentials in the table should be multiplied by -1.

---

* W. M. Latimer, *The Oxidation States of the Elements and Their Potentials in Aqueous Solutions*, 2nd ed. Englewood Cliffs, NJ: Prentice-Hall, 1952.

---

electron-transfer reactions. For example, the standard potential for silver(I) ($+0.799$ V) is more positive than that for copper(II) ($+0.337$ V). We therefore conclude that a piece of copper immersed in a silver(I) solution will cause the reduction of silver ion and the oxidation of the copper. On the other hand, we expect no reaction if we place a piece of silver in a copper(II) solution.

Unlike the data in Table 15-1, standard potentials in Appendix 4 are arranged alphabetically by element to make it easier to locate data for a given electrode reaction.

Example 15-3

Calculate the electrode potential of a silver electrode immersed in a 0.0500 M solution of NaCl using (a) $E^0_{Ag+} = 0.799$ V, and (b) $E^0_{AgCl} = 0.222$ V.

(a) $\qquad\qquad Ag^+ + e^- \rightleftharpoons Ag(s) \qquad E^0 = 0.799$ V

The Ag^+ concentration of this solution is given by

$$[Ag^+] = K_{sp}/[Cl^-] = 1.82 \times 10^{-10}/0.0500 = 3.64 \times 10^{-9} \text{ M}$$

Substituting into the Nernst expression gives

$$E = 0.799 - 0.0592 \log \frac{1}{3.64 \times 10^{-9}} = 0.299 \text{ V}$$

(b) Here we may write

$$E = 0.222 - 0.0592 \log [Cl^-] = 0.222 - 0.0592 \log 0.0500$$
$$= 0.299 \text{ V}$$

15C-6 Limitations to the Use of Standard Electrode Potentials

We shall be using standard electrode potentials throughout the rest of this book to calculate cell potentials and equilibrium constants for redox reactions, as well as to derive data for redox titration curves. You should be aware that such calculations sometimes lead to results that are significantly different from those you would obtain experimentally in the laboratory. There are two main sources of these differences: (1) use of concentrations in place of activities in the Nernst equation, and (2) failure to take into account other equilibria such as dissociation, association, complex formation, and solvolysis.

Feature 15-5
STANDARD ELECTRODE POTENTIALS FOR SYSTEMS
INVOLVING PRECIPITATES OR COMPLEX IONS

In Table 15-1, we find several entries involving Ag(I), including:

$$Ag^+ + e^- \rightleftharpoons Ag(s) \qquad\qquad E^0_{Ag^+} = +0.799 \text{ V}$$
$$AgCl(s) + e^- \rightleftharpoons Ag(s) + Cl^- \qquad E^0_{AgCl} = +0.222 \text{ V}$$
$$Ag(S_2O_3)_2^{3-} + e^- \rightleftharpoons Ag(s) + 2S_2O_3^{2-} \qquad E^0_{Ag(S_2O_3)_2} = +0.17 \text{ V}$$

Each entry gives the potential of a silver electrode in a different environment. Let's see how the three potentials are related.

The Nernst expression for the first half-reaction is

$$E = E^0_{Ag^+} - \frac{0.0592}{1} \log \frac{1}{[Ag^+]}$$

If we replace $[Ag^+]$ with $K_{sp}/[Cl^-]$, we obtain

$$E = E^0_{Ag^+} - \frac{0.0592}{1} \log \frac{[Cl^-]}{K_{sp}} = E^0_{Ag^+} + 0.0592 \log K_{sp} - 0.0592 \log [Cl^-]$$

By definition, the standard potential for the second half-reaction is the potential where $[Cl^-] = 1.00$. That is, when $[Cl^-] = 1.00$, $E = E^0_{AgCl}$. Substituting these values gives

$$E^0_{AgCl} = E^0_{Ag^+} + 0.0592 \log 1.82 \times 10^{-10} - 0.0592 \log (1.00)$$
$$= 0.799 + (-0.577) - 0.000 = 0.222 \text{ V}$$

Figure 15-A illustrates the definition of the standard electrode potential for the Ag/AgCl electrode.

If we proceed in the same way, we obtain for the third equilibrium

$$E^0_{Ag(S_2O_3)_2^{3-}} = E^0_{Ag^+} - 0.0592 \log K_f$$

where K_f is the formation constant for the complex. That is,

$$K_f = \frac{[Ag(S_2O_3)_2^{3-}]}{[Ag^+][S_2O_3^{2-}]^2}$$

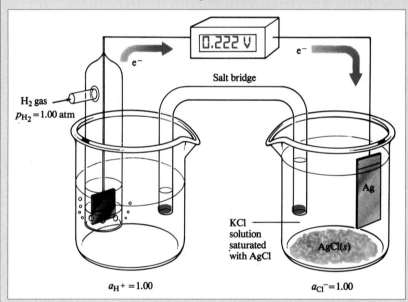

Figure 15-A   Definition of the standard electrode potential for an Ag/AgCl electrode.

Use of Concentrations Instead of Activities

Most analytical oxidation/reduction reactions are carried out in solutions that have such high ionic strengths that activity coefficients cannot be obtained via the Debye-Hückel equation (Equation 8-5, Section 8B-2). Significant errors may result, however, if concentrations are used in the Nernst equation rather than activities. For example, the standard potential for the half-reaction

$$Fe^{3+} + e^- \rightleftharpoons Fe^{2+} \qquad E^0 = +0.771 \text{ V}$$

Feature 15-6

WHY ARE THERE TWO ELECTRODE POTENTIALS
FOR Br_2 IN TABLE 15-1?

In Table 15-1, we find the following data for Br_2:

$$Br_2(aq) + 2e^- \rightleftharpoons 2Br^- \qquad E^0 = +1.087 \text{ V}$$
$$Br_2(l) + 2e^- \rightleftharpoons 2Br^- \qquad E^0 = +1.065 \text{ V}$$

The second standard potential applies only to a solution that is saturated with Br_2. It does *not* apply to undersaturated solutions of the element. You should use 1.065 V to calculate the electrode potential of a 0.0100 M solution of KBr that is saturated with Br_2 and in contact with an excess of the liquid element. In such a case,

$$E = 1.065 - \frac{0.0592}{2} \log [Br^-]^2 = 1.065 - \frac{0.0592}{2} \log (0.0100)^2$$

$$= 1.065 - \frac{0.0592}{2}(-4.00) = 1.183 \text{ V}$$

In this calculation, no term for Br_2 appears in the logarithmic term because it is a pure liquid that is present in excess.

The standard electrode potential shown in the first entry for $Br_2(aq)$ is a *hypothetical* standard potential because the solubility of Br_2 at 25°C is only about 0.18 M. Thus, the recorded value of 1.087 V is based on a system that—in terms of our definition of E^0—cannot be prepared experimentally. Nevertheless, the hypothetical potential does permit us to calculate electrode potentials for solutions that are *undersaturated* in Br_2. For example, if we wish to calculate the electrode potential for a solution that was 0.0100 M in KBr and 0.00100 M in Br_2 we would write

$$E = 1.087 - \frac{0.0592}{2} \log \frac{[Br^-]^2}{[Br_2(aq)]} = 1.087 - \frac{0.0592}{2} \log \frac{(0.0100)^2}{0.00100}$$

$$= 1.087 - \frac{0.0592}{2} \log 0.100 = 1.117 \text{ V}$$

is +0.771 V. When the potential of a platinum electrode, immersed in a solution that is 10^{-4} M in iron(III) ion, iron(II) ion, and perchloric acid, is measured against a standard hydrogen electrode, a reading of close to +0.77 V is obtained as predicted by theory. If, however, perchloric acid is added to this mixture until the acid concentration is 0.1 M, the potential is found to decrease to about +0.75 V. This difference occurs because the activity coefficient of iron(III) is considerably smaller than that of iron(II) (0.4 vs. 0.18) at the high ionic strength of the 0.1 M perchloric acid medium (see Table 8-1). As a result, the ratio of activities of the two species ($[Fe^{2+}]/[Fe^{3+}]$) in the Nernst equation is greater than unity, a condition that leads to a decrease in the electrode potential. In 1 M $HClO_4$, the electrode potential is still smaller (~ 0.73 V).

Effect of Other Equilibria

The application of standard electrode potential data to many systems of interest in analytical chemistry is further complicated by association, dissociation, complex formation, and solvolysis equilibria involving the species that appear in the Nernst equation. These phenomena can be taken into account only if their existence is known and appropriate equilibrium constants are available. Usually, neither of these requirements is met and significant discrepancies arise as a consequence. For example, the presence of 1 M hydrochloric acid in the iron(II)/iron(III) mixture we have just discussed leads to a measured potential of $+0.70$ V; in 1 M sulfuric acid, a potential of $+0.68$ V is observed; and in a 2 M phosphoric acid, the potential is $+0.46$ V. In each of these cases, the iron(II)/iron(III) activity ratio is larger because the complexes of iron(III) with chloride, sulfate, and phosphate ions are more stable than those of iron(II); thus the ratio of the *species* concentrations, $[Fe^{2+}]/[Fe^{3+}]$, in the Nernst equation is greater than unity and the measured potential is less than the standard potential. If formation constants for these complexes were available, it would be possible to make appropriate corrections. Unfortunately, such data are often not available, or if they are, they are not very reliable.

Formal Potentials

Formal potentials are empirically derived potentials that compensate for the types of activity and competing equilibria effects that we have just described. The formal potential $E^{0'}$ of a system is the potential of the half-cell with respect to the standard hydrogen electrode measured under conditions such that the ratio of *analytical concentrations* of reactants and products as they appear in the Nernst equation is exactly unity and the concentrations of other species in the system are all carefully specified. For example, the formal potential for the half-reaction

$$Ag^+ + e^- \rightleftharpoons Ag(s) \qquad E^{0'} = 0.792 \text{ V}$$

in 1.00 M $HClO_4$ could be obtained by measuring the potential of the cell shown in Figure 15-10. Here, the cathode is a silver electrode immersed in a solution that is 1.00 M in $AgNO_3$ and 1.00 M in $HClO_4$; the anode is a standard hydrogen electrode. This cell has a potential of 0.792 V, which is the formal potential of the Ag/Ag^+ couple in 1.00 M $HClO_4$. Note that the standard potential for this couple is 0.799 V.

Formal potentials for many half-reactions are listed in Appendix 4. Substitution of formal potentials for standard electrode potentials in the Nernst equation yields better agreement between calculated and experimental results—provided, of course, that the electrolyte concentration of the solution approximates that for which the formal potential is applicable. It is not surprising that attempts to apply formal potentials to systems that differ substantially in type and in concentration of electrolyte can result in errors that are larger than those associated with the use of standard electrode potentials. In this book, we use whichever is appropriate.

> A formal potential is the electrode potential when the ratio of analytical concentrations of reactants and products of a half-reaction is exactly 1.00 and the molar concentrations of any other solutes are specified.

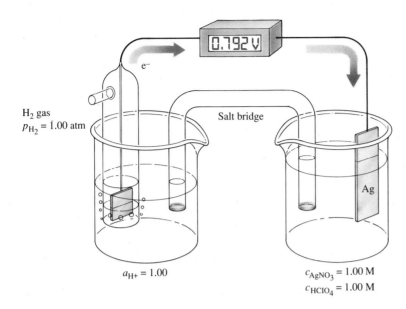

H_2 gas
$p_{H_2} = 1.00$ atm

Salt bridge

e^-

Ag

$a_{H+} = 1.00$

$c_{AgNO_3} = 1.00$ M
$c_{HClO_4} = 1.00$ M

Figure 15-10
Definition of the formal potential for
Ag^+/Ag couple in 1 M $HClO_4$.

15D   QUESTIONS AND PROBLEMS†

15-1. Briefly describe or define
  *(a) oxidation.
  (b) oxidizing agent.
  *(c) salt bridge.
  (d) liquid junction.
  *(e) Nernst equation.

15-2. Briefly describe or define
  *(a) electrode potential.
  (b) formal potential.
  *(c) standard electrode potential.
  (d) liquid-junction potential.
  *(e) oxidation potential.

15-3. Make a clear distinction between
  *(a) reduction and reducing agent.
  (b) a galvanic cell and an electrolytic cell.
  *(c) the anode and the cathode of an electrochemical cell.
  (d) a reversible electrochemical cell and an irreversible electrochemical cell.
  *(e) the standard electrode potential and formal potential.

*15-4. The following entries are found in a table of standard electrode potentials:

$$I_2(s) + 2e^- \rightleftharpoons 2I^- \qquad E^0 = 0.5355 \text{ V}$$

$$I_2(aq) + 2e^- \rightleftharpoons 2I^- \qquad E^0 = 0.615 \text{ V}$$

What is the significance of the difference between these two?

*15-5. Why is it necessary to bubble hydrogen through the electrolyte in a hydrogen electrode?

15-6. The standard electrode potential for the reduction of Ni^{2+} to Ni is -0.25 V. Would the potential of a nickel electrode immersed in a 1.00 M NaOH solution saturated with $Ni(OH)_2$ be more negative than $E^0_{Ni^{2+}}$ or less? Explain.

*15-7. Write balanced net-ionic equations for the following reactions. Supply H^+ and/or H_2O, as needed, to obtain balance.
  *(a) $Fe^{3+} + Sn^{2+} \rightarrow Fe^{2+} + Sn^{4+}$
  (b) $Cr(s) + Ag^+ \rightarrow Cr^{3+} + Ag(s)$
  *(c) $NO_3^- + Cu(s) \rightarrow NO_2(g) + Cu^{2+}$
  (d) $MnO_4^- + H_2SO_3 \rightarrow Mn^{2+} + SO_4^{2-}$
  *(e) $Ti^{3+} + Fe(CN)_6^{3-} \rightarrow TiO^{2+} + Fe(CN)_6^{4-}$
  (f) $H_2O_2 + Ce^{4+} \rightarrow O_2(g) + Ce^{3+}$
  *(g) $Ag(s) + I^- + Sn^{4+} \rightarrow AgI(s) + Sn^{2+}$
  (h) $UO_2^{2+} + Zn(s) \rightarrow U^{4+} + Zn^{2+}$
  *(i) $HNO_2 + MnO_4^- \rightarrow NO_3^- + Mn^{2+}$
  (j) $H_2NNH_2 + IO_3^- + Cl^- \rightarrow N_2(g) + ICl_2^-$

*15-8. Identify the oxidizing agent and the reducing agent on the left side of each equation in 15-7; write a balanced equation for each half-reaction.

---

† Note: Numerical data are molar analytical concentrations where the full formula of a species is provided. Molar equilibrium concentrations are supplied for species displayed as ions.

15-9. Write balanced net-ionic equations for the following reactions. Supply H^+, OH^-, and/or H_2O, as needed, to obtain balance.

*(a) $MnO_4^- + VO^{2+} \rightarrow Mn^{2+} + V(OH)_4^+$

(b) $I_2 + H_2S(g) \rightarrow I^- + S(s)$

*(c) $Cr_2O_7^{2-} + U^{4+} \rightarrow Cr^{3+} + UO_2^{2+}$

(d) $Cl^- + MnO_2(s) \rightarrow Cl_2(g) + Mn^{2+}$

*(e) $IO_3^- + I^- \rightarrow I_2(aq)$

(f) $IO_3^- + I^- + Cl^- \rightarrow ICl_2^-$

*(g) $HPO_3^{2-} + MnO_4^- + OH^- \rightarrow PO_4^{3-} + MnO_4^{2-}$

(h) $SCN^- + BrO_3^- \rightarrow Br^- + SO_4^{2-} + HCN$

*(i) $V^{2+} + V(OH)_4^+ \rightarrow VO^{2+}$

(j) $MnO_4^- + Mn^{2+} + OH^- \rightarrow MnO_2(s)$

*15-10. Identify the oxidizing agent and the reducing agent on the left side of each equation in 15-9; write a balanced equation for each half-reaction.

*15-11. Consider the following oxidation/reduction reactions:

$$AgBr(s) + V^{2+} \rightarrow Ag(s) + V^{3+} + Br^-$$

$$Tl^{3+} + 2Fe(CN)_6^{4-} \rightarrow Tl^+ + 2Fe(CN)_6^{3-}$$

$$2V^{3+} + Zr(s) \rightarrow 2V^{2+} + Zn^{2+}$$

$$Fe(CN)_6^{3-} + Ag(s) + Br^- \rightarrow Fe(CN)_6^{4-} + AgBr(s)$$

$$S_2O_8^{2-} + Tl^+ \rightarrow 2SO_4^{2-} + Tl^{3+}$$

(a) Write each net process in terms of two balanced half-reactions.

(b) Express each half-reaction as a reduction.

(c) Arrange the half-reactions in (b) in order of decreasing effectiveness as electron acceptors.

15-12. Consider the following oxidation/reduction reactions:

$$2H^+ + Sn(s) \rightarrow H_2(g) + Sn^{2+}$$

$$Ag^+ + Fe^{2+} \rightarrow Ag(s) + Fe^{3+}$$

$$Sn^{4+} + H_2(g) \rightarrow Sn^{2+} + 2H^+$$

$$2Fe^{3+} + Sn^{2+} \rightarrow 2Fe^{2+} + Sn^{4+}$$

$$Sn^{2+} + Co(s) \rightarrow Sn(s) + Co^{2+}$$

(a) Write each net process in terms of two balanced half-reactions.

(b) Express each half-reaction as a reduction.

(c) Arrange the half-reactions in (b) in order of decreasing effectiveness as electron acceptors.

*15-13. Calculate the electrode potential of a copper electrode immersed in

(a) 0.0440 M $Cu(NO_3)_2$.

(b) 0.0750 M in NaCl and saturated with CuCl.

(c) 0.0400 M in NaOH and saturated with $Cu(OH)_2$.

(d) 0.0250 M in $Cu(NH_3)_4^{2+}$ and 0.128 M in NH_3. (β_4 for $Cu(NH_3)_4^{2+} = 5.62 \times 10^{11}$)

(e) a solution in which the molar analytical concentration of $Cu(NO_3)_2$ is 4.00×10^{-3} M, that for H_2Y^{2-} is 2.90×10^{-2} M (Y = EDTA), and the pH is fixed at 4.00.

15-14. Calculate the electrode potential of a zinc electrode

immersed in

(a) 0.0600 M $Zn(NO_3)_2$.

(b) 0.01000 M in NaOH and saturated with $Zn(OH)_2$.

(c) 0.0100 M in $Zn(NH_3)_4^{2+}$ and 0.250 M in NH_3. (β_4 for $Zn(NH_3)_4^{2+} = 7.76 \times 10^8$)

(d) a solution in which the molar analytical concentration of $Zn(NO_3)_2$ is 5.00×10^{-3} M, that for H_2Y^{2-} is 0.0445 M, and the pH is fixed at 9.00.

15-15. Use activities to calculate the electrode potential of a hydrogen electrode in which the electrolyte is 0.0100 M HCl and the activity of H_2 is 1.00 atm.

*15-16. Calculate the electrode potential of a platinum electrode immersed in a solution that is

(a) 0.0263 M in K_2PtCl_4 and 0.1492 M in KCl.

(b) 0.0750 M in $Sn(SO_4)_2$ and 2.50×10^{-3} M in $SnSO_4$.

(c) buffered to a pH of 6.00 and saturated with $H_2(g)$ at 1.00 atm.

(d) 0.0353 M in $VOSO_4$, 0.0586 M in $V_2(SO_4)_3$, and 0.100 M in $HClO_4$.

(e) prepared by mixing 25.00 mL of 0.0918 M $SnCl_2$ with an equal volume of 0.1568 M $FeCl_3$.

(f) prepared by mixing 25.00 mL of 0.0832 M $V(OH)_4^+$ with 50.00 mL of 0.01087 M $V_2(SO_4)_3$ and has a pH of 1.00.

15-17. Calculate the electrode potential of a platinum electrode immersed in a solution that is

(a) 0.0813 M in $K_4Fe(CN)_6$ and 0.00566 M in $K_3Fe(CN)_6$.

(b) 0.0400 M in $FeSO_4$ and 0.00845 M in $Fe_2(SO_4)_3$.

(c) buffered to a pH of 5.55 and saturated with H_2 at 1.00 atm.

(d) 0.1996 M in $V(OH)_4^+$, 0.0789 M in VO^{2+}, and 0.0800 M in $HClO_4$.

(e) prepared by mixing 50.00 mL of 0.0607 M $Ce(SO_4)_2$ with an equal volume of 0.100 M $FeCl_2$. Assume solutions were 1.00 M in H_2SO_4 and use formal potentials.

(f) prepared by mixing 25.00 mL of 0.0832 M $V_2(SO_4)_3$ with 50.00 mL of 0.00628 M $V(OH)_4^+$ and has a pH of 1.00.

*15-18. Indicate whether the following half-cells would act as an anode or a cathode when coupled with a standard hydrogen electrode in a galvanic cell. Calculate the cell potential.

(a) $Ni|Ni^{2+}(0.0943 \text{ M})$

(b) $Ag|AgI(sat'd),KI(0.0922 \text{ M})$

(c) $Pt,O_2(780 \text{ torr})|HCl(1.50 \times 10^{-4} \text{ M})$

(d) $Pt|Sn^{2+}(0.0944 \text{ M}),Sn^{4+}(0.350 \text{ M})$

(e) $Ag|Ag(S_2O_3)_2^{3-}(0.00753 \text{ M}),Na_2S_2O_3(0.1439 \text{ M})$

15-19. Indicate whether the following half-cells behave as an anode or a cathode when coupled with a standard hydrogen electrode in a galvanic cell. Calculate the cell potential.

(a) $Cu|Cu^{2+}(0.0897 \text{ M})$

(b) $Cu|CuI(sat'd),KI(0.1214 \text{ M})$

(c) $Pt,H_2(0.984 \text{ atm})|HCl(1.00 \times 10^{-4} \text{ M})$

(d) $Pt|Fe^{3+}(0.0906 \text{ M}),Fe^{2+}(0.1628 \text{ M})$

(e) $Ag|Ag(CN)_2^-(0.0827 \text{ M}),KCN(0.0699 \text{ M})$

*15-20. The solubility-product constant for Ag_2SO_3 is 1.5×10^{-14}. Calculate E^0 for the process

$$Ag_2SO_3(s) + 2e^- \rightleftharpoons 2Ag(s) + SO_3^{2-}$$

15-21. The solubility-product constant for $Ni_2P_2O_7$ is 1.7×10^{-13}. Calculate E^0 for the process

$$Ni_2P_2O_7(s) + 4e^- \rightleftharpoons 2Ni(s) + P_2O_7^{4-}$$

*15-22. The solubility-product constant for Tl_2S is 6×10^{-22}. Calculate E^0 for the reaction

$$Tl_2S(s) + 2e^- \rightleftharpoons 2Tl(s) + S^{2-}$$

15-23. The solubility product for $Pb_3(AsO_4)_2$ is 4.1×10^{-36}. Calculate E^0 for the reaction

$$Pb_3(AsO_4)_2(s) + 6e^- \rightleftharpoons 3Pb(s) + 2AsO_4^{3-}$$

*15-24. Compute E^0 for the process

$$ZnY^{2-} + 2e^- \rightleftharpoons Zn(s) + Y^{4-}$$

where Y^{4-} is the completely deprotonated anion of EDTA. The formation constant for ZnY^{2-} is 3.2×10^{16}.

*15-25. Given the formation constants

$$Fe^{3+} + Y^{4-} \rightleftharpoons FeY^- \qquad K_f = 1.3 \times 10^{25}$$
$$Fe^{2+} + Y^{4-} \rightleftharpoons FeY^{2-} \qquad K_f = 2.1 \times 10^{14}$$

calculate E^0 for the process

$$FeY^- + e^- \rightleftharpoons FeY^{2-}$$

15-26. Calculate E^0 for the process

$$Cu(NH_3)_4^{2+} + e^- \rightleftharpoons Cu(NH_3)_2^+ + 2NH_3$$

given that

$$Cu^+ + 2NH_3 \rightleftharpoons Cu(NH_3)_2^+ \qquad \beta_2 = 7.2 \times 10^{10}$$
$$Cu^{2+} + 4NH_3 \rightleftharpoons Cu(NH_3)_4^{2+} \qquad \beta_4 = 2.0 \times 10^{13}$$

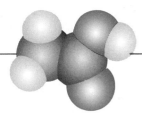

CHAPTER **16**

APPLICATIONS OF STANDARD ELECTRODE POTENTIALS

In this chapter, we show how standard electrode potentials can be used to calculate: (1) thermodynamic cell potentials; (2) equilibrium constants for redox reactions; and (3) data for redox titration curves.

16A THE THERMODYNAMIC POTENTIAL OF ELECTROCHEMICAL CELLS

We can use standard electrode potentials and the Nernst equation to calculate the potential obtainable from a galvanic cell or the potential required to operate an electrolytic cell. The calculated potentials (sometimes called *thermodynamic potentials*) are theoretical in the sense that they refer to cells in which there is no current. Additional factors must be taken into account if there is a current.

The potential of an electrochemical cell E_{cell} is the difference between the *electrode potential* of the cathode $E_{cathode}$ and the *electrode potential* of the anode E_{anode}. That is,

It is important to note that $E_{cathode}$ and E_{anode} in Equation 16-1 are *electrode potentials* as defined at the beginning of Section 15C-2.

$$E_{cell} = E_{cathode} - E_{anode} \qquad (16\text{-}1)$$

where $E_{cathode}$ and E_{anode} are the electrode potentials of the cathode and anode, respectively.

Example 16-1

Calculate the thermodynamic potential of the following cell

$$Cu|Cu^{2+}(0.0200 \text{ M})\|Ag^+(0.0200 \text{ M})|Ag$$

Note that this cell is the galvanic cell shown in Figure 15-2a.

The two half-reactions and standard potentials are

$$Ag^+ + e^- \rightleftharpoons Ag(s) \qquad E^0 = 0.799 \text{ V} \qquad \textbf{(16-2)}$$

$$Cu^{2+} + 2e^- \rightleftharpoons Cu(s) \qquad E^0 = 0.337 \text{ V} \qquad \textbf{(16-3)}$$

The electrode potentials are

$$E_{Ag^+} = 0.799 - 0.0592 \log \frac{1}{0.0200} = 0.6984 \text{ V}$$

$$E_{Cu^{2+}} = 0.337 - \frac{0.0592}{2} \log \frac{1}{0.0200} = 0.2867 \text{ V}$$

We see from the cell diagram that the silver electrode is the cathode and the copper electrode is the anode. Therefore, application of Equation 16-1 gives

$$E_{cell} = E_{Ag^+} - E_{Cu^{2+}} = 0.6984 - 0.2867 = +0.412 \text{ V}$$

Gustav Robert Kirchhoff (1824–1877) was a German physicist who made many important contributions to physics and chemistry. In addition to his work in spectroscopy, he is known for Kirchhoff's laws of electrical circuits. These laws are shown in the form of equations on the stamp above.

Example 16-2

Calculate the thermodynamic potential for the cell

$$Ag|Ag^+(0.0200 \text{ M})\|Cu^{2+}(0.0200 \text{ M})|Cu$$

The electrode potentials for the two half-reactions are identical to the electrode potentials calculated in Example 16-1. That is,

$$E_{Ag^+} = 0.6984 \text{ V} \qquad \text{and} \qquad E_{Cu^{2+}} = 0.2867 \text{ V}$$

In contrast to the previous example, however, the silver electrode is the anode and the copper electrode the cathode. Substituting these electrode potentials into Equation 16-1 gives

$$E_{cell} = E_{Cu^{2+}} - E_{Ag^+} = 0.2867 - 0.6984 = -0.412 \text{ V}$$

In order to develop a current in the electrolytic cell in Example 16-2, it is necessary to apply a potential that is more negative than −0.412 V. (See Figure 15-2b.)

Examples 16-1 and 16-2 illustrate an important fact. The cell potential computed from Equation 16-1 is positive for a galvanic cell and negative for an electrolytic cell.

If you obtain a positive cell potential in a calculation, you know a cell is galvanic; if it is negative, the cell is electrolytic. In the latter case, to develop a current in the cell you must use a voltage source with a potential greater than the calculated cell potential.

Example 16-3

Calculate the potential of the following cell and indicate whether it is galvanic or electrolytic (see Figure 16-1).

$$Pt|UO_2^{2+}(0.0150 \text{ M}), U^{4+}(0.200 \text{ M}), H^+(0.0300 \text{ M})\|$$
$$Fe^{2+}(0.0100 \text{ M}), Fe^{3+}(0.0250 \text{ M})|Pt$$

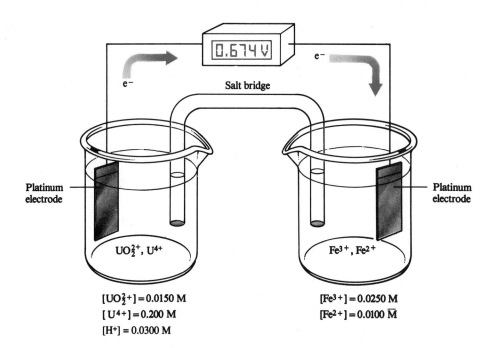

Figure 16-1
Cell for Example 16-3.

The two half-reactions are

$$Fe^{3+} + e^- \rightleftharpoons Fe^{2+} \qquad E^0 = +0.771 \text{ V}$$
$$UO_2^{2+} + 4H^+ + 2e^- \rightleftharpoons U^{4+} + 2H_2O \qquad E^0 = +0.334 \text{ V}$$

The *electrode potential* for the cathode is

$$E_{cathode} = 0.771 - 0.0592 \log \frac{[Fe^{2+}]}{[Fe^{3+}]}$$

$$= 0.771 - 0.0592 \log \frac{0.0100}{0.0250} = 0.771 - (-0.0236)$$

$$= 0.7946 \text{ V}$$

The *electrode potential* for the anode is

$$E_{anode} = 0.334 - \frac{0.0592}{2} \log \frac{[U^{4+}]}{[UO_2^{2+}][H^+]^4}$$

$$= 0.334 - \frac{0.0592}{2} \log \frac{0.200}{(0.0150)(0.0300)^4}$$

$$= 0.334 - 0.2136 = 0.1204 \text{ V}$$

and

$$E_{cell} = E_{cathode} - E_{anode} = 0.7946 - 0.1204 = +0.674 \text{ V}$$

The positive sign means that the cell is galvanic.

Example 16-4

Calculate the theoretical potential for the cell

$$Ag|AgCl(sat'd),HCl(0.0200 \text{ M})|H_2(0.800 \text{ atm}),Pt$$

Note that this cell does not require two compartments (nor a salt bridge) because molecular H_2 has little tendency to react directly with the low concentration of Ag^+ in the electrolyte solution. This is an example of a cell *without liquid junction* (see Figure 16-2).

The two half-reactions and their corresponding standard electrode potentials are (Table 15-1)

$$2H^+ + 2e^- \rightleftharpoons H_2(g) \qquad E^0 = 0.000 \text{ V}$$
$$AgCl(s) + e^- \rightleftharpoons Ag(s) + Cl^- \qquad E^0 = 0.222 \text{ V}$$

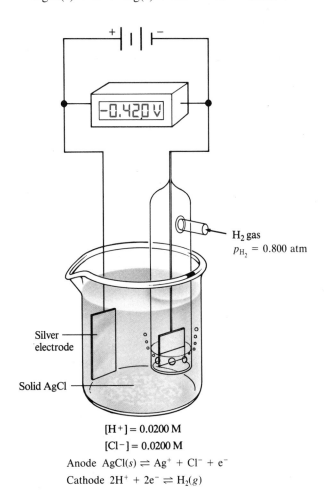

$$[H^+] = 0.0200 \text{ M}$$
$$[Cl^-] = 0.0200 \text{ M}$$

Anode $AgCl(s) \rightleftharpoons Ag^+ + Cl^- + e^-$

Cathode $2H^+ + 2e^- \rightleftharpoons H_2(g)$

Figure 16-2

Cell without liquid junction for Example 16-4.

The two electrode potentials are

$$E_{cathode} = 0.000 - \frac{0.0592}{2} \log \frac{0.800}{(0.0200)^2} = -0.0977 \text{ V}$$

$$E_{anode} = 0.222 - \frac{0.0592}{1} \log 0.0200 = 0.3226 \text{ V}$$

The cell diagram specifies the silver electrode as the anode and the hydrogen electrode as the cathode. Thus,

$$E_{cell} = -0.0977 - 0.3226 = -0.420 \text{ V}$$

The negative sign indicates that the cell reaction

$$2H^+ + 2Ag(s) + 2Cl^- \rightleftharpoons H_2(g) + 2AgCl(s)$$

is nonspontaneous, and thus the cell is electrolytic. It would thus require an external power source for operation in this way.

16B   CALCULATION OF REDOX EQUILIBRIUM CONSTANTS

Let us again consider the equilibrium that is established when a piece of copper is immersed in a solution containing a dilute solution of silver nitrate:

$$Cu(s) + 2Ag^+ \rightleftharpoons Cu^{2+} + 2Ag(s) \tag{16-4}$$

The equilibrium constant for this reaction is

$$K_{eq} = \frac{[Cu^{2+}]}{[Ag^+]^2} \tag{16-5}$$

As we showed in Section 15A-2, this reaction can be carried out in the galvanic cell

$$Cu|Cu^{2+}(xM)\|Ag^+(yM)|Ag$$

A sketch of a cell similar to this is shown in Figure 15-2a. Its cell potential at any instant is given by Equation 16-1:

$$E_{cell} = E_{cathode} - E_{anode} = E_{Ag^+} - E_{Cu^{2+}}$$

As the reaction proceeds, the concentration of Cu(II) ions increases and the concentration of Ag(I) ions decreases. These changes make the potential of the copper electrode more positive and that of the silver electrode less positive. As shown in Figure 15-6, the net effect of these changes is a continuous decrease in the potential of the cell as it discharges. Ultimately the concentrations of Cu(II) and Ag(I) attain their equilibrium values as determined by Equation 16-5, and the current ceases. Under these conditions, *the potential of the cell becomes zero. Thus, at chemical equilibrium*, we may write

$$E_{cell} = 0 = E_{cathode} - E_{anode} = E_{Ag^+} - E_{Cu^{2+}}$$

or

$$E_{cathode} = E_{anode} = E_{Ag^+} = E_{Cu^{2+}} \qquad \text{(16-6)}$$

We can generalize Equation 16-6 by stating that *at equilibrium, the electrode potentials for all half-reactions in an oxidation/reduction system are equal.* This generalization applies regardless of the number of half-reactions present in the system because interactions among *all* must take place until the electrode potentials are identical. For example if we have four oxidation/reduction couples in a solution, interaction among all four takes place until

$$E_{Ox_1} = E_{Ox_2} = E_{Ox_3} = E_{Ox_4} \qquad \text{(16-7)}$$

where E_{Ox_1}, E_{Ox_2}, E_{Ox_3}, and E_{Ox_4} are the electrode potentials for the four half-reactions.

Let us return to the reaction shown in Equation 16-4, and substitute Nernst expressions for the two electrode potentials in Equation 16-6, which gives

$$E^0_{Ag^+} - \frac{0.0592}{2} \log \frac{1}{[Ag^+]^2} = E^0_{Cu^{2+}} - \frac{0.0592}{2} \log \frac{1}{[Cu^{2+}]} \qquad \text{(16-8)}$$

It is important to note that we applied the Nernst equation to the silver half-reaction as it appears in the balanced equation:

$$2Ag^+ + 2e^- \rightleftharpoons 2Ag(s) \qquad E^0 = 0.799 \text{ V}$$

Rearrangement of Equation 16-8 gives

$$\begin{aligned} E^0_{Ag^+} - E^0_{Cu^{2+}} &= \frac{0.0592}{2} \log \frac{1}{[Ag^+]^2} - \frac{0.0592}{2} \log \frac{1}{[Cu^{2+}]} \\ &= \frac{0.0592}{2} \log \frac{1}{[Ag^+]^2} + \frac{0.0592}{2} \log \frac{[Cu^{2+}]}{1} \end{aligned}$$

Finally, then,

$$\frac{2(E^0_{Ag^+} - E^0_{Cu^{2+}})}{0.0592} = \log \frac{[Cu^{2+}]}{[Ag^+]^2} = \log K_{eq} \qquad \text{(16-9)}$$

The concentration terms in Equation 16-9 are *equilibrium concentrations; the ratio [Cu²⁺]/[Ag⁺]² in the logarithmic term is therefore the equilibrium constant for the reaction.*

Remember that *when redox systems are at equilibrium, the electrode potentials of all systems that are present are identical.* This generality applies whether the reactions take place directly in solution or indirectly in a galvanic cell.

Example 16-5

Calculate the equilibrium constant for the reaction shown in Equation 16-4. Substituting numerical values into Equation 16-9 yields

$$\log K_{eq} = \log \frac{[Cu^{2+}]}{[Ag^+]^2} = \frac{2(0.799 - 0.337)}{0.0592}$$

$$= 15.61$$

$$K_{eq} = \text{antilog } 15.61 = 4.1 \times 10^{15}$$

To make calculations like those shown in Example 16-5, you should follow the rounding rule for antilogs that is given on page 73.

Example 16-6

Calculate the equilibrium constant for the reaction

$$2Fe^{3+} + 3I^- \rightleftharpoons 2Fe^{2+} + I_3^-$$

In Appendix 4, we find

$$2Fe^{3+} + 2e^- \rightleftharpoons 2Fe^{3+} \qquad 0.771 \text{ V}$$

$$I_3^- + 2e^- \rightleftharpoons 3I^- \qquad 0.536 \text{ V}$$

We have multiplied the first half-reaction by 2 so that the number of moles of Fe^{3+} and Fe^{2+} are the same as in the balanced overall equation. We write the Nernst equation for Fe^{3+} based on the half-reaction for a two-electron transfer. That is,

$$E_{Fe^{3+}} = E^0_{Fe^{3+}} - \frac{0.0592}{2} \log \frac{[Fe^{2+}]^2}{[Fe^{3+}]^2}$$

and

$$E_{I_3^-} = E^0_{I_3^-} - \frac{0.0592}{2} \log \frac{[I^-]^3}{[I_3^-]}$$

At equilibrium, the electrode potentials are equal and

$$E_{Fe^{3+}} = E_{I_3^-}$$

$$E^0_{Fe^{3+}} - \frac{0.0592}{2} \log \frac{[Fe^{2+}]^2}{[Fe^{3+}]^2} = E^0_{I_3^-} - \frac{0.0592}{2} \log \frac{[I^-]^3}{[I_3^-]}$$

This equation rearranges to

$$\frac{2(E^0_{Fe^{3+}} - E^0_{I_3^-})}{0.0592} = \log \frac{[Fe^{2+}]^2}{[Fe^{3+}]^2} + \log \frac{[I_3^-]}{[I^-]^3}$$

Notice that we have changed the sign of the second logarithmic term by inverting the fraction. Further arrangement gives

$$\log \frac{[Fe^{2+}]^2[I_3^-]}{[Fe^{3+}]^2[I^-]^3} = \frac{2(E^0_{Fe^{3+}} - E^0_{I_3^-})}{0.0592}$$

Recall, however, that here the concentration terms are *equilibrium concentrations* and

$$\log K_{eq} = \frac{2(E^0_{Fe^{3+}} - E^0_{I_3^-})}{0.0592} = \frac{2(0.771 - 0.536)}{0.0592} = 7.939$$

$$K_{eq} = \text{antilog } 7.94 = 8.7 \times 10^7$$

We round the answer to two figures because $\log K_{eq}$ contains only 2 significant figures (the two to the right of the decimal point).

Example 16-7

Calculate the equilibrium constant for the reaction

$$2MnO_4^- + 3Mn^{2+} + 2H_2O \rightleftharpoons 5MnO_2(s) + 4H^+$$

In Appendix 4 we find

$$2MnO_4^- + 8H^+ + 6e^- \rightleftharpoons 2MnO_2(s) + 4H_2O \qquad E^0 = +1.695 \text{ V}$$
$$3MnO_2(s) + 12H^+ + 6e^- \rightleftharpoons 3Mn^{2+} + 6H_2O \qquad E^0 = +1.23 \text{ V}$$

Again we have multiplied both equations by integers so that the number of electrons are equal. When this system is at equilibrium,

$$E_{MnO_4^-} = E_{MnO_2}$$

$$1.695 - \frac{0.0592}{6} \log \frac{1}{[MnO_4^-]^2[H^+]^8} = 1.23 - \frac{0.0592}{6} \log \frac{[Mn^{2+}]^3}{[H^+]^{12}}$$

Inverting the log term on the right and rearranging leads to

$$\frac{6(1.695 - 1.23)}{0.0592} = \log \frac{[H^+]^{12}}{[MnO_4^-]^2[Mn^{2+}]^3[H^+]^8}$$

$$47.1 = \log \frac{[H^+]^4}{[MnO_4^-]^2[Mn^{2+}]^3} = \log K_{eq}$$

$$K_{eq} = \text{antilog } 47.1 = 1 \times 10^{47}$$

Note that the final result has only one significant figure.

16C REDOX TITRATION CURVES

Because most redox indicators respond to changes in electrode potential, the vertical axis in oxidation/reduction titration curves is generally an electrode potential instead of the p-functions that were used for precipitation, complex-formation, and neutralization titration curves. We have seen in Chapter 15 that there is a logarithmic relationship between electrode potential and concentration of the analyte or titrant. As a result, redox titration curves are similar in appearance to those for other types of titrations.

16C-1 Electrode Potentials During Redox Titrations

Let us now consider the redox titration of iron(II) with a standard solution of cerium(IV). The titration is described by the equation

$$Fe^{2+} + Ce^{4+} \rightleftharpoons Fe^{3+} + Ce^{3+}$$

This reaction is rapid and reversible, so that the system is at equilibrium at all times throughout the titration. Consequently, the electrode potentials for the two half-reactions are always identical (Equation 16-7); that is,

$$E_{Ce^{4+}} = E_{Fe^{3+}} = E_{system}$$

where we have termed E_{system} as *the potential of the system*. If a redox indicator has been added to this solution, the ratio of the concentrations of its oxidized and reduced forms must adjust so that the electrode potential for the indicator, E_{In}, is also equal to the system potential; thus, employing Equation 16-7, we may write

$$E_{In} = E_{Ce^{4+}} = E_{Fe^{3+}} = E_{system}$$

We can calculate the electrode potential of a system from standard potential data. For the reaction under consideration, the titration mixture is treated as if it were part of the hypothetical cell

$$SHE \| Ce^{4+}, Ce^{3+}, Fe^{3+}, Fe^{2+} | Pt$$

where SHE symbolizes the standard hydrogen electrode. The potential of the platinum electrode with respect to the standard hydrogen electrode is determined by the tendencies of iron(III) and cerium(IV) to accept electrons—in other words, the tendencies of the following half-reactions to occur determine the potential of the platinum electrode:

$$Fe^{3+} + e^- \rightleftharpoons Fe^{2+}$$
$$Ce^{4+} + e^- \rightleftharpoons Ce^{3+}$$

At equilibrium, the concentration ratios of the oxidized and reduced forms of the two species are such that their attraction for electrons (and thus their electrode potentials) are identical. Note that these concentration ratios vary continuously throughout the titration, as must E_{system}. End points are determined from the characteristic variation in E_{system} that occurs during the titration.

Because $E_{system} = E_{Fe^{3+}} = E_{Ce^{4+}}$, data for a titration curve can be obtained by applying the Nernst equation for *either* the cerium(IV) half-reaction or the iron(III) half-reaction. It turns out, however, that one or the other will be more convenient, depending on the stage of the titration. For example, the iron(III) potential is easier to compute in the region short of the equivalence point because here the species concentrations of iron(II) and iron(III) are appreciable and are approximately equal to their analytical concentrations. In contrast, the concentration of cerium(IV), which is negligible prior to the equivalence point because of the large excess of iron(II), can be obtained at this stage only by calculations based on the equilibrium constant for the reaction. Beyond the equivalence point, the analytical concentrations of cerium(IV) and cerium(III) are computed directly from the volumetric data, while that for iron(II) is not. In this region, then, the cerium(IV) electrode potential is easier to use. Equivalence-point potentials are derived by the method shown in the next section.

> Most end points in oxidation/reduction titrations are based on the rapid changes in E_{system} that occur at or near chemical equivalence.
>
> Before the equivalence point, E_{system} calculations are easiest to make using the Nernst equation for the analyte. Beyond the equivalence point, the Nernst equation for the reagent is used.

Equivalence-Point Potentials

At the equivalence point, the concentrations of cerium(IV) and iron(II) are extremely small and cannot be obtained from the stoichiometry of the reaction. Fortunately, equivalence-point potentials can be calculated from the concentration ratios of the two reactant species and the two product species at chemical equivalence.

At the equivalence point in the titration of iron(II) with cerium(IV), the potential of the system E_{eq} is given by both

$$E_{eq} = E^0_{Ce^{4+}} - \frac{0.0592}{1} \log \frac{[Ce^{3+}]}{[Ce^{4+}]}$$

and

$$E_{eq} = E^0_{Fe^{3+}} - \frac{0.0592}{1} \log \frac{[Fe^{2+}]}{[Fe^{3+}]}$$

Adding these two expressions gives

$$2E_{eq} = E^0_{Ce^{4+}} + E^0_{Fe^{3+}} - \frac{0.0592}{1} \log \frac{[Ce^{3+}][Fe^{2+}]}{[Ce^{4+}][Fe^{3+}]} \qquad \textbf{(16-10)}$$

> The concentration quotient in Equation 16-10 is *not* the usual ratio of product concentrations and reactant concentrations that appears in equilibrium-constant expressions.

The definition of equivalence point requires that

$$[Fe^{3+}] = [Ce^{3+}]$$

$$[Fe^{2+}] = [Ce^{4+}]$$

When we substitute these equalities into Equation 16-10, the concentration quotient becomes unity and the logarithmic term becomes zero:

$$2E_{eq} = E^0_{Ce^{4+}} + E^0_{Fe^{3+}} - \frac{0.0592}{1} \log \frac{[Ce^{3+}][Ce^{4+}]}{[Ce^{4+}][Ce^{3+}]} = E^0_{Ce^{4+}} + E^0_{Fe^{3+}}$$

$$E_{eq} = \frac{E^0_{Ce^{4+}} + E^0_{Fe^{3+}}}{2}$$

(16-11)

Example 16-8 illustrates how the equivalence-point potential is derived for a more complex reaction.

Example 16-8

Derive an expression for the equivalence-point potential in the titration of 0.0500 M UO_2^{2+} with 0.1000 M Ce^{4+}. Assume both solutions are 1.0 M in H_2SO_4.

$$U^{4+} + 2Ce^{4+} + 2H_2O \rightleftharpoons UO_2^{2+} + 2Ce^{3+} + 4H^+$$

In Appendix 4 we find

$$UO_2^{2+} + 4H^+ + 2e^- \rightleftharpoons U^{4+} + 2H_2O \qquad E^0 = 0.334 \text{ V}$$
$$Ce^{4+} + e^- \rightleftharpoons Ce^{3+} \qquad E^{0'} = 1.44 \text{ V}$$

Here we use the formal potential for Ce^{4+} in 1.0 M H_2SO_4.

Proceeding as in the cerium(IV)/iron(II) equivalence-point calculation we write

$$E_{eq} = E^0_{UO_2^{2+}} - \frac{0.0592}{2} \log \frac{[U^{4+}]}{[UO_2^{2+}][H^+]^4}$$

$$E_{eq} = E^{0'}_{Ce^{4+}} - 0.0592 \log \frac{[Ce^{3+}]}{[Ce^{4+}]}$$

In order to combine the log terms we must multiply the first equation by 2 to give

$$2E_{eq} = 2E^0_{UO_2^{2+}} - 0.0592 \log \frac{[U^{4+}]}{[UO_2^{2+}][H^+]^4}$$

Adding this to the previous equation leads to

$$3E_{eq} = 2E^0_{UO_2^{2+}} + E_{Ce^{4+}} - 0.0592 \log \frac{[U^{4+}][Ce^{3+}]}{[UO_2^{2+}][Ce^{4+}][H^+]^4}$$

But at equivalence

$$[U^{4+}] = [Ce^{4+}]/2$$

and

$$[UO_2^{2+}] = [Ce^{3+}]/2$$

Substituting these equations gives, after rearranging,

$$E_{eq} = \frac{2E^0_{UO_2^{2+}} + E^{0'}_{Ce^{4+}}}{3} - \frac{0.0592}{3} \log \frac{2[\cancel{Ce^{4+}}][\cancel{Ce^{3+}}]}{2[\cancel{Ce^{3+}}][\cancel{Ce^{4+}}][H^+]^4}$$

$$= \frac{2E^0_{UO_2^{2+}} + E^{0'}_{Ce^{4+}}}{3} - \frac{0.0592}{3} \log \frac{1}{[H^+]^4}$$

We see that the equivalence-point potential is pH dependent in this titration.

16C-2 Derivation of Titration Curves

Let us first consider the titration of 50.00 mL of 0.0500 M Fe^{2+} with 0.1000 M Ce^{4+} in a medium that is 1.0 M in H_2SO_4 at all times. Formal potential data for both half-cell processes are available in Appendix 4 and are used for these calculations. That is,

$$Ce^{4+} + e^- \rightleftharpoons Ce^{3+} \qquad E^{0'} = 1.44 \text{ V } (1 \text{ M } H_2SO_4)$$

$$Fe^{3+} + e^- \rightleftharpoons Fe^{2+} \qquad E^{0'} = 0.68 \text{ V } (1 \text{ M } H_2SO_4)$$

Initial Potential

The solution contains no cerium species at the outset. In all likelihood, there is a small but unknown amount of Fe^{3+} present due to air oxidation of Fe^{2+}. In any event, we lack sufficient information to calculate an initial potential.

Why is it impossible to calculate the potential of the system before titrant is added?

Potential After the Addition of 5.00 mL of Cerium(IV)

With the first addition of oxidant, the solution has appreciable and readily calculated concentrations of Fe^{3+}, Fe^{2+}, and Ce^{3+}; that of the fourth, Ce^{4+}, is vanishingly small. Therefore, it is more convenient to use the concentrations of the two iron species to calculate the electrode potential of the system.

Remember: The equation for this reaction is

$$Fe^{2+} + Ce^{4+} \rightleftharpoons Fe^{3+} + Ce^{3+}$$

The equilibrium concentration of Fe(III) is equal to the difference between its analytical concentration and the equilibrium concentration of the unreacted Ce(IV):

$$[Fe^{3+}] = \frac{5.00 \times 0.100}{50.00 + 5.00} - [Ce^{4+}] = \frac{0.500}{55.00} - [Ce^{4+}]$$

Similarly, the Fe^{2+} concentration is given by its molarity plus the equilibrium concentration of unreacted $[Ce^{4+}]$:

$$[Fe^{2+}] = \frac{50.00 \times 0.0500 - 5.00 \times 0.1000}{55.00} + [Ce^{4+}] = \frac{2.00}{55.00} + [Ce^{4+}]$$

Generally, redox reactions used in titrimetry are sufficiently complete that the equilibrium concentration of one of the species (in this case $[Ce^{4+}]$) is vanishingly small with respect to the other species present in the solution. Thus, the foregoing two equations can be simplified to

$$[Fe^{3+}] \cong \frac{0.500}{55.00} \quad \text{and} \quad [Fe^{2+}] \cong \frac{2.00}{55.00}$$

Substitution for $[Fe^{2+}]$ and $[Fe^{3+}]$ in the Nernst equation gives

$$E_{system} = +0.68 - \frac{0.0592}{1} \log \frac{2.00/55.00}{0.500/55.00} = 0.64 \text{ V}$$

Note that the volumes in the numerator and denominator cancel, which indicates that the potential is independent of dilution. This independence holds until the solution becomes so dilute that the two assumptions made in the calculation become invalid.

It is worth emphasizing again that the use of the Nernst equation for the Ce(IV)/Ce(III) system would yield the same value for E_{system}, but to do so would require computing $[Ce^{4+}]$ from the equilibrium constant for the reaction.

We calculate additional potentials needed to define the titration curve short of the equivalence point in a similar way. Such data are given in Table 16-1. You may want to confirm one or two of these values.

Equivalence-Point Potential

Substitution of the two formal potentials into Equation 16-11 yields

$$E_{eq} = \frac{E^{0'}_{Ce^{4+}} + E^{0'}_{Fe^{3+}}}{2} = \frac{1.44 + 0.68}{2} = 1.06 \text{ V}$$

Potential After the Addition of 25.10 mL of Cerium(IV)

The molar concentrations of Ce(III), Ce(IV), and Fe(III) are relatively easy to compute at this point but the Fe(II) concentration is not. Therefore, E_{system}

Table 16-1
ELECTRODE POTENTIAL VERSUS SHE IN
TITRATIONS WITH 0.1000 M Ce^{4+}

Reagent Volume, mL	Potential, V vs. SHE		
	50.00 mL of 0.05000 M Fe^{2+}		50.00 mL of 0.02500 M U^{4+}*
5.00	0.64		0.316
15.00	0.69		0.339
20.00	0.72		0.352
24.00	0.76		0.375
24.90	0.82	Equivalence point	0.405
25.00	1.06 ←		→ 0.703
25.10	1.30		1.30
26.00	1.36		1.36
30.00	1.40		1.40

* H_2SO_4 concentration is such that $[H^+] = 1.0$ throughout.

computations based on the cerium half-reaction are more convenient. The concentrations of the two cerium ion species are

$$[Ce^{3+}] = \frac{25.00 \times 0.100}{75.10} - [Fe^{2+}] \cong \frac{2.500}{75.10}$$

$$[Ce^{4+}] = \frac{25.10 \times 0.1000 - 50.00 \times 0.0500}{75.10} + [Fe^{2+}] \cong \frac{0.010}{75.10}$$

Here, the iron(II) concentration is negligible with respect to the analytical concentrations of the two cerium species. Substitution into the Nernst equation for the cerium couple gives

$$E = +1.44 - \frac{0.0592}{1} \log \frac{[Ce^{3+}]}{[Ce^{4+}]} = +1.44 - \frac{0.0592}{1} \log \frac{2.500/75.10}{0.010/75.10}$$

$$= +1.30 \text{ V}$$

The additional postequivalence potentials in Table 16-1 were derived in a similar fashion.

The titration curve of iron(II) with cerium(IV) appears as *A* in Figure 16-3. This plot closely resembles the curves encountered in neutralization, precipitation, and complex-formation titrations with the equivalence point being signaled by a rapid change in the ordinate function. A titration involving 0.00500 M iron(II) and 0.01000 M cerium(IV) yields a curve that, for all practical purposes, is identical to the one we have derived, since the electrode potential of the system is independent of dilution.

In contrast to other titration curves that we have thus far encountered, oxidation/reduction curves are *independent* of the concentration of the reactants except when the solution is very dilute.

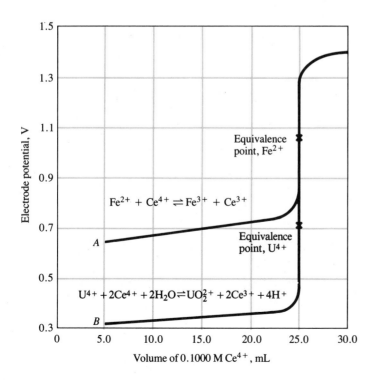

Figure 16-3
Titration curves for 0.1000 M Ce^{4+} titration. *A*: Titration of 50.00 mL of 0.05000 M Fe^{2+}. *B*: Titration of 50.00 mL of 0.02500 M U^{4+}.

Example 16-9

Derive a titration curve for the reaction of 50.00 mL of 0.02500 M U^{4+} with 0.1000 M Ce^{4+}. The solution is 1.0 M in H_2SO_4 throughout the titration (for the sake of simplicity, assume that $[H^+]$ for this solution is also about 1.0 M).

The analytical reaction is

$$U^{4+} + 2H_2O + 2Ce^{4+} \rightleftharpoons UO_2^{2+} + 2Ce^{3+} + 4H^+$$

and in Appendix 4 we find

$$Ce^{4+} + e^- \rightleftharpoons Ce^{3+} \qquad E^{0\prime} = +1.44 \text{ V}$$

$$UO_2^{2+} + 4H^+ + 2e^- \rightleftharpoons U^{4+} + 2H_2O \qquad E^0 = +0.334 \text{ V}$$

Potential After Adding 5.00 mL of Ce^{4+}

$$\text{original amount } U^{4+} = 50.00 \text{ mL } U^{4+} \times 0.02500 \frac{\text{mmol } U^{4+}}{\text{mL } U^{4+}}$$

$$= 1.250 \text{ mmol } U^{4+}$$

$$\text{amount } Ce^{4+} \text{ added} = 5.00 \text{ mL } Ce^{4+} \times 0.1000 \frac{\text{mmol } Ce^{4+}}{\text{mL } Ce^{4+}}$$

$$= 0.5000 \text{ mmol } Ce^{4+}$$

$$\text{amount } UO_2^{2+} \text{ formed} = 0.5000 \text{ mmol } Ce^{4+} \times \frac{1 \text{ mmol } UO_2^{2+}}{2 \text{ mmol } Ce^{4+}}$$

$$= 0.2500 \text{ mmol } UO_2^{2+}$$

$$\text{amount } U^{4+} \text{ remaining} = 1.250 \text{ mmol } U^{4+} - 0.2500 \text{ mmol } UO_2^{2+}$$

$$= \times \frac{1 \text{ mmol } U^{4+}}{\text{mmol } UO_2^{2+}}$$

$$= 1.000 \text{ mmol } U^{4+}$$

$$\text{total volume of solution} = (50.00 + 5.00) \text{ mL} = 55.00 \text{ mL}$$

Applying the Nernst equation for UO_2^{2+}, we obtain

$$E = 0.334 - \frac{0.0592}{2} \log \frac{[U^{4+}]}{[UO_2^{2+}][H^+]^4} = 0.334 - \frac{0.0592}{2} \log \frac{[U^{4+}]}{[UO_2^{2+}](1.00)^4}$$

Substituting concentrations of the two uranium species gives

$$E = 0.334 - \frac{0.0592}{2} \log \frac{1.000 \text{ mmol } U^{4+}/\text{55.00 mL}}{0.2500 \text{ mmol } UO_2^{2+}/\text{55.00 mL}} = 0.316 \text{ V}$$

Other preequivalence-point data, calculated in the same way, are given in Table 16-1.

Equivalence-Point Potential

Following the procedure shown in Example 16-8, we obtain

$$E_{eq} = \frac{(2E^0_{UO_2^{2+}} + E^{0'}_{Ce^{4+}})}{3} - \frac{0.0592}{3} \log \frac{1}{[H^+]^4}$$

Substituting gives

$$E_{eq} = \frac{2 \times 0.334 + 1.44}{3} + \frac{0.0592}{3} \log \frac{1}{(1.00)^4} = 0.782 \text{ V}$$

Potential After Adding 25.10 mL of Ce^{4+}

$$\text{original amount } U^{4+} = 50.00 \text{ mL } U^{4+} \times 0.02500 \frac{\text{mmol } U^{4+}}{\text{mL } U^{4+}}$$

$$= 1.250 \text{ mmol } U^{4+}$$

$$\text{amount } Ce^{4+} \text{ added} = 25.10 \text{ mL } Ce^{4+} \times 0.1000 \frac{\text{mmol } Ce^{4+}}{\text{mL } Ce^{4+}}$$

$$= 2.510 \text{ mmol } Ce^{4+}$$

$$\text{amount } Ce^{3+} \text{ formed} = 1.250 \text{ mmol } U^{4+} \times \frac{2 \text{ mmol } Ce^{3+}}{\text{mmol } U^{4+}}$$

$$= 2.500 \text{ mmol } Ce^{3+}$$

$$\text{amount } Ce^{4+} \text{ remaining} = 2.510 \text{ mmol } Ce^{4+} - 2.500 \text{ mmol } Ce^{3+}$$

$$= \times \frac{1 \text{ mmol } Ce^{4+}}{\text{mmol } Ce^{3+}}$$

$$= 0.010 \text{ mmol } Ce^{4+}$$

$$\text{volume of solution} = 75.10 \text{ mL}$$

$$[Ce^{3+}] = 2.500 \text{ mmol } Ce^{3+}/75.10 \text{ mL}$$

$$[Ce^{4+}] = 0.010 \text{ mmol } Ce^{4+}/75.10 \text{ mL}$$

Substituting into the expression for the formal potential gives

$$E = 1.44 - 0.0592 \log \frac{2.500/75.10}{0.010/75.10} = 1.30 \text{ V}$$

Table 16-1 contains other post-equivalence-point data obtained in this same way.

The data in the third column of Table 16-1 are plotted as curve *B* in Figure 16-3. The two curves in the figure are seen to be identical for volumes greater than 25.10 mL because the concentrations of the two cerium species are identical in this region. It is also interesting that the curve for iron(II) is symmetric about the equivalence point, while the curve for uranium(IV) is not symmetric. In

Redox titration curves are symmetric when the reactants combine in a 1:1 ratio. Otherwise, they are asymmetric.

general, symmetric curves are only obtained when the analyte and titrant react in a 1:1 molar ratio.

16C-3 The Effect of System Variables on Redox Titration Curves

In earlier chapters we considered the effects of reactant concentrations and completeness of the reaction on titration curves. Here, we describe the effects of these variables on oxidation/reduction titration curves.

Reactant Concentration

As we have just seen, E_{system} for an oxidation/reduction titration is ordinarily independent of dilution. Consequently, titration curves for oxidation/reduction reactions are usually independent of analyte and reagent concentrations. This independence is in distinct contrast to what was observed in the other types of titration curves we have encountered.

Completeness of the Reaction

The change in E_{system} in the equivalence-point region of an oxidation/reduction titration becomes larger as the reaction becomes more complete. This effect is demonstrated by the two curves in Figure 16-3. The equilibrium constant for the reaction of cerium(IV) with iron(II) is 7×10^{12} while that for U(IV) is 2×10^{37}. The effect of reaction completeness is further demonstrated in Figure 16-4, which shows curves for the titration of a hypothetical reductant with a standard electrode potential of 0.20 V with several hypothetical oxidants with standard potentials ranging from 0.40 to 1.20 V; the corresponding equilibrium constants lie between about 2×10^3 and 8×10^{16}. Clearly, the greatest change in potential of the system is associated with the reaction that is most complete. In this respect,

Figure 16-4

Effect of titrant electrode potential on reaction completeness. The standard electrode potential for the analyte (E_A^0) is 0.200 V; starting with curve A, standard electrode potentials for the titrant (E_T^0) are 1.20, 1.00, 0.80, 0.60, and 0.40 V, respectively. Both analyte and titrant undergo a one-electron change.

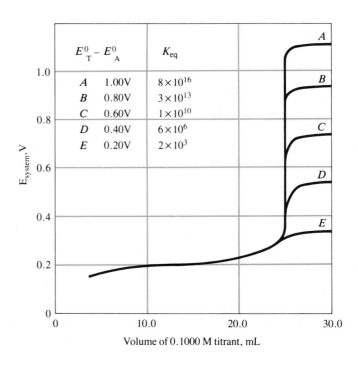

$E_T^0 - E_A^0$		K_{eq}
A	1.00V	8×10^{16}
B	0.80V	3×10^{13}
C	0.60V	1×10^{10}
D	0.40V	6×10^6
E	0.20V	2×10^3

Volume of 0.1000 M titrant, mL

oxidation/reduction titration curves are similar to those involving other types of reactions.

16C-4   Indicators for Oxidation/Reduction Titrations

Two types of chemical indicators are used in oxidation/reduction titrations. *Specific indicators* react with one of the participants in the titration to produce a color. True *oxidation/reduction indicators*, on the other hand, respond to the potential of the system rather than to the appearance or disappearance of a particular species during the course of the titration.

Specific Indicators

Perhaps the best known specific indicator for an oxidation/reduction titration is starch, which forms a deep blue complex with iodine. The appearance or disappearance of this complex serves as a sensitive indicator for titrations in which iodine is either a reactant or a product.

Thiocyanate ion acts as a specific indicator in titrations in which iron(III) is a participant. For example, the disappearance of the red color of the iron(III)/thiocyanate complex signals the end point in the titration of iron(III) with standard titanium(III).

Oxidation/Reduction Indicators

True oxidation/reduction indicators respond to changes in the electrode potential of a system. They are substantially more versatile than specific indicators.

The half-reaction for a typical oxidation/reduction indicator can be written as

$$In_{ox} + ne^- \rightleftharpoons In_{red}$$

If the indicator reaction is reversible, we may write

$$E = E^0 - \frac{0.0592}{n} \log \frac{[In_{red}]}{[In_{ox}]} \qquad \textbf{(16-12)}$$

Typically, a color change from the oxidized form of the indicator to the reduced form requires a change in their concentration ratio from about

$$\frac{[In_{red}]}{[In_{ox}]} \leq \frac{1}{10}$$

to

$$\frac{[In_{red}]}{[In_{ox}]} \geq \frac{10}{1}$$

Thus, the color change of an oxidation/reduction indicator requires about a 100-fold change in the concentration ratio of the two forms. The potential change associated with this transition can be established by substituting these boundary

Protons are involved in the reduction of many indicators. For these, the transition range will also be pH dependent.

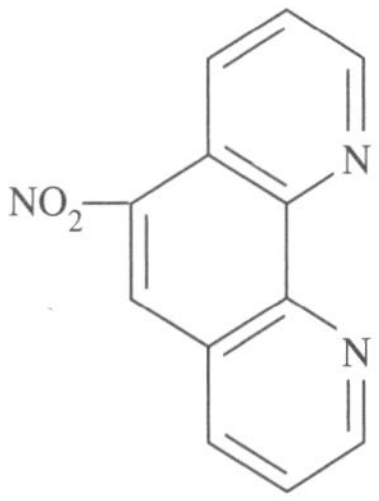

5-Nitro-1,10-phenanthroline

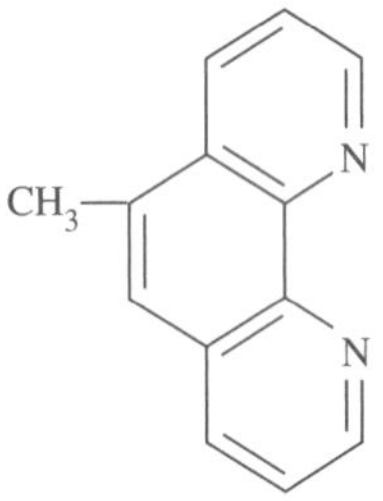

5-Methyl-1,10-phenanthroline

Feature 16-1
A TYPICAL REDOX INDICATOR

A class of organic compounds known as 1,10-phenanthrolines (or orthophenanthrolines) form stable complexes with iron(II) and certain other ions. The parent compound has a pair of nitrogen atoms located in such positions that each can form a covalent bond with the iron(II) ion. Three 1,10-phenanthroline molecules combine with each iron ion to yield a complex with the structure

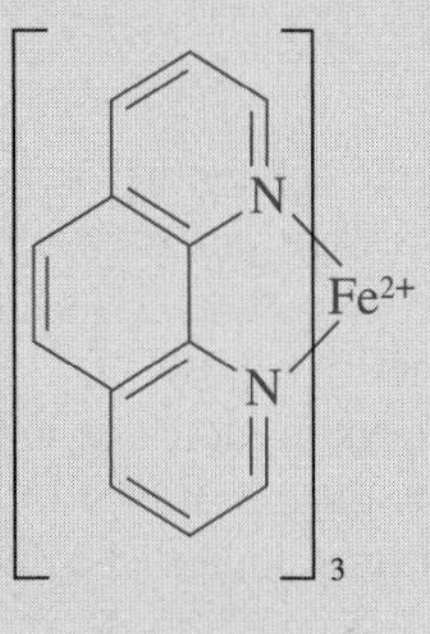

This complex, which is sometimes called "ferroin," is conveniently formulated as $(Phen)_3Fe^{2+}$.

The complexed iron in the ferroin undergoes a reversible oxidation/reduction reaction that can be written

$$(\text{Phen})_3\text{Fe}^{3+} + \text{e}^- \rightleftharpoons (\text{Phen})_3\text{Fe}^{2+} \qquad E^0 = +1.06 \text{ V}$$
$$\underset{\text{pale blue}}{\phantom{(\text{Phen})_3\text{Fe}^{3+}}} \qquad\qquad \underset{\text{red}}{\phantom{(\text{Phen})_3\text{Fe}^{2+}}}$$

In practice, the color of the oxidized form is so slight as to go undetected, and the color change associated with this reduction is from nearly colorless to red. Because of the difference in color intensity, the end point is usually taken when only about 10% of the indicator is in the iron(II) form. The transition potential is thus approximately $+1.11$ V in 1 M sulfuric acid.

Of all the oxidation/reduction indicators, ferroin approaches most closely the ideal substance. It reacts rapidly and reversibly, its color change is sharp, and its solutions are stable and easily prepared. In contrast to many indicators, the oxidized form of ferroin is remarkably inert toward strong oxidizing agents. At temperatures above 60°C, solutions of ferroin decompose.

A number of substituted phenanthrolines have been investigated for their indicator properties, and some have proved to be as useful as the parent compound. Among these, the 5-nitro and 5-methyl derivatives are noteworthy, with transition potentials of $+1.25$ V and $+1.02$ V, respectively.

values into Equation 16-12, which gives

$$E = E^0 \pm \frac{0.0592}{n} \qquad \textbf{(16-13)}$$

For a typical indicator to undergo a useful transition in color, the titrant must cause a change of $0.118/n$ V in the potential of the system. For many indicators, n is 2, so a change of 0.059 V is sufficient.

The potential about which a color transition will occur depends on the standard potential for the particular indicator system. The indicators shown in Table 16-2 have transition potentials ranging from $+1.25$ V to about $+0.3$ V. Turning again to Figure 16-3, we see that all of the indicators, except the first and the last, will provide a satisfactory color change for titrations involving titrant A. In contrast, use of reagent D would succeed only if indigo tetrasulfonate (or an indicator with a similar transition range) were used.

End points for many oxidation/reduction titrations can also be readily observed by making the solution of the analyte solution part of the cell

reference electrode‖analyte solution|Pt

A plot of the potential of this cell as a function of titrant volume will yield curves similar to those in Figures 16-3 and 16-4, and end points can be evaluated graphically. We consider potentiometric titrations in Section 18G.

Challenge: Derive an equation for the effect of pH on the transition range of an indicator for which the half-reaction is $In_{ox} + 2H^+ + 2e^- \rightleftharpoons InH_{2\,red}$.

Reference electrodes are described in Section 18B.

Table 16-2
SELECTED OXIDATION/REDUCTION INDICATORS*

| Indicator | Color | | Transition Potential, V | Conditions |
	Oxidized	Reduced		
5-Nitro-1, 10-phenanthroline iron(II) complex	Pale blue	Red-violet	+1.25	1 M H_2SO_4
2,3'-Diphenylamine dicarboxylic acid	Blue-violet	Colorless	+1.12	7–10 M H_2SO_4
1,10-Phenanthroline iron(II) complex	Pale blue	Red	+1.11	1 M H_2SO_4
5-Methyl 1,10-phenanthroline iron(II) complex	Pale blue	Red	+1.02	1 M H_2SO_4
Erioglaucin A	Blue-red	Yellow-green	+0.98	0.5 M H_2SO_4
Diphenylamine sulfonic acid	Red-violet	Colorless	+0.85	Dilute acid
Diphenylamine	Violet	Colorless	+0.76	Dilute acid
p-Ethoxychrysoidine	Yellow	Red	+0.76	Dilute acid
Methylene blue	Blue	Colorless	+0.53	1 M acid
Indigo tetrasulfonate	Blue	Colorless	+0.36	1 M acid
Phenosafranine	Red	Colorless	+0.28	1 M acid

* Data in part from I. M. Kolthoff and V. A. Stenger, *Volumetric Analysis*, 2nd ed., Vol. 1, p. 140. New York: Interscience, 1942.

16D QUESTIONS AND PROBLEMS

*16-1. Briefly define the electrode potential of a system.

16-2. For an oxidation/reduction titration, briefly distinguish between
   *(a) equilibrium and equivalence.
   (b) a true oxidation/reduction indicator and a specific indicator.

16-3. What is unique about the condition of equilibrium in an oxidation/reduction reaction?

*16-4. Why can the curve for an oxidation/reduction titration be generated through use of the standard electrode potential for either the titrant or the reactant in the Nernst equation?

16-5. How does calculation of the electrode potential of the system at the equivalence point differ from that for any other point of an oxidation/reduction titration?

*16-6. Under what circumstance is the curve for an oxidation/reduction titration asymmetric about the equivalence point?

*16-7. Calculate the theoretical potential of the following cells. Indicate whether the cell, as written, is galvanic or electrolytic.
   (a) $Pb|Pb^{2+}(0.1393 \text{ M})||Cd^{2+}(0.0511 \text{ M})|Cd$
   (b) $Zn|Zn^{2+}(0.0364 \text{ M})||$
      $Tl^{3+}(9.06 \times 10^{-3} \text{ M}),Tl^{+}(0.0620 \text{ M})|Pt$
   (c) $Pt,H_2(765 \text{ torr})|HCl(1.00 \times 10^{-4} \text{ M})||$
      $Ni^{2+}(0.0214 \text{ M})|Ni$
   (d) $Pb|PbI_2(\text{sat'd}),I^-(0.0120 \text{ M})||Hg^{2+}(4.59 \times 10^{-3} \text{ M})|Hg$
   (e) $Pt,H_2(1.00 \text{ atm})|NH_3(0.438 \text{ M}),NH_4^+(0.379 \text{ M})||SHE$
   (f) $Pt|TiO^{2+}(0.0790 \text{ M}),Ti^{3+}(0.00918 \text{ M}),H^+(1.47 \times 10^{-2} \text{ M})||VO^{2+}(0.1340 \text{ M}),V^{3+}(0.0784 \text{ M}),H^+(0.0538 \text{ M})|Pt$

16-8. Calculate the theoretical cell potential of the following cells. Indicate whether each cell, as written, is galvanic or electrolytic.
   (a) $Zn|Zn^{2+}(0.0955 \text{ M})||Co^{2+}(6.78 \times 10^{-3} \text{ M})|Co$
   (b) $Pt|Fe^{3+}(0.1310 \text{ M}),Fe^{2+}(0.0681 \text{ M})||$
      $Hg^{2+}(0.0671 \text{ M})|Hg$
   (c) $Ag|Ag^+(0.1544 \text{ M})||H^+(0.0794 \text{ M})|O_2(1.12 \text{ atm}),Pt$
   (d) $Cu|Cu^{2+}(0.0601 \text{ M})||I^-(0.1350 \text{ M}),AgI(\text{sat'd})|Ag$
   (e) $SHE||HCOOH(0.1302 \text{ M}),HCOO^-(0.0764 \text{ M})|H_2$
      $(1.00 \text{ atm}),Pt$
   (f) $Pt|UO_2^{2+}(7.93 \times 10^{-3} \text{ M}),U^{4+}(6.37 \times 10^{-2} \text{ M}),H^+$
      $(1.16 \times 10^{-3} \text{ M})||$
      $Fe^{3+}(0.003876 \text{ M}),Fe^{2+}(0.1134 \text{ M})|Pt$

16-9. Calculate the potential of
   *(a) a galvanic cell consisting of a lead electrode immersed in 0.0848 M Pb^{2+} and a zinc electrode in contact with 0.1364 M Zn^{2+}.
   (b) a galvanic cell with two platinum electrodes, the one immersed in a solution that is 0.0301 M in Fe^{3+} and 0.0760 M in Fe^{2+}, the other in a solution that is 0.00309 M in $Fe(CN)_6^{4-}$ and 0.1564 M in $Fe(CN)_6^{3-}$.

*(c) a galvanic cell consisting of a standard hydrogen electrode and a platinum electrode immersed in a solution that is 1.46×10^{-3} M in TiO^{2+}, 0.02723 M in Ti^{3+}, and buffered to a pH of 3.00.
   (d) an electrolytic cell consisting of a cobalt electrode immersed in 0.0767 M Co^{2+} and a platinum electrode in contact with 0.200 M HCl; oxygen is evolved from the latter electrode at a pressure of 1.00 atm.
   *(e) an electrolytic cell consisting of two silver electrodes, the one in contact with 0.2058 M Ag^+, the other immersed in 0.0791 M KBr that is saturated with AgBr.
   (f) an electrolytic cell with two platinum electrodes, the one immersed in a solution that is 0.1523 M in I^- and 0.0364 M in I_3^-, the other in 0.02105 M Fe^{3+} and 0.1037 M in Fe^{2+}.

*16-10. Use the shorthand notation (page 263) to describe the cells in 16-9. Each cell is supplied with a salt bridge to provide electrical contact between the solutions in the two cell compartments.

16-11. Generate equilibrium-constant expressions for the following reactions. Calculate numerical values for K_{eq}.
   *(a) $Fe^{3+} + V^{2+} \rightleftharpoons Fe^{2+} + V^{3+}$
   (b) $Fe(CN)_6^{3-} + Cr^{2+} \rightleftharpoons Fe(CN)_6^{4-} + Cr^{3+}$
   *(c) $2V(OH)_4^+ + U^{4+} \rightleftharpoons 2VO^{2+} + UO_2^{2+} + 4H_2O$
   (d) $Tl^{3+} + 2Fe^{2+} \rightleftharpoons Tl^+ + 2Fe^{3+}$
   *(e) $2Ce^{4+} + H_3AsO_3 + H_2O \rightleftharpoons$
      $2Ce^{3+} + H_3AsO_4 + 2H^+ \text{ (1 M } HClO_4)$
   (f) $2V(OH)_4^+ + H_2SO_3 \rightleftharpoons SO_4^{2-} + 2VO^{2+} + 5H_2O$
   *(g) $VO^{2+} + V^{2+} + 2H^+ \rightleftharpoons 2V^{3+} + H_2O$
   (h) $TiO^{2+} + Ti^{2+} + 2H^+ \rightleftharpoons 2Ti^{3+} + H_2O$

*16-12. Calculate the electrode potential of the system at the equivalence point for each of the reactions in 16-11. Use 0.100 M where a value for $[H^+]$ is needed and is not otherwise specified.

*16-13. Select an indicator from Table 16-2 that might be suitable for each of the titrations in 16-11. Write ''none'' if no indicator is suitable.

16-14. Construct curves for the following titrations. Calculate potentials after the addition of 10.00, 25.00, 49.00, 49.90, 50.00, 50.10, 51.00, and 60.00 mL of the reagent. Where necessary, assume that $[H^+] = 1.00$ throughout.
   *(a) 50.00 mL of 0.1000 M V^{2+} with 0.05000 M Sn^{4+}
   (b) 50.00 mL of 0.1000 M $Fe(CN)_6^{3-}$ with 0.1000 M Cr^{2+}
   *(c) 50.00 mL of 0.1000 M $Fe(CN)_6^{4-}$ with 0.05000 M Tl^{3+}
   (d) 50.00 mL of 0.1000 M Fe^{3+} with 0.05000 M Sn^{2+}
   *(e) 50.00 mL of 0.05000 M U^{4+} with 0.02000 M MnO_4^-

---

* Answers to the asterisked problems are given in the answers section at the back of the book.

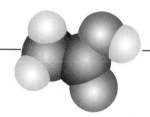

CHAPTER 17

APPLICATIONS OF OXIDATION/REDUCTION TITRATIONS

This chapter is concerned with the preparation of standard solutions of oxidants and reductants and with their applications in analytical chemistry. In addition, auxiliary reagents that convert an analyte to a single oxidation state are described.[1]

17A AUXILIARY OXIDIZING AND REDUCING REAGENTS

The analyte in an oxidation/reduction titration must be in a single oxidation state at the outset. Often, however, the steps that precede the titration (dissolution of the sample and separation of interferences) convert the analyte to a mixture of oxidation states. For example, the solution formed when an iron-containing sample is dissolved usually contains a mixture of iron(II) and iron(III) ions. If we choose to use a standard oxidant for the determination of iron, we must first treat the sample solution with an auxiliary reducing agent; if, on the other hand we plan to titrate with a standard reductant, pretreatment with an auxiliary oxidizing reagent will be needed.[2]

To be useful as a preoxidant or a prereductant, a reagent must react quantitatively with the analyte. In addition, excesses of the reagent must be readily removed because such excesses will inevitably react with standard solution.

[1] For further reading on redox titrimetry, see J. A. Goldman and V. A. Stenger, in *Treatise on Analytical Chemistry*, I. M. Kolthoff and P. J. Elving, Eds., Part I, Vol. 11, Chapter 119. New York: Wiley, 1975; and I. M. Kolthoff and R. Belcher, *Volumetric Analysis*, Vol. 2. New York: Interscience, 1957.

[2] For a brief summary of auxiliary reagents, see J. A. Goldman and V. A. Stenger, in *Treatise on Analytical Chemistry*, I. M. Kolthoff and P. J. Elving, Eds., Part I, Vol. 11, pp. 7204–7206. New York: Wiley, 1975.

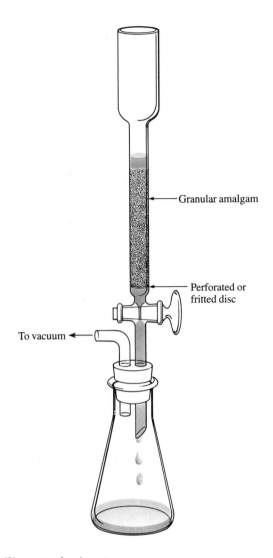

Granular amalgam

Perforated or
fritted disc

To vacuum ◄

Figure 17-1
A Jones reductor.

17A-1 Auxiliary Reducing Reagents

A number of metals are good reducing agents and have been used for the prereduction of analytes. Included among these are zinc, aluminum, cadmium, lead, nickel, copper, and silver (in the presence of chloride ion). Sticks or coils of the metal can be immersed directly in the analyte solution. After reduction is judged complete, the solid is removed manually and rinsed with water. Filtration of the analyte solution is needed to remove granular or powdered forms of the metal. An alternative to filtration is the use of a *reductor*, such as that shown in Figure 17-1.[3] Here, the finely divided metal is held in a vertical glass tube through which the solution is drawn under a mild vacuum. The metal in a reductor is ordinarily sufficient for hundreds of reductions.

A typical *Jones reductor* has a diameter of about 2 cm and holds a 40- to 50-cm column of amalgamated zinc. Amalgamation is accomplished by allowing

[3] For a discussion of reductors, see F. Hecht, in *Treatise on Analytical Chemistry*, I. M. Kolthoff and P. J. Elving, Eds., Part I, Vol. 11, pp. 6703–6707. New York: Wiley, 1975.

zinc granules to stand briefly in a solution of mercury(II) chloride:

$$2Zn(s) + Hg^{2+} \rightarrow Zn^{2+} + Zn(Hg)(s)$$

Zinc amalgam is nearly as effective a reducing agent as the pure metal and has the important virtue of inhibiting the reduction of hydrogen ions by zinc, a parasitic reaction that not only needlessly uses up the reducing agent but also contaminates the sample solution with zinc(II) ions. Solutions that are quite acidic can be passed through a Jones reductor without significant hydrogen formation.

Table 17-1 lists the principal applications of the Jones reductor. Also included in this table are reductions that can be accomplished with a *Walden reductor*, in which granular metallic silver held in a narrow glass column is the reductant. Silver is not a good reducing agent unless chloride or some other ion that forms a silver salt of low solubility is present. For this reason, prereductions with a Walden reductor are generally carried out from hydrochloric acid solutions of the analyte. The coating of silver chloride produced on the metal is removed periodically by dipping a zinc rod into the solution that covers the packing.

Table 17-1 suggests that the Walden reductor is somewhat more selective in its action than is the Jones reductor.

17A-2 Auxiliary Oxidizing Reagents

Sodium Bismuthate

Sodium bismuthate is a powerful oxidizing agent capable of converting manganese(II) quantitatively to permanganate ion. This bismuth salt is a sparingly soluble solid with a formula that is usually written as $NaBiO_3$, although its exact composition is somewhat uncertain. Oxidations are performed by suspending the bismuthate in the analyte solution and boiling for a brief period. The unused reagent is then removed by filtration. The half-reaction for the reduction of

Table 17-1
USES OF THE WALDEN REDUCTOR AND THE JONES REDUCTOR*

Walden $Ag(s) + Cl^- \rightarrow AgCl(s) + e^-$	Jones $Zn(Hg)(s) \rightarrow Zn^{2+} + Hg + 2\,e^-$
$Fe^{3+} + e^- \rightarrow Fe^{2+}$	$Fe^{3+} + e^- \rightleftharpoons Fe^{2+}$
$Cu^{2+} + e^- \rightarrow Cu^+$	$Cu^{2+} + 2e^- \rightleftharpoons Cu(s)$
$H_2MoO_4 + 2H^+ + e^- \rightarrow MoO_2^+ + 2H_2O$	$H_2MoO_4 + 6H^+ + 3e^- \rightleftharpoons Mo^{3+} + 4H_2O$
$UO_2^{2+} + 4H^+ + 2e^- \rightarrow U^{4+} + 2H_2O$	$UO_2^{2+} + 4H^+ + 2e^- \rightleftharpoons U^{4+} + 2H_2O$
	$UO_2^{2+} + 4H^+ + 3e^- \rightleftharpoons U^{3+} + 2H_2O$†
$V(OH)_4^+ + 2H^+ + e^- \rightarrow VO^{2+} + 3H_2O$	$V(OH)_4^+ + 4H^+ + 3e^- \rightleftharpoons V^{2+} + 4H_2O$
TiO^{2+} not reduced	$TiO^{2+} + 2H^+ + e^- \rightleftharpoons Ti^{3+} + H_2O$
Cr^{3+} not reduced	$Cr^{3+} + e^- \rightleftharpoons Cr^{2+}$

* From I. M. Kolthoff and R. Belcher, *Volumetric Analysis,* Vol. 3, p. 12. New York: Interscience, 1957. With permission.

† A mixture of oxidation states is obtained. The Jones reductor may still be used for the analysis of uranium, however, because any U^{3+} formed can be converted to U^{4+} by shaking the solution with air for a few minutes.

sodium bismuthate can be written as

$$NaBiO_3(s) + 4H^+ + 2e^- \rightleftharpoons BiO^+ + Na^+ + 2H_2O$$

Ammonium Peroxydisulfate

Ammonium peroxydisulfate (also called ammonium persulfate), $(NH_4)_2S_2O_8$, is another powerful oxidizing agent. In acidic solution, it converts chromium(III) to dichromate, cerium(III) to cerium(IV), and manganese(II) to permanganate. The half-reaction is

$$S_2O_8^{2-} + 2e^- \rightleftharpoons 2SO_4^{2-} \qquad E^0 = 2.01 \text{ V}$$

The oxidations are catalyzed by traces of silver ion. The excess reagent is readily decomposed by a brief period of boiling:

$$2S_2O_8^{2-} + 2H_2O \rightarrow 4SO_4^{2-} + O_2(g) + 4H^+$$

Sodium Peroxide and Hydrogen Peroxide

Peroxide is a convenient oxidizing agent either as the solid sodium salt or as a dilute solution of the acid. The half-reaction for hydrogen peroxide in acidic solution is

$$H_2O_2 + 2H^+ + 2e^- \rightleftharpoons 2H_2O \qquad E^0 = 1.78 \text{ V}$$

After oxidation is complete, the solution is freed of excess reagent by boiling:

$$2H_2O_2 \rightarrow 2H_2O + O_2(g)$$

17B   APPLICATION OF STANDARD REDUCTANTS

Standard solutions of most reducing agents tend to react with atmospheric oxygen. For this reason, reductants are seldom used for the direct titration of oxidizing analytes; indirect methods are used instead. The two most common indirect methods, which are discussed in the paragraphs that follow, are based on standard solutions of iron(II) ions and standard solutions of sodium thiosulfate.

17B-1 Iron(II) Solutions

Solutions of iron(II) are readily prepared from iron(II) ammonium sulfate, $Fe(NH_4)_2(SO_4)_2 \cdot 6H_2O$ (Mohr's salt), or from the closely related iron(II) ethylenediamine sulfate, $FeC_2H_4(NH_3)_2(SO_4)_2 \cdot 4H_2O$ (Oesper's salt). Air-oxidation of iron(II) takes place rapidly in neutral solutions but is inhibited in the presence of acids, with the most stable preparations being about 0.5 M in H_2SO_4. Such solutions are stable for no longer than one day—if for that long.

Numerous oxidizing agents are conveniently determined by treatment of the analyte solution with a measured excess of standard iron(II) followed by immediate titration of the excess with a standard solution of potassium dichromate or cerium(IV) (Sections 17C-1 and 17C-2). Just before or after the analyte is titrated,

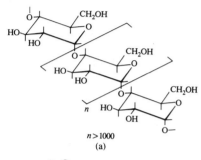
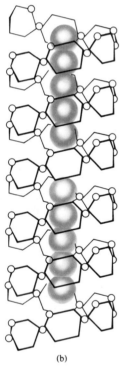

Figure 17-2
Thousands of glucose molecules polymerize to form huge molecules of β-amylose as shown schematically in (a). Molecules of β-amylose tend to assume a helical structure. The iodine species I_5^- as shown in (b) is incorporated into the amylose helix. For further details see R. C. Teitelbaum, S. L. Ruby, and T. J. Marks, *J. Amer. Chem. Soc.,* **1980,** *102,* 3322.

n > 1000
(a)

(b)

the volumetric ratio between the standard oxidant and the iron(II) solution is established by titrating two or three aliquots of the latter with the former.

This procedure has been applied to the determination of organic peroxides; hydroxylamine; chromium(VI); cerium(IV); molybdenum(VI); nitrate, chlorate, and perchlorate ions; and numerous other oxidants (for example, see Problems 17-37 and 17-39).

17B-2 Sodium Thiosulfate

Thiosulfate ion is a moderately strong reducing agent that has been widely used to determine oxidizing agents by an indirect procedure that involves iodine as an intermediate. With iodine, thiosulfate ion is oxidized quantitatively to tetrathionate ion, the half-reaction being

> In its reaction with iodine, each thiosulfate ion gains one electron.

$$2S_2O_3^{2-} \rightleftharpoons S_4O_6^{2-} + 2e^-$$

The quantitative aspect of the reaction with iodine is unique. Other oxidants oxidize the tetrathionate ion, wholly or in part, to sulfate ion.

The scheme used to determine oxidizing agents involves adding an unmeasured excess of potassium iodide to a slightly acidic solution of the analyte. Reduction of the analyte produces a stoichiometrically equivalent amount of iodine. The liberated iodine is then titrated with a standard solution of sodium thiosulfate, $Na_2S_2O_3$, one of the few reducing agents that is stable toward air-oxidation. An example of this procedure is the determination of sodium hypochlorite in bleaches. The reactions are

> Sodium thiosulfate is one of the few reducing agents that is not oxidized by air.

$$OCl^- + 2I^- + 2H^+ \rightarrow Cl^- + I_2 + H_2O \qquad \text{(unmeasured excess KI)}$$
$$I_2 + 2S_2O_3^{2-} \rightarrow 2I^- + S_4O_6^{2-} \qquad\qquad \textbf{(17-1)}$$

The quantitative conversion of thiosulfate ion to tetrathionate ion shown in Equation 17-1 requires a pH smaller than 7. If strongly acidic solutions must be titrated, air-oxidation of the excess iodide must be prevented by blanketing the solution with an inert gas, such as carbon dioxide or nitrogen.

End Points in Iodine/Thiosulfate Titrations

A solution that is about 5×10^{-6} M in I_2 has a discernible color, which corresponds to less than one drop of a 0.05 M iodine solution in 100 mL. Thus, provided the analyte solution is colorless, the disappearance of the iodine color can serve as the indicator in titrations with sodium thiosulfate.

More commonly, titrations involving iodine are performed with a suspension of starch as an indicator. The deep blue color that develops in the presence of iodine is believed to arise from the absorption of iodine into the helical chain of β-amylose, a macro-molecular component of most starches. The closely related α-amylose forms a red adduct with iodine. This reaction is not readily reversible and is thus undesirable. So-called *soluble starch*, which is available from commercial sources, consists principally of β-amylose, the alpha fraction having been removed; indicator solutions are readily prepared from this product.

Aqueous starch suspensions decompose within a few days, primarily because of bacterial action. The decomposition products tend to interfere with the indicator properties of the preparation and may also be oxidized by iodine. The rate of decomposition can be inhibited by preparing and storing the indicator under sterile conditions and by adding mercury(II) iodide or chloroform as a bacteriostat. Perhaps the simplest alternative is to prepare a fresh suspension of the indicator, which requires only a few minutes, on the day it is to be used.

Starch decomposes irreversibly in solutions containing large concentrations of iodine. Therefore, in titrating solutions of iodine with sodium thiosulfate ion, as in the indirect determination of oxidants, addition of the indicator is delayed until the titration is nearly complete as shown by color change from deep red-brown to faint yellow. The indicator can be introduced at the outset when thiosulfate solutions are being titrated directly with iodine.

> Starch undergoes decomposition in solutions with high I_2 concentrations. In titrations of excess I_2 with $Na_2S_2O_3$, addition of the indicator must be deferred until most of the I_2 has been reduced.

The Stability of Sodium Thiosulfate Solutions

Although sodium thiosulfate solutions are resistant to air-oxidation, they do tend to decompose to give sulfur and hydrogen sulfite ion:

$$S_2O_3^{2-} + H^+ \rightleftharpoons HSO_3^- + S(s)$$

Variables that influence the rate of this reaction include pH, the presence of microorganisms, the concentration of the solution, the presence of copper(II) ions, and exposure to sunlight. These variables may cause the concentration of a thiosulfate solution to change by several percent over a period of a few weeks. On the other hand, proper attention to detail will yield solutions that need only occasional restandardization.

The rate of the decomposition reaction increases markedly as the solution becomes acidic.

The most important single cause for the instability of neutral or slightly basic thiosulfate solutions is bacteria that metabolize thiosulfate ion to sulfite and sulfate ions as well as to elemental sulfur. To minimize this problem, standard solutions of the reagent are prepared under reasonably sterile conditions. Bacterial activity appears to be at a minimum at a pH between 9 and 10, which accounts, at least in part, for the reagent's greater stability in slightly basic solutions. The presence of a bactericide, such as chloroform, sodium benzoate, or mercury(II) iodide, also slows decomposition.

> When sodium thiosulfate is added to a strongly acidic medium, a cloudiness develops almost immediately as a consequence of the precipitation of elemental sulfur. Even in neutral solution, this reaction proceeds at such a rate that standard sodium thiosulfate must be restandardized periodically.

The Standardization of Thiosulfate Solutions

Potassium iodate is an excellent primary standard for thiosulfate solutions. In this application, weighed amounts of the primary-standard-grade reagent are dissolved in water containing an excess of potassium iodide. When this mixture is acidified with a strong acid, the reaction

$$IO_3^- + 5I^- + 6H^+ \rightleftharpoons 3I_2 + 2H_2O$$

occurs instantaneously. The liberated iodine is then titrated with the thiosulfate solution. The stoichiometry of the reactions is

$$1 \text{ mol } IO_3^- \equiv 3 \text{ mol } I_2 \equiv 6 \text{ mol } S_2O_3^{2-}$$

Example 17-1

A solution of sodium thiosulfate was standardized by dissolving 0.1210 g KIO_3 (214.00 g/mol) in water, adding a large excess of KI, and acidifying with HCl. The liberated iodine required 41.64 mL of the thiosulfate solution to decolorize the blue starch/iodine complex. Calculate the molarity of the $Na_2S_2O_3$.

$$\text{amount } Na_2S_2O_3 = 0.1210 \text{ g } KIO_3 \times \frac{1 \text{ mmol } KIO_3}{0.21400 \text{ g } KIO_3} \times \frac{6 \text{ mmol } Na_2S_2O_3}{\text{mmol } KIO_3}$$

$$= 3.3925 \text{ mmol } Na_2S_2O_3$$

$$c_{Na_2S_2O_3} = \frac{3.3925 \text{ mmol } Na_2S_2O_3}{41.64 \text{ mL } Na_2S_2O_3} = 0.08147 \text{ M}$$

Other primary standards for sodium thiosulfate are potassium dichromate, potassium bromate, potassium hydrogen iodate, potassium ferricyanide, and metallic copper. All these compounds liberate stoichiometric amounts of iodine when treated with excess potassium iodide.

Applications of Sodium Thiosulfate Solutions

Numerous substances can be determined by the indirect method involving titration with sodium thiosulfate; typical applications are summarized in Table 17-2.

17C APPLICATIONS OF STANDARD OXIDANTS

Table 17-3 summarizes the properties of five of the most widely used volumetric oxidizing reagents. Note that the standard potentials for these reagents vary from 0.5 to 1.5 V. The choice among them depends on the strength of the analyte as

Table 17-2
SOME APPLICATIONS OF SODIUM THIOSULFATE AS A REDUCTANT

Analyte	Half-Reaction	Special Conditions
IO_4^-	$IO_4^- + 8H^+ + 7e^- \rightleftharpoons \frac{1}{2}I_2 + 4H_2O$	Acidic solution
	$IO_4^- + 2H^+ + 2e^- \rightleftharpoons IO_3^- + H_2O$	Neutral solution
IO_3^-	$IO_3^- + 6H^+ + 5e^- \rightleftharpoons \frac{1}{2}I_2 + 3H_2O$	Strong acid
BrO_3^-, ClO_3^-	$XO_3^- + 6H^+ + 6e^- \rightleftharpoons X^- + 3H_2O$	Strong acid
Br_2, Cl_2	$X_2 + 2I^- \rightleftharpoons I_2 + 2X^-$	
NO_2^-	$HNO_2 + H^+ + e^- \rightleftharpoons NO(g) + H_2O$	
Cu^{2+}	$Cu^{2+} + I^- + e^- \rightleftharpoons CuI(s)$	
O_2	$O_2 + 4Mn(OH)_2(s) + 2H_2O \rightleftharpoons 4Mn(OH)_3(s)$	Basic solution
	$Mn(OH)_3(s) + 3H^+ + e^- \rightleftharpoons Mn^{2+} + 3H_2O$	Acidic solution
O_3	$O_3(g) + 2H^+ + 2e^- \rightleftharpoons O_2(g) + H_2O$	
Organic peroxide	$ROOH + 2H^+ + 2e^- \rightleftharpoons ROH + H_2O$	

Table 17-3

SOME COMMON OXIDANTS USED AS STANDARD SOLUTIONS

Reagent and Formula	Reduction Product	Standard Potential, V	Standardized with	Indicator*	Stability†
Potassium permanganate, $KMnO_4$	Mn^{2+}	1.51‡	$Na_2C_2O_4$, Fe, As_2O_3	MnO_4^-	(b)
Potassium bromate, $KBrO_3$	Br^-	1.44‡	$KBrO_3$	(1)	(a)
Cerium(IV), Ce^{4+}	Ce^{3+}	1.44‡	$Na_2C_2O_4$, Fe, As_2O_3	(2)	(a)
Potassium dichromate, $K_2Cr_2O_7$	Cr^{3+}	1.33‡	$K_2Cr_2O_7$, Fe	(3)	(a)
Iodine, I_2	I^-	0.536‡	$BaS_2O_3 \cdot H_2O$, $Na_2S_2O_3$	starch	(c)

* (1) α-Naphthoflavone; (2) 1,10-phenanthroline iron(II) complex (ferroin); (3) diphenylamine sulfonic acid.

† (a) Indefinitely stable; (b) moderately stable, requires periodic standardization; (c) somewhat unstable, requires frequent standardization.

‡ $E^{0'}$ in 1 M H_2SO_4.

a reducing agent, the rate of reaction between oxidant and analyte, the stability of the standard oxidant solutions, the cost, and the availability of a satisfactory indicator.

17C-1 The Strong Oxidants—Potassium Permanganate and Cerium(IV)

Solutions of permanganate ion and cerium(IV) ion are strong oxidizing reagents whose applications closely parallel one another. Half-reactions for the two are

$$MnO_4^- + 8H^+ + 5e^- \rightleftharpoons Mn^{2+} + 4H_2O \qquad E^0 = 1.51 \text{ V}$$
$$Ce^{4+} + e^- \rightleftharpoons Ce^{3+} \qquad E^{0'} = 1.44 \text{ V } (1 \text{ M } H_2SO_4)$$

The formal potential shown for the reduction of cerium(IV) is for solutions that are 1 M in sulfuric acid. In 1 M perchloric acid and 1 M nitric acid, the potentials are 1.70 and 1.61 V, respectively. Solutions of cerium(IV) in the latter two acids are not very stable and thus find limited application.

The half-reaction shown for permanganate ion occurs only in solutions that are 0.1 M or greater in strong acid. In less acidic media, the product may be Mn(III), Mn(IV), or Mn(VI), depending on conditions.

Comparison of the Two Reagents

For all practical purposes, the oxidizing strengths of permanganate and cerium(IV) solutions are comparable. Solutions of cerium(IV) in sulfuric acid, however, are stable indefinitely, whereas permanganate solutions decompose slowly and thus require occasional restandardization. Furthermore, cerium(IV) solutions in sulfuric acid do not oxidize chloride ion and thus can be used to titrate hydrochloric acid solutions of analytes; in contrast, permanganate ion cannot be used with hydrochloric acid solutions unless special precautions are taken to prevent the slow oxidation of chloride ion that leads to overconsumption of the standard reagent. A further advantage of cerium(IV) is that a primary-standard-grade salt of the reagent is available, thus making possible the direct preparation of standard solutions.

Despite these advantages of cerium solutions over permanganate solutions, the latter are more widely used. One reason is the color of permanganate solutions, which is intense enough to serve as an indicator in titrations. Another reason for the popularity of permanganate solutions is their modest cost. The cost of 1 L of 0.02 M solution is about $0.08, whereas 1 L of cerium(IV) of comparable strength costs about $2.20 ($4.40 if reagent of primary-standard grade is employed). Another disadvantage of cerium(IV) solutions is their tendency to form precipitates of basic salts in solutions that are less than 0.1 M in strong acid.

End Points

A useful property of a potassium permanganate solution is its intense purple color, which is sufficient to serve as an indicator for most titrations. As little as 0.01 to 0.02 mL of a 0.02 M solution imparts a perceptible color to 100 mL of water. If the permanganate solution is very dilute, diphenylamine sulfonic acid or the 1,10-phenanthroline complex of iron(II) (Table 16-2) provides a sharper end point.

The permanganate end point is not permanent because excess permanganate ions react slowly with the relatively large concentration of manganese(II) ions present at the end point:

$$2MnO_4^- + 3Mn^{2+} + 2H_2O \rightarrow 5MnO_2(s) + 4H^+$$

The equilibrium constant for this reaction is about 10^{47}, which indicates that the *equilibrium* concentration of permanganate ion is vanishingly small even in highly acidic media. Fortunately, the rate at which this equilibrum is approached is so slow that the end point fades only gradually over a period of perhaps 30 s.

Solutions of cerium(IV) are yellow-orange, but the color is not intense enough to act as an indicator in titrations. Several oxidation/reduction indicators are available for titrations with standard solutions of cerium(IV). The most widely used of these is the iron(II) complex of 1,10-phenanthroline or one of its substituted derivatives (Table 16-2).

The Preparation and Stability of Standard Solutions

Aqueous solutions of permanganate are not entirely stable because the ion tends to oxidize water:

$$4MnO_4^- + 2H_2O \rightarrow 4MnO_2(s) + 3O_2(g) + 4OH^-$$

Although the equilibrium constant for this reaction indicates that the products are favored, permanganate solutions, when properly prepared, are reasonably stable because the decomposition reaction is slow. It is catalyzed by light, heat, acids, bases, manganese(II), and manganese dioxide.

Moderately stable solutions of permanganate ion can be prepared if the effects of these catalysts, particularly manganese dioxide, are minimized. Manganese dioxide is a contaminant in even the best grade of solid potassium permanganate. Furthermore, this compound forms in freshly prepared solutions of the reagent as a consequence of the reaction of permanganate ion with organic matter and dust present in the water used to prepare the solution. Removal of manganese dioxide by filtration before standardization markedly improves the stability of standard permanganate solutions. Before filtration, the reagent solution is allowed to stand for about 24 hours or is heated for a brief period to hasten oxidation of the organic species generally present in small amounts in distilled and deionized water. Paper cannot be used for filtering because permanganate ion reacts with it to form additional manganese dioxide.

Permanganate solutions are moderately stable provided they are freed of manganese dioxide and stored in a dark container.

Standardized permanganate solutions should be stored in the dark. Filtration and restandardization are required if any solid is detected in the solution or on the walls of the storage bottle. In any event, restandardization every one or two weeks is a good precautionary measure.

Solutions containing excess standard permanganate should never be heated because they decompose by oxidizing water. This decomposition cannot be compensated for with a blank. On the other hand, it is possible to titrate hot, acidic solutions of reductants with permanganate without error, provided the reagent is added slowly enough so that large excesses do not accumulate.

Example 17-2

Describe how you would prepare 2.0 L of an approximately 0.010 M solution of $KMnO_4$ (158.03 g/mol).

$$\text{mass } KMnO_4 \text{ needed} = 2.0 \; \cancel{L} \times 0.010 \; \frac{\text{mol } \cancel{KMnO_4}}{\cancel{L}} \times \frac{158.03 \text{ g } KMnO_4}{\text{mol } \cancel{KMnO_4}}$$

$$= 3.16 \text{ g}$$

Dissolve about 3.2 g of $KMnO_4$ in a little water. After solution is complete add water to bring the volume to about 2.0 L. Heat the solution to boiling for a brief period, and let stand until it is cool. Filter through a glass-filtering crucible and store in a clean dark bottle.

The most widely used compounds for the preparation of solutions of cerium(IV) are listed in Table 17-4. Primary-standard-grade cerium ammonium nitrate is available commercially and can be used to prepare standard solutions of the cation directly by weight. More commonly, less expensive reagent-grade cerium(IV) ammonium nitrate or ceric hydroxide is used to prepare solutions that are subsequently standardized. In either case, the reagent is dissolved in a solution that is at least 0.1 M in sulfuric acid to prevent the precipitation of basic salts.

Sulfuric acid solutions of cerium(IV) are remarkably stable and can be stored for months or heated at 100°C for prolonged periods without change in concentration.

Table 17-4
ANALYTICALLY USEFUL CERIUM(IV) COMPOUNDS

Name	Formula	Molar Mass
Cerium(IV) ammonium nitrate	$Ce(NO_3)_4 \cdot 2NH_4NO_3$	548.2
Cerium(IV) ammonium sulfate	$Ce(SO_4)_2 \cdot 2(NH_4)_2SO_4 \cdot 2H_2O$	632.6
Cerium(IV) hydroxide	$Ce(OH)_4$	208.1
Cerium(IV) hydrogen sulfate	$Ce(HSO_4)_4$	528.4

Primary Standards

Several excellent primary standards are available for the standardization of solutions of potassium permanganate and cerium(IV).

Sodium Oxalate. Sodium oxalate is widely used to standardize permanganate and cerium(IV) solutions. In acidic solutions, the oxalate ion is converted to the undissociated acid. Thus, its reaction with the permanganate ion can be depicted as

$$2MnO_4^- + 5H_2C_2O_4 + 6H^+ \rightarrow 2Mn^{2+} + 10CO_2(g) + 8H_2O$$

The reaction between Ce^{4+} and $H_2C_2O_4$ is

$$2Ce^{4+} + H_2C_2O_4 \rightarrow 2Ce^{3+} + 2H^+ + 2CO_2$$

The same oxidation products are formed with cerium(IV).

The reaction between permanganate ion and oxalic acid is complex and proceeds slowly even at elevated temperature unless manganese(II) is present as a catalyst. Thus, when the first few milliliters of standard permanganate are added to a hot solution of oxalic acid, several seconds are required before the color of the permanganate ion disappears. As the concentration of manganese(II) builds up, however, the reaction proceeds more and more rapidly as a result of autocatalysis (see color plate 17).

Autocatalysis is a type of catalysis in which the product of a reaction catalyzes the reaction. This phenomenon causes the rate of the reaction to increase with time.

It has been found that when solutions of sodium oxalate are titrated at 60°C to 90°C, the consumption of permanganate is from 0.1% to 0.4% less than theoretical, probably due to the air-oxidation of a fraction of the oxalic acid. This small error can be avoided by adding 90% to 95% of the required permanganate to a cool solution of the oxalate. After the added permanganate is completely consumed (as indicated by the disappearance of color), the solution is heated to about 60°C and titrated to a pink color that persists for about 30 s. The disadvantage of this procedure is that it requires prior knowledge of the approximate concentration of the permanganate solution so that a proper initial volume of it can be added. For most purposes, direct titration of the hot oxalic acid solution provides adequate data (usually 0.2% to 0.3% high). If greater accuracy is required, a direct titration of the hot solution of one portion of the primary standard can be followed by titration of two or three portions in which the solution is not heated until the end.

Cerium(IV) standardizations against sodium oxalate are usually performed at 50°C in a hydrochloric acid solution containing iodine monochloride as a catalyst.

Example 17-3

You wish to standardize the solution in Example 17-2 against pure $Na_2C_2O_4$ (134.00 g/mol). If you want to use between 30 and 45 mL of the reagent for the standardization, what range of masses of the primary standard should you weigh out?

For a 30-mL titration:

$$\text{amount KMnO}_4 = 30 \; \cancel{\text{mL KMnO}_4} \times 0.010 \; \frac{\text{mmol KMnO}_4}{\cancel{\text{mL KMnO}_4}}$$

$$= 0.30 \; \text{mmol KMnO}_4$$

$$\text{mass Na}_2\text{C}_2\text{O}_4 = 0.30 \; \cancel{\text{mmol KMnO}_4} \times \frac{5 \; \cancel{\text{mmol Na}_2\text{C}_2\text{O}_4}}{2 \; \cancel{\text{mmol KMnO}_4}} \times 0.134 \; \frac{\text{g Na}_2\text{C}_2\text{O}_4}{\cancel{\text{mmol Na}_2\text{C}_2\text{O}_4}}$$

$$= 0.101 \; \text{g Na}_2\text{C}_2\text{O}_4$$

Proceeding in the same way, we find for a 45-mL titration:

$$\text{mass Na}_2\text{C}_2\text{O}_4 = 45 \times 0.010 \times \frac{5}{2} \times 0.134 = 0.151 \; \text{g Na}_2\text{C}_2\text{O}_4$$

Thus, you should weigh between 0.10 and 0.15 g samples of the primary standard.

Table 17-5
SOME APPLICATIONS OF POTASSIUM PERMANGANATE AND CERIUM(IV) IN ACID SOLUTION

Analyte	Half-Reaction	Special Conditions
Sn	$Sn^{2+} \rightleftharpoons Sn^{4+} + 2e^-$	Prereduction with Zn
H_2O_2	$H_2O_2 \rightleftharpoons O_2(g) + 2H^+ + 2e^-$	
Fe	$Fe^{2+} \rightleftharpoons Fe^{3+} + e^-$	Prereduction with $SnCl_2$ or with Jones or Walden reductor
$Fe(CN)_6^{4-}$	$Fe(CN)_6^{4-} \rightleftharpoons Fe(CN)_6^{3-} + e^-$	
V	$VO^{2+} + 3H_2O \rightleftharpoons V(OH)_4^+ + 2H^+ + e^-$	Prereduction with Bi amalgam or SO_2
Mo	$Mo^{3+} + 4H_2O \rightleftharpoons MoO_4^{2-} + 8H^+ + 3e^-$	Prereduction with Jones reductor
W	$W^{3+} + 4H_2O \rightleftharpoons WO_4^{2-} + 8H^+ + 3e^-$	Prereduction with Zn or Cd
U	$U^{4+} + 2H_2O \rightleftharpoons UO_2^{2+} + 4H^+ + 2e^-$	Prereduction with Jones reductor
Ti	$Ti^{3+} + H_2O \rightleftharpoons TiO^{2+} + 2H^+ + e^-$	Prereduction with Jones reductor
$H_2C_2O_4$	$H_2C_2O_4 \rightleftharpoons 2CO_2 + 2H^+ + 2e^-$	
Mg, Ca, Zn, Co, Pb, Ag	$H_2C_2O_4 \rightleftharpoons 2CO_2 + 2H^+ + 2e^-$	Sparingly soluble metal oxalates filtered, washed, and dissolved in acid; liberated oxalic acid titrated
HNO_2	$HNO_2 + H_2O \rightleftharpoons NO_3^- + 3H^+ + 2e^-$	15-min reaction time; excess $KMnO_4$ back-titrated
K	$K_2NaCo(NO_2)_6 + 6H_2O \rightleftharpoons$ $Co^{2+} + 6NO_3^- + 12H^+ + 2K^+ + Na^+ + 11e^-$	Precipitated as $K_2NaCo(NO_2)_6$; filtered and dissolved in $KMnO_4$; excess $KMnO_4$ back-titrated
Na	$U^{4+} + 2H_2O \rightleftharpoons UO_2^{2+} + 4H^+ + 2e^-$	Precipitated as $NaZn(UO_2)_3(OAc)_9$; filtered, washed, dissolved; U determined as above

Example 17-4

Exactly 33.31 mL of the solution in Example 17-2 were required to titrate a 0.1278-g sample of primary-standard $Na_2C_2O_4$. What was the molarity of the $KMnO_4$ reagent?

$$\text{amount } Na_2C_2O_4 = 0.1278 \text{ g } Na_2C_2O_4 \times \frac{1 \text{ mmol } Na_2C_2O_4}{0.13400 \text{ g } Na_2C_2O_4}$$

$$= 0.95373 \text{ mmol } Na_2C_2O_4$$

$$c_{KMnO_4} = 0.95373 \text{ mmol } Na_2C_2O_4 \times \frac{2 \text{ mmol } KMnO_4}{5 \text{ mmol } Na_2C_2O_4} \times \frac{1}{33.31 \text{ mL } KMnO_4}$$

$$= 0.01145 \text{ M}$$

Applications of Potassium Permanganate and Cerium(IV) Solutions

Table 17-5 lists some of the many applications of permanganate and cerium(IV) solutions to the volumetric determination of inorganic species. Both reagents have also been applied to the determination of organic compounds with oxidizable functional groups.

Solutions of $KMnO_4$ and Ce^{4+} can also be standardized with electrolytic iron wire or with potassium iodide.

Example 17-5

Aqueous solutions containing approximately 3% (w/w) H_2O_2 are sold in drug stores as a disinfectant. Propose a method for determining the peroxide content of such a preparation using the standard solution described in Examples 17-3 and 17-4. Assume that you wish to use between 30 and 45 mL of the reagent for a titration. The reaction is

$$5H_2O_2 + 2MnO_4^- + 6H^+ \rightarrow 5O_2 + 2Mn^{2+} + 8H_2O$$

The amount of $KMnO_4$ in 35 to 45 mL of the reagent is between

$$\text{amount } KMnO_4 = 35 \text{ mL } KMnO_4 \times 0.01145 \frac{\text{mmol } KMnO_4}{\text{mL } KMnO_4}$$

$$= 0.401 \text{ mmol } KMnO_4$$

and

$$\text{amount } KMnO_4 = 45 \times 0.01145 = 0.515 \text{ mmol } KMnO_4$$

The amounts of H_2O_2 consumed by these amounts of $KMnO_4$ are

$$\text{amount } H_2O_2 = 0.401 \text{ mmol } KMnO_4 \times \frac{5 \text{ mmol } H_2O_2}{2 \text{ mmol } MnO_4^-} = 1.00 \text{ mmol } H_2O_2$$

and

$$\text{amount } H_2O_2 = 0.515 \times \frac{5}{2} = 1.29 \text{ mmol } H_2O_2$$

Therefore, we need to take samples that contain from 1.00 to 1.29 mmol H_2O_2.

$$\text{mass sample} = 1.00 \text{ mmol } H_2O_2 \times 0.03401 \frac{g\,H_2O_2}{mmol\,H_2O_2} \times \frac{100 \text{ g sample}}{3\,g\,H_2O_2}$$

$$= 1.1 \text{ g sample}$$

to

$$\text{mass sample} = 1.29 \times 0.03401 \times \frac{100}{3} = 1.5 \text{ g sample}$$

Thus we could weigh out from 1.1 to 1.5 g samples. These should be diluted to perhaps 75 to 100 mL with water and made slightly acidic with dilute H_2SO_4 before titration.

17C-2 Potassium Dichromate

In its analytical applications, dichromate ion is reduced to green chromium(III) ion:

$$Cr_2O_7^{2-} + 14H^+ + 6e^- \rightleftharpoons 2Cr^{3+} + 7H_2O \qquad E^0 = 1.33 \text{ V}$$

Dichromate titrations are generally carried out in solutions that are about 1 M in hydrochloric or sulfuric acid. In these media, the formal potential for the half-reaction is 1.0 to 1.1 V.

Potassium dichromate solutions are indefinitely stable, can be boiled without decomposition, and do not react with hydrochloric acid. Moreover, primary-standard reagent is available commercially and at a modest cost. The disadvantages of potassium dichromate compared with cerium(IV) and permanganate ion are its lower electrode potential and the slowness of its reaction with certain reducing agents.

The Preparation, Properties, and Uses of Dichromate Solutions

For most purposes, reagent-grade potassium dichromate is sufficiently pure to permit the direct preparation of standard solutions; the solid simply dried at 150°C to 200°C before being weighed.

The orange color of a dichromate solution is not intense enough for use in end-point detection. However, diphenylamine sulfonic acid (Table 16-2) is an excellent indicator for titrations with this reagent. The oxidized form of the indicator is violet, and its reduced form is essentially colorless; thus, the color change observed in a direct titration is from the green of chromium(III) to violet.

Applications of Potassium Dichromate Solutions

The principal use of dichromate is for the volumetric titration of iron(II) based on the reaction

$$Cr_2O_7^{2-} + 6Fe^{2+} + 14H^+ \rightarrow 2Cr^{3+} + 6Fe^{3+} + 7H_2O$$

Often, this titration is performed in the presence of moderate concentrations of hydrochloric acid.

The reaction of dichromate with iron(II) has been widely used for the indirect determination of a variety of oxidizing agents. In these applications, a measured excess of an iron(II) solution is added to an acidic solution of the analyte. The excess iron(II) is then back-titrated with standard potassium dichromate (Section 17B-1). Standardization of the iron(II) solution by titration with the dichromate is performed concurrently with the analysis because solutions of iron(II) tend to be air-oxidized. This method has been applied to the determination of nitrate, chlorate, permanganate, and dichromate ions as well as organic peroxides and several other oxidizing agents.

Standard solutions of $K_2Cr_2O_7$ have the great advantage that they are indefinitely stable and do not oxidize HCl. Furthermore, primary-standard grade is inexpensive and readily available commercially.

Example 17-6

A 5.00-mL sample of brandy was diluted to 1.000 L in a volumetric flask. The ethanol (C_2H_5OH) in a 25.00-mL aliquot of the diluted solution was distilled into 50.00 mL of 0.02000 M $K_2Cr_2O_7$, and oxidized to acetic acid with heating. Reaction:

$$3C_2H_5OH + 2Cr_2O_7^{2-} + 16H^+ \rightarrow 4Cr^{3+} + 3CH_3COOH + 11H_2O$$

After cooling, 20.00 mL of 0.1253 M Fe^{2+} were pipetted into the flask. The excess Fe^{2+} was then titrated with 7.46 mL of the standard $K_2Cr_2O_7$ to a diphenylamine sulfonic acid end point. Calculate the percent (w/v) C_2H_5OH (46.07 g/mol) in the brandy.

$$\text{total amount } K_2Cr_2O_7 = (50.00 + 7.46) \text{ mL } K_2Cr_2O_7 \times 0.02000 \frac{\text{mmol } K_2Cr_2O_7}{\text{mL } K_2Cr_2O_7}$$

$$= 1.1492 \text{ mmol } K_2Cr_2O_7$$

$$\text{amount } K_2Cr_2O_7 \text{ consumed by } Fe^{2+} = 20.00 \text{ mL } Fe^{2+} \times 0.1253 \frac{\text{mmol } Fe^{2+}}{\text{mL } Fe^{2+}}$$

$$\times \frac{1 \text{ mmol } K_2Cr_2O_7}{6 \text{ mmol } Fe^{2+}}$$

$$= 0.41767 \text{ mmol } K_2Cr_2O_7$$

$$\text{amount } K_2Cr_2O_7 \text{ consumed by } C_2H_5OH = (1.1492 - 0.41767) \text{ mmol } K_2Cr_2O_7$$

$$= 0.73153 \text{ mmol } K_2Cr_2O_7$$

$$\text{mass } C_2H_5OH = 0.73153 \text{ mmol } K_2Cr_2O_7 \times \frac{3 \text{ mmol } C_2H_5OH}{2 \text{ mmol } K_2Cr_2O_7}$$

$$\times 0.04607 \frac{\text{g } C_2H_5OH}{\text{mmol } C_2H_5OH}$$

$$= 0.050552 \text{ g } C_2H_5OH$$

$$\text{percent } C_2H_5OH = \frac{0.050552 \text{ g } C_2H_5OH}{5.00 \text{ mL sample} \times 25.00 \text{ mL}/1000 \text{ mL}} \times 100\%$$

$$= 40.44\% = 40.4\% \ C_2H_5OH$$

17C-3 Iodine

Solutions of iodine are weak oxidizing agents that are used for the determination of strong reductants. The most accurate description of the half-reaction for iodine in these applications is

$$I_3^- + 2e^- \rightleftharpoons 3I^- \qquad E^0 = 0.536 \text{ V}$$

where I_3^- is the triiodide ion.

Standard iodine solutions have relatively limited application compared with the other oxidants we have described because of the significantly smaller electrode potential of the I_3^-/I^- couple. Occasionally, however, this low potential is advantageous because it imparts a degree of selectivity that makes possible the determination of strong reducing agents in the presence of weak ones. An important advantage of iodine is the availability of a sensitive and reversible indicator for the titrations. On the other hand, iodine solutions lack stability and must be restandardized regularly.

The Preparation and Properties of Iodine Solutions

Iodine is not very soluble in water (~ 0.001 M). In order to obtain solutions having analytically useful concentrations of the element, iodine is ordinarily dissolved in moderately concentrated solutions of potassium iodide. In this medium, iodine is reasonably soluble as a consequence of the reaction

Solutions prepared by dissolving iodine in a concentrated solution of potassium iodide are properly called *triiodide solutions*. In practice, however, they are often termed *iodine solutions* because this terminology accounts for the stoichiometric behavior of these solutions ($I_2 + 2e^- \rightarrow 2I^-$).

$$I_2(s) + I^- \rightleftharpoons I_3^- \qquad K = 7.1 \times 10^2$$

Iodine dissolves only slowly in solutions of potassium iodide, particularly if the iodide concentration is low. To ensure complete solution, the iodine is always dissolved in a small volume of concentrated potassium iodide, care being taken to avoid dilution of the concentrated solution until the last trace of solid iodine has disappeared. Otherwise, the molarity of the diluted solution will gradually increase with time. This problem can be avoided by filtering the solution through a sintered glass crucible before standardization.

Iodine solutions lack stability for several reasons, one being the volatility of the solute. Losses of iodine from an open vessel occur in a relatively short time even in the presence of an excess of iodide ion. In addition, iodine slowly attacks most organic materials. Consequently, cork or rubber stoppers are never used to close containers of the reagent, and precautions must be taken to protect standard solutions from contact with organic dusts and fumes.

Air-oxidation of iodide ion also causes changes in the molarity of an iodine solution:

$$4I^- + O_2(g) + 4H^+ \rightarrow 2I_2 + 2H_2O$$

In contrast to the other effects, this reaction causes the molarity of the iodine to increase. Air-oxidation is promoted by acids, heat, and light.

The Standardization and Application of Iodine Solutions

Iodine solutions can be standardized against anhydrous sodium thiosulfate or barium thiosulfate monohydrate, both of which are available commercially. The reaction between iodine and sodium thiosulfate is discussed in detail in Section 17B-2. Often solutions of iodine are standardized against solutions of sodium thiosulfate that have in turn been standardized against potassium iodate or potassium dichromate (Section 17B-2).

Table 17-6 summarizes methods that use iodine as an oxidizing agent.

17C-4   Potassium Bromate as a Source of Bromine

Primary-standard potassium bromate is available from commercial sources and can be used directly to prepare standard solutions that are stable indefinitely. Direct titrations with potassium bromate are relatively few. Instead the reagent is a convenient and widely used stable source of bromine.[4] In this application, an unmeasured excess of potassium bromide is added to an acidic solution of the analyte. Introduction of a measured volume of standard potassium bromate

---

Feature 17-1

THE KARL FISCHER REAGENT FOR DETERMINATION OF WATER

One of the most important titrimetric methods employs a type of standard iodine solution called the *Karl Fischer Reagent* for the determination of water, particularly in organic materials. Karl Fischer Reagent consists of iodine, pyridine, and sulfur dioxide in a $1:3:10$ ratio dissolved in anhydrous methanol. The strength of the reagent is determined by its iodine content. Its reaction with water is described by the equation

$$I_2 + SO_2 + CH_3OH + 3C_5H_5N + H_2O \rightarrow 2C_5H_4NH^+I^- + C_5H_5NH^+SO_4CH_3^-$$

Here, iodine is reduced to iodide and the sulfur dioxide is oxidized to a complex of sulfate ion; water is required for the reaction to proceed. The end point can be detected by eye based on the brown color of the excess reagent, but it is more commonly determined by electroanalytical measurements. Periodic standardization is performed by titration of standard solutions of water in methanol or of primary-standard-grade sodium tartrate dihydrate.

Several instrument manufacturers offer automatic or semiautomatic instruments for determining water in a variety of sample types by Karl Fischer titration. The determination of water with such equipment is one of the commonly used analytical techniques encountered in industrial laboratories.

---

[4] For a discussion of bromate solutions and their applications, see M. R. F. Ashworth, *Titrimetric Organic Analysis,* Part I, pp. 118–130. New York: Interscience, 1964.

Table 17-6
SOME APPLICATIONS OF IODINE SOLUTIONS

Substance Analyzed	Half-Reaction
As	$H_3AsO_3 + H_2O \rightleftharpoons H_3AsO_4 + 2H^+ + 2e^-$
Sb	$H_3SbO_3 + H_2O \rightleftharpoons H_3SbO_4 + 2H^+ + 2e^-$
Sn	$Sn^{2+} \rightleftharpoons Sn^{4+} + 2e^-$
H_2S	$H_2S \rightleftharpoons S(s) + 2H^+ + 2e^-$
SO_2	$SO_3^{2-} + H_2O \rightleftharpoons SO_4^{2-} + 2H^+ + 2e^-$
$S_2O_3^{2-}$	$2\,S_2O_3^{2-} \rightleftharpoons S_4O_6^{2-} + 2e^-$
N_2H_4	$N_2H_4 \rightleftharpoons N_2(g) + 4H^+ + 4e^-$
Ascorbic acid*	$C_6H_8O_6 \rightleftharpoons C_6H_6O_6 + 2H^+ + 2e^-$

* For the structure of ascorbic acid, see Section 29H-3.

results in production of a stoichiometric quantity of bromine:

$$\underset{\substack{\text{standard} \\ \text{solution}}}{BrO_3^-} + \underset{\text{excess}}{5Br^-} + 6H^+ \rightarrow 3Br_2 + 3H_2O$$

1 mol $KBrO_3 \equiv 3$ mol Br_2

This indirect generation circumvents the problems associated with the use of standard bromine solutions, which lack stability.

The primary use of standard potassium bromate is for the determination of organic compounds that react with bromine. Few of these reactions are rapid enough to permit direct titration. Instead, a measured excess of standard bromate is added to the solution that contains the sample plus an excess of potassium bromide. After acidification, the mixture is allowed to stand in a glass-stoppered vessel until reaction of the bromine with the analyte is judged complete. To determine the excess bromine, an unmeasured excess of potassium iodide is introduced to convert the excess bromine to iodine:

$$2I^- + Br_2 \rightarrow I_2 + 2Br^-$$

The liberated iodine is then titrated with standard sodium thiosulfate (Equation 17-1).

Bromine is incorporated into an organic molecule either by substitution or by addition.

Substitution Reactions

Halogen substitution involves the replacement of hydrogen in an aromatic ring by a halogen. Substitution methods have been successfully applied to the determination of aromatic compounds that contain strong ortho-para-directing groups, particularly amines and phenols.

Example 17-7

A 0.2981-g sample of an antibiotic powder containing sulfanilamide was dissolved in HCl and the solution diluted to 100.0 mL. A 20.00-mL aliquot was transferred to a stoppered flask and 25.00 mL of 0.01767 M $KBrO_3$ added. About 10 g of KBr was added to form Br_2, which brominated the sulfanilamide in the sample. After 10 min, an excess of KI was added and the liberated iodine titrated with 12.92 mL of 0.1215 M sodium thiosulfate. The reactions are

$$BrO_3^- + 5Br^- + 6H^+ \rightarrow 3Br_2 + 3H_2O$$

$$Br_2 + 2I^- \rightarrow 2Br^- + I_2 \quad \text{(excess KI)}$$
$$I_2 + 2S_2O_3^{2-} \rightarrow S_4O_6^{2-} + 2I^-$$

Calculate the % $NH_2C_6H_4SO_2NH_2$ (172.21 g/mol) in the powder.

$$\text{total amount } Br_2 = 25.00 \text{ mL } KBrO_3 \times 0.01767 \frac{\text{mmol } KBrO_3}{\text{mL } KBrO_3} \times \frac{3 \text{ mmol } Br_2}{\text{mmol } KBrO_3}$$
$$= 1.32525 \text{ mmol } Br_2$$

We next calculate the amount Br_2 in excess over that required to brominate the analyte.

$$\text{amount excess } Br_2 = \text{amount } I_2$$
$$= 12.92 \text{ mL } Na_2S_2O_3 \times 0.1215 \frac{\text{mmol } Na_2S_2O_3}{\text{mL } Na_2S_2O_3}$$
$$\times \frac{1 \text{ mmol } I_2}{2 \text{ mmol } Na_2S_2O_3}$$
$$= 0.78489 \text{ mmol } Br_2$$

The amount of Br_2 consumed by the sample is given by

$$\text{amount } Br_2 = 1.32525 - 0.78489 = 0.54036 \text{ mmol } Br_2$$
$$\text{mass analyte} = 0.54036 \text{ mmol } Br_2 \times \frac{1 \text{ mmol analyte}}{2 \text{ mmol } Br_2}$$
$$\times 0.17221 \frac{\text{g analyte}}{\text{mmol analyte}}$$
$$= 0.046528 \text{ g analyte}$$
$$\text{percent analyte} = \frac{0.046528 \text{ g analyte}}{0.2891 \text{ g sample} \times 20.00 \text{ mL}/100 \text{ mL}} \times 100\%$$
$$= 80.47\% \text{ sulfanilamide}$$

An important example of the use of a bromine substitution reaction is the determination of 8-hydroxyquinoline:

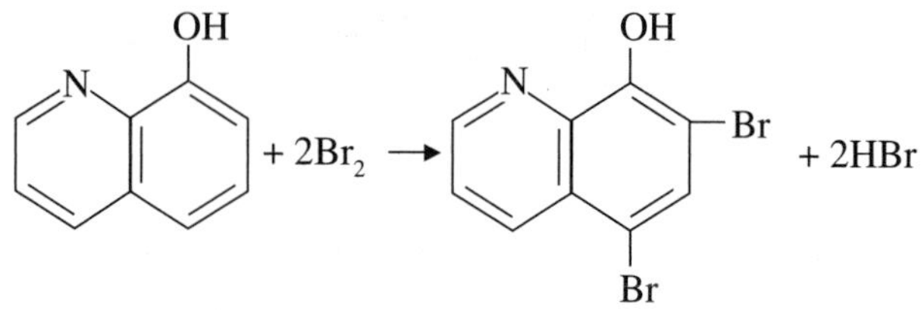

In contrast to most bromine substitutions, this reaction takes place rapidly enough in hydrochloric acid solution to permit direct titration. The titration of 8-hydroxyquinoline with bromine is particularly significant because the former is an excellent precipitating reagent for cations (Section 6D-3). For example, aluminum can be determined according to the sequence

$$Al^{3+} + 3HOC_9H_6N \xrightarrow{\text{pH 4-9}} Al(OC_9H_6N)_3(s) + 3H^+$$

$$Al(OC_9H_6N)_3(s) \xrightarrow{\text{hot 4 M HCl}} 3HOC_9H_6N + Al^{3+}$$

$$3HOC_9H_6N + 6Br_2 \rightarrow 3HOC_9H_4NBr_2 + 6HBr$$

The stoichiometric relationships in this case are

$$1 \text{ mol } Al^{3+} \equiv 3 \text{ mol } HOC_9H_6N \equiv 6 \text{ mol } Br_2 \equiv 2 \text{ mol } KBrO_3$$

Addition Reactions

Addition reactions involve the opening of an olefinic double bond. For example, 1 mol of ethylene reacts with 1 mol of bromine in the reaction

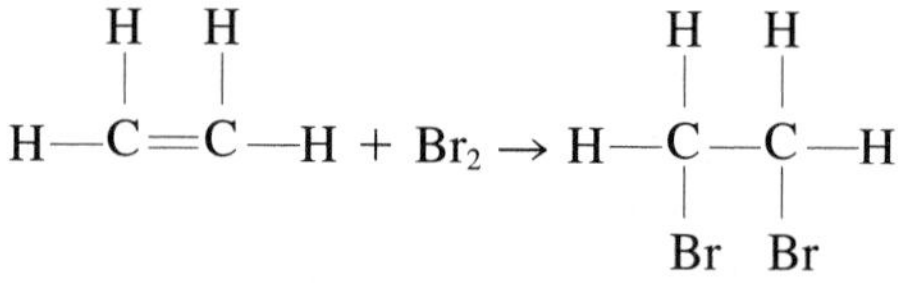

The literature contains numerous references to the use of bromine for the estimation of olefinic unsaturation in fats, oils, and petroleum products.

17D QUESTIONS AND PROBLEMS

*17-1. Write balanced net-ionic equations to describe
  (a) the oxidation of Mn^{2+} to MnO_4^- by ammonium peroxydisulfate.
  (b) the oxidation of Ce^{3+} to Ce^{4+} by sodium bismuthate.
  (c) the oxidation of U^{4+} to UO_2^{2+} by H_2O_2.
  (d) the reaction of $V(OH)_4^+$ in a Walden reductor.
  (e) the titration of H_2O_2 with $KMnO_4$.
  (f) the reaction between KI and ClO_3^- in acidic solution.

17-2. Write balanced net-ionic equations to describe
  (a) the reduction of Fe^{3+} to Fe^{2+} by SO_2.
  (b) the reaction of H_2MoO_4 in a Jones reductor.

(c) the oxidation of HNO_2 by a solution of MnO_4^-.

(d) the reaction of aniline ($C_6H_4NH_2$) with a mixture of $KBrO_3$ and KBr in acidic solution.

(e) the air oxidation of $HAsO_3^{2-}$ to $HAsO_4^{2-}$.

(f) the reaction of KI with HNO_2 in acidic solution.

*17-3. Why is a Walden reductor always used with solutions that contain appreciable concentrations of HCl?

17-4. Why is zinc amalgam preferable to pure zinc in a Jones reductor?

*17-5. Write a balanced net-ionic equation for the reduction of UO_2^{2+} in a Walden reductor.

17-6. Write a balanced net-ionic equation for the reduction of TiO^{2+} in a Jones reductor.

*17-7. Why are standard solutions of reductants less often used for titrations than standard solutions of oxidants?

*17-8. Why are standard $KMnO_4$ solutions seldom used for the titration of solutions containing HCl?

17-9. Why are Ce^{4+} solutions never used for the titration of reductants in basic solutions?

*17-10. Write a balanced net-ionic equation showing why $KMnO_4$ end points fade.

17-11. Why are $KMnO_4$ solutions filtered before they are standardized?

17-12. Why are solutions of $KMnO_4$ and $Na_2S_2O_3$ generally stored in dark reagent bottles?

*17-13. When a solution of $KMnO_4$ was left standing in a buret for 3 h, a brownish ring formed at the surface of the liquid. Write a balanced net-ionic equation to account for this observation.

17-14. What is the primary use of standard $K_2Cr_2O_7$ solutions?

*17-15. Why are iodine solutions prepared by dissolving I_2 in concentrated KI?

17-16. A standard solution of I_2 increased in molarity with standing. Write a balanced net-ionic equation that accounts for the increase.

*17-17. When a solution of $Na_2S_2O_3$ is introduced into a solution of HCl, a cloudiness develops almost immediately. Write a balanced net-ionic equation explaining this phenomenon.

17-18. Suggest a way in which a solution of KIO_3 could be used as a source of known quantities of I_2.

*17-19. Write balanced equations showing how $KBrO_3$ could be used as a primary standard for solutions of $Na_2S_2O_3$.

17-20. Write balanced equations showing how $K_2Cr_2O_7$ could be used as a primary standard for solutions of $Na_2S_2O_3$.

*17-21. Write a balanced net-ionic equation describing the titration of hydrazine (N_2H_4) with standard iodine.

17-22. In the titration of I_2 solutions with $Na_2S_2O_3$, starch indicator is never added until just before chemical equivalence. Why?

17-23. A solution prepared by dissolving a 0.2464-g sample of electrolytic iron wire in acid was passed through a Jones reductor. The iron(II) in the resulting solution required a

39.31-mL titration. Calculate the molar oxidant concentration if the titrant used was

*(a) Ce^{4+} (product: Ce^{3+}).

(b) $Cr_2O_7^{2-}$ (product: Cr^{3+}).

*(c) MnO_4^- (product: Mn^{2+}).

(d) $V(OH)_4^+$ (product: VO^{2+}).

*(e) IO_3^- (product: ICl_2^-).

*17-24. How would you prepare 250.0 mL of 0.03500 M $K_2Cr_2O_7$?

17-25. How would you prepare 2.000 L of 0.02500 M $KBrO_3$?

*17-26. How would you prepare 1.5 L of approximately 0.1 M $KMnO_4$?

17-27. How would you prepare 2.0 L of approximately 0.05 M I_3^-?

*17-28. Titration of 0.1467 g of primary-standard $Na_2C_2O_4$ required 28.85 mL of a potassium permanganate solution. Calculate the molar concentration of $KMnO_4$ in this solution.

17-29. A 0.1809-g sample of pure iron wire was dissolved in acid, reduced to the +2 state, and titrated with 31.33 mL of cerium(IV). Calculate the molar concentration of the Ce^{4+} solution.

*17-30. The iodine produced when an excess of KI was added to a solution containing 0.1518 g of $K_2Cr_2O_7$ required a 46.13-mL titration with $Na_2S_2O_3$. Calculate the molar concentration of the thiosulfate solution.

17-31. A 0.1017-g sample of $KBrO_3$ was dissolved in dilute HCl and treated with an unmeasured excess of KI. The liberated iodine required 39.75 mL of a sodium thiosulfate solution. Calculate the molar concentration of the $Na_2S_2O_3$.

*17-32. The Sb(III) in a 1.080-g ore sample required a 41.67-mL titration with 0.03134 M I_2 [reaction product: Sb(V)]. Express the results of this analysis in terms of (a) percentage of Sb and (b) percentage of stibnite (Sb_2S_3).

17-33. Calculate the percentage of MnO_2 in a mineral specimen if the I_2 liberated by a 0.1344-g sample in the net reaction

$$MnO_2(s) + 4H^+ + 2I^- \rightarrow Mn^{2+} + I_2 + 2H_2O$$

required 32.30 mL of 0.07220 M $Na_2S_2O_3$.

*17-34. Under suitable conditions, thiourea is oxidized to sulfate by solutions of bromate

$$3CS(NH_2)_2 + 4BrO_3^- + 3H_2O \rightarrow$$
$$3CO(NH_2)_2 + 3SO_4^{2-} + 4Br^- + 6H^+$$

A 0.0715-g sample of a material was found to consume 14.1 mL of 0.00833 M $KBrO_3$. What was the percent purity of the thiourea (76.122 g/mol) sample?

*17-35. A 0.7120-g specimen of iron ore was brought into solution and passed through a Jones reductor. Titration of the Fe(II) produced required 39.21 mL of 0.02086 M $KMnO_4$. Express the results of this analysis in terms of percent (a) Fe and (b) Fe_2O_3.

17-36. The Sn in a 0.4352-g mineral specimen was reduced to

the +2 state with Pb and titrated with 29.77 mL of 0.01735 M $K_2Cr_2O_7$. Calculate the results of this analysis in terms of percent (a) Sn and (b) SnO_2.

*17-37. Treatment of hydroxylamine (H_2NOH) with an excess of Fe(II) results in the formation of N_2O and an equivalent amount of Fe(II):

$$2H_2NOH + 4Fe^{3+} \rightarrow N_2O(g) + 4Fe^{2+} + 4H^+ + H_2O$$

Calculate the molar concentration of an H_2NOH solution if the Fe(II) produced by treatment of a 50.00-mL aliquot required 23.61 mL of 0.02170 M $K_2Cr_2O_7$.

17-38. The organic matter in a 0.9280-g sample of burn ointment was eliminated by ashing, following which the solid residue of ZnO was dissolved in acid. Treatment with $(NH_4)_2C_2O_4$ resulted in the formation of the sparingly soluble ZnC_2O_4. The solid was filtered, washed, and then redissolved in dilute acid. The liberated $H_2C_2O_4$ required 37.81 mL of 0.01508 M $KMnO_4$. Calculate the percentage of ZnO in the medication.

*17-39. The $KClO_3$ in a 0.1342-g sample of an explosive was determined by reaction with 50.00 mL of 0.09601 M Fe^{2+}:

$$ClO_3^- + 6Fe^{2+} + 6H^+ \rightarrow Cl^- + 3H_2O + 6Fe^{3+}$$

When the reaction was complete, the excess Fe^{2+} was back-titrated with 12.99 mL of 0.08362 M Ce^{4+}. Calculate the percentage of $KClO_3$ in the sample.

17-40. The tetraethyl lead [$Pb(C_2H_5)_4$] in a 25.00-mL sample of aviation gasoline was shaken with 15.00 mL of 0.02095 M I_2. Reaction:

$$Pb(C_2H_5)_4 + I_2 \rightarrow Pb(C_2H_5)_3I + C_2H_5I$$

After the reaction was complete, the unused I_2 was titrated with 6.09 mL of 0.03465 M $Na_2S_2O_3$. Calculate the weight (in milligrams) of $Pb(C_2H_5)_4$ (323.4 g/mol) in each liter of the gasoline.

*17-41. An 8.13-g sample of an ant-control preparation was decomposed by wet-ashing with H_2SO_4 and HNO_3. The As in the residue was reduced to the trivalent state with hydrazine. After removal of the excess reducing agent, the As(III) required a 23.77-mL titration with 0.02425 M I_2 in a faintly alkaline medium. Express the results of this analysis in terms of percentage of As_2O_3 in the original sample.

17-42. A sample of alkali metal chlorides was analyzed for sodium by dissolving a 0.800-g sample in water and diluting to exactly 500 mL. A 25.0-mL aliquot of this was treated in such a way as to precipitate the sodium as $NaZn(UO_2)_3(OAc)_9 \cdot 6H_2O$. The precipitate was filtered, dissolved in acid, and passed through a lead reductor which converted the uranium to U^{4+}. Oxidation of this to UO_2^{2+} required 19.9 mL of 0.01667 M $K_2Cr_2O_7$. Calculate the percent NaCl in the sample.

*17-43. The ethyl mercaptan concentration in a mixture was deter-

mined by shaking a 1.657-g sample with 50.0 mL of 0.01194 M I_2 in a tightly stoppered flask:

$$2C_2H_5SH + I_2 \rightarrow C_2H_5SSC_2H_5 + 2I^- + 2H^+$$

The excess I_2 was back-titrated with 16.77 mL of 0.01325 M $Na_2S_2O_3$. Calculate the percentage of C_2H_5SH (62.14 g/mol).

17-44. A 4.971-g sample containing the mineral tellurite was dissolved and then treated with 50.00 mL of 0.03114 M $K_2Cr_2O_7$:

$$3TeO_2 + Cr_2O_7^{2-} + 8H^+ \rightarrow 3H_2TeO_4 + 2Cr^{3+} + H_2O$$

Upon completion of the reaction, the excess $Cr_2O_7^{2-}$ required a 10.05-mL back-titration with 0.1135 M Fe^{2+}. Calculate the percentage of TeO_2 in the sample.

*17-45. A sensitive method for I^- in the presence of Cl^- and Br^- entails oxidation of the I^- to IO_3^- with Br_2. The excess Br_2 is then removed by boiling or by reduction with formate ion. The IO_3^- produced is determined by addition of excess I^- and titration of the resulting I_2. A 1.204-g sample of mixed halides was dissolved and analyzed by the foregoing procedure; 20.66 mL of 0.05551 M thiosulfate were required. Calculate the percentage of KI in the sample.

*17-46. A 1.065-g sample of stainless steel was dissolved in HCl (this treatment converts the Cr present to Cr^{3+}) and diluted to 500.0 mL in a volumetric flask. One 50.00-mL aliquot was passed through a Walden reductor and then titrated with 13.72 mL of 0.01920 M $KMnO_4$. A 100.0-mL aliquot was passed through a Jones reductor into 50 mL of approximately 0.10 M Fe^{3+}, which reacted with the Cr^{2+} to form Cr^{3+} and Fe^{2+}. Titration of the resulting solution required 36.43 mL of the $KMnO_4$ solution. Calculate the percentages of Fe and Cr in the alloy.

17-47. A 2.559-g sample containing both Fe and V was dissolved under conditions that converted the elements to Fe(III) and V(V). The solution was diluted to 500.0 mL, and a 50.00-mL aliquot was passed through a Walden reductor and titrated with 17.74 mL of 0.1000 M Ce^{4+}. A second 50.00-mL aliquot was passed through a Jones reductor and required 44.67 mL of the same Ce^{4+} solution. Calculate the percentage of Fe_2O_3 and V_2O_5 in the sample.

*17-48. A 25.0-mL aliquot of a solution containing Tl(I) ion was treated with K_2CrO_4. The Tl_2CrO_4 was filtered, washed free of excess precipitating agent, and dissolved in dilute H_2SO_4. The $Cr_2O_7^{2-}$ produced was titrated with 40.60 mL of 0.1004 M Fe^{2+} solution. What was the mass of Tl in the sample? The reactions are

$$2Tl^+ + CrO_4^{2-} \rightarrow Tl_2CrO_4(s)$$
$$2Tl_2CrO_4(s) + 2H^+ \rightarrow 4Tl^+ + Cr_2O_7^{2-} + H_2O$$
$$Cr_2O_7^{2-} + 6Fe^{2+} + 14H^+ \rightarrow 6Fe^{3+} + 2Cr^{3+} + 7H_2O$$

*17-49. A gas mixture was passed at the rate of 2.50 L/min through a solution of sodium hydroxide for a total of 64.00 min.

The SO_2 in the mixture was retained as sulfite ion

$$SO_2(g) + 2OH^- \rightarrow SO_3^{2-}$$

After acidification with HCl, the sulfite was titrated with 4.98 mL of 0.003125 M KIO_3

$$IO_3^- + 2H_2SO_3 + 2Cl^- \rightarrow ICl_2^- + 2SO_4^{2-} + 2H^+$$

Use 1.20 g/L for the density of the mixture and calculate the concentration of SO_2 in ppm.

17-50. A 24.7-L sample of air drawn from the vicinity of a coking oven was passed over iodine pentoxide at 150°C, where CO was converted to CO_2 and a chemically equivalent quantity of I_2 was produced:

$$I_2O_5(s) + 5CO(g) \rightarrow 5CO_2(g) + I_2(g)$$

The I_2 distilled at this temperature and was collected in a solution of KI. The I_3^- produced was titrated with 7.76 mL of 0.00221 M $Na_2S_2O_3$. Does the air in this space comply with federal regulations that mandate a maximum CO level no greater than 50 ppm? Use 1.20 g/L for the density of the sample.

*17-51. A 30.00-L air sample was passed through an absorption tower containing a solution of Cd^{2+}, where H_2S was retained as CdS. The mixture was acidified and treated with 10.00 mL of 0.01070 M I_2. After the reaction

$$H_2S + I_2 \rightarrow S(s) + 2I^- + 2H^+$$

was complete, the excess iodine was titrated with 12.85 mL of 0.01344 M thiosulfate. Calculate the concentration of H_2S in ppm; use 1.20 g/L for the density of the gas stream.

17-52. A square of photographic film 2.0 cm on an edge was suspended in a 5% solution of $Na_2S_2O_3$ to dissolve the silver halides. After removal and washing of the film, the solution was treated with an excess of Br_2 to oxidize the iodide present to IO_3^- and destroy the excess thiosulfate ion. The solution was then boiled to remove the bromine, and an excess of iodide was added. The liberated iodine was titrated with 13.7 mL of 0.0352 M thiosulfate solution.

(a) Write balanced equations for the reactions involved in the method.

(b) Calculate the milligrams of AgI per square centimeter of film.

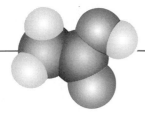

CHAPTER 18

POTENTIOMETRIC METHODS

In Chapter 15, we showed that the potential of an electrode, relative to the standard hydrogen electrode, is determined by the activity of one or more of the species in the solution in which the electrode is immersed. This chapter deals with how electrode potentials are measured and how these data are used to determine the concentration of analytes.[1] Analytical methods that are based on potential measurements are termed *potentiometric methods*.

18A GENERAL PRINCIPLES

In Feature 15-3, we showed that absolute values for individual half-cell potentials cannot be determined in the laboratory. That is, only cell potentials can be obtained experimentally. Figure 18-1 shows a typical cell for potentiometric analysis. This cell can be depicted as

$$\underbrace{\text{reference electrode}}_{E_{\text{ref}}} | \underbrace{\text{salt bridge}}_{E_{\text{j}}} | \underbrace{\text{analyte solution} | \text{indicator electrode}}_{E_{\text{ind}}}$$

The *reference electrode* in this diagram is a half-cell with an accurately known electrode potential, E_{ref}, that is independent of the concentration of the analyte or any other ions in the solution under study. It can be a standard hydrogen electrode but seldom is because a standard hydrogen electrode is somewhat troublesome to maintain and use. By convention, the reference electrode is always treated as the anode in potentiometric measurements. The *indicator electrode*,

Reference electrodes are *always* treated as anodes in this text.

---

[1] For further reading on potentiometric methods, see E. P. Serjeant, *Potentiometry and Potentiometric Titrations*. New York: Wiley, 1984.

which is immersed in a solution of the analyte, develops a potential, E_{ind}, that depends on the activity of the analyte. Most indicator electrodes used in potentiometry are highly selective in their responses. The third component of a potentiometric cell is a salt bridge that prevents the components of the analyte solution from mixing with those of the reference electrode. As noted in Chapter 15, a potential, E_j, develops across the liquid junctions at each end of the salt bridge.

The potential of the cell we have just considered is given by the equation

$$E_{cell} = E_{ind} - E_{ref} + E_j \qquad \textbf{(18-1)}$$

The first term in this equation, E_{ind}, contains the information we seek about the concentration of the analyte. A potentiometric analysis, then, involves measuring a cell potential, correcting this potential for the reference and junction potentials, and computing the analyte concentration from the indicator electrode potential.

The sections that follow deal with the sources of the three potentials shown on the right side of Equation 18-1.

18B   REFERENCE ELECTRODES

The ideal reference electrode has a potential (versus the standard hydrogen electrode) that is accurately known, constant, and completely insensitive to the composition of the analyte solution. In addition, this electrode should be rugged, easy to assemble, and should maintain a constant potential while passing small currents.

18B-1   Calomel Electrodes

A calomel electrode can be represented schematically as

$$Hg \mid Hg_2Cl_2(\text{sat'd}),KCl(x\text{M}) \parallel$$

where x represents the molar concentration of potassium chloride in the solution. Three concentrations of potassium chloride are common, 0.1 M, 1 M, and saturated (about 4.6 M). The saturated calomel electrode (SCE) is the most widely used because it is easily prepared. Its main disadvantage is its somewhat large temperature coefficient. This disadvantage is important only in those rare circumstances where substantial temperature changes occur during a measurement. The *electrode* potential of the saturated calomel electrode is 0.2444 V at 25°C.

The electrode reaction in calomel half-cells is

$$Hg_2Cl_2(s) + 2e^- \rightleftharpoons 2Hg(l) + 2Cl^-(aq)$$

Figure 18-2 illustrates a typical commercial saturated calomel electrode. It consists of a 5- to 15-cm tube that is 0.5 to 1.0 cm in diameter. A mercury/mercury(I) chloride paste in saturated potassium chloride is contained in an inner tube and connected to the saturated potassium chloride solution in the outer tube through a small opening. Contact with the analyte solution is made through a fritted disk, a porous fiber, or a piece of porous vycor (''thirsty glass'') sealed in the end of the outer tubing.

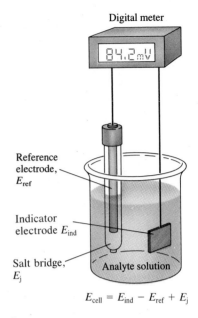

Figure 18-1
A cell for potentiometric analysis.

A hydrogen electrode is seldom used as a reference electrode for day-to-day potentiometric measurements because it is somewhat inconvenient and is also a fire hazard.

The term ''saturated'' in a saturated calomel electrode refers to the KCl concentration. All calomel electrodes are saturated with Hg_2Cl_2 (calomel).

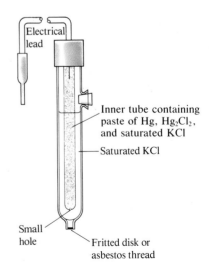

Figure 18-2
Diagram of a typical commercial saturated calomel electrode.

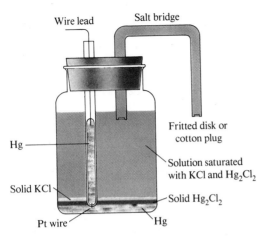

Wire lead   Salt bridge

Fritted disk or
cotton plug

Hg

Solution saturated
with KCl and Hg_2Cl_2

Solid KCl

Solid Hg_2Cl_2

Pt wire   Hg

Half reaction:
$$Hg_2Cl_2(s) + 2\ e^- \rightleftharpoons 2\ Hg + 2\ Cl^-$$

Figure 18-3
A saturated calomel electrode made
from materials readily available in
any laboratory.

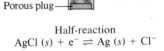

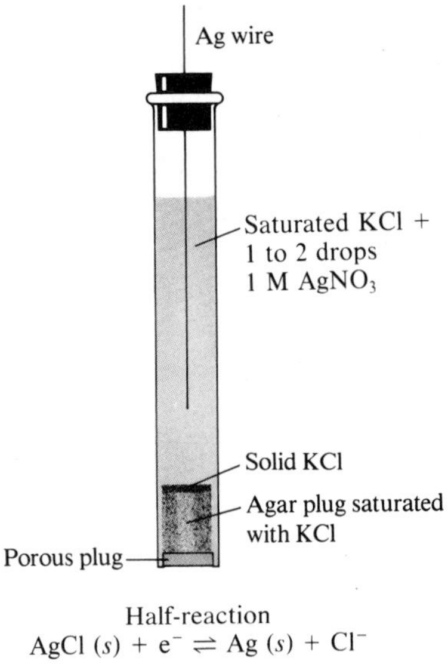

Ag wire

Saturated KCl +
1 to 2 drops
1 M $AgNO_3$

Solid KCl

Agar plug saturated
with KCl

Porous plug

Figure 18-4
An easily constructed Ag/AgCl
electrode.

Half-reaction
$$AgCl\ (s) + e^- \rightleftharpoons Ag\ (s) + Cl^-$$

A salt bridge is readily constructed by filling a U-tube with a conducting gel prepared by heating about 5 g of agar in 100 mL of an aqueous solution containing about 35 g of potassium chloride. When the liquid cools, it sets up into a gel that is a good conductor.

Agar, which is available as translucent flakes, is a heteropolysaccharide that is extracted from certain East Indian seaweed. Solutions of agar in hot water set to a gel on cooling.

At 25°C, the potential of the saturated calomel electrode versus the standard hydrogen electrode is 0.244 V; that for the saturated silver/silver chloride electrode is 0.199 V.

Figure 18-3 shows a saturated calomel electrode that you can easily construct from materials available in your laboratory. A salt bridge (Section 15B) provides electrical contact with the analyte solution.

18B-2 Silver/Silver Chloride Electrodes

A system analogous to the saturated calomel electrode consists of a silver electrode immersed in a solution that is saturated in both potassium chloride and silver chloride:

$$Ag|AgCl(sat'd),KCl(sat'd)\|$$

The half-reaction is

$$AgCl(s) + e^- \rightleftharpoons Ag(s) + Cl^-(aq)$$

The potential of this electrode is 0.199 V at 25°C.

Silver/silver chloride electrodes of various sizes and shapes are on the market. A simple and easily constructed electrode of this type is shown in Figure 18-4.

18C   LIQUID-JUNCTION POTENTIALS

A liquid-junction potential develops across the boundary between two electrolyte solutions that have different compositions. Figure 18-5 shows a very simple liquid junction consisting of a 1 M hydrochloric acid solution that is in contact with a solution that is 0.01 M in that acid. An inert porous barrier, such as a fritted glass plate, prevents the two solutions from mixing. Both hydrogen ions and chloride ions tend to diffuse across this boundary from the more concentrated to the more dilute solution. The driving force for each ion is proportional to the activity difference between the two solutions. In the present example, hydrogen ions are substantially more mobile than chloride ions. Thus, hydrogen ions diffuse more rapidly than chloride ions, and, as shown in the figure, a separation of charge results. The more dilute side of the boundary becomes positively charged because of the more rapid diffusion of hydrogen ions. The concentrated side therefore acquires a negative charge from the excess of slower moving chloride ions. The charge developed tends to counteract the differences in diffusion rates of the two ions so that a condition of equilibrium is attained rapidly. The potential difference resulting from this charge separation may be several hundredths of a volt.

The magnitude of the liquid-junction potential can be minimized by placing a *salt bridge* between the two solutions. The salt bridge is most effective if the mobilities of the negative and positive ions in the bridge are nearly equal and if their concentrations are large. A saturated solution of potassium chloride is good from both standpoints. The net junction potential with such a bridge is typically a few millivolts.

All cells used for potentiometric analyses contain a salt bridge that connects the reference electrode to the analyte solution. As we shall see, uncertainties in the magnitude of the junction potential across this bridge place a fundamental limit on the accuracy of potentiometric methods of analysis.

> The junction potential across a typical salt bridge is a few millivolts.

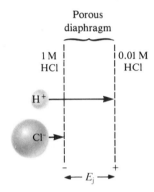

Figure 18-5
Schematic representation of a liquid junction, showing the source of the junction potential E_j. The length of the arrows corresponds to the relative mobility of the two ions.

18D   INDICATOR ELECTRODES

An ideal indicator electrode responds rapidly and reproducibly to changes in the concentration of an analyte ion (or group of ions). Although no indicator electrode is absolutely specific in its response, a few are now available that are remarkably selective. Indicator electrodes are either metallic or membrane.

18D-1   Metallic Indicator Electrodes

It is convenient to classify metallic indicator electrodes as *electrodes of the first kind*, *electrodes of the second kind*, and inert *redox electrodes*.

Electrodes of the First Kind

An electrode of the first kind consists of a pure metal electrode that is in direct equilibrium with its cation in the solution. A single reaction is involved. For example, the equilibrium between a metal X and its cation X^{n+} is

$$X^{n+}(aq) + ne^- \rightleftharpoons X(s)$$

for which

$$E_{ind} = E^0_{X^{n+}} - \frac{0.0592}{n} \log \frac{1}{a_{X^{n+}}} = E^0_{X^{n+}} + \frac{0.0592}{n} \log a_{X^{n+}} \quad \textbf{(18-2)}$$

where E_{ind} is the electrode potential of the metal electrode and $a_{X^{n+}}$ is the activity of the ion (or approximately its molar concentration, $[X^{n+}]$).

We often express the electrode potential of the indicator electrode in terms of the p-function of the cation. Thus, substituting the definition of pX into Equation 18-2 gives

$$E_{ind} = E^0_{X^{n+}} + \frac{0.0592}{n} \log a_{X^{n+}} = E^0_{X^{n+}} - \frac{0.0592}{n} pX \quad \textbf{(18-3)}$$

> Potentiometric analyses provide data in terms of *activities* of analytes in contrast to the other methods we consider in this book, which give the *concentrations* of analytes. Recall that the activity of a species a_X is related to the molar concentration of X by Equation 8-2
>
> $$a_X = \gamma_X[X]$$
>
> where γ_X is the activity coefficient of X, a parameter that varies with the ionic strength of the solution. Because potentiometric data depend on activities, we will not make the usual approximation that $a_X \cong [X]$ in this chapter.

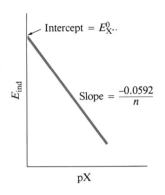

Figure 18-6
A plot of Equation 18-3 for an electrode of the first kind.

Electrode systems of the first kind are not widely used for potentiometric analyses for several reasons. For one, they are not very selective and respond not only to their own cations but also other more easily reduced cations. For example, a copper electrode cannot be used for the determination of copper(II) ions in the presence of silver(I) ions, which are also reduced at the copper surface. Other metal electrodes, such as zinc and cadmium, can only be used in neutral or basic solutions because they dissolve in the presence of acids. Also, other metals are so easily oxidized that their use is restricted to solutions that have been deaerated. Finally, certain harder metals, such as iron, chromium, cobalt, and nickel do not provide reproducible potentials. Moreover, for these electrodes, plots of pX versus activity yield slopes that differ significantly and irregularly from the theoretical ($-0.0592/n$). For these reasons, the only electrode systems of the first kind that have been used are Ag/Ag^+ and Hg/Hg_2^{2+} in neutral solutions and Cu/Cu^{2+}, Zn/Zn^{2+}, Cd/Cd^{2+}, Bi/Bi^{3+}, Tl/Tl^+, and Pb/Pb^{2+} in deaerated solutions.

Electrodes of the Second Kind

Metals not only serve as indicator electrodes for their own cations but also respond to the activities of anions that form sparingly soluble precipitates or stable complexes with such cations. The potential of a silver electrode, for example, correlates reproducibly with the activity of chloride ion in a solution saturated with silver chloride. Here, the electrode reaction can be written as

$$AgCl(s) + e^- \rightleftharpoons Ag(s) + Cl^-(aq) \qquad E^0_{AgCl} = 0.222 \text{ V}$$

The Nernst expression for this process is

$$E_{ind} = E^0_{AgCl} - 0.0592 \log a_{Cl^-} = E^0_{AgCl} + 0.0592 \text{ pCl} \qquad \textbf{(18-4)}$$

Equation 18-4 shows that the potential of a silver electrode is proportional to pCl, the negative logarithm of the chloride ion activity. Thus, in a solution saturated with silver chloride, a silver electrode can serve as an indicator electrode of the second kind for chloride ion. Note that the sign of the log term for an electrode of this type is opposite that for an electrode of the first kind (see Equation 18-3).

Mercury serves as an indicator electrode of the second kind for the EDTA anion Y^{4-}. For example, when a small amount of HgY^{2-} is added to a solution containing Y^{4-}, the half-reaction at a mercury cathode is

$$HgY^{2-} + 2e^- \rightleftharpoons Hg(l) + Y^{4-} \qquad E^0 = 0.21 \text{ V}$$

for which

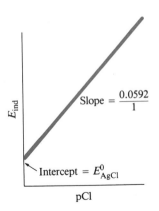

Figure 18-7
A plot of Equation 18-4 for an electrode of the second kind for Cl^-.

$$E_{ind} = 0.21 - \frac{0.0592}{2} \log \frac{a_{Y^{4-}}}{a_{HgY^{2-}}}$$

The formation constant for HgY^{2-} is very large (6.3×10^{21}), and so the concentration of the complex remains essentially constant over a large range of Y^{4-}

concentrations. The Nernst equation for the process can therefore be written as

$$E = K - \frac{0.0592}{2} \log a_{Y^{4-}} = K + \frac{0.0592}{2} pY \qquad (18\text{-}5)$$

where

$$K = 0.21 - \frac{0.0592}{2} \log \frac{1}{a_{HgY^{2-}}}$$

The mercury electrode is thus a valuable electrode of the second kind for EDTA titrations.

Inert Metallic Electrodes for Redox Systems

As noted in Chapter 15, an inert conductor—such as platinum, gold, palladium, or carbon—responds to the potential of redox systems with which it is in contact. For example, the potential of a platinum electrode immersed in a solution containing cerium(III) and cerium(IV) is

$$E_{ind} = E^0_{Ce(IV)} - 0.0592 \log \frac{a_{Ce^{3+}}}{a_{Ce^{4+}}}$$

A platinum electrode is thus a convenient indicator electrode for titrations involving standard cerium(IV) solutions.

18D-2 Membrane Electrodes[2]

For many years, the most convenient method for determining pH has involved measurement of the potential that develops across a thin glass membrane that separates two solutions with different hydrogen ion concentrations. The phenomenon on which the measurement is based was first reported in 1906 and by now has been extensively studied by many investigators. As a result, the sensitivity and selectivity of glass membranes toward hydrogen ions are reasonably well understood. Furthermore, this understanding has led to the development of other types of membranes that respond selectively to more than two dozen other ions.

Membrane electrodes are sometimes called *p-ion electrodes* because the data obtained from them are usually presented as p-functions, such as pH, pCa, or pNO_3. In this section, we consider several types of p-ion membranes. In addition, a gas-sensing probe based on a noncrystalline membrane is described.

It is important to note at the outset of this discussion that membrane electrodes are *fundamentally different* from metal electrodes both in design and in principle. We shall use the glass electrode for pH measurements to illustrate these differences.

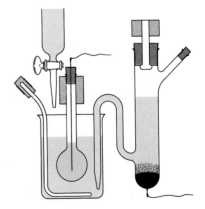

Figure 18-8
The first practical glass electrode of Haber and Klemensiewicz, *Z. Phys. Chem.*, **1909**, *65*, 385.

---

[2] Some suggested sources for additional information on this topic are *Ion-Selective Electrodes in Analytical Chemistry*, H. Freiser, Ed. New York: Plenum Press, 1978; J. Vesely, D. Weiss, and K. Stulik, *Analysis with Ion-Selective Electrodes*. New York: Wiley, 1979; J. Koryta, *Ions, Electrodes, and Membranes*, 2nd ed. New York: Wiley, 1991.

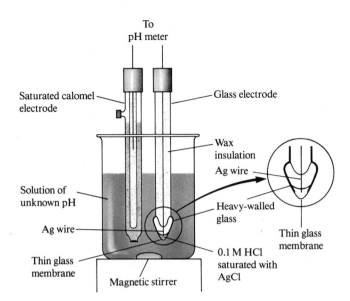

Figure 18-9
Typical electrode system for measuring pH.

The membrane of a typical glass electrode (with a thickness of 0.03 to 0.1 mm) has an electrical resistance of 50 to 500 MΩ.

The external reference electrode may be a Ag/AgCl electrode rather than the SCE shown in Figure 18-9.

18D-3 The Glass Electrode for pH Measurements

Figure 18-9 shows a typical *cell* for measuring pH. The cell consists of a glass indicator electrode and a saturated calomel reference electrode immersed in the solution whose pH is sought. The indicator electrode consists of a thin, pH-sensitive glass membrane sealed onto one end of a heavy-walled glass or plastic tube. A small volume of dilute hydrochloric acid saturated with silver chloride is contained in the tube (the inner solution in some electrodes is a buffer containing chloride ion). A silver wire in this solution forms a silver/silver chloride reference electrode, which is connected to one of the terminals of a potential-measuring device. The calomel electrode is connected to the other terminal.

Figure 18-9 and the schematic representation of this cell in Figure 18-10 show that a glass-electrode system contains *two* reference electrodes: (1) the *external* calomel electrode and (2) the *internal* silver/silver chloride electrode. While the internal reference electrode is a part of the glass electorde, *it is not the pH-sensing element*. Instead, *it is the thin glass membrane at the tip of the electrode that responds to pH*.

The Composition and Structure of Glass Membranes

Much systematic investigation has been devoted to the effects of glass composition on the sensitivity of membranes to protons and other cations, and a number

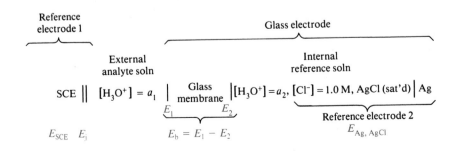

Figure 18-10
Diagram of a glass/calomel cell for measurement of pH.

of formulations are now used for the manufacture of electrodes. Corning 015 glass, which has been widely used for membranes, consists of approximately 22% Na_2O, 6% CaO, and 72% SiO_2. This membrane shows excellent specificity toward hydrogen ions up to a pH of about 9. At higher pH values, however, the glass becomes somewhat responsive to sodium, as well as to other singly charged cations. Other glass formulations are now in use in which sodium and calcium ions are replaced to various degrees by barium and lithium ions. These membranes have superior selectivity and lifetime.

As shown in Figure 18-11, a silicate glass used for membranes consists of an infinite three-dimensional network of SiO_4^{4-} groups in which each silicon is bonded to four oxygens and each oxygen is shared by two silicons. Within the interstices of this structure are sufficient cations to balance the negative charge of the silicate groups. Singly charged cations, such as sodium and lithium, are mobile in the lattice and are responsible for the electrical conduction within the membrane.

The two surfaces of a glass membrane must be hydrated before it will function as a pH electrode. Nonhygroscopic glasses show no pH function. Even hygroscopic glasses lose their pH sensitivity after dehydration by storage over a desiccant. The effect is reversible, however, and the response of a glass electrode is restored by soaking it in water.

The hydration of a pH-sensitive glass membrane involves an ion-exchange reaction between singly charged cations in the interstices of the glass lattice and protons from the solution. The process involves univalent cations exclusively because di- and trivalent cations are too strongly held within the silicate structure to exchange with ions in the solution. Typically, then, the ion-exchange reaction can be written as

$$\underset{\text{soln}}{H^+} + \underset{\text{glass}}{Na^+Gl^-} \rightleftharpoons \underset{\text{soln}}{Na^+} + \underset{\text{glass}}{H^+Gl^-} \qquad \textbf{(18-6)}$$

Oxygen atoms attached to only one silicon atom are negatively charged Gl^- sites. The equilibrium constant for this process is so large that the surfaces of a hydrated glass membrane ordinarily consist entirely of silicic acid (H^+Gl^-). An exception

> Glasses that absorb water are said to be *hygroscopic*.

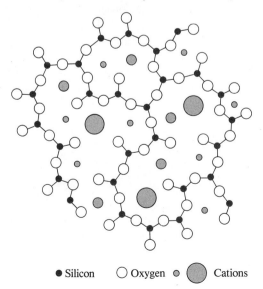

● Silicon   ○ Oxygen   ○ ◯ Cations

Figure 18-11

Cross-sectional view of a silicate glass structure. In addition to the three Si—O bonds shown, each silicon is bonded to an additional oxygen atom, either above or below the plane of the paper. (Adapted with permission from G. A. Perley, *Anal. Chem.*, **1949**, *21*, 395. Copyright 1949 American Chemical Society.)

to this situation exists in highly alkaline media, where the hydrogen ion concentration is extremely small and the sodium ion concentration is large; here, a significant fraction of the sites are occupied by sodium ions.

Membrane Potentials

The lower part of Figure 18-10 shows four potentials that develop in a cell when pH is being determined with a glass electrode. Two of these $E_{Ag,AgCl}$ and E_{SCE}, are reference-electrode potentials that remain constant. There is a third potential across the salt bridge that separates the calomel electrode from the analyte solution. This junction and its associated *junction potential*, E_j, are found in all cells used for the potentiometric measurement of ion concentration. The fourth and most important potential shown in Figure 18-10 is the *boundary potential*, E_b, *which varies with the pH of the analyte solution.* The two reference electrodes simply provide electrical contacts with the solutions so that changes in the boundary potential can be measured.

The Boundary Potential

In Figure 18-10, the boundary potential is shown as being made up of two potentials, E_1 and E_2, which develop at the two surfaces of the glass membrane. The source of these two potentials is the charge that develops as a consequence of the reactions

$$\underset{\text{glass}_1}{H^+Gl^-(s)} \rightleftharpoons \underset{\text{soln}_1}{H^+(aq)} + \underset{\text{glass}_1}{Gl^-(s)} \tag{18-7}$$

$$\underset{\text{glass}_2}{H^+Gl^-(s)} \rightleftharpoons \underset{\text{soln}_2}{H^+(aq)} + \underset{\text{glass}_2}{Gl^-(s)} \tag{18-8}$$

where subscript 1 refers to the interface between the exterior of the glass and the analyte solution and subscript 2 refers to the interface between the internal solution and the interior of the glass. These two reactions cause the two glass surfaces to be negatively charged with respect to the solutions in which they are immersed, thus giving rise to the two potentials E_1 and E_2 shown in Figure 18-10. The positions of the two equilibria that cause the two potentials to develop are determined by the hydrogen ion concentrations in the solutions on the two sides of the membrane. Where these positions differ, the surface at which the greater dissociation has occurred is negative with respect to the other surface. This difference in charge is the boundary potential E_b, which is related to the activities of hydrogen ion in each of the solutions by the Nernst-like equation

$$E_b = E_1 - E_2 = 0.0592 \log \frac{a_1}{a_2} \tag{18-9}$$

where a_1 is the activity of the analyte solution and a_2 is that of the internal solution. For a glass pH electrode, the hydrogen ion activity of the internal solution is held constant so that Equation 18-9 simplifies to

$$E_b = L' + 0.0592 \log a_1 = L' - 0.0592 \text{ pH} \tag{18-10}$$

where

$$L' = -0.0592 \log a_2$$

The boundary potential is then a measure of the hydrogen ion activity of the external solution.

The Potential of the Glass Electrode

The potential of a glass indicator electrode E_{ind} has three components: (1) the boundary potential, given by Equation 18-9, (2) the potential of the internal Ag/AgCl reference electrode, and (3) a small asymmetry potential, E_{asy}, which changes slowly with time and exists because the two surfaces of the membrane are not exactly alike. In equation form, we may write

$$E_{ind} = E_b + E_{Ag/AgCl} + E_{asy}$$

Substitution of Equation 18-10 for E_b gives

$$E_{ind} = L' + 0.0592 \log a_1 + E_{Ag/AgCl} + E_{asy}$$

or

$$E_{ind} = L + 0.0592 \log a_1 = L - 0.0592 \text{ pH} \qquad \text{(18-11)}$$

where L is a combination of the three constant terms. Compare Equations 18-11 and 18-3. Although these two equations are similar in form, remember that the sources of the potential of the electrodes that they describe *are totally different*.

The Alkaline Error

Glass electrodes respond to the concentration of both hydrogen ion and alkali metal ions in basic solution. The magnitude of this *alkaline error* for four different glass membranes is shown in Figure 18-12 (curves C to F). These curves refer to solutions in which the sodium ion concentration was held constant at 1 M while the pH was varied. Note that the error is negative (that is, the measured pH values are lower than the true values), which suggests that the electrode is responding to sodium ions as well as to protons. This observation is confirmed by data obtained for solutions containing different sodium ion concentrations. Thus at pH 12, the electrode with a Corning 015 membrane (curve C in Figure 18-12) registered a pH of 11.3 when immersed in a solution having a sodium ion concentration of 1 M but 11.7 in a solution that was 0.1 M in this ion. All singly charged cations induce an alkaline error whose magnitude depends on both the cation in question and the composition of the glass membrane.

The alkaline error can be satisfactorily explained by assuming an exchange equilibrium between the hydrogen ions on the glass surface and the cations in solution. This process is simply the reverse of that shown in Equation 18-6:

$$\underset{\text{glass}}{H^+Gl^-} + \underset{\text{soln}}{B^+} \rightleftharpoons \underset{\text{glass}}{B^+Gl^-} + \underset{\text{soln}}{H^+}$$

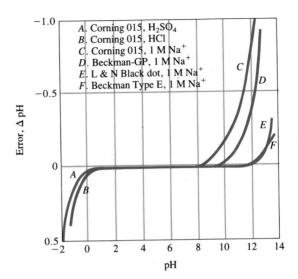

Figure 18-12
Acid and alkaline error for selected glass electrodes at 25°C. (From R. G. Bates, *Determination of pH*, 2nd ed., p. 365. New York: Wiley, 1973. With permission.)

where B^+ represents some singly charged cation, such as sodium ion. In alkaline solutions, the activity of the sodium ions relative to that of the hydrogen ions becomes so large that the electrode responds to both species.

The Acid Error

As shown in Figure 18-12, the typical glass electrode exhibits an error, opposite in sign to the alkaline error, in solution of pH less than about 0.5; pH readings tend to be too high in this region. The magnitude of the error depends on a variety of factors and is generally not very reproducible. The causes of the acid error are not well understood.

18D-4 Glass Electrodes for Cations Other than Protons

The alkaline error in early glass electrodes led to investigations concerning the effect of glass composition on the magnitude of this error. One consequence has been the development of glasses for which the alkaline error is negligible below about pH 12 (see curves *E* and *F*, Figure 18-12). Other studies have led to the discovery of glass compositions that permit the determination of cations other than hydrogen. Incorporation of Al_2O_3 or B_2O_3 in the glass has the desired effect. Glass electrodes that permit the direct potentiometric measurement of such singly charged species as Na^+, K^+, NH_4^+, Rb^+, Cs^+, Li^+, and Ag^+ have been developed. Some of these glasses are reasonably selective toward particular singly charged cations. Glass electrodes for Na^+, Li^+, NH_4^+, and total concentration of univalent cations are now available from commercial sources.

18D-5 Liquid-Membrane Electrodes

A liquid-membrane electrode develops a potential across the interface between the solution containing the analyte and a liquid-ion exchanger that selectively bonds with the analyte ion. These electrodes have been developed for the direct potentiometric measurement of numerous polyvalent cations as well as certain anions.

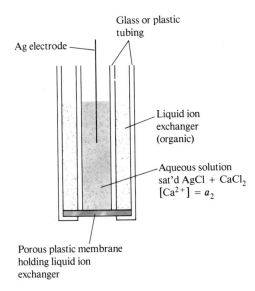

Figure 18-13

Diagram of a liquid-membrane electrode for Ca^{2+}.

Figure 18-13 is a schematic of a liquid-membrane electrode for calcium. It consists of a conducting membrane that selectively bonds calcium ions, an internal solution containing a fixed concentration of calcium chloride, and a silver electrode that is coated with silver chloride to form an internal reference electrode. Notice the similarities between the liquid membrane electrode and the glass electrode as shown in Figure 18-14. The active membrane ingredient is an ion exchanger that consists of a calcium dialkyl phosphate that is nearly insoluble in water. In the electrode shown in Figures 18-13 and 18-14, the ion exchanger is dissolved in an immiscible organic liquid that is forced by gravity into the pores of a hydrophobic porous disk. This disk then serves as the membrane that separates the internal solution from the analyte solution. In a more recent design, the ion exchanger is immobilized in a tough polyvinyl chloride gel cemented to the end of a tube that holds the internal solution and reference electrode. In either design, a dissociation equilibrium develops at each membrane interface that is analogous to Equations 18-7 and 18-8:

> Hydrophobic means water hating. The hydrophobic disk is porous toward organic liquids but repels water.

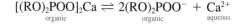

$$\underset{\text{organic}}{[(RO)_2POO]_2Ca} \rightleftharpoons \underset{\text{organic}}{2(RO)_2POO^-} + \underset{\text{aqueous}}{Ca^{2+}}$$

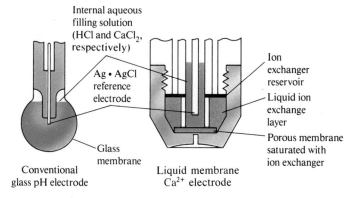

Figure 18-14

Comparison of a liquid-membrane calcium ion electrode with a glass electrode. (Courtesy of Orion Research, Boston, MA.)

where R is a high-molecular-weight aliphatic group. As with the glass electrode, a potential develops across the membrane when the extent of dissociation of the ion exchanger at one surface differs from that at the other surface. This potential is a result of differences in the calcium ion activity of the internal and external solutions. The relationship between the membrane potential and the calcium ion activities is given by an equation that is similar to Equation 18-9:

$$E_b = E_1 - E_2 = \frac{0.0592}{2} \log \frac{a_1}{a_2} \tag{18-12}$$

where a_1 and a_2 are the activities of calcium ion in the external and internal solutions, respectively. Since the calcium ion activity of the internal solution is constant,

$$E_b = N + \frac{0.0592}{2} \log a_1 = N - \frac{0.0592}{2} \text{pCa} \tag{18-13}$$

where N is a constant (compare Equations 18-13 and 18-10). Note that a 2 appears in the denominator of the coefficient of the log term because calcium is divalent.

Figure 18-14 compares the structural features of a glass-membrane electrode and a commercially available liquid-membrane electrode for calcium ion. The sensitivity of the liquid-membrane electrode for calcium ion is reported to be 50 times greater than that for magnesium ion and 1000 times greater than that for sodium or potassium ions. Calcium ion activities as low as 5×10^{-7} M can be measured. Performance of the electrode is independent of pH in the range between 5.5 and 11. At lower pH levels, hydrogen ions undoubtedly replace some of the calcium ions on the exchanger; the electrode then becomes sensitive to pH as well as to pCa.

The calcium ion liquid-membrane electrode is a valuable tool for physiological investigations because this ion plays important roles in such processes as nerve conduction, bone formation, muscle contraction, cardiac expansion and contraction, renal tubular function, and perhaps hypertension. Most of these processes are more influenced by the *activity* than the concentration of the calcium ion; activity, of course, is the parameter measured by the membrane electrode. Thus the calcium ion electrode (and the potassium ion electrode and others) is an important tool for the study of physiological processes.

A liquid-membrane electrode specific for potassium ion is also of great value for physiologists because the transport of neural signals appears to involve movement of this ion across nerve membranes. Investigation of this process requires an electrode that can detect small concentrations of potassium ion in media that contain much larger concentrations of sodium ion. Several liquid-membrane electrodes show promise in meeting this requirement. One is based on the antibiotic valinomycin, a cyclic ether that has a strong affinity for potassium ion. Of equal importance is the observation that a liquid membrane consisting of valinomycin in diphenyl ether is about 10^4 times as responsive to potassium ion as to sodium ion.[3] Figure 18-15 is a photomicrograph of a tiny electrode used for determining the potassium content of a single cell.

Ion-selective microelectrodes can be used to make measurements of ion activities within a living organism.

[3] M. S. Frant and J. W. Ross Jr., *Science*, **1970**, *167*, 987.

Table 18-1 lists some liquid-membrane electrodes available from commercial sources. The anion-sensitive electrodes shown make use of a solution containing an anion-exchange resin in an organic solvent. Liquid-membrane electrodes in which the exchange liquid is held in a polyvinyl chloride gel have been developed for Ca^{2+}, K^+, NO_3^-, and BF_4^-. These have the appearance of crystalline electrodes, which are considered in the following section.

18D-6 Crystalline-Membrane Electrodes

Considerable work has been devoted to the development of solid membranes that are selective toward anions in the same way that some glasses respond to cations. We have seen that anionic sites on a glass surface account for the selectivity of a membrane toward certain cations. By analogy, a membrane with cationic sites might be expected to respond selectively toward anions.

Membranes prepared from cast pellets of silver halides have been used successfully in electrodes for the selective determination of chloride, bromide, and iodide ions. In addition, an electrode based on a polycrystalline Ag_2S membrane is offered by one manufacturer for the determination of sulfide ion. In both types of membranes, silver ions are sufficiently mobile to conduct electricity through the solid medium. Mixtures of PbS, CdS, and CuS with Ag_2S provide membranes that are selective for Pb^{2+}, Cd^{2+}, and Cu^{2+}, respectively. Silver ion must be present in these membranes to conduct electricity because divalent ions are immobile in crystals. The potential that develops across crystalline solid-state electrodes is described by a relationship similar to Equation 18-12.

A crystalline electrode for fluoride ion is available from commercial sources. The membrane consists of a slice of a single crystal of lanthanum fluoride that has been doped with europium(II) fluoride to improve its conductivity. The membrane, supported between a reference solution and the solution to be measured, shows a theoretical response to changes in fluoride ion activity from 10^0 to 10^{-6} M. The electrode is selective for fluoride ion over other common anions by several orders of magnitude; only hydroxide ion appears to offer serious interference. Some solid-state electrodes available from commercial sources are listed in Table 18-2.

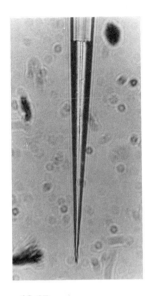

Figure 18-15

Photograph of a potassium liquid-ion exchanger microelectrode with 125 μm of ion exchanger inside the tip. The magnification of the original photo was 400×. (From J. L. Walker, *Anal. Chem.*, **1971**, *43(3)N*, 91A. Reproduced by permission of the American Chemical Society.)

Table 18-1
CHARACTERISTICS OF LIQUID-MEMBRANE ELECTRODES*

Analyte Ion	Concentration Range, M	Major Interferences
Ca^{2+}	10^0 to 5×10^{-7}	Pb^{2+}, Fe^{2+}, Ni^{2+}, Hg^{2+}, Sr^{2+}
Cl^-	10^0 to 5×10^{-6}	I^-, OH^-, SO_4^{2-}
NO_3^-	10^0 to 7×10^{-6}	ClO_4^-, I^-, ClO_3^-, CN^-, Br^-
ClO_4^-	10^0 to 7×10^{-6}	I^-, ClO_3^-, CN^-, Br^-
K^+	10^0 to 1×10^{-6}	Cs^+, NH_4^+, Tl^+
Water hardness (Ca^{2+} + Mg^{2+})	10^0 to 6×10^{-6}	Cu^{2+}, Zn^{2+}, Ni^{2+}, Sr^{2+}, Fe^{2+}, Ba^{2+}

* From *Orion Guide to Ion Analysis*. Boston, MA: Orion Research, 1992.

Table 18-2
CHARACTERISTICS OF SOLID-STATE CRYSTALLINE ELECTRODES*

Analyte Ion	Concentration Range, M	Major Interferences
Br^-	Ag^+: 10^0 to 5×10^{-6}	CN^-, I^-, S^{2-}
Cd^{2+}	Ag^+: 10^{-1} to 1×10^{-7}	Fe^{2+}, Pb^{2+}, Hg^{2+}, Ag^+, Cu^{2+}
Cl^-	Ag^+: 10^0 to 5×10^{-5}	CN^-, I^-, Br^-, S^{2-}
Cu^{2+}	Ag^+: 10^{-1} to 1×10^{-8}	Hg^{2+}, Ag^+, Cd^{2+}
CN^-	Ag^+: 10^{-2} to 1×10^{-6}	S^{2-}
F^-	Ag^+: Sat'd to 1×10^{-6}	OH^-
I^-	Ag^+: 10^0 to 5×10^{-8}	
Pb^{2+}	Ag^+: 10^{-1} to 1×10^{-6}	Hg^{2+}, Ag^+, Cu^{2+}
Ag^+/S^{2-}	Ag^+: 10^0 to 1×10^{-7}	Hg^{2+}
	S^{2-}: 10^0 to 1×10^{-7}	
SCN^-	Ag^+: 10^0 to 5×10^{-6}	I^-, Br^-, CN^-, S^{2-}

* From *Orion Guide to Ion Analysis*. Boston, MA: Orion Research, 1992.

18D-7 Gas-Sensing Probes

Figure 18-16 illustrates the essential features of a potentiometric gas-sensing probe, which consists of a tube containing a reference electrode, a specific ion electrode, and an electrolyte solution. A thin, replaceable, gas-permeable membrane attached to one end of the tube serves as a barrier between the internal and analyte solutions. As can be seen from Figure 18-16, this device is a complete electrochemical cell and is more properly referred to as a probe rather than an electrode. Gas-sensing probes have found widespread use for determining dissolved gases in water and other solvents.

> A gas-sensing probe is a galvanic *cell* whose potential is related to the concentration of a gas in a solution.

Membrane Composition

A *microporous membrane* is fabricated from a hydrophobic polymer. As the name implies, the membrane is highly porous (the average pore size is less than

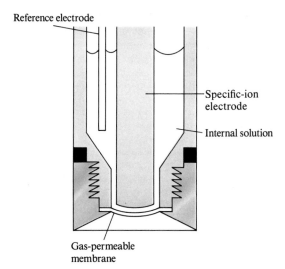

Figure 18-16
Diagram of a gas-sensing probe.

1 μm) and allows the free passage of gases; at the same time, the water-repellent polymer prevents water and solute ions from entering the pores. The thickness of the membrane is about 0.1 mm.

The Mechanism of Response

Using carbon dioxide as an example, we can represent the transfer of gas to the internal solution by the following set of equations:

$$CO_2(aq) \rightleftharpoons CO_2(g)$$

analyte   membrane
solution   pores

$$CO_2(g) \rightleftharpoons CO_2(aq)$$

membrane   internal
pores   solution

$$CO_2(aq) + 2H_2O \rightleftharpoons HCO_3^- + H_3O^+$$

internal   internal
solution   solution

The last equilibrium causes the pH of the internal surface film to change. This change is then detected by the internal glass/calomel electrode system. A description of the overall process is obtained by adding the equations for the three equilibria to give

$$CO_2(aq) + 2H_2O \rightleftharpoons H_3O^+ + HCO_3^-$$

analyte   internal
solution   solution

It can be shown that the potential of the internal cell is given by

$$E_{cell} = L + 0.0592 \log[CO_2(aq)]_{ext} \qquad \textbf{(18-14)}$$

where L is a constant. Thus, the potential between the glass electrode and the reference electrode in the internal solution is determined by the CO_2 concentration in the external solution. *Note that no electrode comes in direct contact with the analyte solution.* Therefore, these devices are gas-sensing *cells*, or *probes*, rather than gas-sensing electrodes. Nevertheless, they continue to be called electrodes in some literature and many advertising brochures.

The only species that interfere are other dissolved gases that permeate the membrane and then affect the pH of the internal solution. The specificity of gas probes depends only on the permeability of the gas membrane. Gas-sensing probes for CO_2, NO_2, H_2S, SO_2, HF, HCN, and NH_3 are now available from commercial sources.

Although often referred to as gas-sensing electrodes, these devices are complete electrochemical cells and should be called gas-sensing probes.

18E   INSTRUMENTS FOR THE MEASUREMENT OF CELL POTENTIALS

Most cells containing a membrane electrode have very high electrical resistance (as much as 10^8 ohms or more). In order to measure potentials of such high-resistance circuits accurately, it is necessary that the voltmeter have an electrical resistance that is several orders of magnitude greater than the resistance of the cell being measured. If the meter resistance is too low, current is drawn from the cell, which has the effect of lowering its output potential, thus creating a

Feature 18-1

POINT-OF-CARE TESTING: BLOOD GASES AND BLOOD ELECTROLYTES WITH PORTABLE INSTRUMENTATION

Modern medicine relies heavily on analytical measurements for diagnosis and treatment in emergency rooms, operating rooms, and intensive care units. Prompt reporting of blood gas and blood electrolyte concentrations is especially important to physicians in these areas. There is often insufficient time to transport blood samples to the clinical laboratory, perform required analyses, and transmit the results back to the bedside. In this feature, we describe an automated instrument, designed specifically to be located near the patient. This instrument determines oxygen, carbon dioxide, sodium, potassium, and calcium concentrations as well as pH and hematocrit in whole blood.[1] The analytes, measurement ranges, and measurement resolutions of the instrument are presented in the following table.

Analyte	Range	Resolution
p_{O_2}	0 to 760 mm Hg	1 mm Hg
p_{CO_2}	5 to 99 mm Hg	1 mm Hg
Na^+	100 to 200 mM	0.1 mM
K^+	0.1 to 9.9 mM	0.01 mM
Ca^{2+}	0.1 to 5.0 mM	0.01 mM
pH	6.80 to 7.80	0.01
Hematocrit	15 to 52%	1%

Most of the analytes (p_{CO_2}, Na^+, K^+, Ca^{2+}, and pH) are determined by potentiometric measurements using membrane-based ion-selective electrode technology; the hematocrit is measured by electrolytic conductivity detection and p_{O_2} is determined with a Clark voltammetric sensor.[2]

The central component of the blood analyzer is the miniature sensor array depicted in Figure 18-A. The individual sensors are located along a narrow flow channel that is connected to the sample/calibration inlet, a reference solution inlet to provide a fresh contact between the sensors and each sample, and a waste output port. The sensor array is incorporated into a disposable cartridge whose internal components are shown in Figure 18-B. In addition to the sensor array, the cartridge contains two standard solutions for calibration of the sensors, the reference solution, a waste reservoir, a distribution valve for routing solutions to the sensors, and a peristaltic pump to move the solutions.

The instrument is calibrated when the distribution valve shown in Figure 18-B is switched to one of two positions corresponding to the two calibration solutions. The peristaltic pump is actuated to draw fresh reference solution and then the calibration solution into the sensor chamber. Following a brief

Hematocrit is the ratio of the volume of red blood cells to the total volume of a blood sample expressed as a percent.

[1] M. E. Meyerhoff, *Clin. Chem.*, **1990**, *36*(8), 1567.
[2] D. A. Skoog, D. M. West, and F. J. Holler, *Fundamentals of Analytical Chemistry*, 6th ed., pp. 487–488. Philadelphia: Saunders College Publishing, 1992.

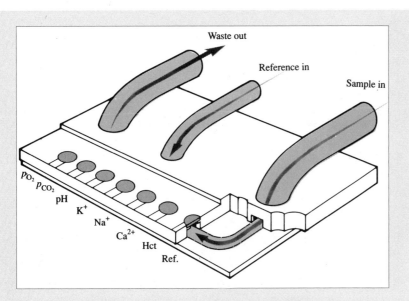

Figure 18-A   Miniature sensor array. (Courtesy of Mallincrodt Sensor Systems, Inc., Ann Arbor, MI.)

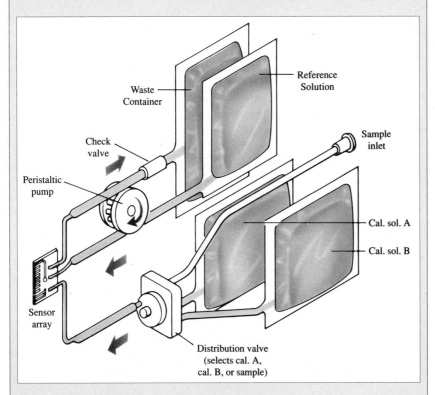

Figure 18-B   Disposable cartridge for blood gas monitor. (Courtesy of Mallincrodt Sensor Systems, Inc., Ann Arbor, MI.)

equilibration period, the sensors are read by the instrument console and stored in the memory of its microcomputer. For measurements on patient samples, 200 μL of whole blood is injected into the sample inlet, the distribution valve switches to the sample line, the sample is pumped into the sensor chamber, and the sensors are read as before. The results of the analysis are then printed and/or stored on a floppy disk for later retrieval. All instrument functions are under the control of the microcomputer.

The analysis time from the time of injection of the sample is 92 seconds. Calibrations are carried out automatically on both calibration solutions every hour, and single-solution calibrations are run every two minutes. The calibration solutions are sealed in gas-impermeable collapsible plastic bags so that there is no contamination from atmospheric gases. Three hundred analyses can be made with each cartridge over a period of seven days. Accuracy and precision of the instrument on whole blood samples are comparable to those of dedicated instruments in the clinical laboratory.

This feature shows how modern ion-selective electrode technology coupled with microcomputer control of the measurement process and data reporting can be used to provide essential measurements of analyte concentrations in whole blood. The measurements are rapid, accurate, and precise, and they are available to the physician at the bedside of the patient.

negative error. This effect is shown in Figure 18-17, which is a plot of the relative error in potential reading as a function of the ratio of the resistance of a meter to the resistance of a cell. When the meter and the cell have the same resistance a relative error of about -50% results. When this ratio is 10, the error is about -9%. When it is 1000 the error is less than 0.1% relative.

Numerous high-resistance, direct-reading digital voltmeters with internal resistances of 10^{11} to 10^{12} ohms are now on the market. These meters are commonly called *pH meters* but could more properly be referred to as *pIon meters* or *ion meters* since they are frequently used for the measurement of concentrations of other ions as well.

The readout of ion meters is either digital or analog (in the latter, a needle sweeps a range on a scale from 0 to 14 pH units). Some meters are capable of precision on the order of 0.001 to 0.005 pH unit. Seldom is it possible to measure pH with a comparable degree of *accuracy*. Inaccuracies of ±0.02 to ±0.03 pH unit are typical.

18F   DIRECT POTENTIOMETRIC MEASUREMENTS

Direct potentiometric measurements provide a rapid and convenient method for the determination of the activity of a variety of cations and anions. Here, the potential of the indicator electrode in contact with the analyte solution is compared with that developed when the electrode is immersed in one or more solutions of known analyte concentration. If the response of the electrode is specific for the analyte, as it often is, no preliminary separation steps are required. Direct potentiometric measurements are also readily adapted to applications requiring continuous and automatic recording of analytical data.

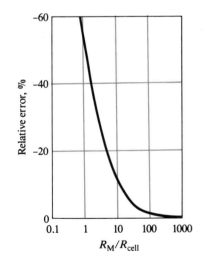

Figure 18-17

Relative error in a cell-potential measurement as a function of the ratio of the electrical resistance of the meter R_M to the resistance of the cell R_{cell}.

18F-1 The Sign Convention and Equations for Direct Potentiometry

The sign convention for potentiometry is consistent with the convention described in Chapter 15 for standard electrode potential.[4] In this convention, the indicator electrode is *always* treated as the *cathode* and the reference electrode as the *anode*.[5] For direct potentiometric measurements, the potential of a cell can then be expressed in terms of the potentials developed by the indicator electrode, the reference electrode, and a junction potential:

$$E_{cell} = E_{ind} - E_{ref} + E_j \tag{18-15}$$

In Section 18D we have described the response of various types of indicator electrodes to analyte activities. For the cation X^{n+} at 25°C, the electrode response takes the general *Nernstian* form

$$E_{ind} = L - \frac{0.0592}{n}pX = L + \frac{0.0592}{n}\log a_X \tag{18-16}$$

where L is a constant and a_X is the activity of the cation. For metallic indicator electrodes, L is ordinarily the standard electrode potential; for membrane electrodes, L is the summation of several constants, including the time-dependent asymmetry potential of uncertain magnitude.

Substitution of Equation 18-16 into Equation 18-15 yields with rearrangement

$$pX = -\log a_X = -\frac{E_{cell} - (E_j - E_{ref} + L)}{0.0592/n} \tag{18-17}$$

The constant terms in parentheses can be combined to give a new constant K.

$$pX = -\log a_X = -\frac{E_{cell} - K}{0.0592/n} = -\frac{n(E_{cell} - K)}{0.0592} \tag{18-18}$$

For an anion A^{n-}, the sign of Equation 18-17 is reversed:

$$pA = \frac{E_{cell} - K}{0.0592/n} = \frac{n(E_{cell} - K)}{0.0592} \tag{18-19}$$

All direct potentiometric methods are based on Equation 18-18 or 18-19. The difference in sign in the two equations has a subtle but important consequence

---

[4] According to Bates, the convention being described here has been endorsed by standardizing groups in the United States and Great Britain as well as IUPAC. See R. G. Bates, in *Treatise on Analytical Chemistry*, 2nd ed., I. M. Kolthoff and P. J. Elving, Eds., Part I, Vol. 1, pp. 831–832. New York: Wiley, 1978.

[5] In effect, the sign convention for electrode potentials described in Section 15C-3 also designates the indicator electrode as the cathode by stipulating that half-reactions always be written as reductions; the standard hydrogen electrode, which is the reference electrode in this case, is then the anode.

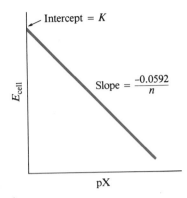

Figure 18-18

A plot of Equation 18-20 for cationic electrodes.

in the way that ion-selective electrodes are connected to pH meters and pIon meters. When the two equations are solved for E_{cell}, we find that for cations

$$E_{cell} = K - \frac{0.0592}{n}\,pX \qquad \textbf{(18-20)}$$

and for anions

$$E_{cell} = K + \frac{0.0592}{n}\,pA \qquad \textbf{(18-21)}$$

Equation 18-20 shows that for a cation-selective electrode, an increase in pX results in a *decrease* in E_{cell} (Figure 18-18). Thus, when a high-resistance voltmeter is connected to the cell in the usual way, with the indicator electrode attached to the positive terminal, the meter reading decreases as pX increases. To eliminate this problem, instrument manufacturers generally reverse the leads so that cation-sensitive electrodes are connected to the *negative* terminal of the voltage measuring device. Meter readings then increase with increases of pX. Anion-selective electrodes, on the other hand, are connected to the *positive* terminal of the meter so that increases in pA also yield larger readings (Figure 18-19).

18F-2   The Electrode-Calibration Method

As we have seen from our discussions in Section 18D, the constant K in Equations 18-18 and 18-19 is made up of several constants, at least one of which, the junction potential, cannot be computed from theory nor measured directly. Thus, before these equations can be used for the determination of pX or pA, K must be evaluated *experimentally* with a standard solution of the analyte.

In the electrode-calibration method, K in Equations 18-18 and 18-19 is determined by measuring E_{cell} for one or more standard solutions of known pX or pA. The assumption is then made that K does not change when the standard is replaced by the analyte solution. The calibration is ordinarily performed at the time pX or pA for the unknown is determined. Recalibration of membrane electrodes may be required if measurements extend over several hours because of slow changes in the asymmetry potential.

The electrode-calibration method offers the advantages of simplicity, speed, and applicability to the continuous monitoring of pX or pA. It suffers, however, from a somewhat limited accuracy because of uncertainties in junction potentials.

Inherent Error in the Electrode-Calibration Procedure

A serious disadvantage of the electrode-calibration method is the inherent error that results from the assumption that K in Equations 18-18 and 18-19 remains constant after calibration. This assumption can seldom, if ever, be exactly true because the electrolyte composition of the unknown almost inevitably differs from that of the solution employed for calibration. The junction potential term contained in K varies slightly as a consequence, even when a salt bridge is used. This error is frequently on the order of 1 mV or more. Unfortunately, because of the nature of the potential/activity relationship, such an uncertainty has an

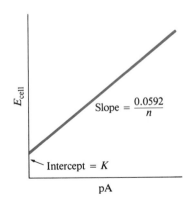

Figure 18-19

A plot of Equation 18-21 for anionic electrodes.

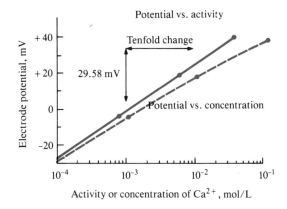

Figure 18-20
Response of a liquid-membrane electrode to variations in the concentration and activity of calcium ion. (Courtesy of Orion Research, Boston, MA.)

amplified effect on the inherent accuracy of the analysis. The relative error in analyte concentration can be estimated from the following equation:[6]

$$\text{percent relative error} = \frac{\Delta a_1}{a_1} \times 100\% = 3.89 \times 10^3 n\, \Delta K\% \approx 4000\, n\, \Delta K\%$$

The quantity $\Delta a_1/a_1$ is the relative error in a_1 associated with an absolute uncertainty ΔK in K. If, for example, ΔK is ± 0.001 V, a relative error in activity of about $\pm 4n\%$ can be expected. *It is important to appreciate that this error is characteristic of all measurements involving cells that contain a salt bridge and that this error cannot be eliminated by even the most careful measurements of cell potentials or the most sensitive and precise measuring devices.*

Activity Versus Concentration

Electrode response is related to analyte activity rather than analyte concentration. We are usually interested in concentration, however, and the determination of this quantity from a potentiometric measurement requires activity coefficient data. Activity coefficients are seldom available because the ionic strength of the solution is either unknown or else is so large that the Debye-Hückel equation is not applicable.

The difference between activity and concentration is illustrated by Figure 18-20, in which the response of a calcium ion electrode is plotted against a logarithmic function of calcium chloride *concentration*. The nonlinearity is due to the increase in ionic strength—and the consequent decrease in the activity of calcium ion—with increasing electrolyte concentration. The upper curve is obtained when these concentrations are converted to activities. This straight line has the theoretical slope of $0.0592/2 = 0.0296$.

Activity coefficients for singly charged species are less affected by changes in ionic strength than are the coefficients for ions with multiple charges. Thus, the effect shown in Figure 18-20 is less pronounced for electrodes that respond to H^+, Na^+, and other univalent ions.

---

[6] For a derivation of this equation see D. A. Skoog, D. M. West, and F. J. Holler, *Fundamentals of Analytical Chemistry*, 6th ed., p. 426. Philadelphia: Saunders College Publishing, 1992.

In potentiometric pH measurements, the pH of the standard buffer used for calibration is generally based on the activity of hydrogen ions. Thus, the results are also on an activity scale. If the unknown sample has a high ionic strength, the hydrogen ion *concentration* will differ appreciably from the activity measured.

An obvious way to convert potentiometric measurements from activity to concentration is to make use of an empirical calibration curve, such as the lower plot in Figure 18-20. For this approach to be successful, it is necessary to make the ionic composition of the standards essentially the same as that of the analyte solution. Matching the ionic strength of standards to that of samples is often difficult, particularly for samples that are chemically complex.

Where electrolyte concentrations are not too great, it is often useful to swamp both samples and standards with a measured excess of an inert electrolyte. The added effect of the electrolyte from the sample matrix becomes negligible under these circumstances, and the empirical calibration curve yields results in terms of concentration. This approach has been used, for example, in the potentiometric determination of fluoride ion in drinking water. Both samples and standards are diluted with a solution that contains sodium chloride, an acetate buffer, and a citrate buffer; the diluent is sufficiently concentrated so that the samples and standards have essentially identical ionic strengths. This method provides a rapid means for measuring fluoride concentrations in the parts-per-million range with an accuracy of about 5% relative.

18F-3 The Standard-Addition Method

The standard-addition method involves determining the potential of the electrode system before and after a measured volume of a standard has been added to a known volume of the analyte solution. Often an excess of an electrolyte is incorporated into the analyte solution at the outset to prevent any major shift in ionic strength that might accompany the addition of standard. As always, it is necessary to assume that the junction potential remains constant during the two measurements.

Example 18-1

A cell consisting of a saturated calomel electrode and a lead ion electrode developed a potential of -0.4706 V when immersed in 50.00 mL of a sample. A 5.00-mL addition of standard 0.02000 M lead solution caused the potential to shift to -0.4490 V. Calculate the molar concentration of lead in the sample.

We shall assume that the activity of Pb^{2+} is approximately equal to $[Pb^{2+}]$ and apply Equation 18-18. Thus,

$$pPb = -\log[Pb^{2+}] = -\frac{E'_{cell} - K}{0.0592/2}$$

where E'_{cell} is the initial measured potential (-0.4706 V).

After the standard solution is added, the potential becomes E''_{cell} (-0.4490 V), and

$$-\log \frac{50.00 \times [Pb^{2+}] + 5.00 \times 0.0200}{50.00 + 5.00} = -\frac{E''_{cell} - K}{0.0592/2}$$

$$-\log(0.9091[Pb^{2+}] + 1.818 \times 10^{-3}) = -\frac{E''_{cell} - K}{0.0592/2}$$

Subtracting this equation from the first leads to

$$-\log \frac{[Pb^{2+}]}{0.9091[Pb^{2+}] + 1.818 \times 10^{-3}} = \frac{2(E''_{cell} - E'_{cell})}{0.0592}$$

$$= \frac{2[-0.4490 - (-0.4706)]}{0.0592} = 0.7297$$

$$\frac{[Pb^{2+}]}{0.9091[Pb^{2+}] + 1.818 \times 10^{-3}} = \text{antilog}\,(-0.7297) = 0.1863$$

$$[Pb^{2+}] = 4.08 \times 10^{-4}\,M$$

18F-4 Potentiometric pH Measurements with a Glass Electrode[7]

The glass electrode is unquestionably the most important indicator electrode for hydrogen ion.[8] It is convenient to use and subject to few of the interferences that affect other pH-sensing electrodes.

The glass/calomel electrode system is a remarkably versatile tool for the measurement of pH under many conditions. It can be used without interference in solutions containing strong oxidants, strong reductants, proteins, and gases; the pH of viscous or even semisolid fluids can be determined. Electrodes for special applications are available. Included among these are small electrodes for pH measurements in one drop (or less) of solution, in a tooth cavity, or in the sweat on the skin; microelectrodes that permit the measurement of pH inside a living cell; rugged electrodes for insertion in a flowing liquid stream to provide a continuous monitoring of pH; and small electrodes that can be swallowed to measure the acidity of the stomach contents (the calomel electrode is kept in the mouth).

Errors That Affect pH Measurements with the Glass Electrode

The ubiquity of the pH meter and the general applicability of the glass electrode tend to lull the chemist into the attitude that any measurement obtained with such equipment is surely correct. The reader must be alert to the fact that there

[7] For a detailed discussion of potentiometric pH measurements, see R. G. Bates, *Determination of pH*, 2nd ed. New York: Wiley, 1973.
[8] A new type of solid-state pH electrode has recently appeared on the market. This electrode is an ion-sensitive field-effect transistor whose output potential is described by an equation similar to Equation 18-11. The advantages of this type of electrode over glass electrodes are reported to be small size, ruggedness, rapid response, and low output impedance. For a description of transitor-type electrodes, see T. Matsuo and M. Esashi, *Sensors and Actuators*, **1981**, *1*, 77–97.

are distinct limitations to the electrode, some of which were discussed in earlier sections.

1. *The alkaline error*. The ordinary glass electrode becomes somewhat sensitive to alkali metal ions and gives low readings at pH values greater than 9.
2. *The acid error*. Values registered by the glass electrode tend to be somewhat high when the pH is less than about 0.5.
3. *Dehydration*. Dehydration may cause erratic electrode performance.
4. *Errors in low ionic strength solutions*. It has been found that significant errors (as much as 1 or 2 pH units) may occur when the pH of samples of low ionic strength, such as lake or stream water, is measured with a glass/calomel electrode system.[9] The prime source of such errors has been shown to be nonreproducible junction potentials, which apparently result from partial clogging of the fritted plug or porous fiber that is used to restrict the flow of liquid from the salt bridge into the analyte solution. In order to overcome this problem, free diffusion junctions (FDJ) of various types have been designed, and one is produced commercially.
5. *Variation in junction potential*. A fundamental source of uncertainty for which a correction cannot be applied is the variation in junction-potential resulting from differences in the composition of the standard and the unknown solution.
6. *Error in the pH of the standard buffer*. Any inaccuracies in the preparation of the buffer used for calibration or any changes in its composition during storage cause an error in subsequent pH measurements. The action of bacteria on organic buffer components is a common cause for deterioration.

The Operational Definition of pH

The utility of pH as a measure of the acidity and alkalinity of aqueous media, the wide availability of commercial glass electrodes, and the relatively recent proliferation of inexpensive solid-state pH meters have made the potentiometric measurement of pH perhaps the most common analytical technique in all of science. It is thus extremely important that pH be defined in a manner that is easily duplicated at various times and in various laboratories throughout the world. To meet this requirement, it is necessary to define pH in operational terms—that is, by the way the measurement is made. Only then will the pH measured by one worker be the same as that by another.

An operational definition of pH is recommended by the National Institute of Standards and Technology (NIST) and by the IUPAC. This definition is based on the direct calibration of a pH meter with carefully prescribed standard buffers followed by potentiometric determination of the pH of unknown solutions.

Consider, for example, the glass/calomel system in Figure 18-9. When these electrodes are immersed in a standard buffer, Equation 18-18 applies and we can write

$$pH_S = -\frac{E_S - K}{0.0592}$$

where E_S is the cell potential when the electrodes are immersed in the buffer.

> Particular care must be taken in measuring the pH of approximately neutral unbuffered solutions, such as samples from lakes and streams.

> Perhaps the most common analytical instrumental technique is the measurement of pH.

> By definition, pH is what you measure with a glass electrode and a pH meter. It is approximately equal to the theoretical definition of $pH = -\log a_{H^+}$.

> An operational definition of a quantity defines the quantity in terms of how it is measured.

[9] See W. Davison and C. Woof, *Anal. Chem.*, **1985**, *57*, 2567; T. R. Harbinson and W. Davison, *Anal. Chem.*, **1987**, *59*, 2450; A. Kopelove, S. Franklin, and G. M. Miller, *Amer. Lab.*, **1989** (6), 40.

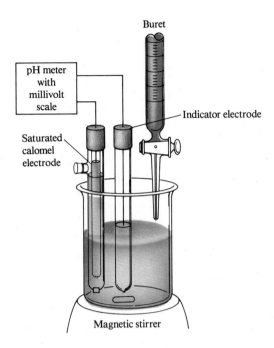

Buret

pH meter
with
millivolt
scale

Indicator electrode

Saturated
calomel
electrode

Magnetic stirrer

Figure 18-21
Apparatus for a potentiometric titration.

Similarly, if the cell potential is E_U when the electrodes are immersed in a solution of unknown pH, we have

$$pH_U = -\frac{E_U - K}{0.0592}$$

By subtracting the first equation from the second and solving for pH_U, we find

$$pH_U = pH_S - \frac{(E_U - E_S)}{0.0592} \qquad \textbf{(18-22)}$$

Equation 18-22 has been adopted throughout the world as the *operational definition of pH*.

Workers at the NIST and elsewhere have used cells without liquid junctions to study primary-standard buffers extensively. Some of the properties of these buffers are presented and discussed in detail elsewhere.[10] For general use, the buffers can be prepared from relatively inexpensive laboratory reagents. For careful work, certified buffers can be purchased from the NIST.

18G   POTENTIOMETRIC TITRATIONS

A *potentiometric titration* involves measurement of the potential of a suitable indicator electrode as a function of titrant volume. The information provided by

[10] R. G. Bates, *Determination of pH*, 2nd ed., Chapter 4. New York: Wiley, 1973.

a potentiometric titration is not the same as that obtained from a direct potentiometric measurement. For example, the direct measurement of 0.100 M solutions of hydrochloric and acetic acids would yield two substantially different hydrogen ion concentrations because the latter is only partially dissociated. In contrast, the potentiometric titration of equal volumes of the two acids would require the same amount of standard base because both solutes have the same number of titratable protons.

Potentiometric titrations provide data that are more reliable than data from titrations that use chemical indicators and are particularly useful with colored or turbid solutions and for detecting the presence of unsuspected species. These titrations are also readily automated. Manual potentiometric titrations, on the other hand, suffer from the disadvantage of being more time-consuming than those involving indicators.

Figure 18-21 illustrates a typical apparatus for performing a manual potentiometric titration. Its use involves measuring and recording the cell potential (in units of millivolts or pH, as appropriate) after each addition of reagent. The titrant is added in large increments at the outset and in smaller and smaller increments as the end point is approached (as indicated by larger changes in response per unit volume).

Automatic *titrators* for carrying out potentiometric titrations are available from several manufacturers. Color plate 18 shows an example of a typical automatic titrator.

18G-1 End-Point Detection

Several methods can be used to determine the end point of a potentiometric titration. The most straightforward involves a direct plot of potential as a function of reagent volume, as in Figure 18-22a; the midpoint in the steeply rising portion of the curve is estimated visually and taken as the end point. A second approach to end-point detection is to plot the change in potential per unit volume of titrant (that is, $\Delta E/\Delta V$) as a function of the average volume V. As shown in Figure 18-22b, a curve is obtained with a maximum that corresponds to the end point.

Figure 18-22c shows that the second derivative for the titration data changes sign at the end point. This change is used as the analytical signal in some automatic titrators.

Figure 18-22

Titration of 2.433 mmol of chloride ion with 0.1000 M silver nitrate. (a) Titration curve. (b) First-derivative curve. (c) Second-derivative curve.

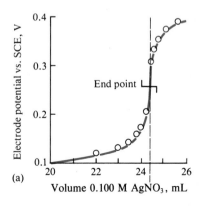

(a)

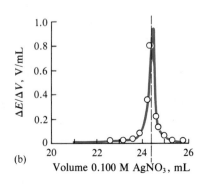

(b)

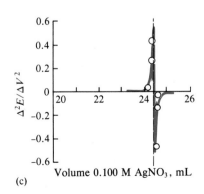

(c)

18H   QUESTIONS AND PROBLEMS

Note: Numerical data are molar analytical concentrations where the full formula of a species is provided. Molar equilibrium concentrations are supplied for species displayed as ions. Unless otherwise noted, neglect the effects of liquid-junction potentials in your calculations.

18-1. Briefly describe or define
 *(a) indicator electrode.
 (b) reference electrode.
 *(c) electrode of the first kind.
 (d) electrode of the second kind.

18-2. Briefly describe or define
 *(a) liquid-junction potential.
 (b) boundary potential.
 (c) asymmetry potential.

*18-3. Describe how a mercury electrode could function as
 (a) an electrode of the first kind for Hg(II).
 (b) an electrode of the second kind for EDTA.

18-4. What is meant by Nernstian behavior in a membrane electrode?

*18-5. Describe the source of pH-dependence in a glass membrane electrode.

18-6. Why is it necessary for the glass in the membrane of a pH-sensitive electrode to be appreciably hygroscopic?

*18-7. List four sources of uncertainty in the measurement of pH with a glass electrode.

18-8. What experimental factor places a limit on the number of significant figures in the response of a membrane electrode?

*18-9. Describe alkaline error in the measurement of pH. Under what circumstances is this error appreciable? How are pH data affected by alkaline error?

18-10. How does a gas-sensing probe differ from other membrane electrodes?

*18-11. How does information supplied by a direct potentiometric measurement of pH differ from that obtained from a potentiometric acid/base titration?

*18-12. (a) Calculate E^0 for the process

$$AgIO_3(s) + e^- \rightleftharpoons Ag(s) + IO_3^-$$

 (b) Use the shorthand notation to describe a cell consisting of a saturated calomel electrode as anode and a silver cathode that could be used to measure pIO_3.
 (c) Develop an equation that relates the potential of the cell in (b) to pIO_3.
 (d) Calculate pIO_3 if the cell in (b) has a potential of 0.294 V.

18-13. (a) Calculate E^0 for the process

$$PbI_2(s) + e^- \rightleftharpoons Pb(s) + 2I^-$$

 (b) Use the shorthand notation to describe a cell con-

sisting of a saturated calomel electrode as anode and a lead cathode that could be used for the measurement of pI.
 (c) Generate an equation that relates the potential of this cell to pI.
 (d) Calculate pI if this cell has a potential of -0.348 V.

18-14. Use the shorthand notation to describe a cell consisting of a saturated calomel electrode as anode and a silver cathode for the measurement of
 *(a) pSCN.   *(b) pI.   *(c) pSO_3.   *(d) pPO_4.

*18-15. Generate an equation that relates pAnion to E_{cell} for each of the cells in Problem 18-14. (For Ag_2SO_3, $K_{sp} = 1.5 \times 10^{-14}$; for Ag_3PO_4, $K_{sp} = 1.3 \times 10^{-20}$.)

18-16. Calculate
 *(a) pSCN if the cell in 18-14(a) has a potential of 0.122 V.
 (b) pI if the cell in 18-14(b) has a potential of -211 mV.
 *(c) pSO_3 if the cell in 18-14(c) has a potential of 300 mV.
 (d) pPO_4 if the cell in 18-14(d) has a potential of 0.244 V.

*18-17. The cell

$$SCE \| Ag_2CrO_4(sat'd), CrO_4^{2-}(x\ M) | Ag$$

is employed for the determination of $pCrO_4$. Calculate $pCrO_4$ when the cell potential is 0.402 V.

18-18. Calculate the potential of the cell

$$SCE \| aqueous\ solution | Hg$$

when the aqueous solution is
 (a) 7.40×10^{-3} M Hg^{2+}.
 (b) 7.40×10^{-3} Hg_2^{2+}.
 (c) $Hg_2SO_4(sat'd), SO_4^{2-}(0.0250\ M)$. For Hg_2SO_4, $K_{sp} = 7.4 \times 10^{-7}$.
 (d) $Hg^{2+}(2.00 \times 10^{-3}), OAc^-(0.100\ M)$.

$$Hg^{2+} + 2OAc^- \rightleftharpoons Hg(OAc)_2(aq)$$
$$K_f = 2.7 \times 10^8$$

*18-19. The standard potential for the reduction of the EDTA complex of Hg(II) is

$$HgY^{2-} + 2e^- \rightleftharpoons Hg(l) + Y^{4-} \qquad E^0 = 0.210\ V$$

Calculate the potential of the cell

$$SCE \| HgY^{4-}(2.00 \times 10^{-4}\ M), Z | Hg$$

---

* Answers to the asterisked problems are given in the answers section at the back of the book.

when Z is
(a) $H^+(1.00 \times 10^{-4}$ M),EDTA(0.0200 M).
(b) $H^+(1.00 \times 10^{-8}$ M),EDTA(0.0200 M).

18-20. Calculate the potential of the cell described in 18-19 when Z is
(a) $H^+(1.00 \times 10^{-6}$ M),EDTA(0.0100 M).
(b) $H^+(1.00 \times 10^{-10}$ M),EDTA(0.0100 M).

*18-21. The cell

$$SCE\|H^+(a = x)|\text{glass electrode}$$

has a potential of 0.2094 V when the solution in the right-hand compartment is a buffer of pH 4.006. The following potentials are obtained when the buffer is replaced with unknowns: (a) -0.3011 V and (b) $+0.1163$ V. Calculate the pH and the hydrogen ion activity of each unknown. (c) Assuming an uncertainty of ±0.002 V in the junction potential, what is the range of hydrogen ion activities within which the true value might be expected to lie?

18-22. The following cell

$$SCE\|MgA_2(a_{Mg^{2+}} = 9.62 \times 10^{-3})|$$
$$\text{membrane electrode for } Mg^{2+}$$

has a potential of 0.367 V.
(a) When the solution of known magnesium activity is replaced with an unknown solution, the potential is $+0.544$ V. What is the pMg of this unknown solution?
(b) Assuming an uncertainty of ±0.002 V in the junction potential, what is the range of Mg^{2+} activities within which the true value might be expected?

*18-23. The cell

$$SCE\|CdA_2(\text{sat'd}),A^-(0.0250 \text{ M})|Cd$$

has a potential of -0.721 V. Calculate the solubility product of CdA_2, neglecting the junction potential.

18-24. The cell

$$SCE\|HA(0.250 \text{ M}),NaA(0.180 \text{ M})|H_2(1.00 \text{ atm}),Pt$$

has a potential of -0.797 V. Calculate the dissociation constant of HA, neglecting the junction potential.

*18-25. A 0.3798-g sample of a purified organic acid was dissolved in water and titrated potentiometrically. A plot of the data revealed a single end point after 27.22 mL of 0.1025 M NaOH had been introduced. Calculate the molecular mass of the acid.

18-26. How could you use the data obtained in Problem 18-25 to evaluate K_{HA} for the unknown acid?

18-27. A 40.00-mL aliquot of 0.05000 M HNO_2 is diluted to 75.00 mL and titrated with 0.0800 M Ce^{4+}. The pH of the solution is maintained at 1.00 throughout the titration; the formal potential of the cerium system is 1.44 V.
*(a) Calculate the potential of the indicator electrode with respect to a saturated calomel reference electrode after the addition of 5.00, 10.00, 15.00, 25.00, 40.00, 49.00, 50.00, 51.00, and 60.00 mL of cerium(IV).
(b) Draw a titration curve for these data.

18-28. Calculate the potential of a silver cathode versus the saturated calomel electrode after the addition of 5.00, 15.00, 25.00, 30.00, 35.00, 39.00, 40.00, 41.00, 45.00, and 50.00 mL of 0.1000 M $AgNO_3$ to 50.00 mL of 0.0800 M KSeCN. Construct a titration curve from these data. (K_{sp} for AgSeCN = 4.20×10^{-16}.)

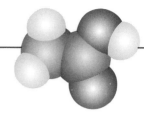

ELECTROGRAVIMETRIC AND COULOMETRIC METHODS

In this chapter we describe two related electroanalytical methods: *electrogravimetry* and *coulometry*. Each is based on an electrolysis that is carried out for a sufficient length of time to ensure complete oxidation or reduction of the analyte to a single product of known composition. In electrogravimetric methods, the product is weighed as a deposit on one of the electrodes (*the working electrode*). In coulometric procedures, the quantity of electrical charge needed to complete the electrolysis is measured.[1]

Electrogravimetry and coulometry are moderately sensitive and among the most accurate and precise techniques available to the chemist. Like gravimetry, these methods require no preliminary calibration against standards because the functional relationship between the quantity measured and the analyte concentration can be derived from theory and atomic-mass data.

> Electrogravimetry and coulometry often have relative errors of a few parts per thousand.

19A THE EFFECT OF CURRENT ON CELL POTENTIALS

Electrogravimetry and coulometry differ from potentiometry in that they require a significant current throughout the analytical process. In contrast, potentiometric measurements are performed under conditions of essentially zero current. When there is a current in an electrochemical cell, the cell potential is no longer simply the difference between the electrode potentials of the cathode and the anode (the thermodynamic potential). Two additional phenomena, *IR drop* and *polarization*, require application of potentials greater than the thermodynamic potential to operate an electrolytic cell and result in the development of potentials smaller than theoretical in a galvanic cell. Before proceeding, we must examine these phenomena in some detail.

---

[1] For further information concerning the methods in this chapter, see J. A. Plambeck, *Electroanalytical Chemistry*. New York: Wiley, 1982; *Laboratory Techniques in Electroanalytical Chemistry*, P. T. Kissinger and W. R. Heineman, Eds. New York: Marcel Dekker, 1984.

19A-1 Ohmic Potential; IR Drop

Electrochemical cells, like metallic conductors, resist the flow charge. In both types of conduction, Ohm's law describes the effect of this resistance. That is,

$$E = IR \tag{19-1}$$

where E is the potential difference in volts across the cell resistance, R is the cell resistance in ohms, and I is the current in the cell in amperes. The product on the right-hand side of this equation is called the *ohmic potential* or the *IR drop* of the cell. In order to develop a current of I amperes in an electrolytic cell, it is necessary to apply an external potential that is IR volts larger than the thermodynamic potential E_{cell}. Thus, we may write

$$E_{ext} = E_{cell} + IR \tag{19-2}$$

where E_{ext} is the externally applied voltage, and E_{cell} is the thermodynamic potential of the cell calculated by the techniques discussed in Section 15C. Note that E_{ext} and I are positive for an electrolytic process.

Figure 19-1 illustrates the effect of a current on the measured potential of an electrochemical cell. The cell is shown in Figure 15-2, page 261, where the electrode system consists of an Ag/Ag^+ couple and a Cu/Cu^{2+} couple joined by a salt bridge. Here, the internal resistance of the cell is represented as a resistor, R_{cell}, with a resistance of 40 Ω. There is no detectable current in the circuit of Figure 19-1a because the switch is open. Furthermore, the digital voltmeter has such a high resistance (>10 MΩ) that it draws no significant current from the cell. The measured potential of the cell under this circumstance is the thermodynamic potential of 0.412 V (see Example 16-1, page 284).

In Figure 19-1b, the switch is shown closed so that there is a current in the variable load resistor R_L and the current meter. The magnitude of this current can be varied by changing the position of the contact L. When its position is such that the current is 2.00 mA, the IR drop in the cell is 2.00 mA $\times$ 40.0 Ω = 0.0800 V. The measured potential across the cell is then 0.412 V − 0.080 V = 0.332 V.

In Figure 19-1c, an external voltage source is applied to the cell in the opposite sense to E_{cell}; that is, the positive terminal of the external source is connected to the positive terminal of the cell through the load resistor. By adjusting the load resistor, the current can be reversed in the cell so that the galvanic process in the cell is reversed, and an electrolytic process occurs. When the load resistor is adjusted so that the current is 2.00 mA, the externally applied voltage E_{ext} is larger than E_{cell} by **IR** = 0.0800 V, so

$$E_{ext} = E_{cell} + IR = 0.412 \text{ V} + 0.080 \text{ V} = 0.492 \text{ V}$$

19A-2 Polarization Effects

Equation 19-2 can be rearranged to give

$$I = \frac{1}{R}E_{ext} - \frac{1}{R}E_{cell} \tag{19-3}$$

André Marie Ampère (1775–1836), French mathematician and physicist, was the first to apply mathematics to the study of electrical current. Consistent with Benjamin Franklin's definitions of positive and negative charge, Ampère defined a positive current to be the direction of flow of positive charge. Although we now know that negative electrons carry current in metals, Ampère's definition has survived to the present. The unit of current, the ampere, is named in his honor.

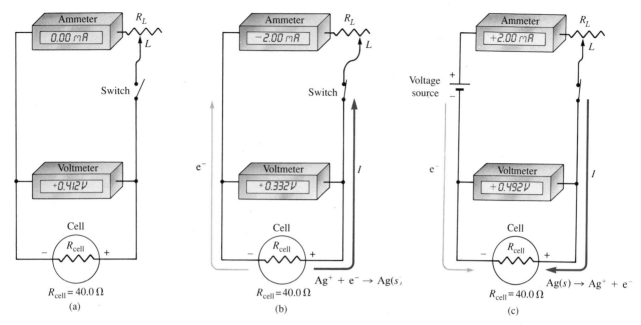

Figure 19-1

Effect of IR drop in cell-on-cell potential. (a) current and $IR \rightarrow 0$; (b) galvanic cell with $R_{cell} = 40.0 \ \Omega$; (c) electrolytic cell with $R_{cell} = 40.0 \ \Omega$.

For small currents and brief periods of time, E_{cell} remains relatively constant during an electrolysis. According to Equation 19-3, a plot of current in an electrochemical cell as a function of the externally applied potential E_{ext} should be a straight line with a slope equal to the reciprocal of the cell resistance R and intercept equal to $-E_{cell}/R$. As shown in Figure 19-2, an experimental plot of current as a function of E_{ext} is in fact linear with small currents. As the externally applied voltage increases and the electrolysis proceeds, the current deviates significantly from linearity, however. Note that if we decrease E_{ext} so that $E_{ext} = E_{cell}$, $I = 0$ as shown in the figure. If E_{ext} is decreased further so that it is negative, the process in the cell is galvanic, and the cell reaction proceeds as shown in the figure. As before, when E_{ext} becomes sufficiently negative the curve deviates significantly from linearity.

When cells exhibit nonlinear current-voltage behavior, they are said to be *polarized*, and the degree of polarization is described by the *overvoltage*, or *overpotential*, which is symbolized by Π in Figure 19-2. Note that polarization requires the application of a potential to an electrolytic cell that is greater than theoretical to give a current of the expected magnitude. Thus, the overpotential required to achieve a current of about 5.4 mA in the galvanic process occurring below the voltage axis is about -0.04 V. That is, the overvoltage is -0.04 V. Similarly, the electrolytic process requires a voltage 0.04 V larger than theoretical to produce a current of 4.8 mA, so the overvoltage is 0.04 V. Thus, when overvoltage is taken into account, Equation 19-2 becomes

> Overvoltage is the potential difference between the theoretical cell potential and the measured cell potential at a given level of current.

$$E_{ext} = E_{cell} + IR + \Pi \qquad \textbf{(19-4)}$$

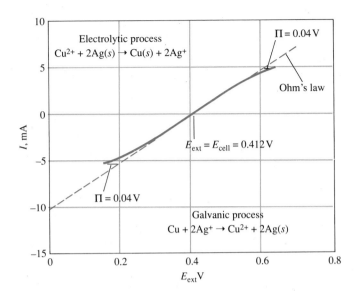

Figure 19-2
The influence of an externally-applied voltage on the current in an electrochemical cell.

Factors that influence polarization: (1) the size, shape, and composition of the electrode; (2) the composition of the electrolyte solution; (3) the temperature and the stirring rate; (4) the current level; and (5) the physical state of the species involved in the cell reaction.

Reactants are transported to or away from an electrode by (1) diffusion, (2) migration, and (3) convection.

Polarization is an electrode phenomenon that may affect either or both of the electrodes in a cell. The degree of polarization of an electrode varies widely. In some instances it approaches zero, but in others, it can be so nearly complete that the current in the cell becomes independent of potential. Polarization phenomena are conveniently divided into two categories: *concentration polarization* and *kinetic polarization*.

Concentration Polarization

Electron transfer between a reactive species in a solution and an electrode can take place only from a thin film of solution located immediately adjacent to the surface of the electrode; this film is only a fraction of a nanometer in thickness and contains a limited number of reactive ions or molecules. In order for there to be a steady current in a cell, this film must be continuously replenished with reactant from the bulk of the solution. That is, as reactant ions or molecules are consumed by the electrochemical reaction, more must be transported into the surface film at a rate that is sufficient to maintain the current. For example, in order to have a current of 2.0 mA in the electrolytic cell described in Figure 15-2b, it is necessary to transport copper ions to the cathode surface at a rate of about 1×10^{-8} mol/s or 6×10^{15} copper ions per second. (Similarly, silver ions must be removed from the surface film of the anode at a rate of 2×10^{-8} mol/s.)

Concentration polarization occurs when reactant species do not arrive at the surface of the electrode or product species do not leave the surface of the electrode fast enough to maintain the desired current. When this happens the current is limited to values less than that predicted by Equation 19-3.

Reactants are transported to the surface of an electrode by three mechanisms: (1) *diffusion*, (2) *migration*, and (3) *convection*. Products are removed from electrode surfaces in the same ways. We will focus on mass-transport processes to the cathode, but our discussion applies equally to anodes.

Diffusion. When there is a concentration difference between two regions of a solution, ions or molecules move from the more concentrated region to the more dilute. This process is called *diffusion* and ultimately leads to a disappearance of the concentration difference. The rate of diffusion is directly proportional to the concentration difference. For example, when copper ions are deposited at a cathode by a current, as illustrated in Figure 19-3, the concentration of Cu^{2+} at the electrode surface $[Cu^{2+}]_0$ is very small. The difference between the concentration at the surface and the concentration in the bulk solution $[Cu^{2+}]$ creates a concentration gradient that causes copper ions to diffuse from the bulk of the solution to the surface film. The rate of diffusion is given by

$$\text{rate of diffusion to cathode surface} = k'([Cu^{2+}] - [Cu^{2+}]_0) \qquad \textbf{(19-5)}$$

where $[Cu^{2+}]$ is the reactant concentration in the bulk of the solution, $[Cu^{2+}]_0$ is its equilibrium concentration at the surface of the cathode, and k' is a proportionality constant. The value of $[Cu^{2+}]_0$ at any instant is fixed by *the potential of the electrode* and can be calculated from the Nernst equation. In the present example, we find the surface copper ion concentration from the relationship

$$E_{\text{cathode}} = E^0_{Cu^{2+}} - \frac{0.0592}{2} \log \frac{1}{[Cu^{2+}]_0}$$

where E_{cathode} is the potential applied to the cathode. As the applied potential becomes more and more negative, $[Cu^{2+}]_0$ becomes smaller and smaller. The result is that the rate of diffusion and the current become correspondingly larger.

Migration. The process by which ions move under the influence of an electric field is called *migration*. This process, shown schematically in Figure 19-4, is the primary cause of mass transfer in the bulk of the solution in a cell. The rate at which ions migrate to or away from an electrode surface generally increases as the electrode potential increases. This charge movement constitutes a current, which also increases with potential.

Convection. Reactants can also be transferred to or from an electrode by mechanical means. Forced *convection*, such as stirring or agitation, will tend to decrease the thickness of the diffusion layer at the surface of an electrode and thus decrease concentration polarization. Natural convection resulting from temperature or density differences can also contribute to the transport of molecules to and from an electrode.

The Importance of Concentration Polarization. As noted earlier, concentration polarization sets in when the effects of diffusion, migration, and convection are insufficient to transport a reactant to or from an electrode surface at a rate that produces a current of the magnitude given by Equation 19-3. Concentration polarization requires applied potentials that are larger than theoretical to maintain a given current in an electrolytic cell (Figure 19-1c). Similarly, the phenomenon causes a galvanic cell potential to be smaller than the value predicted on the basis of the theoretical potential and the *IR* drop (Figure 19-1b).

Concentration polarization is important in several electroanalytical methods. In some applications, its effects are undesirable, and steps are taken to eliminate it. In others, it is essential to the analytical method, and every effort is made to promote its occurrence.

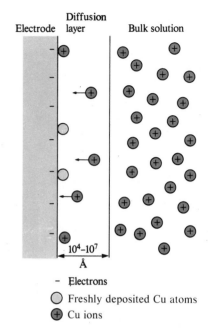

Figure 19-3
Concentration changes at the surface of a cathode. As Cu^{2+} ions are reduced to Cu atoms at the electrode surface, the concentration of Cu^{2+} at the surface becomes very small. Ions then diffuse from the bulk of the solution to the surface as a result of the concentration gradient.

Diffusion is a process in which ions or molecules move from a more concentrated part of a solution to a more dilute.

Convection is the mechanical transport of ions or molecules through a solution as a result of stirring, vibration, or temperature gradients.

The experimental variables that influence the degree of concentration polarization are (1) reactant concentration; (2) total electrolyte concentration; (3) mechanical agitation; and (4) electrode size.

The *current in a kinetically polarized cell is governed by the rate of electron transfer* rather than the rate of mass transfer.

Kinetic Polarization

In kinetic polarization, the magnitude of the current is limited by the rate of one or both of the electrode reactions—that is, the rate of electron transfer between the reactants and the electrodes. In order to offset kinetic polarization, an additional potential, or overvoltage, is required to overcome the energy barrier to the half-reaction.

Kinetic polarization is most pronounced for electrode processes that yield gaseous products and is often negligible for reactions that involve the deposition or solution of a metal. Kinetic effects usually decrease with increasing temperature and decreasing current density. These effects also depend on the composition of the electrode and are most pronounced with softer metals, such as lead, zinc, and particularly mercury. The magnitude of overvoltage effects cannot be predicted from present theory and can only be estimated from empirical information in the literature.[2] In common with *IR* drop, overvoltage effects cause the potential of a galvanic cell to be smaller than theoretical and to require potentials greater than theoretical to operate an electrolytic cell at a desired current.

Current density is defined as current per unit area of an electrode in amperes per square centimeter (A/cm^2) of electrode surface.

The overvoltages associated with the formation of hydrogen and oxygen are often 1 V or more and are of considerable importance because these molecules are frequently produced in electrochemical reactions. Of particular interest is the high overvoltage of hydrogen on such metals as copper, zinc, lead, and mercury. These metals and several others can be deposited without interference from hydrogen evolution. In theory, it is not possible to deposit zinc from a neutral aqueous solution because hydrogen forms at a potential that is considerably less than that required for zinc deposition. In fact, zinc can be deposited on a copper electrode with no significant hydrogen formation because the rate at which the gas forms on both zinc and copper is negligible, as shown by the high hydrogen overvoltage associated with these metals.

Kinetic polarization is most commonly encountered when the reactant or product in an electrochemical cell is gas.

19B THE POTENTIAL SELECTIVITY OF ELECTROLYTIC METHODS

In principle, electrolytic methods offer a reasonably selective means for separating and determining a number of ions. The feasibility of and theoretical conditions for accomplishing a given separation can be readily derived from the standard electrode potentials of the species of interest.

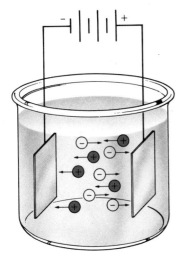

Figure 19-4

Migration is the movement of ions through a solution as a result of electrostatic attraction between the ions and the electrodes.

Example 19-1

Is a quantitative separation of Cu^{2+} and Pb^{2+} by electrolytic deposition feasible in principle? If so, what range of cathode potentials (vs. SCE) can be used? Assume that the sample solution is initially 0.1000 M in each ion and that quantitative removal of an ion is realized when only 1 part in 10,000 remains undeposited.

---

[2] Overvoltage data for various gaseous species on different electrode surfaces have been compiled by J. A. Page, in *Handbook of Analytical Chemistry*, L. Meites, Ed., pp. **5**–184 to **5**–186. New York: McGraw-Hill, 1963.

In Appendix 1, we find

$$Cu^{2+} + 2e^- \rightleftharpoons Cu(s) \qquad E^0 = 0.337 \text{ V}$$
$$Pb^{2+} + 2e^- \rightleftharpoons Pb(s) \qquad E^0 = -0.126 \text{ V}$$

It is apparent that copper will begin to deposit before lead. Let us first calculate the cathode potential required to decrease the Cu^{2+} concentration to 10^{-4} of its original concentration (that is, to 1.00×10^{-5} M). Substituting into the Nernst equation, we obtain

$$E = 0.337 - \frac{0.0592}{2} \log \frac{1}{1.00 \times 10^{-5}} = 0.189 \text{ V}$$

Feature 19-1
OVERVOLTAGE AND THE LEAD/ACID BATTERY

If it were not for the high overvoltage of hydrogen on lead and lead oxide electrodes, the lead/acid storage batteries found in automobiles and trucks would not operate because of hydrogen formation at the cathode both during charging and use. Certain trace metals in the system lower this overvoltage and eventually lead to gassing, or hydrogen formation, which limits the lifetime of the battery. The basic difference between a battery with a 48-month warranty and a 72-month warranty is the concentration of these trace metals in the system.

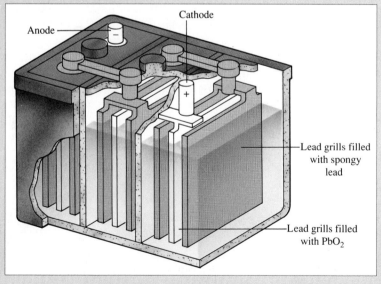

Figure 19-A   The lead storage battery.

Similarly, we can derive the cathode potential at which lead begins to deposit:

$$E = -0.126 - \frac{0.0592}{2} \log \frac{1}{0.100} = -0.156 \text{ V}$$

Therefore, if the cathode potential is maintained between 0.189 V and -0.156 V (vs. SHE), in theory a quantitative separation should occur. To convert these potentials to potentials relative to a saturated calomel electrode, we treat the reference electrode as the anode and write

$$E_{cell} = E_{cathode} - E_{SCE} = 0.192 - 0.244 = -0.052 \text{ V}$$

and

$$E_{cell} = -0.156 - 0.244 = -0.400 \text{ V}$$

Therefore the cathode potential should be kept between -0.052 V and -0.400 V versus the SCE.

Calculations such as the foregoing make it possible to compute the differences in standard electrode potentials theoretically needed to determine one ion without interference from another; these differences range from about 0.04 V for triply charged ions to approximately 0.24 V for singly charged species.

These theoretical separation limits can be approached only by maintaining the potential of the *working electrode* (usually the cathode, at which a metal deposits) at a level required by theory. The potential of this electrode can *be controlled only by variation of the potential applied to the cell*, however. From Equation 19-4, it is evident that variations in the applied potential (E_{cell}) affect not only the cathode potential but also the anode potential, the *IR* drop, and the overpotential. As a consequence, the only *practical* way of achieving separation of species whose electrode potentials differ by a few tenths of a volt is to *measure* the cathode potential continuously against a reference electrode whose potential is known; the applied cell potential can then be adjusted to maintain the cathode potential at the desired level. An analysis performed in this way is called a *controlled-cathode-potential electrolysis*. Controlled-potential methods are discussed in Sections 19C-2 and 19D-4.

19C ELECTROGRAVIMETRIC METHODS OF ANALYSIS

Electrolytic precipitation has been used for over a century for the gravimetric determination of metals. In most applications, the metal is deposited on a weighed platinum cathode, and the increase in mass is determined. Important exceptions to this procedure include the anodic deposition of lead as lead dioxide on platinum and of chloride as silver chloride on silver.

There are two general types of electrogravimetric methods. In the first, no control of the potential of the working electrode is exercised, and the applied cell potential is held at a more or less constant level that provides a large enough current to complete the electrolysis in a reasonable length of time. The second type of electrogravimetric method is the *potentiostatic method*. This procedure is also called the *controlled-cathode-potential* or the *controlled-anode potential method*, depending on whether the working electrode is a cathode or an anode.

> A working electrode is the electrode at which the analytical reaction occurs.

> A potentiostatic method is an electrolytic procedure in which the potential of the working electrode is maintained at a constant level versus a reference electrode, such as a saturated calomel electrode.

19C-1 Electrogravimetry Without Potential Control of the Working Electrode

Electrolytic procedures in which no effort is made to control the potential of the working electrode make use of simple and inexpensive equipment and require little operator attention. In these procedures, the potential applied across the cell is maintained at a more or less constant level throughout the electrolysis.

Instrumentation

As shown in Figure 19-5, the apparatus for an analytical electrodeposition without cathode potential control consists of a suitable cell and a direct-current power supply. The dc power supply usually consists of an ac rectifier, although a 6-V storage battery can be used instead. The voltage applied to the cell is controlled by the rheostat, R. An ammeter and a voltmeter indicate the approximate current and applied voltage. In performing an analytical electrolysis with this apparatus, the applied voltage is adjusted with rheostat R to give a current of several tenths of an ampere. The voltage is then maintained at about the initial level until the deposition is judged to be complete.

Electrolysis Cells

Figure 19-5 shows a typical cell for the deposition of a metal on a solid electrode. Ordinarily, the working electrode is a platinum gauze cylinder 2 or 3 cm in diameter and perhaps 6 cm in length. Often, as shown in the figure, the anode takes the form of a solid platinum stirring paddle that is located inside and connected to the cathode through the external circuit.

The Physical Properties of Electrolytic Precipitates

Ideally, an electrolytically deposited metal should be strongly adherent, dense, and smooth so that it can be washed, dried, and weighed without mechanical loss or reaction with the atmosphere. Good metallic deposits are fine-grained and have a metallic luster; spongy, powdery, or flaky precipitates are likely to be less pure and less adherent.

The principal factors that influence the physical characteristics of deposits are current density, temperature, and the presence of complexing agents. Ordinarily, the best deposits are formed at current densities that are less than 0.1 A/cm^2. Stirring generally improves the quality of a deposit. The effects of temperature are unpredictable and must be determined empirically.

Many metals form smoother and more adherent films when deposited from solutions in which their ions exist primarily as complexes. Cyanide and ammonia complexes often provide the best deposits. The reasons for this effect are not obvious.

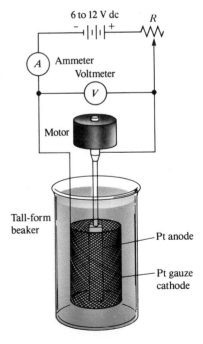

Figure 19-5

Apparatus for the electrodeposition of metals without cathode-potential control.

Applications

In practice, electrolysis at a constant cell potential is limited to the separation of easily reduced cations from those that are more difficult to reduce than hydrogen ion or nitrate ion. The reason for this limitation is illustrated in Figure 19-6, which shows the changes of current, IR drop, and cathode potential during an electrolysis in the cell in Figure 19-5. The analyte here is copper(II) ions in a solution containing an excess of acid. Initially, R is adjusted so that the potential applied to the cell is about -2.5 V, which, as shown in Figure 19-6a, leads to a current of about 1.5 A. The electrolytic deposition of copper is then completed at this applied potential.

As shown in Figure 19-6b, the IR drop decreases continually as the reaction proceeds. The reason for this decrease is primarily concentration polarization at the cathode, which limits the rate at which copper ions are brought to the electrode surface and thus the current. From Equation 19-4, it is apparent that the decrease in IR must be offset by an increase in the cathode overpotential since E_{ext} is constant.

Ultimately, the decrease in current and the increase in cathode potential is slowed at point B by the reduction of hydrogen ions. Because the solution contains a large excess of acid, the current is now no longer limited by concentration polarization, and codeposition of copper and hydrogen takes place simultaneously until the remainder of the copper ions are deposited. Under these conditions, the cathode is said to be *depolarized* by hydrogen ions.

A depolarizer is a chemical that is easily reduced (or oxidized) and stabilizes the potential of a working electrode by minimizing concentration polarization.

Consider now the fate of some metal ion, such as lead(II), which begins to deposit at point A on the cathode potential curve. Clearly it would codeposit well before copper deposition was complete, and an interference would result. In

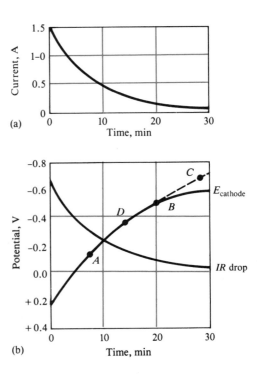

Figure 19-6

(a) Current, (b) IR drop, and cathode-potential change during the electrolytic deposition of copper at a constant-applied-cell potential.

contrast, a metal ion, such as cobalt(II), which reacts at a cathode potential corresponding to point C on the curve, would not interfere because depolarization by hydrogen gas formation prevents the cathode from reaching this potential.

Codeposition of hydrogen during electrolysis often leads to formation of nonadherent deposits, which are unsatisfactory for analytical purposes. This problem can be resolved by introducing another species that is reduced at a less negative potential than hydrogen ion and that does not adversely affect the physical properties of the deposit. One such cathode depolarizer is nitrate ion, its reaction being

$$NO_3^- + 10H^+ + 8e^- \rightleftharpoons NH_4^+ + 3H_2O \qquad \text{(19-6)}$$

Electrolytic methods performed without electrode-potential control, while limited by their lack of selectivity, do have several applications of practical importance. Table 19-1 lists the common elements that are often determined by this procedure.

19C-2 Constant-Cathode-Potential Gravimetry

In the discussion that follows, we shall assume that the working electrode is a cathode at which an analyte is deposited as a metal. However, the discussion is readily extended to an anodic working electrode where nonmetallic deposits are formed.

Instrumentation

In order to separate species with electrode potentials that differ by only a few tenths of a volt, it is necessary to use a more elaborate approach than the one just described. Such techniques are required because concentration polarization at the cathode, if unchecked, causes the potential of that electrode to become so negative that codeposition of the other species present begins before the analyte is completely deposited (Figure 19-6). A large negative drift in the cathode potential can be avoided by employing a three-electrode system, such as that shown in Figure 19-7.

Table 19-1
TYPICAL APPLICATIONS OF ELECTROGRAVIMETRIC
METHODS WITHOUT POTENTIAL CONTROL

Analyte	Weighed as	Cathode	Anode	Conditions
Ag^+	Ag	Pt	Pt	Alkaline CN^- solution
Br^-	AgBr (on anode)	Pt	Ag	
Cd^{2+}	Cd	Cu on Pt	Pt	Alkaline CN^- solution
Cu^{2+}	Cu	Pt	Pt	H_2SO_4/HNO_3 solution
Mn^{2+}	MnO_2 (on anode)	Pt	Pt dish	HCOOH/HCOONa solution
Ni^{2+}	Ni	Cu on Pt	Pt	Ammoniacal solution
Pb^{2+}	PbO_2 (on anode)	Pt	Pt	Strong HNO_3 solution
Zn^{2+}	Zn	Cu on Pt	Pt	Acidic citrate solution

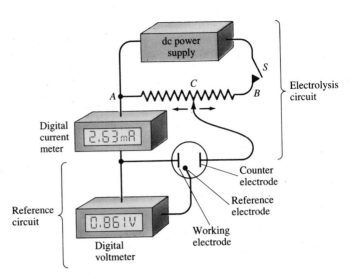

Figure 19-7

Apparatus for controlled-potential electrolysis. Contact C is adjusted as necessary to maintain the working electrode (cathode in this example) at a constant potential. The current in the reference-electrode circuit is essentially zero at all times.

A counter electrode has no effect on the reaction at the working electrode. It simply serves to feed electrons to the working cathode.

A potentiostat is an instrument that maintains the working electrode potential at a constant value.

The controlled-potential apparatus shown in Figure 19-7 has two independent electrical circuits that share a common electrode, the *working electrode* at which the analyte is deposited. The *electrolysis circuit* consists of a dc source, a potential divider (ACB) that permits continuous variation in the potential applied across the working electrode, a *counter electrode*, and a current meter. The *control circuit* is made up of a reference electrode (often a saturated calomel electrode), a high-resistance digital voltmeter, and the working electrode. The electrical resistance of the control circuit is so large that the electrolysis circuit supplies essentially all of the current for the electrolysis.

The purpose of the control circuit is to monitor continuously the potential between the working electrode and the reference electrode. When this potential reaches a level at which codeposition of an interfering species is about to begin, the potential across the working and counter electrodes is decreased by moving contact C to the left. Since the potential of the counter electrode remains constant during this change, the cathode potential becomes smaller, thus preventing interference from codeposition.

The current and the applied, or cell, voltage changes that occur in a typical constant-cathode-potential electrolysis are depicted in Figure 19-8. Note that the applied cell potential has to be decreased continuously throughout the electrolysis. Manual adjustment of the potential is difficult (particularly at the outset) and, above all, time consuming. To avoid such waste of operator time, controlled-cathode-potential electrolyses are generally performed with automated instruments called *potentiostats*, which maintain a constant cathode potential electronically.

Applications

The controlled-cathode-potential method is a potent tool for separating and determining metallic species that have standard potentials that differ by only a few tenths of a volt. An example illustrating the power of the method involves determining copper, bismuth, lead, cadmium, zinc, and tin in mixtures by successive deposition of the metals on a weighed platinum cathode. The first three

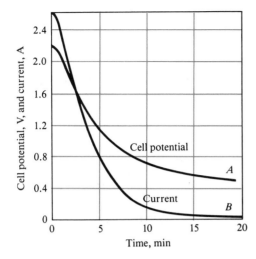

Figure 19-8
Changes in cell potential (*A*) and current (*B*) during a controlled-cathode-potential deposition of copper. The cathode is maintained at -0.36 V (versus SCE) throughout the experiment. (Data from J. J. Lingane, *Anal. Chem. Acta.*, **1948**, *2*, 590. With permission.)

elements are deposited from a nearly neutral solution containing tartrate ion to complex the tin(IV) and prevent its deposition. Copper is first reduced quantitatively by maintaining the cathode potential at -0.2 V with respect to a saturated calomel electrode. After being weighed, the copper-plated cathode is returned to the solution, and bismuth is removed at a potential of -0.4 V. Lead is then deposited quantitatively by increasing the cathode potential to -0.6 V. When lead deposition is complete, the solution is made strongly ammoniacal, and cadmium and zinc are deposited successively at -1.2 and -1.5 V. Finally, the solution is acidified in order to decompose the tin/tartrate complex by the formation of undissociated tartaric acid. Tin is then deposited at a cathode potential of -0.65 V. A fresh cathode must be used here because the zinc redissolves under these conditions. A procedure such as this is particularly attractive for use with a potentiostat because little operator time is required for the complete analysis.

Table 19-2 lists some other separations performed by the controlled-cathode-potential method.

19D   COULOMETRIC METHODS OF ANALYSIS

Coulometric methods are performed by measuring the quantity of electrical charge (electrons) required to convert a sample of an analyte quantitatively to a different oxidation state. Coulometric and gravimetric methods share the common advantage that the proportionality constant between the quantity measured and the analyte mass is derived from accurately known physical constants, thus eliminating the need for calibration standards. In contrast to gravimetric methods, coulometric procedures are usually rapid and do not require that the product of the electrochemical reaction be a weighable solid. Coulometric methods are as accurate as conventional gravimetric and volumetric procedures and in addition are readily automated.[3]

---

[3] For additional information about coulometric methods, see E. Bishop, in *Comprehensive Analytical Chemistry*, C. L. Wilson and D. W. Wilson, Eds., Vol. 11D. New York: Elsevier Scientific, 1975; J. A. Plambeck, *Electroanalytical Chemistry*, Chapter 12. New York: Wiley, 1982; D. J. Curran, in *Laboratory Techniques in Electroanalytical Chemistry*, P. T. Kissinger and W. R. Heineman, Eds., pp. 539–568. New York: Marcel Dekker, 1984.

Table 19-2
**SOME APPLICATIONS OF CONTROLLED-
CATHODE-POTENTIAL ELECTROLYSIS**

Element Determined	Other Elements That Can Be Present
Ag	Cu and heavy metals
Cu	Bi, Sb, Pb, Sn, Ni, Cd, Zn
Bi	Cu, Pb, Zn, Sb, Cd, Sn
Sb	Pb, Sn
Sn	Cd, Zn, Mn, Fe
Pb	Cd, Sn, Ni, Zn, Mn, Al, Fe
Cd	Zn
Ni	Zn, Al, Fe

19D-1 The Quantity of Electrical Charge

Units for the quantity of charge include the coulomb (C) and the faraday (F). *The coulomb is the quantity of electrical charge transported by a constant current of one ampere in one second.* Thus, the number of coulombs (Q) resulting from a constant current of I amperes operated for t seconds is

> 1 coulomb = 1 ampere · second = 1 A · s

$$Q = It \tag{19-7}$$

For a variable current i, the number of coulombs is given by the integral

$$Q = \int_0^t i \, dt \tag{19-8}$$

> A faraday of charge is equivalent to one mole of electrons or 6.022×10^{23} electrons.

The faraday is the quantity of charge that corresponds to one mole or 6.022×10^{23} electrons. The faraday also equals 96,485 C. As shown in Example 19-2, we can use these definitions to calculate the weight of a chemical species that is formed at an electrode by a current of known magnitude.

Example 19-2

A constant current of 0.800 A is used to deposit copper at the cathode and oxygen at the anode of an electrolytic cell. Calculate the number of grams of each product formed in 15.2 min, assuming no other redox reaction.

The two half-reactions are

$$Cu^{2+} + 2e^- \rightleftharpoons Cu(s)$$

$$2H_2O \rightleftharpoons 4e^- + O_2(g) + 4H^+$$

Thus 1 mol of copper is equivalent to 2 mol of electrons and 1 mol of oxygen corresponds to 4 mol of electrons.

Substituting into Equation 19-7 yields

$$Q = 0.800 \text{ A} \times 15.2 \text{ min} \times 60 \text{ s/min} = 729.6 \text{ A} \cdot \text{s} = 729.6 \text{ C}$$

$$\text{no. } F = \frac{729.6 \text{ C}}{96,485 \text{ C/}F} = 7.56 \times 10^{-3}\, F \equiv 7.56 \times 10^{-3} \text{ mol of electrons}$$

The masses of Cu and O_2 are given by

$$\text{mass Cu} = 7.56 \times 10^{-3} \text{ mol } e^- \times \frac{1 \text{ mol Cu}}{2 \text{ mol } e^-} \times \frac{63.55 \text{ g Cu}}{1 \text{ mol Cu}} = 0.240 \text{ g Cu}$$

$$\text{mass } O_2 = 7.56 \times 10^{-3} \text{ mol } e^- \times \frac{1 \text{ mol } O_2}{4 \text{ mol } e^-} \times \frac{32.00 \text{ g } O_2}{1 \text{ mol } O_2} = 0.0605 \text{ g } O_2$$

19D-2 Types of Coulometric Methods

Two methods based on measuring the quantity of charge have been developed: *potentiostatic coulometry* and *amperostatic coulometry*, or *coulometric titrimetry*. Potentiostat methods are performed in much the same way as controlled-potential gravimetric methods, with the potential of the working electrode being maintained at a constant level relative to a reference electrode throughout the electrolysis. Here, however, the electrolysis current is recorded as a function of time to give a curve similar to curve B in Figure 19-8. The analysis is then completed by integrating the current-time curve to obtain the number of coulombs and thus the number of faradays of charge consumed or produced by the analyte.

Coulometric titrations are similar to other titrimetric methods in that analyses are based on measuring the combining capacity of the analyte with a standard reagent. In the coulometric procedure, the reagent is electrons and the standard solution is a constant current of known magnitude. Electrons are added to the analyte (via the direct current) or to some species that immediately reacts with the analyte until an end point is reached. At that point the electrolysis is discontinued. The amount of analyte is determined from the magnitude of the current and the time required to complete the titration. The magnitude of the current in amperes is analogous to the molarity of a standard solution, and the time measurement is analogous to the volume measurement in conventional titrimetry.

19D-3 Current-Efficiency Requirements

A fundamental requirement for all coulometric methods is 100% current efficiency; that is, each faraday of electricity must bring about one equivalent of chemical change in the analyte. Note that 100% current efficiency can be achieved without direct participation of the analyte in electron transfer at an electrode. For example, chloride ions are readily determined by potentiostatic coulometry or by a coulometric titration by generating silver ions at a silver anode. These ions then react with the analyte to form a precipitate or deposit of silver chloride. The quantity of electricity required to complete the silver chloride formation serves as the analytical parameter. In this instance, 100% current efficiency is

Michael Faraday (1791–1867) was one of the foremost chemists and physicists of his time. Among his most important discoveries were Faraday's laws of electrolysis. Faraday, a simple man who lacked mathematical sophistication, was a superb experimentalist and an inspiring teacher and lecturer. The quantity of charge equal to a mole of electrons is named in his honor.

Amperostatic coulometry is also called coulometric titrimetry.

Electrons are the reagent in a coulometric titration.

One equivalent of chemical change is the change that corresponds to 1 mol of electrons. Thus, for the two half-reactions in Example 19-2, one equivalent of chemical change involves production of $\frac{1}{2}$ mol of Cu or $\frac{1}{4}$ mol of O_2.

realized because the number of moles of electrons is exactly equal to the number of moles of chloride ion in the sample despite the fact that these ions do not react directly at the electrode surface.

19D-4 Controlled-Potential Coulometry

In controlled-potential coulometry, the potential of the working electrode is maintained at a constant level such that only the analyte is responsible for the conduction of charge across the electrode/solution interface. The number of coulombs required to convert the analyte to its reaction product is then determined by recording and integrating the current-versus-time curve during the electrolysis.

Instrumentation

The instrumentation for potentiostatic coulometry consists of an electrolysis cell, a potentiostat, and a device for determining the number of coulombs consumed by the analyte.

Cells. Figure 19-9 illustrates two types of cells that are used for potentiostatic coulometry. The first consists of a platinum-gauze working electrode, a platinum-wire counter electrode, and a saturated calomel reference electrode. The counter electrode is separated from the analyte solution by a salt bridge that usually contains the same electrolyte as the solution being analyzed. This bridge is needed to prevent the reaction products formed at the counter electrode from diffusing into the analyte solution and interfering. For example, hydrogen gas is a common product at a cathodic counter electrode. Unless this species is physically isolated from the analyte solution by the bridge, it will react directly with many of the analytes that are determined by oxidation at the working anode.

The second type of cell, shown in Figure 19-9b, is a mercury-pool type. A mercury cathode is particularly useful for separating easily reduced elements as a preliminary step in an analysis. In addition, however, it has found considerable use for the coulometric determination of several metallic cations that form metals that are soluble in mercury. In these applications, little or no hydrogen evolution occurs even at high applied potentials because of the large overvoltage effects. A coulometric cell such as that shown in Figure 19-9b is also useful for the coulometric determination of certain types of organic compounds.

Potentiostats and Coulometers. For controlled-potential coulometry, a potentiostat similar in design to that shown in Figure 19-7 is required. Generally, however, the potentiostat is automated and is equipped with a recorder that provides a plot of current as a function of time, as shown in Figure 19-10. An electronic current integrator then gives the number of coulombs necessary to complete the reaction.

Applications of Controlled-Potential Coulometry

Controlled-potential coulometric methods have been applied to the determination of some 55 elements in inorganic compounds.[4] Mercury appears to be favored

---

[4] For a summary of the applications, see J. E. Harrar, in *Electroanalytical Chemistry*, A. J. Bard, Ed., Vol. 8. New York: Marcel Dekker, 1975; E. Bishop, in *Comprehensive Analytical Chemistry*, C. L. Wilson and D. W. Wilson, Eds., Vol. 11D, Chapter XV. New York: Elsevier, 1975.

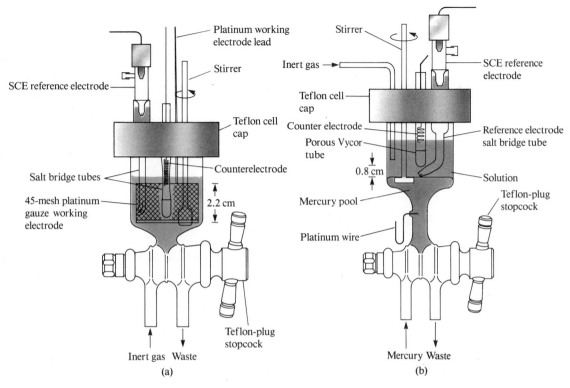

Figure 19-9

Electrolysis cells for potentiostatic coulometry. Working electrode: (a) platinum-gauze, (b) mercury-pool. (Reprinted with permission from J. E. Harrar and C. L. Pomernacki, *Anal. Chem.*, **1973**, *45*, 57. Copyright 1973 American Chemical Society.)

as the cathode, and methods for the deposition at this electrode of more than two dozen metals have been described. The method has found widespread use in the nuclear-energy field for the relatively interference-free determination of uranium and plutonium.

Controlled-potential coulometry also offers possibilities for the electrolytic determination (and synthesis) of organic compounds. For example, trichloroacetic acid and picric acid are quantitatively reduced at a mercury cathode whose potential is suitably controlled:

$$Cl_3CCOO^- + H^+ + 2e^- \rightarrow Cl_2HCCOO^- + Cl^-$$

$$NO_2 \underset{NO_2}{\overset{OH}{\bigcirc}} NO_2 + 18H^+ + 18e^- \rightarrow NH_2 \underset{NH_2}{\overset{OH}{\bigcirc}} NH_2 + 6H_2O$$

Coulometric measurements permit the determination of these compounds with a relative error of a few tenths of a percent.

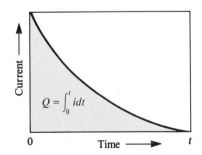

Figure 19-10

For a current that varies with time, the quantity of charge Q in a time t is the shaded area under the curve.

Constant current generators are called amperostats or *galvanostats.*

Auxiliary reagents are essential in coulometric titrations.

19D-5 Coulometric Titrations[5]

Coulometric titrations are carried out with a constant-current source called an *amperostat*, which senses decreases in current in a cell and responds by increasing the potential applied to the cell until the current is restored to its original level. Because of the effects of concentration polarization, 100% current efficiency with respect to the analyte can be maintained only by having present in large excess an auxiliary reagent that is oxidized or reduced at the electrode to give a product that reacts with the analyte. As an example, consider the coulometric titration of iron(II) at a platinum anode. At the beginning of the titration, the primary anodic reaction is

$$Fe^{2+} \rightleftharpoons Fe^{3+} + e^-$$

As the concentration of iron(II) decreases, however, the requirement of a constant current results in an increase in the applied cell potential. Because of concentration polarization, this increase in potential causes the anode potential to increase to the point where the decomposition of water becomes a competing process:

$$2H_2O \rightleftharpoons O_2(g) + 4H^+ + 4e^-$$

The quantity of electricity required to complete the oxidation of iron(II) then exceeds that demanded by theory, and the current efficiency is less than 100%. The lowered current efficiency is prevented, however, by introducing at the outset an unmeasured quantity of cerium(III), which is oxidized at a lower potential than is water:

$$Ce^{3+} \rightleftharpoons Ce^{4+} + e^-$$

With stirring, the cerium(IV) produced is rapidly transported from the surface of the electrode to the bulk of the solution, where it oxidizes an equivalent amount of iron(II):

$$Ce^{4+} + Fe^{2+} \rightleftharpoons Ce^{3+} + Fe^{3+}$$

The net effect is an electrochemical oxidation of iron(II) with 100% current efficiency, even though only a fraction of that species is directly oxidized at the electrode surface.

End Points in Coulometric Titrations

Coulometric titrations, like their volumetric counterparts, require a means for determining when the reaction between analyte and reagent is complete. Gener-

---

[5] For further details on this technique, see D. J. Curran, in *Laboratory Techniques in Electroanalytical Chemistry*, P. T. Kissinger and W. R. Heineman, Eds., Chapter 20. New York: Marcel Dekker, 1984.

ally, the end points described in the chapters on volumetric methods are applicable to coulometric titrations as well. Thus, for the titration of iron(II) just described, an oxidation/reduction indicator, such as 1,10-phenanthroline, can be used; as an alternative, the end point can be established potentiometrically. Similarly, an adsorption indicator or a potentiometric end point can be employed in the coulometric titration of chloride ion by the silver ions generated at a silver anode.

Instrumentation

As shown in Figure 19-11, the equipment required for a coulometric titration includes a source of constant current of one to several hundred milliamperes, a titration vessel, a switch, an electric timer, and a device for monitoring current. Movement of the switch to position 1 simultaneously starts the timer and initiates a current in the titration cell. When the switch is moved to position 2, the electrolysis and the timing are discontinued. With the switch in this position, however, electricity continues to be drawn from the source and passes through a dummy resistor R_D that has about the same electrical resistance as the cell. This arrangement ensures continuous operation of the source, which aids in maintaining the current at a constant level.

The constant-current source for a coulometric titration is often an *amperostat*, an electronic device capable of maintaining a current of 200 mA or more that is constant to a few hundredths percent. Amperostats are available from several instrument manufacturers. A much less expensive source that produces reasonably constant currents can be constructed from several heavy-duty batteries connected in series to give an output potential of 100 to 300 V. This output is applied to the cell that is in series with a resistor whose resistance is large relative to the resistance of the cell. Small changes in the cell conductance then have a negligible effect on the current in the resistor and cell.

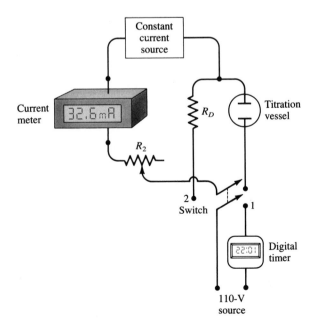

Figure 19-11
Diagram of a coulometric titration apparatus.

An ordinary motor-driven electric clock is unsatisfactory for the measurement of the electrolysis time because the rotor of such a device tends to coast when stopped and lag when started. Digital electronic timers eliminate this problem.

Cells for Coulometric Titrations. Figure 19-12 shows a typical coulometric titration cell consisting of a generator electrode at which the reagent is produced and a counter electrode to complete the circuit. The generator electrode—ordinarily a platinum rectangle, a coil of wire, or a gauze cylinder—has a relatively large surface area to minimize polarization effects. The counter electrode is usually isolated from the reaction medium by a sintered disk or some other porous medium in order to prevent interference by the reaction products from this electrode.

A Comparison of Coulometric and Conventional Titrations

The various components of the titrator in Figure 19-11 have their counterparts in the reagents and apparatus required for a volumetric titration. The constant-current source of known magnitude serves the same function as the standard solution in a volumetric method. The electronic timer and switch correspond to the buret and stopcock, respectively. Electricity is passed through the cell for relatively long periods of time at the outset of a coulometric titration, but the time intervals are made smaller and smaller as chemical equivalence is approached. Note that these steps are analogous to the operation of a buret in a conventional titration.

A coulometric titration offers several significant advantages over a conventional volumetric procedure. Principal among these is the elimination of the problems associated with the preparation, standardization, and storage of standard solutions. This advantage is particularly significant with labile reagents such as chlorine, bromine, and titanium(III) ion, which are sufficiently unstable in aqueous solution to seriously limit their use as volumetric reagents. In contrast, their utilization

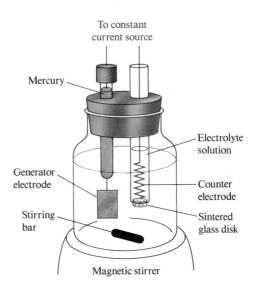

Figure 19-12
A typical coulometric titration cell.

in a coulometric determination is straightforward because they are consumed as soon as they are generated.

Coulometric methods also excel when small amounts of analyte have to be titrated because tiny quantities of reagent are generated with ease and accuracy through the proper choice of current. In contrast, the use of very dilute solutions and the accurate measurement of small volumes are inconvenient at best.

A further advantage of coulometric titrimetry is that a single constant-current source provides reagents for precipitation, complex formation, neutralization, or oxidation/reduction titrations. Finally, coulometric titrations are more readily automated since currents are easier to control than is control of liquid flow.

The current-time measurements required for a coulometric titration are as accurate as or more accurate than the comparable volume/molarity measurements of a conventional volumetric method, particularly where small quantities of reagent are involved. When the accuracy of a titration is limited by the sensitivity of the end point, the two titration methods have comparable accuracies.

> Coulometric methods are as accurate and precise as comparable volumetric methods.

Applications of Coulometric Titrations

Coulometric titrations have been developed for all types of volumetric reactions.[6] Selected applications are described in this section.

Neutralization Titrations. Hydroxide ion can be generated at the surface of a platinum cathode immersed in a solution containing the analyte acid:

$$2H_2O + 2e^- \rightleftharpoons 2OH^- + H_2(g)$$

The platinum anode must be isolated by a diaphragm to eliminate potential interference from the hydrogen ions produced by anodic oxidation of water. As a convenient alternative, a silver wire can be substituted for the platinum anode, provided chloride or bromide ions are added to the analyte solution. The anode reaction then becomes

$$Ag(s) + Br^- \rightleftharpoons AgBr(s) + e^-$$

Silver bromide does not interfere with the neutralization reaction.

Coulometric titrations of acids are much less susceptible to the carbonate error encountered in volumetric methods (Section 12A-3). The only measure required to avoid this error is to remove the carbon dioxide by boiling the solvent or by bubbling an inert gas, such as nitrogen, through the solution for a brief period.

Hydrogen ions generated at the surface of a platinum anode can be used for the coulometric titration of strong as well as weak bases:

$$2H_2O \rightleftharpoons O_2 + 4H^+ + 4e^-$$

---

[6] For additional information, see E. Bishop, in *Comprehensive Analytical Chemistry*, C. L. Wilson and D. W. Wilson, Eds., Vol. 11D, Chapters XVIII to XXIV. New York: Elsevier, 1975; J. T. Stock, *Anal. Chem.*, **1984**, *56*, 1R and **1980**, *52*, 1R.

Table 19-3

SUMMARY OF COULOMETRIC TITRATIONS INVOLVING NEUTRALIZATION, PRECIPITATION, AND COMPLEX-FORMATION REACTIONS

Species Determined •	Generator Electrode Reaction	Secondary Analytical Reaction
Acids	$2H_2O + 2e^- \rightleftharpoons 2OH^- + H_2$	$OH^- + H^+ \rightleftharpoons H_2O$
Bases	$H_2O \rightleftharpoons 2H^+ + \frac{1}{2}O_2 + 2e^-$	$H^+ + OH^- \rightleftharpoons H_2O$
Cl^-, Br^-, I^-	$Ag \rightleftharpoons Ag^+ + e^-$	$Ag^+ + Cl^- \rightleftharpoons AgCl(s)$, etc.
Mercaptans	$Ag \rightleftharpoons Ag^+ + e^-$	$Ag^+ + RSH \rightleftharpoons AgSR(s) + H^+$
Cl^-, Br^-, I^-	$2Hg \rightleftharpoons Hg_2^{2+} + 2e^-$	$Hg_2^{2+} + 2Cl^- \rightleftharpoons Hg_2Cl_2(s)$, etc.
Zn^{2+}	$Fe(CN)_6^{3-} + e^- \rightleftharpoons Fe(CN)_6^{4-}$	$3Zn^{2+} + 2K^+ + 2Fe(CN)_6^{4-} \rightleftharpoons$ $K_2Zn_3[Fe(CN)_6]_2(s)$
$Ca^{2+}, Cu^{2+}, Zn^{2+}, Pb^{2+}$	See Equation 19-9.	$HY^{3-} + Ca^{2+} \rightleftharpoons CaY^{2-} + H^+$, etc.

Here, the cathode must be isolated from the analyte solution to prevent interference from hydroxide ion.

Precipitation and Complex-Formation Reactions. Coulometric titrations with EDTA are carried out by reduction of the ammine mercury(II) EDTA chelate at a mercury cathode:

$$HgNH_3Y^{2-} + NH_4^+ + 2e^- \rightleftharpoons Hg(l) + 2NH_3 + HY^{3-} \qquad (19\text{-}9)$$

Because the mercury chelate is more stable than the corresponding complexes of such cations as calcium, zinc, lead, or copper, complexation of these ions occurs only after the ligand has been freed by the electrode process.

Table 19-4

SUMMARY OF COULOMETRIC TITRATIONS INVOLVING OXIDATION/REDUCTION REACTIONS

Reagent	Generator Electrode Reaction	Substance Determined
Br_2	$2Br^- \rightleftharpoons Br_2 + 2e^-$	As(III), Sb(III), U(IV), Tl(I), I^-, SCN^-, NH_3, N_2H_4, NH_2OH, phenol, aniline, mustard gas, mercaptans, 8-hydroxyquinoline, olefins
Cl_2	$2Cl^- \rightleftharpoons Cl_2 + 2e^-$	As(III), I^-, styrene, fatty acids
I_2	$2I^- \rightleftharpoons I_2 + 2e^-$	As(III), Sb(III), $S_2O_3^{2-}$, H_2S, ascorbic acid
Ce^{4+}	$Ce^{3+} \rightleftharpoons Ce^{4+} + e^-$	Fe(II), Ti(III), U(IV), As(III), I^-, $Fe(CN)_6^{4-}$
Mn^{3+}	$Mn^{2+} \rightleftharpoons Mn^{3+} + e^-$	$H_2C_2O_4$, Fe(II), As(III)
Ag^{2+}	$Ag^+ \rightleftharpoons Ag^{2+} + e^-$	Ce(III), V(IV), $H_2C_2O_4$, As(III)
Fe^{2+}	$Fe^{3+} + e^- \rightleftharpoons Fe^{2+}$	Cr(VI), Mn(VII), V(V), Ce(IV)
Ti^{3+}	$TiO^{2+} + 2H^+ + e^- \rightleftharpoons Ti^{3+} + H_2O$	Fe(III), V(V), Ce(IV), U(VI)
$CuCl_3^{2-}$	$Cu^{2+} + 3Cl^- + e^- \rightleftharpoons CuCl_3^{2-}$	V(V), Cr(VI), IO_3^-
U^{4+}	$UO_2^{2+} + 4H^+ + 2e^- \rightleftharpoons U^{4+} + 2H_2O$	Cr(VI), Ce(IV)

As shown in Table 19-3, several precipitating reagents can be generated coulometrically. The most widely used of these is silver ion, which is generated at a silver anode.

Oxidation/Reduction Titrations. Table 19-4 reveals that a variety of redox reagents can be generated coulometrically. Of particular interest is bromine, the coulometric generation of which forms the basis for a large number of coulometric methods. Of interest as well are reagents not ordinarily encountered in conventional volumetric analysis owing to the instability of their solutions; silver(II), manganese(III), and the chloride complex of copper(I) are examples.

Automatic Coulometric Titrators

A number of instrument manufacturers offer automatic coulometric titrators, most of which employ a potentiometric end point. Some of these instruments are multipurpose and can be used for the determination of a variety of species. Others are designed for a single type of analysis. Examples of instruments designed for a single type of analysis are chloride titrators, in which silver ion is generated coulometrically; sulfur dioxide monitors, where anodically generated bromine oxidizes the analyte to sulfate ions; carbon dioxide monitors, in which the gas, absorbed in monoethanolamine, is titrated with coulometrically generated base; and water titrators, in which Karl Fischer Reagent (see Feature 17-1, page 321) is generated electrolytically.

19E QUESTIONS AND PROBLEMS

Note: Numerical data are molar analytical concentrations in which the full formula of a species is provided. Molar equilibrium concentrations are supplied for species displayed as ions.

19-1. Briefly distinguish between
  *(a) concentration polarization and kinetic polarization.
  (b) an amperostat and a potentiostat.
  *(c) a coulomb and a faraday.
  (d) a working electrode and a counter electrode.
  *(e) the electrolysis circuit and the control circuit for controlled-potential methods.

19-2. Briefly define
  *(a) current density.
  (b) ohmic potential.
  *(c) coulometric titration.
  (d) controlled-potential electrolysis.
  *(e) current efficiency.
  (f) overvoltage.

*19-3. Describe three mechanisms responsible for the transport of dissolved species to and from an electrode surface.

19-4. How does the existence of a current affect the potential of an electrochemical cell?

*19-5. How do concentration polarization and kinetic polarization resemble one another? How do they differ?

19-6. What experimental variables affect concentration polarization in an electrochemical cell?

*19-7. Describe conditions that favor kinetic polarization in an electrochemical cell.

19-8. How do electrogravimetric and coulometric methods differ from potentiometric methods?

*19-9. Identify three factors that influence the physical characteristics of an electrolytic deposit.

19-10. What is the purpose of a cathode depolarizer?

*19-11. What is the function of (a) an amperostat and (b) a potentiostat?

19-12. Differentiate between amperostatic coulometry and potentiostatic coulometry.

*19-13. Why is it ordinarily necessary to isolate the working electrode from the counter electrode in a controlled-potential coulometric analysis?

19-14. Why is an auxiliary reagent always required in a coulometric titration?

19-15. Calculate the number of cations reduced at the surface of a cathode during each second that an electrochemical cell is operated at 0.020 A and the ions involved are
  (a) univalent.
  *(b) divalent.
  (c) trivalent.

19-16. Calculate the theoretical potential needed to initiate the deposition of
  *(a) copper from a solution that is 0.150 M in Cu^{2+} and

buffered to a pH of 3.00. Oxygen is evolved at the anode at 1.00 atm.

(b) tin from a solution that is 0.120 M in Sn^{2+} and buffered to a pH of 4.00. Oxygen is evolved at the anode at 770 torr.

*(c) silver bromide on a silver anode from a solution that is 0.0864 M in Br^- and buffered to a pH of 3.40. Hydrogen is evolved at the cathode at 765 torr.

(d) Tl_2O_3 from a solution that is 4.00×10^{-3} M in Tl^+ and buffered to a pH of 8.00. The solution is also made 0.100 M in Cu^{2+}, which acts as a cathode depolarizer. For the process

$$Tl_2O_3(s) + 3H_2O + 4e^- \rightleftharpoons 2Tl^+ + 6OH^-$$

$$E^0 = 0.020 \text{ V}$$

*19-17. Calculate the initial potential needed for a current of -0.078 A in the cell

$$Co|Co^{2+}(6.40 \times 10^{-2} \text{ M})\|Zn^{2+}(3.75 \times 10^{-3} \text{ M})|Zn$$

if this cell has a resistance of 5.00 Ω.

19-18. The cell

$$Sn|Sn^{2+}(8.22 \times 10^{-4} \text{ M})\|Cd^{2+}(7.50 \times 10^{-2} \text{ M})|Cd$$

has a resistance of 3.95 Ω. Calculate the initial potential that will be needed for a current of -0.072 A in this cell.

*19-19. Copper is to be deposited from a solution that is 0.200 M in Cu(II) and is buffered to a pH of 4.00. Oxygen is evolved from the anode at a partial pressure of 740 torr. The cell has a resistance of 3.60 Ω; the temperature is 25°C. Calculate

(a) the theoretical potential needed to initiate deposition of copper from this solution.

(b) the *IR* drop associated with a current of -0.10 A in this cell.

(c) the initial potential, given that the overvoltage of oxygen is 0.50 V under these conditions.

(d) the potential of the cell when $[Cu^{2+}]$ is 8.00×10^{-6}, assuming that *IR* drop and O_2 overvoltage remain unchanged.

19-20. Nickel is to be deposited on a platinum cathode (area = 120 cm^2) from a solution that is 0.200 M in Ni^{2+} and buffered to a pH of 2.00. Oxygen is evolved at a partial pressure of 1.00 atm at a platinum anode with an area of 80 cm^2. The cell has a resistance of 3.15 Ω; the temperature is 25°C. Calculate

(a) the theoretical potential needed to initiate the deposition of nickel.

(b) the *IR* drop for a current of -1.10 A.

(c) the current density at the anode and the cathode.

(d) the initial applied potential, given that the overvoltage of oxygen on platinum is approximately 0.52 V under these conditions.

(e) the applied potential when the nickel concentration has decreased to 2.00×10^{-4} M (assume that all variables other than $[Ni^{2+}]$ remain constant).

*19-21. Silver is to be deposited from a solution that is 0.150 M in $Ag(CN)_2^-$, 0.320 M in KCN, and buffered to a pH of 10.00. Oxygen is evolved at the anode at a partial pressure of 1.00 atm. The cell has a resistance of 2.90 Ω; the temperature is 25°C. Calculate

(a) the theoretical potential needed to initiate deposition of silver from this solution.

(b) the *IR* drop associated with a current of -0.12 A.

(c) the initial applied potential, given that the O_2 overvoltage is 0.80 V.

(d) the applied potential when $[Ag(CN)_2^-]$ is 1.00×10^{-5} M, assuming no changes in *IR* drop and O_2 overvoltage.

19-22. A solution is 0.150 M in Co^{2+} and 0.0750 M in Cd^{2+}. Calculate

(a) the Co^{2+} concentration in the solution as the first cadmium starts to deposit.

(b) the cathode potential needed to lower the Co^{2+} concentration to 1×10^{-5} M.

*19-23. A solution is 0.0500 M in BiO^+ and 0.0400 M in Co^{2+} and has a pH of 2.50.

(a) What is the concentration of the more readily reduced cation at the onset of deposition of the less reducible one?

(b) What is the potential of the cathode when the concentration of the more easily reduced species is 1.00×10^{-6} M?

19-24. Electrogravimetric analysis involving control of the cathode potential is proposed as a means for separating Bi^{3+} and Sn^{2+} in a solution that is 0.200 M in each ion and buffered to pH 1.50.

(a) Calculate the theoretical cathode potential at the start of deposition of the more readily reduced ion.

(b) Calculate the residual concentration of the more readily reduced species at the outset of the deposition of the less readily reduced species.

(c) Propose a range (vs. SCE), if such exists, within which the cathode potential should be maintained; consider a residual concentration less than 10^{-6} M as constituting quantitative removal.

*19-25. Halide ions can be deposited on a silver anode via the reaction

$$Ag(s) + X^- \rightleftharpoons AgX(s) + e^-$$

(a) If 1.00×10^{-5} M is used as the criterion for quantitative removal, is it theoretically feasible to separate Br^- from I^- through control of the anode potential in a solution that is initially 0.250 M in each ion?

(b) Is a separation of Cl^- and I^- theoretically feasible in a solution that is initially 0.250 M in each ion?

(c) If a separation is feasible in either (a) or (b), what range of anode potential (versus SCE) should be used?

19-26. A solution is 0.100 M in each of two reducible cations, A and B. Removal of the more reducible species (A) is deemed complete when [A] has been decreased to 1.00×10^{-5} M. What minimum difference in standard electrode potentials will permit the isolation of A without interference from B when

	A is	B is
*(a)	univalent	univalent
(b)	divalent	univalent
*(c)	trivalent	univalent
(d)	univalent	divalent
*(e)	divalent	divalent
(f)	trivalent	divalent
*(g)	univalent	trivalent
(h)	divalent	trivalent
*(i)	trivalent	trivalent

*19-27. Calculate the time needed for a constant current of 0.961 A to deposit 0.500 g of Co(II) as
   (a) elemental cobalt on the surface of a cathode.
   (b) Co_3O_4 on an anode.

19-28. Calculate the time needed for a constant current of 1.20 A to deposit 0.500 g of
   (a) Tl(III) as the element on a cathode.
   (b) Tl(I) as Tl_2O_3 on an anode.
   (c) Tl(I) as the element on a cathode.

*19-29. A 0.1516-g sample of a purified organic acid was neutralized by the hydroxide ion produced in 5 min and 24 s by a constant current of 0.401 A. Calculate the equivalent weight of the acid (the weight of acid that contains 1 mol of titratable protons).

19-30. The CN^- concentration of 10.0 mL of a plating solution was determined by titration with electrogenerated hydrogen ion to a methyl orange end point. A color change occurred after 3 min and 22 s with a current of 43.4 mA. Calculate the number of grams of NaCN per liter of solution.

*19-31. An excess of $HgNH_3Y^{2-}$ was introduced to 25.00 mL of well water. Express the hardness of the water in terms of ppm $CaCO_3$ if the EDTA needed for the titration was generated at a mercury cathode (Equation 19-9) in 2.02 min by a constant current of 31.6 mA.

19-32. Electrolytically generated I_2 was used to determine the amount of H_2S in 100.0 mL of brackish water. Following addition of excess KI, a titration required a constant current of 36.32 mA for 10.12 min. The reaction was

$$H_2S + I_2 \rightleftharpoons S(s) + 2H^+ + 2I^-$$

Express the results of the analysis in terms of ppm H_2S.

*19-33. The nitrobenzene in 210 mg of an organic mixture was reduced to phenylhydroxylamine at a constant potential of −0.96 V (vs. SCE) applied to a mercury cathode:

$$C_6H_5NO_2 + 4H^+ + 4e^- \rightleftharpoons C_6H_5NHOH + H_2O$$

The sample was dissolved in 100 mL of methanol; after electrolysis for 30 min, the reaction was judged complete. An electronic coulometer in series with the cell indicated

that the reduction required 26.74 C. Calculate the percentage of $C_6H_5NO_2$ in the sample.

19-34. The phenol content of water downstream from a coking furnace was determined by coulometric analysis. A 100-mL sample was rendered slightly acidic, and an excess of KBr was introduced. To produce Br_2 for the reaction

$$C_6H_5OH + 3Br_2 \rightleftharpoons Br_3C_6H_2OH(s) + 3HBr$$

a steady current of 0.0313 A for 7 min and 33 s was required. Express the results of this analysis in terms of parts of C_6H_5OH per million parts of water. (Assume that the density of water is 1.00 g/mL.)

*19-35. At a potential of −1.0 V (vs. SCE), CCl_4 in methanol is reduced to $CHCl_3$ at a mercury cathode:

$$2CCl_4 + 2H^+ + 2e^- + 2Hg(l) \rightleftharpoons 2CHCl_3 + Hg_2Cl_2(s)$$

At −1.80 V, the $CHCl_3$ further reacts to give CH_4:

$$2CHCl_3 + 6H^+ + 6e^- + 6Hg(l) \rightleftharpoons 2CH_4 + 3Hg_2Cl_2(s)$$

A 0.750-g sample containing CCl_4, $CHCl_3$, and inert organic species was dissolved in methanol and electrolyzed at −1.0 V until the current approached zero. A coulometer indicated that 11.63 C was required to complete the reaction. The potential of the cathode was adjusted to −1.8 V. Completion of the titration at this potential required an additional 68.6 C. Calculate the percentage of CCl_4 and $CHCl_3$ in the mixture.

19-36. A 0.1309-g sample containing only $CHCl_3$ and CH_2Cl_2 was dissolved in methanol and electrolyzed in a cell containing a mercury cathode; the potential of the cathode was held constant at −1.80 V (vs. SCE). Both compounds were reduced to CH_4 (see 19-35 for the reaction). Calculate the percentage of $CHCl_3$ and CH_2Cl_2 if 306.7 C was required to complete the reduction.

*19-37. Traces of $C_6H_5NH_2$ can be determined by reaction with an excess of electrolytically generated Br_2:

The polarity of the working electrode is then reversed, and the excess Br_2 is determined by a coulometric titration involving the generation of Cu(I):

$$Br_2 + 2Cu^+ \rightarrow 2Br^- + 2Cu^{2+}$$

Suitable quantities of KBr and $CuSO_4$ were added to a 25.0-mL sample containing aniline. Calculate the number of micrograms of $C_6H_5NH_2$ in the sample from the data:

Working Electrode Functioning As	Generation Time with a Constant Current of 1.51 mA, min
anodect	3.76
cathode	0.270

19-38. Quinone can be reduced to hydroquinone with an excess of electrolytically generated Sn(II):

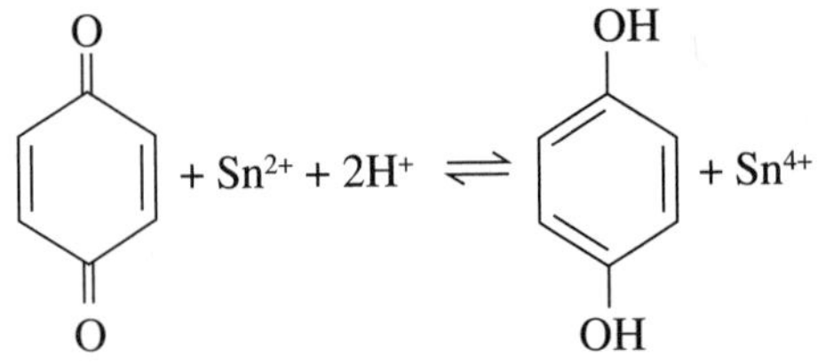

The polarity of the working electrode is then reversed, and the excess Sn(II) is oxidized with Br_2 generated in a coulometric titration:

$$Sn^{2+} + Br_2 \rightleftharpoons Sn^{4+} + 2Br^-$$

Appropriate quantities of $SnCl_4$ and KBr were added to a 50.0-mL sample. Calculate the weight of $C_6H_4O_2$ in the sample from the data:

Working Electrode Functioning As	Generation Time with a Constant Current of 1.062 mA, min
cathode	8.34
anodect	0.691

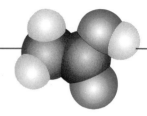

CHAPTER **20**

AN INTRODUCTION TO SPECTROSCOPIC METHODS OF ANALYSIS

Spectroscopic methods of analysis are based on measurement of the electromagnetic radiation produced or absorbed by analytes.[1] *Emission methods* make use of radiation given off when an analyte is exposed to thermal or electrical energy. *Fluorescence methods* are also based on radiation emitted by an analyte. With fluorescence, however, the emitted radiation is generated by exposing the sample to a beam of electromagnetic radiation from a lamp. *Absorption methods*, in contrast, are based on the decrease in power (*attenuation*) of a beam of electromagnetic radiation as a consequence of its interaction with the analyte.

> A beam of radiation is *attenuated* when its intensity or power is decreased.

Spectroscopic methods are also classified according to region of the electromagnetic spectrum. These regions include X-ray, ultraviolet, visible, infrared, microwave, and radio-frequency. Historically, early spectroscopic methods were restricted to use of visible radiation, and for this reason were termed *optical methods*. However, because of similar instrumentation, this terminology has been extended to include methods employing ultraviolet and infrared radiation despite the fact that the human optic nerve is sensitive to neither of these types of radiation. In this book we focus largely on optical methods that are based on ultraviolet/visible radiation, although occasional reference is made to infrared spectroscopic methods.

> Other analytically useful types of electromagnetic radiation include X-ray, infrared, microwave, and radio-frequency radiation.

> Optical spectroscopy involves the study of the emission, fluorescence, and absorption of ultraviolet, visible, and infrared radiation.

---

[1] For further study, see: E. J. Meehan, in *Treatise on Analytical Chemistry*, 2nd ed., Part I, Vol. 7, Chapters 1–3, P. J. Elving, E. J. Meehan, and I. M. Kolthoff, Eds. New York: Wiley, 1981; J. D. Ingle Jr. and S. R. Crouch, *Analytical Spectroscopy*. Englewood Cliffs, NJ: Prentice-Hall, 1988; J. E. Crooks, *The Spectrum in Chemistry*. New York: Academic Press, 1978.

20A   PROPERTIES OF ELECTROMAGNETIC RADIATION

Electromagnetic radiation is a form of energy that is transmitted through space at enormous velocities. Many of the properties of electromagnetic radiation are conveniently described by treating the radiation as sinusoidal waves that have such properties as wavelength, frequency, velocity, and amplitude. In contrast to other wave phenomena—for example, sound—electromagnetic radiation requires no supporting medium for its transmission; thus, it readily passes through a vacuum.

The wave model fails to account for phenomena associated with the absorption and emission of radiant energy. For these processes, electromagnetic radiation must be treated as a stream of discrete particles or wave packets of energy called *photons* or *quanta*. The energy of a photon is proportional to the frequency of the radiation. These dual views of radiation as particles and as waves are not mutually exclusive but complementary. Indeed, this duality applies to the behavior of streams of electrons as well as other elementary particles such as protons and is completely rationalized by wave mechanics.

20A-1 Wave Properties of Electromagnetic Radiation

For many purposes electromagnetic radiation is conveniently pictured as an electric field that undergoes sinusoidal oscillations in space. Figure 20-1 is a two-dimensional representation of a beam of monochromatic (single wavelength) radiation. The electric field is represented as a vector whose length is proportional to the field strength. The abscissa in this plot is either time as the radiation passes a fixed point in space or distance at a fixed time. Note that the direction in which the field oscillates is perpendicular to the direction in which the radiation is being propagated.

A wave in which the direction of displacement is perpendicular to the direction of propagation, as in Figure 20-1, is called a *transverse* wave.

Wave Parameters

In Figure 20-1, the *amplitude A* of the sinusoidal wave is defined as the length of the electrical vector at the maximum in the wave. The time required for the passage of successive maxima (or minima) through a fixed point in space is

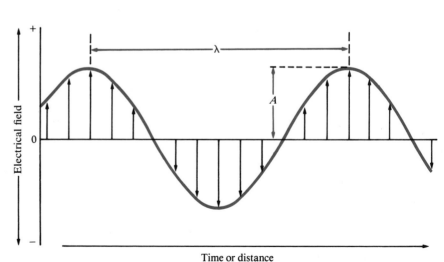

Figure 20-1

Representation of a beam of monochromatic radiation of wavelength λ and amplitude A. The arrows represent the electrical vector of the radiation.

called the *period p* of the radiation. The *frequency ν* is the number of oscillations of the field per second and is equal to $1/p$.

It is important to realize that *frequency* is determined by the source and remains *constant* regardless of the medium traversed by the radiation. In contrast, the *velocity* of propagation v_i of the wave front through a medium is *dependent* on both the medium and the frequency; the subscript i is employed to indicate this frequency dependence. Another parameter of interest is the *wavelength λ_i*, which is the linear distance between successive maxima or minima of a wave (see Figure 20-1). Multiplication of the frequency in waves per second by the wavelength in centimeters gives the velocity v_i of propagation in centimeters per second:

$$v_i = \nu\lambda_i \tag{20-1}$$

Velocity of Radiation

In a vacuum, the velocity at which radiation is propagated becomes independent of wavelength and is at its maximum. This velocity, which is given the symbol c, has been determined to be 2.99792×10^{10} cm/s. The velocity of radiation in air differs only slightly from c (it is about 0.03% less). Thus, for a vacuum, or for air, the velocity of light is conveniently rounded to

$$c = \nu\lambda \quad = 3.00 \times 10^{10} \text{ cm/s} = 3.00 \times 10^{10} \text{ m/s} \tag{20-2}$$

Radiation is propagated at a rate that is less than c in a medium containing matter because the electromagnetic field of the radiation interacts with the electrons in the atoms or molecules of the medium and is thus slowed. Since the radiant frequency is invariant and fixed by the source, the *wavelength of radiation must decrease* as it passes from a vacuum to a medium containing matter (Equation 20-1). This effect is illustrated in Figure 20-2 for a beam of visible radiation. Note that the wavelength shortens nearly 200 nm, or more than 30%, as the radiation passes from air into glass; a reverse change occurs when the radiation again enters air.

The *wavenumber, $\bar{\nu}$*, is yet another way of describing electromagnetic radiation. It is defined as the number of waves per centimeter and is equal to $1/\lambda$. By definition, $\bar{\nu}$ has units of cm^{-1}.

Radiant Power and Intensity

The *power, P*, is the energy of a beam that reaches a given area per second; the *intensity* is the power-per-unit solid angle. Both quantities are related to the amplitude of the radiation (see Figure 20-1). Although it is not strictly correct to do so, power and intensity are frequently used interchangeably.

20A-2 Particle Properties of Electromagnetic Radiation

Some interactions with matter require that we treat electromagnetic radiation as discrete packets of energy, called *photons* or *quanta*. The energy of a photon

---

The unit of frequency is the *hertz*, Hz, which corresponds to one cycle per second. That is, 1 Hz = 1 s^{-1}.

The frequency of a beam of electromagnetic radiation never changes.

WAVELENGTH UNITS FOR VARIOUS SPECTRAL REGIONS

Region	Unit	Definition
X-ray	Angstrom unit, Å	10^{-10} m
Ultraviolet/ visible	Nanometer, nm	10^{-9} m^0
Infrared	Micrometer, μm	10^{-6} m^0

To three significant figures, Equation 20-2 is equally applicable in air or vacuum.

The refractive index, η, of a medium is a measure of its interaction with radiation, and is defined by $\eta = c/v_i$. For example, the refractive index of water at room temperature is 1.33, which means that radiation passes through water at a rate of $c/1.33$ or 2.26×10^{10} cm/s. In other words, light travels 1.33 times as fast in vacuum as it does in water.

The velocity and the wavelength of radiation become proportionally smaller as the radiation passes from a vacuum, or from air, to a denser medium, while the frequency remains constant.

The *wavenumber, $\bar{\nu}$*, in cm^{-1} is generally used to describe infrared radiation. The most useful part of the infrared spectrum for the detection and determination of organic species is from 2.5 to 15 μm in wavelength, which corresponds to a wavenumber range of 4000 to 667 cm^{-1}. Wavenumber is used by infrared spectroscopists because the parameter is a measure of the frequency of molecular vibrations brought about by absorption of infrared radiation.

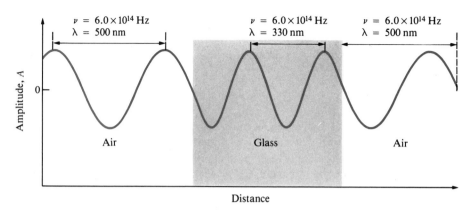

Figure 20-2
Change in wavelength as radiation
passes from air into a dense glass and
back to air.

Equations 20-3 and 20-4 give the
energy of radiation in SI units of
joules, where one joule (J) is the work
done by a force of one newton (N) acting over a distance of one meter.

Both frequency and wavenumber are proportional to the energy of a photon.

On occasion we speak of a "mole of pho-
tons," meaning 6.02×10^{23} packets of radiation.

REGIONS OF THE OPTICAL SPECTRUM

Region	Wavelength Range
UV	180–380 nm
Visible	380–780 nm
Near-IR	0.78–2.5 μm
Mid-IR	2.5–50 μm

Recall the order of the colors in the spectrum by the mnemonic *ROY G BIV*.

*R*ed
*O*range
*Y*ellow
*G*reen
*B*lue
*I*ndigo
*V*iolet

depends on the frequency of the radiation and is given by

$$E = h\nu \tag{20-3}$$

where h is the Planck constant (6.63×10^{-34} J s). We can also relate the energy of radiation to wavelength and wavenumber:

$$E = \frac{hc}{\lambda} = hc\bar{\nu} \tag{20-4}$$

Note that the wavenumber, in common with frequency, is directly proportional to energy.

20B   THE ELECTROMAGNETIC SPECTRUM

The electromagnetic spectrum covers an enormous range of wavelengths (or energies). An X-ray photon ($\lambda \sim 10^{-10}$ m), for example, is approximately 10,000 times as energetic as one that is emitted by an incandescent tungsten wire ($\lambda \sim 10^{-6}$ m) and 10^{11} times as energetic as a photon in the radio-frequency range ($\lambda \sim 10^{5}$ m).

The major divisions of the electromagnetic spectrum are depicted in the color plate located inside the front cover of this book. Note that both the frequency and the wavelength scales are logarithmic; note also that the region to which the human eye is perceptive (the *visible spectrum*) is but a minute part of the whole spectrum. Such diverse radiations as gamma rays or radio waves differ from visible light only in the matter of frequency and, hence, energy.

Figure 20-3 shows the regions of the magnetic spectrum that are used for spectroscopic analyses. Also shown are the types of atomic and molecular transitions that are responsible for absorption and emission in each region. Note that the wavelength and wavenumber scales are logarithmic.

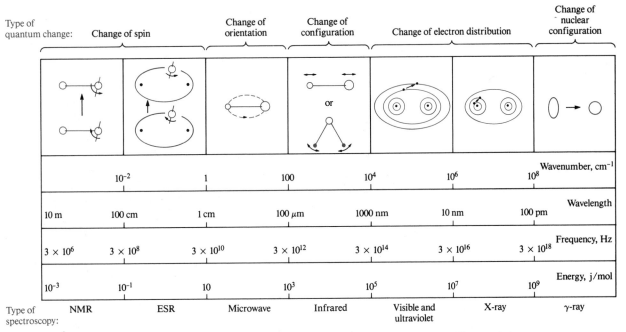

| Change of spin | Change of orientation | Change of configuration | Change of electron distribution | Change of nuclear configuration |

| Wavenumber, cm^{-1} |

10^{-2} 1   100 10^4 10^6 10^8

| Wavelength |

10 m   100 cm   1 cm   100 μm   1000 nm   10 nm   100 pm

| Frequency, Hz |

3×10^6 3×10^8 3×10^{10} 3×10^{12} 3×10^{14} 3×10^{16} 3×10^{18}

| Energy, j/mol |

10^{-3} 10^{-1} 10 10^3 10^5 10^7 10^9

Type of spectroscopy: NMR   ESR   Microwave   Infrared   Visible and ultraviolet   X-ray γ-ray

Figure 20-3

The regions of the electromagnetic spectrum. (From C. N. Banwell, *Fundamentals of Molecular Spectroscopy*, 3rd ed., p. 7. New York: McGraw-Hill, 1983.)

20C COMPONENTS OF INSTRUMENTS FOR OPTICAL SPECTROSCOPY

Most optical spectroscopic instruments are made up of five components, including: (1) a stable source of radiant energy; (2) a wavelength selector that isolates a limited region of the spectrum for measurement; (3) one or more sample containers; (4) a radiation detector, which converts radiant energy to a measurable signal (usually electrical); and (5) a signal processor and readout that displays the transduced signal on a meter scale, an oscilloscope face, a digital meter, or a recorder chart. Figure 20-4 illustrates the three ways these components are configured for carrying out optical spectroscopic measurements. As can be seen in the figure, components (3), (4), and (5) have similar configurations for each type of measurement.

The first two instrumental configurations, which are used for the measurement of absorption and fluorescence, require an external source of radiant energy. For absorption, the beam from the source passes directly through the sample into the wavelength selector (in some instruments, the position of the sample and selector is reversed). In fluorescence measurements, the source induces the sample to emit characteristic radiation, which is measured at an angle (usually 90 deg) with respect to the source beam.

Emission spectroscopy differs from the other types in that no external radiation source is required; the sample itself is the emitter (see Figure 20-4c). In emission spectroscopy, the sample in an external container is fed into an arc, a spark, or a

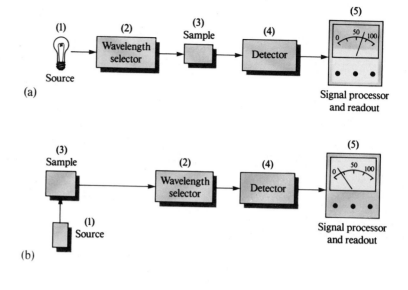

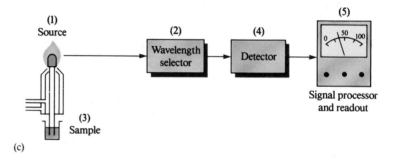

Figure 20-4

Components of various types of instruments for optical spectroscopy: (a) absorption spectroscopy; (b) fluorescence and scattering spectroscopy; (c) emission spectroscopy.

flame. The source then both contains the sample and causes it to emit characteristic radiation.

20C-1 The Transmittance of Optical Materials

The cells, windows, lenses, and wavelength dispersing elements in an optical instrument must be transparent in the wavelength region that has been selected. Figure 20-5 shows the usable wavelength ranges for several construction materials that find use for the ultraviolet, visible, and infrared regions of the spectrum. Ordinary silicate glass is completely adequate for the visible region and has the considerable advantage of low cost. Below about 380 nm, glass begins to absorb and fused silica or quartz must be substituted.

20C-2 Radiation Sources

To be suitable for spectroscopic studies, a source must generate a beam of radiation with sufficient power for easy detection and measurement. In addition, its output power should be stable for reasonable periods. Typically, the radiant power of a source varies exponentially with the potential of the electrical supply. Thus, a regulated power source is often needed to provide the required stability.

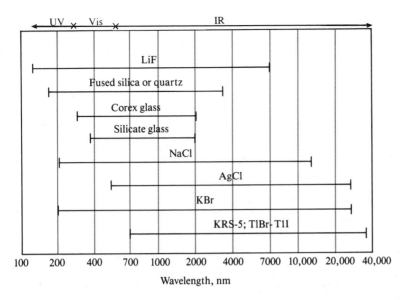

Figure 20-5

Transmittance range for various optical materials.

Spectroscopic sources are of two types: *continuous sources*, which emit radiation that changes in intensity only slowly as a function of wavelength, and *line sources*, which emit a limited number of bands of radiation, each of which spans a very limited range of wavelengths. Table 20-1 lists the most widely used continuous spectroscopic sources. Details concerning line sources are found in Section 24D-2.

Continuous Sources of Visible Radiation

An ordinary tungsten filament lamp provides a continuous spectrum from 320 to 2500 nm (Figure 20-6b). Its intensity varies as the fourth power of voltage. For this reason, an electronically stabilized voltage supply is usually required. An ordinary tungsten-filament lamp is generally operated at a temperature of 2900 K, which produces useful radiation for the wavelength region between 350 and 2500 nm.

Tungsten/halogen lamps contain a small quantity of iodine within the quartz envelope that houses the filament. Quartz allows the filament to be operated at

A continuous source provides radiation of all wavelengths within a particular spectral region.

Table 20-1

CONTINUOUS SOURCES FOR OPTICAL SPECTROSCOPY

Source	Wavelength Region, nm	Type of Spectroscopy
Xenon lamp	250–600	Molecular fluorescence
H_2 and D_2 lamps	160–380	UV molecular absorption
Tungsten/halogen lamp	240–2500	UV/vis/near-IR molecular absorption
Tungsten lamp	350–2200	Vis/near-IR molecular absorption
Nernst glower	400–20,000	IR molecular absorption
Nichrome wire	750–20,000	IR molecular absorption
Globar	1200–40,000	IR molecular absorption

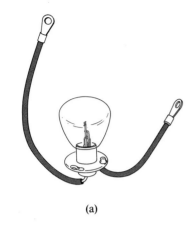

(a)

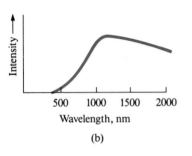

(b)

Figure 20-6

(a) A tungsten lamp. (b) Its emission spectrum.

a temperature of about 3500 K, which leads to higher intensities and extends the range of the lamp well into the ultraviolet. The lifetime of a tungsten/halogen lamp is more than double that of an ordinary tungsten lamp because the life of the latter is limited by sublimation of the tungsten from the filament. In the presence of iodine, the sublimed tungsten reacts to give gaseous WI_2 molecules, which then diffuse back to the hot filament, where they decompose and redeposit as tungsten atoms. Tungsten/halogen lamps are finding ever-increasing use in modern spectroscopic instruments because of their extended wavelength range, greater intensity, and longer life.

Continuous Sources of Ultraviolet Radiation

Deuterium (and also hydrogen) lamps are used to provide continuous radiation in the range of 160 to 380 nm and are the most common sources for ultraviolet spectroscopy. A deuterium lamp consists of a cylindrical tube with a quartz window from which the radiation exits (see Figure 20-7). The tube contains deuterium at a low pressure. Excitation is carried out by applying about 40 V between a heated oxide-coated electrode and a metal electrode. A regulated power supply is required to produce radiation of constant intensity. The emission spectrum for a typical deuterium lamp is shown in Figure 20-8.

Continuous Sources of Infrared Radiation

Continuous infrared radiation is obtained from heated inert solids. A *Globar* source consists of a 5- by 50-mm silicon carbide rod. Radiation in the region from 1 to 40 μm is emitted when the Globar is heated to about 1500°C by the passage of electricity.

A *Nernst glower* is a cylinder of zirconium and yttrium oxides having typical dimensions of 2 by 20 mm; it emits infrared radiation when heated to a high temperature by an electric current. Electrically heated spirals of nichrome wire also serve as infrared sources.

20C-3 Wavelength Selectors

Spectroscopic instruments are ordinarily equipped with one or more devices to restrict the radiation being measured to a narrow band that is absorbed or emitted by the analyte. Such devices greatly enhance both the selectivity and the sensitivity of an instrument. In addition, for absorption measurements, narrow bands increase the likelihood that the instrument will respond linearly to analyte concentration. Two general types of wavelength selectors, *monochromators* and *filters*, are used to provide narrow bands of radiation. Monochromators have the advantage that the output wavelength can be varied continuously over a considerable spectral range. Filters offer the advantages of simplicity, ruggedness, and cheapness. We will largely restrict our discussion in this book to monochromators because they are currently the more common type of wavelength selector found in modern spectroscopic instruments.

Monochromators

Monochromators are of two general types, one of which employs a grating to disperse radiation into its component wavelengths; the other uses a prism for

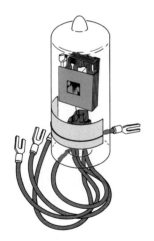

Figure 20-7

A deuterium lamp.

this purpose. Currently, nearly all commercially available monochromators are based on gratings, and we shall limit our discussion to this type.

Figure 20-9a shows the design of a typical grating monochromator. Radiation from a source enters the monochromator via a narrow rectangular opening or *slit*. The radiation is then collimated by a concave mirror, a device that produces a parallel beam that then strikes the surface of a *reflection grating*. For illustrative purposes, the radiation is shown as consisting of just two wavelengths, λ_1 and λ_2 where λ_1 is longer than λ_2. The pathway of the longer radiation after it is reflected from the grating is shown by dashed lines; the solid lines show the path of the shorter wavelength. Note that the shorter wavelength radiation λ_2 is reflected off the grating at a sharper angle than is λ_1. That is, *angular dispersion* of the radiation takes place. Angular dispersion results from diffraction, which occurs at the reflective surface (see Feature 20-1). The two wavelengths are focused by another concave mirror onto the *focal plane* of the monochromator, where they appear as two images of the entrance slit, one for λ_1 and the other for λ_2 (Figure 20-9b). By rotating the grating, either λ_1 or λ_2 can be focused on the exit slit of the monochromator.

If a detector is located at the exit slit of the monochromator shown in Figure 20-9a and the grating is rotated so that one of the lines shown (say, λ_1) is *scanned* across the slit from $\lambda_1 - \Delta\lambda$ to $\lambda_1 + \Delta\lambda$, where $\Delta\lambda$ is a small wavelength difference, the output of the detector takes the Gaussian shape shown in Figure 20-10. The *effective bandwidth* of the monochromator, which is defined in the figure, depends on the size and quality of the dispersing element, the slit widths, and the focal length of the monochromator. A high-quality monochromator will exhibit an effective bandwidth of a few tenths of a nanometer or less in the ultraviolet and visible regions. The effective bandwidth of a monochromator that is satisfactory for most quantitative applications is from about 1 to 20 nm.

Many monochromators are equipped with adjustable slits to permit some control over the bandwidth. A narrow slit decreases the effective bandwidth but

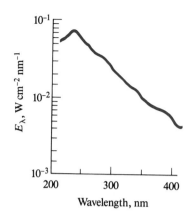

Figure 20-8
Output from a deuterium lamp.

A parallel beam of radiation is one that does not diverge or widen appreciably after leaving its source.

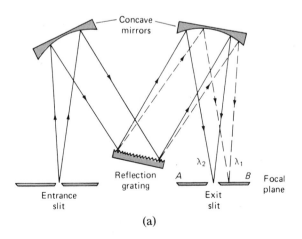

(a)

(b)

Figure 20-9
(a) Overhead view of a grating monochromator ($\lambda_1 > \lambda_2$). (b) View of the focal plane.

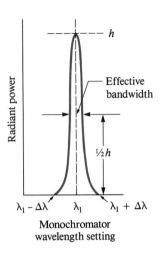

Figure 20-10
Output of an exit slit as monochromator is scanned from $\lambda_1 - \Delta\lambda$ to $\lambda_1 + \Delta\lambda$.

Feature 20-1
DISPERSION BY REFLECTION GRATINGS

Reflection gratings consist of a flat reflecting aluminum surface a few centimeters in length that contains a large number of parallel and closely spaced grooves. The surface is covered with a coating of transparent silica to prevent corrosion of the metal. A magnified cross section of a few of the grooves is shown in Figure 20-A. For ultraviolet/visible radiation a grating typically has 1200 to 1400 grooves/mm. For infrared radiation, 10 to 200 grooves/mm is common.

In the grating shown in the figure, parallel beams of monochromatic radiation labeled 1 and 2 strike two of the broad faces at an incident angle i to the *grating normal*. Maximum constructive interference is shown as occurring at the angle r. Beam 2 travels a greater distance than beam 1 and the difference in their paths is equal to $\overline{CB} + \overline{BD}$ (shown as a broader colored line in the figure). For constructive interference to occur, this difference must equal $\mathbf{n}\lambda$:

$$\mathbf{n}\lambda = \overline{CB} + \overline{BD}$$

where $\mathbf{n}$, a small whole number, is called the diffraction *order*. Note, however, that angle CAB is equal to angle i and that angle DAB is identical to angle r. Therefore, from trigonometry,

$$\overline{CB} = d \sin i$$

where d is the spacing between the reflecting surfaces. It is also seen that

$$\overline{BD} = d \sin r$$

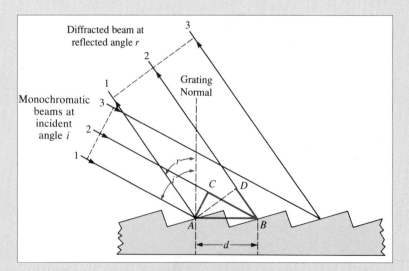

Figure 20-A Mechanisms of diffraction from an echellette-type grating.

Substitution of the last two expressions into the first gives the condition for constructive interference:

$$\mathbf{n}\lambda = d(\sin i + \sin r)$$

This equation suggests that several values of λ exist for a given diffraction angle r. Thus, if a first-order line ($\mathbf{n} = 1$) of 900 nm is found at r, second-order (450 nm) and third-order (300 nm) lines also appear at this angle. Ordinarily, the first-order line is the most intense; indeed, it is possible to design gratings that concentrate as much as 90% of the incident intensity in this order. The higher-order lines can generally be removed by filters. For example, glass, which absorbs radiation below 350 nm, eliminates the high-order spectra associated with first-order radiation in most of the visible region.

also diminishes the power of the emergent beam. Thus, the minimum practical bandwidth may be limited by the sensitivity of the detector. For qualitative analysis, narrow slits and minimum effective bandwidths are required if a spectrum consists of narrow peaks. For quantitative work, on the other hand, wider slits permit operation of the detector system at lower amplification, which in turn provides greater reproducibility of response.

Filters

Figure 20-11 shows the transmission characteristics of a typical *absorption filter* and a typical *interference filter*. The vertical axis in these plots is the percentage of radiation of a given wavelength that is transmitted by the device. Filters are generally characterized by the wavelength at which maximum transmission occurs and by their effective bandwidths. For example, the interference filter shown in the figure would be described as having a transmittance maximum of 66% at 450 nm and an effective bandwidth of about 10 nm.

   The most common type of absorption filter is a piece of colored glass mounted in a holder. Filters of this type have effective bandwidths of 50 to 250 nm and usually have lower peak transmittances than do interference filters. They are significantly less expensive than their interference counterparts. As their name implies, interference filters remove most of the radiation in a beam by destructive interference. They generally have effective bandwidths of 5 to 20 nm.

20C-4 Radiation Detectors and Transducers

A *detector* is a device that indicates the existence of some physical phenomenon. Familiar examples of detectors include photographic film for indicating the presence of electromagnetic or radioactive radiation, the pointer of a balance for detecting mass differences, and the mercury level in a thermometer for detecting temperature changes. The human eye is also a detector; it converts visible radiation into an electrical signal that is passed to the brain via the chain of neurons in the optic nerve.

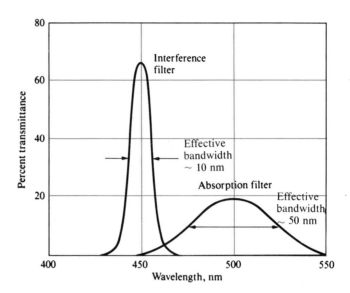

Figure 20-11
Output from two types of filters.

A transducer is a type of detector that converts various types of chemical and physical quantities to electrical voltage, charge, or current.

A *transducer* is a special type of detector that converts signals, such as light intensity, pH, mass, and temperature into *electrical* signals that can be subsequently amplified, manipulated, and finally converted into numbers proportional to the magnitude of the original signal.

Properties of Radiation Transducers

The ideal transducer for electromagnetic radiation responds rapidly to low levels of radiant energy over a broad wavelength range. In addition, it produces an electrical signal that is easily amplified and has a relatively low noise level. Finally, it is essential that the electrical signal produced by the transducer be directly proportional to the power of the beam P:

$$G = KP + K' \qquad (20\text{-}5)$$

where G is the electrical response of the detector in units of current, resistance, or potential. The proportionality constant K measures the sensitivity of the detector in terms of electrical response per unit of radiant power. Many detectors exhibit a small constant response, known as a *dark current* K', even when no radiation impinges on their surfaces. Instruments with detectors that have a significant dark-current response are ordinarily equipped with a compensating circuit that permits subtraction of a signal proportional to the dark current to reduce K' to zero. Thus under ordinary circumstances, we can write

A dark current is a current produced by a photoelectric detector in the absence of light.

$$G = KP \qquad (20\text{-}6)$$

Types of Transducers

As shown in Table 20-2, we encounter two general types of transducers: one type responds to photons, the other to heat. All photon detectors are based on the interaction of radiation with a reactive surface to produce electrons (photo-

Feature 20-2

SIGNALS, NOISE, AND THE SIGNAL-TO-NOISE RATIO

The output of an analytical instrument fluctuates in a random way. These fluctuations limit the precision of the instrument and are the net result of a large number of uncontrolled random variables in the instrument and in the chemical system under study. An example of such a variable is the random arrival of photons at the photocathode of a photomultiplier tube. The term *noise* is used to describe these fluctuations, and each uncontrolled variable is a noise source. The term comes from audio and electronic engineering, where undesirable signal fluctuations appear to the ear as static, or noise. The average value of the output of an electronic device is called the *signal*, and the standard deviation of the signal is a measure of the noise.

An important figure of merit for analytical instruments, stereos, compact-disk players, and many other types of electronic devices is the *signal-to-noise ratio* (S/N). The signal-to-noise ratio is usually defined as the ratio of the average value of the output signal to its standard deviation. The signal-to-noise behavior of an absorption spectrophotometer is illustrated in the

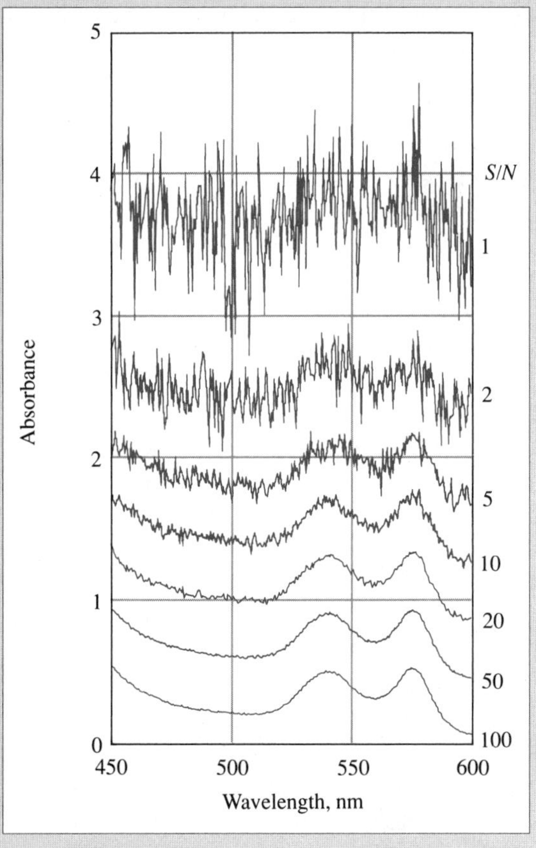

Generally, the output from analytical instruments fluctuates in a random way as a consequence of the operation of a large number of uncontrolled variables. These fluctuations, which limit the sensitivity of an instrument, are called *noise*. The terminology is derived from radio engineering, where the presence of unwanted signal fluctuations is recognizable to the ear as static, or noise.

Common causes for noise include vibration, pickup from 60-Hz lines, temperature variations, frequency or voltage fluctuations in the power supply, and the random arrival of photons at the detector.

Figure 20-B   Absorption spectra of hemoglobin with identical signal levels but different amounts of noise.

spectra of hemoglobin shown in Figure 20-B. The spectrum at the bottom of the figure has $S/N = 100$, and you can easily pick out the peaks at 540 nm and 580 nm. As the S/N degrades to about two in the second spectrum from the top of the figure, the peaks are barely visible. Somewhere between $S/N = 2$ and $S/N = 1$, the peaks disappear altogether into the noise and are impossible to identify. As modern instruments have become computerized and controlled by sophisticated electronic circuits, various methods have been developed to increase the signal-to-noise ratio of instrument outputs. These methods include analog filtering, lock-in amplification, boxcar averaging, smoothing, and Fourier transformation.[1]

[1] D. A. Skoog and J. J. Leary, *Principles of Instrumental Analysis,* Chapter 4. Philadelphia: Saunders College Publishing, 1992.

emission) or to promote electrons to energy states in which they can conduct electricity (photoconduction). Only ultraviolet, visible, and near-infrared radiation have sufficient energy to cause these processes to occur; thus photon detectors are limited to wavelengths shorter than about 2 μm.

Generally, infrared radiation is detected by measuring the temperature rise of a blackened material located in the path of the beam or by measuring the increase in electrical conductivity of a photoconducting material when it absorbs infrared radiation. Because the temperature changes resulting from the absorption of the infrared energy are minute, close control of the ambient temperature is required if large errors are to be avoided. It is usually the detector system that limits the sensitivity and precision of an infrared instrument.

Photon Detectors

Widely used types of photon detectors include: (1) phototubes, (2) photomultiplier tubes, (3) silicon photodiodes, and (4) photovoltaic cells.

Table 20-2
DETECTORS FOR SPECTROSCOPY

Type	Wavelength Range, nm
Photon Detectors	
Phototubes	150–1000
Photomultiplier tubes	150–1000
Silicon diodes	350–1100
Photoconductors	750–3000
Photovoltaic cells	380–780
Heat Detectors	
Thermocouples	600–20,000
Bolometers	600–20,000
Pneumatic cells	600–40,000
Pyroelectric cells	1000–20,000

Phototubes. As shown in Figure 20-12, a phototube consists of a semicylindrical cathode and a wire anode sealed inside an evacuated transparent envelope. The concave surface of the cathode supports a layer of photoemissive material, such as an alkali metal or metal oxide, that tends to emit electrons when it is irradiated. When a potential is applied across the electrodes, the emitted *photoelectrons* flow to the wire anode, producing a current (a *photocurrent*) that is readily amplified and displayed or recorded.

> Photoelectrons are electrons that are ejected from a photosensitive surface by electromagnetic radiation.

The number of electrons ejected from a photoemissive surface is directly proportional to the radiant power of the beam striking that surface. With an applied potential of about 90 V, all these electrons reach the anode to give a current that is also proportional to radiant power. Phototubes frequently produce a small dark current in the absence of radiation (Equation 20-5) that results from thermally induced electron emission.

Photomultiplier Tubes. The *photomultiplier tube* (PMT), shown schematically in Figure 20-13, is similar in construction to the phototube but is significantly more sensitive. The composition of its cathode surface is similar to that of a phototube, with electrons being emitted upon exposure to radiation. The emitted electrons are accelerated toward a *dynode* (labeled 1 in the figure) maintained at a potential 90 V more positive than the cathode. Each accelerated photoelectron that strikes the dynode surface produces several additional electrons, all of which are then accelerated to dynode 2, which is 90 V more positive than dynode 1. Again, electron amplification occurs. By the time this process has been repeated at each of the remaining dynodes, 10^6 to 10^7 electrons have been produced for each photon. This cascade of electrons is finally collected at the anode. The resulting current is then further amplified electronically and measured.

> One of the major advantages of photomultipliers is their automatic internal amplification. About 10^6 to 10^7 electrons arrive at the anode for each photon that strikes the photocathode of a photomultiplier tube.

> With modern electronic instrumentation, it is possible to detect the arrival of individual photons at the photocathode of a PMT. The current pulses produced when photons strike the cathode are counted, and the accumulated count is a measure of the intensity of the electromagnetic radiation impinging on the PMT. Photon counting is advantageous when light intensity, or the rate at which photons arrive at the photocathode, is low.

Silicon Photodiodes. Crystalline silicon is a *semiconductor*—that is, a material whose electrical conductivity is less than that of a metal but greater than that of an electrical insulator. Silicon is a Group IV element and thus has four valence electrons. In a silicon crystal, each of these electrons is combined with electrons from four other silicon atoms to form four covalent bonds. At room temperature, sufficient thermal agitation occurs in this structure to liberate an occasional

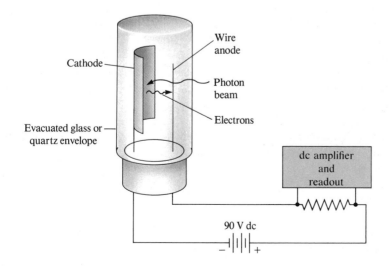

Cathode

Wire anode

Photon beam

Electrons

Evacuated glass or quartz envelope

dc amplifier and readout

90 V dc

Figure 20-12

A phototube and accessory circuit. The photocurrent induced by the radiation causes a potential difference across the resistor; this potential is then amplified to drive a meter or recorder.

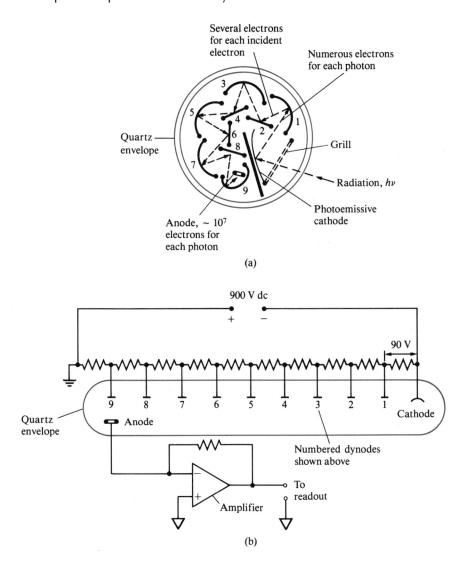

Figure 20-13

Diagram of a photomultiplier tube. (a) Cross section of the tube. (b) Electrical circuit.

electron from its bonded state, leaving it free to move throughout the crystal. Thermal excitation of an electron leaves behind a positively charged region termed a *hole*, which, like the electron, is also mobile. The mechanism of hole movement is stepwise, with a bound electron from a neighboring silicon atom jumping into the electron-deficient region (the hole) and thereby creating another positive hole in its wake. Conduction in a semiconductor involves the movement of electrons and holes in opposite directions.

The conductivity of silicon can be greatly enhanced by *doping*, a process whereby a tiny, controlled amount (approximately 1 ppm) of a Group V or Group III element is distributed homogenously throughout a silicon crystal. For example, when a crystal is doped with a Group V element, such as arsenic, four out of five of the valence electrons of the dopant form covalent bonds with four silicon atoms leaving one electron free to contribute to the conductivity of the crystal, as shown in Figure 20-14. In contrast, when the silicon is doped with a Group III element, such as gallium, which has but three valence electrons, an excess of holes develops, which also enhances conductivity (see Figure 20-15). A semiconductor containing unbonded electrons (*n*egative charges) is termed an *n-type* semiconductor, and one containing an excess of holes (*p*ositive charges) is a *p-type*. In an *n*-type semiconductor, electrons are the *majority carrier*; in a *p*-type, holes are the majority carrier.

Present silicon technology makes it possible to fabricate what is called a *pn junction* or a *pn diode*, which is conductive in one direction and not in the other. Figure 20-16a is a schematic of a silicon diode. The *pn* junction is shown as a dashed line through the middle of the crystal. Electrical wires are attached to both ends of the device. Figure 20-16b shows the junction in its conduction mode, wherein the positive terminal of a dc source is connected to the *p* region and the negative terminal to the *n* region (the diode is said to be *forward-biased* under these conditions). The excess electrons in the *n* region and the positive holes in the *p* region move toward the junction, where they combine and annihilate each other. The negative terminal of the source injects new electrons into the *n* region, which can continue the conduction process. The positive terminal extracts electrons from the *p* region, thus creating new holes that are free to migrate toward the *pn* junction.

Figure 20-16c illustrates the behavior of a silicon diode under *reverse biasing*. Here, the majority carriers are drawn away from the junction, leaving a nonconductive *depletion layer*. The conductance under reverse bias is only about 10^{-6} to 10^{-8} of that under forward biasing; thus, a silicon diode will pass a current in one direction but not in the other.

A reverse-biased silicon diode can serve as a radiation detector because ultraviolet and visible photons are sufficiently energetic to create additional electrons and holes when they strike the depletion layer of a *pn* junction. The resulting increase in conductivity is readily measured and is directly proportional to radiant power. A silicon-diode detector is more sensitive than a simple vacuum phototube but less sensitive than a photomultiplier tube.

Diode-Array Detectors. Silicon photodiodes have acquired notable importance recently because 1000 or more can be fabricated side by side on a single small silicon chip (the width of individual diodes is about 0.02 mm). With one or two of these *diode-array detectors* placed along the length of the focal plane of a monochromator, all wavelengths can be monitored simultaneously, thus making high-speed spectroscopy possible. Multichannel instruments based on diode arrays are discussed in Section 22A-1 and depicted in color plate 19.

Photovoltaic Cells. A photovoltaic cell (or photocell), the simplest of all radiation transducers, consists of a flat copper or iron electrode on which is deposited a layer of a semiconducting material, such as selenium or copper(I) oxide. The outer surface of the semiconductor is coated with a thin, transparent film of gold, silver, or lead, which serves as the second, or collector, electrode. When radiation is absorbed on the surface of the semiconductor, electrons and holes are formed and migrate in opposite directions, thus creating a current. If the two electrodes are connected through a low-resistance external circuit, the current produced is directly proportional to the power of the incident beam. The currents are large enough (10 to 100 μA) to be measured with a simple microammeter without amplification.

The typical photovoltaic cell has maximum sensitivity at about 550 nm, with the response falling off to perhaps 10% of maximum at 350 and 750 nm. The

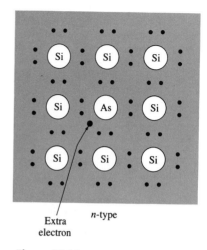

Figure 20-14

Two-dimensional representation of *n*-type silicon showing ''impurity'' atom.

In electronics, a bias is a dc voltage that is inserted in series with a circuit element.

A device in which an electrical current passes more readily in one direction than the other is termed a *rectifier*. Rectifiers are used to convert alternating currents to direct currents.

A silicon photodiode is a reverse-biased silicon diode that is used for measuring radiant power.

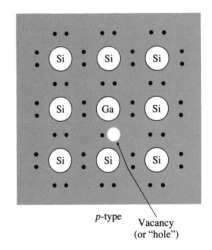

Figure 20-15

Two-dimensional representation of *p*-type silicon showing ''impurity'' atom.

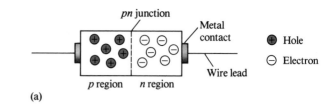

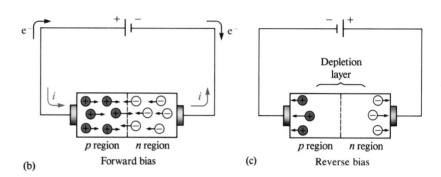

Figure 20-16

(a) Schematic of a silicon diode.
(b) Flow of electricity under forward bias. (c) Formation of depletion layer, which prevents flow of electricity under reverse bias.

photocell constitutes a rugged, low-cost detector of visible radiation that has the advantage of not requiring an external power source. It is not, however, as sensitive as other detectors and in addition suffers from *fatigue*, which causes its current output to decrease gradually with continued illumination. Despite these disadvantages, photovoltaic cells are quite useful for simple, portable, low-cost filter instruments.

Heat Detectors

The convenient photon detectors discussed in the previous section cannot be used to measure infrared radiation because photons of these frequencies lack the energy to cause photoemission of electrons; as a consequence, thermal detectors must be used. Unfortunately, the performance characteristics of thermal detectors are definitely inferior to those of phototubes, photomultiplier tubes, silicon diodes, or photovoltaic cells.

A thermal detector consists of a tiny blackened surface that absorbs infrared radiation and increases in temperature as a consequence. The temperature rise is converted to an electrical signal that is amplified and measured. Under the best of circumstances, the temperature changes involved are tiny, amounting to a few thousandths of a degree Celsius. The difficulty of measurement is compounded by thermal radiation from the surroundings, which is always a potential source of uncertainty. To minimize this background radiation, or noise, thermal detectors are housed in a vacuum and are carefully shielded from their surroundings. To further reduce the effects of this external noise, the beam from the source is chopped by a rotating disk inserted between source and detector. Chopping produces a beam that fluctuates regularly from zero intensity to a maximum. The transducer converts this periodic radiation signal to an alternating electrical current that can be amplified and separated from the dc signal resulting from the background radiation. Despite all these measures, infrared measurements are significantly less precise than measurements of ultraviolet and visible radiation. As shown in Table 20-2, four types of heat detectors are used for infrared spectroscopy.

Signal Processors and Readouts

A signal processor is ordinarily an electronic device that amplifies the electrical signal from the detector; in addition, it may alter the signal from dc to ac (or the reverse), change the phase of the signal, and filter it to remove unwanted components. The signal processor may also be called on to perform such mathematical operations on the signal as differentiation, integration, or conversion to a logarithm.

Several types of readout devices are found in modern instruments. Digital meters, scales of potentiometers, recorders, cathode-ray tubes, and monitors of microcomputers are some examples.

20C-5 Sample Containers

Sample containers, which are usually called *cells* or *cuvettes*, must have windows fabricated from a material that is transparent in the spectral region of interest. Thus, as shown in Figure 20-5, quartz or fused silica is required for the ultraviolet region (below 350 nm) and may be used in the visible region and to about 3000 nm in the infrared. Because of its lower cost, silicate glass is ordinarily used for the region between 375 and 2000 nm. Plastic containers have also found application in the visible region. The most common window material for infrared studies is crystalline sodium chloride.

The best cells have windows that are normal to the direction of the beam to minimize reflection losses. The most common cell length for studies in the ultraviolet and visible regions is 1 cm; matched, calibrated cells of this size are available from several commercial sources. Other path lengths, from shorter than 0.1 cm to 10 cm, can also be purchased. Transparent spacers for shortening the path length of 1-cm cells to 0.1 cm are also available. Some typical cells are shown in Figure 20-17.

Avoid touching or scratching the windows of cuvettes.

Figure 20-17
Typical examples of commercially available cells.

For reasons of economy, cylindrical cells are sometimes encountered. Particular care must be taken to duplicate the position of such cells with respect to the beam; otherwise variations in path length and reflection loss at the curved surfaces can cause significant error.

The quality of spectroscopic data is critically dependent on the way the matched cells are used and maintained. Fingerprints, grease, or other deposits on the walls markedly alter the transmission characteristics of a cell. Thus, thorough cleaning before and after use is imperative, and care must be taken to avoid touching the windows after cleaning is complete. Matched cells should never be dried by heating in an oven or over a flame because this may cause physical damage or a change in path length. Matched cells should be calibrated against each other regularly with an absorbing solution.

20D   QUESTIONS AND PROBLEMS

20-1. Describe the differences between the following and list any particular advantages possessed by one over the other:
  (a) filters and monochromators as wavelength selectors.
  ***(b)** photovoltaic cells and phototubes as detectors for electromagnetic radiation.
  (c) phototubes and photomultiplier tubes.
  ***(d)** photon and heat detectors.

20-2. Define the term *effective bandwidth of a filter*.

***20-3.** Why are photomultiplier tubes unsuited for the detection of infrared radiation?

20-4. Why do quantitative and qualitative analyses often require different monochromator slit widths?

***20-5.** Why is iodine sometimes introduced into a tungsten lamp?

20-6. Calculate the frequency in hertz of
  ***(a)** an X-ray beam with a wavelength of 2.65 Å.
  (b) an emission line for copper at 211.0 nm.
  ***(c)** the line at 694.3 nm produced by a ruby laser.
  (d) the output of a CO_2 laser at 10.6 μm.
  ***(e)** an infrared absorption peak at 19.6 μm.
  (f) a microwave beam at 1.86 cm.

20-7. Calculate the wavelength in centimeters of
  ***(a)** an airport tower transmitting at 118.6 MHz.
  (b) a VOR (radio navigation aid) transmitting at 114.10 kHz.

***(c)** an NMR signal at 105 MHz.
  (d) an infrared absorption peak with a wavenumber of 1210 cm^{-1}.

20-8. A typical simple infrared spectrophotometer covers a wavelength range from 3 to 15 μm. Express its range (a) in wavenumbers and (b) in hertz.

***20-9.** A sophisticated ultraviolet/visible/near-IR instrument has a wavelength range of 185 to 3000 nm. What are its wavenumber and frequency ranges?

20-10. Calculate the frequency in hertz and the energy in joules of an X-ray photon with a wavelength of 2.70 Å.

***20-11.** Calculate the wavelength in cm and the energy in joules associated with a signal at 220 mHz.

20-12. Calculate the wavelength of
  ***(a)** the sodium line at 589 nm in an aqueous solution with a refractive index of 1.35.
  (b) the output of a ruby laser at 694.3 nm when it is passing through a piece of quartz that has a refractive index of 1.55.

20-13. Calculate the velocity, frequency, and wavelength of a potassium emission line at 404.5 nm as light from this source passes through an optical cell whose refractive index η_D is 1.43.

* Answers to the asterisked problems are given in the answers section at the back of the book.

CHAPTER 21

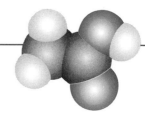

THEORY OF MOLECULAR ABSORPTION SPECTROSCOPY

Every molecular analyte is capable of absorbing certain characteristic wavelengths of electromagnetic radiation. In this process, the energy of the radiation is temporarily transferred to the molecule and the intensity of the radiation is decreased as a consequence. When absorption occurs the radiation is said to be *attenuated*. In this chapter, we first discuss the relationship between the degree of attenuation of radiation and the concentration of an absorbing analyte. We then turn to a consideration of the mechanisms by which various types of molecular species absorb ultraviolet, visible, and infrared radiation.

> In spectroscopy, *attenuate* means to decrease the energy per unit area of a beam of radiation.

21A TERMS EMPLOYED IN ABSORPTION SPECTROSCOPY

Table 21-1 lists the common terms and symbols used in absorption spectroscopy. This nomenclature is recommended by the American Society for Testing Materials as well as by the American Chemical Society. The third column contains alternative symbols you may encounter in the older literature. Because a standard nomenclature is highly desirable in order to avoid ambiguities, we urge you to learn and use the recommended terms and symbols and avoid the older terms.

21A-1 Transmittance

Figure 21-1 depicts the attenuation of a collimated beam of monochromatic radiation before and after it has passed through a layer of solution with a thickness of b cm and a concentration c of an absorbing species. As a consequence of interactions between the photons and absorbing particles, the power of the beam decreases from P_0 to P. The *transmittance* T of the solution is defined as the

> Percent transmittance $= \%T = \dfrac{P}{P_0} \times 100\%$

Table 21-1

IMPORTANT TERMS AND SYMBOLS EMPLOYED
IN ABSORPTION MEASUREMENT

Term and Symbol*	Definition	Alternative Name and Symbol
Radiant power P, P_0	Energy of radiation (in ergs) impinging on a 1-cm² area of a detector per second	Radiation intensity I, I_0
Absorbance A	$\log \dfrac{P_0}{P}$	Optical density D; extinction E
Transmittance T	$\dfrac{P}{P_0}$	Transmission T
Path length of radiation† b	—	l, d
Absorptivity† a	$\dfrac{A}{bc}$	Extinction coefficient k
Molar absorptivity‡ ε	$\dfrac{A}{bc}$	Molar extinction coefficient

* Terminology recommended by the American Chemical Society (*Anal. Chem.*, **1990**, *62*, 91).
† c may be expressed in g/L or in other specified concentration units; b may be expressed in cm or in other units of length.
‡ c is expressed in mol/L; b is expressed in cm.

fraction of incident radiation transmitted by the solution:

$$T = P/P_0 \tag{21-1}$$

Transmittance is often expressed as a percentage.

21A-2 Absorbance

The absorbance of a solution is defined by the equation

$$A = -\log_{10} T = \log \frac{P_0}{P} \tag{21-2}$$

Note that the absorbance of a solution increases as the transmittance decreases.

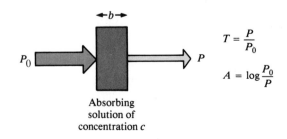

Figure 21-1

Attenuation of a beam of radiation by an absorbing solution.

21A-3 Experimental Transmittances and Absorbances

Ordinarily, transmittance and absorbance, as defined by Equations 21-1 and 21-2, cannot be measured in the laboratory because the solution to be studied must be held in some sort of container. Interaction between the radiation and the container walls is inevitable, with losses in power occurring at each interface as a result of reflection and possibly absorption (see Figure 21-3). Reflection losses are substantial. For example, it can be shown that about 8.5% of a beam of yellow light is lost by reflection when it passes through a glass cell containing water, such as shown in Figure 21-3. In addition to such losses, scattering by large molecules or inhomogeneities in the solvent may cause attenuation of the beam as it traverses the solution.

To compensate for these effects, the power of the beam transmitted through a cell containing the analyte solution is compared with one which traverses an identical cell containing only the solvent for the sample. An experimental absorbance that closely approximates the true absorbance for the solution is thus obtained; that is,

$$A = \log \frac{P_0}{P} \cong \log \frac{P_{solvent}}{P_{solution}} \qquad \textbf{(21-3)}$$

The terms P_0 and P, when used henceforth, refer to the power of radiation after it has passed through cells containing the solvent and the analyte, respectively.

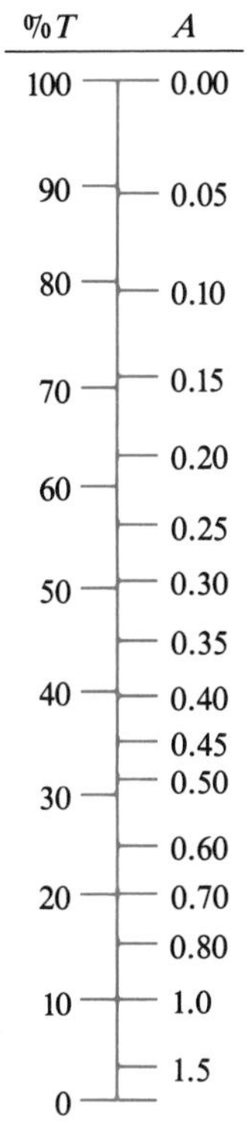

21B BEER'S LAW

According to Beer's law, absorbance is linearly related to the concentration of the absorbing species, c, and the path length, b, of the radiation in the absorbing medium. That is,

$$A = \log(P_0/P) = abc \qquad \textbf{(21-4)}$$

where a is a proportionality constant called the *absorptivity*. Because absorbance

%T	A
100	0.00
90	0.05
80	0.10
70	0.15
	0.20
60	0.25
50	0.30
	0.35
40	0.40
	0.45
30	0.50
	0.60
20	0.70
	0.80
10	1.0
	1.5
0	

Figure 21-2

A spectrophotometer readout. The readout scales on some spectrophotometers are linear in percent transmittance. The absorbance scales must then be logarithmic.

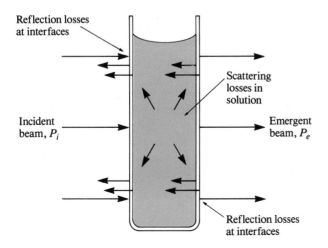

Figure 21-3

Reflection and scattering losses.

The molar absorptivity of a species in an absorption peak is often used to characterize the species. Peak molar absorptivities for many organic compounds range from 10 or less to 10,000 or more. Some transition metal complexes have molar absorptivities of 10,000 to 50,000. High molar absorptivities are desirable for quantitative analysis because they lead to greater analytical sensitivity.

is a unitless quantity, the absorptivity must have units that cancel the units of b and c.[1]

When concentration in Equation 21-4 is expressed in moles per liter and b is in centimeters, the proportionality constant is called the *molar absorptivity* and is given the special symbol ε. Thus,

$$A = \varepsilon bc \qquad \text{where } C = \frac{mol}{L} \qquad (21\text{-}5)$$

where ε has the units of $\text{L cm}^{-1} \text{mol}^{-1}$.

Example 21-1

A 7.50×10^{-5} M solution of potassium permanganate has a transmittance of 36.4% when measured in a 1.05-cm cell at a wavelength of 525 nm. Calculate (a) the absorbance of this solution; (b) the molar absorptivity of $KMnO_4$.

(a) $A = -\log T = -\log 0.364 = -(-0.4389) = 0.439$

(b) Rearranging Equation 21-5

$$\varepsilon = A/bc = 0.4389/(1.05 \text{ cm} \times 7.50 \times 10^{-5} \text{ mol/L})$$

$$= 5.57 \times 10^3 \text{ cm}^{-1} \text{mol}^{-1} \text{L}$$

21B-1 Application of Beer's Law to Mixtures

Beer's law also applies to solutions containing more than one kind of absorbing substance. Provided no interaction occurs among the various species, the total absorbance for a multicomponent system is the sum of the individual absorbances:

$$A_{\text{total}} = A_1 + A_2 + \cdots + A_n = \varepsilon_1 bc_1 + \varepsilon_2 bc_2 + \cdots + \varepsilon_n bc_n \qquad (21\text{-}6)$$

where the subscripts refer to absorbing components $1, 2, \cdots, n$.

21B-2 Limitations to the Applicability of Beer's Law

The linear relationship between absorbance and path length at a fixed concentration is a generalization for which there are few if any exceptions. On the other hand, deviations from the direct proportionality between absorbance and concentration (when b is constant) are frequently observed. Some of these deviations are fundamental and represent real limitations to the law. Others occur as a consequence of the manner in which the absorbance measurements are made (*instrumental deviations*) or as a result of chemical changes associated with concentration changes (*chemical deviations*).

---

[1] For a derivation of Beer's law see D. J. Swinehart, *J. Chem. Educ.*, **1972**, *32*, 333.

Real Limitations to Beer's Law

Beer's law is successful in describing the absorption behavior of dilute solutions only and in this sense is a *limiting law*. At high concentrations (usually >0.01 M), the average distances between ions or molecules of the absorbing species are diminished to the point where each particle affects the charge distribution of its neighbors. This interaction can alter their ability to absorb a given wavelength of radiation. Because the extent of interaction depends on concentration, the occurrence of this phenomenon causes deviations from the linear relationship between absorbance and concentration. A similar effect is sometimes encountered in dilute solutions of absorbers that contain high concentrations of other species, particularly electrolytes. The close proximity of ions to the absorber alters the molar absorptivity of the latter by electrostatic interactions, which lead to departures from Beer's law.

> Real limitations to Beer's law are encountered only in relatively concentrated solutions of the analyte or in concentrated electrolyte solutions.

Chemical Deviations

Deviations from Beer's law appear when the absorbing species undergoes association, dissociation, or reaction with the solvent to give products that absorb differently than the analyte. As shown in Example 21-2, the extent of such departures can be predicted from the molar absorptivities of the absorbing species and the equilibrium constants for the equilibria that are involved.

> Chemical deviations from Beer's law are encountered when the absorbing species participates in a concentration-dependent equilibrium such as a dissociation or association reaction.

Example 21-2

Solutions containing various concentrations of the acidic indicator HIn ($K_a = 1.42 \times 10^{-5}$) were prepared in 0.1 M HCl and 0.1 M NaOH. In both media, a linear relationship between absorbance and concentration was observed at 430 and 570 nm. From the magnitude of the acid dissociation constant, it is apparent that, for all practical purposes, the indicator is entirely in the undissociated form (HIn) in the HCl solution and completely dissociated as In^- in NaOH. The molar absorptivities at the two wavelengths were found to be

	ε_{430}	ε_{570}
HIn (HCl solution)	6.30×10^2	7.12×10^3
In^- (NaOH solution)	2.06×10^4	9.60×10^2

Derive absorbance data (1.00-cm cell) at the two wavelengths for unbuffered solutions with indicator concentrations ranging from 2×10^{-5} to 16×10^{-5} M. Plot the data.

Let us calculate the concentration of HIn and In^- in an unbuffered 2.00×10^{-5} M solution of the indicator. From the equation for the dissociation reaction, it is apparent that

$$[H^+] = [In^-]$$

Furthermore

$$[In^-] + [HIn] = 2.00 \times 10^{-5} \text{ M}$$

Substitution of these relationships into the expression for K_a gives

$$\frac{[In^-]^2}{2.00 \times 10^{-5} - [In^-]} = 1.42 \times 10^{-5}$$

Rearrangement yields the quadratic expression

$$[In^-]^2 + (1.42 \times 10^{-5})[In^-] - 2.84 \times 10^{-10} = 0$$

which can be solved to give

$$[In^-] = 1.12 \times 10^{-5}$$
$$[HIn] = 2.00 \times 10^{-5} - 1.12 \times 10^{-5} = 0.88 \times 10^{-5}$$

The absorbances at the two wavelengths are found by substituting into Equation 21-6

$$A_{430} = (6.30 \times 10^2) \times 1.00 \times (0.88 \times 10^{-5}) + (2.06 \times 10^4)$$
$$\times 1.00 \times (1.12 \times 10^{-5})$$
$$= 0.236$$
$$A_{570} = 7.12 \times 10^3 \times 1.00 \times 0.88 \times 10^{-5} + 9.60 \times 10^2$$
$$\times 1.00 \times 1.12 \times 10^{-5}$$
$$= 0.073$$

The following data were derived in a similar way and are plotted in Figure 21-4.

c_{HIn}, M	[HIn]	[In⁻]	A_{430}	A_{570}
2.00×10^{-5}	0.88×10^{-5}	1.12×10^{-5}	0.236	0.073
4.00×10^{-5}	2.22×10^{-5}	1.78×10^{-5}	0.381	0.175
8.00×10^{-5}	5.27×10^{-5}	2.73×10^{-5}	0.596	0.401
12.00×10^{-5}	8.52×10^{-5}	3.48×10^{-5}	0.771	0.640
16.00×10^{-5}	11.9×10^{-5}	4.11×10^{-5}	0.922	0.887

The data for the plot in Figure 21-4 were computed by the method shown in Example 21-2 and illustrate chemical deviations from Beer's law. Note that the direction of curvature for the one plot is opposite that for the other.

Instrumental Deviations with Polychromatic Radiation

Beer's law is also a limiting law in the sense that it applies only to absorbance measurements with monochromatic radiation. Truly monochromatic sources, such as lasers, are not practical for routine analytical measurements. Instead, a polychromatic continuous source is employed in conjunction with a grating or a filter that isolates a more or less symmetric band of wavelengths around the wavelength to be used (Figures 20-10 and 20-11). The derivation in Feature 21-1 illustrates how such a source may lead to deviations from Beer's law.

Deviations from Beer's law often occur when polychromatic radiation is used to measure absorbance.

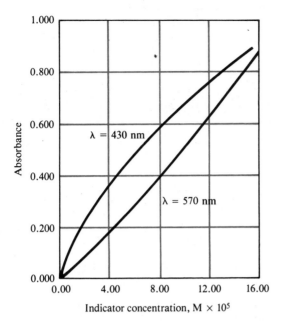

Figure 21-4

Chemical deviations from Beer's law for unbuffered solutions of the indicator HIn. For data, see Example 21-2.

It is an experimental fact that deviations from Beer's law resulting from the use of a polychromatic beam are not appreciable, provided the radiation used does not encompass a spectral region in which the absorber exhibits large changes in absorbance as a function of wavelength. This observation is illustrated in Figure 21-5.

Most molecular absorption peaks in the ultraviolet/visible spectral region are sufficiently broad (see Figure 21-11, for example) that close adherence to Beer's law can be expected if measurements are made at a peak maximum. In the infrared region, however, peaks are generally so narrow (see Figure 21-15) that departures from Beer's law are the rule. Nevertheless, quantitative analysis with infrared radiation is feasible although calibration plots are usually curved and require a larger number of data points to precisely define the relationship between absorbance and concentration.

Atomic absorption peaks in the ultraviolet/visible region are also so narrow (0.002 to 0.005 nm) that quantitative determination of the elements by ordinary absorption spectroscopy is impossible (unless a very special type of monochromator is employed). The problem created by the limited width of atomic absorption lines was solved in the 1950s by the use of line sources that emit radiation having bandwidths even narrower than the width of absorption peaks. For example, if sodium is to be determined, a sodium vapor lamp is used that emits lines that are narrower than sodium absorption peaks and at the same wavelengths. With this type of source, linear plots of absorbance versus concentration are observed. Atomic absorption spectroscopy is a powerful tool for elemental analysis and is discussed in detail in Chapter 24.

Instrumental Deviations in the Presence of Stray Radiation

The radiation employed for absorbance measurements is usually contaminated with small amounts of *stray* radiation due to instrumental imperfections. Stray

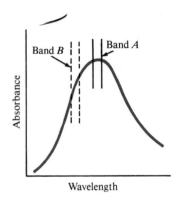

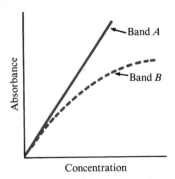

Figure 21-5

The effect of polychromatic radiation on Beer's law. Band *A* shows little deviation because ε does not change greatly throughout the band. Band *B* shows marked deviation because ε undergoes significant changes in this region.

Feature 21-1

DEMONSTRATION OF AN INSTRUMENTAL DEVIATION FROM BEER'S LAW

Consider a beam made up of just two wavelengths λ' and λ'', and assume that Beer's law applies strictly to each individually. With this assumption, we can write for radiation λ'

$$A' = \log \frac{P_0'}{P'} = \varepsilon' bc$$

$$\frac{P_0'}{P'} = 10^{\varepsilon' bc} \qquad \text{and} \qquad P' = P_0' 10^{-\varepsilon' bc}$$

Similarly, for λ''

$$\frac{P_0''}{P''} = 10^{\varepsilon'' bc} \qquad \text{and} \qquad P'' = P_0'' 10^{-\varepsilon'' bc}$$

When an absorbance measurement is made with radiation composed of both wavelengths, the power of the beam emerging from the solution is given by $P' + P''$ and that of the beam emerging from the solvent by $P_0' + P_0''$. Therefore, the measured absorbance is

$$A_m = \log \left(\frac{P_0' + P_0''}{P' + P''} \right)$$

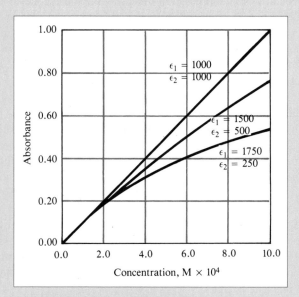

Figure 21-A
Deviations from Beer's law with polychromatic light. The absorber has the indicated molar absorptivities at the two wavelengths λ' and λ''.

which can be rewritten as

$$A_m = \log\left(\frac{P_0' + P_0''}{P_0'10^{-\varepsilon' bc} + P_0''10^{-\varepsilon'' bc}}\right)$$
$$= \log(P_0' + P_0'') - \log(P_0'10^{-\varepsilon' bc} + P_0''10^{-\varepsilon'' bc})$$

Now, when $\varepsilon' = \varepsilon''$, this equation simplifies to

$$A_m = \varepsilon' bc$$

and Beer's law is followed. As shown in Figure 21-A, however, the relationship between A_m and concentration is no longer linear when the molar absorptivities differ. Moreover, departures from linearity become greater as the difference between ε' and ε'' increases. When this treatment is expanded to include additional wavelengths, the effect remains the same.

radiation is the result of scattering and reflection off the surfaces of gratings, lenses, filters, and windows. It often differs greatly in wavelength from the principal radiation and, in addition, may not have passed through the sample or solvent.

When measurements are made in the presence of stray radiation, the observed absorbance is given by

$$A' = \log\frac{P_0 + P_s}{P + P_s}$$

where P_s is the power of the stray radiation. Figure 21-6 shows a plot of A' versus concentration for various levels of P_s relative to P_0.

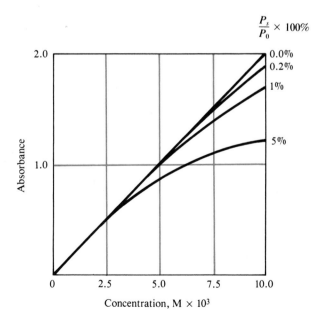

Figure 21-6

Apparent deviation from Beer's law caused by various amounts of stray radiation.

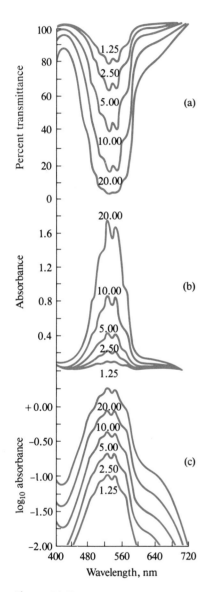

Figure 21-7

Methods for plotting spectral data. The numbers for the curves indicate ppm of $KMnO_4$ in the solution; $b = 2.00$ cm. (From M. G. Mellon, *Analytical Absorption Spectroscopy*, pp. 104–106. New York: Wiley, 1950. With permission.)

Feature 21-2
WHY IS A RED SOLUTION RED?

A solution, such as $FeSCN^{2+}$, is red not because the complex adds red radiation to the solvent. Instead it absorbs green from the incoming white radiation and transmits the red component unaltered. Thus in a colorimetric analysis of iron based on its thiocyanate complex, the maximum change in absorbance with concentration occurs with green radiation; the absorbance change with red radiation is negligible. In general, then, the radiation used for a colorimetric analysis should be the complementary color of the analyte solution. The table below shows this relationship for various parts of the visible spectrum.

THE VISIBLE SPECTRUM

Wavelength Region, nm	Color	Complementary Color
400–435	Violet	Yellow–green
435–480	Blue	Yellow
480–490	Blue–green	Orange
490–500	Green–blue	Red
500–560	Green	Purple
560–580	Yellow-green	Violet
580–595	Yellow	Blue
595–650	Orange	Blue–green
650–750	Red	Green–blue

Note that the instrumental deviations illustrated in Figures 21-5 and 21-6 result in absorbances that are smaller than theoretical. It can be shown that instrumental deviations always lead to negative absorbance errors.[2]

21B-3 Absorption Spectra

An *absorption spectrum* is a plot of percent transmittance, absorbance, log absorbance, or molar absorptivity of an analyte as a function of wavelength or wavenumber (or occasionally frequency). Figure 21-7b shows that the greatest difference between curves occurs in the region of high absorbance. When percent transmittance is plotted, the reverse is true. A plot with log A as ordinate leads to loss of spectral detail but is convenient for comparing solutions of different concentrations since the curves are displaced equally along the vertical axis for equal multiples of concentration.

---

[2] E. J. Meehan, in *Treatise on Analytical Chemistry*, 2nd ed., P. J. Elving, E. J. Meehan, and I. M. Kolthoff, Eds., Part I, Vol. 7, pp. 71–79. New York: Interscience, 1981.

21C THEORY OF MOLECULAR ABSORPTION

According to quantum theory, every molecular species has a unique set of energy states, the lowest of which is the *ground state*. At room temperature, most molecules are present in their ground state. When a photon of radiation passes near a molecule, absorption becomes probable if (and only if) the energy of the photon matches *exactly* the energy difference between the ground state and one of the higher energy states of the molecule. Under these circumstances, the energy of the photon is transferred to the molecule, converting it to the higher energy state, which is termed an *excited state*. Excitation of a species M to its excited state M* can be depicted by the equation

$$M + h\nu \rightarrow M^*$$

After a brief period (10^{-8} to 10^{-9} s), the excited species *relaxes* to its original, or ground, state, most commonly by transferring its excess energy to other atoms or molecules in the medium. This process, which causes a minute increase in temperature of the surroundings, is described by the equation

$$M^* \rightarrow M + \text{heat}$$

Relaxation may also occur by *photochemical decomposition* of M* to form new species or by the *fluorescent* or *phosphorescent* reemission of radiation. It is important to note that the lifetime of M* is so very short that its concentration at any instant is ordinarily negligible. Furthermore, the amount of thermal energy released during relaxation is usually so small as to be undetectable. Thus, absorption measurements have the advantage of creating minimal disturbance of the system under study.

> The lowest energy state of an atom or molecule is called its ground state.

> Excitation is a process in which a chemical species absorbs thermal, electrical, or radiant energy and is promoted to a higher energy state called an *excited state*.

> Relaxation is a process in which an excited species gives up its excess energy and returns to a lower energy state.

21C-1 Types of Molecular Transitions

Molecules undergo three types of quantized transitions when excited by absorbing ultraviolet, visible, and infrared radiation. For ultraviolet and visible radiation, excitation involves promoting an electron residing in a low energy molecular or atomic orbital to a higher energy orbital. In order for this transition to occur, the energy, $h\nu$, of the photon must be exactly the same as the energy difference between the two orbital energies. The transition of an electron between two orbitals is termed an *electronic transition* and the absorption process is called *electronic absorption*.

In addition to electronic transitions, molecules exhibit two other types of radiation-induced transitions, namely *vibrational transitions* and *rotational transitions*. Vibrational transitions come about because molecules have a multitude of quantized energy levels (or *vibrational states*) associated with the bonds that hold the molecule together.

To get an idea of the nature of vibrational states, picture a bond as a flexible spring with atoms attached at both ends. In Figure 21-8a, two types of stretching vibration are shown. With each vibration, atoms first approach and then move away from one another. The potential energy of the system at any instant depends on the extent to which the spring-like bond is stretched or compressed. For an ordinary spring, the energy of the system varies continuously and reaches a

> An electronic transition involves transition of an electron from one molecular energy state to another.

maximum when the spring is fully stretched or fully compressed. In contrast, the energy of a spring system with atomic dimensions can assume only certain discrete energies called *vibrational energy levels*. Figure 21-8b shows other types of molecular vibrations. The energy associated with each of these vibrational states usually differs from others and from the energies associated with stretching vibrations.

In addition to quantized vibrational states, a molecule has a host of quantized rotational states that are associated with the rotational motion of a molecule around its center of mass.

The overall energy E associated with an orbital of a molecule is then given by

$$E = E_{\text{electronic}} + E_{\text{vibrational}} + E_{\text{rotational}} \qquad (21\text{-}7)$$

where $E_{\text{electronic}}$ is the energy associated with the electrons in the various outer orbitals of the molecule, and $E_{\text{vibrational}}$ is the energy of the molecule as a whole due to interatomic vibrations. The term $E_{\text{rotational}}$ accounts for the energy associated with rotation of the molecule about its center of mass.

21C-2 Energy-Level Diagrams for Molecules

Figure 21-9 is a partial energy-level diagram that depicts typical processes that occur when a polyatomic species absorbs infrared, visible, and ultraviolet radiation. The energies, E_1 and E_2, of two of the several electronically excited states of a molecule are shown relative to the energy of its ground state E_0. In addition, the relative energies of a few of the many vibrational states associated with each electronic state are indicated by the lighter horizontal lines labeled 1, 2, 3, and 4 in Figure 21-9 (the lowest vibrational levels are labeled 0). Note that the differences in energy among the vibrational states are significantly smaller than

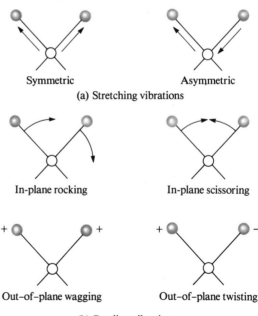

Symmetric   Asymmetric

(a) Stretching vibrations

In-plane rocking   In-plane scissoring

Out–of–plane wagging   Out–of–plane twisting

(b) Bending vibrations

Figure 21-8

Types of molecular vibrations. The plus sign indicates motion from the page toward the reader; the minus sign indicates motion away from the reader.

among energy levels of the electronic states (typically, an order of magnitude smaller).

Although they are not shown in Figure 21-9, a set of rotational energy states is superimposed on each of the vibrational states shown in the energy diagram. The energy differences among these states are smaller than those among vibrational states by an order of magnitude.

$$\Delta E_{electronic} \cong 10\Delta E_{vibrational} \cong 100\Delta E_{rotational}$$

Infrared Absorption

Infrared radiation generally is not sufficiently energetic to cause electronic transitions but can induce transitions in the vibrational and rotational states associated with the *ground electronic state* of the molecule. Four of these transitions are depicted in the lower left part of Figure 21-9. For absorption to occur, the analyte must be irradiated with frequencies corresponding exactly to the energies indicated by the lengths of the four arrows.

Absorption of Ultraviolet and Visible Radiation

The center arrows in Figure 21-9 suggest that the molecules under consideration absorb visible radiation of five wavelengths, thereby promoting electrons to the five vibrational levels of the excited electronic level E_1. Ultraviolet photons that are more energetic are required to produce the absorption indicated by the five arrows to the right.

As suggested by Figure 21-9, molecular absorption in the ultraviolet and visible regions consists of absorption *bands* made up of closely spaced lines. (A real molecule has many more vibrational energy levels than shown here; thus the typical absorption band consists of a multitude of lines.) In a solution, the absorbing species are surrounded by solvent, and the band nature of molecular absorption often becomes blurred because collisions tend to spread the energies of the quantum states, thus giving smooth and continuous absorption peaks.

Figure 21-10 shows visible absorption spectra for 1,2,4,5-tetrazine that were obtained under three different conditions. The upper spectrum is for the compound in the vapor state. Here, the individual tetrazine molecules are sufficiently separated from one another to vibrate and rotate freely, and many individual absorption peaks resulting from transitions among the various vibrational and rotational states are clearly evident. In the condensed state and in solutions, however, freedom to rotate is largely lost, and lines due to differences in rotational energy levels are obliterated. Furthermore, in the presence of solvent molecules, energies of the various vibrational levels are modified in an irregular way. Thus, the energy in a given electronic state in an assemblage of molecules takes on the more or less Gaussian distribution shown in Figure 21-10b. Here the tetrazine was in a nonpolar solvent (hexane), and only discrete peaks corresponding to electronic transitions are evident. In a polar solvent, such as water, the electronic peaks blend together to give a single smooth absorption peak (Figure 21-10c).

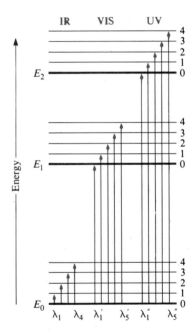

Figure 21-9
Energy-level diagram showing some of the energy changes that occur during absorption of infrared (IR), visible (VIS), and ultraviolet (UV) radiation. Note that with some molecules, a transition from E_0 to both E_1 and E_2 may require UV radiation of different wavelengths. With other molecules, transitions to both E_1 and E_2 may take place as a result of absorption of visible radiation.

21C-3   Molecular Species that Absorb Ultraviolet and Visible Radiation

Absorption measurements in the visible and ultraviolet regions of the spectrum provide qualitative and quantitative information about organic, inorganic, and biochemical molecules.

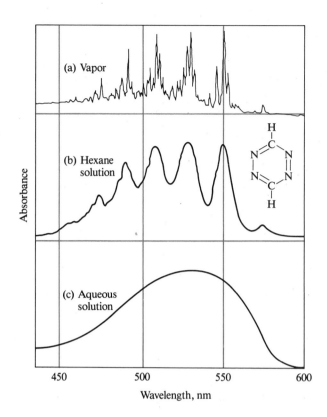

Figure 21-10
Typical ultraviolet absorption spectra.
The compound is 1,2,4,5-tetrazine.
(From S. F. Mason, *J. Chem. Soc.*,
1959, 1265. With permission.)

Absorption by Organic Compounds

Two types of electrons are responsible for the absorption of ultraviolet and visible
radiation by organic molecules: (1) shared electrons that participate directly in
bond formation and are thus associated with more than one atom and (2) unshared
outer electrons that are largely localized about such atoms as oxygen, the halogens,
sulfur, and nitrogen.

The wavelength at which an organic molecule absorbs depends on how tightly
its several electrons are bound. The shared electrons in such single bonds as
carbon/carbon or carbon/hydrogen are so firmly held that absorption occurs only
in a region of the ultraviolet spectrum ($\lambda < 180$ nm) where the components of
air also absorb (this region is known as the *vacuum ultraviolet*). The experimental
difficulty of making measurements in the vacuum ultraviolet limits the use of
this region for analytical purposes.

Electrons involved in double and triple bonds of organic molecules are more
loosely held and are therefore more easily excited by radiation; thus, species
with unsaturated bonds generally exhibit useful absorption peaks in the readily
accessible ultraviolet region (>180 nm). Unsaturated organic functional groups
that absorb in the ultraviolet or visible regions are known as *chromophores*.
Table 21-2 lists common chromophores and the approximate wavelengths at
which they absorb. The data for position and peak intensity can only serve as a
rough guide for identification purposes, since both are influenced by solvent
effects as well as other structural details of the molecule. Moreover, conjugation
between two (or more) chromophores tends to cause shifts in peak maxima to

Chromophores are functional groups
that absorb radiation.

Table 21-2
ABSORPTION CHARACTERISTICS OF SOME
COMMON ORGANIC CHROMOPHORES

Chromophore	Example	Solvent	λ_{max}, nm	ε_{max}
Alkene	$C_6H_{13}CH{=}CH_2$	n-Heptane	177	13,000
Conjugated alkene	$CH_2{=}CHCH{=}CH_2$	n-Heptane	217	21,000
Alkyne	$C_5H_{11}C{\equiv}C{-}CH_3$	n-Heptane	178	10,000
			196	2000
			225	160
Carbonyl	$\overset{\displaystyle CH_3 O}{\underset{\displaystyle CH_3CCH_3}{CH_3 \parallel}}$	n-Hexane	186	1000
			280	16
	$\overset{\displaystyle CH_3 O}{\underset{\displaystyle CH_3CH}{CH_3 \parallel}}$	n-Hexane	180	Large
			293	12
Carboxyl	$\overset{\displaystyle CH_3 O}{\underset{\displaystyle CH_3COH}{CH_3 \parallel}}$	Ethanol	204	41
Amido	$\overset{\displaystyle CH_3 O}{\underset{\displaystyle CH_3CNH_2}{CH_3 \parallel}}$	Water	214	60
Azo	$CH_3N{=}NCH_3$	Ethanol	339	5
Nitro	CH_3NO_2	Isooctane	280	22
Nitroso	C_4H_9NO	Ethyl ether	300	100
			665	20
Nitrate	$C_2H_5ONO_2$	Dioxane	270	12
Aromatic	Benzene	n-Hexane	204	7900
			256	200

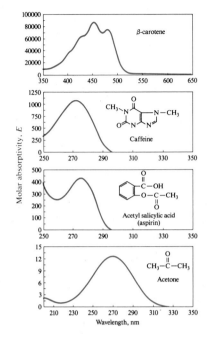

Figure 21-11

Ultraviolet spectra for typical organic
compounds.

longer wavelengths. Typical spectra for organic compounds are shown in Figure
21-11.

The unshared electrons in such elements as sulfur, bromine, and iodine are
less strongly held than the shared electrons of a saturated bond. Organic molecules
incorporating these elements frequently exhibit useful peaks in the ultraviolet
region as a result.

Absorption by Inorganic Species

In general, the ions and complexes of elements in the first two transition series
absorb broad bands of visible radiation in at least one of their oxidation states
and are, as a consequence, colored (see, for example, Figure 21-12). Here,
absorption involves transitions between filled and unfilled d-orbitals with energies
that depend on the ligands bonded to the metal ions. The energy differences
between these d-orbitals (and thus the position of the corresponding absorption
peak) depend on the position of the element in the periodic table, its oxidation
state, and the nature of the ligand bonded to it.

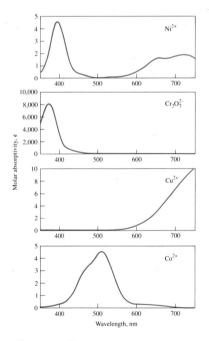

Figure 21-12

Absorption spectra of aqueous solutions of transition metal ions.

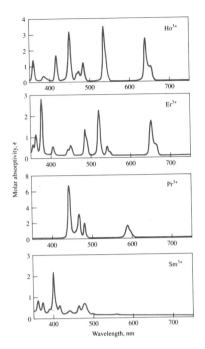

Figure 21-13

Absorption spectra of aqueous solutions of rare earth ions.

Charge-transfer spectra result when the absorption of a photon promotes an electron from ligand to metal or from metal to ligand in transition metal complexes.

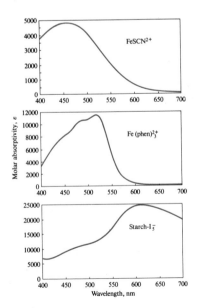

Figure 21-14

Charge-transfer spectra.

Absorption spectra of ions of the lanthanide and actinide transitions series differ substantially from those shown in Figure 21-12. The electrons responsible for absorption by these elements ($4f$ and $5f$, respectively) are shielded from external influences by electrons that occupy orbitals with larger principal quantum numbers. As a result, the bands tend to be narrow and relatively unaffected by the species bonded by the outer electrons (see Figure 21-13).

Charge-Transfer Absorption

For quantitative purposes, *charge-transfer absorption* is particularly important because molar absorptivities are unusually large ($\varepsilon_{max} > 10,000$), a circumstance that leads to high sensitivity. Many inorganic and organic complexes exhibit this type of absorption and are therefore called *charge-transfer complexes*.

A charge-transfer complex consists of an electron-donor group bonded to an electron acceptor. When this product absorbs radiation, an electron from the donor is transferred to an orbital that is largely associated with the acceptor. The excited state is thus the product of a kind of internal oxidation/reduction process. This behavior differs from that of an organic chromophore in which the excited electron is in a *molecular* orbital that is shared by two or more atoms.

Familiar examples of charge-transfer complexes include the phenolic complex of iron(III), the 1,10-phenanthroline complex of iron(II), the iodide complex of molecular iodine, and the ferro/ferricyanide complex responsible for the color of Prussian blue. The red color of the iron(III)/thiocyanate complex is a further example of charge-transfer absorption. Absorption of a photon results in the transfer of an electron from the thiocyanate ion to an orbital that is largely associated with the iron(III) ion. The product is an excited species involving predominantly iron(II) and the thiocyanate radical SCN. As with other types of electronic excitation, the electron in this complex ordinarily returns to its original state after a brief period. Occasionally, however, an excited complex may dissociate and produce photochemical oxidation/reduction products.

In most charge-transfer complexes involving a metal ion, the metal serves as the electron acceptor. Exceptions are the 1,10-phenanthroline complexes of iron(II) (Section 16C-1) and copper(I), where the ligand is the acceptor and the metal ion the donor. A few other examples of this type of complex are known.

21C-4 Molecular Species That Absorb Infrared Radiation

With the exception of a few homonuclear compounds, such as O_2, Cl_2, and N_2, all molecules, organic and inorganic, absorb infrared radiation. Thus, infrared spectroscopy is one of the most generally applicable of all analytical methods. As we noted in Section 21C-2, absorption of infrared radiation involves transitions among vibrational energy levels of the lowest electronic energy levels of molecules (see the left-hand side of Figure 21-9). The number of ways a molecule can vibrate is related to the number of bonds it contains and thus the number of atoms making up the molecule. The number of vibrations is large, for even a simple molecule. For example, *n*-butanal ($CH_3CH_2CH_2CHO$) has 33 vibrational modes, most differing from each other in energy. Not all these vibrational modes produce infrared peaks; nevertheless, as shown in Figure 21-15, the spectrum of *n*-butanal is complex.

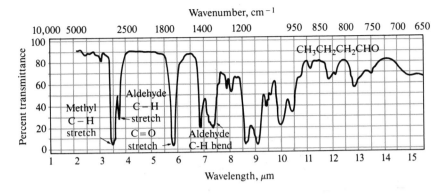

Figure 21-15
Infrared spectrum for *n*-butanal (*n*-butyraldehyde). Note that transmittance rather than absorbance is plotted.

Infrared absorption occurs not only with organic molecules but with simple inorganic compounds such as CO_2, CO, H_2S, NO_2, and SO_2 as well as covalently bonded metal complexes of various types.

21D   QUESTIONS AND PROBLEMS

21-1. Briefly describe or define
  *(**a**) chromophore.
  (**b**) an electronic transition of a molecule.
  *(**c**) monochromatic radiation.
  (**d**) the ground state of a molecule.
  *(**e**) absorption spectrum.
  (**f**) vacuum ultraviolet.

21-2. What is the relationship between
  *(**a**) absorbance and transmittance?
  (**b**) absorptivity a and molar absorptivity ε?

21-3. How do electronic transitions resemble vibrational transitions? How do they differ?

*(**21-4.**) Identify factors that cause the Beer's law relationship to depart from linearity.

21-5. What type of transition is associated with molecular absorption of
  (**a**) infrared radiation?
  (**b**) ultraviolet/visible radiation?

*(**21-6.**) Why is a solution of $Cu(NH_3)_4^{2+}$ blue?

21-7. What is the mechanism of charge-transfer absorption? Why is this type of absorption of interest in analytical chemistry?

21-8. What are the units for absorptivity when the path length is given in centimeters and the concentration is expressed in
  *(**a**) parts per million?
  (**b**) micrograms per liter?
  *(**c**) weight-volume percent?
  (**d**) grams per liter?

21-9. Evaluate the missing quantities in the table shown below. Where needed, use 200 for the molar mass of the analyte.

21-10. Express the following absorbances in terms of percent transmittance.
  *(**a**) 0.0510   (**d**) 0.261
  (**b**) 0.918   *(**e**) 0.485
  *(**c**) 0.379   (**f**) 0.702

	A	% T	ε L mol⁻¹ cm⁻¹	a cm⁻¹ ppm⁻¹	b cm	c M	ppm
*(**a**)	0.172		4.23×10^3		1.00		
(**b**)		44.9		0.0258		1.35×10^{-4}	
*(**c**)	0.520		7.95×10^3		1.00		
(**d**)		39.6		0.0912			1.76
*(**e**)			3.73×10^3		0.100	1.71×10^{-3}	
(**f**)		83.6			1.00	8.07×10^{-6}	
*(**g**)	0.798				1.50		33.6
(**h**)		11.1	1.35×10^4			7.07×10^{-5}	
*(**i**)		5.23	9.78×10^3				5.24
(**j**)	0.179				1.00	7.19×10^{-5}	

21-11. Convert the accompanying transmittance data to absorbances.

*(a) 25.5% (d) 3.58%
(b) 0.567 *(e) 0.085
*(c) 32.8% (f) 53.8%

*21-12. Calculate the percent transmittance of solutions that have twice the absorbance of the solutions in 21-10.

*21-13. Calculate the absorbances of solutions with half the transmittance of those in 21-11.

*21-14. A solution containing 4.48 ppm $KMnO_4$ has a transmittance of 0.309 in a 1.00-cm cell at 520 nm. Calculate the molar absorptivity of $KMnO_4$.

21-15. Beryllium(II) forms a complex with acetylacetone (molar mass = 207.2). Calculate the molar absorptivity of the complex, given that a 1.34 ppm solution of Be(II) has a transmittance of 55.7% when measured in a 1.00-cm cell at 295 nm, the wavelength of maximum absorption.

*21-16. At 580 nm, the wavelength of its maximum absorption, the complex $FeSCN^{2+}$ has a molar absorptivity of 7.00×10^3 L cm^{-1} mol^{-1}. Calculate

(a) the absorbance of a 2.50×10^{-5} M solution of the complex at 580 nm in a 1.00-cm cell.

(b) the absorbance of a solution in which the concentration of the complex is twice that in (a).

(c) the transmittance of the solutions described in (a) and (b).

(d) the absorbance of a solution that has half the transmittance of that described in (a).

*21-17. A 2.50-ml aliquot of a solution that contains 3.8 ppm iron(III) is treated with an appropriate excess of KSCN and diluted to 50.0 mL. What is the absorbance of the resulting solution at 580 nm in a 2.50-cm cell? (See 21-16 for absorptivity data.)

21-18. A solution containing the complex formed between Bi(III) and thiourea has a molar absorptivity of 9.32×10^3 L cm^{-1} mol^{-1} at 470 nm.

(a) What is the absorbance of a 6.24×10^{-5} M solution of the complex at 470 nm in a 1.00-cm cell?

(b) What is the percent transmittance of the solution described in (a)?

(c) What is the molar concentration of the complex in a solution that has the absorbance described in (a) when measured at 470 nm in a 5.00-cm cell?

*21-19. The complex formed between Cu(I) and 1,10-phenanthroline has a molar absorptivity of 6.94×10^3 L cm^{-1} mol^{-1} at 435 nm, the wavelength of maximum absorption. Calculate

(a) the absorbance of an 8.50×10^{-5} M solution of the complex when measured in a 1.00-cm cell at 435 nm.

(b) the percent transmittance of the solution in (a).

(c) the concentration of a solution that in a 5.00-cm cell has the same absorbance as the solution in (a).

(d) the path length through a 3.40×10^{-5} M solution of the complex that is needed for an absorbance that is the same as the solution in (a).

21-20. The equilibrium constant for the conjugate acid/base pair

$$HIn + H_2O \rightleftharpoons H_3O^+ + In^-$$

is 8.00×10^{-5}. From the additional information

Species	Absorption Maximum, nm	Molar Absorptivity	
		430 nm	600 nm
HIn	430	8.04×10^3	1.23×10^3
In^-	600	0.775×10^3	6.96×10^3

*(a) calculate the absorbance at 430 nm and 600 nm for the following indicator concentrations: 3.00×10^{-4} M, 2.00×10^{-4} M, 1.00×10^{-4} M, 0.500×10^{-4} M, and 0.250×10^{-4} M.

(b) plot absorbance as a function of indicator concentration.

21-21. The equilibrium constant for the reaction

$$2CrO_4^{2-} + 2H^+ \rightleftharpoons Cr_2O_7^{2-} + H_2O$$

is 4.2×10^{14}. The molar absorptivities for the two principal species in a solution of $K_2Cr_2O_7$ are

λ	$\varepsilon_1(CrO_4^{2-})$	$\varepsilon_2(Cr_2O_7^{2-})$
345	1.84×10^3	10.7×10^2
370	4.81×10^3	7.28×10^2
400	1.88×10^3	1.89×10^2

Four solutions were prepared by dissolving 4.00×10^{-4}, 3.00×10^{-4}, 2.00×10^{-4}, and 1.00×10^{-4} moles of $K_2Cr_2O_7$ in water and diluting to 1.00 L with a pH 5.60 buffer. Derive theoretical absorbance values (1.00-cm cells) for each solution and plot the data for (a) 345 nm; (b) 370 nm; (c) 400 nm.

CHAPTER 22

APPLICATIONS OF MOLECULAR ABSORPTION SPECTROSCOPY

Molecular spectroscopy based on ultraviolet, visible, and infrared radiation is widely used for the identification and determination of a variety of inorganic, organic, and biochemical species.[1] Molecular ultraviolet/visible absorption spectroscopy is employed primarily for quantitative analysis and is one of the most common methods used in chemical and clinical laboratories. Infrared absorption spectroscopy is a powerful tool for determining the structure of both inorganic and organic compounds. In addition, it is now assuming an important role in the quantitative determination of environmental pollutants.

In this chapter we first describe instruments for molecular absorption spectroscopy with emphasis on those that make use of ultraviolet/visible radiation. We then turn to their applications for determining inorganic, organic, and biochemical compounds.

22A INSTRUMENTS FOR OPTICAL ABSORPTION MEASUREMENTS

The optical components described in Section 20C have been combined in various ways to produce two types of instruments that are used for absorption measurements with ultraviolet, visible, or infrared radiation: *spectrophotometers* and

---

[1] For more detailed treatment of absorption spectroscopy, see E. J. Meehan, in *Treatise on Analytical Chemistry*, 2nd ed., P. J. Elving, E. J. Meehan, and I. M. Kolthoff, Eds., Part I, Vol. 7, Chapter 2. New York: Wiley, 1981; *Techniques in Visible and Ultraviolet Spectrometry*, C. Burgess and A. Knowles, Eds., Vol. 1. New York: Chapman and Hall, 1981; and J. D. Ingle Jr. and S. R. Crouch, *Spectrochemical Analysis*, Chapters 2, 3, and 13. Englewood Cliffs, NJ: Prentice-Hall, 1988.

Photometers use filters to provide wavelength selection. Spectrophotometers use a grating or prism for this purpose.

photometers. Spectrophotometers employ a grating or a prism monochromator to provide a narrow band of radiation for measurements. Photometers, in contrast, use an absorption filter or an interference filter for this purpose. Spectrophotometers offer the considerable advantage that the wavelength used can be varied continuously, thus making it possible to record entire absorption spectra. Photometers have the advantages of simplicity, ruggedness, and low cost. We shall focus most of our attention on spectrophotometers.

22A-1 Ultraviolet/Visible Spectrophotometers

Several dozen models of spectrophotometers are marketed by various instrument manufacturers. Some are designed for the visible region alone. Others have ranges that extend from 180 to 200 nm in the ultraviolet through the entire visible region to 750 to 800 nm. A few cover the ultraviolet/visible region as well as the near-infrared to approximately 3000 nm.

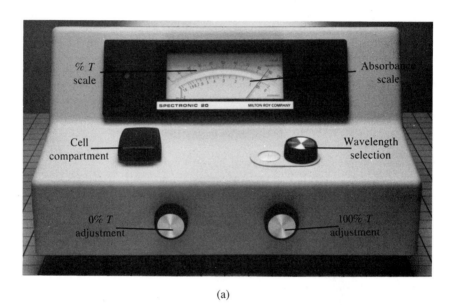

(a)

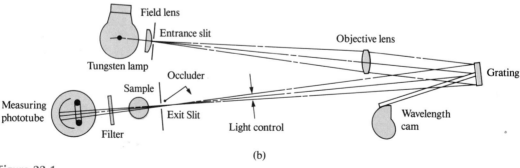

(b)

Figure 22-1

(a) The Spectronic 20 spectrophotometer. (b) Its optical diagram. (Courtesy of Milton Roy Company, Analytical Products Division, Rochester, NY.)

Single-Beam Instruments

Figure 22-1 shows the design of a simple and inexpensive spectrophotometer, the Spectronic 20, which is designed for the visible region of the spectrum. The original version of this instrument first appeared on the market in the mid-1950s, and the modified version shown in the figure is still being manufactured and widely sold. More of these instruments are currently in use throughout the world than any other single spectrophotometer model.

The Spectronic 20 is equipped with a tungsten-filament light source operated by a stabilized power supply. The intensity of radiation from the lamp is sufficiently constant to produce reproducible absorbance data. Radiation from the source passes through a fixed slit to the surface of a reflection grating. The diffracted radiation then passes through an exit slit to the cylindrical sample or reference cell and then to a phototube. The amplified electrical signal from the detector powers a meter with a $5\frac{1}{2}$-inch scale that is linear in percent transmittance; the face of the scale is also scribed with a logarithmic scale that is calibrated in absorbance.

The Spectronic 20 is equipped with an *occluder*, which is a vane that automatically falls between the beam and the detector whenever the cuvette is removed from its holder. The light control device consists of a V-shaped aperture that is moved in and out of the beam to control the intensity of the beam falling on the phototube (see Figure 22-2).

To obtain a percent transmittance reading, the pointer of the meter is first zeroed with the sample compartment empty so that the occluder blocks the beam, and no radiation reaches the detector. This process is called the *0% T calibration*, or *adjustment*. A cell containing the blank is then inserted into the cell holder and the pointer is brought to the *100% T* mark by adjusting the position of the light control aperture and thus the amount of light reaching the detector. This adjustment is called the *100% T calibration*, or *adjustment*. Finally, the sample is placed in the cell compartment, and the percent transmittance or the absorbance is read directly from the scale of the meter.

The spectral range of the Spectronic 20 is 340 to 625 nm (an accessory phototube extends the range to 950 nm). Other specifications for the instrument include an effective bandwidth of 20 nm and a wavelength accuracy of ± 2.5 nm.

Single-beam instruments, such as the Spectronic 20, are well suited for quantitative absorption measurements at a single wavelength. Here, simplicity of instrumentation, low cost, and ease of maintenance offer distinct advantages.

Several instrument manufacturers offer single-beam instruments for both ultraviolet and visible measurements. The lower wavelength extremes for these instruments range from 190 to 210 nm and the upper from 800 to 1000 nm. All are equipped with interchangeable tungsten and deuterium or hydrogen lamps. Most employ photomultiplier tubes for detectors and gratings for dispersing radiation. Some are equipped with digital readout devices; others employ large meters. Prices for these instruments range from $2000 to $8000.

Double-Beam Instruments

Many modern photometers and spectrophotometers are based on a double-beam design. Figure 22-3b illustrates a double-beam-in-space instrument in which two beams are formed in space by a V-shaped mirror called a beam splitter. One beam passes through the reference solution to a photodetector, and the second

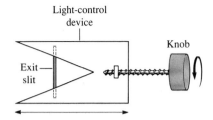

Figure 22-2

End view of the exit slit of the Spectronic 20 spectrophotometer pictured in Figure 22-1 showing the light-control device.

The 0% T and 100% T adjustments should be made immediately before each transmittance or absorbance measurement.

To obtain reproducible transmittance measurements it is essential that the radiant power of the source remain constant during the time that the 100% T adjustment is made and the % T is read from the meter.

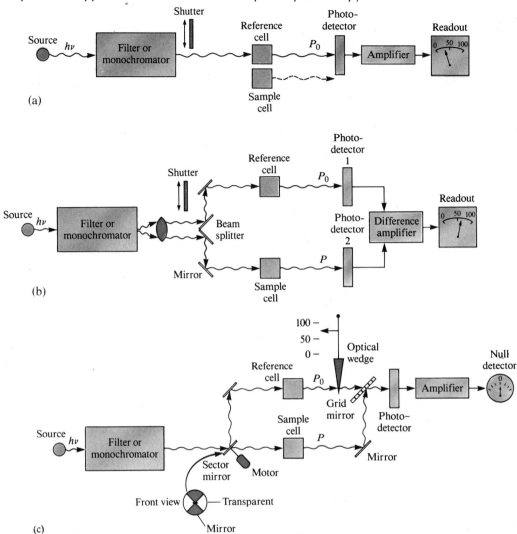

Figure 22-3

Instrument designs for photometers and spectrophotometers: (a) single-beam instrument; (b) double-beam instrument with beams separated in space; (c) double-beam instrument with beams separated in time.

Double-beam instruments are either double-beam-in-space or double-beam-in-time.

simultaneously traverses the sample to a second, matched photodetector. The two outputs are amplified, and their ratio (or the log of their ratio) is determined electronically and displayed on the readout device or the chart of a recorder. With manual instruments, the measurement is a two-step operation: first, the zero adjustment is made with a shutter in place between selector and beam splitter. Second, the shutter is opened and the transmittance or absorbance is read directly from the meter.

The second type of double-beam instrument is illustrated in Figure 22-3c. Here the beams are separated in time by a rotating sector mirror that directs the entire beam from the monochromator first through the reference cell and then through the sample cell. The pulses of radiation are recombined by another sector mirror, which transmits one pulse to the detector and reflects the other. As shown by the insert labeled "front view" in Figure 22-3c, the motor-driven sector mirror is made up of pie-shaped segments, half of which are mirrored and half of which are transparent. The mirrored sections are held in place by blackened metal frames that periodically interrupt the beam and prevent its reaching the detector. The detector circuit is programmed to use these periods to perform the 0% T adjustment.

The instrument shown in Figure 22-3c is a null type in which the beam passing through the solvent is attenuated until its intensity just matches that of the beam passing through the sample. Attenuation is accomplished in this design with an *optical wedge*, whose transmission decreases linearly along its length. Thus, the null point is reached by moving the wedge in the beam until the two electrical pulses are indicated as identical by the null detector. The transmittance (or absorbance) is then read directly from the pointer attached to the wedge.

Double-beam instruments offer the advantage that they compensate for all but the most short-term fluctuations in the radiant output of the source as well as for drift in the detector and amplifier. They also compensate for wide variations in source intensity with wavelength (see Figures 20-6b and 20-8). Furthermore, the double-beam design is well suited for the continuous recording of transmittance or absorbance spectra. Consequently, most modern ultraviolet and visible recording instruments are double-beam (usually in-time). Many infrared spectrophotometers are also based on this design.

The light source and the detector electronics do not need to be as stable in double-beam instruments as they do for single-beam instruments because in the former the 0% T and 100% T adjustments are made nearly simultaneously with the measurement of T.

Multichannel Instruments

Multichannel, or diode array, spectrometers are products of modern optoelectronic technology that makes it possible to record an entire ultraviolet or visible spectrum rapidly. The heart of these instruments is an array of several hundred silicon diode detectors (page 399) that are fabricated side by side on a single silicon chip as shown in Figure 22-4. Typically, the chips are 1 to 6 cm in length, and the widths of the individual diodes are 0.015 to 0.050 mm. The chip also contains a capacitor and an electronic switch for each diode. A computer-driven shift register sequentially closes each switch momentarily, which causes each capacitor to be charged to -5 V. Radiation impinging on the diode surface causes partial discharge of its capacitor. This lost charge is replaced during the next switching cycle. The resulting charging currents, which are proportional to radiant power, are amplified, digitized, and stored in computer memory. With one or two of these diode arrays placed along the length of the focal plane of a grating monochromator, all wavelengths can be monitored simultaneously and data for an entire spectrum collected and stored in a second or less.

Photos of diode-array detectors are shown in color plate 19.

Figure 22-5 shows an optical diagram of a typical multichannel ultraviolet/visible spectrophotometer. Because of the few optical components, the radiation throughput is much higher than that of traditional spectrophotometers. As a result, a single deuterium lamp can serve as a source for not only the ultraviolet but also the visible region (up to 820 nm). After passing through the solvent or analyte solution, the radiation is focused on an entrance slit and then passes onto the surface of a reflection grating. The detector is a diode array made up of 316 elements each with a dimension of 18 × 0.5 mm. The dispersion of the grating

The detector in the HP 8452A spectrometer shown in Figure 22-5 consists of 316 individual diode detectors, each coupled to its own capacitor and on–off switch, and all formed on a single silicon chip.

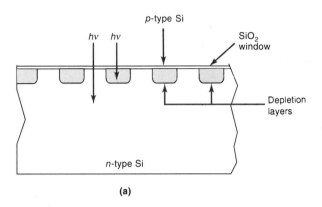

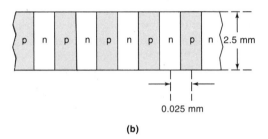

Figure 22-4

A reverse-biased linear diode-array detector: (a) cross section; (b) top view.

and the size of the diode elements are such that a resolution of 2 nm is realized throughout the entire spectral region. Because the system contains no moving parts, the wavelength reproducibility from scan to scan is exceedingly high.

A single scan from 200 to 820 nm with this instrument requires 0.1 s. In order to improve the precision of the measurements, however, spectra are generally scanned for a second or more with the data being collected in computer memory for subsequent averaging. With such short exposure times, photodecomposition of samples is minimized despite the location of the sample between the source and the monochromator. The stability of the source and electronic system is such that the solvent signal (100% T) needs to be observed and stored only every five to ten minutes.

The spectrophotometer shown in Figure 22-5 is designed to be interfaced with most personal computer systems. The instrument (without the computer) costs about $7000 to $9000.

22A-2 Infrared Spectrophotometers

Two types of spectrometers are encountered in infrared spectroscopy: dispersive and Fourier transform.

Dispersive Instruments

Dispersive infrared instruments differ from the double-beam spectrophotometer shown in Figure 22-3c in the location of the cell compartment with respect to the monochromator. In most ultraviolet/visible instruments, cells are located between the monochromator and the detector in order to avoid photochemical decomposition, which may occur if samples are exposed to the full power of the

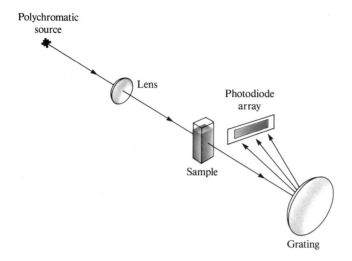

Figure 22-5
Diagram of a multichannel spectrophotometer based on a grating and photodiode detector.

source. Infrared radiation, in contrast, is not sufficiently energetic to bring about photodecomposition; thus the cell compartment can be located between the source and the monochromator. This arrangement is advantageous because any scattered radiation generated in the cell compartment is largely removed by the monochromator.

As shown in Section 20C, the components of infrared instruments differ considerably in detail from those in ultraviolet and visible instruments. Thus, infrared sources are heated solids rather than deuterium or tungsten lamps, infrared gratings are much coarser than those required for ultraviolet/visible radiation, and infrared detectors respond to heat rather than photons. Furthermore, the optical components of infrared instruments are constructed from polished salts, such as sodium chloride or potassium bromide.

Fourier Transform Instruments

When Fourier transform infrared spectrometers first appeared on the market in the early 1970s, they were bulky, expensive (more than $100,000), and required frequent mechanical adjustments. For these reasons, their use was limited to special applications in which their unique characteristics (great speed, high resolution, high sensitivity, and unparalleled wavelength precision and accuracy) were essential. By now Fourier transform spectrometers have been reduced to benchtop size and have become reliable and easy to maintain. Furthermore, the price of simpler models has been reduced to a point where they are competitive with all but the simplest dispersive instruments. For these reasons Fourier transform infrared instruments are largely displacing dispersive instruments in most laboratories.

Fourier transform instruments contain no dispersing element, and all wavelengths are detected and measured simultaneously. In order to obtain radiant power as a function of wavelength it is necessary to modulate the source signal in such a way that it can be decoded by a Fourier transformation, a mathematical operation that requires a high-speed computer. The theory of Fourier transform measurements is beyond the scope of this book.[2]

Except for the least expensive spectrometers, modern infrared instruments are of the Fourier transform type.

---

[2] For discussions of the principles of Fourier-transform spectroscopy, see W. D. Perkins, *J. Chem. Educ.*, **1986**, *63*, A5, A196; and L. Glasser, *J. Chem. Educ.*, **1987**, *64*, A228, A260, A296.

22A-3 Photometers

Photometers that employ ultraviolet, visible, or infrared radiation are marketed for various purposes. Their most common applications are in monitoring liquids and gases in industrial plants, where their simplicity, ruggedness, and ease of maintenance are advantages. Photometers also find many applications where portability is an important requirement. Finally, photometers are often used as detectors in liquid chromatography because of their compactness and simplicity. Where high spectral purity is not important (and often it is not), the accuracy and precision of measurements made with a photometer can approach those made with a spectrophotometer. The disadvantages of photometers are their lesser versatility, their inability to generate entire spectra, and their generally wider effective bandwidths.

Feature 22-1

INFRARED PHOTOMETERS FOR ROUTINE DETERMINATION
OF ATMOSPHERIC POLLUTANTS

Figure 22-A is an optical diagram of a portable infrared photometer that has been designed for the routine quantitative determination of organic pollutants in the atmosphere. The source is a ceramic rod wound with a nichrome wire, and the transducer is a pyroelectric detector. A variety of interference filters that transmit in the region of 3000 to 750 cm^{-1} or 3.3 to 13 μm are available; each is designed for the determination of a specific compound. The filters are readily interchangeable. Gaseous samples are introduced into the cell by means of a battery-operated pump. The path length of the cell as shown is 0.5 m; a series of reflecting mirrors (not shown) permits increases in the cell length to 20.5 m in increments of 0.5 m. This feature greatly enhances the concentration range of the instrument.

The photometer is reported to be sensitive to a few tenths of a part per million of such substances as acrylonitrile, chlorinated hydrocarbons, carbon monoxide, phosgene, and hydrogen cyanide.

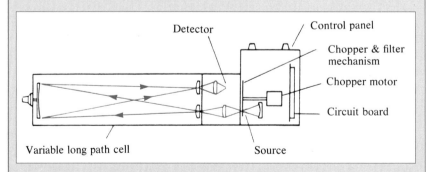

Figure 22-A
A portable infrared gas analyzer for air pollutants. (Courtesy of Wilks Scientific Corporation.)

22B   ULTRAVIOLET/VISIBLE SPECTROPHOTOMETRY AND PHOTOMETRY

Qualitative applications of ultraviolet/visible spectrophotometry are limited because the spectra of most compounds in solution consist of one or, at most, a few broad peaks with no fine structure that would be required for unambiguous identification. In contrast, the method is one of the most powerful and widely used tools for quantitative analysis.[3] Important characteristics of quantitative ultraviolet/visible spectrophotometry include wide applicability to organic, inorganic, and biochemical systems; good sensitivity; detection limits of 10^{-4} to 10^{-7} M; moderate to high selectivity; reasonable accuracy and precision (relative errors in the 1% to 3% range and with special techniques, as low as a few tenths of a percent); and speed and convenience. In addition, spectrophotometric methods are readily automated.

The most important applications of ultraviolet/visible spectrophotometry are quantitative, whereas for infrared spectrophotometry, they are qualitative.

Many automated analytical instruments are based on photometric and spectrophotometric measurements.

22B-1   Scope

The applications of molecular spectrophotometric methods are numerous and varied. The reader can obtain a notion of the scope of spectrophotometry by consulting the series of review articles published biennially in *Analytical Chemistry*,[4] as well as monographs on the subject.[5]

Application to Absorbing Species

Table 21-2 lists many common organic chromophoric groups. Spectrophotometric determination of organic compounds containing one or more of these groups is thus potentially feasible; many such applications can be found in the literature.

A number of inorganic species also absorb. We have noted that many ions of the transition metals and their complexes are colored in solution and can thus be determined by spectrophotometric measurement. In addition, a number of other species including nitrite, nitrate, and chromate ions, the oxides of nitrogen, the elemental halogens, and ozone show characteristic absorption peaks.

Applications to Nonabsorbing Species

Many nonabsorbing analytes, both inorganic and organic, can be determined photometrically by causing them to react with chromophoric reagents to produce products that absorb strongly in the ultraviolet and visible regions. The successful application of these color-forming reagents usually requires that their reaction with the analyte be forced to near completion.

---

[3] For a wealth of detailed, practical information on spectrophotometric practices, see *Techniques in Visible and Ultraviolet Spectrometry*, Vol. I, *Standards in Absorption Spectroscopy*, C. Burgess and A. Knowles, Eds. London: Chapman and Hall, 1981; and J. R. Edisbury, *Practical Hints on Absorption Spectrometry*. New York: Plenum Press, 1968.

[4] L. G. Hargis and J. A. Howell, *Anal. Chem.*, **1984**, *56*, 225R; **1986**, *58*, 108R; **1988**, *60*, 131R; **1990**, *62*, 155R; **1992**, *64*, 66R.

[5] See, for example, E. B. Sandell and H. Onishi, *Photometric Determination of Traces of Metals*, 4th ed. New York: Wiley, 1978; *Colorimetric Determination of Nonmetals*, 2nd ed., D. F. Boltz, Ed. New York: Interscience, 1978; F. D. Snell, *Photometric and Fluorometric Methods of Analysis*. New York: Wiley, 1978.

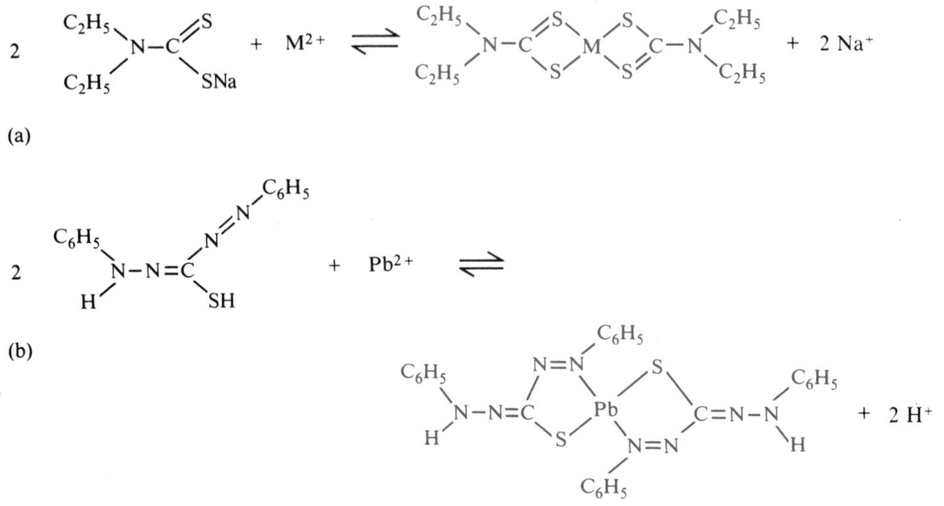

(a)

(b)

Figure 22-6
Typical chelating reagents for absorption. (a) Diethyldithiocarbamate. (b) Diphenylthiocarbazone.

Hundreds of chelating reagents have been developed for determining the concentration of various inorganic cations.

Typical inorganic reagents include the following: thiocyanate ion for iron, cobalt, and molybdenum; the anion of hydrogen peroxide for titanium, vanadium, and chromium; and iodide ion for bismuth, palladium, and tellurium. Of even greater importance are organic chelating reagents that form stable colored complexes with cations. Common examples include diethyldithiocarbamate for copper, diphenylthiocarbazone for lead, 1,10-phenanthroline for iron, and dimethylglyoxime for nickel. Figure 22-6 shows the color-forming reaction for the first two of these reagents. The structure of the 1,10-phenanthroline complex of iron(II) is shown on page 302; the reaction of nickel(II) with dimethylglyoxime to form a red precipitate is given on page 113. In the application of this reagent to the photometric determination of nickel, an aqueous solution of the cation is extracted with the chelating agent dissolved in an immiscible organic solvent. The absorbance of the bright red organic solution of the complex serves as a measure of the cation concentration.

Other reagents are available that react with organic functional groups to produce colors that are useful for quantitative analysis. For example, the red color of the 1:1 complexes that form between low-molecular-weight aliphatic alcohols and cerium(IV) can be used for the quantitative estimation of such alcohols.

22B-2 Procedural Details

A first step in any photometric or spectrophotometric method is the development of conditions that yield a reproducible relationship (preferably linear) between absorbance and analyte concentration.

Wavelength Selection

In order to realize maximum sensitivity, spectrophotometric absorbance measurements are ordinarily made at a wavelength corresponding to an absorption peak because the sensitivity (change in absorbance per unit of concentration) is greatest at this point. In addition, the absorption curve is often flat at a maximum, which leads to good adherence to Beer's law (see Figure 21-5) and less uncertainty from failure to reproduce precisely the wavelength setting of the instrument.

Ordinarily use a peak wavelength for quantitative spectrophotometric measurements.

Variables That Influence Absorbance

Common variables that influence the absorption spectrum of a substance include the type of solvent, the pH of the solution, the temperature, high electrolyte concentrations, and the presence of interfering substances. The effects of these variables must be known and conditions for the analysis chosen such that the absorbance of the analyte will not be materially affected by small, uncontrolled variations in their magnitudes.

Cleaning and Handling of Cells

Accurate spectrophotometric analysis requires the use of high quality, matched cells. These should be regularly calibrated against one another to detect differences that can arise from scratches, etching, and wear. Equally important is the use of proper cell cleaning techniques.

Avoid touching the optical windows of cells because their transmission characteristics can be altered by fingerprints.

Never dry cells in an oven.

Determination of the Relationship Between Absorbance and Concentration

The calibration standards for a photometric or a spectrophotometric analysis should approximate as closely as possible the overall composition of the actual samples, and should encompass a reasonable range of analyte concentrations. Seldom, if ever, is it safe to assume adherence to Beer's law and only use a single standard to determine the molar absorptivity; it is even less prudent to base the results of an analysis on a literature value for the molar absorptivity.

The measured absorbance of a given solution will usually vary somewhat from instrument to instrument. Thus, determination should *never* be based on molar absorptivities found in the literature.

The Standard Addition Method

The difficulties that attend production of standards with an overall composition closely resembling that of the sample can be formidable, if not insurmountable. Under such circumstances, the *standard addition* approach may prove useful. Here, a known amount of analyte is introduced to a second aliquot of the sample. Provided Beer's law is obeyed (and this must be confirmed experimentally), the difference in absorbance is used to calculate the analyte concentration of the sample. For further details on the standard addition method, see page 467.

Example 22-1

A 2.00-mL urine specimen was treated with reagents to generate color with phosphate, following which the sample was diluted to 100 mL. Photometric measurement for the phosphate in a 25.0-mL aliquot yielded an absorbance of 0.428. Addition of 1.00 mL of a solution containing 0.0500 mg of phosphate to a second 25.0-mL aliquot resulted in an absorbance of 0.517. Use these data to calculate the milligrams of phosphate in each milliliter of the specimen.

The absorbance of the second measurement must be corrected for dilution.

Thus,

$$\text{corrected absorbance} = 0.517 \times \frac{26.0\ \text{mL}}{25.0\ \text{mL}} = 0.538$$

$$\text{absorbance caused by } 0.0500\ \text{mg phosphate} = 0.538 - 0.428 = 0.110$$

$$\text{weight phosphate in } \frac{25.0\ \text{mL}}{100\ \text{mL}} \text{ of specimen} = \frac{0.428}{0.110} \times 0.050\ \text{mg} = 0.195\ \text{mg}$$

Finally, then,

$$\text{concentration of phosphate} = \frac{100\ \text{mL}}{25.0\ \text{mL}} \times 0.195\ \text{mg} \times \frac{1}{2.00\ \text{mL}} = 0.390\ \frac{\text{mg}}{\text{mL}}$$

Analysis of Mixtures

The total absorbance of a solution at any given wavelength is equal to the sum of the absorbances of the individual components in the solution (Equation 21-6). This relationship makes it possible in principle to determine the concentrations of the individual components of a mixture even if there is total overlap in their spectra. For example, Figure 22-7 shows the spectrum of a solution containing a mixture of species M and species N as well as absorption spectra for the individual components. Clearly, no wavelength exists at which the absorbance is due to just one of these components. To analyze the mixture, molar absorptivities for M and N are first determined at wavelengths λ_1 and λ_2 with enough standards to be sure that Beer's law is obeyed over an absorbance range that encompasses the absorbance of the sample. Note that the wavelengths selected are ones at which the two spectra differ significantly. Thus, at λ_1, the molar absorptivity of component M is much larger than that for component N. The reverse is true for λ_2. To complete the analysis, the absorbance of the mixture is determined at the same two wavelengths. Example 22-2 demonstrates how the composition of the mixture is derived from data of this kind.

Example 22-2

Palladium(II) and gold(III) can be analyzed simultaneously through reaction with methiomeprazine ($C_{19}H_{24}N_2S_2$). The absorption maximum for the Pd complex occurs at 480 nm, while that for the Au complex is at 635 nm. Molar absorptivity data at these wavelengths are

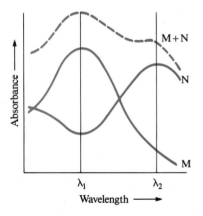

Figure 22-7
Absorption spectrum of a two-component mixture (M + N), with spectra of the individual components.

	Molar Absorptivity, ε	
	480 nm	**635 nm**
Pd complex	3.55×10^3	5.64×10^2
Au complex	2.96×10^3	1.45×10^4

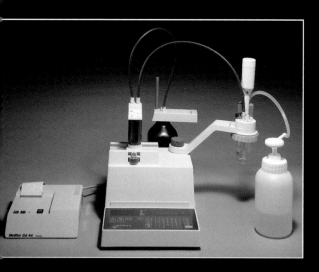

Color Plate 18

Computer-controlled automatic potentiometric titrator. The sample is pipetted into the sample cup. The titrator adds titrant, measures the pH of the titration mixture as titrant is added, and locates the equivalence point. The results of the titration are printed to provide a permanent record of the analysis.

Photo courtesy of Mettler–Toledo, Inc., Hightstown, NJ.

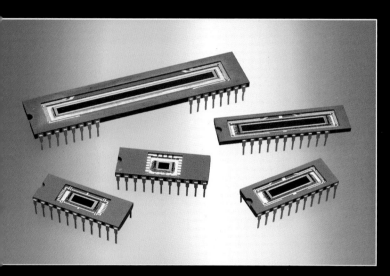

Color Plate 19

Photodiode arrays of various sizes. The arrays contain 256, 512, 1024, 2048, and 4096 diodes (Section 20C-4).

Photo courtesy of EG&G Reticon, Sunnyvale, CA.

Color Plate 20

Laminar flow burner for atomic absorption spectroscopy (Section 24B-1).

Photo courtesy of Varian Instruments, Sunnyvale, CA.

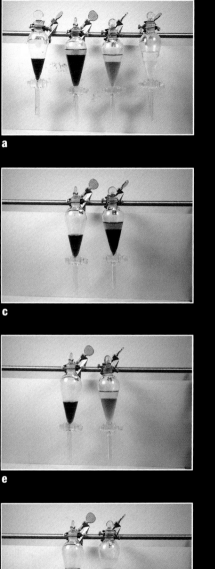

a

b

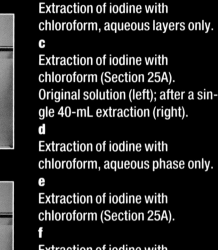

c

d

e

f

g

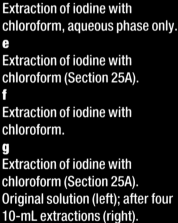

h

Color Plate 21

a
Extraction of iodine with chloroform (Section 25A). Left to right: iodine solution; after single 40-mL extraction; after two 20-mL extractions; after four 10-mL extractions.

b
Extraction of iodine with chloroform, aqueous layers only.

c
Extraction of iodine with chloroform (Section 25A). Original solution (left); after a single 40-mL extraction (right).

d
Extraction of iodine with chloroform, aqueous phase only.

e
Extraction of iodine with chloroform (Section 25A).

f
Extraction of iodine with chloroform.

g
Extraction of iodine with chloroform (Section 25A). Original solution (left); after four 10-mL extractions (right).

h
Extraction of iodine with chloroform, aqueous phase only.

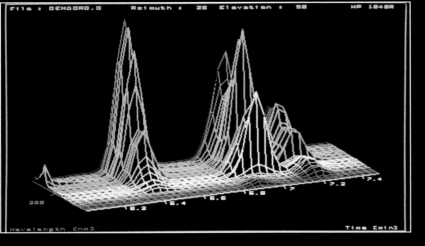

Color Plate 22

High-performance liquid chromatogram (Section 27B) obtained with a multichannel diode array spectrometer (Section 22A-1) as detector. The chromatogram is displayed on the terminal of a computer, which controls the instrument. Retention time is displayed horizontally, and various detected wavelengths are presented in different colors.

Photo courtesy of Hewlett–Packard Company, Palo Alto, CA.

Color Plate 23

Electronic analytical balance with 0.1-mg precision (Section 28B-2).

Photo courtesy of Mettler–Toledo, Inc., Hightstown, NJ.

Color Plate 24

Stamps from various countries celebrating scientists and events of importance to analytical chemistry. The numbers following each citation refer to the Scott Standard Postage Stamp Catalogue. Stamps Commemorating **(1)** the Centennial of the Metric Convention of 1875, France, #1475; **(2)** T. W. Richards, Sweden, #1104; **(3)** Svante Arrhenius, Sweden, #549; **(4)** Cato Guldberg and Peter Waage, Norway, #453; **(5)** Walther Nernst, Sweden, #655; **(6)** Allessandro Volta, Italy, #527; **(7)** André Marie Ampère, Monaco, #1001; **(8)** Gustav Robert Kirchhoff, Germany, #9N345; **(9)** spectroscopic analysis, Vatican, #655; **(10)** Jöns Jacob Berzelius, Sweden, #1293. **(11)** Stamp honoring Archer J. P. Martin and Richard L. M. Synge for their work in chromatography, Great Britain, #808. **(12)** Stamp depicting the visible spectrum and atomic lines, Canada, #613. **(13)** Stamp announcing an international spectroscopy colloquium in Madrid, Spain, #1570.

These stamps are from the private collection of Professor C. M. Lang.

A 25.0-mL sample was treated with an excess of methiomeprazine and subsequently diluted to 50.0 mL. Calculate the molar concentrations of Pd(II), c_{Pd}, and Au(III), c_{Au}, in the sample if the diluted solution had an absorbance of 0.533 at 480 nm and 0.590 at 635 nm when measured in a 1.00-cm cell.

At 480 nm

$$0.533 = (3.55 \times 10^3)(1.00) \times c_{Pd} + (2.96 \times 10^3)(1.00) \times c_{Au}$$

or

$$c_{Pd} = \frac{0.533 - 2.96 \times 10^3 \times c_{Au}}{3.55 \times 10^3}$$

At 635 nm

$$0.590 = (5.64 \times 10^2)(1.00) \times c_{Pd} + (1.45 \times 10^4)(1.00) \times c_{Au}$$

Substitution for c_{Pd} in this expression gives

$$0.590 = \frac{5.64 \times 10^2(0.533 - 2.96 \times 10^3 \times c_{Au})}{3.55 \times 10^3} + 1.45 \times 10^4 \times c_{Au}$$

$$= 0.0847 - 4.70 \times 10^2 \times c_{Au} + 1.45 \times 10^4 \times c_{Au}$$

$$c_{Au} = \frac{0.590 - 0.0847}{1.403 \times 10^4} = 3.60 \times 10^{-5}\ M$$

and

$$c_{Pd} = \frac{0.533 - (2.96 \times 10^3)(3.60 \times 10^{-5})}{3.55 \times 10^3} = 1.20 \times 10^{-4}\ M$$

Since the analysis involved a two-fold dilution, the concentrations of Pd(II) and Au(III) in the original sample were 7.20×10^{-5} and 2.40×10^{-4} M, respectively.

Mixtures containing more than two absorbing species can be analyzed, in principle at least, if one additional absorbance measurement is made for each added component. The uncertainties in the resulting data become greater, however, as the number of measurements increases. Some of the newer computerized spectrophotometers are capable of minimizing these uncertainties by overdetermining the system; that is, these instruments use many more data points than unknowns and effectively match the entire spectrum of the unknown as closely as possible by deriving synthetic spectra for various concentrations of the components. The derived spectra are then compared with that of the analyte until a close match is found. The spectrum for standard solutions of each component is required, of course.

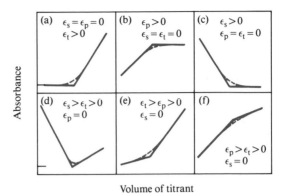

Figure 22-8
Typical photometric titration curves. Molar absorptivities of the substance titrated, product, and titrant are ϵ_s, ϵ_p, and ϵ_t.

22B-3 Spectrophotometric and Photometric Titrations

Spectrophotometric and photometric measurements are useful for locating the equivalence points of titrations.[6] This application of absorption measurements requires that one or more of the reactants or products absorb radiation or that an absorbing indicator be present.

Titration Curves

A photometric titration curve is a plot of absorbance (corrected for volume change) as a function of titrant volume. If conditions are chosen properly, the curve consists of two straight-line regions with different slopes, one occurring at the outset of the titration and the other located well beyond the equivalence-point region; the end point is taken as the intersection of extrapolated linear portions of the two lines.

Figure 22-8 shows typical photometric titration curves. Figure 22-8a is the curve for the titration of a nonabsorbing species with an absorbing titrant that is decolorized by the reaction. An example is the titration of thiosulfate ion with triiodide ion. The titration curve for the formation of an absorbing product from colorless reactants is shown in Figure 22-8b. An example is the titration of iodide ion with a standard solution of iodate ion to form triiodide. The remaining plots in Figure 22-8 illustrate the type of curve obtained with various combinations of absorbing analytes, titrants, and products.

In order to obtain titration curves with linear portions that can be extrapolated, the absorbing system(s) must obey Beer's law. Furthermore, absorbances must be corrected for volume changes by multiplying the observed absorbance by $(V + v)/V$, where V is the original volume of the solution and v is the volume of added titrant.

Instrumentation

Spectrophotometric and photometric titrations are ordinarily performed with an instrument that has been modified so that the titration vessel is held in the light

---

[6] For further information, see J. B. Headridge, *Photometric Titrations*. New York: Pergamon Press, 1961.

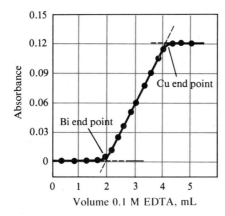

Figure 22-9
Photometric titration curve at 745 nm for 100 mL of a solution that was 2.0×10^{-3} M in Bi^{3+} and Cu^{2+}. The titrant was 0.1 M EDTA. (A. L. Underwood, *Anal. Chem.*, **1954**, *26*, 1322. With permission of the American Chemical Society.)

path. After the instrument is set to a suitable wavelength the 0% *T* adjustment is made in the usual way. With radiation passing through the analyte solution to the detector, the instrument is then adjusted to a convenient absorbance reading by varying the source intensity or the detector sensitivity. Ordinarily, no attempt is made to measure the true absorbance since relative values are adequate for end-point detection. Titration data are then collected without alteration of the instrument settings. The power of the radiation source and the response of the detector must remain constant during a photometric titration.

Applications of Spectrophotometric and Photometric Titrations

Spectrophotometric and photometric titrations often provide more accurate results than a direct photometric determination because the data from several measurements are pooled in determining the end point. Furthermore, the presence of other absorbing species may not interfere since only a change in absorbance is being measured.

One advantage of a photometric end point is that the experimental data are taken well away from the equivalence-point region. Consequently, the equilibrium constants of the reactions need not be as favorable as those required for a titration that depends on observations near the equivalence point (for example, potentiometric or indicator end points). For the same reason, more dilute solutions may be titrated.

The photometric end point has been applied to all types of reactions.[7] For example, most standard oxidizing agents have characteristic absorption spectra and thus produce photometrically detectable end points. Although standard acids or bases do not absorb, the introduction of acid/base indicators permits photometric neutralization titrations. The photometric end point has also been used to great advantage in titrations with EDTA and other complexing agents. Figure 22-9 illustrates the application of this technique to the successive titration of bismuth(III) and copper(II). At 745 nm, the cations, the reagent, and the bismuth complex formed in the first part of the titration do not absorb but the copper complex does. Thus, the solution exhibits no absorbance until essentially all the

---

[7] See, for example, the review by A. L. Underwood in *Advances in Analytical Chemistry and Instrumentation*, C. N. Reilley, Ed., Vol. 3, pp. 31–104. New York: Interscience, 1964.

Figure 22-10

Infrared spectrum for *n*-butanal (*n*-butyraldehyde). Note that transmittance rather than absorbance is plotted. [*Catalog of Selected Infrared Spectral Data*, Serial No. 225, Thermodynamics Research Center Data Project, Thermodynamics Research Center, Texas A&M University, College Station, TX (loose-leaf data sheets extant, 1964).]

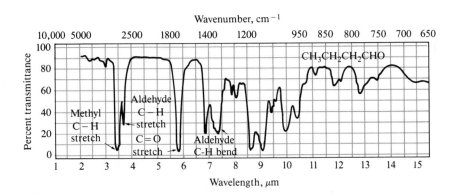

bismuth has been titrated. With the first formation of the copper complex, an increase in absorbance occurs. The increase continues until the copper equivalence point is reached. Additional reagent causes no further absorbance change. Clearly, two well-defined end points result.

22C APPLICATIONS OF INFRARED ABSORPTION SPECTROSCOPY

Infrared spectroscopy has been employed for both qualitative and quantitative analysis. Its qualitative applications are by far the more important of the two.

22C-1 Qualitative Applications

Infrared spectra are unique.

Infrared absorption spectroscopy is one of the most powerful and important tools available to the chemist for identifying and determining the structure of organic, inorganic, and biochemical species. As we noted earlier all molecular species absorb infrared radiation with the exception of a handful of homonuclear species, such as molecular hydrogen, oxygen, and nitrogen. Furthermore, the spectra of even relatively simple compounds are complex and provide numerous maxima and minima that are useful for identification purposes (see Figure 22-10). Indeed the infrared spectrum in the range of 2.5 to 15 μm of a compound provides a unique fingerprint, which is readily distinguished from absorption patterns of all other compounds; only optical isomers have identical spectra. Because techniques for the identification of organic compounds from their infrared spectra are usually treated in detail in organic laboratory courses, we will not attempt to deal with the subject in this book.[8]

22C-2 Quantitative Applications

Instrumental deviations from Beer's law are encountered in infrared photometry and spectrophotometry because peaks are frequently narrow and wide-slit widths must often be used because of the low sensitivity of infrared detectors.

Quantitative applications of infrared spectroscopy are much more limited than are such applications with ultraviolet/visible radiation because of the low molar absorptivities and narrowness of infrared peaks and instrumental difficulties in measuring transmittances accurately.

---

[8] For a description of these techniques, see R. M. Silverstein, G. C. Bassler, and T. C. Morrill, *Spectrometric Identification of Organic Compounds*, 5th ed., Chapter 3. New York: Wiley, 1991.

Table 22-1

EXAMPLES OF INFRARED VAPOR ANALYSIS FOR OSHA COMPLIANCE

Compound	Allowable Exposure, ppm*	Wavelength, μm	Minimum Detectable Concentration, ppm†
Carbon disulfide	4	4.54	0.5
Chloroprene	10	11.4	4
Diborane	0.1	3.9	0.05
Ethylenediamine	10	13.0	0.4
Hydrogen cyanide	4.7‡	3.04	0.4
Methyl mercaptan	0.5	3.38	0.4
Nitrobenzene	1	11.8	0.2
Pyridine	5	14.2	0.2
Sulfur dioxide	2	8.6	0.5
Vinyl chloride	1	10.9	0.3

Courtesy of The Foxboro Company, Foxboro, MA 02035.
* 1992 OSHA exposure limits for 8-hr time-weighted average.
† For 20.25-m cell.
‡ Short-term exposure limit; 15-min time-weighted average that shall not be exceeded at any time during the work day.

Absorbance Measurements

The use of matched cuvettes for solvent and analyte is seldom practical for infrared measurements because of the difficulty in obtaining cells with identical transmission characteristics. Part of this difficulty results from degradation of the transparency of infrared cell windows (typically polished sodium chloride) with use due to attack by traces of moisture in the atmosphere and in samples. Furthermore, path lengths are hard to reproduce because infrared cells are often less than 1 mm thick. Such narrow cells are required to permit the transmission of measurable intensities of infrared radiation through pure samples or through very concentrated solutions of the analyte. Measurements of dilute analyte solutions, as is done in ultraviolet or visible spectroscopy, are frequently precluded by the lack of good solvents that transmit over appreciable regions of the infrared spectrum.

For these reasons, a reference absorber is often dispensed with entirely in infrared work, and the intensity of the radiation passing through the sample is simply compared with that of the unobstructed beam; alternatively, a salt plate may be placed in the reference beam. Either way, the resulting transmittance is ordinarily less than 100%, even in regions of the spectrum where the sample is totally transparent. This effect is readily seen by examining the spectrum in Figure 22-10.

Typical Applications

Infrared spectrophotometry offers the potential for determining an unusually large number of substances because nearly all molecular species absorb in the region. Moreover, the uniqueness of an infrared spectrum provides a degree of specificity that is matched or exceeded by relatively few other analytical methods. This specificity has particular application to the analysis of mixtures of organic compounds with similar structures.

Figure 22-11
Experimental curves relating relative concentration uncertainties to absorbance for two spectrophotometers. Data obtained with (a) a Spectronic 20, a low-cost instrument, and (b) a Cary 118, a research-quality instrument. (From W. E. Harris and B. Kratochvil, *An Introduciton to Chemical Analysis*, p. 384. Philadelphia: Saunders College Publishing, 1981. With permission.)

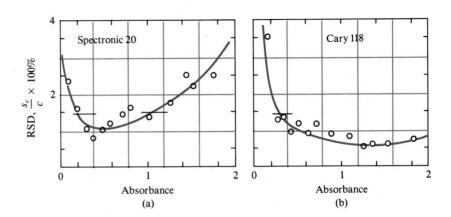

The recent proliferation of government regulations on atmospheric contaminants has demanded the development of sensitive, rapid, and highly specific methods for a variety of chemical compounds. Infrared absorption procedures appear to meet this need better than any other single analytical tool.

Table 22-1 illustrates the variety of atmospheric pollutants that can be determined with the simple, portable filter photometer shown in Figure 22-A; a separate interference filter is used for each analyte species. Of the more than 400 chemicals for which maximum tolerable limits have been set by the Occupational Safety and Health Administration (OSHA), more than half have absorption characteristics that make them amenable to determination by infrared photometry or spectrophotometry. Obviously, peak overlaps are to be expected with so many compounds absorbing; nevertheless, the method does provide a moderately high degree of selectivity.

22D ERRORS IN SPECTROPHOTOMETRIC ANALYSES[9]

In most of the analytical methods we have considered so far, the relative error in the result can be directly related to the relative instrumental error associated with the measurement. Thus, in a volumetric analysis, a relative error of 0.03 mL in a 10-mL titration will lead to a relative uncertainty in the concentration of the analyte of 0.03 mL/10 mL or 3 ppt. This straightforward relationship between instrumental error and concentration error is not found in spectrophotometric methods as shown by the experimental data plotted in Figure 22-11. To obtain these plots, a series of standard solutions was analyzed and the relative standard deviation of the results computed for various concentrations of the analyte. In the figures, these relative standard deviations are plotted as a function of absorbance for two spectrophotometers. One was a Spectronic 20 like the one shown in Figure 22-1; the other was a Cary 118 instrument, which is an expensive, research-quality spectrophotometer.

It is evident from both plots that concentration measurements at absorbances lower than about 0.1 are not very reliable and should be avoided. The reason for the poor precision in this region can be understood by writing Beer's law in

---

[9] For a detailed discussion of the errors associated with spectrophotometric measurements in the visible and ultraviolet regions, see L. D. Rathman, S. R. Crouch, and J. D. Ingle Jr., *Anal. Chem.*, **1972**, *44*, 1375.

the form

$$A = abc = \log \frac{P_0}{P} = \log P_0 - \log P$$

Note that concentration is directly proportional to the *difference* between two measured quantities, $\log P_0$ and $\log P$. At low concentrations, $\log P$ is nearly as large as $\log P_0$, and A is a small difference between two large numbers. Therefore, the relative uncertainties in A and c are large.

Note in Figure 22-11a that large errors are also encountered with the Spectronic 20 when the measured absorbance is above about 1.2. Here, the power of the beam is so very low after it has passed through the analyte solution that it cannot be measured accurately. As shown in Figure 22-11b, this type of error is much less severe with a highly sensitive research-quality instrument.

22E QUESTIONS AND PROBLEMS

22-1. Describe the differences between the following and list any particular advantages possessed by one over the other:
(a) spectrophotometers and photometers.
(b) single-beam and double-beam instruments for absorbance measurements.
(c) conventional and diode-array spectrophotometers.

*22-2. What minimum requirement is needed to obtain reproducible results with a single-beam spectrophotometer?

22-3. What is the purpose of (a) the 0% T adjustment and (b) the 100% T adjustment of a spectrophotometer?

*22-4. What experimental variables must be controlled to assure reproducible absorbance data?

22-5. What advantage can be claimed for the standard addition method? What minimum condition is needed for the successful application of this method?

*22-6. The molar absorptivity for the complex formed between bismuth(III) and thiourea is 9.32×10^3 L cm^{-1} mol^{-1} at 470 nm. Calculate the range of permissible concentrations for the complex if the absorbance is to be no less than 0.15 nor greater than 0.80 when the measurements are made in 1.00-cm cells.

22-7. The molar absorptivity for aqueous solutions of phenol at 211 nm is 6.17×10^3 L cm^{-1} mol^{-1}. Calculate the permissible range of phenol concentrations that can be used if the transmittance is to be less than 80% and greater than 5% when the measurements are made in 1.00-cm cells.

*22-8. The logarithm of the molar absorptivity for acetone in ethanol is 2.75 at 366 nm. Calculate the range of acetone concentrations that can be used if the transmittance is to be greater than 10% and less than 90% with a 1.50-cm cell.

22-9. The logarithm of the molar absorptivity of phenol in aqueous solution is 3.812 at 211 nm. Calculate the range of phenol concentrations that can be used if the absorbance is to be greater than 0.100 and less than 2.000 with a 1.25-cm cell.

22-10. A photometer with a linear response to radiation gave a reading of 685 mV with a blank in the light path and 179 mV when the blank was replaced by an absorbing solution. Calculate
*(a) the percent transmittance and absorbance of the absorbing solution.
(b) the expected transmittance if the concentration of absorber is one half that of the original solution.
*(c) the transmittance to be expected if the light path through the original solution is doubled.

22-11. A portable photometer with a linear response to radiation registered 73.6 μA with a blank solution in the light path. Replacement of the blank with an absorbing solution yielded a response of 24.9 μA. Calculate
(a) the percent transmittance of the sample solution.
*(b) the absorbance of the sample solution.
(c) the transmittance to be expected for a solution in which the concentration of the absorber is one third that of the original sample solution.
*(d) the transmittance to be expected for a solution that has twice the concentration of the sample solution.

22-12. Sketch a photometric titration curve for the titration of Sn^{2+} with MnO_4^-. What color radiation should be used for this titration? Explain.

22-13. Iron(III) reacts with SCN$^-$ to form the red complex, $FeSCN^{2+}$. Sketch a photometric titration curve for Fe(III) with SCN$^-$ ion when a photometer with a green filter is used to collect data. Why is a green filter used?

*22-14. Ethylenediaminetetraacetic acid abstracts bismuth(III)

---

* Answers to the asterisked problems are given in the answers section at the back of the book.

from its thiourea complex:

$$Bi(tu)_6^{3+} + H_2Y^{2-} \rightarrow BiY^- + 6tu + 2H^+$$

where tu is the thiourea molecule, $(NH_2)_2CS$. Predict the shape of a photometric titration curve based on this process, given that the Bi(III)/thiourea complex is the only species in the system that absorbs at 465 nm, the wavelength selected for the analysis.

22-15. The accompanying data (1.00-cm cells) were obtained for the spectrophotometric titration of 10.00 mL of Pd(II) with 2.44×10^{-4} M Nitroso R:[10]

Volume of Nitroso R, mL	A_{500}
0	0
1.00	0.147
2.00	0.271
3.00	0.375
4.00	0.371
5.00	0.347
6.00	0.325
7.00	0.306
8.00	0.289

Calculate the concentration of the Pd(II) solution, given that the ligand-to-cation ratio in the colored product is 2:1.

*22-16. A 4.97-g petroleum specimen was decomposed by wet-ashing and subsequently diluted to 500 mL in a volumetric flask. Cobalt was determined by treating 25.00-mL aliquots of this diluted solution as follows:

	Reagent Volume		
Co(II), 3.00 ppm	Ligand	H_2O	Absorbance
0.00	20.00	5.00	0.398
5.00	20.00	0.00	0.510

Assume that the Co(II)/ligand chelate obeys Beer's law, and calculate the percentage of cobalt in the original sample.

22-17. Iron(III) forms a complex with thiocyanate ion that has the formula $FeSCN^{2+}$. The complex has an absorption maximum at 580 nm. A specimen of well water was assayed according to the scheme given in Table 22-2. Calculate the concentration of iron in parts per million.

22-18. A. J. Mukhedkar and N. V. Deshpande[11] report on a simultaneous determination for cobalt and nickel based on absorption by their 8-quinolinol complexes. Molar absorptivities are $\varepsilon_{Co} = 3529$ and $\varepsilon_{Ni} = 3228$ at 365 nm and $\varepsilon_{Co} = 428.9$ and $\varepsilon_{Ni} = 0$ at 700 nm. Calculate the concentration of nickel and cobalt in each of the following solutions (1.00-cm cells):

Solution	A_{700}	A_{365}
*1	0.0235	0.617
2	0.0714	0.755
*3	0.0945	0.920
4	0.0147	0.592

22-19. Molar absorptivity data for the cobalt and nickel complexes with 2,3-quinoxalinedithiol are $\varepsilon_{Co} = 36,400$ and $\varepsilon_{Ni} = 5520$ at 510 nm and $\varepsilon_{Co} = 1240$ and $\varepsilon_{Ni} = 17,500$ at 656 nm. A 0.425-g sample was dissolved and diluted to 50.0 mL. A 25.0-mL aliquot was treated to eliminate interferences; after addition of 2,3-quinoxalinedithiol, the volume was adjusted to 50.0 mL. This solution had an absorbance of 0.446 at 510 nm and 0.326 at 656 nm in a 1.00-cm cell. Calculate the parts per million of cobalt and nickel in the sample.

22-20. The indicator HIn has an acid dissociation constant of 4.80×10^{-6} at ordinary temperatures. The accompanying absorbance data are for 8.00×10^{-5} M solutions of the indicator measured in 1.00-cm cells in strongly acidic and strongly alkaline media.

Table 22-2

		Volumes, mL				
Sample	Sample Volume	Oxidizing Reagent	Fe(II) 2.75 ppm	KSCN 0.050 M	H_2O	Absorbance, 580 nm (1.00-cm cells)
1	50.00	5.00	5.00	20.00	20.00	0.549
2	50.00	5.00	0.00	20.00	25.00	0.231

[10] O. W. Rollins and M. M. Oldham, *Anal. Chem.*, **1971**, *43*, 262.
[11] *Anal. Chem.*, **1963**, *35*, 47.

	Absorbance	
λ, nm	pH 1.00	pH 13.00
420	0.535	0.050
445	0.657	0.068
450	0.658	0.076
455	0.656	0.085
470	0.614	0.116
510	0.353	0.223
550	0.119	0.324
570	0.068	0.352
585	0.044	0.360
595	0.032	0.361
610	0.019	0.355
650	0.014	0.284

Estimate the wavelength at which absorption by the indicator becomes independent of pH (that is, the isosbestic point).

22-21. Calculate the absorbance (1.00-cm cells) at 450 nm of a solution in which the total molar concentration of the indicator described in 22-20 is 8.00×10^{-5} and the pH is *(a) 4.92, (b) 5.46, *(c) 5.93, (d) 6.16.

22-22. What is the absorbance at 595 nm (1.00-cm cells) of a solution that is 1.25×10^{-4} M in the indicator of 22-20 and has a pH of *(a) 5.30, (b) 5.70, *(c) 6.10?

22-23. Several buffer solutions were made 1.00×10^{-4} M in the indicator of 22-20. Absorbance data (1.00-cm cells) are

Solution	A_{450}	A_{595}
*A	0.344	0.310
B	0.508	0.212
*C	0.653	0.136
D	0.220	0.380

Calculate the pH of each solution.

22-24. Construct an absorption spectrum for an 8.00×10^{-5} M solution of the indicator of 22-20 when measurements are made with 1.00-cm cells and

*(a) $\dfrac{[\text{HIn}]}{[\text{In}^-]} = 3.00$.

(b) $\dfrac{[\text{HIn}]}{[\text{In}^-]} = 1.00$.

(c) $\dfrac{[\text{HIn}]}{[\text{In}^-]} = \dfrac{1}{3.00}$.

22-25. Solutions of P and Q individually obey Beer's law over a large concentration range. Spectral data for these species in 1.00-cm cells are

	Absorbance	
λ, nm	8.55×10^{-5} M P	2.37×10^{-4} M Q
400	0.078	0.500
420	0.087	0.592
440	0.096	0.599
460	0.102	0.590
480	0.106	0.564
500	0.110	0.515
520	0.113	0.433
540	0.116	0.343
560	0.126	0.255
580	0.170	0.170
600	0.264	0.100
620	0.326	0.055
640	0.359	0.030
660	0.373	0.030
680	0.370	0.035
700	0.346	0.063

(a) Plot an absorption spectrum for a solution that is 8.55×10^{-5} M in P and 2.37×10^{-4} M in Q.

(b) Calculate the absorbance (1.00-cm cells) at 440 nm of a solution that is 4.00×10^{-5} M in P and 3.60×10^{-4} M in Q.

(c) Calculate the absorbance (1.00-cm cells) at 620 nm for a solution that is 1.61×10^{-4} M in P and 7.35×10^{-4} M in Q.

22-26. Use the data in 22-25 to calculate the molar concentration of P and Q in each of the following solutions:

A_{400}	A_{620}		A_{400}	A_{620}
*(a) 0.357	0.803.	(d)	0.910	0.338.
(b) 0.830	0.448.	*(e)	0.480	0.825.
*(c) 0.248	0.333.	(f)	0.194	0.315.

*22-27. A standard solution was put through appropriate dilutions to give the concentrations of iron shown below. The iron (II)-1,10, phenanthroline complex was then developed in 25.0-mL aliquots of these solutions, and then each was diluted to 50.0 mL. The following absorbances (1.00-cm cells) were recorded at 510 nm:

Fe(II) Concentration in Original Solutions, ppm	A_{510}
4.00	0.160
10.0	0.390
16.0	0.630
24.0	0.950
32.0	1.260
40.0	1.580

(a) Sketch a calibration curve from these data.

(b) Use the method of least squares to derive an equation relating absorbance and the concentration of iron(II).

*(c) Calculate the standard deviation of the slope and intercept.

22-28. The method developed 22-27 was used for the routine determination of iron in 25.0-mL aliquots of groundwater. Express the concentration (as ppm Fe) in samples that yielded the accompanying absorbance data (1.00-cm cell). Calculate the relative standard deviation of the result. Repeat the calculation assuming the absorbance data are means of three measurements.

*(a) 0.143   *(c) 0.068   *(e) 1.512
 (b) 0.675   (d) 1.009   (f) 0.546

CHAPTER 23

MOLECULAR FLUORESCENCE SPECTROSCOPY

Fluorescence is an analytically important emission process in which atoms or molecules are excited by the absorption of electromagnetic radiation. The excited species then relax to the ground state, giving up their excess energy as photons. One of the most attractive features of fluorescence methods is their inherent sensitivity, which is often one to three orders of magnitude better than that for absorption spectroscopy. Another advantage of fluorescence methods is their large linear ranges, which are often significantly greater than those encountered in absorption spectroscopy. Fluorescence methods are, however, much less widely applicable than absorption methods because of the relatively limited number of chemical systems that can be made to fluoresce. In this chapter we consider the theory and applications of molecular fluorescence.

Fluorescence emission occurs in 10^{-5} s or less. In contrast, phosphorescence may go on for several minutes or even hours. Fluorescence is much more widely used for analyses than phosphorescence.

23A THEORY OF MOLECULAR FLUORESCENCE

Figure 23-1 is the partial energy diagram for a hypothetical molecular species. Three electronic energy states are shown, E_0, E_1, and E_2, where E_0 is the ground state and E_1 and E_2 are electronic excited states. Each of the electronic states is shown as having four excited vibrational states. Irradiation of this species with a band of radiation made up of wavelengths λ_1 to λ_5 (Figure 23-1a) results in the momentary population of the five vibrational levels of the first excited electronic state E_1. Similarly, when the molecules are irradiated with a more energetic band radiation made up of shorter wavelengths λ_1' through λ_5', the five vibrational levels of the higher energy electronic state E_2 become briefly populated.

23A-1 Relaxation Processes

Generally, the lifetime of an excited species is brief because there are several ways an excited atom or molecule can give up its excess energy and relax to its

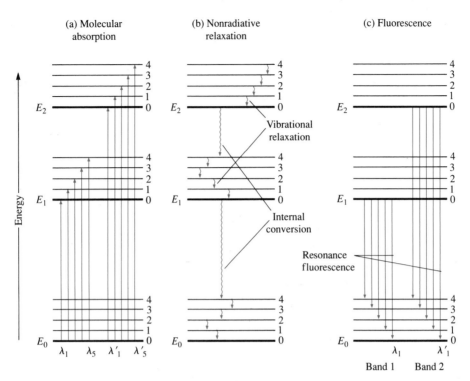

Figure 23-1

Energy-level diagram showing some of the energy changes that occur during (a) absorption, (b) nonradiative relaxation, and (c) fluorescence by a molecular species.

Vibrational relaxation takes 10^{-15} s or less. Electronic relaxation takes 10^{-5} to 10^{-9} s.

ground state. Two of the most important of these mechanisms, nonradiative relaxation and fluorescent relaxation, are illustrated in Figure 23-1b and c.

Two types of nonradiative relaxation are shown in Figure 23-1b. *Vibrational deactivation*, or *relaxation*, depicted by the short wavy arrows between vibrational energy levels, takes place during collisions between excited molecules and molecules of the solvent. During these collisions, the excess vibrational energy is transferred to solvent molecules in a series of steps as indicated in the figure. The gain in vibrational energy of the solvent is reflected in a slight increase in the temperature of the medium. Vibrational relaxation is such an efficient process that the average lifetime of an excited *vibrational* state is only about 10^{-15} s.

Nonradiative relaxation between the lowest vibrational level of an excited electronic state and the upper vibrational level of another electronic state can also occur. This type of relaxation, which is sometimes called *internal conversion*, is depicted by the two longer wavy arrows in Figure 23-1b. Internal conversion is much less efficient than vibrational relaxation so that the average lifetime of an electronic excited state is between 10^{-5} and 10^{-9} s. The mechanisms by which this type of relaxation occurs are not fully understood, but the net effect is again a tiny rise in the temperature of the medium.

Figure 23-1c depicts another relaxation process: fluorescence. Note that bands of radiation are produced when molecules fluoresce because the electronically excited molecules can relax to any of the several vibrational states of the ground electronic state. Like molecular absorption bands, molecular fluorescence bands are made up of a multitude of closely spaced lines that are often difficult to resolve.

Fluorescence bands consist of a host of closely spaced lines.

Resonance Lines and the Stokes Shift

Note that the lines that terminate the two fluorescence bands on the short-wavelength, or high-energy, side (λ_1 and λ_1') are identical in energy to the two lines labeled λ_1 and λ_1' in the absorption diagram in Figure 23-1a. These lines are termed *resonance lines* because the absorption and fluorescence wavelengths are identical. Note also that molecular fluorescence bands are made up largely of lines that have longer wavelengths and thus lower energies than the band of absorbed radiation responsible for their excitation. This shift to longer wavelengths is sometimes called the *Stokes shift*.

Resonance fluorescence has an identical wavelength to the radiation that caused the fluorescence.

To develop a better understanding of Stokes shifts, let us consider what occurs when the molecule under consideration is irradiated by a single wavelength λ_5'. As shown in Figure 23-1a, absorption of this radiation promotes an electron into vibrational level 4 of the second excited electronic state E_2. In 10^{-15} s or less, vibrational relaxation to the zero vibrational level of E_2 occurs (Figure 23-1b). At this point, further relaxation can follow either the nonradiative route depicted by the longer wavy line in Figure 23-1b or the radiative route shown in Figure 23-1c. If the radiative route is followed, relaxation to any of the several vibrational levels of the ground state takes place, giving a band (band 2) of emitted wavelengths as shown. All of these lines have lower energies, or longer wavelengths, than the excitation line, λ_5'.

Stokes-shifted fluorescence is longer in wavelength than the radiation that caused the fluorescence.

Quantum yield = $\Phi = \dfrac{R_f}{R_f + R_r}$ where R_f is the rate of fluorescent relaxation and R_r is the rate of radiationless relaxation.

Let us now turn to those molecules in excited state E_2 that undergo internal conversion to electronic state E_1. As before, further relaxation can take a nonradiative or a radiative route to the ground state. If the latter occurs, band 1 of fluorescence is produced. Note that here the Stokes shift is from the more energetic band 2 to the less energetic band 1. Note also that band 1 can be produced not only by the mechanism just described but also by the absorption of radiation of wavelengths λ_1 through λ_5 (Figure 23-1).

Relationship Between Excitation Spectra and Fluorescence Spectra

Because the energy differences between vibrational states is about the same for both ground and excited states, the absorption, or *excitation spectrum* and the fluorescence spectrum for a compound often appear as approximate mirror images of one another with overlap occurring at the resonance line. This effect is demonstrated by the spectra appearing in Figure 23-2.

23A-2 Fluorescent Species

As shown in Figure 23-1, fluorescence is one of several mechanisms by which a molecule returns to the ground state after it has been excited by absorption of radiation. Thus all absorbing molecules have the potential to fluoresce. Most do not, however, because their structure provides radiationless pathways by which relaxation can occur *at greater rate* than fluorescent emission.

The *quantum yield* Φ of molecular fluorescence is simply the ratio of the number of molecules that fluoresce to the total number of excited molecules (or the ratio of photons emitted to photons absorbed). Highly fluorescent molecules,

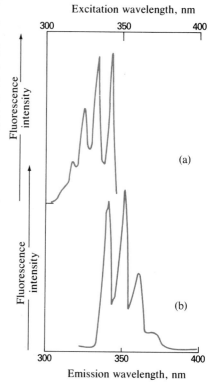

Figure 23-2
Fluorescence spectra for 1 ppm anthracene in alcohol: (a) excitation spectrum; (b) emission spectrum.

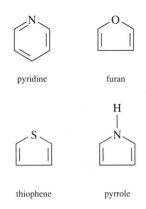

Figure 23-3

Typical aromatic molecules that do not fluoresce.

Many aromatic compounds fluoresce.

Rigid molecules or complexes tend to fluoresce.

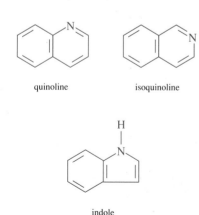

Figure 23-4

Typical aromatic molecules that fluoresce.

such as fluorescein, have quantum efficiencies that approach unity under some conditions. Nonfluorescent species have efficiencies that are essentially zero.

Fluorescence and Structure

Compounds containing aromatic rings give the most intense and most useful molecular fluorescent emission. While certain aliphatic and alicyclic carbonyl compounds as well as highly conjugated double-bonded structures also fluoresce, their numbers are small in comparison with the number of fluorescent compounds containing aromatic systems.

Most unsubstituted aromatic hydrocarbons fluoresce in solution, with the quantum efficiency increasing with the number of rings and their degree of condensation. The simplest heterocyclics, such as pyridine, furan, thiophene, and pyrrole (see Figure 23-3), do not exhibit molecular fluorescence, but fused-ring structures containing these rings often do (see Figure 23-4).

Substitution on an aromatic ring causes shifts in the wavelength of absorption maxima and corresponding changes in the fluorescence peaks. In addition, substitution frequently affects the fluorescence efficiency. These effects are demonstrated by the data in Table 23-1.

The Effect of Structural Rigidity

It is found experimentally that fluorescence is particularly favored in rigid molecules. For example, under similar conditions of measurement, the quantum efficiency of fluorene is nearly 1.0 whereas that of biphenyl is about 0.2. The difference in behavior appears to be largely a result of the increased rigidity furnished by the bridging methylene group in fluorene (see Figure 23-5). This rigidity lowers the rate of nonradiative relaxation to the point where relaxation by fluorescence has time to occur. Many similar examples can be cited. In addition, enhanced emission frequently results when fluorescing dyes are adsorbed on a solid surface; here again, the added rigidity provided by the solid may account for the observed effect.

The influence of rigidity has also been invoked to account for the increase in fluorescence of certain organic chelating agents when they are complexed with a metal ion. For example, the fluorescence intensity of 8-hydroxyquinoline is much less than that of the zinc complex (see Figure 23-6).

Temperature and Solvent Effects

In most molecules, the quantum efficiency of fluorescence decreases with increasing temperature because the increased frequency of collision at elevated temperatures improves the probability of collisional relaxation. A decrease in solvent viscosity leads to the same result.

23B EFFECT OF CONCENTRATION ON FLUORESCENCE INTENSITY

The power of fluorescent radiation F is proportional to the radiant power of the excitation beam absorbed by the system:

$$F = K'(P_0 - P) \tag{23-1}$$

where P_0 is the power of the beam incident on the solution and P is its power after it traverses a length b of the medium. The constant K' depends on the quantum efficiency of the fluorescence. Moreover, as shown in Feature 23-1, at constant P_0

$$F = Kc \qquad (23\text{-}2)$$

where c is the concentration of the fluorescent species and K is a proportionality constant. Thus, a plot of the fluorescent power of a solution versus the concentration of the emitting species should be, and ordinarily is, linear at low concentrations. When c becomes great enough that the absorbance is larger than about 0.05 (or the transmittance is smaller than about 90%), linearity is lost and F lies below an extrapolation of the straight-line plot (see Figure 23-7). This effect is a result of *self-quenching* in which analyte molecules absorb the fluorescence produced by other analyte molecules. Indeed, at very high concentrations, F reaches a maximum and then begins to decrease with increasing concentration.

Table 23-1

EFFECT OF SUBSTITUTION ON THE FLUORESCENCE OF BENZENE DERIVATIVES*

Compound	Relative Intensity of Fluorescence
Benzene	10
Toluene	17
Propylbenzene	17
Fluorobenzene	10
Chlorobenzene	7
Bromobenzene	5
Iodobenzene	0
Phenol	18
Phenolate ion	10
Anisole	20
Aniline	20
Anilinium ion	0
Benzoic acid	3
Benzonitrile	20
Nitrobenzene	0

* In ethanol solution. Taken from W. West, *Chemical Applications of Spectroscopy* (*Techniques of Organic Chemistry*, Vol. IX, p. 730). New York: Interscience, 1956. Reprinted by permission of John Wiley & Sons.

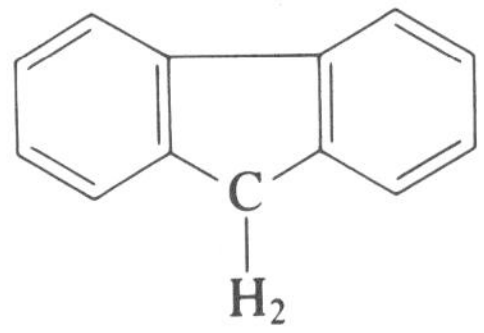

Feature 23-1
HOW IS FLUORESCENCE RELATED TO CONCENTRATION?

In order to relate F to the concentration c of the fluorescing particle, we write Beer's law in the form

$$\frac{P}{P_0} = 10^{-\varepsilon bc}$$

where ε is the molar absorptivity of the fluorescing species and εbc is the absorbance A. By substituting this equation into Equation 23-1, we obtain

$$F = K'P_0(1 - 10^{-\varepsilon bc})$$

A power expansion of the exponential term in this equation leads to

$$F = K'P_0\left[2.3\varepsilon bc - \frac{(-2.3\varepsilon bc)^2}{2!} - \frac{(-2.3\varepsilon bc)^3}{3!} - \cdots\right]$$

Provided $\varepsilon bc = A < 0.05$, all the subsequent terms in the brackets are small with respect to the first and we can write

$$F = 2.3K'\varepsilon bcP_0 \qquad (23\text{-}3)$$

or, at constant P_0

$$F = Kc$$

(which is Equation 23-2).

fluorene
$\Phi \to 1$

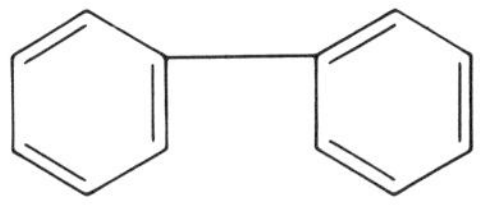

biphenyl
$\Phi = 0.2$

Figure 23-5
Effect of rigidity on quantum yield.

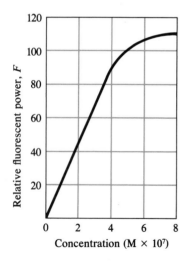

Figure 23-6
Effect of rigidity on fluorescence.

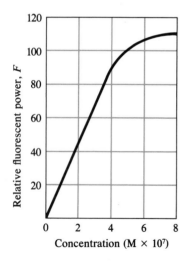

Figure 23-7
Calibration curve for the spectrofluor-ometric determination of tryptophan in soluble proteins from the lens of a mammalian eye.

23C FLUORESCENCE INSTRUMENTS

Figure 23-8 shows a typical configuration for the components of *fluorometers* and *spectrofluorometers*. These components are identical to the ones described in Section 20C for ultraviolet/visible spectroscopy. A fluorometer, like a photometer, employs filters for wavelength selection. Most spectrofluorometers, in contrast, employ a filter to limit the excitation radiation and a grating monochromator to disperse the fluorescent radiation from the sample.

As shown in Figure 23-8, fluorescence instruments are usually double-beam in design in order to compensate for fluctuations in the power of the source. The beam directed toward the sample first passes through a primary filter or a primary monochromator, which transmits radiation that excites fluorescence but excludes or limits radiation that corresponds to the fluorescence wavelengths. Fluorescence radiation is propagated from the sample in all directions but is most conveniently observed at right angles to the excitation beam; at other angles, increased scattering from the solution and the cell walls may cause large errors in the intensity measurement. The emitted radiation reaches a photoelectric detector after passing through the secondary filter or monochromator, which isolates a fluorescence peak for measurement.

The reference beam passes through an attenuator to decrease its power to approximately that of the fluorescence radiation (the power reduction is usually by a factor of 100 or more). The signals from the reference and sample phototubes are then processed by a difference amplifier whose output is displayed on a meter or recorder.

The sophistication, performance characteristics, and cost of fluorometers and spectrofluorometers differ as widely as do the corresponding instruments for absorption measurements. In many regards, filter-type instruments are better suited for quantitative analytical work than are the more elaborate instruments based on monochromators. Generally, fluorometers are more sensitive than spectrofluorometers because filters have a higher radiation throughput than do monochromators. In addition, source and detector can be positioned closer to the sample in the simpler instrument, a factor that enhances sensitivity.

23D APPLICATIONS OF FLUORESCENCE METHODS

Fluorescence methods are generally one to three orders of magnitude more sensitive than methods based on absorption because the sensitivity of the former can be enhanced either by increasing the power of the excitation beam (Equation 23-3) or by amplifying the detector signal. Note that neither of these options improves the sensitivity of methods based on absorption because the concentration-related parameter in Beer's law is the logarithm of a ratio

$$c = k \log(P_0/P)$$

where $k = 1/ab$. Increasing the power P_0 increases P proportionately and thus has no effect on sensitivity. Similarly, increasing the amplification of the detector signal affects the two measured quantities in an identical way and leads to no improvement.

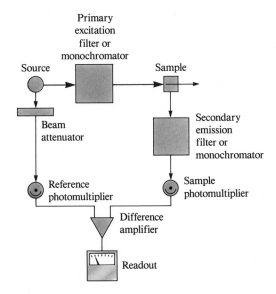

Figure 23-8

Components of a fluorometer or a spectrofluorometer.

23D-1 Methods for Inorganic Species

There are two types of inorganic fluorometric methods. Direct methods are based on the reaction of the analyte with a chelating agent to form a complex that fluoresces. In contrast, indirect methods depend on the diminution, or *quenching*, of fluorescence of a reagent as a result of its reaction with the analyte. Quenching is used primarily for the determination of anions.

The most successful fluorometric reagents for the determination of cations are aromatic compounds with two or more donor functional groups that permit chelate formation with the metal ion. A few typical fluorometric reagents and their applications are found in Table 23-2 and Figure 23-9. With most of these reagents, the cation is extracted into a solution of the reagent in an immiscible organic solvent, such as chloroform. The fluorescence of the organic solution is then measured.[1]

Nonradiative relaxation of transition-metal chelates is so efficient that fluorescence of these species is seldom encountered. It is worth noting that most transition metals absorb in the ultraviolet or visible region, whereas nontransition metal ions do not. For this reason, fluorometry often complements spectrophotometry as a method for the determination of cations.

23D-2 Methods for Organic and Biochemical Species

The number of applications of fluorometric methods to organic problems is impressive. Weissler and White have summarized the most important of these in several tables.[2] More than 100 entries are found under the heading "Organic

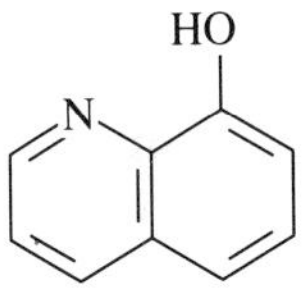

HO

8-hydroxyquinoline
(reagent for Al, Be, and
other metal ions)

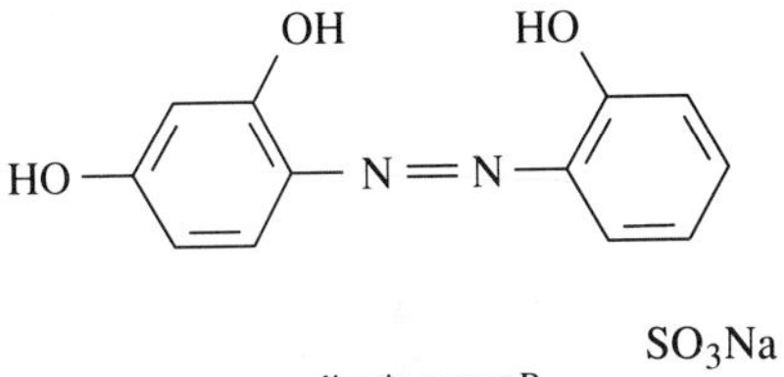

OH   HO

N = N

SO₃Na

alizarin garnet R
(reagent for Al, F⁻)

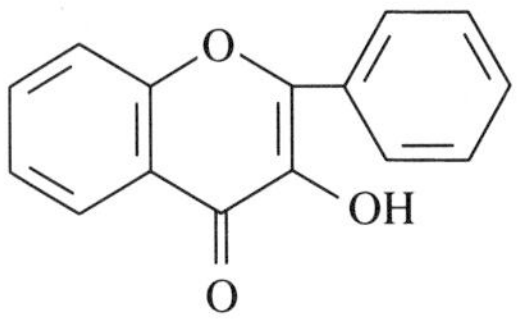

O

OH
O

flavanol
(reagent for Zr and Sn)

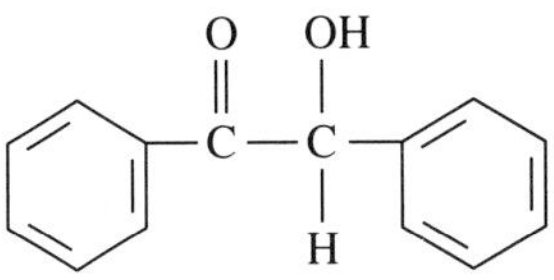

O   OH
C — C
H

benzoin
(reagent for B, Zn, Ge, and Si)

Figure 23-9

Some fluorometric reagents for cations.

[1] For a more complete summary of fluorescent chelating reagents, see A. Weissler and C. E. White, in *Handbook of Analytical Chemistry*, L. Meites, Ed., pp. **6**–178 to **6**–181. New York: McGraw-Hill, 1963.
[2] A. Weissler and C. E. White, in *Handbook of Analytical Chemistry*, L. Meites, Ed., pp. **6**–182 to **6**–196. New York: McGraw-Hill, 1963.

Table 23-2
SELECTED FLUOROMETRIC METHODS FOR INORGANIC SPECIES*

| Ion | Reagent | Wavelength, nm | | Sensitivity, $\mu g/mL$ | Interference |
		Absorption	Fluorescence		
Al^{3+}	Alizarin garnet R	470	500	0.007	Be, Co, Cr, Cu, F^-, NO_3^-, Ni, PO_4^{3-}, Th, Zr
F^-	Al complex of Alizarin garnet R (quenching)	470	500	0.001	Be, Co, Cr, Cu, Fe, Ni, PO_4^{3-}, Th, Zr
$B_4O_7^{2-}$	Benzoin	370	450	0.04	Be, Sb
Cd^{2+}	2-(o-Hydroxyphenyl)-benzoxazole	365	Blue	2.001	NH_3
Li^+	8-Hydroxyquinoline	370	580	0.2	Mg
Sn^{4+}	Flavanol	400	470	0.1	F^-, PO_4^{3-}, Zr
Zn^{2+}	Benzoin	—	Green	10.001	B, Be, Sb, colored ions

* From *Handbook of Analytical Chemistry*, L. Meites, Ed., pp. **6**–178 to **6**–181. New York: McGraw-Hill, 1963.

and General Biochemical Substances," including such diverse compounds as adenine, anthranilic acid, aromatic polycyclic hydrocarbons, cysteine, guanidine, indole, naphthols, certain nerve gases, proteins, salicylic acid, skatole, tryptophan, uric acid, and warfarin. Some 50 medicinal agents that can be determined fluorometrically are listed. Included among these are adrenaline, alkylmorphine, chloroquin, digitalis principles, lysergic acid diethylamide (LSD), penicillin, phenobarbital, procaine, and reserpine. Methods for the analysis of ten steroids and an equal number of enzymes and coenzymes are also listed in these tables. Some of the plant products listed are chlorophyll, ergot alkaloids, rauwolfia serpentian alkaloids, flavonoids, and rotenone.

Without question, the most important application of fluorometry is in the analysis of food products, pharmaceuticals, clinical samples, and natural products. The sensitivity and selectivity of the method make it a particularly valuable tool in these fields.

23E QUESTIONS AND PROBLEMS

23-1. Briefly describe or define
  *(a) fluorescence.
  (b) vibrational relaxation.
  *(c) internal conversion.
  (d) resonance fluorescence.
  *(e) Stokes shift.
  (f) quantum yield.
  *(g) self-quenching.

23-2. How do vibrational relaxation and electronic relaxation resemble one another? How do they differ?

*23-3. Why do some absorbing compounds fluoresce and others do not?

23-4. What structural features appear to favor fluorescence?

*23-5. Describe the components of a fluorometer.

23-6. Why are most fluorescence instruments double-beam in design?

*23-7. Why are fluorometers often more useful than spectrofluorometers for quantitative analysis?

23-8. Which of the compounds that follow would you expect to have a greater fluorescence quantum yield? Explain.

(a)

phenolphthalein

fluorescein

(b)

o,o'-dihydroxyazobenzene

bis(o-hydroxyphenyl)hydrazine

*23-9. Briefly explain why
   (a) fluorescent emission ordinarily occurs at wavelengths that are longer than that of the excitation radiation.
   (b) fluorescence measurements have the capability of greater sensitivity than absorbance measurements.

*23-10. The quinine in an antimalarial tablet was dissolved in sufficient 0.10 M HCl to give 500 mL of solution. A 15.00-mL aliquot was then diluted to 100.0 mL with the acid. The fluorescent intensity for the diluted sample at 347.5 nm provided a reading of 288 on an arbitrary scale. A standard 100 ppm quinine solution registered 180 when measured under conditions identical to those for the diluted sample. Calculate the milligrams of quinine in the tablet.

23-11. The analysis in Problem 23-10 was modified to make use of a standard addition. As before, a tablet was dissolved in sufficient 0.10 M HCl to give 1.000 L. Dilution of a 20.00-mL aliquot to 100 mL gave a solution that gave a reading of 540 at 347.5 nm. A second 20.00-mL aliquot was mixed with 10.0 mL of 50.0 ppm quinine solution before dilution to 100 mL. The fluorescent intensity of

this solution was 600. Calculate the milligrams of quinine in the tablet.

*23-12. Warfarin ($C_{19}H_{16}O_4$), the active ingredient in rodenticides, will fluoresce at 385 nm. A 0.842-g sample was brought into solution and diluted to 1.000 L. The fluorescence of this solution was found to be 596 (arbitrary units) at 385 nm. Mixture of a 25.00-mL aliquot of the sample with 5.00 mL of standard 0.250 ppm warfarin yielded a solution with an intensity of 670. Calculate the percentage of warfarin in the sample.

23-13. The reduced form of nicotinamide adenine dinucleotide (NADH) is an important and highly fluorescent coenzyme. It has an absorption maximum of 340 nm and an emission maximum at 465 nm. Standard solutions of NADH gave the following fluorescence intensities:

Concn NADH, μmol/L	Relative Intensity, I_r
0.200	4.52
0.400	9.01
0.600	13.71
0.800	17.91

   (a) Construct a calibration curve for NADH.
   (b) An unknown exhibits a relative fluorescence of 12.16. Use the plot from (a) to determine the concentration of NADH.
   (c) Derive a least-squares equation for the plot in part (a).
   *(d) Calculate the standard deviation of the slope and the intercept for the curve.
   *(e) A sample exhibits a relative fluorescence of 12.16. Calculate the concentration of NADH.
   *(f) Calculate the relative standard deviation for the result in (e).
   *(g) Calculate the relative standard deviation for the result in (e) if the reading of 12.16 was the mean of three measurements.

23-14. To four 10.0-mL aliquots of a water sample were added 0.00, 1.00, 2.00, and 3.00 mL of a standard NaF solution containing 10.0 ppb F^-. Exactly 5.00 mL of a solution containing an excess of Al-acid Alizarin Garnet R complex, a strongly fluorescing complex, were added to each and the solutions were diluted to 50.0 mL. The fluorescent intensities of the four solutions are given below. A blank solution gave a meter reading of 89.2.

mL Sample	mL of Std F^-	Meter Reading
5.00	0.00	68.2
5.00	1.00	55.3
5.00	2.00	41.3
5.00	3.00	28.8

(a) Explain the chemistry of the analytical method.

(b) Plot the data.

(c) Calculate the ppb F^- in the sample.

(d) By least-squares analysis derive an equation relating the decrease in fluorescence to the volume of standard reagent.

(e) Calculate the ppb F^- based upon the relationship derived in part (d).

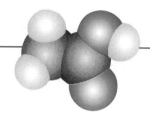

CHAPTER **24**

ATOMIC SPECTROSCOPY

Atomic spectroscopy is used for the qualitative and quantitative determination of perhaps 70 elements. Sensitivities of atomic methods lie typically in the parts-per-million to parts-per-billion range. Additional virtues of these methods are speed, convenience, unusually high selectivity, and moderate instrument costs.[1]

Spectroscopic determination of atomic species can only be performed on a gaseous medium in which the individual atoms (or sometimes elementary ions such as Fe^+, Mg^+, or Al^+) are well separated from one another. Consequently, the first step in all atomic spectroscopic procedures is *atomization*, a process in which the sample is volatilized and decomposed to produce an atomic gas. The efficiency and reproducibility of the atomization step in large measure determine the method's sensitivity, precision, and accuracy; that is, atomization is by far the most critical step in atomic spectroscopy.

> Atomization is a process in which a sample is converted into gaseous atoms or elementary ions.

As shown in Table 24-1, several methods are used to atomize samples for atomic spectroscopic studies. The most widely used of these is flame atomization, which is discussed in some detail in this chapter. Note that flame atomized samples produce atomic absorption, emission, and fluorescence spectra. We shall also consider briefly three other atomization methods listed in Table 24-1: electro-thermal, inductively coupled plasma, and direct current plasma.

24A   SOURCES OF ATOMIC SPECTRA

Atomic emission, absorption, and fluorescence spectra are much simpler than the corresponding molecular spectra because there are no vibrational and rotational states. Thus atomic emission, absorption, and fluorescence spectra are made up of a limited number of narrow peaks, or lines.

---

[1] References that deal with the theory and applications of atomic spectroscopy include C. Th. J. Alkemade *et al., Metal Vapors in Flames.* Elmsford, NY: Pergamon Press, 1982; B. Magyar, *Guidelines to Planning Atomic Spectrometric Analysis.* New York: Elsevier, 1982; J. D. Ingle Jr. and S. R. Crouch, *Spectrochemical Analysis,* Chapters 7–11. Englewood Cliffs, NJ: Prentice-Hall, 1988.

Table 24-1
CLASSIFICATION OF ATOMIC SPECTRAL METHODS

Atomization Method	Typical Atomization Temperature, °C	Basis for Method	Common Name and Abbreviation of Method
Flame	1700–3150	Absorption	Atomic absorption spectroscopy, AAS
		Emission	Atomic emission spectroscopy, AES
		Fluorescence	Atomic fluorescence spectroscopy, AFS
Electrothermal	1200–3000	Absorption	Electrothermal atomic absorption spectroscopy
		Fluorescence	Electrothermal atomic fluorescence spectroscopy
Inductively coupled argon plasma	6000–8000	Emission	Inductively coupled plasma spectroscopy, ICP
		Fluorescence	Inductively coupled plasma fluorescence spectroscopy
Direct-current argon plasma	6000–10,000	Emission	DC plasma spectroscopy, DCP
Electric arc	4000–5000	Emission	Arc-source emission spectroscopy
Electric spark	40,000(?)	Emission	Spark-source emission spectroscopy

24A-1 Emission Spectra

Figure 24-1 is a partial energy level diagram for atomic sodium showing the source of three of its most prominent emission lines. These lines are generated by heating gaseous sodium to 2000°C to 3000°C in a flame. The heat promotes the single outer electrons of the atoms from their ground state $3s$ orbitals to $3p$, $4p$, or $5p$ excited state orbitals. After a microsecond or less, the excited atoms relax to the ground state and give up their energy as photons of visible or ultraviolet radiation. As shown at the right of the figure, the wavelengths of the emitted radiation are 590, 330, and 285 nm.

24A-2 Absorption Spectra

Figure 24-2a shows three of several absorption peaks for sodium vapor. The source of these peaks is indicated in the partial energy diagram shown in Figure 24-2b. Here, absorption of radiation of 285, 330, and 590 nm excites the single outer electron of sodium from its ground-state $3s$ energy level to the excited $3p$, $4p$, and $5p$ orbitals, respectively. After a few microseconds, the excited atoms relax to their ground state by transferring their excess energy to other atoms or molecules in the medium. Alternatively, relaxation may take the form of fluorescence.

The absorption and emission spectra for sodium are relatively simple and consist of perhaps 40 peaks. For elements that have several outer electrons that can be excited, absorption spectra may be much more complex and consist of hundreds of peaks.

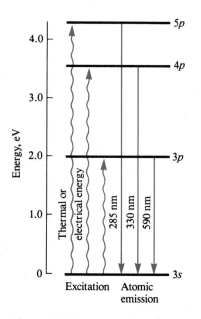

Figure 24-1
Source of three emission lines for sodium.

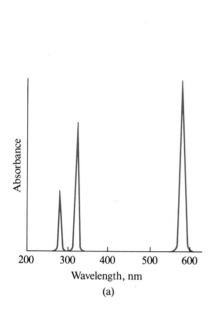

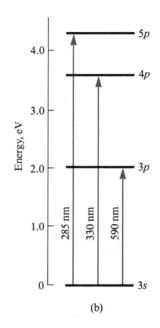

Figure 24-2
(a) Partial absorption spectrum for sodium vapor. (b) Electronic transitions, responsible for the lines in (a).

24B ATOMIC SPECTROSCOPY BASED ON FLAME ATOMIZATION

As shown in Table 24-1, three types of atomic spectroscopy are based on flame atomization: (1) atomic absorption spectroscopy (AAS), (2) atomic emission spectroscopy (AES), and (3) atomic fluorescence spectroscopy (AFS). We shall consider the first two of these methods.

In flame atomization, a solution of the analyte (usually aqueous) is converted to a mist, or *nebulized*, and carried into the flame by a flow of gaseous oxidant or fuel. Emission and absorption spectra are generated in the resulting hot, gaseous medium.

24B-1 Flame Atomizers

A flame atomizer consists of a pneumatic nebulizer, which converts the sample solution into a mist, or *aerosol*, that is then fed into a burner. One such nebulizer is the concentric tube type, shown in Figure 24-3, in which the liquid sample is sucked through a capillary tube by a high-pressure stream of gas flowing around the tip of the tube (the *Bernoulli effect*). This process of liquid transport is called *aspiration*. The high-velocity gas breaks the liquid up into fine droplets of various sizes, which are then carried into the flame. Cross-flow nebulizers are also employed in which the high-pressure gas flows across a capillary tip at right angles. In most atomizers, the high-pressure gas is the oxidant, with the aerosol-containing oxidant being mixed subsequently with the fuel.

Figure 24-4 is a diagram of a typical commercial laminar flow burner that employs a concentric tube nebulizer. The aerosol is mixed with fuel and flows past a series of baffles that remove all but the finest droplets. As a result of the baffles, the majority of the sample collects in the bottom of the mixing chamber, where it is drained to a waste container. The aerosol, oxidant, and fuel are then

Atomic *p* orbitals are in fact split into two energy levels that differ only slightly in energy. The energy difference between the two levels is small enough that the emission appears to be a single line, as suggested by Figure 24-1. With a very high resolution spectrometer each of these lines appears as a doublet.

Note that the wavelengths of the absorption and emission peaks for sodium are all identical.

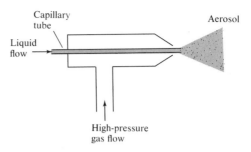

Figure 24-3
A concentric tube nebulizer.

passed through a slotted burner, which provides a flame that is usually 5 or 10 cm in length.

Laminar flow burners provide a relatively quiet flame and a long path length. These properties tend to enhance sensitivity and reproducibility. The mixing chamber in this type of burner contains a potentially explosive mixture, which can be ignited by flashback if the flow rates are not sufficient. Note that the burner in Figure 24-4 is equipped with pressure relief vents for this reason.

24B-2 Properties of Flames

When a nebulized sample is carried into a flame, the solvent evaporates in the *primary combustion zone* of the flame, which is located just above the tip of the burner (Figure 24-5). The resulting finely divided solid particles are carried to a region in the center of the flame called the *interzonal region*. Here, in this hottest part of the flame, gaseous atoms and elementary ions are formed from the solid particles. Excitation of atomic emission spectra also takes place in this region. Finally, the atoms and ions are carried to the outer edge, or *secondary combustion zone*, where oxidation may occur before the atomization products disperse into the atmosphere. Because the velocity of the fuel/oxidant mixture

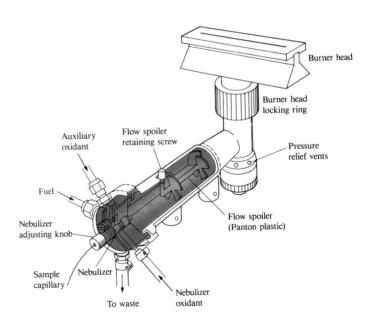

Figure 24-4
A laminar-flow burner. (Courtesy of Perkin-Elmer Corporation, Norwalk, CT.)

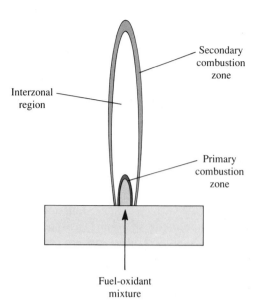

Secondary
combustion
zone

Interzonal
region

Primary
combustion
zone

Fuel-oxidant
mixture

Figure 24-5
Regions in a flame.

through the flame is high, only a fraction of the sample undergoes all these processes; indeed, a flame is not a very efficient atomizer.

Types of Flames Used in Atomic Spectroscopy

Table 24-2 lists the common fuels and oxidants employed in flame spectroscopy and the approximate range of temperatures realized with each of these mixtures. Note that temperatures of 1700°C to 2400°C are obtained with the various fuels when air serves as the oxidant. At these temperatures, only species that are easily excited, such as the alkali and alkaline earth metals, produce usable emission spectra. For heavy-metal species, which are less readily excited, oxygen or nitrous oxide must be employed as the oxidant. With the common fuels these oxidants produce temperatures of 2500°C to 3100°C.

The Effects of Flame Temperature

Both emission and absorption spectra are affected in a complex way by variations in flame temperature. Higher temperatures increase the total atom population of the flame and thus the sensitivity. With certain elements, however, this increase in atom population is more than offset by the loss of atoms as a result of ionization.

Flame temperature also determines the relative number of excited and unexcited atoms in a flame. In an air/acetylene flame, for example, the ratio of excited to unexcited magnesium atoms can be computed to be about 10^{-8}, whereas in an oxygen/acetylene flame, which is about 700°C hotter, this ratio is about 10^{-6}. Control of temperature is thus of prime importance in flame emission methods. For example, with a 2500°C flame, a temperature increase of 10°C causes the number of sodium atoms in the excited $3p$ state to increase by about 3%. In contrast, the corresponding *decrease* in the much larger number of ground-state atoms is only about 0.002%. Therefore, emission methods, based as they are on the population of *excited atoms*, require much closer control of flame temperature

> Nebulize means to reduce to a fine spray.

Modern flame-atomic-absorption instruments use laminar flow burners almost exclusively.

Table 24-2

FLAMES USED IN ATOMIC SPECTROSCOPY

Fuel and Oxidant	Temperature, °C
Gas/Air	1700–1900
Gas/O_2	2700–2800
H_2/Air	2000–2100
H_2/O_2	2550–2700
*C_2H_2/Air	2100–2400
*C_2H_2/O_2	3050–3150
*C_2H_2/N_2O	2600–2800

* Acetylene

The width of atomic emission peaks is about 10^{-3} nm.

than do absorption procedures, in which the analytical signal depends on the number of *unexcited atoms*.

The number of unexcited atoms in a typical flame exceeds the number of excited ones by a factor of 10^3 to 10^{10} more. This discrepancy suggests that absorption methods should be significantly more sensitive than emission methods. In fact, however, several other variables also influence sensitivity, and the two methods tend to complement each other in this regard. Table 24-3 illustrates this point.

Absorption and Emission Spectra in Flames

Both atomic and molecular emission and absorption spectra are found when a sample is atomized in a flame. A typical flame-emission spectrum is shown in Figure 24-6. Atomic emissions in this spectrum consist of narrow peaks, or lines, such as that for sodium at about 330 nm, potassium at approximately 404 nm, and calcium at 423 nm. Also present are broad emission bands that result from excitation of molecular species such as MgOH, MgO, CaOH, and OH. Here, vibrational transitions superimposed on electronic transitions produce closely spaced lines that are not completely resolved by the spectrometer.

Atomic absorption spectra are seldom recorded because to do so requires a sophisticated monochromator that produces radiation with unusually narrow band widths. Such spectra have much the same appearance as Figure 24-6 with both atomic and molecular absorption peaks being present. The vertical axis here would be absorbance rather than relative power.

Ionization in Flames

All elements ionize to some degree in a flame, which leads to a mixture of atoms, ions, and electrons in the hot medium. For example, when a sample containing barium is atomized, the equilibrium

$$Ba \rightleftharpoons Ba^+ + e^-$$

is established in the interzonal region of the flame. The position of this equilibrium

Table 24-3

COMPARISON OF DETECTION LIMITS FOR VARIOUS ELEMENTS BY FLAME ABSORPTION METHODS AND FLAME EMISSION METHODS

Flame Emission More Sensitive	Sensitivity About the Same	Flame Absorption More Sensitive
Al, Ba, Ca, Eu, Ga, Ho, In, K, La, Li, Lu, Na, Nd, Pr, Rb, Re, Ru, Sm, Sr, Tb, Tl, Tm, W, Yb	Cr, Cu, Dy, Er, Gd, Ge, Mn, Mo, Nb, Pd, Rh, Sc, Ta, Ti, V, Y, Zr	Ag, As, Au, B, Be, Bi, Cd, Co, Fe, Hg, Ir, Mg, Ni, Pb, Pt, Sb, Se, Si, Sn, Te, Zn

Adapted with permission from E. E. Pickett and S. R. Koirtyohann, *Anal. Chem.*, **1969**, *41* (14), 42A. Copyright 1969 American Chemical Society.

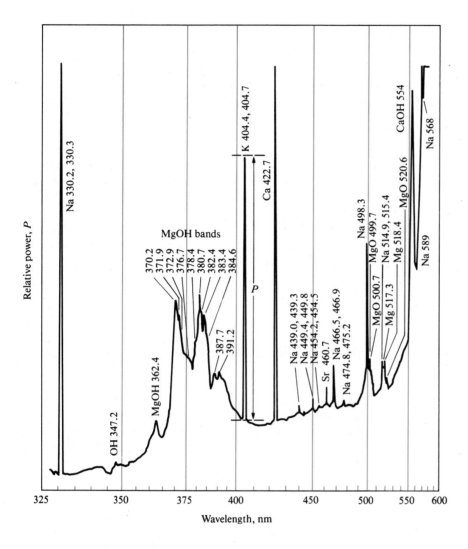

Figure 24-6

Emission spectrum of a brine obtained with an oxyhydrogen flame. (R. Hermann and C. T. J. Alkemade, *Chemical Analysis by Flame Photometry*, 2nd ed., p. 484. New York: Interscience, 1963. With permission.)

depends on the temperature of the flame and the total concentration of barium as well as on the concentration of the electrons produced from the ionization of *all other elements* present in the sample. At the temperatures of the hottest flames (>3000 K), nearly half of the barium exists in ionic form. The emission and absorption spectra of Ba and Ba^+ are, however, totally different from one another. Thus, in a high-temperature flame, two spectra for barium appear, one for the atom and one for its ion. For this reason (and others), control of the flame temperatures is important in flame spectroscopy.

Ionization of an atomic species in a flame is an equilibrium process that can be treated by the law of mass action.

The spectrum of an atom is entirely different from that of its ions.

24B-3   Flame-Atomic-Absorption Spectroscopy

Flame-atomic-absorption spectroscopy is currently the most widely used of all the atomic methods listed in Table 24-1 because of its simplicity, effectiveness, and relatively low cost. The general use of this technique by chemists for elemental analysis began in the early 1950s and grew explosively after that. The reason that atomic absorption methods were not widely used until that time was directly related to problems created by the very narrow widths of atomic absorption lines.

Line Widths

The natural width of an atomic absorption or an atomic emission line is on the order of 10^{-5} nm. Two effects, however, cause line widths to be broadened by a factor of 100 (or more). As we will see, line broadening is an important consideration in the design of atomic absorption instruments.

Doppler Broadening. Doppler broadening results from the rapid motion of atoms as they emit radiation. The wavelength of radiation received by the detector is slightly shorter when the emitter is moving toward the detector and slightly longer when the emitter is moving away from the detector. This difference is a manifestation of the well-known Doppler shift (see Figure 24-7). The net effect is an increase in the width of the emission line. For precisely the same reason, the Doppler effect also causes broadening of absorption lines. This type of broadening becomes more pronounced as the flame temperature increases because of the consequent increased rate of motion of the atoms.

> Both Doppler broadening and pressure broadening are temperature dependent.

Pressure Broadening. Pressure broadening arises from collisions among atoms that result in slight variations in their ground-state energies and thus slight energy differences between ground and excited states. Like Doppler broadening, pressure broadening becomes greater with increases in temperature. Therefore, broader absorption and emission peaks are always encountered at elevated temperatures.

The Effect of Narrow Line Widths on Absorbance Measurements. Since transition energies for atomic absorption lines are unique for each element, analytical methods based on atomic absorption are highly selective. The narrow lines do, however, create a problem in quantitative analysis that is not often encountered in molecular absorption.

> The widths of atomic absorption peaks are much less than the effective bandwidths of most monochromators.

No ordinary monochromator is capable of yielding a band of radiation as narrow as the peak width of a typical atomic absorption line (0.002 to 0.005

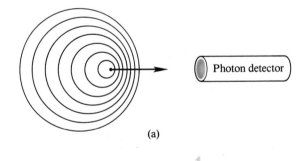

(a)

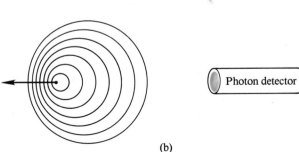

(b)

Figure 24-7

Cause of Doppler broadening. (a) When an atom moves toward a photon detector and emits radiation, the detector sees wave crests more often and detects radiation of higher frequency. (b) When an atom moves away from a photon detector and emits radiation, the detector sees crests less frequently and thus detects radiation of lower frequency.

nm). As a result, the use of radiation that has been isolated from a continuous source by a monochromator inevitably causes instrumental departures from Beer's law (see the discussion of instrument deviations from Beer's law in Section 21B-2). In addition, since the fraction of radiation absorbed from such a beam is small, the detector receives a signal that is less attenuated (that is, $P \rightarrow P_0$) and the sensitivity of the measurement is reduced. This effect is illustrated by the lower curve in Figure 21-5.

The problem created by narrow absorption peaks was surmounted in the mid-1950s by using radiation from a source that emits not only a *line of the same wavelength* as the one selected for absorption measurements but also a line that is *narrower*. For example, a mercury vapor lamp is selected as the external radiation source for the determination of mercury. Gaseous mercury atoms electrically excited in such a lamp return to the ground state by *emitting* radiation with wavelengths that are identical to the wavelengths *absorbed* by analyte mercury atoms in the flame. Since the lamp is operated at a temperature lower than that of the flame, the Doppler and pressure broadening of the emission lines from the lamp are less than the corresponding broadening of the analyte absorption peaks in the hot flame that holds the sample. The effective bandwidths of the lines emitted by the lamp are, therefore, significantly less than the corresponding bandwidths of the absorption peaks for the analyte in the flame.

Figure 24-8 illustrates the strategy generally used in atomic absorption methods. Figure 24-8a shows four narrow *emission* lines from a typical atomic lamp source. The figure also illustrates how one of these lines is isolated by a filter or monochromator. Figure 24-8b shows the *flame absorption spectrum* for the analyte between the wavelengths λ_1 and λ_2; note that the width of the absorption peak in the flame is significantly greater than the width of the emission line from the lamp. As shown in Figure 24-8c, the intensity of the incident beam P_0 has been decreased to P by passage through the sample. Since the bandwidth of the emission line from the lamp is now significantly less than the bandwidth of the absorption peak in the flame, $\log P_0/P$ is likely to be linearly related to concentration.

Line Sources

Two types of lamps are used in atomic absorption instruments: *hollow-cathode lamps* and *electrodeless-discharge lamps*.

Hollow-Cathode Lamps. The most useful radiation source for atomic absorption spectroscopy is the *hollow-cathode lamp*, shown schematically in Figure 24-9. It consists of a tungsten anode and a cylindrical cathode sealed in a glass tube containing an inert gas, such as argon, at a pressure of 1 to 5 torr. The cathode is either fabricated from the analyte metal or else serves as a support for a coating of that metal.

The application of a potential of about 300 V across the electrodes causes ionization of the argon and generation of a current of 5 to 10 mA as the argon cations and electrons migrate to the two electrodes. If the potential is sufficiently large, the argon cations strike the cathode with sufficient energy to dislodge some of the metal atoms and thereby produce an atomic cloud; this process is called *sputtering*. Those sputtered metal atoms that are in an excited state emit their characteristic wavelengths as they return to the ground state. It is important to

Sputtering is a process in which atoms or ions are ejected from a surface by a beam of charged particles.

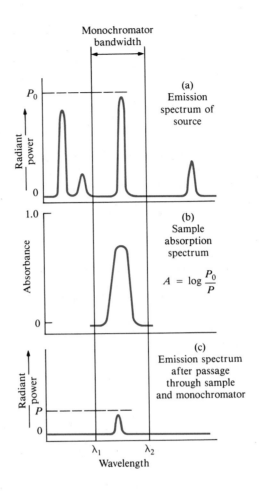

Figure 24-8
Atomic absorption of a resonance line.

recall that the atoms producing emission lines in the lamp are at a significantly lower temperature than the analyte atoms in the flame. Thus the emission lines from the lamp are broadened less than the absorption peaks in the flame. The sputtered metal atoms in a lamp eventually diffuse back to the cathode surface (or to the walls of the lamp) and are deposited.

Hollow-cathode lamps for about 40 elements are available from commercial sources. Some are fitted with a cathode containing more than one element; such lamps provide spectral lines for the determination of several species. The development of the hollow-cathode lamp is widely regarded as the single most important event in the evolution of atomic absorption spectroscopy.

Hollow-cathode lamps made atomic absorption spectroscopy practical.

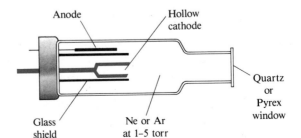

Figure 24-9
Diagram of a hollow-cathode lamp.

Electrodeless-Discharge Lamps. Electrodeless-discharge lamps are useful sources of atomic line spectra and provide radiant intensities that are usually one to two orders of magnitude greater than their hollow-cathode counterparts. A typical lamp is constructed from a sealed quartz tube containing an inert gas, such as argon, at a pressure of a few torr and a small quantity of the analyte metal (or its salt). The lamp contains no electrode but instead is energized by an intense field of radio-frequency or microwave radiation. The argon ionizes in this field, and the ions are accelerated by the high-frequency component of the field until they gain sufficient energy to excite (by collision) the atoms of the metal whose spectrum is sought.

Electrodeless-discharge lamps are available commercially for several elements. Their performance does not appear to be as reliable as that of the hollow-cathode lamp.

Source Modulation

In an atomic absorption measurement, it is necessary to discriminate between radiation from the source and that from the flame. Much of the latter is eliminated by the monochromator, which is always located between the flame and the detector. The thermal excitation of a fraction of the analyte atoms in the flame, however, produces radiation of the wavelength at which the monochromator is set. Because such radiation is not removed, it acts as a potential source of interference.

The effect of analyte emission is overcome by *modulating* the output from the hollow-cathode lamp so that its intensity fluctuates at a constant frequency. The detector thus receives an alternating signal from the hollow-cathode lamp and a continuous signal from the flame and converts these signals into the corresponding types of electric current. A relatively simple electronic system then eliminates the unmodulated dc signal produced by the flame and passes the ac signal from the source to an amplifier and finally to the readout device.

> Modulation is the deliberate and systematic variation of some property of a signal such as frequency, amplitude, or wavelength. In AAS, the output of the source is modulated from continuous to intermittent.

Modulation is most often accomplished by interposing a motor-driven circular chopper *between the source and the flame* (Figure 24-10). Segments of the metal chopper have been removed so that radiation passes through the device half of the time and is reflected the other half. Rotation of the chopper at a constant speed causes the beam reaching the flame to vary periodically from zero intensity to some maximum and then back to zero. As an alternative, the power supply for the source can be designed for intermittent (or ac) operation.

Beam choppers are widely used in spectroscopy to produce modulated beams of radiation.

Instruments for Atomic Absorption Spectroscopy

An atomic absorption instrument contains the same basic components as an instrument designed for molecular absorption measurements: a source, a sample container (here, a flame reservoir), a wavelength selector, and a detector/readout system. Both single- and double-beam instruments are offered by numerous manufacturers. The range of sophistication and the cost (upward from a few thousand dollars) are both substantial.

Photometers. As a minimum, an instrument for atomic absorption spectroscopy must be capable of providing a sufficiently narrow bandwidth to isolate the line chosen for a measurement from other lines that may interfere with or diminish the sensitivity of the method. A photometer equipped with a hollow-cathode

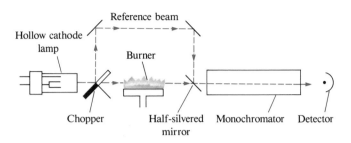

Figure 24-10

Optical paths in a double-beam atomic absorption spectrophotometer.

source and filters is satisfactory for measuring concentrations of the alkali metals, which have only a few widely spaced resonance lines in the visible region. A more versatile photometer is sold with readily interchangeable interference filters and lamps. A separate filter and lamp are used for each element. Satisfactory results for the determination of 22 metals are claimed.

Spectrophotometers. Most atomic absorption measurements are made with instruments equipped with an ultraviolet/visible grating monochromator. Figure 24-10 is a schematic of a typical double-beam instrument. Radiation from the hollow-cathode lamp is chopped and mechanically split into two beams, one of which passes through the flame and the other around the flame. A half-silvered mirror returns both beams to a single path, by which they pass alternately through the monochromator and to the detector. The signal processor then separates the ac signal generated by the chopped light source from the dc signal produced by the flame. The logarithm of the ratio between the reference and sample components of the ac signal is then computed and sent to the readout device for display as absorbance.

Interferences

Spectral interferences are encountered in AAS due to scattering by potential products in the flame and absorption by molecular species in the flame.

Two types of interference are encountered in atomic absorption methods. *Spectral interferences* occur when particulate matter from the atomization scatters the incident radiation from the source or when the absorption of an interfering species is so close to the analyte wavelength that overlap of absorption peaks occurs. *Chemical interferences* result from various chemical processes that occur during atomization and alter the absorption characteristics of the analyte.

Spectral Interferences. Interference due to overlapping lines is rare because the emission lines of hollow-cathode sources are so very narrow. Nevertheless, such an interference can occur if the separation between two lines is on the order of 0.01 nm. For example, a vanadium line at 308.211 nm will interfere in an analysis based on the aluminum absorption line at 308.215 nm. The interference is readily avoided, however, by selecting a different aluminum line (309.27 nm, for example.)

Spectral interferences also result from the presence of either *molecular* combustion products that exhibit broad-band absorption or particulate products that scatter radiation. Both diminish the power of the transmitted beam and lead to positive analytical errors. Where the source of these products is the fuel/oxidant mixture alone, corrections are readily obtained from absorbance measurements made with a blank aspirated into the flame.

A much more troublesome problem is encountered when the source of absorption or scattering originates in the sample matrix. In this type of interference, the power of the transmitted beam P is attenuated by the matrix components, but the incident beam power P_0 is not; a positive error in absorbance and thus concentration results. For example, a potential matrix interference due to absorption occurs in the determination of barium in alkaline earth mixtures. The wavelength of the barium line used for atomic absorption analysis appears in the center of a broad absorption *band* for molecular CaOH; interference by calcium in a barium analysis results. The net effect is readily eliminated by substituting nitrous oxide for air as the oxidant; the higher temperature decomposes the CaOH and eliminates the absorption band.

Spectral interference due to scattering by products of atomization often occurs when concentrated solutions containing elements such as titanium, zirconium, and tungsten—all of which form stable oxides—are aspirated into the flame. Metal oxide particles with diameters greater than the wavelength of light appear to be formed and cause scattering of the incident beam.

Fortunately, spectral interferences by matrix products are not widely encountered with flame atomization and usually can be avoided by variations in such analytical parameters as temperature and fuel-to-oxidant ratio. Alternatively, if the source of interference is known, an excess of the interferent can be added to both sample and standards. Provided the excess is large with respect to the concentration from the sample matrix, the contribution of the latter will become insignificant. The added substance is sometimes called a *radiation buffer*.

> A radiation buffer is a substance that is added in large excess to both samples and standards in atomic spectroscopy to swamp the effect of matrix species and thus minimize interference.

Chemical Interferences. Chemical interferences can frequently be minimized by a suitable choice of operating conditions. Perhaps the most common type of chemical interference is by anions that form compounds of low volatility with the analyte and thus decrease the rate at which it is atomized. Low results are the consequence. An example is the decrease in calcium absorbance observed with increasing concentrations of sulfate or phosphate ions, which form nonvolatile compounds with calcium ion.

> Chemical interferences in AAS are caused by reactions between the analyte and the interferents that decrease the analyte concentration in the flame.

Interferences due to the formation of species of low volatility can often be eliminated or moderated by use of higher temperatures. Alternatively, *releasing agents*, which are cations that react preferentially with the interference and prevent its interaction with the analyte, can be introduced. For example, the addition of excess strontium or lanthanum ion minimizes interference by phosphate in the determination of calcium. Here, the strontium or lanthanum replaces the analyte in the nonvolatile compound formed with the interferent.

> Releasing agents are cations that react selectively with anions and thus prevent their interfering in the determination of a cationic analyte.

Protective agents prevent interference by preferentially forming stable but volatile species with the analyte. Three common reagents for this purpose are EDTA, 8-hydroxyquinoline, and APDC (the ammonium salt of 1-pyrrolidine-carbodithioic acid). For example, the presence of EDTA has been shown to eliminate interference by silicon, phosphate, and sulfate in the determination of calcium.

> Protective agents are reagents that form stable volatile complexes with an analyte and thus prevent interference by anions that form nonvolatile compounds with the analyte.

Ionization Effects. The ionization of atoms and molecules is usually inconsequential in combustion mixtures that involve air as the oxidant. In high-temperature flames, however, where oxygen or nitrous oxide serves as the oxidant, ionization becomes appreciable, and a significant concentration of free electrons exists as a consequence of the equilibrium

$$M \rightleftharpoons M^+ + e^-$$

where M represents a neutral atom or molecule and M^+ is its ion. Ordinarily, the spectrum of M^+ is quite different from that of M, so that ionization of the analyte ions leads to low results. It is important to appreciate that treating the ionization process as an equilibrium—with free electrons as one of the products—implies that the degree of ionization of an analyte atom is strongly influenced by the presence of other ionizable metals in the flame. Thus, if the medium contains not only species M but species B as well, and if B ionizes according to the equation

$$B \rightleftharpoons B^+ + e^-$$

then the degree of ionization of M is decreased by the mass-action effect of the electrons formed from B. The errors caused by analyte ionization can frequently be eliminated by addition of an *ionization suppressor*, which provides a relatively high concentration of electrons to the flame; suppression of analyte ionization results. Potassium salts are frequently used as ionization suppressors because of the low ionization energy of this element.

Quantitative Analysis by Atomic Absorption Spectroscopy

Atomic absorption spectroscopy provides a sensitive means of determining more than 60 elements. The method is well suited for routine measurements by relatively unskilled operators.

Region of Flame for Quantitative Measurements. Figure 24-11 shows the absorbance of three elements as a function of distance above the burner tip. For magnesium and silver, the initial rise in absorbance is a consequence of the longer exposure to the heat, which leads to a greater concentration of atoms in the radiation path. The absorbance for magnesium, however, reaches a maximum near the center of the flame and then falls off as oxidation to magnesium oxide takes place. This effect is not seen with silver because this element is much more resistant to oxidation. For chromium, which forms very stable oxides, maximum absorbance lies immediately above the burner tip. For this element, oxide formation begins as soon as chromium atoms are formed.

It is clear from Figure 24-11 that the part of a flame to be used in an analysis must vary from element to element and furthermore that the position of the flame with respect to the source must be reproduced closely during calibration and analysis. Generally the flame position is adjusted to yield a maximum absorbance reading.

Calibration. Quantitative atomic absorption methods are usually based on calibration curves, which in principle are linear. Departures from linearity occur, however, and analyses should *never* be based on the measurement of a single standard with the assumption that Beer's law is being followed. In addition, the production of an atomic vapor involves sufficient uncontrollable variables to warrant measuring the absorbance of at least one standard solution each time an analysis is performed. Any deviation of the standard from its original calibration value can then be applied as a correction to the analytical results.

Standard-Addition Method. The standard-addition method is extensively used in atomic absorption spectroscopy. In this procedure, two or more aliquots of the sample are transferred to volumetric flasks. One is diluted to volume directly, and a known amount of analyte is introduced into the other before dilution to the same volume. The absorbance of each is measured (several different standard additions are recommended if the method is unfamiliar). If a linear relationship exists between absorbance and concentration (and this must be verified experimentally), the following relationships apply:

$$A_x = \frac{kV_x c_x}{V_T}$$

$$A_T = \frac{kV_x c_x}{V_T} + \frac{kV_s c_s}{V_T} \qquad (24\text{-}1)$$

where V_x and c_x are the volume and concentration of the analyte solution, V_s and c_s are volume and concentration of the standard, and V_T is the total volume; A_x and A_T are the absorbances of the sample alone and the sample plus standard, respectively. These two equations are readily combined to give

$$c_x = \left(\frac{A_x}{A_T - A_x}\right) \frac{c_s V_s}{V_x}$$

If several standard additions have been made, A_T in Equation 24-1 can be plotted against c_s as in Figure 24-12. The resulting straight line can then be extrapolated to $A_T = 0$. Substituting this relationship into Equation 24-1 and rearranging gives

$$c_x = \frac{-c_s (V_s)_0}{V_x}$$

where $(V_s)_0$ is the extrapolated volume of standard. Use of the standard-addition method tends to compensate for variations caused by physical and chemical interferences in the analyte solution.

Detection Limits and Accuracy. The second column in Table 24-4 shows detection limits for a number of common elements by flame atomic absorption and also by flame emission methods. Under usual conditions, the relative error of flame absorption analysis is of the order of 1% to 2%. With special precautions, this figure can be lowered to a few tenths of one percent.

24B-4 Flame Emission Spectroscopy

Atomic emission spectroscopy employing a flame (also called flame emission spectroscopy or flame photometry) has found widespread application in elemental analysis. Its most important uses are in the determination of sodium, potassium, lithium, and calcium, particularly in biological fluids and tissues. Because of its convenience, speed, and relative freedom from interferences, flame emission

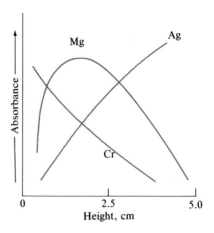

Figure 24-11
Flame-absorbance profile for three elements.

The standard-addition method often compensates for interferences in the sample matrix.

Atomic emission spectroscopy can provide both qualitative and quantitative information about an analyte.

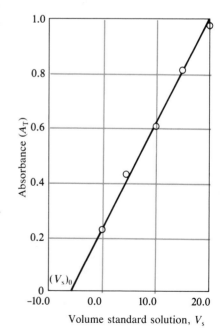

Figure 24-12
A plot of Equation 24-1 for standard-addition data.

Table 24-4

DETECTION LIMITS OF ATOMIC SPECTROSCOPY METHODS FOR SELECTED ELEMENTS*

Element	Absorption, Flame†	Absorption, Electrothermal‡	Emission, Flame†	Emission, ICP†,§
Al	30	0.005	5	2
As	100	0.02	0.0005	40
Ca	1	0.02	0.1	0.02
Cd	1	0.0001	800	2
Cr	3	0.01	4	0.3
Cu	2	0.002	10	0.1
Fe	5	0.005	30	0.3
Hg	500	0.1	0.0004	1
Mg	0.1	0.00002	5	0.05
Mn	2	0.0002	5	0.06
Mo	30	0.005	100	0.2
Na	2	0.0002	0.1	0.2
Ni	5	0.02	20	0.4
Pb	10	0.002	100	2
Sn	20	0.1	300	30
V	20	0.1	10	0.2
Zn	2	0.00005	0.0005	2

* All values in nanograms/milliliter = 10^{-3} μg/mL = 10^{-3} ppm.
† From V. A. Fassel and R. N. Kniseley, *Anal. Chem.*, **1974**, *46*, 1111A. With permission. Copyright 1974 American Chemical Society.
‡ From C. W. Fuller, *Electrothermal Atomization for Atomic Absorption Spectroscopy*, pp. 65–83. London: The Chemical Society, 1977. With permission of The Royal Society of Chemistry.
§ ICP = inductively coupled plasma.

Atomic emission and atomic absorption instruments are alike except that no lamp is required for emission measurement.

One of Kirchhoff's early prism spectroscopes. From his paper in *Abhandl. Berlin Akad.*, **1862**, 227.

spectroscopy has become the method of choice for these elements, which are otherwise difficult to determine. The method has also been applied, with various degrees of success, to the determination of perhaps half the elements in the periodic table.

In addition to its quantitative applications, flame emission spectroscopy is also useful for qualitative analysis. Complete spectra are readily recorded; identification of the elements present is then based on the peak wavelengths, which are unique for each element. In this respect, flame emission has a clear advantage over flame absorption, which does not provide complete absorption spectra because of the discontinuous nature of the radiation sources that must be used.

Instruments

Instruments for flame emission work are similar in design to those for flame absorption except that in the former the flame acts as the radiation source; a hollow-cathode lamp and chopper are therefore unnecessary.

For nonroutine analysis, a recording ultraviolet/visible spectrophotometer with a resolution of perhaps 0.05 nm is desirable. Such an instrument can provide

Feature 24-1
AUTOMATED FLAME PHOTOMETERS FOR DETERMINING SODIUM AND POTASSIUM IN BLOOD SERUM

Several instrument manufacturers supply automated flame photometers designed specifically for the determination of sodium and potassium in blood serum and other biological samples. Figure 24-A is a schematic of one of these instruments. The instrument incorporates three separate photometers: one each for sodium and potassium and one for lithium, which serves as the *internal standard* for the analysis. Each photometer contains an interference filter that transmits an emission line of only one of the three elements. A fixed amount of lithium is added to each of the standards and samples to provide a reference signal from the lithium transducer. The magnitudes of the signals from the sodium and potassium transducers are then each divided by the magnitude of the reference signal from the lithium transducer. This scheme improves the accuracy of the analysis because the emission intensities of the lines from the three elements are affected in the same way by experimental variables such as flame temperature, fuel flow rate, and background radiation. The technique works only if there is no lithium in the sample other than the amount that is added as the internal standard.

Some instruments of this type are highly automated. They withdraw samples from a turntable, dialyze the samples to remove protein and particulates, dilute the samples with the lithium internal standard, and aspirate them into a flame. The photometers calibrate themselves automatically after every few samples and make over 100 simultaneous sodium and potassium determina-

Gustav Robert Kirchhoff (1824–1877) was a German physicist who, along with his colleague, chemist Robert Wilhelm Bunsen (1811–1899), discovered spectroscopic analysis. Bunsen's burner was used to atomize samples of the elements and Kirchhoff's prism spectroscope was used to analyze the light given off by the incandescent samples. Using this technique, they were able to identify several new elements and demonstrate the presence of a number of elements in the sun by analyzing sunlight. This discovery is depicted on this Vatican stamp. The sun's corona, one of Kirchhoff's spectroscopes, and the absorption spectrum of hydrogen are also illustrated.

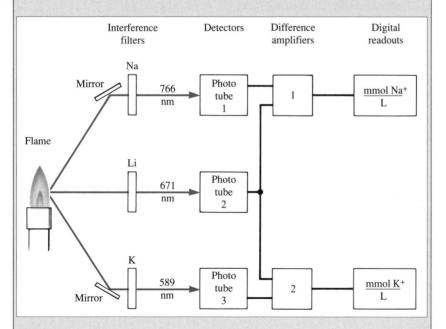

Figure 24-A
A three-channel photometer for monitoring Na^+ and K^+ in blood serums. The center channel is for monitoring the internal standard line for Li^+.

tions per hour. Such devices are extremely useful in the clinical laboratory because they perform analyses on urine, plasma, or serum samples with excellent accuracy and precision.

complete emission spectra that are useful for the identification of the elements present in a sample. Figure 24-6 is an example of a typical flame emission spectrum excited by an oxyhydrogen flame. The sample was a brine, and spectral lines and bands for several elements are identified. The figure illustrates how background corrections are made with a recorded spectrum.

Simple filter photometers often suffice for routine determinations of the alkali and alkaline earth metals. A low-temperature flame is employed to prevent excitation of most other metals. As a consequence, the spectra are simple, and interference filters can be used to isolate the desired emission line.

Interferences

The interferences encountered in flame emission spectroscopy have the same sources as those encountered in atomic absorption methods (Section 24B-3); the severity of any given interference often differs for the two procedures, however.

Analytical Techniques

The analytical techniques for flame emission spectroscopy are similar to those described for atomic absorption spectroscopy. Both calibration curves and the standard-addition method are employed. In addition, internal standards can be used to compensate for flame variables.

24C ATOMIC SPECTROSCOPY WITH ELECTROTHERMAL ATOMIZERS

Electrothermal atomizers, which first appeared on the market in about 1970, generally provide enhanced sensitivity because the entire sample is atomized in a short period, and the average residence time of the atoms in the optical path is a second or more.[2] Electrothermal atomizers are used for atomic absorption and atomic fluorescence measurements, but have not been generally applied for emission work. They are, however, beginning to be used for vaporizing samples in inductively coupled emission spectroscopy.

In electrothermal atomizers, a few microliters of sample are first evaporated at a low temperature and then ashed at a somewhat higher temperature in an electrically heated graphite tube or cup. After ashing, the current is rapidly increased to several hundred amperes, which causes the temperature to soar to perhaps 2000°C to 3000°C; atomization of the sample occurs in a period from a few milliseconds to a few seconds. The absorption or fluorescence of the atomized particles is then monitored in the region immediately above the heated surface.

---

[2] For detailed discussions of electrothermal atomizers see S. R. Koirtyohann and M. L. Kaiser, *Anal. Chem.*, **1982**, *54*, 1515A; C. W. Fuller, *Electrothermal Atomization for Atomic Absorption Spectroscopy.* London: The Chemical Society, 1978; A. Varma, *CRC Handbook of Furnace Atomic Absorption Spectroscopy.* Boca Raton, FL: CRC Press, 1989.

24C-1 Atomizer Designs

Figure 24-13a is a cross-sectional view of a commercial electrothermal atomizer. Atomization occurs in a cylindrical graphite tube that is open at both ends and has a central hole for introduction of sample by means of a micropipette. The tube is about 5 cm long and has an internal diameter of somewhat less than 1 cm. The interchangeable graphite tube fits snugly into a pair of cylindrical graphite electrical contacts located at the two ends of the tube. These contacts are held in a water-cooled metal housing. Two inert gas streams are provided. The external stream prevents the entrance of outside air and a consequent incineration of the tube. The internal stream flows into the two ends of the tube and out the central sample port. This stream not only excludes air but also serves to carry away vapors generated from the sample matrix during the first two heating stages.

Figure 24-13b illustrates the so-called L'vov platform, which is often used in graphite furnaces such as that shown in (a). The platform is also graphite and is located beneath the sample entrance port. The sample is evaporated and ashed on this platform in the usual way. When the tube temperature is raised rapidly, however, atomization is delayed since the sample is no longer directly on the furnace wall. As a consequence, atomization occurs in an environment in which the temperature is not changing so rapidly. More reproducible peaks are obtained as a consequence.

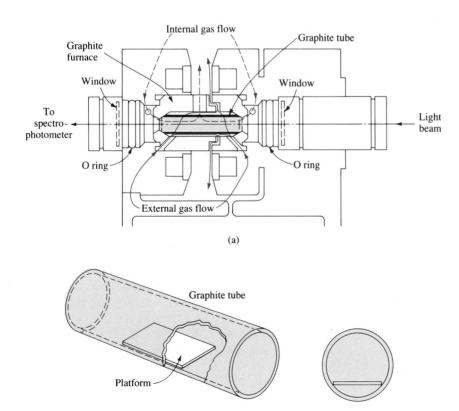

(a)

(b)

Figure 24-13

(a) Cross-sectional view of a graphite furnace. (b) The L'vov platform and its position in the graphite furnace. (a) Courtesy of the Perkin-Elmer Corp., Norwalk, CT; (b) Reprinted with permission from: W. Slavin, *Anal. Chem.*, **1982**, *54*, 689A. Copyright 1982 American Chemical Society.

Feature 24-2
MERCURY AND ITS DETERMINATION BY COLD-VAPOR
ATOMIC ABSORPTION SPECTROPHOTOMETRY

People's fascination with mercury began when prehistoric cave dwellers discovered the mineral cinnabar (HgS) and used it as a red pigment. Our first written record of the element came from Aristotle who described it as ''liquid silver'' in the fourth century B.C. Today there are thousands of uses of mercury and its compounds in medicine, metallurgy, electronics, agriculture, and many other fields. Because it is a liquid metal at room temperature, mercury is used to make flexible and efficient electrical contacts in scientific, industrial, and household applications. Thermostats, silent light switches, and fluorescent light bulbs are but a few examples of electrical applications.

A useful property of metallic mercury is that it forms amalgams with other metals, which have a host of uses. For example, metallic sodium is produced as an amalgam by electrolysis of molten sodium chloride. Dentists use a 50% amalgam with an alloy of silver for fillings.

The toxicological effects of mercury have been known for many years. The bizarre behavior of the Mad Hatter in Lewis Carroll's *Alice in Wonderland* was a result of the effects of mercury and mercury compounds on the Hatter's brain. Mercury absorbed through the skin and lungs destroys brain cells, which are not regenerated. Hatters of the nineteenth century used mercury compounds in processing fur to make felt hats. These workers and workers in other industries have suffered the debilitating symptoms of mercurialism such as loosening of teeth, tremors, muscle spasms, personality changes, depression, irritability, and nervousness.

The toxicity of mercury is complicated by its tendency to form both inorganic and organic compounds. Inorganic mercury is relatively insoluble in body tissues and fluids, so it is expelled from the body about ten times faster than organic mercury. Organic mercury, usually in the form of alkyl compounds, such as methyl mercury, is somewhat soluble in fatty tissues such as the liver. Methyl mercury accumulates to toxic levels and is expelled from the body quite slowly.

Mercury concentrates in the environment as illustrated in Figure 24-B. Inorganic mercury is converted to organic mercury by anaerobic bacteria in sludge deposited at the bottom of lakes, streams, and other bodies of water. Small aquatic animals consume the organic mercury and are in turn eaten by larger life forms. As the mercury moves up the food chain from microbes, to shrimp, to fish, and ultimately to larger animals such as swordfish, the mercury becomes ever more concentrated. Some sea creatures such as oysters may concentrate mercury by a factor of 100,000. At the top of the food chain, the concentration of mercury reaches levels as high as 20 ppm. The Food and Drug Administration has set a legal limit of 1.0 ppm methyl mercury in fish for human consumption. As a result, mercury levels in some areas threaten local fishing industries.

Analytical methods for the determination of mercury play an important role in monitoring the safety of food and water supplies. One of the most

Figure 24-B
Biological concentration of mercury in the environment.

useful methods is based on the atomic absorption by metallic mercury vapor of 253.7 nm radiation. Figure 24-C shows an apparatus that is used to determine mercury by atomic absorption at room temperature.[1]

A sample suspected of containing mercury is decomposed in a hot mixture of nitric acid and sulfuric acid, which converts the mercury to mercury(II) salts. These are reduced to the metal with a mixture of hydroxylamine sulfate and tin(II) sulfate. Air is then pumped through the solution to carry the resulting mercury-containing vapor through the drying tube and into the observation cell. Water vapor is trapped by Drierite in the drying tube so that only mercury vapor and air pass through the cell. The monochromator of the atomic absorption spectrophotometer is tuned to a band around 254 nm so that radiation from the 253.7-nm line of the mercury hollow cathode lamp passes through the quartz windows of the observation cell, which is placed in the light path of the instrument. The absorbance is directly proportional to the concentration of mercury in the cell, which is in turn proportional to the concentration of mercury in the sample. Solutions of known mercury concentration are treated in a similar way to calibrate the apparatus. The method depends on the low solubility of mercury in the reaction mixture and its appreciable vapor pressure, which is 2×10^{-3} torr at 25°C. The detection limit of the method is less than 1 ppb, and it is used to determine mercury in foods, metals, ores, and environmental samples. The method has

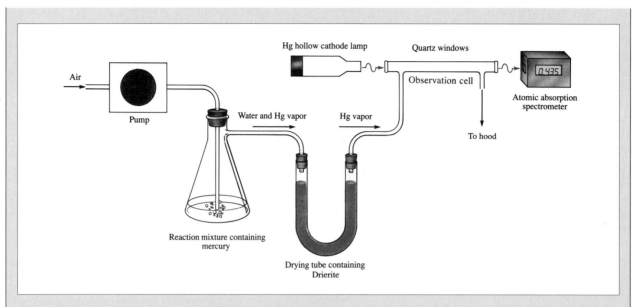

Figure 24-C
Apparatus for cold-vapor atomic absorption determination of mercury.

the advantages of sensitivity, simplicity, and room-temperature operation. Automated mercury analysis systems that are capable of handling 20 to 40 samples per hour with detection limits of 0.2 ppb are now being marketed.

---

[1] W. R. Hatch and W. L. Ott, *Anal. Chem.*, **1968**, *40*(14), 2085.

24C-2 Output Signal

At a wavelength at which absorbance occurs, the detector output rises to a maximum after a few seconds of ignition followed by a rapid decay back to zero as the atomization products escape into the surroundings. The change is rapid enough (often < 1 s) to require a high-speed data acquisition system. Quantitative analyses are usually based on peak height although peak area has also been used.

Figure 24-14 shows typical output signals from an atomic absorption spectrophotometer equipped with a carbon rod atomizer. The series of peaks on the right show the absorbance at the wavelength of a lead peak as a function of time when a 2-μL sample of a canned orange juice was atomized. During both drying and ashing, peaks are produced, probably due to particulate ignition products. The three peaks on the left are for lead standards employed for calibration. The sample peak on the far right indicates a lead concentration of 0.1 μg/mL of juice.

Applications of Electrothermal Atomizers

Electrothermal atomizers offer the advantage of unusually high sensitivity for small volumes of sample. Typically, sample volumes between 0.5 and 10 μL are sufficient; under these circumstances, absolute detection limits typically lie in the range of 10^{-10} to 10^{-13} g of analyte.

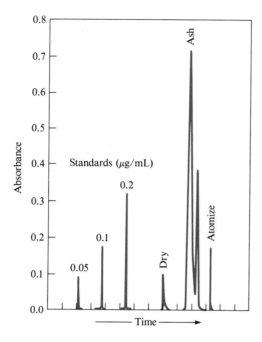

Figure 24-14

Typical output from a spectrophotometer equipped with an electrothermal atomizer. The time for drying and ashing are 20 s and 60 s, respectively. (Courtesy of Varian Instrument Division, Palo Alto, CA.)

The relative precision of electrothermal methods is generally in the range of 5% to 10% compared with the 1% or better that can be expected for flame or plasma atomization. Furthermore, furnace methods are slow—typically requiring several minutes per element. Still another disadvantage is that chemical interference effects are often more severe with electrothermal atomization than with flame atomization. A final disadvantage is the analytical range is low, being usually less than two orders of magnitude. Consequently, electrothermal atomization is ordinarily applied only when flame or plasma atomization provides inadequate detection limits.

24D   ATOMIC EMISSION METHODS BASED ON PLASMA SOURCES

Plasma atomizers, which became available commercially in the mid-1970s, offer several advantages over flame atomizers.[3] Plasma atomization has been used for both thermal emission and fluorescence spectroscopy. It has not been widely used as an atomizer for atomic absorption methods.

By definition, a plasma is a conducting gaseous mixture containing a significant concentration of cations and electrons. In the argon plasma employed for emission analyses, argon ions and electrons are the principal conducting species, although

> A plasma is a gas containing relatively high concentration of cations and electrons.

---

[3] For a detailed discussion of the various plasma sources, see R. D. Sacks, in *Treatise on Analytical Chemistry*, 2nd ed., P. J. Elving, E. J. Meehan, and I. M. Kolthoff, Eds., Part I, Vol. 7, pp. 516–526. New York: Wiley, 1981; J. D. Ingle Jr. and S. R. Crouch, *Spectrochemical Analysis*, Chapter 8. Englewood Cliffs, NJ: Prentice-Hall, 1988.

cations from the sample also contribute. Argon ions, once formed in a plasma, are capable of absorbing sufficient power from an external source to maintain the temperature at a level at which further ionization sustains the plasma indefinitely; temperatures as great as 10,000 K are encountered. Three power sources have been employed in argon plasma spectroscopy. One is a dc electrical source capable of maintaining a current of several amperes between electrodes immersed in the argon plasma. The second and third are powerful radio-frequency and microwave-frequency generators through which the argon flows. Of the three, the radio-frequency, or *inductively coupled plasma* (ICP), source appears to offer the greatest advantage in terms of sensitivity and freedom from interference. On the other hand, the *dc plasma source* (DCP) has the virtues of simplicity and lower cost.

24D-1 The Inductively Coupled Plasma Source

Figure 24-15a is a schematic of an inductively coupled plasma source. It consists of three concentric quartz tubes through which streams of argon flow at a total rate of between 11 and 17 L/min. The diameter of the largest tube is about 2.5 cm. Surrounding the top of this tube is a water-cooled induction coil powered by a radio-frequency generator capable of producing 2 kW of energy at about 27 MHz. Ionization of the flowing argon is initiated by a spark from a Tesla coil. The resulting ions, and their associated electrons, then interact with the fluctuating magnetic field (labeled H in Figure 24-15) produced by the induction coil I. This interaction causes the ions and electrons within the coil to flow in the closed annular paths depicted in the figure; ohmic heating is the consequence of their resistance to this movement.

During the 1980s, low-flow, low-power torches appeared on the market. Currently, one manufacturer offers only this type of torch, whereas other companies offer them as an option. Typically, these torches require a total argon flow of lower than 10 L/min and require less than 800 W of radio-frequency power.

The temperature of this plasma is high enough to require that it be thermally isolated from the quartz cylinder. Isolation is achieved by flowing argon tangentially around the walls of the tube, as indicated by the arrows in Figure 24-15a. The tangential flow cools the inside walls of the central tube and centers the plasma radially.

Sample Injection

The sample is carried into the hot plasma at the head of the tubes by argon flowing at about 1 L/min through the central quartz tube. The sample can be an aerosol, a thermally generated vapor, or a fine powder.

Plasma Appearance and Spectra

The typical plasma has a very intense, brilliant white, nontransparent core topped by a flame-like tail. The core, which extends a few millimeters above the tube, produces a spectral continuum on which is superimposed the atomic spectrum for argon. The continuum apparently results when argon and other ions recombine with electrons. In the region 10 to 30 mm above the core, the continuum fades

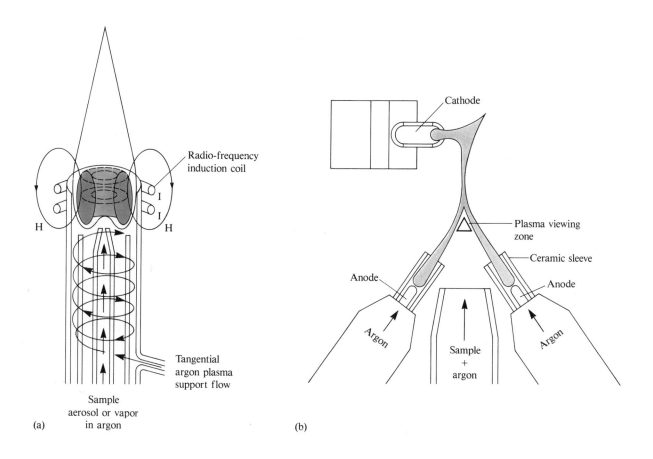

(a)

(b)

Radio-frequency induction coil

Tangential argon plasma support flow

Sample aerosol or vapor in argon

H   H

I
I

Cathode

Plasma viewing zone

Ceramic sleeve

Anode   Anode

Argon   Sample + argon   Argon

Figure 24-15

Plasma sources. (a) Inductively coupled plasma source. (b) A three-electrode dc plasma jet. (a) From V. A. Fassel, *Science*, **1978**, *202*, 185. With permission. Copyright 1978 by the American Association for the Advancement of Science; (b) Courtesy of Applied Research Laboratories, Inc., Sunland, CA.

and the plasma is optically transparent. Spectral observations are generally made 15 to 20 mm above the induction coil. Here, the background radiation is remarkably free of argon lines and is well suited to spectral measurements. Many of the most sensitive analyte lines in this region of the plasma are from ions such as Ca^+, Cd^+, Cr^+, and Mn^+.

Analyte Atomization and Ionization

By the time the sample atoms reach the observation point in the plasma, they have had a residence time of about 2 ms at temperatures ranging from 6000 to 8000 K. These times and temperatures are two to three times as great as those attainable in the hottest combustion flames (acetylene/nitrous oxide). As a consequence, atomization is more nearly complete, and fewer chemical interferences are encountered. Surprisingly, ionization interference effects are small or nonexistent, perhaps because the large concentration of electrons from the ionization of the argon maintains a more or less constant electron concentration in the plasma.

Several other advantages are associated with the plasma source. First, atomization occurs in a chemically inert environment, which should also enhance the

Plasma temperatures are substantially higher than those in a flame.

lifetime of the analyte. In addition, and in contrast to flame sources, the temperature cross section of the plasma is relatively uniform. As a consequence, calibration curves tend to remain linear over several orders of magnitude of concentration.

24D-2 The Direct-Current Argon Plasma Source

Direct-current plasma jets were first described in the 1920s and were systematically investigated as sources for emission spectroscopy for more than two decades. It was not until recently, however, that a source of this type was designed that provides data reproducible enough to successfully compete with flame and inductively coupled plasma sources.

Figure 24-15b is a diagram of a commercially available dc plasma source that is well suited to the excitation of emission spectra. This plasma-jet source consists of three electrodes arranged in an inverted Y configuration. A graphite anode is located in each arm of the Y, and a tungsten cathode is located at the inverted base. Argon flows from the two anode blocks toward the cathode. The plasma jet is formed when the cathode is momentarily brought into contact with the anodes. Ionization of the argon occurs, and the current that develops (≈ 14 A) generates additional ions to sustain itself indefinitely. The temperature is perhaps 10,000 K in the arc core and 5000 K in the viewing region. The sample is aspirated into the area between the two arms of the Y, where it is atomized, excited, and its spectrum viewed.

Spectra produced by the plasma jet tend to have fewer lines than those produced by the inductively coupled plasma, and the lines formed in the former are largely from atoms rather than ions. Sensitivities achieved with the dc plasma jet appear to range from an order of magnitude lower to about the same as those obtainable with the inductively coupled plasma. The reproducibilities of the two systems are similar. Significantly less argon is required for the dc plasma, and the auxiliary power supply is simpler and less expensive. On the other hand, the graphite electrodes must be replaced every few hours, whereas the inductively coupled plasma source requires little or no maintenance.

24D-3 Instruments for Plasma Spectroscopy

Several manufacturers offer instruments for plasma emission spectroscopy. In general, these consist of a high-quality grating spectrophotometer for the ultraviolet and visible regions and a photomultiplier detector. Many are automated so that an entire spectrum can be scanned sequentially. Others have several photomultiplier tubes located in the focal plane so that the lines for several elements (two dozen or more) can be monitored simultaneously. Such instruments are very expensive.

24D-4 Quantitative Applications of Plasma Sources

Unquestionably, inductively coupled and dc plasma sources yield significantly better quantitative analytical data than other emission sources. The excellence of these results stems from the high stability, low noise, low background, and freedom from interferences of the sources when operated under appropriate experimental conditions. The performance of the inductively coupled plasma source is somewhat better than that of the dc plasma source in terms of detection limits. The latter, however, is less expensive to purchase and operate and is entirely adequate for many applications.

In general, the detection limits with the inductively coupled plasma source appear comparable to or better than those of other atomic spectral procedures. Table 24-4 compares the sensitivity of several of these methods.

24E   QUESTIONS AND PROBLEMS

*24-1. Describe the basic differences between atomic emission and atomic absorption spectroscopy.

24-2. Define
　*(a) atomization.
　(b) pressure broadening.
　*(c) Doppler broadening.
　(d) turbulent-flow nebulizer.
　*(e) plasma.
　(f) hollow-cathode lamp.
　*(g) sputtering.
　(h) ionization suppressor.
　*(i) spectral interference.
　(j) chemical interference.
　*(k) radiation buffer.
　(l) releasing agent.
　*(m) protective agent.

*24-3. Why is atomic emission more sensitive to flame instability than atomic absorption?

24-4. Why is source modulation used in atomic absorption spectroscopy?

*24-5. In a hydrogen/oxygen flame, an atomic absorption peak for iron decreased in the presence of large concentrations of sulfate ion.
　(a) Suggest an explanation for this observation.
　(b) Suggest three possible methods for overcoming the potential interference of sulfate in a quantitative determination of iron.

24-6. Why are the lines from a hollow-cathode lamp generally narrower than the lines emitted by atoms in a flame?

*24-7. In the concentration range from 500 to 2000 ppm of U, a linear relationship is found between absorbance at 351.5 nm and concentration. At lower concentrations, the relationship becomes nonlinear unless about 2000 ppm of an alkali metal salt is introduced. Explain.

24-8. What is the purpose of an internal standard in flame emission methods?

*24-9. A 5.00-mL sample of blood was treated with trichloroacetic acid to precipitate proteins. After centrifugation, the resulting solution was brought to pH 3 and extracted with two 5-mL portions of methyl isobutyl ketone containing

the organic lead-complexing agent APCD. The extract was aspirated directly into an air/acetylene flame and yielded an absorbance of 0.502 at 283.3 nm. Five-milliliter aliquots of standard solutions containing 0.400 and 0.600 ppm of lead were treated in the same way and yielded absorbances of 0.396 and 0.599. Calculate the parts per million of lead in the sample, assuming that Beer's law is followed.

24-10. The sodium in a series of cement samples was determined by flame emission spectroscopy. The flame photometer was calibrated with a series of standards containing 0, 20.0, 40.0, 60.0, and 80.0 μg Na_2O per milliliter. The instrument readings for these solutions were 3.1, 21.5, 40.9, 57.1, and 77.3.
　(a) Plot the data.
　(b) Derive a least-squares line for the data.
　(c) Calculate standard deviations for the slope and the intercept for the line in (b).
　(d) The following data were obtained for replicate 1.00-g samples of cement dissolved in HCl and diluted to 100.0 mL after neutralization.

		Emission Reading		
	Blank	Sample A	Sample B	Sample C
Replicate 1	5.1	28.6	40.7	73.1
Replicate 2	4.8	28.2	41.2	72.1
Replicate 3	4.9	28.9	40.2	spilled

Calculate the % Na_2O in each sample. What are the absolute and relative standard deviations for the average of each determination?

*24-11. The chromium in an aqueous sample was determined by pipetting 10.0 mL of the unknown into each of five 50.0-mL volumetric flasks. Various volumes of a standard containing 12.2 ppm Cr were added to the flasks, and the solutions were then diluted to volume.

Unknown, mL	Standard, mL	Absorbance
10.0	0.0	0.201
10.0	10.0	0.292
10.0	20.0	0.378
10.0	30.0	0.467
10.0	40.0	0.554

(a) Plot absorbance as a function of volume of standard V_s.

(b) Derive an expression relating absorbance to the concentrations of standard and unknown (c_s and c_x) and the volumes of the standards and unknown (V_s and V_x) as well as the volume to which the solutions were diluted (V_T).

(c) Derive expressions for the slope and the intercept of the straight line obtained in (a) in terms of the variables listed in (b).

(d) Show that the concentration of the analyte is given by the relationship $c_x = bc_s/mV_x$, where m and b are the slope and the intercept of the straight line in (a).

(e) Determine values for m and b by the method of least squares.

(f) Calculate the standard deviation for the slope and the intercept in (e).

(g) Calculate the ppm Cr in the sample using the relationship given in (d).

CHAPTER 25

ANALYTICAL SEPARATIONS BY EXTRACTION AND ION EXCHANGE

An interference in a chemical analysis arises whenever a species in the sample matrix either produces a signal that is indistinguishable from that of the analyte or, alternatively, attenuates the analyte signal. Few if any analytical signals are so specific as to be totally free of interference. As a consequence, most analytical methods require one or more preliminary steps to eliminate the effects of interferents.

Two general methods are available for dealing with interferences. The first makes use of a *masking agent* to immobilize or chemically bind the interferent in a form in which it no longer contributes to or attenuates the signal from the analyte.[1] Clearly, a masking agent must not affect the behavior of the analyte significantly. An example of masking is the use of fluoride ion to prevent iron(III) from interfering in the iodometric determination of copper(II). In this analysis, the analyte solution is treated with an excess of iodide ion, which reacts with copper(II) ions to give a stoichiometric amount of iodine. The liberated iodine is then titrated with a standard solution of sodium thiosulfate. Iron(III) interferes in this determination because it also oxidizes iodide ion to some extent. The interference is avoided by introducing fluoride ion in excess. Here, masking results from the strong tendency of fluoride ions to complex iron(III) but not copper(II). The consequence is a decrease in the electrode potential of the iron(III) system to the point where only copper(II) ions from the sample oxidize iodide to iodine. We noted another example of masking in Section 14B-8, where cyanide ion is used to mask heavy metal ions and prevent their interference in the titration of magnesium and calcium ions with EDTA.

> The sample matrix is the medium containing an analyte of interest.

> An interferent is a chemical species that causes a systematic error in an analysis by enhancing or attenuating the analytical signal.

> A masking agent is a reagent that chemically binds an interferent and prevents it from causing errors in an analysis.

---

[1] For a monograph on masking agents, see D. D. Perrin, *Masking and Demasking Reactions.* New York: Wiley-Interscience, 1970.

The second approach for avoiding the effect of an interferent involves separating the analyte and the interferent as separate phases. The classical way of performing this type of separation is based on precipitating the analyte selectively with an appropriate chemical reagent such as hydrogen sulfide (Section 7C) or any of several organic precipitating agents (Section 6D-2). Another method of removing the analyte as a separate phase involves electrolysis at controlled electrode potential as described in Section 19C-2. A third method of separation is based on converting the analyte to a gaseous phase that can then be isolated by distillation.

In this chapter we consider two other methods of isolating an analyte from interferences: *extraction* and *ion exchange*. In Chapters 26 and 27 we describe various types of chromatographic separations. These methods are also based on distributing the analyte and interference between two phases.

Chemical species are generally separated by converting them to different phases that can then be mechanically isolated.

25A SEPARATION BY EXTRACTION

The extent to which solutes, both inorganic and organic, distribute themselves between two immiscible liquids differs enormously, and these differences have been used for decades to accomplish separations of chemical species. This section considers applications of the distribution phenomenon to analytical separations.

25A-1 Theory

The partition of a solute between two immiscible phases is an equilibrium phenomenon that is governed by the *distribution law*. If the solute species A is allowed to distribute itself between water and an organic phase, the resulting equilibrium may be written as

$$A_{aq} \rightleftharpoons A_{org}$$

where the subscripts refer to the aqueous and the organic phases, respectively. Ideally, the ratio of activities for A in the two phases will be constant and independent of the total quantity of A; that is, at any given temperature,

$$K = \frac{(a_A)_{org}}{(a_A)_{aq}} \cong \frac{[A_{org}]}{[A_{aq}]} \qquad (25\text{-}1)$$

where $(a_A)_{org}$ and $(a_A)_{aq}$ are the activities of A in each of the phases and the bracketed terms are molar concentrations of A. The equilibrium constant K is known as the *partition*, or *distribution*, *coefficient*. As with many other equilibria, molar concentrations can often be substituted for activities without serious error. Generally, the numerical value for K approximates the ratio of the solubility of A in each solvent.[2]

---

[2] If the solute exists in different states of aggregation in the two solvents, the equilibrium becomes

$$x(A_y)_{aq} \rightleftharpoons y(A_x)_{org}$$

and the partition coefficient is then

$$K = \frac{[(A_x)_{org}]^y}{[(A_y)_{aq}]^x}$$

Partition coefficients are useful because they permit the calculation of the concentration of an analyte remaining in a solution after a certain number of extractions. They also provide guidance as to the most efficient way to perform an extractive separation. Thus, it is readily shown (see Feature 25-1) that for the simple system described by Equation 25-1, the concentration of A remaining in an aqueous solution after n extractions with an organic solvent is given by the equation

$$[A_{aq}]_n = \left(\frac{V_{aq}}{V_{org}K + V_{aq}}\right)^n [A_{aq}]_0 \qquad (25\text{-}2)$$

where $[A_{aq}]_n$ is the concentration of A remaining in the aqueous solution after extracting V_{aq} mL of the solution having an original concentration of $[A_{aq}]_0$ with n portions of the organic solvent, each having a volume of V_{org}. Example 25-1 illustrates how this equation can be used to decide the most efficient way to perform an extraction.

It is always better to use several small portions of solvent to extract a sample than to extract with one large portion.

Example 25-1

The distribution coefficient for iodine between an organic solvent and H_2O is 85. Calculate the concentration of I_2 remaining in the aqueous layer after extraction of 50.0 mL of 1.00×10^{-3} M I_2 with the following quantities of the organic solvent: (a) 50.0 mL; (b) two 25.0-mL portions; (c) five 10.0-mL portions.

Substitution into Equation 25-2 gives

(a) $\quad [I_{2aq}] = \left(\dfrac{50.0}{50.0 \times 85 + 50.0}\right)^1 \times 1.00 \times 10^{-3} = 1.16 \times 10^{-5}$

(b) $\quad [I_{2aq}] = \left(\dfrac{50.0}{25.0 \times 85 + 50.0}\right)^2 \times 1.00 \times 10^{-3} = 5.28 \times 10^{-7}$

(c) $\quad [I_{2aq}] = \left(\dfrac{50.0}{10.0 \times 85 + 50.0}\right)^5 \times 1.00 \times 10^{-3} = 5.29 \times 10^{-10}$

Figure 25-1 shows that the improved efficiency of multiple extractions falls off rapidly as a total fixed volume is subdivided into smaller and smaller portions. As we see, little is gained by dividing the extracting solvent into more than five or six portions.

25A-2 Applications

An extraction is frequently more attractive than a precipitation method for separating inorganic species. The processes of equilibration and separation of phases in a separatory funnel are less tedious and time consuming than conventional precipitation, filtration, and washing. Moreover, difficulties associated with coprecipitation are avoided. Finally, and in contrast to the precipitation process, extraction procedures are ideally suited for the isolation of trace quantities of analytes.

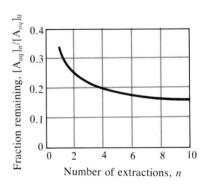

Figure 25-1

Plot of Equation 25-4, with $K = 2$ and $V_{aq} = 100$ mL. The total volume of organic solvent is also 100 mL: thus, $V_{org} = 100/n$.

The Extractive Separation of Metal Ions As Chelates

Many organic chelating agents are weak acids that react with metal ions to give uncharged complexes that are highly soluble in organic solvents such as ethers, hydrocarbons, ketones, and chlorinated species (including chloroform and carbon tetrachloride).[3] On the other hand, most metal chelates are nearly insoluble in water. Similarly, the chelating agents themselves are often quite soluble in organic solvents but of limited solubility in water.

Feature 25-1
DERIVATION OF EQUATION 25-2

Consider a simple system that is described by Equation 25-1. Suppose x_0 mmol of the solute A in V_{aq} mL of aqueous solution is extracted with V_{org} mL of an immiscible organic solvent. At equilibrium, x_1 mmol of A will remain in the aqueous layer, and $(x_0 - x_1)$ mmol will have been transferred to the organic layer. The concentrations of A in the two layers will then be

$$[A_{aq}] = \frac{x_1}{V_{aq}}$$

and

$$[A_{org}] = \frac{(x_0 - x_1)}{V_{org}}$$

Substitution of these quantities into Equation 25-1 and rearrangement gives

$$x_1 = \left(\frac{V_{aq}}{V_{org}K + V_{aq}}\right)x_0$$

Similarly, the number of millimoles, x_2, remaining after a second extraction with the same volume of solvent will be

$$x_2 = \left(\frac{V_{aq}}{V_{org}K + V_{aq}}\right)x_1$$

Substitution of the previous equation into this expression gives

$$x_2 = \left(\frac{V_{aq}}{V_{org}K + V_{aq}}\right)^2 x_0$$

By the same argument, the number of millimoles, x_n, that remain after

---

[3] The use of chlorinated solvents is decreasing because of concerns about their health effects and their possible role in the depletion of the ozone layer.

n extractions will be given by the expression

$$x_n = \left(\frac{V_{aq}}{V_{org}K + V_{aq}}\right)^n x_0$$

Finally, this equation can be written in terms of the initial and final concentrations of x in the aqueous layer by substituting the relationships

$$x_n = [A_{aq}]_n V_{aq} \quad \text{and} \quad x_0 = [A_{aq}]_0 V_{aq}$$

Thus,

$$[A_{aq}]_n = \left(\frac{V_{aq}}{V_{org}K + V_{aq}}\right)^n [A_{aq}]_0$$

which is Equation 25-2.

Figure 25-2 shows the equilibria that develop when an aqueous solution of a divalent cation, such as zinc(II), is extracted with an organic solution containing a large excess of 8-hydroxyquinoline (see Section 6D-3 for the structure and reactions of this chelating agent). Four equilibria are shown. The first involves distribution of the 8-hydroxyquinoline, HQ, between the organic and aqueous layers. The second is the acid dissociation of the HQ to give H^+ and Q^- ions in the aqueous layer. The third equilibrium is the complex-formation reaction giving MQ_2. The fourth is distribution of the chelate between the two solvents (but for the fourth equilibrium, MQ_2 would precipitate out of the aqueous solution). The overall equilibrium is the sum of these four reactions or

$$2HQ(org) + M^{2+}(aq) \rightleftharpoons MQ_2(org) + 2H^+(aq)$$

The equilibrium constant for this reaction is

$$K' = \frac{[MQ_2(org)][H^+(aq)]^2}{[HQ(org)]^2[M^{2+}(aq)]}$$

Ordinarily, HQ is present in the organic layer in large excess with respect to M^{2+} in the aqueous phase so that $[HQ(org)]$ remains essentially constant during the extraction. Therefore, the equilibrium-constant expression can be simplified to

$$K'[HQ(org)]^2 = K = \frac{[MQ_2(org)][H^+(aq)]^2}{[M^{2+}(aq)]}$$

or

$$\frac{[MQ_2(org)]}{[M^{2+}(aq)]} = \frac{K}{[H^+(aq)]^2}$$

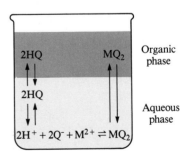

Figure 25-2
Equilibria in the extraction of an aqueous cation M^{2+} into an immiscible organic solvent containing 8-hydroxyquinoline.

Thus, we see that the ratio of concentration of the metal species in the two layers is inversely proportional to the square of the hydrogen ion concentration of the aqueous layer. Equilibrium constants K vary widely from metal ion to metal ion, and these differences make it possible to selectively extract one cation from another by buffering the aqueous solution at a level where one is extracted nearly completely and the second remains largely in the aqueous phase.

Several useful extractive separations with 8-hydroxyquinoline have been developed. Furthermore, numerous chelating agents that behave in a similar way are described in the literature.[4] As a consequence, pH-controlled extractions provide a powerful method for separating metallic ions.

Extraction of Metal Chlorides and Nitrates

A number of inorganic species can be separated by extraction with suitable solvents. For example, a single ether extraction of a 6 M hydrochloric acid solution will cause better than 50% of several ions to be transferred to the organic medium; included among these are iron(III), antimony(V), titanium(III), gold(III), molybdenum(VI), and tin(IV). Other ions, such as aluminum(III) and the divalent cations of cobalt, lead, manganese, and nickel are not extracted.

Uranium(VI) can be separated from such elements as lead and thorium by ether extraction of a solution that is 1.5 M in nitric acid and saturated with ammonium nitrate. Bismuth and iron(III) are also extracted to some extent from this medium.

25B   SEPARATION BY ION EXCHANGE

Ion exchange is a process by which ions held on a porous, essentially insoluble solid are exchanged for ions in a solution that is brought in contact with the solid. The ion-exchange properties of clays and zeolites have been recognized and studied for more than a century. Synthetic ion-exchange resins were first produced in 1935 and have since found widespread application in water softening, water deionization, solution purification, and ion separation.

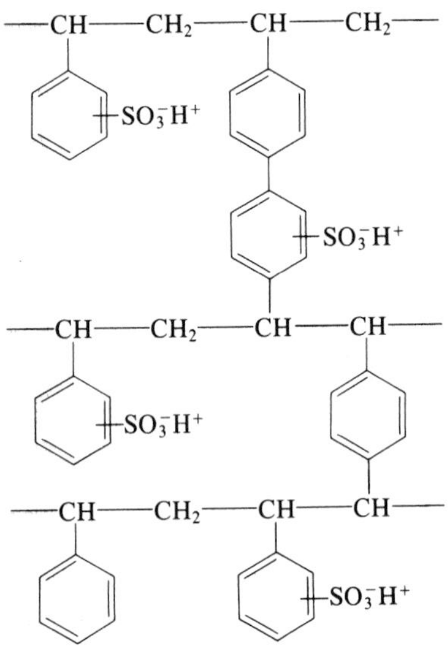

25B-1   Ion-Exchange Resins

Synthetic ion-exchange resins are high-molecular-weight polymers that contain large numbers of an ionic functional group per molecule. Cation-exchange resins contain acidic groups, while anion-exchange resins have basic groups. Strong-acid type exchangers have sulfonic acid groups ($-SO_3^-H^+$) attached to the polymeric matrix (see Figure 25-3) and have wider application than weak-acid type exchangers, which owe their action to carboxylic acid ($-COOH$) groups. Similarly, strong-base anion exchangers contain quaternary amine [$-N(CH_3)_3^+OH^-$] groups, while weak-base types contain secondary or tertiary amines.

Cation exchange is illustrated by the equilibrium

$$x\underset{\text{solid}}{RSO_3^-H^+} + \underset{\text{soln}}{M^{x+}} \rightleftharpoons \underset{\text{solid}}{(RSO_3^-)_xM^{x+}} + \underset{\text{soln}}{xH^+}$$

Figure 25-3

Structure of a cross-linked polystyrene ion-exchange resin. Similar resins are used in which the $-SO_3^-H^+$ group is replaced by $-COO^-H^+$, $-NH_3^+OH^-$, and $-N(CH_3)_3^+OH^-$ groups.

[4] For example, see G. H. Morrison, H. Freiser, and J. F. Cosgrove, in *Handbook of Analytical Chemistry*, L. Meites, Ed., p. **10**–5. New York: McGraw-Hill, 1963.

where M^{x+} represents a cation and R represents *that part* of a resin molecule that contains one sulfonic acid group. The analogous equilibrium involving a strong-base anion exchanger and an anion A^{x-} is

$$xRN(CH_3)_3^+OH^- + A^{x-} \rightleftharpoons [RN(CH_3)_3^+]_x A^{x-} + xOH^-$$

25B-2 Ion-Exchange Equilibria

Ion-exchange equilibria can be treated by the law of mass action. For example, when a dilute solution containing calcium ions is passed through a column packed with a sulfonic acid resin, the following equilibrium is established:

$$Ca^{2+}(aq) + 2H^+(res) \rightleftharpoons Ca^{2+}(res) + 2H^+(aq)$$

for which an equilibrium constant K is given by

$$K = \frac{[Ca^{2+}(res)][H^+(aq)]^2}{[Ca^{2+}(aq)][H^+(res)]^2} \qquad \textbf{(25-3)}$$

As usual, the bracketed terms are molar concentrations (strictly, activities) of the species in the two phases. Note that $[Ca^{2+}(res)]$ and $[H^+(res)]$ are molar concentrations of the two ions *in the solid phase*. In contrast to most solids, however, these concentrations can vary from zero to some maximum value when all of the negative sites on the resin are occupied by only one species.

Ion-exchange separations are ordinarily performed under conditions in which one ion predominates in *both* phases. Thus, in the removal of calcium ions from a dilute and somewhat acidic solution, the calcium ion concentration will be much smaller than that of hydrogen ion in both the aqueous and resin phases; that is,

$$[Ca^{2+}(res)] \ll [H^+(res)]$$

and

$$[Ca^{2+}(aq)] \ll [H^+(aq)]$$

As a consequence, the hydrogen ion concentration is essentially constant in both phases. Thus, Equation 25-3 can be rearranged to

$$\frac{[Ca^{2+}(res)]}{[Ca^{2+}(aq)]} = K\frac{[H^+(res)]^2}{[H^+(aq)]^2} = K_D \qquad \textbf{(25-4)}$$

where K_D is a distribution constant analogous to the constant that governs an extraction equilibrium (Equation 25-1). Note that K_D in Equation 25-4 represents the affinity of the resin for calcium ion relative to another ion (here, H^+). In general, where K_D for an ion is large, a strong tendency for the stationary phase to retain that ion exists; where K_D is small, the opposite is true. Selection of a common reference ion (such as H^+) permits a comparison of distribution ratios for various ions on a given type of resin. Such experiments reveal that polyvalent ions are much more strongly retained than singly charged species. Within a given charge group, differences that exist among values for K_D appear to be related to the size of the hydrated ion, as well as other properties. Thus, for a typical sulfonated cation-exchange resin, values of K_D for univalent ions decrease in the order $Ag^+ > Cs^+ > Rb^+ > K^+ > NH_4^+ > Na^+ > H^+ > Li^+$. For divalent cations the order is $Ba^{2+} > Pb^{2+} > Sr^{2+} > Ca^{2+} > Ni^{2+} > Cd^{2+} > Cu^{2+} > Co^{2+} > Zn^{2+} > Mg^{2+} > UO_2^{2+}$.

25B-3 Applications

Ion-exchange resins are used to eliminate ions that would otherwise interfere with an analysis. For example, iron(III), aluminum(III), as well as many other cations, tend to coprecipitate with barium sulfate during the determination of sulfate ion. Passage of a solution containing sulfate through a cation-exchange resin results in the retention of these ions and the release of an equivalent number of hydrogen ions. Sulfate ions pass freely through the column and can be precipitated as barium sulfate from the effluent.

Another valuable application of ion-exchange resins involves the concentration of ions from a very dilute solution. Thus, traces of metallic elements in large volumes of natural waters can be collected on a cation-exchange column and subsequently liberated from the resin by treatment with a small volume of an acidic solution; the result is a considerably more concentrated solution for analysis.

The total salt content of a sample can be determined by titrating the hydrogen ion released as an aliquot of sample passes through a cation exchanger in the acidic form. Similarly, a standard hydrochloric acid solution can be prepared by diluting to known volume the effluent resulting from treatment of a cation-exchange resin with a known weight of sodium chloride. Substitution of an anion-exchange resin in its hydroxide form will permit the preparation of a standard base solution.

As shown in Section 27B-4 ion-exchange resins are particularly useful for the chromatographic separation of both inorganic and organic ionic species.

25C   QUESTIONS AND PROBLEMS

*25-1. What is a masking agent and how does it function?

25-2. How do strong and weak acid synthetic ion-exchange resins differ in structure?

*25-3. The distribution coefficient for X between n-hexane and water is 9.6. Calculate the concentration of X remaining in the aqueous phase after 50.0 mL of 0.150 M X is treated by extraction with the following quantities of n-hexane: (a) one 40.0-mL portion, (b) two 20.0-mL portions, (c) four 10.0-mL portions, and (d) eight 5.00-mL portions.

25-4. The distribution coefficient for Z between n-hexane and water is 6.25. Calculate the percent of Z remaining in 25.0 mL of water that was originally 0.0600 M in Z after extraction with the following volumes of n-hexane: (a) one 25.0-mL portion, (b) two 12.5-mL portions, (c) five 5.00-mL portions, and (d) ten 2.50-mL portions.

*25-5. What volume of n-hexane is required to decrease the concentration of X in 25-3 to 1.00×10^{-4} M if 25.0 mL of 0.0500 M X are extracted with (a) 25.0-mL portions, (b) 10.0-mL portions, and (c) 2.0-mL portions?

25-6. What volume of n-hexane is required to decrease the concentration of Z in 25-4 to 1.00×10^{-5} M if 40.0 mL of 0.0200 M Z are extracted with (a) 50.0-mL portions of n-hexane, (b) 25.0-mL portions, (c) 10.0-mL portions?

*25-7. What minimum distribution coefficient is needed to permit removal of 99% of a solute from 50.0 mL of water with (a) two 25.0-mL extractions with toluene and (b) five 10.0-mL extractions with toluene?

25-8. If 30.0 mL of water that is 0.0500 M in Q are to be extracted with four 10.0-mL portions of an immiscible organic solvent, what is the minimum distribution coefficient that allows transfer of all but the following percentages of the solute to the organic layer: *(a) 1.00×10^{-4}, (b) 1.00×10^{-2}, (c) 1.00×10^{-3}?

*25-9. A 0.150 M aqueous solution of the weak organic acid HA was prepared from the pure compound, and three 50.0-mL aliquots were transferred to 100-mL volumetric flasks. Solution 1 was diluted to 100 mL with 1.0 M $HClO_4$, solution 2 was diluted to the mark with 1.0 M NaOH, and solution 3 was diluted to the mark with water. A 25.0-mL aliquot of each was extracted with 25.0 mL of *n*-hexane. The extract from solution 2 contained no detectable trace of A-containing species, indicating that A^- is not soluble in the organic solvent. The extract from solution 1 contained no ClO_4^- or $HClO_4$ but was found to be 0.0454 M in HA (by extraction with standard NaOH and back-titration with standard HCl). The extract from solution 3 was found to be 0.0225 M in HA. Assume that HA does not associate or dissociate in the organic solvent, and calculate

(a) the distribution ratio for HA between the two solvents.

(b) the concentration of the *species* HA and A^- in aqueous solution 3 after extraction.

(c) the dissociation constant of HA in water.

25-10. To determine the equilibrium constant for the reaction

$$I_2 + 2SCN^- \rightleftharpoons I(SCN)_2^- + I^-$$

25.0 mL of a 0.0100 M aqueous solution of I_2 was extracted with 10.0 mL of $CHCl_3$. After extraction, spectrophotometric measurements revealed that the I_2 concentration *of the aqueous layer* was 1.12×10^{-4} M. An aqueous solution that was 0.0100 M in I_2 and 0.100 M in KSCN was then prepared. After extraction of 25.0 mL of this solution with 10.0 mL of $CHCl_4$, the concentration of I_2 *in the $CHCl_3$ layer* was found from spectrophotometric measurement to be 1.02×10^{-3} M.

(a) What is the distribution coefficient for I_2 between $CHCl_3$ and H_2O?

(b) What is the formation constant for $I(SCN)_2^-$?

*25-11. The total cation content of natural water is often determined by exchanging the cations for hydrogen ions on a strong-acid ion-exchange resin. A 25.0-mL sample of a natural water was diluted to 100 mL with distilled water, and 2.0 g of a cation-exchange resin was added. After stirring, the mixture was filtered and the solid remaining on the filter paper was washed with three 15.0-mL portions of water. The filtrate and washings required 15.3 mL of 0.0202 M NaOH to give a bromocresol green end point.

(a) Calculate the number of milliequivalents of cation present in exactly 1 L of sample. (Here, the equivalent weight of a cation is its formula weight divided by its charge.)

(b) Report the results in terms of milligrams of $CaCO_3$ per liter.

25-12. An organic acid was isolated and purified by recrystallization of its barium salt. To determine the equivalent weight of the acid, a 0.393-g sample of the salt was dissolved in about 100 mL of water. The solution was passed through a strong-acid ion-exchange resin, and the column was then washed with water; the eluate and washings were titrated with 18.1 mL of 0.1006 M NaOH to a phenolphthalein end point.

(a) Calculate the equivalent weight of the organic acid.

(b) A potentiometric titration curve of the solution resulting when a second sample was treated in the same way revealed two end points, one at pH 5 and the other at pH 9. Calculate the molecular weight of the acid.

*25-13. Describe the preparation of exactly 2 L of 0.1500 M HCl from primary-standard grade NaCl using a cation-exchange resin.

25-14. An aqueous solution containing $MgCl_2$ and HCl was analyzed by first titrating a 25.00-mL aliquot to a bromocresol green end point with 18.96 mL of 0.02762 M NaOH. A 10.00-mL aliquot was then diluted to 50.00 mL with distilled water and passed through a strong-acid ion-exchange resin. The eluate and washings required 36.54 mL of the NaOH solution to reach the same end point. Report the molar concentrations of HCl and $MgCl_2$ in the sample.

CHAPTER 26

AN INTRODUCTION TO CHROMATOGRAPHIC METHODS

Chromatography was invented by the Russian botanist Mikhail Tswett shortly after the turn of this century. He passed solutions containing plant pigments, such as chlorophylls and xanthophylls, through glass columns packed with finely divided calcium carbonate. The separated species appeared as colored bands on the column, which accounts for the name he chose for the method (Greek: *chroma* meaning "color" and *graphein* meaning "to write").

Chromatography is a technique in which the components of a mixture are separated based on the rates at which they are carried through a stationary phase by a gaseous or liquid mobile phase.

Planar and column chromatography are based on the same types of equilibria.

Chromatography is an analytical method that is widely used for the separation, identification, and determination of the chemical components in complex mixtures. No other separation method is as powerful and generally applicable as is chromatography.[1]

26A   A GENERAL DESCRIPTION OF CHROMATOGRAPHY

The term *chromatography* is difficult to define rigorously because the name has been applied to such a variety of systems and techniques. All of these methods, however, have in common the use of a *stationary phase* and a *mobile phase*. Components of a mixture are carried through the stationary phase by the flow of a gaseous or liquid mobile phase, separations being based on differences in migration rates among the sample components.

26A-1 Classification of Chromatographic Methods

Chromatographic methods are of two types. In *column chromatography*, the stationary phase is held in a narrow tube and the mobile phase is forced through the tube under pressure or by gravity. In *planar chromatography*, the stationary

---

[1] General references on chromatography include: *Chromatography: Fundamentals and Applications of Chromatography and Electrophotometric Methods*, *Part A Fundamentals*, *Part B Applications*, E. Heftmann, Ed. New York: Elsevier, 1983; P. Sewell and B. Clarke, *Chromatographic Separations*. New York: Wiley, 1988; *Chromatographic Theory and Basic Principles*, J. A. Jonsson, Ed. New York: Marcel Dekker, 1987; J. C. Giddings, *Unified Separation Science*. New York: Wiley, 1991.

phase is supported on a flat plate or in the pores of a paper. Here the mobile phase moves through the stationary phase by capillary action or under the influence of gravity. We will deal with column chromatography only.

As shown in the first column of Table 26-1, chromatographic methods fall into three categories based on the nature of the mobile phase. The three types of phases include liquids, gases, and supercritical fluids. The second column of the table reveals that there are five types of liquid chromatography and three types of gas chromatography that differ in the nature of the stationary phase and the types of equilibria between phases.

> Liquid chromatography can be performed in columns and on planar surfaces, but gas chromatography is restricted to column procedures.

26A-2 Elution Chromatography

Figure 26-1 shows how two components, A and B, are resolved on a column by *elution chromatography*. Elution involves washing a solute through a column by additions of fresh solvent. A single portion of the sample dissolved in the mobile phase is introduced at the head of the column (at time t_0 in Figure 26-1), where components A and B distribute themselves between the two phases. Introduction of additional mobile phase (the *eluent*) forces the dissolved portion of the sample down the column, where further partition between the mobile phase and fresh portions of the stationary phase occurs (time t_1). Partitioning between the fresh solvent and the stationary phase takes place simultaneously at the original site of the sample.

Further additions of solvent carry solute molecules down the column in a continuous series of transfers between the two phases. Because solute movement can occur only in the mobile phase, the average *rate* at which a solute migrates *depends on the fraction of time it spends in that phase.* This fraction is small

> Elution is a process in which solutes are washed through a stationary phase by the movement of a mobile phase.

> An eluent is a solvent used to carry the components of a mixture through a stationary phase.

Table 26-1
CLASSIFICATION OF COLUMN CHROMATOGRAPHIC METHODS

General Classification	Specific Method	Stationary Phase	Type of Equilibrium
Liquid chromatography (LC) (mobile phase: liquid)	Liquid-liquid, or partition	Liquid adsorbed on a solid	Partition between immiscible liquids
	Liquid-bonded phase	Organic species bonded to a solid surface	Partition between liquid and bonded surface
	Liquid-solid, or adsorption	Solid	Adsorption
	Ion exchange	Ion-exchange resin	Ion exchange
	Size exclusion	Liquid in interstices of a polymeric solid	Partition/sieving
Gas chromatography (GC) (mobile phase: gas)	Gas-liquid	Liquid adsorbed on a solid	Partition between gas and liquid
	Gas-bonded phase	Organic species bonded to a solid surface	Partition between liquid and bonded surface
	Gas-solid	Solid	Adsorption
Supercritical-fluid chromatography (SFC) (mobile phase: supercritical fluid)		Organic species bonded to a solid surface	Partition between supercritical fluid and bonded surface

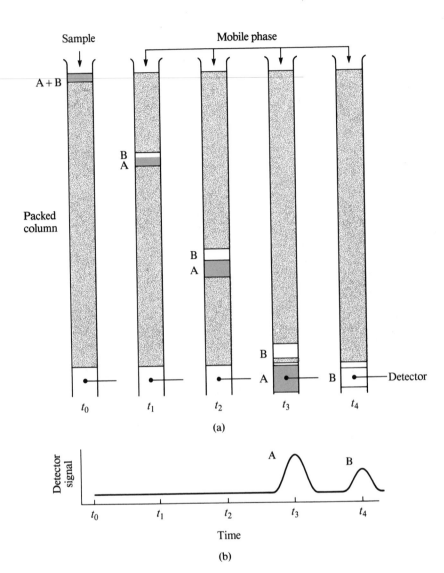

Figure 26-1

(a) Diagram showing the separation of a mixture of components A and B by column elution chromatography. (b) The output of the signal detector at the various stages of elution shown in (a).

for solutes that are strongly retained by the stationary phase (component B in Figure 26-1, for example) and large where retention in the mobile phase is more likely (component A). Ideally, the resulting differences in rates cause the components in a mixture to separate into *bands*, or *zones*, along the length of the column (see Figure 26-2). Isolation of the separated species is then accomplished by passing a sufficient quantity of mobile phase through the column to cause the individual bands to pass out the end (to be *eluted* from the column), where they can be collected (times t_3 and t_4 in Figure 26-1).

Chromatograms

> A chromatogram is a plot of some function of solute concentration versus elution time or elution volume.

If a detector that responds to solute concentration is placed at the end of the column and its signal is plotted as a function of time (or of volume of added mobile phase), a series of symmetric peaks is obtained, as shown in the lower part of Figure 26-1. Such a plot, called a *chromatogram*, is useful for both

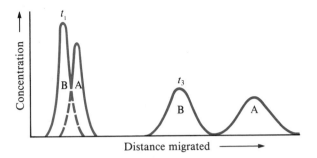

Figure 26-2
Concentration profiles of solute bands A and B at two different times in their migration down the column in Figure 26-1. The times t_1 and t_3 are indicated in Figure 26-1.

qualitative and quantitative analysis. The positions of the peaks on the time axis can be used to identify the components of the sample; the areas under the peaks provide a quantitative measure of the amount of each species.

The Effects of Relative Migration Rates and Band Broadening on Resolution

Figure 26-2 shows concentration profiles for the bands containing solutes A and B on the column in Figure 26-1 at time t_1 and at a later time t_3.[2] Because B is more strongly retained by the stationary phase than is A, B lags during the migration. Clearly, the distance between the two increases as they move down the column. At the same time, however, broadening of both bands takes place, which lowers the efficiency of the column as a separating device. While band broadening is inevitable, conditions can often be found where it occurs more slowly than band separation. Thus, as shown in Figure 26-2, a clean separation of species is possible provided the column is sufficiently long.

Several chemical and physical variables influence the rates of band separation and band broadening. As a consequence, improved separations can often be realized by the control of variables that either (1) increase the rate of band separation, or (2) decrease the rate of band spreading. These alternatives are illustrated in Figure 26-3.

The variables that influence the relative rates at which solutes migrate through a stationary phase are described in the next section. Following this discussion, we turn to those factors that play a part in zone broadening.

26B MIGRATION RATES OF SOLUTES

The effectiveness of a chromatographic column in separating two solutes depends in part on the relative rates at which the two species are eluted. These rates are in turn determined by the partition ratios of the solutes between the two phases.

26B-1 Partition Ratios in Chromatography

All chromatographic separations are based on differences in the extent to which solutes are partitioned between the mobile and the stationary phase. For the solute

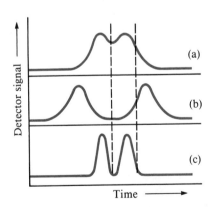

Figure 26-3
Two-component chromatograms illustrating two methods of improving separation: (a) original chromatogram with overlapping peaks; improvement brought about by (b) an increase in band separation; and (c) a decrease in bandwidth.

---

[2] Note that the relative positions of the bands for A and B in the concentration profile in Figure 26-2 appear to be reversed from their positions in the lower part of Figure 26-1. The difference is that the abscissa is distance along the column in Figure 26-2 but time in Figure 26-1. Thus, in Figure 26-1, the *front* of a peak lies to the left and the *tail* to the right; in Figure 26-2, the reverse is true.

species A, the equilibrium involved is described by the equation

$$A_{mobile} \rightleftharpoons A_{stationary}$$

The equilibrium constant K for this reaction is called a *partition ratio*, or *partition coefficient*, and is defined as

$$K = \frac{c_S}{c_M} \tag{26-1}$$

where c_S is the molar analytical concentration of a solute in the stationary phase and c_M is its analytical concentration in the mobile phase. Ideally, the partition ratio is constant over a wide range of solute concentrations; that is, c_S is directly proportional to c_M.

> The retention time t_R is the time between injection of a sample and the appearance of a solute peak at the detector of a chromatographic column.

> The dead time t_M is the time it takes for an unretained species to pass through a column.

26B-2 Retention Time

Figure 26-4 is a simple chromatogram made up of just two peaks. The small peak on the left is for a species that is *not* retained by the stationary phase. The time t_M after sample injection for this peak to appear is sometimes called the *dead time*. The dead time provides a measure of the average rate of migration of the mobile phase and is an important parameter in identifying analyte peaks. Often the sample or the mobile phase will contain an unretained species. When they do not, such a species may be added to aid in peak identification. The larger peak on the right in Figure 26-4 is that of an analyte species. The time required for this peak to reach the detector after sample injection is called the *retention time* and is given the symbol t_R.

The average linear rate of solute migration, $\bar{v}$, is

$$\bar{v} = \frac{L}{t_R} \tag{26-2}$$

where L is the length of the column packing. Similarly, the average linear velocity, u, of the molecules of the mobile phase is

$$u = \frac{L}{t_M} \tag{26-3}$$

Figure 26-4

A typical chromatogram for a two-component mixture. The small peak on the left represents a solute that is not retained on the column and so reaches the detector almost immediately after elution is started. Thus its retention time t_M is approximately equal to the time required for a molecule of the mobile phase to pass through the column.

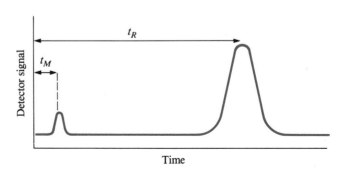

26B-3 The Relationship Between Migration Rate and Partition Ratio

In order to relate the rate of migration of a solute to its partition ratio, we express the rate as a fraction of the velocity of the mobile phase:

$$\bar{v} = u \times \text{fraction of time solute spends in mobile phase}$$

This fraction, however, equals the average number of moles of solute in the mobile phase at any instant divided by the total number of moles of solute in the column:

$$\bar{v} = u \times \frac{\text{moles of solute in mobile phase}}{\text{total moles of solute}}$$

The total number of moles of solute in the mobile phase is equal to the molar concentration, c_M, of the solute in that phase multiplied by its volume, V_M. Similarly, the number of moles of solute in the stationary phase is given by the product of the concentration, c_S, of the solute in the stationary phase and its volume, V_S. Therefore,

$$\bar{v} = u \times \frac{c_M V_M}{c_M V_M + c_S V_S} = u \times \frac{1}{1 + c_S V_S / c_M V_M}$$

Substitution of Equation 26-1 into this equation gives an expression for the rate of solute migration as a function of its partition ratio as well as a function of the volumes of the stationary and mobile phases:

$$\bar{v} = u \times \frac{1}{1 + K V_S / V_M} \tag{26-4}$$

The two volumes can be estimated from the method by which the column is prepared.

26B-4 The Capacity Factor

The *capacity factor* is an important experimental parameter that is widely used to describe the migration rates of solutes on columns. For a solute A, the capacity factor k'_A is defined as

$$k'_A = \frac{K_A V_S}{V_M} \tag{26-5}$$

where K_A is the partition ratio for the species A. Substitution of Equation 26-5 into 26-4 yields

$$\bar{v} = u \times \frac{1}{1 + k'_A} \tag{26-6}$$

Michael Tswett (1872–1919), a Russian botanist, discovered the basic principles of column chromatography. He separated plant pigments by eluting a mixture of the pigments on a column of calcium carbonate. The various pigments separated into colored bands; hence the name chromatography.

In order to show how k'_A can be derived from a chromatogram, we substitute Equations 26-2 and 26-3 into Equation 26-6:

$$\frac{L}{t_R} = \frac{L}{t_M} \times \frac{1}{1 + k'_A} \tag{26-7}$$

This equation rearranges to

$$k'_A = \frac{t_R - t_M}{t_M} \tag{26-8}$$

As shown in Figure 26-4, t_R and t_M are readily obtained from a chromatogram. When the capacity factor for a solute is much less than unity, elution occurs so rapidly that accurate determination of the retention times is difficult. When the capacity factor is larger than perhaps 20 to 30, elution times become inordinately long. Ideally, separations are performed under conditions in which the capacity factors for the solutes in a mixture are in the range between 1 and 5.

The capacity factors in gas chromatography can be varied by changing the temperature and the column packing. In liquid chromatography, capacity factors can often be manipulated to give better separations by varying the composition of the mobile phase and the stationary phase.

> Ideally, the capacity factor for analytes in a sample is between 1 and 5.

26B-5 Relative Migration Rates: The Selectivity Factor

> The selectivity factor for two analytes in a column provides a measure of how well the column will separate the two.

The *selectivity factor* α of a column for the two species A and B is defined as

$$\alpha = \frac{K_B}{K_A} \tag{26-9}$$

where K_B is the partition ratio for the more strongly retained species B and K_A is the constant for the less strongly held or more rapidly eluted species A. According to this definition, α *is always greater than unity*.

Substitution of Equation 26-5 and the analogous equation for solute B into Equation 26-9 provides after rearrangement a relationship between the selectivity factor for two solutes and their capacity factors:

$$\alpha = \frac{k'_B}{k'_A} \tag{26-10}$$

where k'_B and k'_A are the capacity factors for B and A, respectively. Substitution of Equation 26-8 for the two solutes into Equation 26-10 gives an expression that permits the determination of α from an experimental chromatogram:

$$\alpha = \frac{(t_R)_B - t_M}{(t_R)_A - t_M}$$ **(26-11)**

In Section 26D-1 we show how we use the selectivity factor to compute the resolving power of a column.

26C   THE EFFICIENCY OF CHROMATOGRAPHIC COLUMNS

The efficiency of a chromatographic column refers to the amount of band broadening that occurs when a compound passes through the column. Before defining column efficiency in more quantitative terms, let us examine the reasons that bands become broader as they move down a column.

26C-1 The Rate Theory of Chromatography

The *rate theory* of chromatography describes the shapes and breadths of elution peaks in quantitative terms based on a random-walk mechanism for the migration of molecules through a column. A detailed discussion of the rate theory is beyond the scope of this book.[3] We can, however, give a qualitative picture of why bands broaden and what variables improve column efficiency.

If you examine the chromatograms shown in this and the next chapter, you will see that the elution peaks look very much like the Gaussian or normal error curves that you encountered in Chapters 4 and 5. As shown in Section 4B, normal error curves are rationalized by assuming that the uncertainty associated with any single measurement is the summation of a much larger number of small, individually undetectable and random uncertainties, each of which has an equal probability of being positive or negative. In a similar way, the typical Gaussian shape of a chromatographic band can be attributed to the additive combination of the random motions of the myriad molecules making up a band as it moves down the column.

It is instructive to consider a single solute molecule as it undergoes many thousands of transfers between the stationary and the mobile phases during elution. Residence time in either phase is highly irregular. Transfer from one phase to the other requires energy, and the molecule must acquire this energy from its surroundings. Thus, the residence time in a given phase may be transitory for some molecules and relatively long for others. Recall that movement down the column can occur *only while the molecule is in the mobile phase*. As a consequence, certain particles travel rapidly by virtue of their accidental inclusion in the mobile phase for a majority of the time, whereas others lag because they happen to be incorporated in the stationary phase for a greater-than-average length of time. The result of these random individual processes is a symmetric spread of velocities around the mean value, which represents the behavior of the average analyte molecule.

Some chromatographic peaks are nonideal and exhibit *tailing* or *fronting*. In the former case the tail of the peak, appearing to the right on the chromatogram, is drawn out while the front is steepened. With fronting, the reverse is the case. A common cause of tailing and fronting is a nonlinear distribution coefficient. Fronting also occurs when the sample introduced onto a column is too large. Distortions of this kind are undesirable because they lead to poorer separations and less reproducible elution times. In our discussion we assume that tailing and fronting are minimal.

---

[3] See J. J. Hawkes, *J. Chem. Educ.*, **1983**, *60*, 393.

Diffusion also contributes to band broadening. Recall from Section 19A-2 that molecules tend to diffuse from a more concentrated part of a solution to a more dilute, the rate being proportional to the concentration difference. In the center of a chromatographic band, the concentration of a species is high, while at the two edges the concentration approaches zero. Therefore, molecules tend to migrate to either edge of the band. Band broadening is the result. Note that half of the diffusion will be in the direction of flow and the other half in an opposed direction.

Another cause of band broadening is shown in Figure 26-5, which shows that individual molecules in the mobile phase follow paths of different lengths as they traverse the column. Thus, their arrival time at the detector differs, and bands are broader as a consequence.

The breadth of a band increases as it moves down the column because more time is allowed for spreading by these various mechanisms to occur. Thus, zone breadth is directly related to residence time in the column and inversely related to the flow velocity of the mobile phase.

26C-2 A Quantitative Description of Column Efficiency

Two related terms are widely used as quantitative measures of the efficiency of chromatographic columns: (1) *plate height H* and (2) *number of theoretical plates N*. The two are related by the equation

$$N = L/H \qquad (26\text{-}12)$$

> The plate height H is also known as the height equivalent of a theoretical plate (HETP).

where L is the length (usually in centimeters) of the column packing. Feature 26-1 describes how these measures of column efficiency got their names.

The efficiency of chromatographic columns increases as the number of plates becomes greater and as the plate height becomes smaller. Enormous differences

> The efficiency of a column is great when H is small and N is large.

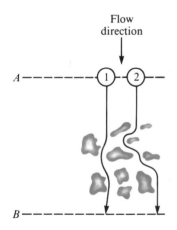

Figure 26-5

Typical pathways of two molecules during elution. Note that the distance traveled by molecule 2 is greater than that traveled by molecule 1. Thus, molecule 2 will arrive at B later than molecule 1.

Feature 26-1
WHAT IS THE SOURCE OF THE TERMS PLATE AND PLATE HEIGHT?

The 1952 Nobel Prize for chemistry was awarded to two Englishmen, A. J. P. Martin and R. L. M. Synge, for their work in the development of modern chromatography. In their theoretical studies, they adapted a model that was first developed in the early 1920s to describe separations on fractional distillation columns. Fractionating columns, which were first used in the petroleum industry for separating closely related hydrocarbons, consisted of numerous interconnected bubble-cap plates (see Figure 26-A) at which vapor-liquid equilibria were established when the column was operated under reflux conditions.

Martin and Synge treated a chromatographic column as if it were made up of a series of contiguous bubble-cap-like plates, within which equilibrium conditions always prevail. This plate model successfully accounts for the Gaussian shape of chromatographic peaks as well as for factors that influence differences in solute-migration rates. The plate model is totally incapable of accounting for zone broadening, however, because of its basic assumption

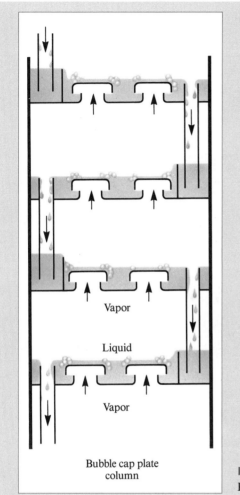

Figure 26-A
Plates in a fractionating column.

that equilibrium conditions prevail throughout a column during elution. This assumption can never be valid in the dynamic state that exists in a chromatographic column, where phases are moving past one another at such a pace that sufficient time is not available for equilibration.

in efficiencies are encountered in columns, depending on their type and the kinds of mobile and stationary phases they contain. Efficiencies in terms of plate numbers can vary from a few hundred to several hundred thousand; plate heights from a few tenths to one thousandth of a centimeter or smaller are not uncommon.

Definition of Plate Height, *H*

In Section 4C-2, we pointed out that the breadth of a Gaussian curve is described by the standard deviation σ and the variance σ^2. Because chromatographic bands are also Gaussian and because the efficiency of a column is reflected in the breadth of chromatographic peaks, the variance per unit length of column is used by chromatographers as a measure of column efficiency. That is, the column

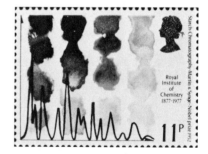

Stamp in honor of biochemists Archer J. P. Martin (b. 1910) and Richard L. M. Synge (b. 1914) who were awarded the 1952 Nobel Prize in chemistry for their contributions to the development of modern chromatography.

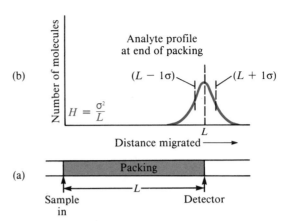

Figure 26-6
Definition of plate height $H = \sigma^2/L$.

efficiency H is defined as

$$H = \frac{\sigma^2}{L} \qquad (26\text{-}13)$$

This definition of column efficiency is illustrated in Figure 26-6a, which shows a column that has a packing L cm in length. Above this schematic is a plot showing the distribution of molecules along the length of the column at the moment the analyte peak reaches the end of the packing (that is, at the retention time t_R). The curve is Gaussian, and the locations of $L + 1\sigma$ and $L - 1\sigma$ are indicated as broken vertical lines. Note that L carries units of centimeters and σ^2 units of centimeters squared; thus H represents a linear distance in centimeters as well (Equation 26-13). In fact, the plate height can be thought of as the length of column that contains a fraction of the analyte that lies between L and $L - \sigma$. Because the area under a normal error curve bounded by $\pm\sigma$ is about 68% of the total area (page 62), the plate height, as defined, contains 34% of the analyte.

Experimental Determination of the Number of Plates in a Column

The number of theoretical plates, N, and the plate height, H, are widely used in the literature and by instrument manufacturers as measures of column performance. Figure 26-7 shows how N can be determined from a chromatogram. Here, the retention time of a peak t_R and the width of the peak at its base W (in units of time) are measured. It can be shown (see Feature 26-2) that the number of

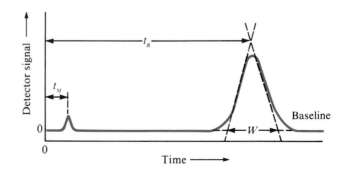

Figure 26-7
Determination of the standard deviation τ from a chromatographic peak: $W = 4\tau$.

plates can then be computed by the simple relationship

$$N = 16 \left(\frac{t_R}{W} \right)^2 \qquad \text{(26-14)}$$

To obtain H, the length of the column L is measured and Equation 26-12 applied.

Feature 26-2
DERIVATION OF EQUATION 26-14

The variance of the peak shown in Figure 26-7 has units of seconds squared because the abscissa is time in seconds (or sometimes in minutes). This time-based variance is usually designated as τ^2 to distinguish it from σ^2, which has units of centimeters squared. The two standard deviations τ and σ are related by

$$\tau = \frac{\sigma}{L/t_R} \qquad \text{(26-15)}$$

where L/t_R is the average linear velocity of the solute in centimeters per second.

Figure 26-7 illustrates a simple means for approximating τ from an experimental chromatogram. Tangents at the inflection points on the two sides of the chromatographic peak are extended to form a triangle with the base line. The area of this triangle can be shown to be approximately 96% of the total area under the peak. In Section 5B-2 it was shown that about 96% of the area under a Gaussian peak is included within plus or minus two standard deviations ($\pm 2\sigma$) of its maximum. Thus, the intercepts shown in Figure 26-7 occur at approximately $\pm 2\tau$ from the maximum, and $W = 4\tau$, where W is the magnitude of the base of the triangle. Substituting this relationship into Equation 26-15 and rearranging yields

$$\sigma = \frac{LW}{4t_R}$$

Substitution of this equation for σ into Equation 26-13 gives

$$H = \frac{LW^2}{16t_R^2} \qquad \text{(26-16)}$$

To obtain N, we substitute into Equation 26-12 and rearrange to get

$$N = 16 \left(\frac{t_R}{W} \right)^2$$

Thus, N can be calculated from two time measurements, t_R and W; to obtain H, the length of the column packing L must also be known.

26C-3 Variables That Affect Column Efficiency

Band broadening, and thus loss of column efficiency, is the consequence of the finite rate at which several mass-transfer processes occur during migration of a solute down a column. Some of the variables that affect these rates are controllable and can be exploited to improve separations.

The Effect of Mobile-Phase Flow Rate

The extent of band broadening depends on the length of time the mobile phase is in contact with the stationary phase. Therefore, as shown in Figure 26-8, column efficiency depends on the rate of flow of the mobile phase. Note that for both liquid and gas-liquid chromatography, minimum plate heights (or maximum efficiencies) occur at relatively low flow rates. Maximum efficiency for liquid chromatography occurs at rates that are well below those for gas chromatography. In fact, the minima for liquid chromatography are often so low that they are not observed under ordinary operating conditions.

Generally, liquid chromatograms are obtained at lower flow rates than gas chromatograms. Furthermore, as shown in Figure 26-8, plate heights for liquid chromatographic columns are an order of magnitude or more smaller than those of gas chromatographic columns. Offsetting this advantage, however, is the difference in lengths of the two types of columns. Gas chromatographic columns can be 50 m or more in length. Liquid columns, on the other hand, can be no longer than 25 to 50 cm because of the high pressure drops along their lengths. As a consequence, the number of plates, N, in a gas chromatographic column may exceed that in a liquid chromatographic column by a factor of several hundred.

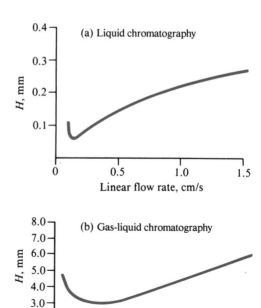

Figure 26-8

Effect of mobile-phase flow rate on plate height for (a) liquid chromatography and (b) gas chromatography.

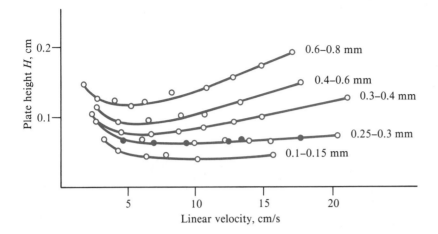

Figure 26-9
Effect of particle size on plate height.
The numbers to the right are particle
diameters. (From J. Boheman and
J. H. Purnell, in *Gas Chromatography
1958*, D. H. Desty, Ed. New York:
Academic Press, 1958. With permis-
sion of Butterworths, Stoneham, MA.)

Other Variables

It has been found that column efficiency can be increased by decreasing the
particle size of column packings, by employing thinner layers of the immobilized
film (where the stationary phase is a liquid adsorbed on a solid), and by lowering
the mobile-phase viscosity. Increases in temperature also reduce band broadening
under most circumstances. Figure 26-9 illustrates how particle size affects plate
heights of gas-liquid chromatographic columns.

26D   COLUMN RESOLUTION

The *resolution* R_s of a column provides a quantitative measure of its ability to
separate two analytes. The significance of this term is illustrated in Figure 26-10,
which consists of chromatograms for species A and B on three columns with
different resolving powers. The resolution of each column is defined as

$$R_s = \frac{2\Delta Z}{W_A + W_B} = \frac{2[(t_R)_B - (t_R)_A]}{W_A + W_B} \tag{26-17}$$

where all of the terms on the right side are as defined in the figure.

It is evident from the figure that a resolution of 1.5 gives an essentially complete
separation of A and B, whereas a resolution of 0.75 does not. At a resolution of
1.0, zone A contains about 4% B and zone B contains about 4% A. At a resolution
of 1.5, the overlap is about 0.3%. The resolution for a given stationary phase
can be improved by lengthening the column, thus increasing the number of plates.
An adverse consequence of the added plates, however, is an increase in the time
required for the resolution.

26D-1 Effect of Capacity and Selectivity Factors on Resolution

A useful equation is readily derived that relates the resolution of a column to
the number of plates it contains, as well as to the capacity and selectivity factors

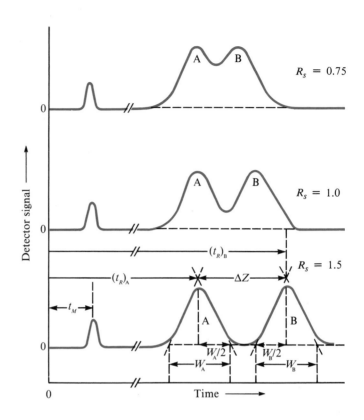

Figure 26-10
Separation at three resolutions: $R_s = 2\Delta Z/(W_A + W_B)$.

of a pair of solutes on the column. It can be shown[4] that for the two solutes A and B in Figure 26-10, the resolution is given by the equation

$$R_s = \frac{\sqrt{N}}{4}\left(\frac{\alpha - 1}{\alpha}\right)\left(\frac{k'_B}{1 + k'_B}\right) \quad \text{(26-18)}$$

where k'_B is the capacity factor of the slower-moving species and α is the selectivity factor. This equation can be rearranged to give the number of plates needed to realize a given resolution:

$$N = 16R_s^2\left(\frac{\alpha}{\alpha - 1}\right)^2\left(\frac{1 + k'_B}{k'_B}\right)^2 \quad \text{(26-19)}$$

26D-2 Effect of Resolution on Retention Time

As mentioned earlier, the goal in chromatography is the highest possible resolution in the shortest possible elapsed time. Unfortunately, these goals tend to be incompatible; consequently, a compromise between the two is usually necessary. The

---

[4] See D. A. Skoog and J. J. Leary, *Principles of Instrumental Analysis*, 4th ed., pp. 592–593. Philadelphia: Saunders College Publishing, 1992.

time $(t_R)_B$ required to elute the two species in Figure 26-10 with a resolution of R_s is given by

$$(t_R)_B = \frac{16 R_s^2 H}{u} \left(\frac{\alpha}{\alpha - 1} \right)^2 \frac{(1 + k_B')^3}{(k_B')^2} \qquad \textbf{(26-20)}$$

where u is the linear velocity of the mobile phase.

Example 26-1

Substances A and B have retention times of 16.40 and 17.63 min, respectively, on a 30.0-cm column. An unretained species passes through the column in 1.30 min. The peak widths (at base) for A and B are 1.11 and 1.21 min, respectively. Calculate (a) column resolution, (b) average number of plates in the column, (c) plate height, (d) length of column required to achieve a resolution of 1.5, and (e) time required to elute substance B on the longer column.

(a) Employing Equation 26-17, we find

$$R_s = 2(17.63 - 16.40)/(1.11 + 1.21) = 1.06$$

(b) Equation 26-14 permits computation of N:

$$N = 16 \left(\frac{16.40}{1.11} \right)^2 = 3493 \quad \text{and} \quad N = 16 \left(\frac{17.63}{1.21} \right)^2 = 3397$$

$$N_{av} = (3493 + 3397)/2 = 3445 = 3.4 \times 10^3$$

(c)

$$H = L/N = 30.0/3445 = 8.7 \times 10^{-3} \text{ cm}$$

(d) k' and α do not change greatly with increasing N and L. Thus, substituting N_1 and N_2 into Equation 26-19 and dividing one of the resulting equations by the other yields

$$\frac{(R_s)_1}{(R_s)_2} = \frac{\sqrt{N_1}}{\sqrt{N_2}}$$

where the subscripts 1 and 2 refer to the original and to the longer column, respectively. Substituting the appropriate values for N_1, $(R_s)_1$, and $(R_s)_2$ gives

$$\frac{1.06}{1.5} = \frac{\sqrt{3445}}{\sqrt{N_2}}$$

$$N_2 = 3445 \left(\frac{1.5}{1.06} \right)^2 = 6.9 \times 10^3$$

But

$$L = NH = 6.9 \times 10^3 \times 8.7 \times 10^{-3} = 60 \text{ cm}$$

(e) Substituting $(R_s)_1$ and $(R_s)_2$ into Equation 26-20 and dividing yields

$$\frac{(t_R)_1}{(t_R)_2} = \frac{(R_s)_1^2}{(R_s)_2^2} = \frac{17.63}{(t_R)_2} = \frac{(1.06)^2}{(1.5)^2}$$

$$(t_R)_2 = 35 \text{ min}$$

Thus, to obtain the improved resolution, the separation time must be doubled.

26E   APPLICATIONS OF CHROMATOGRAPHY

Chromatography is a powerful and versatile tool for separating closely related chemical species. In addition, it can be employed for the qualitative identification and quantitative determination of separated species.

26E-1   Qualitative Analysis

Chromatography is widely used for recognizing the presence or absence of components in mixtures that contain a limited number of species whose identities are known. For example, 30 or more amino acids in a protein hydrolysate can be detected with a reasonable degree of certainty by means of a chromatogram. On the other hand, because a chromatogram provides but a single piece of information about each species in a mixture (the retention time), the application of the technique to the qualitative analysis of complex samples of unknown composition is limited. This limitation has been largely overcome by linking chromatographic columns directly with ultraviolet, infrared, and mass spectrometers. The resulting *hyphenated* instruments are powerful tools for identifying the components of complex mixtures.

A hyphenated technique is an analytical method in which two instrumental techniques are coupled to produce a more powerful analytical procedure. Examples include gas chromatography/mass spectrometry, liquid chromatography/voltammetry, and inductively coupled plasma/mass spectrometry.

It is important to note that while a chromatogram may not lead to positive identification of the species in a sample, it often provides sure evidence of the *absence* of species. Thus, failure of a sample to produce a peak at the same retention time as a standard obtained under identical conditions is strong evidence that the compound in question is absent (or present at a concentration below the detection limit of the procedure).

26E-2   Quantitative Analysis

Chromatography owes its enormous growth in part to its speed, simplicity, relatively low cost, and wide applicability as a tool for separations. However, it is doubtful that its use would have become so widespread had it not been for the fact that it can also provide quantitative information about separated species.

Quantitative chromatography is based on a comparison of either the height or the area of an analyte peak with that of one or more standards. If conditions are controlled properly, both of these parameters vary linearly with concentration.

Analyses Based on Peak Height

The height of a chromatographic peak is obtained by connecting the base lines on the two sides of the peak by a straight line and measuring the perpendicular distance from this line to the peak. This measurement can ordinarily be made with reasonably high precision and yields accurate results, provided variations in column conditions do not alter peak width during the period required to obtain chromatograms for sample and standards. The variables that must be controlled closely are column temperature, eluent flow rate, and rate of sample injection. In addition, care must be taken to avoid overloading the column. The effect of sample-injection rate is particularly critical for the early peaks of a chromatogram. Relative errors of 5% to 10% due to this cause are not unusual with syringe injection.

Analyses Based on Peak Area

Peak area is independent of broadening effects caused by the variables mentioned in the previous paragraph. From this standpoint, therefore, area is a more satisfactory analytical parameter than peak height. On the other hand, peak heights are more easily measured and, for narrow peaks, more accurately determined.

Most modern chromatographic instruments are equipped with electronic integrators that provide precise measurements of relative peak areas. If such equipment is not available, a manual estimate must be made. A simple method that works well for symmetric peaks of reasonable widths is to multiply peak height by the width at one-half peak height.

Calibration with Standards

The most straightforward method for quantitative chromatographic analyses involves the preparation of a series of standard solutions that approximate the composition of the unknown. Chromatograms for the standards are then obtained, and peak heights or areas are plotted as a function of concentration. A plot of the data should yield a straight line passing through the origin; analyses are based on this plot. Frequent standardization is necessary for highest accuracy.

The Internal-Standard Method

The highest precision for quantitative chromatography is obtained by using internal standards because the uncertainties introduced by sample injection, flow rate, and variations in column conditions are minimized. In this procedure, a carefully measured quantity of an internal-standard is introduced into each standard and sample, and the ratio of analyte peak area (or height) to internal-standard peak area (or height) is the analytical parameter. For this method to be successful, it is necessary that the internal-standard peak be well separated from the peaks of all other components in the sample, but it must appear close to the analyte peak. With a suitable internal standard, precisions of 0.5% to 1% relative are reported.

26F   QUESTIONS AND PROBLEMS

26-1. Define
*(a) elution.
(b) mobile phase.
*(c) stationary phase.
(d) partition ratio.
*(e) retention time.
(f) capacity factor.
*(g) selectivity factor.
(h) plate height.

26-2. List the variables that lead to zone broadening.

*26-3.** What is the difference between gas-liquid and liquid-liquid chromatography?

26-4. What is the difference between liquid-liquid and liquid-solid chromatography?

*26-5.** Describe a method for determining the number of plates in a column.

26-6. Name two general methods for improving the resolution of two substances on a chromatographic column.

*26-7.** The following data are for a liquid chromatographic column:

Length of packing	24.7 cm
Flow rate	0.313 mL/min
V_M	1.37 mL
V_S	0.164 mL

A chromatogram of a mixture of species A, B, C, and D provided the following data:

	Retention Time, min	Width of Peak Base (W), min
Nonretained	3.1	————
A	5.4	0.41
B	13.3	1.07
C	14.1	1.16
D	21.6	1.72

Calculate
(a) the number of plates from each peak.
(b) the mean and the standard deviation for N.
(c) the plate height for the column.

*26-8.** From the data in 26-7, calculate for A, B, C, and D
(a) the capacity factor.
(b) the partition coefficient.

*26-9.** From the data in 26-7, calculate for species B and C
(a) the resolution.
(b) the selectivity factor.
(c) the length of column necessary to separate the two species with a resolution of 1.5.
(d) the time required to separate the two species on the column in (c).

*26-10.** From the data in 26-7, calculate for species C and D
(a) the resolution.

(b) the length of column necessary to separate the two species with a resolution of 1.5.

26-11. The following data were obtained by gas-liquid chromatography on a 40-cm packed column:

Compound	t_R, min	$W_{1/2}$, min
Air	1.9	
Methylcyclohexane	10.0	0.76
Methylcyclohexene	10.9	0.82
Toluene	13.4	1.06

Calculate
(a) an average number of plates from the data.
(b) the standard deviation for the average in (a).
(c) an average plate height for the column.
(d) the capacity factor for each of the three species.

26-12. Refer to 26-11 and calculate the resolution for
(a) methylcyclohexene and methylcyclohexane.
(b) methylcyclohexane and toluene.
(c) methylcyclohexane and toluene.

26-13. If a resolution of 1.5 is desired in separating methylcyclohexane and methylcyclohexene in 26-11,
(a) how many plates are required?
(b) how long must the column be if the same packing is employed?
(c) what is the retention time for methylcyclohexene on the column in 26-11b?

*26-14.** If V_S and V_M for the column in 26-11 are 19.6 and 62.6 mL, respectively, and a nonretained air peak appears after 1.9 min, calculate the
(a) capacity factor for each compound.
(b) partition coefficient for each compound.
(c) selectivity factor for methylcyclohexane and methylcyclohexene.

*26-15.** From distribution studies, species M and N are known to have water/hexane partition coefficients of 6.01 and 6.20 ($K = [M]_{H_2O}/[M]_{hex}$). The two species are to be separated by elution with hexane in a column packed with silica gel containing adsorbed water. The ratio V_S/V_M for the packing is 0.422.
(a) Calculate the capacity factor for each solute.
(b) Calculate the selectivity factor.
(c) How many plates are needed to provide a resolution of 1.5?
(d) How long a column is needed if the plate height of the packing is 2.2×10^{-3} cm?
(e) If a flow rate of 7.10 cm/min is employed, how long will it take to elute the two species?

26-16. Repeat the calculations in 26-15 assuming $K_M = 5.81$ and $K_N = 6.20$.

---

* Answers to the asterisked problems are given in the answers section at the back of the book.

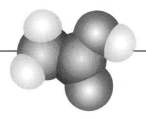

CHAPTER 27

APPLICATIONS OF CHROMATOGRAPHY

In this chapter, we discuss how analytical separations are carried out by gas-liquid chromatography and by high-performance liquid chromatography.

27A GAS-LIQUID CHROMATOGRAPHY

In *gas-liquid chromatography* (GLC), or *gas chromatography* (GC), the components of a vaporized sample are fractionated as a consequence of being partitioned between a mobile *gaseous* phase and a liquid stationary phase held in a column.[1]

27A-1 Apparatus

The basic components of a typical instrument for performing gas chromatography are shown in Figure 27-1 and are described briefly in this section.

Carrier Gas Supply

The gaseous mobile phase must be chemically inert. Helium is the most common mobile phase, although argon, nitrogen, and hydrogen are also used. These gases are available in pressurized tanks. Pressure regulators, gauges, and flow meters are required to control the flow rate of the gas.

Pressures at the column inlet usually range from 10 to 50 psi (lb/in² above room pressure) and provide flow rates of 25 to 50 mL/min. Flow rates are generally measured by a simple soap-bubble meter, such as that shown at the end of the column in Figure 27-1 and in Figure 27-2. A soap film is formed in

> In all types of chromatography, liquid stationary phases are immobilized on solid surfaces by adsorption or by chemical bonding.

Helium is the most common mobile phase in gas-liquid chromatography.

---

[1] For detailed treatment of GLC, see J. Willet, *Gas Chromatography*. New York: Wiley, 1987; *Modern Practice of Gas Chromatography*, 2nd ed., R. L. Grob, Ed. New York: Wiley, 1985; J. A. Perry, *Introduction to Analytical Gas Chromatography*. New York: Dekker, 1981.

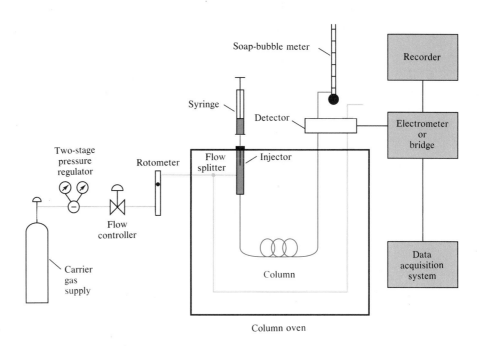

Figure 27-1
Schematic of a gas chromatograph.

Packed columns offer the advantages of larger sample size and ease and convenience of use. Capillary columns have the advantages of higher resolution, shorter analysis time, and higher sensitivity.

Figure 27-2

A soap-bubble flow meter. (Courtesy of Chrompack, Inc., Raritan, NJ.)

the path of the gas when a rubber bulb containing a solution of soap is squeezed; the time required for this film to move between two graduations on the buret is measured and converted to a flow rate.

Sample Injection System

Column efficiency requires that the sample be of a suitable size and be introduced as a ''plug'' of vapor; slow injection or oversized samples cause band spreading and poor resolution. Calibrated microsyringes, such as those shown in Figure 27-3, are used to inject liquid samples through a rubber or silicone diaphragm or septum into a heated sample port located at the head of the column. The sample port (Figure 27-4) is ordinarily about 50°C above the boiling point of the least volatile component of the sample. For ordinary packed analytical columns, sample sizes range from a few tenths of a microliter to 20 μL. Capillary columns require samples that are smaller by a factor of 100 or more. Here, a sample splitter is often needed to deliver only a small known fraction (1 : 100 to 1 : 500) of the injected sample, with the remainder going to waste.

Packed Columns

Both *packed* and *open-tubular* (or *capillary*) columns are used in gas-liquid chromatography. The former, which are discussed in this section, can accommodate larger samples and are generally more convenient to use. The latter, which are described in the next section, are of considerable importance because of their unparalleled resolution.

Present-day packed columns are fabricated from glass or metal tubing; they are typically 2 to 3 m long and have inside diameters of 2 to 4 mm. The tubes are ordinarily formed as coils with diameters of roughly 15 cm to permit convenient thermostating in an oven.

The packing, or support, for a column holds the liquid stationary phase in place, so that the surface area exposed to the mobile phase is as large as possible. The ideal solid packing consists of small, uniform, spherial particles with good mechanical strength and with a specific surface of at least 1 m²/g. In addition, the material should be inert at elevated temperatures and be uniformly wetted by the liquid phase. No substance that meets perfectly all of these criteria is yet available.

The earliest packings for gas chromatography were prepared from naturally occurring diatomaceous earth, which consists of the skeletons of thousands of species of single-celled plants that inhabited ancient lakes and seas (Figure 27-5 is an enlarged photo of a diatom obtained with a scanning electron microscope). These packings are still the most widely used support materials and are often treated chemically with dimethylchlorosilane, which gives a surface layer of methyl groups. This treatment reduces the tendency of the packing to adsorb polar molecules.

The particle size of packings for gas chromatography typically falls in the range of 60 to 80 mesh (250 to 170 μm) or 80 to 100 mesh (170 to 149 μm). The use of smaller particles is not practical because the pressure drop within the column becomes prohibitively high.

Open-Tubular Columns

Open-tubular or capillary columns were first described in the 1950s, when it became apparent from theoretical considerations that such columns should provide separations that were unprecedented in terms of speed and number of theoretical plates.[2] At that time, several investigations demonstrated that columns with as many as 300,000 plates or more were practical. Despite such spectacular results, the use of capillary columns was delayed until the late 1970s because of a number of problems associated with their use. These problems have now become manageable and a number of instrument vendors offer open-tubular equipment for routine use. Capillary columns, which are generally constructed of glass or fused silica, typically have inside diameters of 0.25 to 0.50 mm and lengths of 25 to 50 m. Their inner surfaces are coated with a thin layer of the stationary phase, which may be any of the liquids described in Section 27A-2.

The manufacture of fused-silica columns, which are currently the most widely used capillary columns, is based on techniques developed for the production of optical fibers. Silica capillaries, which have much thinner walls than their glass or metal counterparts, have outside diameters of about 0.3 mm. The tubes are given added strength by an exterior polyimide coating. The resulting columns are quite flexible and strong, and can be bent into coils with diameters of a few inches (Figure 27-6). An important advantage of fused silica columns is their minimal tendency to adsorb analyte molecules.

Column Thermostating

Reproducible retention times require control of the column temperature to within a few tenths of a degree. For this reason, the coiled column is ordinarily housed in a thermostated oven. The optimum temperature depends on the boiling points

---

[2] For a detailed description of open-tubular columns, see M. L. Lee, F. J. Yang, and K. D. Bartle, *Open Tubular Column Gas Chromatography*. New York: Wiley, 1984.

Figure 27-3

A set of microsyringes for sample injection. (Courtesy of Chrompack, Inc., Raritan, NJ.)

In 1987 a column manufacturer reported drawing a fused silica column 1300 m in length with over 2 million theoretical plates.

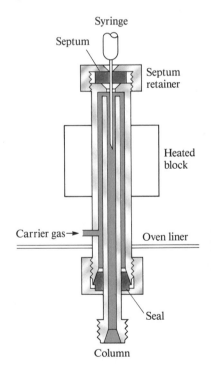

Figure 27-4

A heated sample port. (From H. H. Willard, L. L. Merritt, J. A. Dean, and F. A. Settle, *Instrumental Methods of Analysis*, 7th ed., p. 542. Belmont, CA: Wadsworth, 1988. With permission.)

Figure 27-5

A photomicrograph of a diatom. Magnification 5000×.

> Temperature programming is a technique in which the temperature of a gas-chromatographic column is increased continuously or in steps during elution.

of the sample components. A temperature that is roughly equal to or slightly above the average boiling point of a sample results in a reasonable elution period (2 to 30 min). For samples with a broad boiling range, it may be necessary to employ *temperature programming*, whereby the column temperature is increased either continuously or in steps as the separation proceeds. Figure 27-7c shows the improvement in a chromatogram brought about by temperature programming.

In general, optimum resolution is associated with minimal temperature. Lower temperatures, however, result in longer elution times and hence slower analyses. Figure 27-7 illustrates this principle.

Detectors

Detection devices for gas-liquid chromatography must respond rapidly to minute concentrations of solutes as they exit the column. The solute concentration in the carrier gas at any instant is no more than a few parts per thousand and often is smaller by one or two orders of magnitude. Moreover, the time during which a peak passes the detector is typically one second (or less), which requires that the device be capable of exhibiting its full response during this brief period.

Other desirable properties for a detector include linear response, stability, and uniform response for a wide variety of chemical species or, alternatively, a predictable and selective response toward one or more classes of solutes. No single detector fulfills all of these requirements. Some of the more common detectors are listed in Table 27-1. Two of the most widely used detectors are discussed in the sections that follow.

The Thermal Conductivity Detector. The *thermal conductivity detector*, or *katharometer*, which was one of the earliest detectors for gas chromatography, still finds wide application. This device consists of an electrically heated source whose temperature at constant electric power depends on the thermal conductivity of the surrounding gas. The heated element may be a fine platinum, gold, or tungsten wire (see Figure 27-8) or, alternatively, a small thermistor. The electrical resistance of this element depends on the thermal conductivity of the gas. Twin detectors are ordinarily used, one being located ahead of the sample injection chamber and the other immediately beyond the column; alternatively, the gas stream is split as in Figure 27-1. The detectors are incorporated in two arms of a simple bridge circuit such that the thermal conductivity of the carrier is canceled. In addition, the effects of variations in temperature, pressure, and electric power are minimized. The thermal conductivities of helium and hydrogen are roughly six to ten times greater than those of most organic compounds. Thus, even small amounts of organic species cause relatively large decreases in the thermal conductivity of the column effluent, which results in a marked rise in the temperature of the detector. Detection by thermal conductivity is less satisfactory with carrier gases whose conductivities more closely resemble those of most sample components.

The advantages of the thermal conductivity detector include its simplicity, its large linear dynamic range (about five orders of magnitude), its general response to both organic and inorganic species, and its nondestructive character, which permits collection of solutes after detection. The chief limitation of the katharometer is its relatively low sensitivity ($\sim 10^{-9}$ g/mL carrier). Other detectors exceed this sensitivity by factors of 10^4 to 10^7.

Figure 27-6

A 25-m fused-silica capillary column. (Courtesy of Chrompack, Inc., Raritan, NJ.)

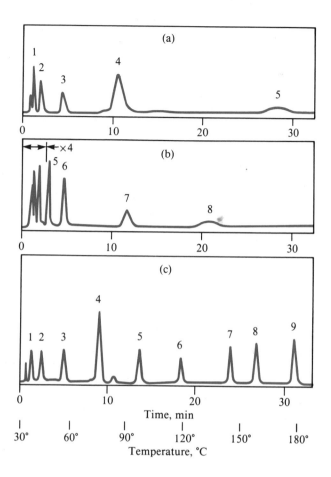

Figure 27-7

Effect of temperature on gas chromatograms. (a) Isothermal at 45°C; (b) isothermal at 145°C; (c) programmed at 30°C to 180°C. (From W. E. Harris and H. W. Habgood, *Programmed Temperature Gas Chromatography*, p. 10. New York: Wiley, 1966. Reprinted with permission.)

The Flame-Ionization Detector. Flame-ionization detectors are perhaps the most widely used of all detectors for gas chromatography. These devices are based on the fact that most organic compounds, when pyrolyzed in a hot flame, produce ionic intermediates that conduct electricity through the flame. Hydrogen is used as the carrier gas with this detector, and the eluent is mixed with oxygen and combusted in a burner equipped with a pair of electrodes (see Figure 27-9). Detection involves monitoring the conductivity of the combustion products. The ionization detector exhibits a high sensitivity ($\sim 10^{-13}$ g/mL), a large linear response, and low noise. It is also rugged and easy to use. A disadvantage of the ionization detector is that it destroys the sample.

Selective Detectors

The effluent from chromatographic columns is often monitored continuously by the selective techniques of spectroscopy or electrochemistry. The resulting so-called *hyphenated methods* (for instance, GC-MS and GC-IR) provide the chemist with powerful tools for identifying the components of complex mixtures.[3]

Hyphenated methods couple the separation capabilities of chromatography with the qualitative and quantitative detection capabilities of electrical or spectral methods.

Table 27-1
SOME GAS
CHROMATOGRAPHY
DETECTORS

Type	Typical Detection Limit
Thermal conductivity	400 pg/mL
Flame ionization	2 pg/s
Electron capture	5 fg/s
Fourier transform IR	0.2 to 40 ng
Mass spectrometer	0.25 to 100 pg

[3] For a review on hyphenated methods, see T. Hirschfield, *Anal. Chem.*, **1980**, *52*, 297A; C. L. Wilkens, *Science*, **1983**, *222*, 251.

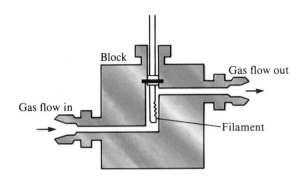

Figure 27-8

A typical thermal conductivity detector. (Courtesy of Varian Instrument Division, Palo Alto, CA.)

Generally, these procedures require large computer memories for storing spectral or electrochemical data for each of the components eluted from the column.

27A-2 Liquid Phases for Gas-Liquid Chromatography

Desirable properties for the immobilized liquid phase in a gas-liquid chromatographic column include: (1) *low volatility* (ideally, the boiling point of the liquid should be at 100°C higher than the maximum operating temperature for the column); (2) *thermal stability*; (3) *chemical inertness*; (4) *solvent characteristics* such that k' and α (Sections 26B-4 and 26B-5) values for the solutes to be resolved fall within a suitable range.

Hundreds of liquids have been proposed as stationary phases in the development of gas-liquid chromatography. By now, only a handful—perhaps a dozen or less—suffice for most applications. The proper choice among these liquids is often crucial to the success of a separation. Qualitative guidelines exist for making this choice, but in the end, the best stationary phase can only be determined in the laboratory.

The retention time for an analyte on a column depends on its partition ratio (Equation 26-1), which in turn is related to the chemical nature of the liquid stationary phase. Clearly, to be useful in gas-liquid chromatography, the immobilized liquid must generate different partition ratios for different sample components. In addition, however, these ratios must not be extremely large or extremely small because the former leads to prohibitively long retention times and the latter results in such short retention times that separations are incomplete.

The polarities of common organic functional groups in increasing order are: aliphatic hydrocarbons < olefins < aromatic hydrocarbons < halides < sulfides < ethers < nitro compounds < esters ≈ aldehydes ≈ ketones < alcohols ≈ amines < sulfones < sulfoxides < amides < carboxylic acids < water.

To have a reasonable residence time in the column, an analyte must show some degree of compatibility (solubility) with the stationary phase. Here, the principle of "like dissolves like" applies, where "like" refers to the polarities of the analyte and the immobilized liquid. Polarity is the electrical field effect in the immediate vicinity of a molecule and is measured by the dipole moment of the species. Polar stationary phases contain functional groups such as —CN, —CO, and —OH. Hydrocarbon-type stationary phases and dialkyl siloxanes are nonpolar, whereas polyester phases are highly polar. Polar analytes include alcohols, acids, and amines; solutes of medium polarity include ethers, ketones, and aldehydes. Saturated hydrocarbons are nonpolar. Generally, the polarity of the stationary phase should match that of the sample components. When the match is good, the order of elution is determined by the boiling point of the eluents.

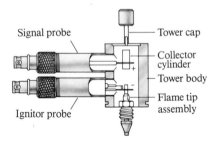

Figure 27-9

A flame-ionization detector. (Courtesy of Varian Instrument Division, Palo Alto, CA.)

Table 27-2 lists the most widely used stationary phases for both packed and open-tubular-column gas chromatography in order of increasing polarity. These six liquids can probably provide satisfactory separations for 90% or more of the samples encountered by the scientist.

Five of the liquids listed in Table 27-2 are polydimethyl siloxanes that have the general structure

$$R - \underset{\underset{R}{|}}{\overset{\overset{R}{|}}{Si}} - O - \left[\underset{\underset{R}{|}}{\overset{\overset{R}{|}}{Si}} - O \right]_n \underset{\underset{R}{|}}{\overset{\overset{R}{|}}{Si}} - R$$

In the first of these, polydimethyl siloxane, the —R groups are all —CH_3, giving a liquid that is relatively nonpolar. In the other polysiloxanes shown in the table, a fraction of the methyl groups are replaced by functional groups such as phenyl (—C_6H_5), cyanopropyl (—C_3H_6CN), and trifluoropropyl (—$C_3H_6CF_3$). The percentage descriptions in each case give the amount of substitution of the named

Table 27-2

SOME COMMON LIQUID STATIONARY PHASES FOR GAS–LIQUID CHROMATOGRAPHY

Stationary Phase	Common Trade Name	Maximum Temperature, °C	Common Applications
Polydimethyl siloxane	OV-1, SE-30	350	General purpose nonpolar phase; hydrocarbons; polynuclear aromatics; drugs; steroids; PCBs
5% Phenyl-polydimethyl siloxane	OV-3, SE-52	350	Fatty acid methyl esters; alkaloids; drugs; halogenated compounds
50% Phenyl-polydimethyl siloxane	OV-17	250	Drugs; steroids; pesticides; glycols
50% Trifluoropropyl-polydimethyl siloxane	OV-210	200	Chlorinated aromatics; nitroaromatic; alkyl substituted benzenes
Polyethylene glycol	Carbowax 20M	250	Free acids; alcohols; ethers; essential oils; glycols
50% Cyanopropyl-polydimethyl siloxane	OV-275	240	Polyunsaturated fatty acids; rosin acids; free acids; alcohols

Figure 27-10

Typical chromatograms from open tubular columns coated with (a) polydimethyl siloxane, (b) 5% phenyl polydimethyl siloxane, (c) 50% phenyl polydimethyl siloxane, (d) 50% trifluoropropylpolydimethyl siloxane, (e) polyethylene glycol, and (f) 50% cyanopropylpolydimethyl siloxane.

group for methyl groups on the polysiloxane backbone. Thus, for example, 5% phenyl-polydimethyl siloxane has a phenyl ring bonded to 5% by number of the silicon atoms in the polymer. These substitutions increase the polarity of the liquids to various degrees.

The fifth entry in Table 27-2 is a polyethylene glycol with the structure

$$HO-CH_2-CH_2-(O-CH_2-CH_2)_n-OH$$

It finds widespread use for separating polar species. Figure 27-10 illustrates typical applications of the phases listed in Table 27-2 for open-tubular columns.

KETONES

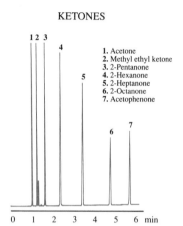

1. Acetone
2. Methyl ethyl ketone
3. 2-Pentanone
4. 2-Hexanone
5. 2-Heptanone
6. 2-Octanone
7. Acetophenone

Column: 30 m × 0.53 mm I.D.
Coating: DB-1, 1.5 μm
Temp: 60°C to 120°C at 10°C/min (FID)
Flow rate: Helium at 26 mL/min

(a)

ALKALOIDS

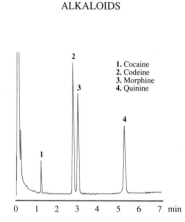

1. Cocaine
2. Codeine
3. Morphine
4. Quinine

Column: 15 m × 0.53 mm I.D.
Coating: DB-5, 1.5 μm
Temp: 200°C to 270°C at 10°C/min (FID)
Flow rate: Helium at 34 mL/min

(b)

STEROIDS

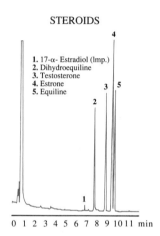

1. 17-α- Estradiol (Imp.)
2. Dihydroequiline
3. Testosterone
4. Estrone
5. Equiline

Column: 30 m × 0.53 mm I.D.
Coating: DB-17, 1.0 μm
Temp: 200°C to 280°C at 10°C/min (FID)
Flow rate: Helium at 12 mL/min

(c)

CHLORINATED AROMATICS

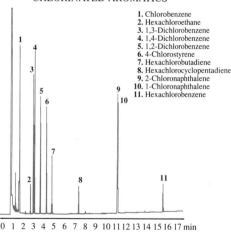

1. Chlorobenzene
2. Hexachloroethane
3. 1,3-Dichlorobenzene
4. 1,4-Dichlorobenzene
5. 1,2-Dichlorobenzene
6. 4-Chlorostyrene
7. Hexachlorobutadiene
8. Hexachlorocyclopentadiene
9. 2-Chloronaphthalene
10. 1-Chloronaphthalene
11. Hexachlorobenzene

Column: 30 m × 0.53 mm I.D.
Coating: DB-210, 1.0 μm
Temp: 85°C to 175°C at 10°C/min (FID)
Flow rate: Helium at 10 mL/min

(d)

BLOOD ALCOHOLS

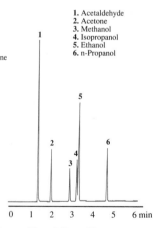

1. Acetaldehyde
2. Acetone
3. Methanol
4. Isopropanol
5. Ethanol
6. n-Propanol

Column: 30 m × 0.53 mm I.D.
Coating: DB-WAX, 1.0 μm
Temp: 40°C (2 min) to 80°C at 10°C/min (FID)
Flow rate: Helium at 7 mL/min

(e)

RAPE SEED OIL

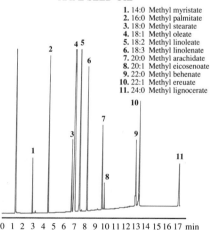

1. 14:0 Methyl myristate
2. 16:0 Methyl palmitate
3. 18:0 Methyl stearate
4. 18:1 Methyl oleate
5. 18:2 Methyl linoleate
6. 18:3 Methyl linolenate
7. 20:0 Methyl arachidate
8. 20:1 Methyl eicosenoate
9. 22:0 Methyl behenate
10. 22:1 Methyl ereuate
11. 24:0 Methyl lignocerate

Column: 30 m × 0.25 mm I.D.
Coating: DB-225, 0.15 μm
Temp: 180°C (2 min) to 220°C at 4°C/min (FID)
Flow rate: Hydrogen at 36.5 cm/sec

(f)

27A-3 Applications of Gas-Liquid Chromatography

Gas-liquid chromatography is applicable to species that are appreciably volatile and thermally stable at temperatures up to a few hundred degrees Celsius. An enormous number of compounds of interest to man possess these qualities. Consequently, gas chromatography has been widely applied to the separation and determination of the components in a variety of sample types. Figure 27-10 shows chromatograms for a few such applications.

27B HIGH-PERFORMANCE LIQUID CHROMATOGRAPHY

Early liquid-chromatographic columns were glass tubes with diameters of perhaps 10 to 50 mm that held 50- to 500-cm lengths of solid particles for the stationary phase. To ensure reasonable flow rates, the particle size of the solid was kept larger than 150 to 200 μm; even then, flow rates were at best a few tenths of a milliliter per minute. Attempts to speed up this classic procedure by application of vacuum or pressure were not effective because increases in flow rates were accompanied by increases in plate heights and accompanying decreases in column efficiency.

Early in the development of liquid chromatography, it was realized that large decreases in plate heights could be expected to accompany decreases in the particle size of packings. It was not until the late 1960s, however, that the technology for producing and using packings with particle diameters as small as 5 to 10 μm was developed. This technology required sophisticated instruments that contrasted markedly with the simple devices that preceded them. The name *high-performance liquid chromatography* (HPLC) is often employed to distinguish these newer procedures from their predecessors, which still find considerable use for preparative purposes.[4]

Figure 27-11 illustrates the improved column efficiency that can be obtained with smaller diameter particles. It is of interest to note that in none of these plots is the minimum shown in Figure 26-9a reached. The reason for this difference is that diffusion in liquids is much slower than in gases; consequently, its effect on plate heights is observed only at extremely low flow rates.

27B-1 Apparatus

Pumping pressures of several hundred atmospheres are required to achieve reasonable flow rates with packings in the 3 to 10 μm size range, which are common in modern liquid chromatography. As a consequence of these high pressures, the equipment for high-performance liquid chromatography tends to be considerably more elaborate and expensive than that encountered in other types of chromatography. Figure 27-12 is a diagram showing the important components of a typical high-performance liquid-chromatographic apparatus.

---

[4] References for HPLC include: L. R. Snyder and J. J. Kirkland, *Introduction to High-Performance Liquid Chromatography*, 2nd ed. New York: Chapman and Hall, 1979; P. Brown and R. A. Hartwick, *High-Performance Liquid Chromatography*. New York: Wiley, 1988; S. Lindsey, *High-Performance Liquid Chromatography*. New York: Wiley, 1987; V. Meyer, *Practical High-Performance Liquid Chromatography*. New York: Wiley, 1988.

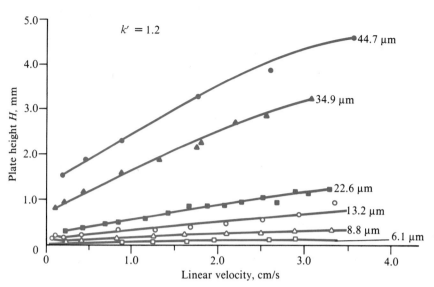

Figure 27-11

Effect of packing particle size and flow rate on plate height in liquid chromatography. Column dimensions: 30 cm × 2.4 mm. Solute: N,N-diethyl-*n*-aminoazobenzene. Mobile phase: mixture of hexane, methylene chloride, isopropyl alcohol. (From R. E. Majors, *J. Chromatogr. Sci.*, **1973**, *11*, 92. With permission.)

Mobile Phase Reservoirs and Solvent Treatment Systems

A modern HPLC apparatus is equipped with one or more glass or stainless steel reservoirs, each of which contains 500 mL or more of a solvent. Provisions are often included to remove dissolved gases and dust from the liquids. The former produce bubbles in the column and thereby cause band spreading; in addition, both bubbles and dust interfere with the performance of detectors. Degassers

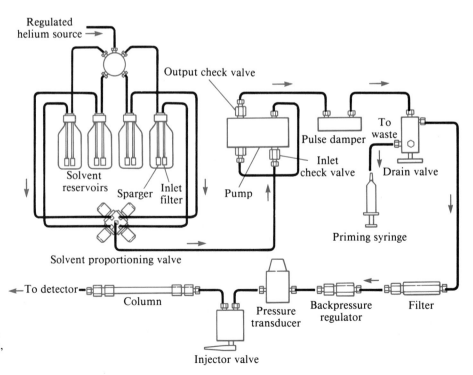

Figure 27-12

Schematic of an apparatus for high-performance liquid chromatography. (Courtesy of Perkin-Elmer, Norwalk, CT.)

may consist of a vacuum pumping system, a distillation system, a device for heating and stirring, or, as shown in Figure 27-12, a system for *sparging*, in which the dissolved gases are swept out of solution by fine bubbles of an inert gas that is not soluble in the mobile phase.

An elution with a single solvent of constant composition is termed *isocratic*. In *gradient elution*, two (and sometimes more) solvent systems that differ significantly in polarity are employed. The ratio of the two solvents is varied in a preprogrammed way, sometimes continuously and sometimes in a series of steps. Gradient elution frequently improves separation efficiency, just as temperature programming helps in gas chromatography. Modern high-performance liquid-chromatographic instruments are often equipped with proportionating valves that introduce liquids from two or more reservoirs at rates that vary continuously (Figure 27-12).

> Sparging is a process in which dissolved gases are swept out of a solvent by bubbles of an inert, insoluble gas.

> An isocratic elution in HPLC is one in which the composition of the solvent remains constant.

> A gradient elution in HPLC is one in which the composition of the solvent is changed continuously or in a series of steps.

Pumping Systems

The requirements for liquid-chromatographic pumps are severe and include: (1) the generation of pressures of up to 6000 psi (lb/in^2); (2) pulse-free output; (3) flow rates ranging from 0.1 to 10 mL/min; (4) flow reproducibilities of 0.5% relative or better; and (5) resistance to corrosion by a variety of solvents.

It should be noted that the high pressures generated by liquid-chromatographic pumps do not constitute an explosion hazard because liquids are not very compressible. Thus, rupture of a component results only in solvent leakage. To be sure, such leakage may constitute a fire hazard.

Sample Injection Systems

The most widely used method of sample introduction in liquid chromatography is based on sampling loops such as that shown in Figure 27-13. These devices are often an integral part of modern liquid chromatography equipment and have interchangeable loops that provide a choice of sample sizes ranging from 5 to 500 μL. The reproducibility of injections with a typical sampling loop is a few tenths of a percent relative.

Liquid Chromatography Columns

Liquid-chromatographic columns are ordinarily constructed from smooth-bore stainless steel tubing, although heavy-walled glass tubing is occasionally encountered. The latter is restricted to pressures that are less than about 600 psi. Hundreds of packing columns differing in size and packing are available from several manufacturers. Costs generally range from $200 to $500.[5]

Analytical Columns

The majority of liquid-chromatographic columns range in length from 10 to 30 cm. Normally, the columns are straight, with added length where needed being gained by coupling two or more columns together. The inside diameter of liquid columns is often 4 to 10 mm; the common particle sizes of packings are 3, 5,

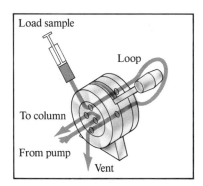

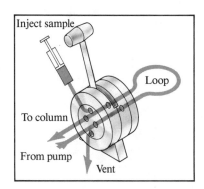

Figure 27-13

A sampling loop for liquid chromatography. (Courtesy of Beckman Instruments, Fullerton, CA.)

---

[5] For descriptions of recent commercially available HPLC columns, see R. E. Majors, *LC-GC*, **1992**, *10*, 188.

Figure 27-14

High-speed isocratic separation. Column dimensions: 4 cm length, 0.4 cm i.d.; Packing: 3 μm sperisorb; Mobile phase: 4.1% ethyl acetate in *n*-hexane. Compounds: (1) *p*-xylene, (2) anisole, (3) benzyl acetate, (4) dioctyl phthalate, (5) dipentyl phthalate, (6) dibutyl phthalate, (7) dipropyl phthalate, (8) diethyl phthalate. (From R. P. W. Scott, *Small Bore Liquid Chromatography Columns: Their Properties and Uses*, p. 156. New York: Wiley, 1984. Reprinted with permission of John Wiley & Sons, Inc.)

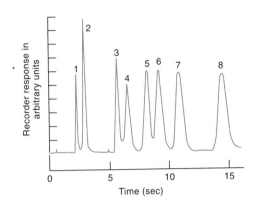

and 10 μm. Perhaps the most common column currently in use is one that is 25 cm in length, 4.6 mm in inside diameter, and packed with 5 μm particles. Columns of this type contain 40,000 to 60,000 plates/meter.

Recently, manufacturers have been producing high-speed, high-performance columns, which have smaller dimensions than those just described.[6] Such columns may have inside diameters that range from 1 to 4.6 mm and may be packed with 3 or 5 μm particles. Often, their lengths are as short as 3 to 7.5 cm. Such columns contain as many as 100,000 plates/meter and have the advantage of speed and minimal solvent consumption. The latter property is of considerable importance because the high-purity solvents required for liquid chromatography are expensive to purchase and to dispose of after use. Figure 27-14 illustrates the speed with which a separation can be performed on this type of column. Here, eight components of diverse type are separated in about 15 s. The column is 4 cm in length and has an inside diameter of 4 mm; it is packed with 3 μm particles.

Guard Columns

Often, a short guard column is introduced before the analytical column to increase the life of the analytical column by removing particulate matter and contaminants from the solvents. In addition, in liquid-liquid chromatography, the guard column serves to saturate the mobile phase with the stationary phase so that losses of this solvent from the analytical column are minimized. The composition of the guard-column packing should be similar to that of the analytical column; the particle size is usually larger, however, to minimize pressure drop.

Column Thermostats

For many applications, close control of column temperature is not necessary and columns are operated at room temperature. Often, however, better chromatograms are obtained by maintaining column temperatures constant to a few tenths of a degree Celsius. Most modern commercial instruments are now equipped with

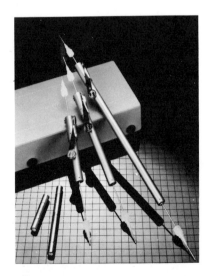

Three HPLC columns. (Courtesy of Chrompak, Inc., Raritan, NJ.)

---

[6] See *Microcolumn High-Performance Liquid Chromatography*, P. Kucera, Ed. New York: Elsevier, 1984; *Small Bore Liquid Chromatography Columns: Their Properties and Uses*, R. P. W. Scott, Ed. New York: Wiley, 1984; M. Novotny, *Anal. Chem.*, **1988**, *60*, 500A.

Table 27-3
PERFORMANCES OF HPLC DETECTORS

HPLC Detector	Commercially Available	Mass LOD (commercial detectors)*	Mass LOD (state of the art)†
Absorbance	Yes‡	100 pg–1 ng	1 pg
Fluorescence	Yes‡	1–10 pg	10 fg
Electrochemical	Yes‡	10 pg–1 ng	100 fg
Refractive index	Yes	100 ng–1 μg	10 ng
Conductivity	Yes	500 pg–1 ng	500 ng
Mass spectrometry	Yes	100 pg–1 ng	1 pg
FT–IR	Yes	1 μg	100 ng
Light scattering	Yes	10 μg	500 ng
Optical activity	No	—	1 ng
Element selective	No	—	10 ng
Photoionization	No	—	1 pg–1 ng

* Mass LOD (limit of detection) is calculated for injected mass that yields a signal equal to five times the σ noise, using a mol wt of 200 g/mol, 10 μL injected for conventional or 1 μL injected for microbore HPLC.
† Same definition as *, but the injected volume is generally smaller.
‡ Commercially available for microbore HPLC also.
(From E. S. Yeung and R. E. Synovec, *Anal. Chem.*, **1986**, *58*, 1238. With permission. Copyright American Chemical Society.)

column heaters that control column temperatures to a few tenths of a degree from near ambient to 150°C. Columns may also be fitted with water jackets fed from a constant temperature bath to give precise temperature control.

Detectors

No highly sensitive, universal detector systems, such as those for gas chromatography, are available for high-performance liquid chromatography. Thus, the system used depends on the nature of the sample. Table 27-3 lists some of the common detectors and their properties.

The most widely used detectors for liquid chromatography are based on absorption of ultraviolet or visible radiation (see Figure 27-15). Both photometers and spectrophotometers, specifically designed for use with chromatographic columns, are available from commercial sources. The former often make use of the 254- and 280-nm lines from a mercury source because many organic functional groups absorb in the region. Deuterium or tungsten filament sources with interference filters also provide a simple means of detecting absorbing species. Some modern instruments are equipped with filter wheels that contain several interference filters, which can be rapidly switched into place. Spectrophotometric detectors are considerably more versatile than photometers and are also widely used in high-performance instruments.

Another detector, which has found considerable application, is based on the changes in the refractive index of the solvent that are caused by analyte molecules. In contrast to most of the other detectors listed in Table 27-3, the refractive index indicator is general rather than selective and responds to the presence of all solutes.

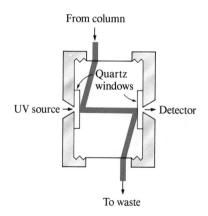

Figure 27-15
A UV detector for HPLC.

27B-2 High-Performance Partition Chromatography

Partition chromatography has become the most widely used of all liquid-chromatographic procedures. This technique can be subdivided into *liquid-liquid* and *liquid-bonded-phase* chromatography. The difference between the two is the method by which the stationary phase is held on the support particles of the packing. With liquid-liquid, retention is by physical adsorption, while with bonded-phase, covalent bonds are involved. Early partition chromatography was exclusively liquid-liquid; now, however, bonded-phase methods predominate, with liquid-liquid separation being relegated to certain special applications.

Bonded-Phase Packings

Most bonded-phase packings are prepared by reaction of an organochlorosilane with the —OH groups formed on the surface of silica particles by hydrolysis in hot, dilute hydrochloric acid. The product is an organosiloxane. The reaction for one such SiOH site on the surface of a particle can be written as

$$
\begin{array}{c}
\quad\quad\quad\quad\quad\quad\quad CH_3 \quad\quad\quad\quad\quad CH_3 \\
\quad\quad\quad\quad\quad\quad\quad | \quad\quad\quad\quad\quad\quad | \\
-\text{Si}-\text{O}-\text{H} + \text{Cl}-\text{Si}-\text{R} \rightarrow -\text{Si}-\text{O}-\text{Si}-\text{R} + \text{HCl} \\
\quad\quad\quad\quad\quad\quad\quad | \quad\quad\quad\quad\quad\quad | \\
\quad\quad\quad\quad\quad\quad\quad CH_3 \quad\quad\quad\quad\quad CH_3
\end{array}
$$

where R is often a straight chain octyl- or octyldecyl-group. Other organic functional groups that have been bonded to silica surfaces include aliphatic amines, ethers, and nitriles, as well as aromatic hydrocarbons. Thus, a variety of polarities for the bonded stationary phase is available.

Bonded-phase packings have the advantge of markedly greater stability than physically held stationary phases. With the latter, periodic recoating of the solid surfaces is required because the stationary phase is gradually dissolved away into the mobile phase. Furthermore, gradient elution is not practical with liquid-liquid packings, again because the immobilized liquid slowly dissolves into the mobile phase. The main disadvantage of bonded-phase packings is their somewhat limited sample capacity.

Normal- and Reversed-Phase Packings

Two types of partition chromatography are encountered. Early work in liquid chromatography was based on highly polar stationary phases such as triethylene glycol or water; a relatively nonpolar solvent such as hexane or *i*-propyl ether then served as the mobile phase. For historic reasons, this type of chromatography is now called *normal-phase chromatography*. In *reversed-phase chromatography*, the stationary phase is nonpolar, often a hydrocarbon, and the mobile phase is a relatively polar solvent (such as water, methanol, or acetonitrile).[7] In normal-phase chromatography, the *least* polar component is eluted first; *increasing the polarity of the mobile phase then decreases the elution time. In contrast, in

> In liquid-liquid partition chromatography, the stationary phase is a solvent that is held in place by adsorption on the surface of packing particles.

> In liquid-bonded-phase partition chromatography, the stationary phase is an organic species that is attached to the surface of the packing particles by chemical bonds.

> In normal-phase partition chromatography, the stationary phase is polar and the mobile phase nonpolar. In reversed-phase partition chromatography, the polarity of these phases is reversed.

---

[7] For a detailed discussion of reversed-phase HPLC, see A. M. Krstulovic and P. R. Brown, *Reversed-Phase High-Performance Liquid Chromatography*. New York: Wiley, 1982.

the reversed-phase method, the *most* polar component elutes first, and *increasing* the mobile phase polarity *increases* the elution time.

It has been estimated that more than three quarters of all HPLC separations are currently performed with reversed-phase, bonded, octyl- or octyldecyl siloxane packings. With such preparations, the long-chain hydrocarbon groups are aligned parallel to one another and perpendicular to the surface of the particle, giving a brushlike, nonpolar, hydrocarbon surface. The mobile phase used with these packings is often an aqueous solution containing various concentrations of such solvents as methanol, acetonitrile, or tetrahydrofuran.

Choice of Mobile and Stationary Phases

Successful partition chromatography requires a proper balance of intermolecular forces among the three participants in the separation process—the solute, the mobile phase, and the stationary phase. These intermolecular forces are described qualitatively in terms of the relative polarity possessed by each of the three reactants (see margin note, page 514 for relative polarities of organic functional groups).

As a rule, most chromatographic separations are achieved by matching the polarity of the analyte to that of the stationary phase; a mobile phase of considerably different polarity is then used. This procedure is generally more successful than one in which the polarities of the analyte and the mobile phase are matched but are different from that of the stationary phase; in this case, the stationary phase often cannot compete successfully for the sample components. Retention times then become too short for practical application. At the other extreme is the situation where the polarities of the analyte and stationary phase are too much alike; in this case, retention times become inordinately long.

Applications

Figures 27-16 and 27-17 illustrate typical applications of bonded-phase partition chromatography. Table 27-4 further illustrates the variety of samples to which the technique is applicable.

27B-3 High-Performance Adsorption Chromatography

All of the pioneering work in chromatography was based on adsorption of analyte species on a solid surface, a method in which the stationary phase is the surface of a finely divided polar solid. With such a packing, the analyte competes with the mobile phase for sites on the surface of the packing, and retention is the result of adsorption forces.

Stationary and Mobile Phases

Finely divided silica and alumina are the only stationary phases that find extensive use for adsorption chromatography. Silica is preferred for most (but not all) applications because of its higher sample capacity and its wider range of useful forms. The adsorption characteristics of the two substances parallel one another. For both, retention times become longer as the polarity of the analyte increases.

In adsorption chromatography, the only variable that affects the partition coefficient of analytes is the composition of the mobile phase (in contrast to partition

> In normal-phase chromatography, the least polar analyte is eluted first. In reversed-phase chromatography, the least polar analyte is eluted last.

> In adsorption chromatography, analyte species are adsorbed onto the surface of a polar packing.

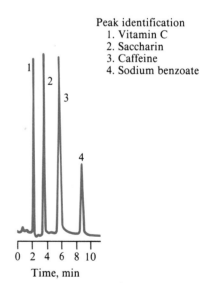

Peak identification
1. Vitamin C
2. Saccharin
3. Caffeine
4. Sodium benzoate

Figure 27-16

Liquid-bonded-phase chromatogram of soft-drink additives. Packing: polar cyano packing. Isocratic elution with 6% HOAc/94% water. (Courtesy of DuPont Instrument Systems, Wilmington, DE.)

Table 27-4
TYPICAL APPLICATIONS OF HIGH-PERFORMANCE
PARTITION CHROMATOGRAPHY

Field	Typical Mixtures
Pharmaceuticals	Antibiotics, sedatives, steroids, analgesics
Biochemicals	Amino acids, proteins, carbohydrates, lipids
Food products	Artificial sweeteners, antioxidants, aflatoxins, additives
Industrial chemicals	Condensed aromatics, surfactants, propellants, dyes
Pollutants	Pesticides, herbicides, phenols, PCBs
Forensic chemistry	Drugs, poisons, blood alcohol, narcotics
Clinical medicine	Bile acids, drug metabolites, urine extracts, estrogens

chromatography, where the polarity of the stationary phase can also be varied). Fortunately, enormous variations in retention and thus resolution accompany variations in the solvent system, and only rarely is a suitable mobile phase unavailable.

Applications of Adsorption Chromatography

Currently, liquid-solid HPLC is used extensively for the separation of relatively nonpolar, water-insoluble organic compounds with molecular weights that are less than about 5000. One unique strength of adsorption chromatography is its ability to resolve isomeric mixtures such as meta and para substituted benzene derivatives.

27B-4 High-Performance Ion-Exchange Chromatography

In Section 25B we described some of the applications of ion-exchange resins to analytical separations. In addition, these materials are useful as stationary phases for liquid chromatography, where they are used to separate charged species.[8] In most cases, conductivity measurements are used for detecting eluents.

Two types of ion chromatography are currently in use: *suppressor-based* and *single-column*. They differ in the method used to prevent the conductivity of the eluting electrolyte from interfering with the measurement of analyte conductivities.

Ion Chromatography Based on Suppressors

Conductivity detectors have many of the properties of the ideal detector. They can be highly sensitive, they are universal for charged species, and as a general rule, they respond in a predictable way to concentration changes. Furthermore, such detectors are simple to operate, inexpensive to construct and maintain, easy to miniaturize, and ordinarily give prolonged, trouble-free service. The only limitation to the use of conductivity detectors, which delayed their general applica-

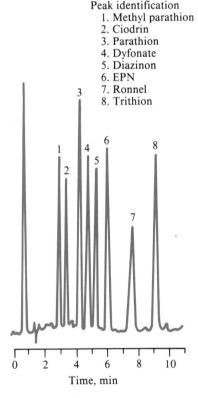

Peak identification
1. Methyl parathion
2. Ciodrin
3. Parathion
4. Dyfonate
5. Diazinon
6. EPN
7. Ronnel
8. Trithion

Figure 27-17
Liquid-bonded-phase chromatogram of organic phosphate insecticides. Packing: nonpolar C_8. Gradient elution: 67% CH_3OH/33% H_2O to 80% CH_3OH/20% H_2O. (Courtesy of IBM Instruments, Inc., Danbury, CT.)

---

[8] For a brief review of ion chromatography, see J. S. Fritz, *Anal. Chem.*, **1987**, *59*, 335A. For a detailed description of the method, see H. Small, *Ion Chromatography*. New York: Plenum Press, 1989; D. T. Gjerde and J. S. Fritz, *Ion Chromatography*, 2nd ed. New York: A. Heuthig, 1987.

tion to ion chromatography until the mid-1970s, was due to the high electrolyte concentrations required to elute most analyte ions in a reasonable time. As a consequence, the conductivity from the mobile-phase components tends to swamp that from the analyte ions, thus greatly reducing the detector sensitivity.

In 1975, the problem created by the high conductance of eluents was solved by the introduction of an *eluent suppressor column* immediately following the ion-exchange column.[9] The suppressor column is packed with a second ion-exchange resin that effectively converts the ions of the eluting solvent to a molecular species of limited ionization without affecting the conductivity due to the analyte ions. For example, when cations are being separated and determined, hydrochloric acid is chosen as the eluting reagent, and the suppressor column is an anion-exchange resin in the hydroxide form. The product of the reaction between the eluent and the suppressor is water:

$$H^+(aq) + Cl^-(aq) + resin^+OH^-(s) \rightarrow resin^+Cl^-(s) + H_2O$$

The analyte cations are not retained by this second column.

For anion separations, the suppressor packing is the acid form of a cation-exchange resin and sodium bicarbonate or carbonate is the eluting agent. The reaction in the suppressor is

$$Na^+(aq) + HCO_3^-(aq) + resin^-H^+(s) \rightarrow resin^-Na^+(s) + H_2CO_3(aq)$$

The largely undissociated carbonic acid does not contribute significantly to the conductivity.

Single-Column Ion Chromatography

Recently, equipment has become available commercially for ion chromatography in which no suppressor column is used. This approach depends on the small differences in conductivity between the eluted sample ions and the prevailing eluent ions. To amplify these differences, low-capacity exchangers that permit elution with dilute eluent solutions are used. Furthermore, eluents of low equivalent conductance are chosen.[10]

Single-column ion chromatography offers the advantage of not requiring special equipment for suppression. However, it is a somewhat less sensitive method for determining anions than suppressor column methods.

Applications

Figures 27-18 and 27-19 illustrate two applications of ion chromatography based on a suppressor column and conductometric detection. In each instance, the ions were present in the parts-per-million range. The method is particularly important for anion analysis.

[9] H. Small, T. S. Stevens, and W. C. Bauman, *Anal. Chem.*, **1975**, *47*, 1801.
[10] See R. M. Becker, *Anal. Chem.*, **1980**, *52*, 1510; J. R. Benson, *Amer. Lab.*, **1985**, (6), 30; T. Jupille, *Amer. Lab.*, **1986**, (5), 114.

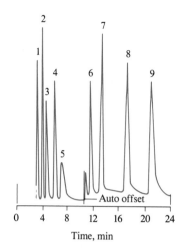

	ppm
1. Li^+	0.5
2. Na^+	2
3. NH_4^+	3
4. K^+	3
5. Morpholine	30
6. Cyclohexylamine	10
7. Mg^{2+}	1
8. Ca^{2+}	2
9. Sr^{2+}	10

Figure 27-18

Ion chromatogram of a mixture of cations. (Courtesy of Dionex, Sunnyvale, CA.)

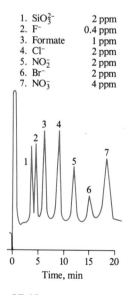

1. SiO_3^{2-}	2 ppm
2. F^-	0.4 ppm
3. Formate	1 ppm
4. Cl^-	2 ppm
5. NO_2^-	2 ppm
6. Br^-	2 ppm
7. NO_3^-	4 ppm

Figure 27-19

Ion chromatogram of a mixture of anions. (Courtesy of Dionex, Sunnyvale, CA.)

Feature 27-1
BUCKYBALLS: THE CHROMATOGRAPHIC SEPARATION
OF FULLERENES

Our ideas about the nature of matter are often profoundly influenced by chance discoveries. No event in recent memory has captured the imagination of both the scientific community and the public as did the serendipitous discovery in 1985 of the soccerball-shaped molecule C_{60}. This molecule, illustrated in Figure 27-A, its cousin C_{70}, and other similar molecules discovered since 1985 are called *fullerenes*, or more commonly, *buckyballs*.[1] The compounds are named in honor of the famous architect, R. Buckminster Fuller, who designed many geodesic dome buildings having the same hexagonal/pentagonal structure as buckyballs. Since their discovery, thousands of research groups throughout the world have studied various chemical and physical properties of these highly stable molecules. They represent a third allotropic form of carbon besides graphite and diamond.

The preparation of buckyballs is almost trivial. When an ac arc is established between two carbon electrodes in a flowing helium atmosphere, the soot that is collected is rich in C_{60} and C_{70}. Although the preparation is easy, the separation and purification of more than a few milligrams of C_{60} proved tedious and expensive. Recently, it was found that relatively large quantities of buckyballs can be separated using size exclusion chromatography.[2] Fuller-

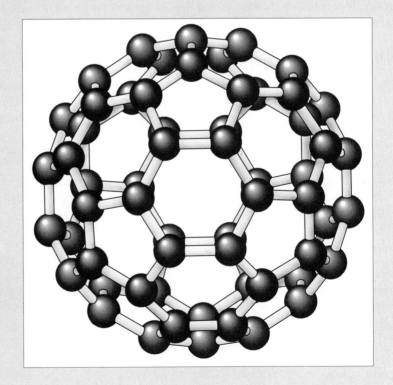

Figure 27-A   Buckminster fullerene, C_{60}.

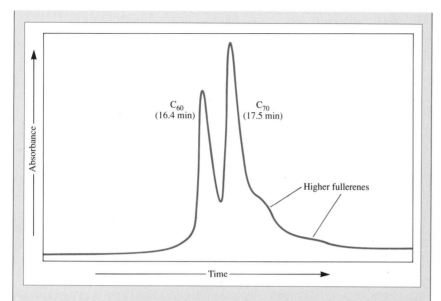

Figure 27-B   Separation of fullerenes.

enes are extracted from soot prepared as mentioned above and injected on a 19 mm × 30 cm 500 Å Ultrastyragel column (Waters Chromatography Division of Millipore Corporation), using toluene as the mobile phase and UV/visible detection following separation. A typical chromatogram is shown in Figure 27-B. The peaks in the chromatogram are labeled with their identities and retention times.

Note that C_{60} elutes before C_{70} and the higher fullerenes. This is contrary to what we expect; the smallest molecule, C_{60}, should be retained more strongly than C_{70} and the higher fullerenes. It has been suggested that the interaction between the solute molecules and the gel is on the surface of the gel rather than in its pores. Since C_{70} and the higher fullerenes have larger surface areas than C_{60}, the higher fullerenes are retained more strongly on the surface of the gel and thus elute after C_{60}. Using automated apparatus, this method of separation may be used to prepare several grams of 99.8% pure C_{60} from 5 to 10 g of C_{60} to C_{70} mixture in a twenty-four hour period. These quantities of C_{60} can then be used to prepare and study the chemistry and physics of derivatives of this interesting and unusual form of carbon.

As an example, C_{60} forms stoichiometric compounds with the alkali metals that have the general formula M_3C_{60}, where M is potassium, rubidium, or cesium. Researchers have found that these compounds are superconductors at temperatures below 30 K. In the future these or similar compounds may help lead the way to practical high-temperature superconductors, which would revolutionize the electrical, electronics, and telecommunications industries and conserve vast amounts of energy.

[1] R. F. Curl and R. E. Smalley, *Scientific American*, **1991**, *265*(4), 54.
[2] M. S. Meier and J. P. Selegue, *J. Org. Chem.*, **1992**, *57*, 1924; A. Gügel and K. Müllen, *J. Chromatogr.*, **1993**, *628*, 23.

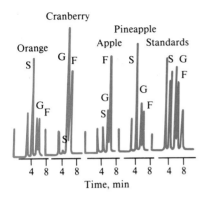

Figure 27-20
Gel-filtration chromatogram for glucose (G), fructose (F), and sucrose (S) in canned juices. (Courtesy of DuPont Instrument Systems, Wilmington, DE.)

In size-exclusion chromatography, fractionation is based on molecular size.

Gel filtration is a type of size-exclusion chromatography in which the packing is hydrophilic. It is used for separating polar species.

Gel permeation is a type of size-exclusion chromatography in which the packing is hydrophobic. It is used to separate nonpolar species.

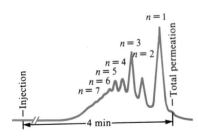

Figure 27-21
Gel-permeation separation of components in an epoxy resin. (Courtesy of DuPont Instrument Systems, Wilmington, DE.)

27B-5 High-Performance Size-Exclusion Chromatography

Size-exclusion, or gel, chromatography is the newest of the liquid-chromatographic procedures. It is a powerful technique that is particularly applicable to high-molecular-weight species.[11]

Packings

Packings for size-exclusion chromatography consist of small ($\sim 10 \ \mu m$) silica or polymer particles containing a network of uniform pores into which solute and solvent molecules can diffuse. While in the pores, molecules are effectively trapped and removed from the flow of the mobile phase. The average residence time of analyte molecules depends on their effective size. Molecules that are significantly larger than the average pore size of the packing are excluded and thus suffer no retention; that is, they travel through the column at the rate of the mobile phase. Molecules that are appreciably smaller than the pores can penetrate throughout the pore maze and are thus entrapped for the greatest time; they are last to be eluted. Between these two extremes are intermediate-size molecules whose average penetration into the pores of the packing depends on their diameters. The fractionation that occurs within this group is directly related to molecular size and, to some extent, molecular shape. Note that size-exclusion separations differ from the other chromatographic procedures in the respect that no chemical or physical interactions between analytes and the stationary phase are involved. Indeed, every effort is made to avoid such interactions because they lead to impaired column efficiencies.

Numerous size-exclusion packings are on the market. Some are hydrophilic for use with aqueous mobile phases; others are hydrophobic and are used with nonpolar organic solvents. Chromatography based on the hydrophilic packings is sometimes called *gel filtration*, while techniques based on hydrophobic packings are termed *gel permeation*. With both types of packings many pore diameters are available. Ordinarily, a given packing will accommodate a 2- to 2.5-decade range of molecular weight. The average molecular weight suitable for a given packing may be as small as a few hundred or as large as several million.

Applications

Figures 27-20 and 27-21 illustrate typical applications of size-exclusion chromatography. In the first, a hydrophilic packing was used to exclude molecular weights greater than 1000. The chromatogram in Figure 27-21 was obtained with a hydrophobic packing in which the eluent was tetrahydrofuran. The sample was a commercial epoxy resin in which each monomer unit had a molecular weight of 280 (n = number of monomer units).

Another important application of size-exclusion chromatography is the rapid determination of the molecular weight or the molecular weight distribution of large polymers or natural products. Here, the elution volumes of the sample are compared with elution volumes for a series of standard compounds that have the same chemical characteristics.

[11] See W. W. Yao, J. J. Kirkland, and D. D. Bly, *Modern Size Exclusion Liquid Chromatography.* New York: Wiley, 1979.

Table 27-5
COMPARISON OF HIGH-PERFORMANCE
LIQUID CHROMATOGRAPHY AND
GAS–LIQUID CHROMATOGRAPHY

Characteristics of both methods

· Efficient, highly selective, widely applicable
· Only small sample required
· May be nondestructive of sample
· Readily adapted to quantitative analysis

Advantages of HPLC

· Can accommodate nonvolatile and thermally unstable samples
· Generally applicable to inorganic ions

Advantages of GLC

· Simple and inexpensive equipment
· Rapid
· Unparalleled resolution (with capillary columns)
· Easily interfaced with mass spectroscopy

27C   COMPARISON OF HIGH-PERFORMANCE LIQUID CHROMATOGRAPHY WITH GAS-LIQUID CHROMATOGRAPHY

Table 27-5 provides a comparison between high-performance liquid chromatography and gas-liquid chromatography. When either is applicable, gas-liquid chromatography offers the advantage of speed and simplicity of equipment. On the other hand, high-performance liquid chromatography is applicable to nonvolatile substances (including inorganic ions) and thermally unstable materials, whereas gas-liquid chromatography is not. Often, the two methods are complementary.

27D   QUESTIONS AND PROBLEMS

27-1. What is a chromatogram?

***27-2.** What are the effects of slow sample injection on a gas chromatogram?

27-3. How do planar and column chromatography differ?

***27-4.** What is meant by temperature programming in gas chromatography?

***27-5.** What is the fundamental difference between N and H as measures of separation efficiencies of columns?

27-6. What variables must be controlled if satisfactory quantitative data are to be obtained from chromatograms?

***27-7.** Describe the physical differences between open tubular and packed columns. What are the advantages and disadvantages of each?

27-8. Define the following terms used in HPLC:

*(a) sparging.
 (b) gradient elution.
*(c) isocratic elution.
 (d) reversed-phase packing.
*(e) stop-flow injection.
 (f) sampling loops.
*(g) bonded-phase packings.
 (h) gel filtration.

27-9. Indicate the order in which the following compounds would be eluted from an HPLC column containing a reversed-phase packing:

*(a) benzene, diethyl ether, n-hexane.
 (b) acetone, dichloroethane, acetamide.

27-10. Indicate the order of elution of the following compounds from a normal-phase packed HPLC column:

*(a) ethyl acetate, acetic acid, dimethylamine.

(b) propylene, hexane, benzene, dichlorobenzene.

*27-11. Describe the fundamental difference between adsorption and partition chromatography.

27-12. Describe the fundamental difference between ion-exchange and size-exclusion chromatography.

*27-13. Describe the difference between gel-filtration and gel-permeation chromatography.

27-14. What types of species can be separated by HPLC but not by GLC?

*27-15. One method for quantitative determination of the concentration of constituents in a sample analyzed by gas chromatography is the *area-normalization method*. Here, complete elution of all of the sample constituents is necessary. The area of each peak is then measured and corrected for differences in detector response to the different eluates. This correction involves dividing the area by an empirically determined correction factor. The concentration of the analyte is found from the ratio of its corrected area to the total corrected area of all peaks. For a chromatogram containing three peaks, the relative areas were found to be 16.4, 45.2, and 30.2 in the order of increasing retention time. Calculate the percentage of each compound if the relative detector responses were 0.60, 0.78, and 0.88, respectively.

27-16. Peak areas and relative detector responses are to be used to determine the concentration of the five species in a sample. The area-normalization method described in 27-15 is to be used. The relative areas for the five gas chromatographic peaks follow. Also shown are the relative responses of the detector. Calculate the percentage of each component in the mixture.

Compound	Relative Peak Area	Relative Detector Response
A	32.5	0.70
B	20.7	0.72
C	60.1	0.75
D	30.2	0.73
E	18.3	0.78

CHAPTER 28

THE CHEMICALS, APPARATUS, AND UNIT OPERATIONS OF ANALYTICAL CHEMISTRY

This chapter is concerned with the practical aspects of the unit operations encountered in an analytical laboratory, as well as with the apparatus and chemicals used in these operations.

28A THE SELECTION AND HANDLING OF REAGENTS AND OTHER CHEMICALS

The purity of reagents has an important bearing on the accuracy attained in any analysis. Therefore, it is essential that the quality of a reagent be consistent with the use for which it is intended.

28A-1 The Classification of Chemicals

Reagent Grade

Reagent-grade chemicals conform to the minimum standards set forth by the Reagent Chemical Committee of the American Chemical Society (ACS)[1] and are used wherever possible in analytical work. Some suppliers label their products with the maximum limits of impurity allowed by the ACS specifications; others print actual assay for the various impurities.

---

[1] Committee on Analytical Reagents, *Reagent Chemicals*, 8th ed. Washington, D.C.: American Chemical Society, 1993.

Primary-Standard Grade

The qualities required of a *primary standard*—in addition to extraordinary purity—are set forth in Section 9A-3. Primary-standard reagents have been carefully analyzed by the supplier, and the assay is printed on the container label. The National Institute of Standards and Technology is an excellent source for primary standards. This agency also provides *reference standards*, which are complex substances that have been exhaustively analyzed.[2]

The National Institute of Standards and Technology (NIST) is the current name of what was formerly the National Bureau of Standards.

Special-Purpose Reagent Chemicals

Chemicals that have been prepared for a specific application are also available. Included among these are solvents for spectrophotometry and high-performance liquid chromatography. Information pertinent to the intended use is supplied with these reagents. Data provided with a spectrophotometric solvent, for example, might include its absorbance at selected wavelengths and its ultraviolet cutoff wavelength.

28A-2 Rules for Handling Reagents and Solutions

High quality in a chemical analysis requires reagents and solutions of established purity. A freshly opened bottle of a reagent-grade chemical ordinarily can be used with confidence; whether this same confidence is justified when the bottle is half empty depends entirely on the way it has been handled after being opened. The following rules should be observed to prevent the accidental contamination of reagents and solutions.

1. Select the best grade of chemical available for analytical work. Whenever possible, pick the smallest bottle that will supply the desired quantity.
2. Replace the top of every container *immediately* after removal of the reagent; do not rely on someone else to do this.
3. Hold the stoppers of reagent bottles between your fingers; never set a stopper on a desk top.
4. *Unless specifically directed otherwise, never return any excess reagent to a bottle.* The money saved by returning excesses is seldom worth the risk of contaminating the entire bottle.
5. Unless directed otherwise, never insert spatulas, spoons, or knives into a bottle that contains a solid chemical. Instead, shake the capped bottle vigorously or tap it gently against a wooden table to break up any encrustation; then pour out the desired quantity. If this does not work, use a clean porcelain spoon.
6. Keep the reagent shelf and the laboratory balance clean and neat. Clean up any spills immediately, even though someone else might be waiting to use the same chemical or reagent.
7. Observe local regulations concerning the disposal of surplus reagents and solutions.[3]

---

[2] United States Department of Commerce, *NIST Standard Reference Materials Catalog, 1992–93, NIST Special Publication 260.* Washington, D.C.: Government Printing Office, 1992.
[3] For recommended methods of disposing of chemical wastes, see *Prudent Practices for Disposal of Chemicals from Laboratories.* Washington, D.C.: National Academy of Science, 1983.

28B   THE MEASUREMENT OF MASS[4]

Highly reliable mass data are ordinarily required at one stage or another in the course of a chemical analysis. An *analytical balance* is used to acquire these data. Less precise but more rugged *laboratory balances* suffice for mass determinations in cases in which demands for reliability are not critical.

28B-1   Types of Analytical Balances

By definition, an *analytical balance* is a weighing instrument with a maximum capacity that ranges from one gram to a few kilograms with a precision of at least one part in 10^5 at maximum capacity. The precision and accuracy of many modern analytical balances exceed one part in 10^6 at full capacity.

The most commonly encountered analytical balances (*macrobalances*) have a maximum capacity ranging between 160 and 200 g; measurements can be made with a standard deviation of ± 0.1 mg. *Semimicroanalytical balances* have a maximum loading of 10 to 30 g with a precision of ± 0.01 mg. A typical *microanalytical balance* has a capacity of 1 to 3 g and has a precision of ± 0.001 mg.

The analytical balance has undergone a dramatic evolution over the past several decades. The traditional, or *equal-arm*, analytical balance had two pans attached to either end of a light-weight beam that pivoted about a knife edge located in the center of the beam. The object to be weighed was placed on one pan; sufficient standard masses were then added to the other pan to restore the beam to its original position. Weighing with such a balance was tedious and time consuming.

The first *single-pan analytical balance* appeared on the market in 1946. The speed and convenience of weighing with this balance were vastly superior to what could be realized with the traditional equal-arm balance. Consequently, this balance rapidly replaced the latter. The single-pan balance is still in use in some laboratories but in turn is being rapidly replaced by the *electronic balance*, which has neither a beam nor a knife edge. The design and operation of the electronic balance are discussed in Section 28B-2. The single-pan balance is described briefly in Section 28B-3.

> An analytical balance has a maximum capacity that ranges from 1 g to several kilograms and a precision at maximum capacity of at least 1 part in 10^5.

> The most common type of analytical balance, a macrobalance, has a maximum load of 160 to 200 g and a precision of 0.1 mg.

> A semimicroanalytical balance has a maximum load of 10 to 30 g and a precision of 0.01 mg.

> A microanalytical balance has a maximum load of 1 to 3 g and a precision of 0.001 mg, or 1 μg.

28B-2   The Electronic Analytical Balance[5]

Figure 28-1 is a diagram that illustrates the principle of an electronic analytical balance. The pan rides above a hollow metal cylinder that is surrounded by a coil and fits over the inner pole of a cylindrical permanent magnet. An electric current in the coil creates a magnetic field that supports or *levitates* the cylinder, the pan and indicator arm, and whatever load is on the pan. The current is adjusted so that the level of the indicator arm is in the null position when the pan is empty. Placing an object on the pan causes the pan and indicator arm to move downward, which increases the amount of light striking the photocell of the null detector. The increased current from the photocell is amplified and fed into the coil, creating a larger magnetic field, which returns the pan to its original null position. A device such as this, in which a small electric current causes a mechani-

---

[4] For the distinction between mass and weight, see Feature 2-1.
[5] For a more detailed discussion, see R. M. Schoonover, *Anal. Chem.*, **1982**, *54*, 973A.

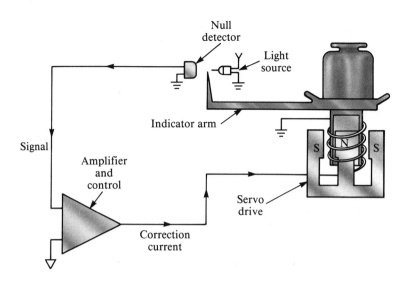

Figure 28-1
Electronic analytical balance. (From R. M. Schoonover, *Anal. Chem.*, **1982**, *54*, 974A. Published 1982 American Chemical Society.)

A servo system is a device in which a small electric signal causes a mechanical system to return to a null position.

A photograph of two electronic balances is shown in color plate 23.

A tare is the mass of an empty sample container. Taring is the process of setting a balance to read zero with the empty container on the pan.

cal system to maintain a null position, is called a *servo system*. The current required to keep the pan and object in the null position is directly proportional to the mass of the object and is readily measured, digitized, and displayed. The calibration of an electronic balance involves the use of a standard mass and adjustment of the current so that the mass of the standard is exhibited on the display. Most electronic balances are equipped with a built-in standard weight, thus making periodic recalibration simple and convenient. With some, periodic recalibration is performed automatically.

Figure 28-2 shows the configurations for two electronic analytical balances. In each, the pan is tethered to a system of constraints known collectively as a *cell*. The cell incorporates several *flexures* that permit limited movement of the pan and prevent torsional forces (resulting from off-center loading) from disturbing the alignment of the balance mechanism. At null, the beam is parallel to the gravitational horizon and each flexure pivot is in a relaxed position.

Figure 28-2a shows an electronic balance with the pan located below the cell. Higher precision is achieved with this arrangement than with the top-loading design shown in Figure 28-2b. Even so, top-loading electronic balances have a precision that equals or exceeds that of the best mechanical balances and additionally provide free access to the pan. Protection from air currents is needed to permit discrimination between small differences in mass (<1 mg). An analytical balance is thus always enclosed in a case equipped with doors to permit the introduction or removal of objects (see Figure 28-2a).

Electronic balances generally feature an automatic *taring control* that causes the display to read zero with a container (such as a boat or weighing bottle) on the pan. Most balances permit taring to 100% of capacity.

Some electronic balances have dual capacities and dual precisions. These features permit the capacity to be decreased from that of a macrobalance to that of a semimicrobalance (30 g) with a concomitant gain in precision to 0.01 mg. Thus, we have two balances in one.

A modern electronic analytical balance provides unprecedented speed and ease of use. For example, one instrument is controlled by touching a single bar at various positions along its length. One position on the bar turns the instrument

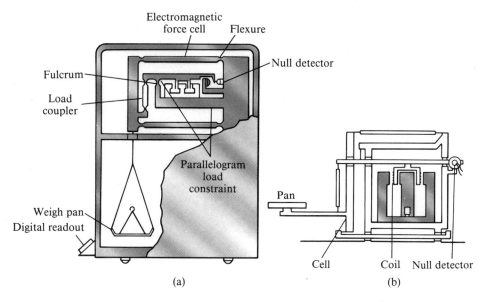

(a)   (b)

Figure 28-2
Electronic analytical balances. (a) Classical configuration with pan beneath the cell. (From R. M. Schoonover, *Anal. Chem.*, **1982**, *54*, 976A. Published 1982 American Chemical Society.) (b) A top-loading design. Note that the mechanism is enclosed in a windowed case. (Reprinted with permission *Amer. Lab.*, **1983**, *15*(3), 72. Copyright 1983 by International Scientific Communications, Inc.)

on or off, another automatically calibrates the balance against a standard weight, and a third zeros the display, either with or without an object on the pan. Reliable mass data are obtainable with little or no instruction or practice.

28B-3 The Single-Pan Mechanical Analytical Balance

Components

Although they differ considerably in appearance and performance characteristics, all mechanical balances—equal-arm as well as single-pan—have several common components. Figure 28-3 is a diagram of a typical single-pan balance. Fundamental to this instrument is a light-weight *beam* that is supported on a planar surface by a prism-shaped knife edge (A). Attached to the left end of the beam is a pan for holding the object to be weighed and a full set of weights held in place by hangers. These weights can be lifted from the beam one at a time by a mechanical arrangement that is controlled by a set of knobs on the exterior of the balance case. The right end of the beam holds a counterweight of such a size as to just balance the pan and weights on the left end of the beam.

A second knife edge (B) is located near the left end of the beam and serves to support a second planar surface, which is located in the inner side of a *stirrup* that couples the pan to the beam. The two knife edges and their planar surfaces are fabricated from extraordinarily hard materials (agate or synthetic sapphire) and form two bearings that permit motion of the beam and pan with a minimum of friction. The performance of a mechanical balance is critically dependent on the perfection of these two bearings.

Single-pan balances are also equipped with a *beam arrest* and a *pan arrest*. The beam arrest is a mechanical device that raises the beam so that the central knife edge no longer touches its bearing surface and simultaneously frees the stirrup from contact with the outer knife edge. The purpose of both arrest mechanisms is to prevent damage to the bearings while objects are being placed on or

To avoid damage to the knife edges and bearing surfaces, the arrest system for a mechanical balance should be engaged at all times other than during actual weighing.

Agate is a fine-grained, milky or grayish quartz. It is extremely hard and can be polished to a sharp edge or a flat surface.

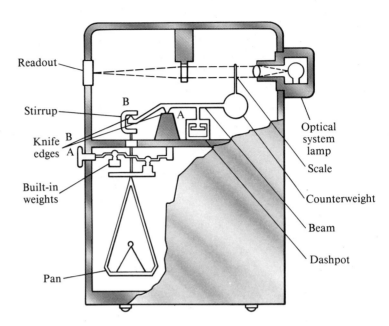

Figure 28-3
Modern single-pan analytical balance.
(From R. M. Schoonover, *Anal.
Chem.*, **1982**, *54*, 973A. Published
1982 American Chemical Society.)

removed from the pan. When engaged, the pan arrest supports most of the mass
of the pan and its contents and thus prevents oscillation. Both arrests are controlled
by a lever mounted on the outside of the balance case and should be engaged
whenever the balance is not in use.

An *air damper* (also known as a *dashpot*) is mounted near the end of the
beam opposite the pan. This device consists of a piston that moves within a
concentric cylinder attached to the balance case. Air in the cylinder undergoes
expansion and contraction as the beam is set in motion; the beam rapidly comes
to rest as a result of this opposition to motion.

Weighing with a Single-Pan Balance

The beam of a properly adjusted balance assumes an essentially horizontal position with no object on the pan and all of the weights in place. When the pan
and beam arrests are disengaged, the beam is free to rotate around the knife
edge. Placing an object on the pan causes the left end of the beam to move
downward. Weights are then removed systematically one by one from the beam
until the imbalance is less than 100 mg. The angle of deflection of the beam with
respect to its original horizontal position is directly proportional to the mass in
milligrams that must be removed to restore the beam to its original horizontal
position. The optical system shown in the upper part of Figure 28-3 measures
the angle of deflection and converts this angle to milligrams. A *reticle*, which is
a small transparent screen mounted on the beam, is scribed with a scale that
reads 0 to 100 mg. A beam of light passes through the scale to an enlarging lens,
which in turn focuses a small part of the enlarged scale onto a frosted glass plate
located on the front of the balance. A vernier makes it possible to read this scale
to the nearest 0.1 mg.

28B-4 Precautions in Using an Analytical Balance

An analytical balance is a delicate instrument that you must handle with care. Consult with your instructor for detailed directions for weighing with your particular model of balance. Observe the following general rules for working with an analytical balance regardless of the make and model.

1. Center the load on the pan as well as possible.
2. Protect the balance from corrosion. Objects to be placed on the pan should be limited to nonreactive metals, nonreactive plastics, and vitreous materials.
3. Observe special precautions (Section 28C-6) for the weighing of liquids.
4. Consult with the instructor if the balance appears to need adjustment.
5. Keep the balance and its case scrupulously clean. A camel's-hair brush is useful for the removal of spilled material or dust.
6. Always allow an object that has been heated to return to room temperature before attempting to weigh it.
7. Use tongs or finger pads to prevent the uptake of moisture by dried objects.
8. For single-pan balances, be certain that you have engaged the arresting mechanisms for the beam whenever the loading is being changed and whenever the balance is not in use.

28B-5 Sources of Error in Weighing

Three common sources of error are encountered in weighing with an analytical balance: (1) buoyancy difference between the object and the weight used for calibration, (2) temperature difference between the object and the ambient air, and (3) static electricity on the object.

Buoyancy Effects[6]

A *buoyancy error* will affect data if the density of the object being weighed differs significantly from that of the standard weight or weights used for calibration of an electronic balance or the set of weights used in a single-pan balance. This error has its origin in the difference in the buoyant force exerted by the medium (air) on the object and on the standard weight used for calibration. Correction for buoyancy is accomplished by means of the equation

> A buoyancy error is the weighing error that develops when the object being weighed has a significantly different density from the weight used to calibrate an electronic balance or the set of weights employed in a single-pan balance.

$$M_1 = M_2 + M_2 \left(\frac{d_{air}}{d_{obj}} - \frac{d_{air}}{d_{wt}} \right) \qquad \textbf{(28-1)}$$

where M_1 is the corrected mass of the object, M_2 is the mass of the standard weight or weights, d_{obj} is the density of the object, d_{wt} is the density of the weight or weights, and d_{air} is the density of the air displaced by them; d_{air} has a value of $0.0012 \ \text{g} \cdot \text{cm}^{-3}$.

The consequences of Equation 28-1 are shown in Figure 28-4, in which the relative error due to buoyancy is plotted against the density of objects weighed in air with a balance that has been calibrated with a weight having a density of

---

[6] For further information, see R. Battino and A. G. Williamson, *J. Chem. Educ.*, **1984**, *64*, 51.

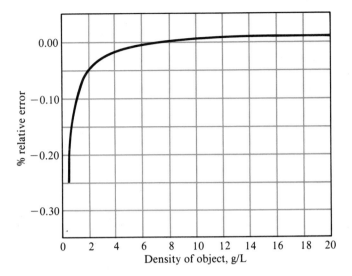

Figure 28-4

Effect of buoyancy on weighing data (density of weights = 8 g·cm⁻³). Plot of relative error as a function of the density of the object weighed.

8 g·cm⁻³. Note that this error is less than 0.1% for objects that have a density of 2 g·cm⁻³ or greater. It is thus seldom necessary to apply a correction to the mass of most solids. The same cannot be said for low-density solids, liquids, or gases, however. For these, the effects of buoyancy are significant and you must apply a correction.

The density of weights used for calibration ranges from 7.8 to 8.4 g · cm⁻³, depending on the manufacturer. Use of 8 g · cm⁻³ is adequate for most purposes. If a greater accuracy is required, the specifications for the balance to be used should be consulted for the necessary density data.

Example 28-1

A bottle had a mass of 7.6504 g empty and 9.9716 g after introduction of an organic liquid with a density of 0.92 g · cm⁻³. The balance was calibrated with a stainless steel weight (d = 8.0 g · cm⁻³). Correct the mass of the sample for the effects of buoyancy.

The apparent mass of the liquid is 9.9716 − 7.6504 = 2.3212 g. The same buoyant force acts on the container during both weighings; thus, we need to consider only the force that acts on the 2.3212 g of liquid. Substitution of 0.0012 g·cm⁻³ for d_{air}, 0.92 g·cm⁻³ for d_{obj}, and 8.0 g·cm⁻³ for d_{wt} in Equation 28-1 gives

$$M_1 = 2.3212 + 2.3212 \left(\frac{0.0012}{0.92} - \frac{0.0012}{8.0} \right) = 2.3239 \text{ g}$$

Failure to correct for buoyancy leads to a relative error of

$$\text{Rel error} = \frac{2.3212 - 2.3239}{2.3239} \times 100\% = -0.12\%$$

Temperature Effects

Attempts to weigh an object whose temperature is different from that of its surroundings will result in a significant error. Failure to allow sufficient time for a heated object to return to room temperature is the most common source of this problem. Errors due to a difference in temperature have two sources. First, convection currents within the balance case exert a buoyant effect on the pan and object. Second, warm air trapped in a closed container weighs less than the same volume at a lower temperature. Both effects cause the apparent mass of the object to be low. This error can amount to as much as 10 or 15 mg for a typical porcelain filtering crucible or a weighing bottle (Figure 28-5). Therefore, heated objects must always be cooled to room temperature before being weighed.

Always allow heated objects to return to room temperature before you attempt to weigh them.

Effect of Static Electricity

A porcelain or glass object will occasionally acquire a static charge that is sufficient to cause a balance to perform erratically; this problem is particularly serious when the relative humidity is low. Spontaneous discharge frequently occurs after a short period. A low-level source of radioactivity (such as a photographer's brush) in the balance case will provide sufficient ions to relieve the charge. Alternatively, the object can be wiped with a faintly damp chamois.

28B-6 Auxiliary Balances

Balances that are less precise than analytical balances find extensive use in the analytical laboratory. These offer the advantages of speed, ruggedness, large capacity, and convenience. They should be used whenever high sensitivity is not required.

Top-loading auxiliary balances are particularly convenient. A sensitive top-loading balance will accommodate 150 to 200 g with a precision of about 1 mg—an order of magnitude less than a macroanalytical balance. Some balances of this type tolerate loads as great as 25,000 g with a precision of ±0.05 g. Most

Use auxiliary laboratory balances for weighings that do not require great accuracy.

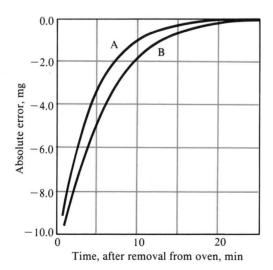

Figure 28-5

Effect of temperature on weighing data. Absolute error in weight as a function of time after the object was removed from a 110°C drying oven. A: porcelain filtering crucible. B: weighing bottle containing about 7.5 g of KCl.

are equipped with a taring device that brings the balance reading to zero with an empty container on the pan. Some are fully automatic, require no manual dialing or weight handling, and provide a digital readout of the mass. Modern top-loading balances are electronic.

A triple-beam balance with a sensitivity less than that of a typical top-loading auxiliary balance is also useful. This is a single-pan balance with three decades of weights that slide along individual calibrated scales. The precision of a triple-beam balance may be one or two orders of magnitude less than that of a top-loading instrument but is adequate for many weighing operations. This type of balance offers the advantages of simplicity, durability, and low cost.

28C   THE EQUIPMENT AND MANIPULATIONS ASSOCIATED WITH WEIGHING

The mass of many solids changes with humidity, owing to their tendency to absorb weighable amounts of moisture. This effect is especially pronounced when a large surface area is exposed, as with a reagent chemical or a sample that has been ground to a fine powder. The first step in a typical analysis, then, involves drying the sample so that the results will not be affected by the humidity of the surrounding atmosphere.

A sample, a precipitate, or a container is brought to *constant mass* by a cycle that involves heating (ordinarily for one hour or more) at an appropriate temperature, cooling, and weighing. This cycle is repeated as many times as needed to obtain successive masses that agree within 0.2 to 0.3 mg of one another. The establishment of constant mass provides some assurance that the chemical or physical processes that occur during the heating (or ignition) are complete.

> Drying or ignition to constant mass is a process in which a solid is cycled through heating, cooling, and weighing steps until its mass becomes constant to within 0.2 to 0.3 mg.

28C-1   Weighing Bottles

Solids are conveniently dried and stored in *weighing bottles*, two common varieties of which are shown in Figure 28-6. The ground-glass portion of the cap-style bottle shown on the left is on the outside and does not come into contact with the contents; this design eliminates the possibility of some of the sample becoming entrained on and subsequently lost from the ground-glass surface.

Plastic weighing bottles are available, and offer the advantage of being more rugged than glass.

28C-2   Desiccators and Desiccants

Oven drying is the most common way of removing moisture from solids. This approach is not appropriate for substances that decompose or for those from which water is not removed at the temperature of the oven.

> A desiccator is a device for drying substances or objects.

Dried materials are stored in *desiccators* while they cool in order to minimize the uptake of moisture. Figure 28-7 shows the components of a typical desiccator. The base section contains a chemical drying agent, such as anhydrous calcium chloride, calcium sulfate (Drierite), anhydrous magnesium perchlorate (Anhydrone or Dehydrite), or phosphorus pentoxide. The ground-glass surfaces are lightly coated with grease.

When you remove or replace the lid of a desiccator use a sliding motion to minimize the likelihood of disturbing the sample. An airtight seal is achieved by slight rotation and downward pressure on the positioned lid.

Figure 28-6
Typical weighing bottles.

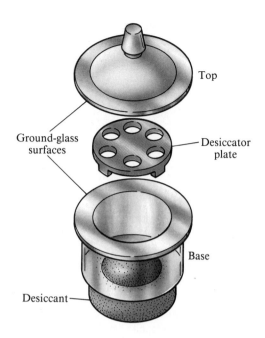

Figure 28-7
Components of a typical desiccator.

When you place a heated object in a desiccator, the increase in pressure as the enclosed air is warmed may be sufficient to break the seal between lid and base. Conversely, if the seal is not broken, the cooling of heated objects can cause development of a partial vacuum. Both of these conditions can cause the contents of the desiccator to be physically lost or contaminated. Although it defeats the purpose of the desiccator somewhat, you should allow some cooling to occur before the lid is seated. It is also helpful to break the seal once or twice during cooling to relieve any excessive vacuum that develops. Finally, you should lock the lid in place with your thumbs as you move the desiccator from one place to another.

Very hygroscopic materials should be stored in containers equipped with snug covers, such as weighing bottles; the covers remain in place during storage in the desiccator. Most other solids can be safely stored uncovered.

28C-3 The Manipulation of Weighing Bottles

Heating at 105°C to 110°C is sufficient to remove the moisture from the surface of most solids. Figure 28-8 depicts the arrangement recommended for drying a sample. The weighing bottle is contained in a labeled beaker with a ribbed cover glass. This arrangement protects the sample from accidental contamination and also allows for the free access of air. Crucibles containing a precipitate that can be freed of moisture by simple drying can be treated in the same way. The beaker containing the weighing bottle or crucible to be dried must be carefully marked to permit identification.

You should avoid handling a dried object with your fingers because detectable amounts of water or oil from your skin may be transferred to the object. This problem is avoided by using tongs, chamois finger cots, clean cotton gloves, or strips of paper to handle dried objects for weighing. Figure 28-9 shows how a weighing bottle is manipulated with a strip of paper.

Figure 28-8
Arrangement for the drying of samples.

Figure 28-9
Method for quantitative transfer of a solid sample. Note the use of paper strips to avoid contact between glass and skin.

28C-4 Weighing by Difference

Weighing by difference is a simple method for determining a series of sample masses. First the bottle and its contents are weighed. One sample is then transferred from the bottle to a container; gentle tapping of the bottle with its top and slight rotation of the bottle provide control over the amount of sample transferred. The bottle and its residual contents are then weighed. The mass of the sample is the difference between the two masses. It is essential that all the solid removed from the weighing bottle be transferred without loss to the container.

28C-5 Weighing Hygroscopic Solids

Hygroscopic substances rapidly absorb moisture from the atmosphere and therefore require special handling. You need a weighing bottle for each sample to be weighed. Place the approximate amount of sample needed in the individual bottles and heat for an appropriate time. When heating is complete, quickly cap the bottles and cool in a desiccator. Weigh one of the bottles after opening it momentarily to relieve any vacuum. Quickly empty the contents of the bottle into its receiving vessel, cap immediately, and weigh the bottle again (along with any solid that did not get transferred). Repeat for each sample and determine the sample masses by difference.

28C-6 Weighing Liquids

The mass of a liquid is always obtained by difference. Liquids that are noncorrosive and relatively nonvolatile can be transferred to previously weighed containers with snugly fitting covers (such as weighing bottles); the mass of the container is subtracted from the total mass.

A volatile or corrosive liquid should be sealed in a weighed glass ampoule. The ampoule is first heated, and then the neck is immersed in the sample; as cooling occurs, the liquid is drawn into the bulb. The ampoule is then inverted and the neck sealed off with a small flame. The ampoule and its contents, along with any glass removed during sealing, are cooled to room temperature and weighed. Finally, the ampoule is transferred to an appropriate container and broken. A volume correction for the glass of the ampoule may be needed if the receiving vessel is a volumetric flask.

28D THE CLEANING AND MARKING OF LABORATORY WARE

A chemical analysis is ordinarily performed in duplicate or triplicate. Thus, each vessel that holds a sample must be marked so that its contents can be positively identified. Flasks, beakers, and some crucibles have small etched areas on which semipermanent markings can be made with a pencil.

Special marking inks are available for porcelain surfaces. The marking is baked permanently into the glaze by heating at a high temperature. A saturated solution of iron(III) chloride, while not as satisfactory as the commercial preparation, can also be used for marking.

Every beaker, flask, or crucible that will contain the sample must be thoroughly cleaned before being used. The apparatus should be washed with a hot detergent solution and then rinsed—initially with large amounts of tap water and finally with several small portions of deionized water.[7] Properly cleaned glassware will be coated with a uniform and unbroken film of water. *It is seldom necessary to dry the interior surface of glassware before use*; drying is ordinarily a waste of time at best and a potential source of contamination at worst.

Figure 28-10
Arrangement for the evaporation of a liquid.

28E   THE EVAPORATION OF LIQUIDS

It is frequently necessary to decrease the volume of a solution that contains a nonvolatile solute. Figure 28-10 illustrates how this operation is performed. The ribbed cover glass permits vapors to escape and protects the remaining solution from accidental contamination. Less satisfactory is the use of glass hooks to provide space between the rim of the beaker and a conventional cover glass.

Evaporation is often difficult to control, due to the tendency of some solutions to overheat locally. The bumping that results can be sufficiently vigorous to cause partial loss of the solution. Careful and gentle heating will minimize the danger of such loss. Where their use is permissible, glass beads will also minimize bumping.

> Bumping is sudden, often violent boiling that tends to spatter solution out of its container.

Some unwanted species can be eliminated during evaporation. For example, chloride and nitrate can be removed from a solution by adding sulfuric acid and evaporating until copious white fumes of sulfur trioxide are observed (this operation must be performed in a hood). Urea is effective in removing nitrate ion and nitrogen oxides from acidic solutions. The removal of ammonium chloride is best accomplished by adding concentrated nitric acid and evaporating the solution to a small volume. Ammonium ion is rapidly oxidized on heating; the solution is then evaporated to dryness.

> Unless you are directed otherwise, do not dry the interior surfaces of glassware or porcelain ware.

Organic constituents can frequently be eliminated from a solution by adding sulfuric acid and heating to the appearance of sulfur trioxide fumes (hood); this process is known as *wet-ashing*. Nitric acid can be added toward the end of heating to hasten oxidation of the last traces of organic matter.

> Wet-ashing is the oxidation of the organic constituents of a sample with oxidizing reagents such as nitric acid, sulfuric acid, hydrogen peroxide, aqueous bromine, or a combination of these reagents.

28F   THE EQUIPMENT AND MANIPULATIONS FOR FILTRATION AND IGNITION

28F-1   Apparatus

Simple Crucibles

Simple crucibles serve only as containers. Porcelain, aluminum oxide, silica, and platinum crucibles maintain constant mass—within the limits of experimental error—and are used principally to convert a precipitate into a suitable weighing form. The solid is first collected on a filter paper. The filter and contents are then transferred to a weighed crucible, and the paper is ignited.

Simple crucibles of nickel, iron, silver, zirconium, and gold are used as containers for the high-temperature fusion of samples that are not soluble in aqueous reagents. Attack by both the atmosphere and the contents may cause these cruci-

---

[7] References to deionized water in this chapter and the one that follows apply equally to distilled water.

bles to suffer mass changes. Moreover, such attacks will contaminate the sample with species derived from the crucible. We select the crucible whose products will offer the least interference in subsequent steps of the analysis.

Filtering Crucibles

Filtering crucibles serve not only as containers but also as filters. A vacuum is used to hasten the filtration; a tight seal between crucible and filtering flask is accomplished with any of several types of rubber adaptors (see Figure 28-11; a complete filtration train is shown in Figure 28-16). Collection of a precipitate with a filtering crucible is frequently less time-consuming than with paper.

Sintered-glass (also called *fritted-glass*) crucibles are manufactured in fine, medium, and coarse porosities (marked *f*, *m*, and *c*). The upper temperature limit for a sintered-glass crucible is ordinarily about 250°C. Porcelain filtering crucibles can tolerate substantially higher temperatures without damage.

A *Gooch crucible* has a perforated bottom that supports a fibrous mat. Asbestos was at one time the filtering medium of choice for a Gooch crucible; current regulations concerning this material have virtually eliminated its use. Small circles of glass matting have now replaced asbestos; they are used in pairs to protect against disintegration during the filtration. Glass mats can tolerate temperatures in excess of 500°C.

Filter Paper

Paper is an important filtering medium. Ashless paper is manufactured from cellulose fibers that have been treated with hydrochloric and hydrofluoric acids to remove metallic impurities and silica; ammonia is then used to neutralize the acids. The residual ammonium salts in many filter papers may be sufficient to affect the analysis for nitrogen by the Kjeldahl method (Section 29C-11).

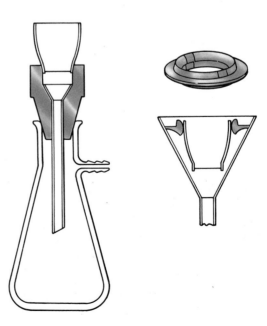

Figure 28-11
Adaptors for filtering crucibles.

All papers tend to pick up moisture from the atmosphere, and ashless paper is no exception. It is thus necessary to destroy the paper by ignition if the precipitate collected on it is to be weighed. Typically, 9- or 11-cm circles of ashless paper leave a residue that has a mass less than 0.1 mg, an amount that is ordinarily negligible. Ashless paper can be obtained in several porosities.

Gelatinous precipitates, such as hydrous iron(III) oxide, clog the pores of any filtering medium. A coarse-porosity ashless paper is most effective for the filtration of such solids, but even here clogging occurs. This problem can be minimized by mixing a dispersion of ashless filter paper with the precipitate prior to filtration. Filter paper pulp is available in tablet form from chemical suppliers; if necessary, the pulp can be prepared by treating a piece of ashless paper with concentrated hydrochloric acid and washing the disintegrated mass free of acid.

Table 28-1 summarizes the characteristics of common filtering media. None satisfies all requirements.

Heating Equipment

Many precipitates can be weighed directly after being brought to constant mass in a low-temperature drying oven. Such an oven is electrically heated and capable of maintaining a constant temperature to within 1°C (or better). The maximum attainable temperature ranges from 140°C to 260°C, depending on make and model; for many precipitates, 110°C is a satisfactory drying temperature. The efficiency of a drying oven is greatly increased by the forced circulation of air. The passage of predried air through an oven designed to operate under a partial vacuum represents an additional improvement.

Microwave laboratory ovens are currently available on the market. Where applicable, these greatly shorten drying cycles. For example, slurry samples that

Table 28-1
COMPARISON OF FILTERING MEDIA FOR GRAVIMETRIC ANALYSES

Characteristic	Paper	Gooch Crucible, Glass Mat	Glass Crucible	Porcelain Crucible	Aluminum Oxide Crucible
Speed of filtration	Slow	Rapid	Rapid	Rapid	Rapid
Convenience and ease of preparation	Troublesome, inconvenient	Convenient	Convenient	Convenient	Convenient
Maximum ignition temperature, °C	None	>500	200–500	1100	1450
Chemical reactivity	Carbon has reducing properties	Inert	Inert	Inert	Inert
Porosity	Many available	Several available	Several available	Several available	Several available
Convenience with gelatinous precipitates	Satisfactory	Unsuitable; filter tends to clog	Unsuitable; filter tends to clog	Unsuitable; filter tends to clog	Unsuitable; filter tends to clog
Cost	Low	Low	High	High	High

require 12 to 16 hours for drying in a conventional oven are reported to be dried within 5 to 6 min in a microwave oven.[8]

An ordinary heat lamp can be used to dry a precipitate that has been collected on ashless paper and to char the paper as well. The process is conveniently completed by ignition at an elevated temperature in a muffle furnace.

Burners are convenient sources of intense heat. The maximum attainable temperature depends on the design of the burner and the combustion properties of the fuel. Of the three common laboratory burners, the Meker provides the highest temperatures, followed by the Tirrill and Bunsen types.

A heavy-duty electric furnace (*muffle furnace*) is capable of maintaining controlled temperatures of 1100°C or higher. Long-handled tongs and heat-resistant gloves are needed for protection when transferring objects to or from such a furnace.

28F-2 The Manipulations Associated with Filtration and Ignition

Preparation of Crucibles

A crucible used to convert a precipitate to a form suitable for weighing must maintain—within the limits of experimental error—a constant mass throughout the drying or ignition. The crucible is first cleaned thoroughly (filtering crucibles are conveniently cleaned by backwashing on a filtration train) and subjected to the same regimen of heating and cooling as will be needed for the precipitate. This process is repeated until constant mass has been achieved, that is, until consecutive weighings differ by 0.3 mg or less.

> Backwashing a filtering crucible is done by turning the crucible upside down in the adaptor (Figure 28-11) and sucking water through the inverted crucible.

Filtration and Washing of Precipitates

The steps involved in filtering an analytical precipitate are *decantation*, *washing*, and *transfer*. In decantation, as much supernatant liquid as possible is passed through the filter while the precipitated solid is kept essentially undisturbed in the beaker where it was formed. This procedure speeds the overall filtration rate by delaying the time at which the pores of the filtering medium become clogged with precipitate. A stirring rod is used to direct the flow of decantate (Figure 28-12). When flow ceases, the drop of liquid at the end of the pouring spout is collected with the stirring rod and returned to the beaker. Wash liquid is next added to the beaker and thoroughly mixed with precipitate. The solid is allowed to settle, following which this liquid is also decanted through the filter. Several such washings may be required, depending on the precipitate. Most washing should be carried out *before* the solid is transferred; this results in a more thoroughly washed precipitate and a more rapid filtration.

> Decantation is the process of pouring a liquid gently so as to not disturb a solid in the bottom of the container.

The transfer process is illustrated in Figures 28-12c and d. The bulk of the precipitate is moved from beaker to filter by suitably directed streams of wash liquid. As in decantation and washing, a stirring rod provides direction for the flow of material to the filtering medium.

The last traces of precipitate that cling to the inside of the beaker are dislodged with a *rubber policeman*, which is a small section of rubber tubing that has been crimped on one end. The policeman is fitted onto the end of a stirring rod and

---

[8] See *Anal. Chem.*, **1986**, *58*, 1424A; E. S. Beary, *Anal. Chem.*, **1988,** *60*, 742.

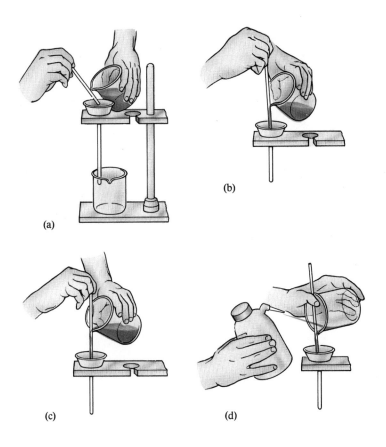

(a)

(b)

(c)

(d)

Figure 28-12
Steps in filtering: (a and b) washing by decantation; (c and d) transfer of the precipitate.

is wetted with wash liquid before use. Any solid collected with it is combined with the main portion on the filter. Small pieces of ashless paper can be used to wipe the last traces of hydrous oxide precipitates from the wall of the beaker; these papers are ignited along with the paper that holds the bulk of the precipitate.

Many precipitates possess the exasperating property of *creeping*, or spreading over a wetted surface against the force of gravity. Filters are never filled to more than three quarters of capacity, owing to the possibility that some of the precipitate could be lost as the result of creeping. The addition of a small amount of nonionic detergent, such as Triton X-100, to the supernatant liquid or to the wash liquid can be helpful in minimizing creeping.

A gelatinous precipitate must be completely washed before it is allowed to dry. These precipitates shrink and develop cracks as they dry. Further additions of wash liquid simply pass through these cracks and accomplish little or no washing.

Creeping is a process in which a solid moves up the side of a wetted container or filter paper.

Do not permit a gelatinous precipitate to dry until it has been washed completely.

28F-3 Directions for the Filtration and Ignition of Precipitates

Preparation of a Filter Paper

Figure 28-13 shows the sequence followed in folding a filter paper and seating it in a 60-deg funnel. The paper is folded exactly in half (a), firmly creased, and folded again (b). A triangular piece from one of the corners is torn off parallel to the second fold (c). The paper is then opened so that the untorn quarter forms

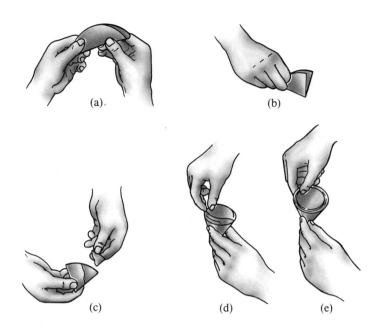

Figure 28-13
Folding and seating a filter paper.

a cone (d). The cone is fitted into the funnel, and the second fold is creased (e). Seating is completed by dampening the cone with water from a wash bottle and *gently* patting it with a finger. There will be no leakage of air between the funnel and a properly seated cone; in addition, the stem of the funnel will be filled with an unbroken column of liquid.

The Transfer of Paper and Precipitate to a Crucible

After filtration and washing have been completed, the filter and its contents must be transferred from the funnel to a crucible that has been brought to constant mass. Ashless paper has very low wet strength and must be handled with care during this transfer. The danger of tearing is lessened considerably if the paper is allowed to dry somewhat before it is removed from the funnel.

Figure 28-14 illustrates the transfer process. The triple-thick portion of the filter paper is drawn across the funnel to flatten the cone along its upper edge (a); the corners are next folded inward (b); the top edge is then folded over (c). Finally, the paper and its contents are eased into the crucible (d) so that the bulk of the precipitate is near the bottom.

Ashing of a Filter Paper

If a heat lamp is to be used, the crucible is placed on a clean, nonreactive surface, such as a wire screen covered with aluminum foil. The lamp is then positioned about 1 cm above the rim of the crucible and turned on. Charring takes place without further attention. The process is accelerated considerably if the paper is moistened with no more than one drop of concentrated ammonium nitrate solution. Elimination of the residual carbon is accomplished with a burner.

Considerably more attention must be paid if a burner is used to ash a filter paper. The burner produces much higher temperatures than a heat lamp. The possibility thus exists for the mechanical loss of precipitate if moisture is expelled

You should have a burner for each crucible. You can tend to the ashing of several filter papers at the same time.

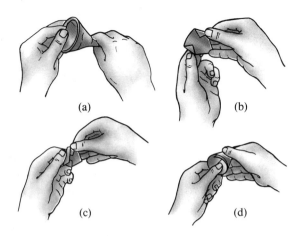

(a) (b)

(c) (d)

Figure 28-14

Transferring a filter paper and precipitate to a crucible.

too rapidly in the initial stages of heating or if the paper bursts into flame. Also, partial reduction of some precipitates can occur through reaction with the hot carbon of the charring paper; such reduction is a serious problem if reoxidation following ashing is inconvenient. These difficulties can be minimized by positioning the crucible as illustrated in Figure 28-15. The tilted position allows for the ready access of air; a clean crucible cover should be available to extinguish any flame that might develop.

Heating is commenced with a small flame. The temperature is gradually increased as moisture is evolved and the paper begins to char. The intensity of heating that can be tolerated can be gauged by the amount of smoke given off. Thin wisps are normal. A significant increase in the amount of smoke indicates that the paper is about to flash and that heating should be temporarily discontinued. Any flame that does appear should be immediately extinguished with a crucible cover. (The cover may become discolored, owing to the condensation of carbonaceous products; these products must ultimately be removed from the cover by ignition to confirm the absence of entrained particles of precipitate.) When no further smoking can be detected, heating is increased to eliminate the residual carbon. Strong heating, as necessary, can then be undertaken.

This sequence ordinarily precedes the final ignition of a precipitate in a muffle furnace, where a reducing atmosphere is equally undesirable.

The Use of Filtering Crucibles

A vacuum filtration train (Figure 28-16) is used where a filtering crucible can be used instead of paper. The trap isolates the filter flask from the source of vacuum.

28F-4 Rules for the Manipulation of Heated Objects

Careful adherence to the following rules will minimize the possibility of accidental loss of a precipitate.

1. Practice unfamiliar manipulations before putting them to use.
2. *Never* place a heated object on the benchtop; instead, place it on a wire gauze or a heat-resistant ceramic plate.

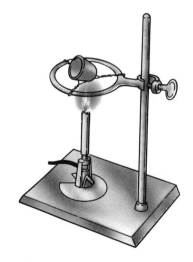

Figure 28-15

Ignition of a precipitate. Proper crucible position for preliminary charring.

Figure 28-16
Train for vacuum filtration.

3. Allow a crucible that has been subjected to the full flame of a burner or to a muffle furnace to cool momentarily (on a wire gauze or ceramic plate) before transferring it to the desiccator.
4. Keep the tongs and forceps used to handle heated objects scrupulously clean. In particular, do not allow the tips to touch the benchtop.

28G   THE MEASUREMENT OF VOLUME

The precise measurement of volume is as important to many analytical methods as is the precise measurement of mass.

28G-1 Units of Volume

The unit of volume is the *liter* (L), defined as one cubic decimeter. The *milliliter* (mL) is one one-thousandth of a liter, or one cubic centimeter (cm^3), and is used where the liter represents an inconveniently large volume unit.

28G-2 The Effect of Temperature on Volume Measurements

The volume occupied by a given mass of liquid varies with temperature as does the device that holds the liquid during measurement. Most volumetric measuring devices are made of glass, however, which fortunately has a small coefficient of expansion. Consequently, variations in the volume of a glass container with temperature need not be considered in ordinary analytical work.

The coefficient of expansion for dilute aqueous solutions (approximately 0.025%/°C) is such that a 5°C change has a measurable effect on the reliability of ordinary volumetric measurements.

Example 28-2

A 40.00-mL sample is taken from an aqueous solution at 5°C; what volume does it occupy at 20°C?

$$V_{20°} = V_{5°} + 0.00025(20 - 5)(40.00) = 40.00 + 0.15 = 40.15 \text{ mL}$$

Volumetric measurements must be referred to some standard temperature; this reference point is ordinarily 20°C. The ambient temperature of most laboratories is sufficiently close to 20°C to eliminate the need for temperature corrections in volume measurements for aqueous solutions. In contrast, the coefficient of expansion for organic liquids may require corrections for temperature differences of 1°C or less.

28G-3 Apparatus for the Precise Measurement of Volume

The reliable measurement of volume is performed with the *pipet*, the *buret*, and the *volumetric flask*.

Volumetric equipment is marked by the manufacturer to indicate not only the manner of calibration (usually TD for "to deliver" or TC for "to contain") but also the temperature at which the calibration strictly applies. Pipets and burets are ordinarily calibrated to deliver specified volumes, whereas volumetric flasks are calibrated on a to-contain basis.

Pipets

Pipets permit the transfer of accurately known volumes from one container to another. Common types are shown in Figure 28-17; information concerning their use is given in Table 28-2. A *volumetric*, or *transfer*, pipet (Figure 28-17a) delivers a single, fixed volume between 0.5 and 200 mL. Many such pipets are color-coded by volume for convenience in identification and sorting. *Measuring* pipets (Figures 28-17b and c) are calibrated in convenient units to permit delivery of any volume up to a maximum capacity ranging from 0.1 to 25 mL.

Volumetric and measuring pipets are filled to a calibration mark at the outset; the manner in which the transfer is completed depends on the particular type. Because an attraction exists between most liquids and glass, a small amount of liquid tends to remain in the tip after the pipet is emptied. This residual liquid is never blown out of a volumetric pipet or from some measuring pipets; it *is* blown out of other types of pipets (Table 28-2).

Hand-held Eppendorf micropipets (Figure 28-17d) deliver adjustable microliter volumes of liquid. With these pipets, a known and adjustable volume of air is

Tolerances, Class A Transfer Pipets

Capacity, mL	Tolerances, mL
0.5	±0.006
1	±0.006
2	±0.006
5	±0.01
10	±0.02
20	±0.03
25	±0.03
50	±0.05
100	±0.08

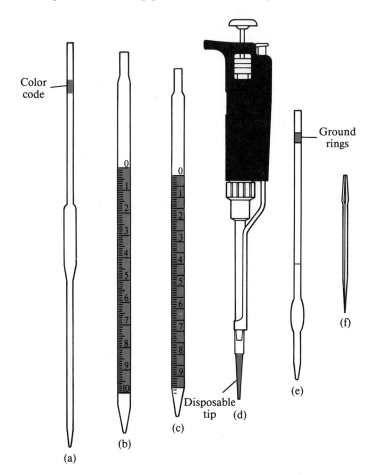

Figure 28-17
Typical pipets: (a) volumetric, (b) Mohr, (c) serological, (d) Eppendorf micropipet, (e) Ostwald-Folin, (f) lambda.

Table 28-2
CHARACTERISTICS OF PIPETS

Name	Type of Calibration*	Function	Available Capacity, mL	Type of Drainage
Volumetric	TD	Delivery of fixed volume	1–200	Free
Mohr	TD	Delivery of variable volume	1–25	To lower calibration line
Serological	TD	Delivery of variable volume	0.1–10	Blow out last drop†
Serological	TD	Delivery of variable volume	0.1–10	To lower calibration line
Ostwald-Folin	TD	Delivery of fixed volume	0.5–10	Blow out last drop†
Lambda	TC	Containment of fixed volume	0.001–2	Wash out with suitable solvent
Lambda	TD	Delivery of fixed volume	0.001–2	Blow out last drop†
Eppendorf	TD	Delivery of variable or fixed volume	0.001–1	Tip emptied by air displacement

* TD: to deliver; TC: to contain.
† A frosted ring near the top of recently manufactured pipets indicates that the last drop is to be blown out.

Range and Precision of Typical Eppendorf Micropipets

Volume Range, μL	Standard Deviation, μL
1–20	<0.04 @ 2 μL
	<0.06 @ 20 μL
10–100	<0.10 @ 15 μL
	<0.15 @ 100 μL
20–200	<0.15 @ 25 μL
	<0.30 @ 200 μL
100–1000	<0.6 @ 250 μL
	<1.3 @ 1000 μL
500–5000	<3 @ 1.0 mL
	<8 @ 5.0 mL

Tolerances, Class A Burets

Volume mL	Tolerances, mL
5	±0.01
10	±0.02
25	±0.03
50	±0.05
100	±0.20

displaced from the plastic disposable tip by depressing the pushbutton on the top of the pipet to a first stop. This button operates a spring-loaded piston that forces air out of the pipet. The volume of displaced air can be varied by a locking digital micrometer adjustment located on the front of the device. The plastic tip is then inserted into the liquid and the pressure on the button released causing liquid to be drawn into the tip. The tip is then placed against the wall of the receiving vessel, and the pushbutton is again depressed to the first stop. After one second, the pushbutton is depressed further to a second stop, which completely empties the tip. The range of volumes and precision of typical pipets of this type are shown in the margin.

Numerous *automatic* pipets are available for situations that call for the repeated delivery of a particular volume. In addition, motorized, computer-controlled microliter pipets are now available (see Figure 28-18). These devices are programmed to function as pipets, dispensers of multiple volumes, burets, and as a means for diluting samples. The volume desired is entered on a keyboard and is displayed on a panel. A motor-driven piston dispenses the liquid. Maximum volumes range from 10 to 2500 μL.

Burets

Burets, like measuring pipets, enable the analyst to deliver any volume up to their maximum capacity. The precision attainable with a buret is substantially greater than with a pipet.

A buret consists of a calibrated tube to hold titrant plus a valve arrangement by which the flow of titrant is controlled. This valve is the principal source of difference among burets. The simplest valve consists of a close-fitting glass bead inside a short length of rubber tubing that connects the buret and its tip; only when the tubing is deformed does liquid flow past the bead.

A buret equipped with a glass stopcock for a valve relies on a lubricant between the ground-glass surfaces of stopcock and barrel for a liquid-tight seal. Some solutions, notably bases, cause a glass stopcock to freeze on long contact; therefore, thorough cleaning is needed after each use. Valves made of Teflon are commonly encountered; these are unaffected by most common reagents and require no lubricant.

Volumetric Flasks

Volumetric flasks are manufactured with capacities ranging from 5 mL to 5 L and are usually calibrated to contain a specified volume when filled to a line etched on the neck. They are used for the preparation of standard solutions and for the dilution of samples to a fixed volume prior to taking aliquots with a pipet. Some are also calibrated on a to-deliver basis; these are readily distinguished by two reference lines on the neck. If delivery of the stated volume is desired, the flask is filled to the upper line. (See Figure 28-19 for a sketch of typical volumetric flasks.)

28G-4   General Considerations Concerning the Use of Volumetric Equipment

Volume markings are blazed on clean volumetric equipment by the manufacturer. An equal degree of cleanliness is needed in the laboratory if these markings are to have their stated meanings. Only clean glass surfaces support a uniform film of liquid. Dirt or oil causes breaks in this film; the existence of breaks is a certain indication of an unclean surface.

Cleaning

A brief soaking in a warm detergent solution is usually sufficient to remove the grease and dirt responsible for water breaks. Prolonged soaking should be avoided because a rough area or ring is likely to develop at a detergent/air interface. This ring cannot be removed and causes a film break that destroys the usefulness of the equipment.

After being cleaned, the apparatus must be thoroughly rinsed with tap water and then with three or four portions of distilled water. It is seldom necessary to dry volumetric ware.

Avoiding Parallax

The top surface of a liquid confined in a narrow tube exhibits a marked curvature, or *meniscus*. It is common practice to use the bottom of the meniscus as the point of reference in calibrating and using volumetric equipment. This minimum can be established more exactly by holding an opaque card or piece of paper behind the graduations (Figure 28-20).

In reading volumes, your eye must be at the level of the liquid surface to avoid an error due to *parallax*, a condition that causes the volume to appear smaller than its actual value if the meniscus is viewed from above and larger if the meniscus is viewed from below (Figure 28-20).

28G-5   Directions for the Use of a Pipet

The following directions pertain specifically to volumetric pipets but can be modified for the use of other types as well.

Liquid is drawn into a pipet through the application of a slight vacuum. *You must never use your mouth for suction because of the possibility of accidentally*

Figure 28-18

A hand-held, battery-operated, motorized pipet. (Courtesy of Rainin Instrument Co., Inc., Woburn, MA.)

Tolerances, Class A Volumetric Flasks

Capacity, mL	Tolerances, mL
5	±0.02
10	±0.02
25	±0.03
50	±0.05
100	±0.08
250	±0.12
500	±0.20
1000	±0.30
2000	±0.50

A meniscus is the curved surface of a liquid at its interface with the atmosphere.

Parallax is the apparent displacement of a liquid level or of a pointer as an observer changes position. It occurs when an object is viewed from a position that is not at a right-angle to the object.

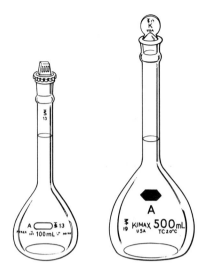

Figure 28-19
Typical volumetric flasks.

An aliquot is a measured fraction of the volume of a liquid sample.

ingesting the liquid being pipetted. Instead, use a rubber suction bulb (Figure 28-21a) or a rubber tube connected to a vacuum source.

Cleaning

Use a rubber bulb to draw detergent solution to a level 2 to 3 cm above the calibration mark of the pipet. Drain this solution and then rinse the pipet with several portions of tap water. Inspect for film breaks; repeat this portion of the cleaning cycle if necessary. Finally, fill the pipet with distilled water to perhaps one third of its capacity and carefully rotate it so that the entire interior surface is wetted. Repeat this rinsing step at least twice more.

Measurement of an Aliquot

Use a rubber bulb to draw a small volume of the liquid to be sampled into the pipet and thoroughly wet the entire interior surface. Repeat with *at least* two additional portions. Then carefully fill the pipet to a level somewhat above the graduation mark (Figure 28-21a). Quickly replace the bulb with your *forefinger* to arrest the outflow of liquid (Figure 28-21b). Make certain there are no bubbles in the bulk of the liquid or foam at the surface. Tilt the pipet slightly from the vertical and wipe the exterior free of adhering liquid (Figure 28-21c). Touch the tip of the pipet to the wall of a glass vessel (*not* the container into which the aliquot is to be transferred), and slowly allow the liquid level to drop by partially releasing your forefinger (Note 1). Halt further flow as the bottom of the meniscus coincides exactly with the graduation mark. Then place the pipet tip well within the receiving vessel and allow the liquid to drain. When free flow ceases, rest the tip against the inner wall of the receiver for a full ten seconds (Figure 28-21d). Finally, withdraw the pipet with a rotating motion to remove any liquid adhering to the tip. *The small volume remaining inside the tip of a volumetric pipet should not be blown or rinsed into the receiving vessel* (Note 2).

Notes on Using a Pipet

1. The liquid can best be held at a constant level if your forefinger is *faintly* moist. Too much moisture makes control impossible.
2. Rinse the pipet thoroughly after use.

Figure 28-20

Method for reading a buret. The eye should be level with the meniscus. The reading shown is 34.39 mL. If viewed from position 1, the reading appears smaller than 34.39 mL; from position 2, it appears larger.

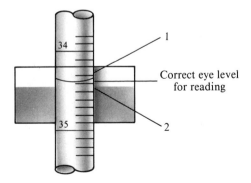

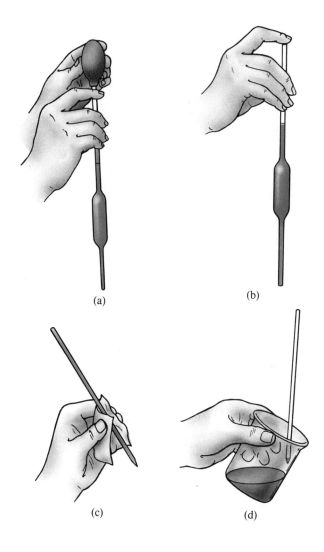

(a)

(b)

(c)

(d)

Figure 28-21
Steps in dispensing an aliquot.

28G-6 Directions for the Use of a Buret

Before it is placed in service, a buret must be scrupulously clean; in addition, its valve must be liquid-tight. In order to avoid leakage, glass stopcocks must be coated with a thin layer of a lubricant. Teflon stopcocks are more convenient because they require no lubricant.

Buret readings should be estimated to the nearest 0.01 mL.

Cleaning

Thoroughly clean the tube of the buret with detergent and a long brush. Rinse thoroughly with tap water and then with distilled water. Inspect for water breaks. Repeat the treatment if necessary.

Lubrication of a Glass Stopcock

Carefully remove all old grease from a glass stopcock and its barrel with a paper towel and dry both parts completely. Lightly grease the stopcock, taking care to

Figure 28-22

Recommended method for manipulation of a buret stopcock.

avoid the area adjacent to the hole. Insert the stopcock into the barrel and rotate it vigorously with slight inward pressure. A proper amount of lubricant has been used when (1) the area of contact between stopcock and barrel appears nearly transparent, (2) the seal is liquid-tight, and (3) no grease has worked its way into the tip (Note 1, 2).

Filling

Make certain the stopcock is closed. Add 5 to 10 mL of the titrant, tilt, and carefully rotate the buret to wet the interior completely. Allow the liquid to drain through the tip. *Repeat this procedure at least two more times.* Then fill the buret well above the zero mark. Free the tip of air bubbles by rapidly rotating the stopcock and permitting small quantities of the titrant to pass. Finally, lower the level of the liquid just to or somewhat below the zero mark. Allow for drainage (~1 min), and then record the initial volume reading, estimating to the nearest 0.01 mL (Note 3).

Titration

Figure 28-22 illustrates the preferred method for the manipulation of a stopcock; when your hand is held as shown, any tendency for lateral movement by the stopcock will be in the direction of firmer seating. Be sure the tip of the buret is well within the titration vessel (ordinarily a flask). Introduce the titrant in increments of about 1 mL. Swirl (or stir) constantly to ensure thorough mixing. Decrease the size of the increments as the titration progresses; add titrant dropwise in the immediate vicinity of the end point (Note 4, 5). When it is judged that only a few more drops are needed, rinse the walls of the container (Note 6). Allow for drainage (at least 30 s) at the completion of the titration. Then record the final volume, again to the nearest 0.01 mL.

Notes

1. The use of silicone lubricants is not recommended; contamination by such preparations is difficult—if not impossible—to remove.
2. As long as the flow of liquid is not impeded, fouling of a buret tip with stopcock grease is not a serious matter. Removal is best accomplished with organic solvents. A stoppage during a titration can be freed by *gentle* warming of the tip with a lighted match.
3. Before a buret is returned to service after reassembly, it is advisable to test for leakage. Simply fill the buret with water and establish that the volume reading does not change with time.
4. When unfamiliar with a particular titration, many chemists prepare an extra sample. No care is lavished on its titration since its functions are to reveal the nature of the end point and to provide a rough estimate of titrant requirements. This deliberate sacrifice of one sample usually results in an overall saving of time.
5. Increments smaller than one drop can be taken by allowing a small volume of titrant to form on the tip of the buret and then touching the tip to the wall of the flask. This partial drop is then combined with the bulk of the liquid as in Note 3.
6. Instead of being rinsed toward the end of a titration, the flask can be tilted

and rotated so that the bulk of the liquid picks up any drops that adhere to the inner surface.

28G-7 Directions for the Use of a Volumetric Flask

Before being put into use, volumetric flasks should be washed with detergent and thoroughly rinsed. Only rarely do they need to be dried. If required, however, drying is best accomplished by clamping the flask in an inverted position. Insertion of a glass tube connected to a vacuum line hastens the process.

Direct Weighing into a Volumetric Flask

The direct preparation of a standard solution requires the introduction of a known mass of solute to a volumetric flask. A powder funnel will minimize the possibility of loss of solid during the transfer. Rinse the funnel thoroughly; collect the washings in the flask.

The foregoing procedure may be inappropriate if heating is needed to dissolve the solute. Instead, weigh the solid into a beaker or flask, add solvent, heat to dissolve the solute, and allow the solution to cool to room temperature. Transfer this solution quantitatively to the volumetric flask, as described in the next section.

The Quantitative Transfer of Liquid to a Volumetric Flask

Insert a funnel into the neck of the volumetric flask; use a stirring rod to direct the flow of liquid from the beaker into the funnel. Tip off the last drop of liquid on the spout of the beaker with the stirring rod. Rinse both the stirring rod and the interior of the beaker with distilled water and transfer the washings to the volumetric flask, as before. Repeat the rinsing process *at least* two more times.

Dilution to the Mark

After the solute has been transferred, fill the flask about half-full and swirl the contents to hasten solution. Add more solvent and again mix well. Bring the liquid level almost to the mark, and allow time for drainage (~1 min); then use a medicine dropper to make such final additions of solvent as are necessary (Note). Firmly stopper the flask, and invert it repeatedly to ensure thorough mixing. Transfer the contents to a storage bottle that either is dry or has been thoroughly rinsed with several small portions of the solution from the flask.

> The solute should be completely dissolved *before* you dilute to the mark.

Note. If, as sometimes happens, the liquid level accidentally exceeds the calibration mark, the solution can be saved by correcting for the excess volume. Use a gummed label to mark the location of the meniscus. After the flask has been emptied, carefully refill to the manufacturer's etched mark with water. Use a buret to determine the additional volume needed to fill the flask so that the meniscus is at the gummed-label mark. This volume must be added to the nominal volume of the flask when calculating the concentration of the solution.

28H   THE CALIBRATION OF VOLUMETRIC WARE

Volumetric glassware is calibrated by measuring the mass of a liquid (usually distilled water) of known density and temperature that is contained in (or delivered by) the device. In carrying out a calibration, a buoyancy correction must be made

(Section 28D-4) since the density of water is quite different from that of the weights.

The calculations associated with calibration, while not difficult, are somewhat involved. The raw mass data are first corrected for buoyancy with Equation 28-1. Next, the volume of the apparatus at the temperature of calibration (T) is obtained by dividing the density of the liquid at that temperature into the corrected mass. Finally, this volume is corrected to the standard temperature of 20°C as in Example 28-2.

Table 28-3 is provided to ease the computational burden of calibration. Corrections for buoyancy with respect to stainless steel or brass weights (the density difference between the two is small enough to be neglected) and for the volume change of water and of glass containers have been incorporated into these data. Multiplication by the appropriate factor from Table 28-3 converts the mass of water at temperature T to (1) the corresponding volume at that temperature or (2) the volume at 20°C.

Table 28-3
VOLUME OCCUPIED BY 1.000 G OF
WATER WEIGHED IN AIR AGAINST
STAINLESS STEEL WEIGHTS*

| Temperature, T °C | Volume, mL | |
	At T	**Corrected to 20°C**
10	1.0013	1.0016
11	1.0014	1.0016
12	1.0015	1.0017
13	1.0016	1.0018
14	1.0018	1.0019
15	1.0019	1.0020
16	1.0021	1.0022
17	1.0022	1.0023
18	1.0024	1.0025
19	1.0026	1.0026
20	1.0028	1.0028
21	1.0030	1.0030
22	1.0033	1.0032
23	1.0035	1.0034
24	1.0037	1.0036
25	1.0040	1.0037
26	1.0043	1.0041
27	1.0045	1.0043
28	1.0048	1.0046
29	1.0051	1.0048
30	1.0054	1.0052

* Corrections for buoyancy (stainless steel weights) and change in container volume have been applied.

Example 28-3

A 25-mL pipet delivers 24.976 g of water weighed against stainless steel weights at 25°C. Use the data in Table 28-3 to calculate the volume delivered by this pipet at 25°C and 20°C.

$$\text{At 25°C: } V = 24.976 \text{ g} \times 1.0040 \text{ mL/g} = 25.08 \text{ mL}$$
$$\text{At 20°C: } V = 24.976 \text{ g} \times 1.0037 \text{ mL/g} = 27.07 \text{ mL}$$

28H-1   General Directions for Calibration Work

All volumetric ware should be painstakingly freed of water breaks before being calibrated. Burets and pipets need not be dry; volumetric flasks should be thoroughly drained and dried at room temperature. The water used for calibration should be in thermal equilibrium with its surroundings. This condition is best established by drawing the water well in advance, noting its temperature at frequent intervals, and waiting until no further changes occur.

Although an analytical balance can be used for calibration, weighings to the nearest milligram are perfectly satisfactory for all but the very smallest volumes. Thus a top-loading balance with a milligram sensitivity is conveniently used. Weighing bottles or small, well-stoppered conical flasks can serve as receivers for the calibration liquid.

Calibration of a Volumetric Pipet

Determine the mass of the empty stoppered receiver to the nearest milligram. Transfer a portion of temperature-equilibrated water to the receiver with the pipet, weigh the receiver and its contents (again, to the nearest milligram), and calculate the mass of water delivered from the difference in these masses. Calculate the volume delivered with the aid of Table 28-3. Repeat the calibration several times; calculate the mean volume delivered and its standard deviation.

Calibration of a Buret

Fill the buret with temperature-equilibrated water and make sure that no air bubbles are trapped in the tip. Allow about one minute for drainage; then lower the liquid level to bring the bottom of the meniscus to the 0.00-mL mark. Touch the tip to the wall of a beaker to remove any adhering drop. Wait ten minutes and recheck the volume; if the stopcock is tight, there should be no perceptible change. During this interval, weigh (to the nearest milligram) a 125-mL conical flask fitted with a rubber stopper.

Once tightness of the stopcock has been established, slowly transfer (at about 10 mL/min) approximately 10 mL of water to the flask. Touch the tip to the wall of the flask. Wait one minute, record the volume (to the nearest 0.01 mL) that was apparently delivered, and refill the buret. Weigh the flask and its contents to the nearest milligram; the difference between this mass and the initial value gives the mass of water delivered. Use Table 28-3 to convert this mass to the true volume. Subtract the apparent volume from the true volume. This difference is the correction that should be applied to the apparent volume to give the true volume. Repeat the calibration until agreement within ±0.02 mL is achieved.

Starting again from the zero mark, repeat the calibration, this time delivering about 20 mL to the receiver. Test the buret at 10-mL intervals over its entire volume. Prepare a plot of the correction to be applied as a function of volume delivered. The correction associated with any interval can be determined from this plot.

Calibration of a Volumetric Flask

Weigh the clean, dry flask to the nearest milligram. Then fill to the mark with equilibrated water and reweigh. Calculate the volume contained with the aid of Table 28-3.

Calibration of a Volumetric Flask Relative to a Pipet

The calibration of a volumetric flask relative to a pipet provides an excellent method for partitioning a sample into aliquots. These directions pertain to a 50-mL pipet and a 500-mL volumetric flask; other combinations are equally convenient.

Carefully transfer ten 50-mL aliquots from the pipet to a dry 500-mL volumetric flask. Mark the location of the meniscus with a gummed label. Cover with a label varnish to ensure permanence. Dilution to the label permits the same pipet to deliver precisely a one-tenth aliquot of the solution in the flask. Note that recalibration is necessary if another pipet is to be used.

28I THE LABORATORY NOTEBOOK

A laboratory notebook is needed to record measurements and observations concerning an analysis. The book should be permanently bound with consecutively numbered pages (if necessary, the pages should be hand-numbered before any entries are made). Save the first few pages for a table of contents that is updated as entries are made.

28I-1 Rules for the Maintenance of a Laboratory Notebook

> Remember that you can discard an experimental measurement *only if you have certain knowledge that you made an experimental error*. Thus, you must carefully record experimental observations in your notebook as soon as they occur.

1. *Record all data and observations directly into the notebook in ink.* Neatness is desirable, but you should not accomplish neatness by transcribing data from a sheet of paper to the notebook or from one notebook to another. The risk of misplacing—or incorrectly transcribing—crucial data and thereby ruining an experiment is unacceptable.
2. Supply each entry or series of entries with a heading or label. A series of weighing data for a set of empty crucibles should carry the heading ''Empty Crucible Mass'' (or something similar), for example, and the mass of each crucible should be identified by the same number or letter used to label the crucible.

> An erroneous entry in a laboratory notebook should never be erased but should be crossed out instead.

3. Date each page of the notebook as it is used.
4. *Never* attempt to erase or obliterate an incorrect entry. Instead, cross it out

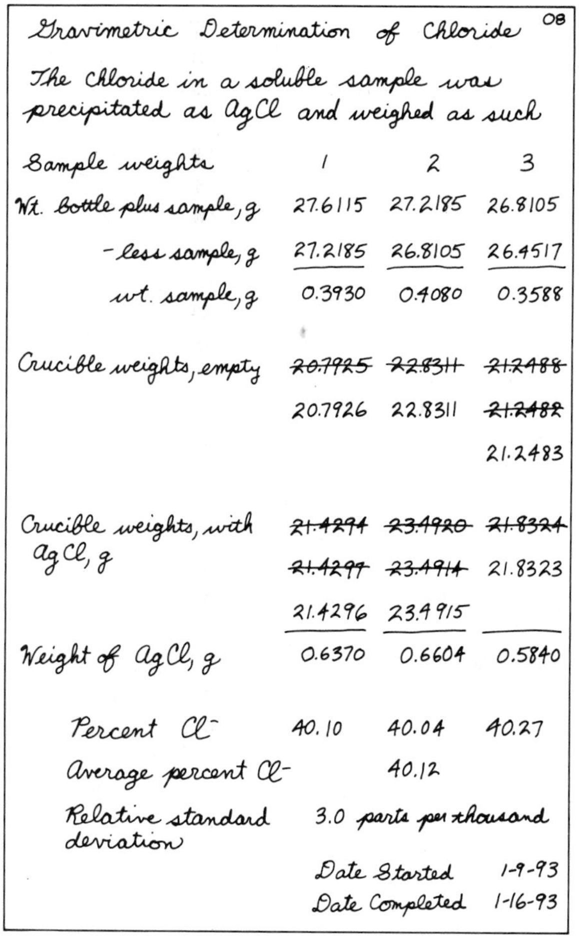

Gravimetric Determination of Chloride   08

The Chloride in a soluble sample was precipitated as AgCl and weighed as such

Sample weights	1	2	3
Wt. Bottle plus sample, g	27.6115	27.2185	26.8105
− less sample, g	27.2185	26.8105	26.4517
wt. sample, g	0.3930	0.4080	0.3588
Crucible weights, empty	~~20.7925~~	~~22.8311~~	~~21.2488~~
	20.7926	22.8311	~~21.2482~~
			21.2483
Crucible weights, with AgCl, g	~~21.4294~~	~~23.4920~~	~~21.8324~~
	~~21.4297~~	~~23.4914~~	21.8323
	21.4296	23.4915	
Weight of AgCl, g	0.6370	0.6604	0.5840
Percent Cl⁻	40.10	40.04	40.27
Average percent Cl⁻		40.12	
Relative standard deviation		3.0 parts per thousand	

Date Started   1-9-93
Date Completed   1-16-93

Figure 28-23
Laboratory notebook data page.

with a single horizontal line and locate the correct entry as nearby as possible. Do not write over incorrect numbers; with time, it may become impossible to distinguish the correct entry from the incorrect one.

5. Never remove a page from the notebook. Draw diagonal lines across any page that is to be disregarded. Provide a brief rationale for disregarding the page.

28I-2 Format

The instructor should be consulted concerning the format to be used in keeping the laboratory notebook.[9] One convention involves using each page consecutively for the recording of data and observations as they occur. The completed analysis is then summarized on the next available page spread (that is, left and right facing pages). As shown in Figure 28-23, the first of these two facing pages should contain the following entries:

[9] See also Howard M. Kanare, *Writing the Laboratory Notebook*. Washington, D.C.: The American Chemical Society, 1985.

1. the title of the experiment ("The Gravimetric Determination of Chloride"),
2. a brief statement of the principles on which the analysis is based,
3. a complete summary of the weighing, volumetric, and/or instrument response data needed to calculate the results,
4. a report of the best value for the set and a statement of its precision.

The second page (facing) should contain the following items:

1. equations for the principal reactions in the analysis,
2. an equation showing how the results were calculated,
3. a summary of observations that appear to bear on the validity of a particular result or the analysis as a whole. *Any such entry must have been originally recorded in the notebook at the time the observation was made.*

28J   SAFETY IN THE LABORATORY

Work in a chemical laboratory necessarily involves a degree of risk; accidents can and do happen. Strict adherence to the following rules will go far toward preventing (or minimizing the effect of) accidents.

1. At the outset, learn the location of the nearest eye fountain, fire blanket, shower, and fire extinguisher. Learn the proper use of each, and do not hesitate to use this equipment should the need arise.
2. **WEAR EYE PROTECTION AT ALL TIMES.** The potential for serious and perhaps permanent eye injury makes it mandatory that adequate eye protection be worn at all times by students, instructors, and visitors. Eye protection should be donned before entering the laboratory and should be used continuously until it is time to leave. Serious eye injuries have occurred to people performing such innocuous tasks as computing or writing in a laboratory notebook; such incidents are usually the result of someone else losing control of an experiment. Regular prescription glasses are not adequate substitutes for eye protection approved by the Occupational Safety and Health Administration (OSHA). Contact lenses should *never* be used in the laboratory because laboratory fumes may react with them and have a harmful effect on your eyes.
3. Most of the chemicals in a laboratory are toxic; some are very toxic, and some—such as concentrated solutions of acids and bases—are highly corrosive. Avoid contact between these liquids and your skin. In the event of such contact, *immediately* flood the affected area with copious quantities of water. If a corrosive solution is spilled on clothing, remove the garment immediately. Time is of the essence; modesty cannot be a matter of concern.
4. **NEVER** perform an unauthorized experiment. Such activity is grounds for disqualification at many institutions.
5. Never work alone in the laboratory; be certain that someone is always within earshot.
6. Never bring food or beverages into the laboratory. Do not drink from laboratory glassware. Do not smoke in the laboratory.
7. Always use a bulb to draw liquids into a pipet; **NEVER** use your mouth to provide suction.

8. Wear adequate foot covering (no sandals). Confine long hair with a net. A laboratory coat or apron will provide some protection and may be required.

9. Be extremely tentative in touching objects that have been heated; hot glass looks just like cold glass.

10. Always fire-polish the ends of freshly cut glass tubing. **NEVER** attempt to force glass tubing through the hole of a stopper. Instead, make sure that both tubing and hole are wet with soapy water. Protect your hands with several layers of towel while inserting glass into a stopper.

11. Use fume hoods whenever toxic or noxious gases are likely to be evolved. Be cautious in testing for odors; use your hand to waft vapors above containers toward your nose.

12. Notify the instructor in the event of an injury.

13. Dispose of solutions and chemicals as instructed. It is illegal to flush solutions containing heavy metal ions or organic liquids down the drain in many localities; alternative arrangements are required for the disposal of such liquids.

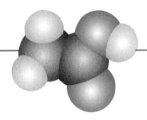

CHAPTER 29

SELECTED METHODS OF ANALYSIS

This chapter contains detailed directions for performing a variety of chemical analyses. The methods have been chosen to introduce you to analytical techniques that are widely used by chemists.

Your chances of success in the laboratory will greatly improve if you take time before beginning any analysis to read carefully and *understand* each step in the method and to develop a plan for how and when you will perform each step. For greatest efficiency, such study and planning should take place *before you enter the laboratory.*

The discussion in this section is aimed at helping you develop efficient work habits in the laboratory and also at providing you with some general information about an analytical chemistry laboratory.

BACKGROUND INFORMATION

Before you start an analysis, you should understand the significance of each step in the procedure in order to avoid the pitfalls and potential sources of error that exist in all analytical methods. Information about these steps can usually be found in preliminary discussion sections of this chapter, in earlier chapters that are referred to in the discussion section, and in the "Notes" that follow many of the procedures in this chapter. If, after reading these materials, you still don't understand the reason for one or more of the steps in the method, consult your instructor *before you begin laboratory work.*

The Accuracy of Measurements

In looking over an analytical procedure, you should decide which measurements must be made with maximum precision, and thus with maximum care, as opposed to those that can be carried out rapidly with little concern for precision. Generally, measurements that appear in the equation used to compute the results must be performed with maximum precision. The remaining measurements can *and should* be made less carefully to conserve time. The words *about* and *approximately* are frequently used to indicate that a measurement does not have to be done carefully. You should not waste time and effort to measure, for example, a volume to ± 0.02 mL when an uncertainty of ± 0.5 mL or even ± 5 mL will have no discernible effect on the results.

In some procedures, a statement such as "weigh three 0.5-g samples to the nearest 0.1 mg" is encountered. Here, samples of perhaps 0.4 to 0.6 g are acceptable, but their weights must be known to the nearest 0.1 mg. The number of significant figures in the specification of a volume or a weight is also a guide as to the care that should be taken in making a measurement. For example, the statement "add 10.00 mL of a solution to the beaker" indicates that you should measure the volume carefully with a buret or a pipet with the aim of limiting the uncertainty to perhaps ± 0.02 mL. In contrast, if the directions read "add 10 mL," the measurement can be made with a graduated cylinder.

Time Utilization

You should study carefully the time requirements of the several unit operations involved in an analysis *before you start work*. Such study will reveal operations that require considerable elapsed, or clock, time but little or no operator time—for example, when a sample is dried in an oven, cooled in a desiccator, or evaporated on a hot plate. The experienced chemist plans to use such periods of waiting to perform other operations or perhaps to begin a new analysis. Some people find it worthwhile to prepare a written time schedule for each laboratory session to avoid periods when no work can be done.

Time planning is also needed to identify places where an analysis can be interrupted for overnight or longer, as well as those operations that must be completed without a break.

Reagents

Directions for the preparation of reagents accompany many of the procedures. Before preparing such reagents, be sure to check to see if they are already prepared and available on a side shelf for general use.

If a reagent is known to pose a hazard, you should plan *in advance of the laboratory period* the steps that you should take to minimize injury or damage. Furthermore, you must acquaint yourself with the rules that apply in your laboratory for the disposal of waste liquids and solids. These rules vary from one part of the country to another and even among laboratories in the same locale.

Water

Some laboratories use deionizers to purify water; others employ stills for this purpose. The terms "distilled water" and "deionized water" are used interchangeably in the directions that follow. Either type is satisfactory for analytical work.

You should use tap water only for preliminary cleaning of glassware. The cleaned equipment should then be rinsed with at least three small portions of distilled or deionized water.

29A GRAVIMETRIC METHODS OF ANALYSIS

General aspects, calculations, and typical applications of gravimetric analysis are discussed in detail in Chapter 6.

29A-1 The Gravimetric Determination of Chloride in a Soluble Sample

Discussion

The chloride content of a soluble salt can be determined by precipitation as silver chloride:

$$Ag^+ + Cl^- \rightarrow AgCl(s)$$

The precipitate is collected in a weighed filtering crucible and washed. Its mass is then determined after it has been dried to constant mass at 110°C.

The solution containing the sample is kept slightly acidic during the precipitation to eliminate possible interference from anions of weak acids (such as CO_3^{2-}) that form sparingly soluble silver salts in a neutral environment. A moderate excess of silver ion is needed to diminish the solubility of silver chloride, but a large excess is avoided to minimize coprecipitation of silver nitrate.

Silver chloride forms first as a colloid and is subsequently coagulated with heat. Nitric acid and the small excess of silver nitrate promote coagulation by providing a moderately high electrolyte concentration. Nitric acid in the wash solution maintains the electrolyte concentration and eliminates the possibility of peptization during the washing step; the acid subsequently decomposes to give volatile products when the precipitate is dried. See Section 6A-2 for additional information concerning the properties and treatment of colloidal precipitates.

In common with other silver halides, finely divided silver chloride undergoes photodecomposition:

$$2AgCl(s) \xrightarrow{h\nu} 2Ag(s) + Cl_2(g)$$

The elemental silver produced in this reaction is responsible for the violet color that develops in the precipitate. In principle, this reaction leads to low results for chloride ion. In practice, however, its effect is negligible provided you avoid direct and prolonged exposure of the precipitate to sunlight.

If photodecomposition of silver chloride occurs before filtration, the additional reaction

$$3Cl_2(aq) + 3H_2O + 5Ag^+ \rightarrow 5AgCl(s) + ClO_3^- + 6H^+$$

tends to cause high results.

Some photodecomposition of silver chloride is inevitable as the analysis is ordinarily performed. It is worthwhile to minimize exposure of the solid to intense sources of light as far as possible.

Because silver nitrate is expensive, any unused reagent should be collected in a storage container. Similarly, precipitated silver chloride should be retained after the analysis is complete.[1]

Procedure

Clean three medium-porosity sintered-glass or porcelain filtering crucibles by allowing about 5 mL of concentrated HNO_3 to stand in each for about 5 min. Use a vacuum (Figure 28-16) to draw the acid through the crucible. Rinse each crucible with three portions of tap water, and then discontinue the vacuum. Next, add about 5 mL of 6 M NH_3 and wait for about 5 min before drawing it through the filter. Finally, rinse each crucible with six to eight portions of distilled or deionized water. Provide each crucible with an identifying mark. Dry the crucibles to constant mass by heating at 110°C while the other steps in the analysis are being carried out. The first drying should be for at least 1 h; subsequent heating periods can be somewhat shorter (30 to 40 min). This process of heating and drying should be repeated until the mass becomes constant to within 0.2 to 0.3 mg.

Transfer the unknown to a weighing bottle and dry it at 110°C (Figure 28-8) for 1 to 2 h; allow the bottle and contents to cool to room temperature in a desiccator. Weigh (to the nearest 0.1 mg) individual samples by difference into 400-mL beakers (Note 1). Dissolve each sample in about 100 mL of distilled water to which 2 to 3 mL of 6 M HNO_3 have been added.

Slowly, and with good stirring, add 0.2 M $AgNO_3$ to each of the cold sample solutions until AgCl is observed to coagulate (Notes 2 and 3); then introduce an additional 3 to 5 mL. Heat almost to boiling, and digest the solids for about 10 min. Add a few drops of $AgNO_3$ to confirm that precipitation is complete. If more precipitate forms, add about 3 mL of $AgNO_3$, digest, and again test for completeness of precipitation. Pour any unused $AgNO_3$ into a waste container (**not** into the original reagent bottle). Cover each beaker, and store in a dark place for at least 2 h—preferably until the next laboratory period.

Read the instructions for filtration in Section 28F-2. Decant the supernatant liquids through weighed filtering crucibles. Wash the precipitates several times (while they are still in the beaker) with a solution consisting of 2 to 5 mL of 6 M HNO_3 per liter of distilled water; decant these washings through the filters. Quantitatively transfer the AgCl from the beakers to the individual crucibles with fine streams of wash solution; use rubber policemen to dislodge any particles that adhere to the walls of the beakers. Continue washing until the filtrates are essentially free of Ag^+ ion (Note 4).

Dry the precipitate at 110°C for at least 1 h. Store the crucibles in a desiccator while they cool. Determine the mass of the crucibles and their

Be sure to label your beakers and crucibles.

To digest means to heat an unstirred precipitate in the *mother liquor*, i.e., the solution from which it is formed.

[1] Silver can be recovered from silver chloride and from surplus reagent by reduction with ascorbic acid; see J. W. Hill and L. Bellows, *J. Chem. Educ.*, **1986**, *63* (4), 357; see also J. P. Rawat and S. Iqbal M. Kamoonpuri, *ibid.*, **1986**, *63* (6), 537 for recovery (as $AgNO_3$) based on ion exchange. See D. D. Perrin, W. L. F. Armarego and D. R. Perrin, *Chemistry International*, **1987**, *9* (1), 3 concerning a potential hazard in the recovery of silver nitrate.

contents. Repeat the cycle of heating, cooling, and weighing until consecutive weighings agree to within 0.2 mg. Calculate the percentage of Cl^- in the sample.

When the analysis is complete, remove the precipitates by gently tapping the crucibles over a piece of glazed paper. Transfer the collected AgCl to a container for silver wastes. Remove the last traces of AgCl by filling the crucibles with 6 M NH_3 and allowing them to stand.

Notes

1. Consult with the instructor concerning an appropriate sample size.
2. Determine the approximate amount of $AgNO_3$ needed by calculating the volume that would be required if the unknown were pure NaCl.
3. Use a separate stirring rod for each sample and leave it in its beaker throughout the determination.
4. To test the washings for Ag^+, collect a small volume in a test tube and add a few drops of HCl. Washing is judged complete when little or no turbidity develops.

29A-2 The Gravimetric Determination of Nickel in Steel

Discussion

The nickel in a steel sample can be precipitated from a slightly alkaline medium with an alcoholic solution of dimethylglyoxime (Section 6D-3). Interference from iron(III) is eliminated by masking with tartaric acid. The product is freed of moisture by drying at 110°C.

The bulky character of nickel dimethylglyoxime limits the mass of nickel that can be accommodated conveniently and thus the sample mass. Care must also be taken to control the excess of alcoholic dimethylglyoxime used. If too much is added, the alcohol concentration becomes sufficient to dissolve appreciable amounts of the nickel dimethylglyoxime, which leads to low results. If the alcohol concentration becomes too low, however, some of the reagent may precipitate and cause a positive error.

Iron(III) forms a highly stable complex with tartrate ion, which prevents it from precipitating as $Fe_2O_3 \cdot xH_2O$ in slightly alkaline solutions.

Preparation of Solutions

(a) *Dimethylglyoxime, 1% (w/v).* Dissolve 10 g of dimethylglyoxime in 1 L of ethanol. (Sufficient for about 50 precipitations.)
(b) *Tartaric acid, 15% (w/v).* Dissolve 225 g of tartaric acid in sufficient water to give 1500 mL of solution. Filter before use if the solution is not clear. (Sufficient for about 50 precipitations.)

Procedure

Clean and mark three medium-porosity sintered-glass crucibles (Note 1); bring them to constant mass by drying at 110°C for at least 1 h.

Weigh (to the nearest 0.1 mg) samples containing between 30 and 35 mg of nickel into individual 400-mL beakers (Note 2). Dissolve each sample in about 50 mL of 6 M HCl with gentle warming (**use a hood**). Carefully add approximately 15 mL of 6 M HNO_3, and boil gently to expel any oxides of nitrogen that may have been produced. Dilute to about 200 mL and heat to boiling. Introduce about 30 mL of 15% tartaric acid and sufficient concentrated NH_3 to produce a faint odor of NH_3 in the vapors over the solutions (Note 3); then add another 1 to 2 mL of NH_3. If the solutions are not clear at this stage, proceed as directed in Note 4. Make the solutions acidic with HCl (no odor of NH_3), heat to 60°C to 80°C, and add about 20 mL of the 1% dimethylglyoxime solution. With good stirring, add 6 M NH_3 until a slight excess exists (faint odor of NH_3) plus an additional 1 to 2 mL. Digest the precipitates for 30 to 60 min, cool for at least 1 h, and filter.

Wash the solids with water until the washings are free of Cl^- (Note 5). Bring the crucibles and their contents to constant mass at 110°C. Report the percentage of nickel in the sample. The dried precipitate has the composition $Ni(C_4H_7O_2N_2)_2$ (288.92 g/mol).

Notes

1. Medium-porosity porcelain filtering crucibles or Gooch crucibles with glass pads can be substituted for sintered-glass crucibles in this determination.
2. Use a separate stirring rod for each sample and leave it in the beaker throughout.
3. The presence or absence of excess NH_3 is readily established by odor; use a waving motion with your hand to waft the vapors toward your nose.
4. If $Fe_2O_3 \cdot xH_2O$ forms on addition of NH_3, acidify the solution with HCl, introduce additional tartaric acid, and neutralize again. Alternatively, remove the solid by filtration. Thorough washing with a hot NH_3/NH_4Cl solution is required; the washings are combined with the solution containing the bulk of the sample.
5. Test the washings for Cl^- by collecting a small portion in a test tube, acidifying with HNO_3, and adding a drop or two of 0.1 M $AgNO_3$. Washing is judged complete when little or no turbidity develops.

29B NEUTRALIZATION TITRATIONS

Neutralization titrations are performed with standard solutions of strong acids or bases. While a single solution (of either acid or base) is sufficient for the titration of a given type of analyte, it is convenient to have standard solutions of both acid and base available in the event a back-titration is needed to locate end points more exactly. The concentration of one solution is established by titration against a primary standard; the concentration of the other is then determined from the acid/base ratio (that is, the volume of acid needed to neutralize 1.000 mL of the base).

29B-1 The Effect of Atmospheric Carbon Dioxide on Neutralization Titrations

Water in equilibrium with the atmosphere is about 1×10^{-5} M in carbonic acid as a consequence of the equilibrium

$$CO_2(g) + H_2O \rightleftharpoons H_2CO_3(aq)$$

At this concentration level, the amount of 0.1 M base consumed by the carbonic acid in a typical titration is negligible. With more dilute reagents (<0.05 M), however, the water used as a solvent for the analyte and in the preparation of reagents must be freed of carbonic acid by boiling for a brief period.

Water that has been purified by distillation rather than by deionization is often supersaturated with carbon dioxide and may thus contain sufficient acid to affect the results of an analysis.[2] The instructions that follow are based on the assumption that the amount of carbon dioxide in the water supply can be neglected without causing serious error. For further discussion on the effects of carbon dioxide in neutralization titrations, see Section 12A-3.

29B-2 Preparation of Indicator Solutions for Neutralization Titrations

Discussion

The theory of acid/base indicators is discussed in Section 10A-2. An indicator exists for virtually any pH range between 1 and 13.[3] Directions follow for the preparation of indicator solutions suitable for most neutralization titrations.

Procedure

Stock solutions ordinarily contain between 0.5 and 1.0 g of indicator per liter. (One liter of indicator is sufficient for hundreds of titrations.)

(a) *Bromocresol green.* Dissolve the sodium salt directly in distilled water.
(b) *Phenolphthalein, thymolphthalein.* Dissolve the solid indicator in a solution consisting of 800 mL ethanol and 200 mL of distilled or deionized water.

---

[2] Water that is to be used for neutralization titrations can be tested by adding 5 drops of phenolphthalein to a 500-mL portion. Less than 0.2 to 0.3 mL of 0.1 M OH$^-$ should suffice to produce the first faint pink color of the indicator. If a larger volume is needed, the water should be boiled and cooled before it is used to prepare standard solutions or to dissolve samples.
[3] See, for example, J. Beukenkemp and W. Rieman III, in *Treatise on Analytical Chemistry*, I. M. Kolthoff and P. J. Elving, Eds., Part I, Vol. 11, pp. 6987–7001. New York: Wiley, 1974.

29B-3 Preparation of Dilute Hydrochloric Acid Solutions

Discussion

The preparation and standardization of acids are considered in Sections 12A-1 and 12A-2.

Procedure

For a 0.1 M solution, add about 8 mL of concentrated HCl to about 1 L of distilled water (Note). Mix thoroughly, and store in a glass-stoppered bottle.

Note

It is advisable to eliminate CO_2 from the water by a preliminary boiling if very dilute solutions (< 0.05 M) are being prepared.

29B-4 Preparation of Carbonate-Free Sodium Hydroxide

Discussion

See Sections 12A-3 and 12A-4 for information concerning the preparation and standardization of bases.

Procedure

If so directed by the instructor, prepare a bottle for protected storage (Figure 12-2; also Note 1). Transfer 1 L of distilled water to the storage bottle (see the Note in Section 29B-3). Decant 4 to 5 mL of 50% NaOH into a small container (Note 2), add it to the water, and *mix thoroughly*. **USE EXTREME CARE IN HANDLING** 50% NaOH, which is highly corrosive. If the reagent comes into contact with your skin, **IMMEDIATELY** flush the area with *copious* amounts of water.

Protect the solution from unnecessary contact with the atmosphere.

Notes

1. A solution of base that will be used up within two weeks can be stored in a tightly capped polyethylene bottle. After each removal of base, squeeze the bottle while tightening the cap to minimize the air space above the reagent. The bottle will become embrittled after extensive use as a container for bases.
2. Be certain that any solid Na_2CO_3 in the 50% NaOH has settled to the bottom of the container and that the decanted liquid is absolutely clear. If necessary, filter the base through a glass mat in a Gooch crucible; collect the clear filtrate in a test tube inserted in the filter flask.

29B-5 The Determination of the Acid/Base Ratio

Discussion

If both acid and base solutions have been prepared, it is useful to determine their volumetric combining ratio. Knowledge of this ratio and the concentration of one solution permits calculation of the molarity of the other.

Procedure

Instructions for placing a buret into service are given in Sections 28G-4 and 28G-6; consult these instructions if necessary. Place a test tube or small beaker over the top of the buret that holds the NaOH solution to minimize contact between the solution and the atmosphere.

Record the initial volumes of acid and base in the burets to the nearest 0.01 mL. Deliver 35 to 40 mL of the acid into a 250-mL conical flask. Touch the tip of the buret to the inside wall of the flask, and rinse down with a little distilled water. Add two drops of phenolphthalein (Note 1) and then sufficient base to render the solution a definite pink. Introduce acid dropwise to discharge the color, and again rinse down the walls of the flask. Carefully add base until the solution again acquires a faint pink hue that persists for at least 30 s (Notes 2 and 3). Record the final buret volumes (again, to nearest 0.01 mL). Repeat the titration. Calculate the acid/base volume ratio. The ratios for duplicate titrations should agree to within 1 to 2 ppt. Perform additional titrations, if necessary, to achieve this order of precision.

Notes

1. The volume ratio can also be determined with an indicator that has an acidic transition range, such as bromocresol green. If the NaOH is contaminated with carbonate, the ratio obtained with this indicator will differ significantly from the value obtained with phenolphthalein. In general, the acid/base ratio should be evaluated with the indicator that is to be used in subsequent titrations.
2. Fractional drops can be formed on the buret tip, touched to the wall of the flask, and then rinsed down with a small amount of water.
3. The phenolphthalein end point fades as CO_2 is absorbed from the atmosphere.

29B-6 Standardization of Hydrochloric Acid Against Sodium Carbonate

Discussion

See Section 12A-2.

Procedure

Dry a quantity of primary-standard Na_2CO_3 for about 2 h at 110°C (Figure 28-8), and cool in a desiccator. Weigh individual 0.20- to 0.25-g samples (to the nearest 0.1 mg) into 250-mL conical flasks, and dissolve each in about 50 mL of distilled water. Introduce 3 drops of bromocresol green, and titrate with HCl until the solution just begins to change from blue to green. Boil the solution for 2 to 3 min, cool to room temperature (Note 1), and complete the titration (Note 2).

Determine an indicator correction by titrating approximately 100 mL of 0.05 M NaCl and 3 drops of indicator. Boil briefly, cool, and complete the titration. Subtract any volume needed for the blank from the titration volumes. Calculate the concentration of the HCl solution.

Notes

1. The indicator should change from green to blue as CO_2 is removed during heating. If no color change occurs, an excess of acid was added originally. This excess can be back-titrated with base, provided the acid/base combining ratio is known; otherwise, the sample must be discarded.
2. It is permissible to back-titrate with base to establish the end point with greater certainty.

29B-7 Standardization of Sodium Hydroxide Against Potassium Hydrogen Phthalate

Discussion

See Section 12A-4.

Procedure

Dry a quantity of primary-standard potassium hydrogen phthalate (KHP) for about 2 h at 110°C (Figure 28-8), and cool in a desiccator. Weigh individual 0.7- to 0.8-g samples (to the nearest 0.1 mg) into 250-mL conical flasks, and dissolve each in 50 to 75 mL of distilled water. Add 2 drops of phenolphthalein; titrate with base until the pink color of the indicator persists for 30 s (Note). Calculate the concentration of the NaOH solution.

Note

It is permissible to back-titrate with acid to establish the end point more precisely. Record the volume used in the back-titration. Use the acid/base ratio to calculate the net volume of base used in the standardization.

29B-8 The Determination of Potassium Hydrogen Phthalate in an Impure Sample

Discussion

The unknown is a mixture of KHP and a neutral salt. This analysis is conveniently performed concurrently with the standardization of the base.

Procedure

Consult with the instructor concerning an appropriate sample size. Then follow the directions in Section 29B-7.

29B-9 The Determination of the Acid Content of Vinegars and Wines

Discussion

The total acid content of a vinegar or a wine is readily determined by titration with a standard base. It is customary to report the acid content of vinegar in terms of acetic acid, the principal acidic constituent, even though other acids are present. Similarly, the acid content of a wine is expressed as percent tartaric acid, even though there are other acids in the sample. Most vinegars contain about 5% acid (w/v) expressed as acetic acid; wines ordinarily contain somewhat under 1% acid (w/v) expressed as tartaric acid.

Procedure

(a) *If the unknown is a vinegar* (Note 1), pipet 25.00 mL into a 250-mL volumetric flask and dilute to the mark with distilled water. Mix thoroughly, and pipet 50.00-mL aliquots into 250-mL conical flasks. Add about 50 mL of water and 2 drops of phenolphthalein (Note 2) to each, and titrate with standard 0.1 M NaOH to the first permanent (~30 s) pink color.

Report the acidity of the vinegar as percent (w/v) CH_3COOH (60.053 g/mol).

(b) *If the unknown is a wine*, pipet 50.00-mL aliquots into 250-mL conical flasks, add about 50 mL of distilled water and 2 drops of phenolphthalein to each (Note 2), and titrate to the first permanent (~ 30 s) pink color.

Express the acidity of the wine sample as percent (w/v) tartaric acid $C_2H_4O_2(COOH)_2$ (150.09 g/mol) (Note 3).

Notes

1. The acidity of bottled vinegar tends to decrease on exposure to air. It is recommended that unknowns be stored in individual vials with snug covers.
2. The amount of indicator used should be increased as necessary to make the color change visible in colored samples.
3. Tartaric acid has two acidic hydrogens, both of which are titrated at a phenolphthalein end point.

29B-10 The Determination of Sodium Carbonate in an Impure Sample

Discussion

The titration of sodium carbonate is discussed in Section 12A-2 in connection with its use as a primary standard; the same considerations apply for the determination of carbonate in an unknown that has no interfering contaminants.

> **Procedure**
>
> Dry the unknown at 110°C for 2 h, and then cool in a desiccator. Consult with the instructor on an appropriate sample size. Then follow the instructions in Section 29B-6.
>
> Report the percentage of Na_2CO_3 in the sample.

29B-11 The Determination of Amine Nitrogen by the Kjeldahl Method

Discussion

These directions are suitable for the Kjeldahl determination of protein in materials such as blood meal, wheat flour, pasta products, dry cereals, and pet foods. A simple modification permits the analysis of unknowns that contain more highly oxidized forms of nitrogen.[4]

In the Kjeldahl method (see Section 12B-1), the organic sample is digested in hot concentrated sulfuric acid, which converts carbon and hydrogen in the sample to CO_2 and H_2O and the amine nitrogen to ammonium sulfate. After cooling, the sulfuric acid is neutralized by the addition of an excess of concentrated sodium hydroxide. The ammonia liberated by this treatment is then distilled into a measured excess of a standard solution of acid; the excess is determined by back titration with standard base.

Figure 29-1 illustrates typical equipment for a Kjeldahl distillation. The long-necked container, which is used for both digestion and distillation, is called a *Kjeldahl flask*. In the apparatus in Figure 29-1a, the base is added slowly by partially opening the stopcock from the NaOH storage vessel; the liberated ammonia is then carried to the receiving flask by steam distillation.

In an alternative method (Figure 29-1b), a dense, concentrated sodium hydroxide solution is carefully poured down the side of the Kjeldahl flask to form a second, lower layer. The flask is then quickly connected to a spray trap and an ordinary condenser before loss of ammonia can occur. Only then are the two layers mixed by gentle swirling of the flask.

Quantitative collection of ammonia requires the tip of the condenser to extend into the liquid in the receiving flask throughout the distillation step. The tip must be removed before heating is discontinued, however. Otherwise, the liquid will be drawn back into the apparatus.

---

[4] See *Official Methods of Analysis*, 15th ed., p. 18. Washington, D.C.: Association of Official Analytical Chemists, 1990.

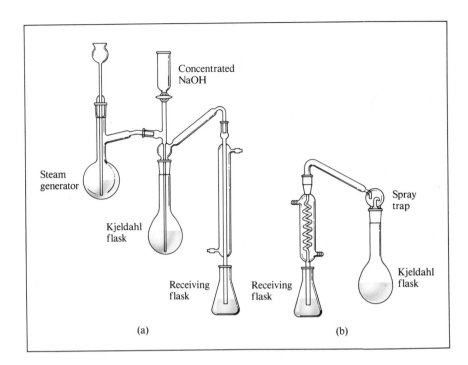

Figure 29-1

Kjeldahl distillation apparatus.

Two methods are commonly used for collecting and determining the ammonia liberated from the sample. In one, the ammonia is distilled into a measured volume of standard acid. After the distillation is complete, the excess acid is back-titrated with standard base. An indicator with an acidic transition range is required because of the acidity of the ammonium ions present at equivalence. A convenient alternative, which requires only one standard solution, involves the collection of the ammonia in an unmeasured excess of boric acid, which retains the ammonia by the reaction

$$H_3BO_3 + NH_3 \rightarrow NH_4^+ + H_2BO_3^-$$

The dihydrogen borate ion produced is a reasonably strong base that can be titrated with a standard solution of hydrochloric acid:

$$H_2BO_3^- + H_3O^+ \rightarrow H_3BO_3 + H_2O$$

At the equivalence point, the solution contains boric acid and ammonium ions; an indicator with an acidic transition interval (such as bromocresol green) is again required.

Procedure

Preparation of Samples

Consult with the instructor on sample size. *If the unknown is powdered* (such as blood meal), weigh samples onto individual 9-cm filter papers (Note 1). Fold the paper around the sample and drop each into a Kjeldahl flask (the paper keeps the samples from clinging to the neck of the flask). *If the unknown is not powdered* (such as breakfast cereals or pasta), the samples can be weighed by difference directly into the Kjeldahl flasks.

Add 25 mL of concentrated H_2SO_4, 10 g of powdered K_2SO_4, and the catalyst (Note 2) to each flask.

Digestion

Clamp the flasks in a slanted position in a hood or vented digestion rack. Heat carefully to boiling. Discontinue heating briefly if foaming becomes excessive; never allow the foam to reach the neck of the flask. Once foaming ceases and the acid is boiling vigorously, the samples can be left unattended; prepare the distillation apparatus during this time. Continue digestion until the solution becomes colorless or faint yellow; 2 to 3 h may be needed for some materials. If necessary, *cautiously* replace the acid lost by evaporation.

When digestion is complete, discontinue heating, and allow the flasks to cool to room temperature; swirl the flasks if the contents show signs of solidifying. Cautiously add 250 mL of water to each flask and again allow the solution to cool to room temperature.

Distillation of Ammonia

Arrange a distillation apparatus similar to that shown in Figure 29-1. Pipet 50.00 mL of standard 0.1 M HCl into the receiver flask (Note 3). Clamp the flask so that the tip of the adapter extends below the surface of the standard acid. Circulate water through the condenser jacket.

Hold the Kjeldahl flask at an angle and gently introduce about 60 mL of 50% (w/v) NaOH solution, taking care to minimize mixing with the solution in the flask. *The concentrated caustic solution is highly corrosive and should be handled with great care* (Note 4). Add several pieces of granulated zinc (Note 5) and a small piece of litmus paper. *Immediately* connect the Kjeldahl flask to the spray trap. Cautiously mix the contents by gentle swirling. The litmus paper should be blue after mixing is complete, indicating that the solution is basic.

Bring the solution to a boil, and distill at a steady rate until one half to one third of the original volume remains. Control the rate of heating to prevent the liquid in the receiver flask from being drawn back into the Kjeldahl flask. After distillation is judged complete, lower the receiver flask to bring the adapter well clear of the liquid. Discontinue heating, disconnect the apparatus, and rinse the inside of the condenser with small portions of distilled water, collecting the washings in the receiver flask. Add 2 drops of bromocresol green to the receiver flask, and titrate the residual HCl with standard 0.1 M NaOH to the color change of the indicator.

Report the percentage of nitrogen and the percentage of protein (Note 6) in the unknown.

Notes

1. If filter paper is used to hold the sample, carry a similar piece through the analysis as a blank. Acid-washed filter paper is frequently contami-

nated with measurable amounts of ammonium ion and should be avoided if possible.

2. Any of the following catalyze the digestion: a crystal of $CuSO_4$, 0.1 g of selenium, 0.2 g of $CuSeO_3$. The catalyst can be omitted, if desired.

3. A modification of this procedure uses about 50 mL of 4% boric acid solution in lieu of the standard HCl in the receiver flask. After distillation is complete, the ammonium borate produced is titrated with standard 0.1 M HCl, with 2 to 3 drops of bromocresol green as indicator.

4. If any sodium hydroxide solution comes into contact with your skin, wash the affected area **IMMEDIATELY** with copious amounts of water.

5. Granulated zinc (10 to 20 mesh) is added to minimize bumping during the distillation; it reacts slowly with the base to give small bubbles of hydrogen that prevent super-heating of the liquid.

6. The percentage of protein in the unknown is calculated by multiplying the % N by an appropriate factor: 5.70 for cereals, 6.25 for meats, and 6.38 for dairy products.

29C PRECIPITATION TITRATIONS

As noted in Chapter 13, most precipitation titrations make use of a standard silver nitrate solution as titrant. Directions follow for the volumetric titration of chloride ion using an adsorption indicator.

29C-1 Preparation of a Standard Silver Nitrate Solution

Procedure

Use a top-loading balance to transfer the approximate mass of $AgNO_3$ to a weighing bottle (Note 1). Dry at 110°C for about 1 h but not much longer (Note 2), and then cool to room temperature in a desiccator. Weigh the bottle and contents (to the nearest 0.1 mg). Transfer the bulk of the $AgNO_3$ to a volumetric flask using a powder funnel. Cap the weighing bottle and reweigh it and any solid that remains. Rinse the powder funnel thoroughly. Dissolve the $AgNO_3$, dilute to the mark with water, and mix well (Note 3). Calculate the molar concentration of this solution.

Notes

1. Consult with the instructor concerning the volume and concentration of $AgNO_3$ to be prepared. The mass of $AgNO_3$ to be taken is as follows:

Silver Ion Concentration, M	Approximate Mass (g) of $AgNO_3$ Needed to Prepare		
	1000 mL	500 mL	250 mL
0.10	16.9	8.5	4.2
0.05	8.5	4.2	2.1
0.02	3.4	1.8	1.0

2. Prolonged heating causes partial decomposition of $AgNO_3$. Some discoloration may occur, even after only 1 h at 110°C; the effect of this decomposition on the purity of the reagent is ordinarily imperceptible.
3. Silver nitrate solutions should be stored in a dark place when not in use.

29C-2 The Determination of Chloride by Titration with an Adsorption Indicator

Discussion

In this titration, the anionic adsorption indicator dichlorofluorescein is used to locate the end point. With the first excess of titrant, the indicator becomes incorporated in the counter-ion layer surrounding the silver chloride and imparts color to the solid (Section 13B-1). In order to obtain a satisfactory color change, it is desirable to maintain the particles of silver chloride in the colloidal state. Dextrin is added to the solution to stabilize the colloid and prevent its coagulation.

Preparation of Solutions

Dichlorofluorescein indicator (sufficient for several hundred titrations). Dissolve 0.2 g of dichlorofluorescein in a solution prepared by mixing 75 mL of ethanol and 25 mL of water.

Procedure

Dry the unknown at 110°C for about 1 h; allow it to return to room temperature in a desiccator. Weigh individual samples (to the nearest 0.1 mg) into individual conical flasks, and dissolve them in appropriate volumes of distilled water (Note 1). To each, add about 0.1 g of dextrin and 5 drops of indicator. Titrate (Note 2) with $AgNO_3$ to the first permanent pink color of silver dichlorofluoresceinate. Report the percentage of Cl^- in the unknown.

Notes

1. Use 0.25-g samples for 0.1 M $AgNO_3$ and about half that amount for 0.05 M reagent. Dissolve the former in about 200 mL of distilled water and the latter in about 100 mL. If 0.02 M $AgNO_3$ is to be used, weigh a 0.4-g sample into a 500-mL volumetric flask, and take 50-mL aliquots for titration.
2. Colloidal AgCl is sensitive to photodecomposition, particularly in the presence of the indicator; attempts to perform the titration in direct sunlight will fail. If photodecomposition appears to be a problem, establish the approximate end point with a rough preliminary titration, and use this information to estimate the volumes of $AgNO_3$ needed for the other samples. For each subsequent sample, add the indicator and dextrin only

after most of the $AgNO_3$ has been added, and then complete the titration without delay.

29D   COMPLEX-FORMATION TITRATIONS WITH EDTA

See Chapter 14 for a discussion of the analytical uses of EDTA as a chelating reagent. Directions follow for a direct titration of magnesium and a determination of the hardness of a natural water.

29D-1   Preparation of Solutions

Procedure

A pH-10 buffer and an indicator solution are needed for these titrations.

(a) *Buffer solution, pH 10.* (Sufficient for 80 to 100 titrations). Dilute 57 mL of concentrated NH_3 and 7 g of NH_4Cl in sufficient distilled water to give 100 mL of solution.
(b) *Eriochrome Black T indicator.* (Sufficient for about 100 titrations.) Dissolve 100 mg of the solid in a solution containing 15 mL of ethanolamine and 5 mL of absolute ethanol. This solution should be freshly prepared every two weeks; refrigeration slows its deterioration.

29D-2   Preparation of Standard 0.01 M EDTA Solution

Discussion

For a description of the properties of reagent-grade $Na_2H_2Y \cdot 2H_2O$ and its use in the direct preparation of standard EDTA solutions, see Section 14B-1.

Procedure

Dry about 4 g of the purified dihydrate $Na_2H_2Y \cdot 2H_2O$ (Note 1) at 80°C for 1 h to remove superficial moisture. Cool to room temperature in a desiccator. Weigh (to the nearest milligram) about 3.8 g into a 1-L volumetric flask (Note 2). Use a powder funnel to ensure quantitative transfer; rinse the funnel well with water before removing it from the flask. Add 600 to 800 mL of water (Note 3) and swirl periodically. Dissolution may take 15 min or longer. When all the solid has dissolved, dilute to the mark with water and mix well (Note 4). In calculating the molarity of the solution, correct the weight of the salt for the 0.3% moisture it ordinarily retains after drying at 80°C.

Notes

1. Directions for the purification of the disodium salt are described by W. J. Blaedel and H. T. Knight, *Anal. Chem.*, **1954**, *26* (4), 741.

2. The solution can be prepared from the anhydrous disodium salt, if desired. The weight taken should be about 3.6 g.
3. Water used in the preparation of standard EDTA solutions must be totally free of polyvalent cations. If any doubt exists concerning its quality, pass the water through a cation-exchange resin before use.
4. As an alternative, an EDTA solution that is approximately 0.01 M can be prepared and standardized by direct titration against a Mg^{2+} solution of known concentration (using the directions in Section 29D-3).

29D-3 The Determination of Magnesium by Direct Titration

Discussion

See Section 14B-7.

Procedure

Submit a clean 500-mL volumetric flask to receive the unknown, dilute to the mark with water, and mix thoroughly. Transfer 50.00-mL aliquots to 250-mL conical flasks, add 1 to 2 mL of pH-10 buffer and 3 to 4 drops of Eriochrome Black T indicator to each. Titrate with 0.01 M EDTA until the color changes from red to pure blue (Notes 1 and 2).

Express the results as parts per million of Mg^{2+} in the sample.

Notes

1. The color change tends to be slow in the vicinity of the end point. Care must be taken to avoid overtitration.
2. Other alkaline earths, if present, are titrated along with the Mg^{2+}; removal of Ca^{2+} and Ba^{2+} can be accomplished with $(NH_4)_2CO_3$. Most polyvalent cations are also titrated. Precipitation as hydroxides or the use of a masking reagent may be needed to eliminate this source of interference.

29D-4 The Determination of Hardness in Water

Discussion

See Section 14B-9.

Procedure

Acidify 100.0-mL aliquots of the sample with a few drops of HCl, and boil gently for a few minutes to eliminate CO_2. Cool, add 3 to 4 drops of methyl red, and neutralize with 0.1 M NaOH. Introduce 2 mL of pH-10 buffer, 3 to 4 drops of Eriochrome Black T, and titrate with standard 0.01 M Na_2H_2Y to a color change from red to pure blue (Note).

Report the results in terms of milligrams of $CaCO_3$ per liter of water.

Note

The color change is sluggish if Mg^{2+} is absent. In this event, add 1 to 2 mL of 0.1 M MgY^{2-} before starting the titration. This reagent is prepared by adding 2.645 g of $MgSO_4 \cdot 7H_2O$ to 3.722 g of $Na_2H_2Y \cdot 2H_2O$ in 50 mL of distilled water. The solution is then made faintly alkaline to phenolphthalein and diluted to 100 mL. A small portion, mixed with pH-10 buffer and a few drops of Eriochrome Black T indicator, should have a dull violet color. A single drop of 0.01 M EDTA solution should cause a color change to blue, while an equal volume of 0.01 Mg^{2+} should cause a change to red. If necessary, adjust the composition with EDTA or with Mg^{2+} until these criteria are met.

29E   TITRATIONS WITH POTASSIUM PERMANGANATE

The properties and uses of potassium permanganate are described in Section 17C-1. Directions follow for the determination of iron in an ore and calcium in a limestone.

29E-1 Preparation of 0.02 M Potassium Permanganate

Discussion

See page 313 for a discussion of the precautions needed in the preparation and storage of permanganate solutions.

Procedure

Dissolve about 3.2 g of $KMnO_4$ in 1 L of distilled water. Keep the solution at a gentle boil for about 1 h. Cover and let stand overnight. Remove MnO_2 by filtration (Note 1) through a fine-porosity filtering crucible (Note 2) or through a Gooch crucible fitted with glass mats. Transfer the solution to a clean glass-stoppered bottle; store in the dark when not in use.

Notes

1. The heating and filtering steps can be omitted if the permanganate solution is standardized and used on the same day.
2. Remove the MnO_2 that collects on the fritted plate with 1 M H_2SO_4 containing a few milliliters of 3% H_2O_2, followed by a rinse with copious quantities of water.

29E-2 Standardization of Potassium Permanganate Solutions

Discussion

See Section 17C-1 for a discussion of primary standards for permanganate solutions. Directions follow for standardization with sodium oxalate.

Procedure

Dry about 1.5 g of primary-standard $Na_2C_2O_4$ at 110°C for at least 1 h. Cool in a desiccator; weigh (to the nearest 0.1 mg) individual 0.2- to 0.3-g samples into 400-mL beakers. Dissolve each in about 250 mL of 1 M H_2SO_4. Heat each solution to 80°C to 90°C, and titrate with $KMnO_4$ while stirring with a thermometer. The pink color imparted by one addition should be permitted to disappear before any further titrant is introduced (Notes 1 and 2). Reheat if the temperature drops below 60°C. Take the first persistent (~30 s) pink color as the end point (Notes 3 and 4). Determine a blank by titrating an equal volume of the 1 M H_2SO_4.

Correct the titration data for the blank, and calculate the concentration of the permanganate solution (Note 5).

Notes

1. Promptly wash any $KMnO_4$ that spatters on the walls of the beaker into the bulk of the liquid with a stream of water.
2. Finely divided MnO_2 will form along with Mn^{2+} if the $KMnO_4$ is added too rapidly and will cause the solution to acquire a faint brown discoloration. Precipitate formation is not a serious problem so long as sufficient oxalate remains to reduce the MnO_2 to Mn^{2+}; the titration is simply discontinued until the brown color disappears. The solution must be free of MnO_2 at the end point.
3. The surface of the permanganate solution rather than the bottom of the meniscus can be used to measure titrant volumes. Alternatively, backlighting with a flashlight or a match will permit reading of the meniscus in the conventional manner.
4. A permanganate solution should not be allowed to stand in a buret any longer than necessary because partial decomposition to MnO_2 may occur. Freshly formed MnO_2 can be removed from a glass surface with 1 M H_2SO_4 containing a small amount of 3% H_2O_2.
5. As noted on page 315, this procedure yields molarities that are a few tenths of a percent low. For more accurate results, introduce from a buret sufficient permanganate to react with 90% to 95% of the oxalate (about 40 mL of 0.02 M $KMnO_4$ for a 0.3-g sample). Let the solution stand until the permanganate color disappears. Then warm to about 60°C and complete the titration, taking the first permanent pink (~ 30 s) as the end point (Notes 3 and 4). Determine a blank by titrating an equal volume of the 1 M H_2SO_4.

29E-3 The Determination of Calcium in a Limestone

Discussion

In common with a number of other cations, calcium is conveniently determined by precipitation with oxalate ion. The solid calcium oxalate is filtered, washed free of excess precipitating reagent, and dissolved in dilute acid. The oxalic acid liberated in this step is then titrated with standard permanganate or some other oxidizing reagent. This method is applicable to samples that contain magnesium and the alkali metals. Most other cations must be absent since they either precipitate or coprecipitate as oxalates and cause positive errors in the analysis.

Factors Affecting the Composition of Calcium Oxalate Precipitates. It is essential that the mole ratio between calcium and oxalate be exactly unity in the precipitate and thus in solution at the time of titration. A number of precautions are needed to ensure this condition. For example, the calcium oxalate formed in a neutral or ammoniacal solution is likely to be contaminated with calcium hydroxide or a basic calcium oxalate, either of which will cause low results. The formation of these compounds is prevented by adding the oxalate to an acidic solution of the sample and slowly forming the desired precipitate by the dropwise addition of ammonia. The coarsely crystalline calcium oxalate that is produced under these conditions is readily filtered. Losses resulting from the solubility of calcium oxalate are negligible above pH 4, provided washing is limited to freeing the precipitate of excess oxalate.

Coprecipitation of sodium oxalate becomes a source of positive error in the determination of calcium whenever the concentration of sodium in the sample exceeds that of calcium. The error from this source can be eliminated by reprecipitation.

Magnesium, if present in high concentration, may also be a source of contamination. An excess of oxalate ion helps prevent this interference through the formation of soluble oxalate complexes with magnesium. Prompt filtration of the calcium oxalate can also help prevent interference because of the pronounced tendency of magnesium oxalate to form supersaturated solutions from which precipitate formation occurs only after an hour or more. These measures do not suffice for samples that contain more magnesium than calcium. Here, reprecipitation of the calcium oxalate becomes necessary.

> Reprecipitation is a method for minimizing coprecipitation errors by dissolving the initial precipitate and then reforming the solid.

The Composition of Limestones. Limestones are composed principally of calcium carbonate; dolomitic limestones contain large amounts of magnesium carbonate as well. Calcium and magnesium silicates are also present in smaller amounts, along with the carbonates and silicates of iron, aluminum, manganese, titanium, sodium, and other metals.

Hydrochloric acid is an effective solvent for most limestones. Only silica, which does not interfere with the analysis, remains undissolved. Some limestones are more readily decomposed after they have been ignited; a few yield only to a carbonate fusion.

The method that follows is remarkably effective for determining calcium in most limestones. Iron and aluminum, in amounts equivalent to that of calcium, do not interfere. Small amounts of manganese and titanium can also be tolerated.

Procedure

Sample Preparation

Dry the unknown for 1 to 2 h at 110°C, and cool in a desiccator. If the material is readily decomposed in acid, weigh 0.25- to 0.30-g samples (to the nearest 0.1 mg) into 250-mL beakers. Add 10 mL of water to each sample and cover with a watch glass. Add 10 mL of concentrated HCl dropwise, taking care to avoid losses due to spattering as the acid is introduced.

Precipitation of Calcium Oxalate

Add 5 drops of saturated bromine water to oxidize any iron in the samples and boil gently (**use a hood**) for 5 min to remove the excess Br_2. Dilute each sample solution to about 50 mL, heat to boiling, and add 100 mL of hot 6% (w/v) $(NH_4)_2C_2O_4$ solution. Add 3 to 4 drops of methyl red, and precipitate CaC_2O_4 by slowly adding 6 M NH_3. As the indicator starts to change color, add the NH_3 at a rate of one drop every 3 to 4 s. Continue until the solutions turn to the intermediate yellow-orange color of the indicator (pH 4.5 to 5.5). Allow the solutions to stand for no more than 30 min (Note) and filter; medium-porosity filtering crucibles or Gooch crucibles with glass mats are satisfactory. Wash the precipitates with several 10-mL portions of cold water. Rinse the outside of the crucibles to remove residual $(NH_4)_2C_2O_4$, and return them to the beakers in which the CaC_2O_4 was formed.

Titration

Add 100 mL of water and 50 mL of 3 M H_2SO_4 to each of the beakers containing the precipitated calcium oxalate and the crucible. Heat to 80°C to 90°C, and titrate with 0.02 M permanganate. The temperature should be above 60°C throughout the titration; reheat if necessary.

Report the percentage of CaO in the unknown.

Note

The period of standing can be longer if the unknown contains no Mg^{2+}.

29E-4 The Determination of Iron in an Ore

Discussion

The common ores of iron are hematite (Fe_2O_3), magnetite (Fe_3O_4), and limonite ($2Fe_2O_3 \cdot 3H_2O$). Steps in the analysis of these ores are (1) dissolution of the sample, (2) reduction of iron to the divalent state, and (3) titration of iron(II) with a standard oxidant.

The Decomposition of Iron Ores. Iron ores often decompose completely in hot concentrated hydrochloric acid. The rate of attack by this reagent is increased by the presence of a small amount of tin(II) chloride. The tendency of iron(II) and iron(III) to form chloro complexes accounts for the effectiveness of hydrochloric acid over nitric or sulfuric acid as a solvent for iron ores.

Many iron ores contain silicates that may not be entirely decomposed by treatment with hydrochloric acid. Incomplete decomposition is indicated by a dark residue that remains after prolonged treatment with the acid. A white residue of hydrated silica, which does not interfere in any way, is indicative of complete decomposition.

The Prereduction of Iron. Because part or all of the iron is in the trivalent state after decomposition of the sample, prereduction to iron(II) must precede titration with the oxidant. Any of the methods described in Section 17A-1 can be used. Perhaps the most satisfactory prereductant for iron is tin(II) chloride:

$$2Fe^{3+} + Sn^{2+} \rightarrow 2Fe^{2+} + Sn^{4+}$$

The only other common species reduced by this reagent are the high oxidation states of arsenic, copper, mercury, molybdenum, tungsten, and vanadium.

The excess reducing agent is eliminated by the addition of mercury(II) chloride:

$$Sn^{2+} + 2HgCl_2 \rightarrow Hg_2Cl_2(s) + Sn^{4+} + 2Cl^-$$

The slightly soluble mercury(I) chloride (Hg_2Cl_2) does not reduce permanganate, nor does the excess mercury(II) chloride ($HgCl_2$) reoxidize iron(II). Care must be taken, however, to prevent the occurrence of the alternative reaction

$$Sn^{2+} + HgCl_2 \rightarrow Hg(l) + Sn^{4+} + 2Cl^-$$

Elemental mercury reacts with permanganate and causes the results of the analysis to be high. The formation of mercury, which is favored by an appreciable excess of tin(II), is prevented by careful control of this excess and by the rapid addition of excess mercury(II) chloride. A proper reduction is indicated by the appearance of a small amount of a silky white precipitate after the addition of mercury(II). Formation of a gray precipitate at this juncture indicates the presence of metallic mercury; the total absence of a precipitate indicates that an insufficient amount of tin(II) chloride was used. In either event, the sample must be discarded.

The Titration of Iron(II). The reaction of iron(II) with permanganate is smooth and rapid. The presence of iron(II) in the reaction mixture, however, *induces* the oxidation of chloride ion by permanganate, a reaction that does not ordinarily proceed rapidly enough to cause serious error. High results are obtained if this parasitic reaction is not controlled. Its effects can be eliminated through removal of the hydrochloric acid by evaporation with sulfuric acid or by introduction of *Zimmermann-Reinhardt reagent*, which contains manganese(II) in a fairly concentrated mixture of sulfuric and phosphoric acids.

The oxidation of chloride ion during a titration is believed to involve a direct reaction between this species and the manganese(III) ions that form as an intermediate in the reduction of permanganate ion by iron(II). The presence of manganese(II) in the Zimmermann-Reinhardt reagent is believed to inhibit the formation of chlorine by decreasing the potential of the manganese(III)/manganese(II) couple. Phosphate ion is believed to exert a similar effect by forming stable manganese(III) complexes. Moreover, phosphate ions react with iron(III) to form nearly colorless complexes so that the yellow color of the iron(III)/chloro complexes does not interfere with the end point.[5]

---

[5] The mechanism by which Zimmermann-Reinhardt reagent acts has been the subject of much study. For a discussion of this work, see H. A. Laitinen, *Chemical Analysis*, pp. 369–372. New York: McGraw-Hill, 1960.

Preparation of Reagents

The following solutions suffice for about 100 titrations.

(a) *Tin(II) chloride, 0.25 M.* Dissolve 60 g of iron-free $SnCl_2 \cdot 2H_2O$ in 100 mL of concentrated HCl; warm if necessary. After the solid has dissolved, dilute to 1 L with distilled water and store in a well-stoppered bottle. Add a few pieces of mossy tin to help preserve the solution.

(b) *Mercury(II) chloride, 5% (w/v).* Dissolve 50 g of $HgCl_2$ in 1 L of distilled water.

(c) *Zimmermann-Reinhardt reagent.* Dissolve 300 g of $MnSO_4 \cdot 4H_2O$ in 1 L of water. Cautiously add 400 mL of concentrated H_2SO_4, 400 mL of 85% H_3PO_4, and dilute to 3 L.

Procedure

Sample Preparation

Dry the ore at 110°C for at least 3 h, and then allow it to cool to room temperature in a desiccator. Consult with the instructor for a sample size that will require from 25 to 40 mL of standard 0.02 M $KMnO_4$. Weigh samples into 500-mL conical flasks. To each, add 10 mL of concentrated HCl and about 3 mL of 0.25 M $SnCl_2$ (Note 1). Cover each flask with a small watch glass or flask cover. Heat the flasks in a hood at just below boiling until the samples are decomposed and the undissolved solid—if any—is pure white (Note 2). Use another 1 to 2 mL of $SnCl_2$ to eliminate any yellow color that may develop as the solutions are heated. Heat a blank consisting of 10 mL of HCl and 3 mL of $SnCl_2$ for the same amount of time.

After the ore has been decomposed, remove the excess Sn(II) by the dropwise addition of 0.02 M $KMnO_4$ until the solutions become faintly yellow. Dilute to about 15 mL. Add sufficient $KMnO_4$ solution to impart a faint pink color to the blank; then decolorize with one drop of the $SnCl_2$ solution.

Take samples and blank individually through subsequent steps to minimize air-oxidation of iron(II).

Reduction of Iron

Heat the sample solution nearly to boiling, and make dropwise additions of 0.25 M $SnCl_2$ until the yellow color just disappears; then add two more drops (Note 3). Cool to room temperature and *rapidly* add 10 mL of 5% $HgCl_2$ solution. A small amount of silky white Hg_2Cl_2 should precipitate (Note 4). The blank should be treated with the $HgCl_2$ solution.

Titration

Following addition of the $HgCl_2$, wait 2 to 3 min. Then add 25 mL of Zimmermann-Reinhardt reagent and 300 mL of water. Titrate *immediately* with standard 0.02 M $KMnO_4$ to the first faint pink that persists for 15 to 20 s. Do not add the $KMnO_4$ rapidly at any time. Correct the titrant volume for the blank.

Report the percentage of Fe_2O_3 in the sample.

Notes

1. The $SnCl_2$ hastens decomposition of the ore by reducing iron(III) oxides to iron(II). Insufficient $SnCl_2$ is indicated by the appearance of yellow iron(III)/chloride complexes.
2. If dark particles persist after the sample has been heated with acid for several hours, filter the solution through ashless paper, wash the residue with 5 to 10 mL of 6 M HCl, and retain the filtrate and washings. Ignite the paper and its contents in a small platinum crucible. Mix 0.5 to 0.7 g of Na_2CO_3 with the residue and heat until a clear melt is obtained. Cool, add 5 mL of water, and then cautiously add a few milliliters of 6 M HCl. Warm the crucible until the melt has dissolved, and combine the contents with the original filtrate. Evaporate the solution to 15 mL and continue the analysis.
3. The solution may not become entirely colorless but instead may acquire a faint yellow-green hue. Further additions of $SnCl_2$ will not alter this color. The addition of too much $SnCl_2$ can be corrected by adding 0.2 M $KMnO_4$ and repeating the reduction.
4. The absence of precipitate indicates that insufficient $SnCl_2$ was used and that the reduction of iron(III) was incomplete. A gray residue indicates the presence of elemental mercury, which reacts with $KMnO_4$. The sample must be discarded in either event.
5. These directions can be used to standardize a permanganate solution against primary standard iron. Weigh (to the nearest 0.1 mg) 0.2-g lengths of electrolytic iron wire into 250-mL conical flasks and dissolve in about 10 mL of concentrated HCl. Dilute each sample to about 75 mL. Then take each individually through the reduction and titration steps.

29F   TITRATIONS WITH IODINE

The oxidizing properties of iodine, the composition and stability of triiodide solutions, and the applications of this reagent in volumetric analysis are discussed in Section 17C-3. Starch is ordinarily employed as an indicator for iodimetric titrations.

29F-1   Preparation of Reagents

Procedure

(a) *Iodine approximately 0.05 M.* Weigh about 40 g of KI into a 100-mL beaker. Add 12.7 g of I_2 and 10 mL of water. Stir for several minutes

(Note 1). Introduce an additional 20 mL of water and stir again for several minutes. Carefully decant the bulk of the liquid into a storage bottle containing 1 L of distilled water. It is essential that any undissolved iodine remain in the beaker (Note 2).

Notes

1. Iodine dissolves slowly in the KI solution. Thorough stirring is needed to hasten the process.
2. Any solid I_2 inadvertently transferred to the storage bottle will cause the concentration of the solution to increase gradually. Filtration through a sintered-glass crucible eliminates this potential source of difficulty.

(b) *Starch indicator*. (Sufficient for about 100 titrations.) Rub 1 g of soluble starch and 15 mL of water into a paste. Dilute to about 500 mL with boiling water, and heat until the mixture is clear. Cool; store in a tightly stoppered bottle. For most titrations, 3 to 5 mL of the indicator are used.

The indicator is readily attacked by airborne organisms and should be freshly prepared every few days.

29F-2 Standardization of Iodine Solutions

Discussion

Arsenic(III) oxide, long a favored primary standard for iodine solutions, is now seldom used because of the elaborate federal regulations governing the use of even small amounts of arsenic-containing compounds. Barium thiosulfate monohydrate and anhydrous sodium thiosulfate have been proposed as alternative standards.[6] Perhaps the most convenient method for determining the concentration of an iodine solution is the titration of aliquots with a sodium thiosulfate solution that has been standardized against pure potassium iodate. Instructions for this method follow.

Preparation of Reagents

(a) *Sodium thiosulfate, 0.1 M*. Follow the directions in Sections 29G-1 and 29G-2 for the preparation and standardization of this solution.
(b) *Starch indicator*. See Section 29F-1(b).

---

[6] W. M. McNevin and O. H. Kriege, *Anal. Chem.*, **1953**, *25* (5), 767; A. A. Woolf, *Anal. Chem.*, **1982**, *54* (12), 2134.

Procedure

Transfer 25.00-mL aliquots of the iodine solution to 250-mL conical flasks, and dilute to about 50 mL. *Take each aliquot individually through subsequent steps.* Introduce approximately 1 mL of 3 M H_2SO_4, and titrate immediately with standard sodium thiosulfate until the solution becomes a faint straw-yellow. Add about 5 mL of starch indicator, and complete the titration, taking as the end point the change in color from blue to colorless (Note).

Note

The blue color of the starch/iodine complex may reappear after the titration has been completed, owing to the air-oxidation of iodide ion.

29F-3 The Determination of Antimony in Stibnite

Discussion

The analysis of stibnite, a common antimony ore, is a typical application of iodimetry and is based on the oxidation of Sb(III) to Sb(V):

$$SbO_3^{3-} + I_2 + H_2O \rightleftharpoons SbO_4^{3-} + 2I^- + 2H^+$$

The position of this equilibrium is strongly dependent on the hydrogen ion concentration. In order to force the reaction to the right, it is common practice to carry out the titration in the presence of an excess of sodium hydrogen carbonate, which consumes the hydrogen ions as they are produced.

Stibnite is an antimony sulfide ore containing silica and other contaminants. Provided the material is free of iron and arsenic, the analysis of stibnite for its antimony content is straightforward. Samples are decomposed in hot concentrated hydrochloric acid to eliminate sulfide as gaseous hydrogen sulfide. Care is needed to prevent loss of volatile antimony(III) chloride during this step. The addition of potassium chloride helps by favoring formation of nonvolatile chloro complexes such as $SbCl_4^-$ and $SbCl_6^{3-}$.

Sparingly soluble basic antimony salts, such as SbOCl, often form when the excess hydrochloric acid is neutralized; these react incompletely with iodine and cause low results. The difficulty is overcome by adding tartaric acid, which forms a soluble complex ($SbOC_4H_4O_6^-$) from which antimony is rapidly oxidized by the reagent.

Procedure

Dry the unknown at 110°C for 1 h, and allow it to cool in a desiccator. Weigh individual samples (Note 1) into 500-mL conical flasks. Introduce about 0.3 g of KCl and 10 mL of concentrated HCl to each flask. Heat the mixtures (**use a hood**) just below boiling until only white or slightly gray residues of SiO_2 remain.

Add 3 g of tartaric acid to each sample and heat for an additional 10 to

15 min. Then, with good swirling, add water (Note 2) from a pipet or buret until the volume is about 100 mL. If reddish Sb_2S_3 forms, discontinue dilution and heat further to eliminate H_2S; add more HCl if necessary.

Add 3 drops of phenolphthalein, and neutralize with 6 M NaOH to the first faint pink of the indicator. Discharge the color by the dropwise addition of 6 M HCl, and then add 1 mL in excess. Introduce 4 to 5 g of $NaHCO_3$, taking care to avoid losses of solution by spattering during the addition. Add 5 mL of starch indicator, rinse down the inside of the flask, and titrate with standard 0.05 M I_2 to the first blue color that persists for 30 s.

Report the percentage of Sb_2S_3 in the unknown.

Notes

1. Samples should contain between 1.5 and 2 mmol of antimony; consult with your instructor for an appropriate sample size. Weighings to the nearest milligram are adequate for samples larger than 1 g.
2. The slow addition of water, with efficient stirring, is essential to prevent the formation of SbOCl.

29G   TITRATIONS WITH SODIUM THIOSULFATE

Numerous methods are based on the reducing properties of iodide ion:

$$2I^- \rightarrow I_2 + 2e^-$$

Iodine, the reaction product, is ordinarily titrated with a standard sodium thiosulfate solution, with starch serving as the indicator:

$$I_2 + 2S_2O_3^{2-} \rightarrow 2I^- + S_4O_6^{2-}$$

A discussion of thiosulfate methods is found in Section 17B-2.

29G-1 Preparation of 0.1 M Sodium Thiosulfate

Procedure

Boil about 1 L of distilled water for 10 to 15 min. Allow the water to cool to room temperature; then add about 25 g of $Na_2S_2O_3 \cdot 5H_2O$ and 0.1 g of Na_2CO_3. Stir until the solid has dissolved. Transfer the solution to a clean glass or plastic bottle, and store in a dark place.

29G-2 Standardization of Sodium Thiosulfate Against Potassium Iodate

Discussion

Solutions of sodium thiosulfate are conveniently standardized by titration of the iodine produced when an unmeasured excess of potassium iodide is added to a known volume of an acidified standard potassium iodate solution. The reaction is

$$IO_3^- + 5I^- + 6H^+ \rightarrow 3I_2 + 3H_2O$$

Note that each mole of iodate results in the production of three moles of iodine. The procedure that follows is based on this reaction.

Preparation of Solutions

(a) *Potassium iodate, 0.0100 M.* Dry about 1.2 g of primary-standard KIO_3 at 110°C for at least 1 h and cool in a desiccator. Weigh (to the nearest 0.1 mg) about 1.1 g into a 500-mL volumetric flask; use a powder funnel to ensure quantitative transfer of the solid. Rinse the funnel well, dissolve the KIO_3 in about 200 mL of distilled water, dilute to the mark, and mix thoroughly.

(b) *Starch indicator.* See Section 29F-1 Procedure (b).

Procedure

Pipet 50.00-mL aliquots of standard iodate solution into 250-mL conical flasks. *Treat each sample individually from this point to minimize error resulting from the air-oxidation of iodide ion.* Introduce 2 g of iodate-free KI, and swirl the flask to hasten solution. Add 2 mL of 6 M HCl, and immediately titrate with thiosulfate until the solution becomes pale yellow. Introduce 5 mL of starch indicator, and titrate with constant stirring to the disappearance of the blue color. Calculate the molarity of the iodine solution.

29G-3 Standardization of Sodium Thiosulfate Against Copper

Discussion

Thiosulfate solutions can also be standardized against pure copper wire or foil. This procedure is advantageous when the solution is to be used for the determination of copper because any systematic error in the method tends to be canceled.

Copper(II) is reduced quantitatively to copper(I) by iodide ion:

$$2Cu^{2+} + 4I^- \rightarrow 2CuI(s) + I_2$$

The importance of CuI formation in forcing this reaction to completion can be seen from the following standard electrode potentials:

$$Cu^{2+} + e^- \rightleftharpoons Cu^+ \qquad E^0 = 0.15 \text{ V}$$

$$I_2 + 2e^- \rightleftharpoons 2I^- \qquad E^0 = 0.54 \text{ V}$$

$$Cu^{2+} + I^- + e^- \rightleftharpoons CuI(s) \qquad E^0 = 0.86 \text{ V}$$

The first two potentials suggest that iodide should have no tendency to reduce copper(II); the formation of CuI, however, favors the reduction. The solution must contain at least 4% excess iodide to force the reaction to completion. Moreover, the pH must be less than 4 to prevent the formation of basic copper species that react slowly and incompletely with iodide ion. The acidity of the solution cannot be greater than about 0.3 M, however, because of the tendency of iodide ion to undergo air-oxidation, a process catalyzed by copper salts. Nitrogen oxides also catalyze the air-oxidation of iodide ion. A common source of these oxides is the nitric acid ordinarily used to dissolve metallic copper and other copper-containing solids. Urea is used to scavenge nitrogen oxides from solutions:

$$(NH_2)_2CO + 2HNO_2 \rightarrow 2N_2(g) + CO_2(g) + 3H_2O$$

The titration of iodine by thiosulfate tends to yield slightly lower results owing to the adsorption of small but measurable quantities of iodine on solid CuI. The adsorbed iodine is released only slowly, even when thiosulfate is in excess; transient and premature end points result. This difficulty is largely overcome by the addition of thiocyanate ion. The sparingly soluble copper(I) thiocyanate replaces part of the copper iodide at the surface of the solid:

$$CuI(s) + SCN^- \rightarrow CuSCN(s) + I^-$$

Accompanying this reaction is the release of the adsorbed iodine, which thus becomes available for titration. The addition of thiocyanate must be delayed until most of the iodine has been titrated to prevent interference from a slow reaction between the two species, possibly

$$2SCN^- + I_2 \rightarrow 2I^- + (SCN)_2$$

Preparation of Solutions

(a) *Urea, 5% (w/v).* Dissolve about 5 g of urea in sufficient water to give 100 mL of solution. Approximately 10 mL will be needed for each titration.

(b) *Starch indicator.* See Section 29F-1 Procedure (b).

Procedure

Use scissors to cut copper wire or foil into 0.20- to 0.25-g portions. Wipe the metal free of dust and grease with a filter paper; do not dry it in an oven. The pieces of copper should be handled with paper strips, cotton gloves, or tweezers to prevent contamination by contact with the skin.

Use a weighed watch glass or weighing bottle to obtain the mass of individual copper samples (to the nearest 0.1 mg) by difference. Transfer each sample to a 250-mL conical flask. Add 5 mL of 6 M HNO_3, cover with a small watch glass, and warm gently (**use a hood**) until the metal has dissolved. Dilute with about 25 mL of distilled water, add 10 mL of 5% (w/v) urea, and boil briefly to eliminate nitrogen oxides. Rinse the watch glass, collecting the rinsings in the flask. Cool.

Add concentrated NH_3 dropwise and with thorough mixing to produce the intensely blue $Cu(NH_3)_4^{2+}$; the solution should smell faintly of ammonia (Note). Make dropwise additions of 3 M H_2SO_4 until the color of the complex just disappears, and then add 2.0 mL of 85% H_3PO_4. Cool to room temperature.

Treat each sample individually from this point on to minimize the air-oxidation of iodide ion. Add 4.0 g of KI to the sample, and titrate immediately with $Na_2S_2O_3$ until the solution becomes pale yellow. Add 5 mL of starch indicator, and continue the titration until the blue color becomes faint. Add 2 g of KSCN; swirl vigorously for 30 s. Complete the titration, using the disappearance of the blue starch/I_2 color as the end point.

Calculate the molarity of the $Na_2S_2O_3$ solution.

Note

Do not sniff vapors directly from the flask; instead, use a waving motion of your hand to waft them toward your nose.

29G-4 The Determination of Copper in Brass

Discussion

The standardization procedure described in Section 29G-3 is readily adapted to the determination of copper in brass, an alloy that also contains appreciable amounts of tin, lead, and zinc (and perhaps minor amounts of nickel and iron). The method is relatively simple and applicable to brasses with less than 2% iron. A weighed sample is treated with nitric acid, which causes the tin to precipitate as a hydrated oxide of uncertain composition. Evaporation with sulfuric acid to the appearance of sulfur trioxide eliminates the excess nitrate, redissolves the tin compound, and possibly causes the formation of lead sulfate. The pH is adjusted through the addition of ammonia, followed by acidification with a measured amount of phosphoric acid. An excess of potassium iodide is added, and the liberated iodine is titrated with standard thiosulfate. See Section 29G-3 for additional discussion.

Procedure

If so directed, free the metal of oils by treatment with an organic solvent; briefly heat in an oven to drive off the solvent. Weigh (to the nearest 0.1 mg) 0.3-g samples into 250-mL conical flasks, and introduce 5 mL of 6 M HNO_3 into each; warm (**use a hood**) until solution is complete. Add 10 mL of concentrated H_2SO_4, and evaporate (**hood**) until copious white fumes of SO_3 are given off. Allow the mixture to cool. Cautiously add 30 mL of distilled water, boil for 1 to 2 min, and cool again.

Follow the instructions in the third and fourth paragraphs of the *Procedure* in Section 29G-3.

Report the percentage of Cu in the sample.

29H   TITRATIONS WITH POTASSIUM BROMATE

Applications of standard bromate solutions to the determination of organic functional groups are described in Section 17C-4. Directions follow for the determination of ascorbic acid in vitamin C tablets.

29H-1 Preparation of Solutions

Procedure

(a) *Potassium bromate, 0.015 M.* Transfer about 1.5 g of reagent-grade potassium bromate to a weighing bottle, and dry at 110°C for at least 1 h. Cool in a desiccator. Weigh approximately 1.3 g (to the nearest 0.1 mg) into a 500-mL volumetric flask; use a powder funnel to ensure quantitative transfer of the solid. Rinse the funnel well, and dissolve the $KBrO_3$ in about 200 mL of distilled water. Dilute to the mark, and mix thoroughly.

   Solid potassium bromate can cause a fire if it comes into contact with damp organic material (such as paper toweling in a waste container). Consult with the instructor concerning the disposal of any excess.

(b) *Sodium thiosulfate, 0.05 M.* Follow the directions in Section 29G-1; use about 12.5 g of $Na_2S_2O_3 \cdot 5H_2O$ per liter of solution.

(c) *Starch indicator.* See Section 29F-1 Procedure (b).

29H-2 Standardization of Sodium Thiosulfate Against Potassium Bromate

Discussion

Iodine is generated by the reaction between a known volume of standard potassium bromate and an unmeasured excess of potassium iodide:

$$BrO_3^- + 6I^- + 6H^+ \rightarrow Br^- + 3I_2 + 3H_2O$$

The iodine produced is titrated with the sodium thiosulfate solution.

Procedure

Pipet 25.00-mL aliquots of the $KBrO_3$ solution into 250-mL conical flasks and rinse the interior wall with distilled water. *Treat each sample individually beyond this point.* Introduce 2 to 3 g of KI and about 5 mL of 3 M H_2SO_4.

Immediately titrate with $Na_2S_2O_3$ until the solution is pale yellow. Add 5 mL of starch indicator, and titrate to the disappearance of the blue color. Calculate the concentration of the thiosulfate solution.

29H-3 The Determination of Ascorbic Acid in Vitamin C Tablets by Titration with Potassium Bromate

Discussion

Ascorbic acid, $C_6H_8O_6$, is cleanly oxidized to dehydroascorbic acid by bromine:

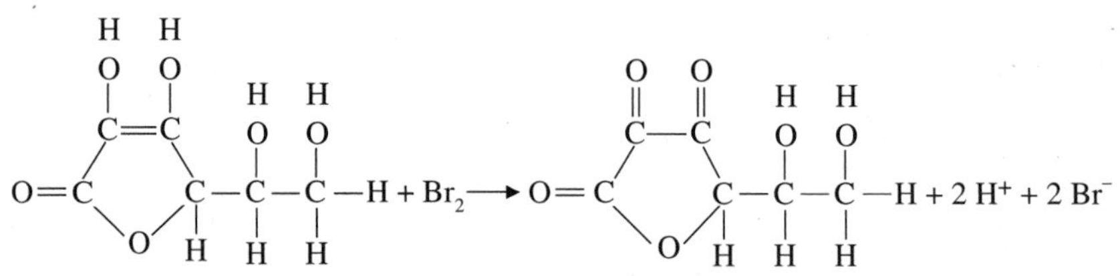

An unmeasured excess of potassium bromide is added to an acidified solution of the sample. The solution is titrated with standard potassium bromate to the first permanent appearance of excess bromine; this excess is then determined iodometrically with standard sodium thiosulfate. The entire titration must be performed without delay to prevent air-oxidation of the ascorbic acid.

Procedure

Weigh (to the nearest milligram) 3 to 5 vitamin C tablets (Note 1). Pulverize them thoroughly in a mortar, and transfer the powder to a dry weighing bottle. Weigh individual 0.40- to 0.50-g samples (to the nearest 0.1 mg) into dry 250-mL conical flasks. *Treat each sample individually beyond this point.* Dissolve the sample (Note 2) in 50 mL of 1.5 M H_2SO_4; then add about 5 g of KBr. Titrate immediately with standard $KBrO_3$ to the first faint yellow due to excess Br_2. Record the volume of $KBrO_3$ used. Add 3 g of KI and 5 mL of starch indicator; back-titrate (Note 3) with standard 0.05 M $Na_2S_2O_3$.

Calculate the average mass (in milligrams) of ascorbic acid (176.13 g/mol) in each tablet.

Notes

1. This method is not applicable to chewable vitamin C tablets.
2. The binder in many vitamin C tablets remains in suspension throughout the analysis. If the binder is starch, the characteristic color of the complex with iodine appears on the addition of KI.
3. The volume of thiosulfate needed for the back-titration seldom exceeds a few milliliters.

29I POTENTIOMETRIC METHODS

Potentiometric measurements provide a highly selective method for the quantitative determination of numerous cations and anions. A discussion of the principles and applications of potentiometric measurements is found in Chapter 18. Detailed instructions are given in this section on the use of potentiometric measurements to locate end points in volumetric titrations. In addition, a procedure for the direct potentiometric determination of fluoride ion in drinking water and in toothpaste is described.

29I-1 General Directions for Performing a Potentiometric Titration

The procedure that follows is applicable to the titrimetric methods described in this section. With the proper choice of indicator electrode, it can also be applied to most of the volumetric methods given in Sections 29B through 29H.

1. Dissolve the sample in 50 to 250 mL of water. Rinse a suitable pair of electrodes with deionized water, and immerse them in the sample solution. Provide magnetic (or mechanical) stirring. Position the buret so that reagent can be delivered without splashing.
2. Connect the electrodes to the meter, commence stirring, and record the initial buret volume and the initial potential (or pH).
3. Record the meter reading and buret volume after each addition of titrant. Introduce fairly large volumes (about 5 mL) at the outset. Withhold a succeeding addition until the meter reading remains constant within 1 to 2 mV (or 0.05 pH unit) for at least 30 s (Note). Judge the volume of reagent to be added by estimating a value for $\Delta E/\Delta V$ after each addition. In the immediate vicinity of the equivalence point, introduce the reagent in 0.1-mL increments. Continue the titration 2 to 3 mL beyond the equivalence point, increasing the volume increments as $\Delta E/\Delta V$ again becomes smaller.

Note. Stirring motors occasionally cause erratic meter readings. You may need to turn off the motor while you are making meter readings.

29I-2 The Potentiometric Titration of Chloride and Iodide in a Mixture

Discussion

Figure 29-2 is an argentometric titration curve for a mixture of iodide and chloride ions. Initial additions of silver nitrate result in formation of silver iodide exclusively because the solubility of that salt is only about 5×10^{-7} that of silver chloride. It can be shown that this solubility difference is great enough so that formation of silver chloride is delayed until all but 7×10^{-5} % of the iodide has precipitated. Thus, short of the equivalence point, the curve is essentially indistinguishable from that for iodide alone. Just beyond the iodide equivalence point, the silver ion concentration is determined by the concentration of chloride ion in the solution, and the titration curve becomes essentially identical to that for chloride ion by itself.

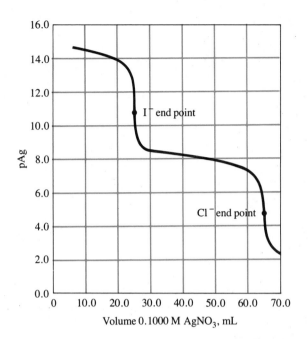

Figure 29-2

Titration curve for 50.00 mL of a solution that is 0.0500 M in I^- and 0.0800 M in Cl^-.

Curves resembling Figure 29-2 can be obtained experimentally by measuring the potential of a silver electrode immersed in the analyte solution. Hence, a chloride/iodide mixture can be analyzed for each of its components. This technique is not as applicable to analyzing iodide/bromide or bromide/chloride mixtures, however, because the solubility differences between the silver salts are not great enough. Thus, the more soluble salt begins to form in significant amounts before precipitation of the less soluble salt is complete. The silver indicator electrode can be a commercial billet type or simply a polished wire. A calomel electrode can be used as reference, although diffusion of chloride ion from the salt bridge may cause the results of the titration to be measurably high. This source of error can be eliminated by placing the calomel electrode in a potassium nitrate solution that is in contact with the analyte solution by means of a KNO_3 salt bridge. Alternatively, the analyte solution can be made slightly acidic with several drops of nitric acid; a glass electrode can then serve as the reference electrode because the pH of the solution and thus its potential will remain essentially constant throughout the titration.

The titration of I^-/Cl^- mixtures demonstrates how a potentiometric titration can have multiple end points. The potential of the silver electrode is proportional to pAg. Thus, a plot of E_{Ag} against titrant volume yields an experimental curve with the same shape as the curve shown in Figure 29-2 (the ordinate units will be different, of course).

In titrating I^-/Cl^- mixtures, the volume of silver nitrate needed to reach the I^- end point is generally somewhat greater than theoretical. This effect is the result of coprecipitation of the more soluble AgCl during formation of the less soluble AgI. An overconsumption of reagent thus occurs in the first part of the titration. The total volume closely approaches the correct amount.

Despite this coprecipitation error, the potentiometric method is useful for the analysis of halide mixtures. With approximately equal quantities of iodide and chloride, relative errors can be kept within 2% relative.

Preparation of Reagents

(a) *Silver nitrate, 0.05 M.* Follow the instructions in Section 29C-1.
(b) *Potassium nitrate salt bridge.* Bend an 8-mm glass tube into a U-shape with arms that are long enough to extend nearly to the bottom of two 100-mL beakers. Heat 50 mL of water to boiling, and stir in 1.8 g of powdered agar; continue to heat and stir until a uniform suspension is formed. Dissolve 12 g of KNO_3 in the hot suspension. Allow the mixture to cool somewhat. Clamp the U-tube with the openings facing up, and use a medicine dropper to fill it with the warm agar suspension. Cool the tube under a cold-water tap to form the gel. When the bridge is not in use, immerse the ends in 2.5 M KNO_3.

Procedure

Obtain the unknown in a clean 250-mL volumetric flask; dilute to the mark with water, and mix well.

Transfer 50.00 mL of the sample to a clean 100-mL beaker, and add a drop or two of concentrated HNO_3. Place about 25 mL of 2.5 M KNO_3 in a second 100-mL beaker, and make contact between the two solutions with the agar salt bridge. Immerse a silver electrode in the analyte solution and a calomel reference electrode in the second beaker. Titrate with $AgNO_3$ as described in Section 29I-1. Use small increments of titrant in the vicinity of the two end points.

Plot the data, and establish end points for the two analyte ions. Plot a theoretical titration curve, assuming the measured concentrations of the two constituents to be correct.

Report the number of milligrams of I^- and Cl^- in the sample or as otherwise instructed.

29I-3 The Potentiometric Determination of Solute Species in a Carbonate Mixture

Discussion

A glass/calomel electrode system can be used to locate end points in neutralization titrations and to estimate dissociation constants. As a preliminary step to the titrations, the electrode system is standardized against one or more buffers of known pH.

The unknown is issued as an aqueous solution prepared from one or perhaps two adjacent members of the following series: $NaHCO_3$, Na_2CO_3, and $NaOH$ (see Section 12B-2). The object is to determine which of these components were used to prepare the unknown as well as the weight percent of each solute.

Most unknowns require a titration with either standard acid or standard base. A few may require separate titrations, one with acid and one with base. The initial pH of the unknown provides guidance concerning the appropriate titrant(s); a study of Figure 12-3 and Table 12-2 may be helpful in interpreting the data.

Preparation of Solutions

Standardized 0.1 M HCl and/or 0.1 M NaOH. Follow the directions in Sections 29B-3 through 29B-7.

Procedure

Obtain the unknown in a clean 250-mL volumetric flask. Dilute to the mark and mix well. Transfer a small amount of the diluted unknown to a beaker, and determine its pH. Titrate a 50.00-mL aliquot with standard acid or standard base (or perhaps both). Use the resulting titration curves to select indicator(s) suitable for end-point detection, and perform duplicate titrations with these.

Identify the solute species in the unknown, and report the weight/volume percent of each. Calculate the approximate dissociation constant that can be obtained for any carbonate-containing species from the titration data. Estimate the ionic strength of the solution and correct the calculated constant to give an approximate thermodynamic dissociation constant.

29I-4 The Direct Potentiometric Determination of Fluoride Ion

Discussion

The solid-state fluoride electrode (Section 18D-6) has found extensive use in the determination of fluoride in a variety of materials. Directions follow for the determination of this ion in drinking water and in toothpaste. A total ionic strength adjustment buffer (TISAB) is used to adjust all unknowns and standards to essentially the same ionic strength; when this reagent is used, the *concentration* of fluoride, rather than its activity, is measured. The pH of the buffer is about 5, a level at which F^- is the predominant fluorine-containing species. The buffer also contains cyclohexylaminedinitrilotetraacetic acid, which forms stable chelates with iron(III) and aluminum(III), thus freeing fluoride ion from its complexes with these cations.

Review Sections 18D and 18F before undertaking these experiments.

Preparation of Solutions

(a) *Total ionic strength adjustment buffer (TISAB)*. This solution is marketed commercially under the trade name TISAB.[7] Sufficient buffer for 15 to 20 determinations can be prepared by mixing (with stirring) 57 mL of glacial acetic acid, 58 g of NaCl, 4 g of cyclohexylaminedinitrilotetra-acetic acid, and 500 mL of distilled water in a 1-L beaker. Cool the contents in a water or ice bath, and carefully add 6 M NaOH to a pH of 5.0 to 5.5. Dilute to 1 L with water, and store in a plastic bottle.

(b) *Standard fluoride solution, 100 ppm*. Dry a quantity of NaF at 110°C for 2 h. Cool in a desiccator; then weigh (to the nearest milligram) 0.22 g into a 1-L volumetric flask. (**CAUTION! NaF IS HIGHLY TOXIC.** *Immediately* wash any skin touched by this compound with copious quantities of water.) Dissolve in water, dilute to the mark, mix well, and store in a plastic bottle. Calculate the exact concentration of fluoride in parts per million.

A standard F^- solution can be purchased from commercial sources.

Procedure

The apparatus for this experiment consists of a solid-state fluoride electrode, a saturated calomel electrode, and a pH meter. A sleeve-type calomel electrode is needed for the toothpaste determination because the measurement is made on a suspension that tends to clog the liquid junction. The sleeve must be loosened momentarily to renew the interface after each series of measurements.

Determination of Fluoride in Drinking Water

Transfer 50.00-mL portions of the water to 100-mL volumetric flasks, and dilute to the mark with TISAB solution.

Prepare a 5-ppm F^- solution by diluting 25.0 mL of the 100-ppm standard to 500 mL in a volumetric flask. Transfer 5.00-, 10.0-, 25.0-, and 50.0-mL aliquots of the 5-ppm solution to 100-mL volumetric flasks, add 50 mL of TISAB solution, and dilute to the mark. (These solutions correspond to 0.5, 1.0, 2.5, and 5.0 ppm F^- in the sample.)

After thorough rinsing and drying with paper tissue, immerse the electrodes in the 0.5-ppm standard. Stir mechanically for 3 min; then record the potential. Repeat with the remaining standards and samples.

Plot the measured potential against the log of the concentration of the standards. Use this plot to determine the concentration in parts per million of fluoride in the unknown.

---

[7] Orion Research, Boston, MA.

> ### Determination of Fluoride in Toothpaste[8]
>
> Weigh (to the nearest milligram) 0.2 g of toothpaste into a 250-mL beaker. Add 50 mL of TISAB solution, and boil for 2 min with good mixing. Cool and then transfer the suspension quantitatively to a 100-mL volumetric flask, dilute to the mark with distilled water, and mix well. Follow the directions for the analysis of drinking water, beginning with the second paragraph.
> Report the parts per million of F^- in the sample.

29J ELECTROGRAVIMETRIC METHODS

A convenient example of an electrogravimetric method of analysis is the simultaneous determination of copper and lead in a sample of brass. Additional information concerning electrogravimetric methods is found in Section 19C.

29J-1 The Electrogravimetric Determination of Copper and Lead in Brass

Discussion

This procedure is based on the deposition of metallic copper on a cathode and of lead as PbO_2 on an anode. First, the hydrous oxide of tin ($SnO_2 \cdot xH_2O$) that forms when the sample is treated with nitric acid must be removed by filtration. Lead dioxide is deposited quantitatively at the anode from a solution with a high nitrate ion concentration; copper is only partially deposited on the cathode under these conditions. It is therefore necessary to eliminate the excess nitrate after deposition of the PbO_2 is complete. Removal is accomplished through the addition of urea:

$$6NO_3^- + 6H^+ + 5(NH_2)_2CO \rightarrow 8N_2(g) + 5CO_2(g) + 13H_2O$$

Copper then deposits quantitatively from the solution after the nitrate ion concentration has been decreased.

> #### Procedure
>
> Preparation of Electrodes
>
> Immerse the platinum electrodes in hot 6 M HNO_3 for about 5 min (Note 1). Wash them thoroughly with distilled water, rinse with several small portions of ethanol, and dry in an oven at 110°C for 2 to 3 min. Cool and weigh both anodes and cathodes to the nearest 0.1 mg.

---

[8] From T. S. Light and C. C. Cappuccino, *J. Chem. Educ.*, **1975**, *52*, 247.

Preparation of Samples

It is not necessary to dry the unknown. Weigh (to the nearest 0.1 mg) 1-g samples into 250-mL beakers. Cover the beakers with watch glasses. Cautiously add about 35 mL of 6 M HNO_3 (**use a hood**). Digest for at least 30 min; add more acid if necessary. Evaporate to about 5 mL but never to dryness (Note 2).

To each sample, add 5 mL of 3 M HNO_3, 25 mL of water, and one quarter of a tablet of filter paper pulp; digest without boiling for about 45 min. Filter off the $SnO_2 \cdot xH_2O$, using a fine-porosity filter paper (Note 3); collect the filtrates in tall-form electrolysis beakers. Use many small washes with hot 0.3 M HNO_3 to remove the last traces of copper; test for completeness with a few drops of NH_3. The final volume of filtrate and washings should be between 100 and 125 mL; either add water or evaporate to attain this volume.

Electrolysis

With the current switch off, attach the cathode to the negative terminal and the anode to the positive terminal of the electrolysis apparatus. Briefly turn on the stirring motor to be sure the electrodes do not touch. Cover the beakers with split watch glasses and commence the electrolysis. Maintain a current of 1.3 A for 35 min.

Rinse the cover glasses and add 10 mL of 3 M H_2SO_4 followed by 5 g of urea to each beaker. Maintain a current of 2 A until the solutions are colorless. To test for completeness of the electrolysis, remove one drop of the solution with a medicine dropper, and mix it with a few drops of NH_3 in a small test tube. If the mixture turns blue, rinse the contents of the tube back into the electrolysis vessel, and continue the electrolysis for an additional 10 min. Repeat the test until no blue $Cu(NH_3)_4^{2+}$ is produced.

When electrolysis is complete, discontinue stirring but leave the current on. Rinse the electrodes thoroughly with water as they emerge from the solution. After rinsing is complete, turn off the electrolysis apparatus (Note 4), disconnect the electrodes, and dip them in acetone. Dry the cathodes for about 3 min and the anodes for about 15 min at 110°C. Allow the electrodes to cool in air, and then weigh them.

Report the percentages of lead (Note 5) and copper in the brass.

Notes

1. Alternatively, grease and organic materials can be removed by heating platinum electrodes to redness in a flame. Electrode surfaces should not be touched with your fingers after cleaning because grease and oil cause nonadherent deposits that can flake off during washing and weighing.
2. Chloride ion must be totally excluded from this determination because it attacks the platinum anode during electrolysis. This reaction is not only destructive but also causes positive errors in the analysis by codepositing platinum with copper on the cathode.
3. If desired, the tin content can be determined gravimetrically by ignition of the $SnO_2 \cdot xH_2O$ to SnO_2.

4. It is important to maintain a potential between the electrodes until they have been removed from the solution and washed. Some copper may redissolve if this precaution is not observed.

5. Experience has shown that a small amount of moisture is retained by the PbO_2 and that better results are obtained if 0.8643 is used instead of 0.8662, the stoichiometric factor.

29K METHODS BASED ON THE ABSORPTION OF RADIATION

Molecular absorption methods are discussed in Chapters 21 and 22. Directions follow for (1) the use of a calibration curve for the determination of iron in water, (2) the use of a standard-addition procedure for the determination of manganese in steel, and (3) a spectrophotometric determination of the pH of a buffer solution.

29K-1 The Cleaning and Handling of Cells

The accuracy of spectrophotometric measurements is critically dependent on the availability of good-quality matched cells. These should be calibrated against one another at regular intervals to detect differences resulting from scratches, etching, and wear. Equally important is the proper cleaning of the exterior sides (the *windows*) just before the cells are inserted into a photometer or spectrophotometer. The preferred method is to wipe the windows with a lens paper soaked in methanol; the methanol is then allowed to evaporate, leaving the windows free of contaminants. It has been shown that this method is far superior to the usual procedure of wiping the windows with a dry lens paper, which tends to leave a residue of lint and a film on the window.[9]

29K-2 The Determination of Iron in a Natural Water

Discussion

The red-orange complex that forms between iron(II) and 1,10-phenanthroline (orthophenanthroline) is useful for determining iron in water supplies. The reagent is a weak base that reacts to form 1,10-phenanthrolinium ion, $PhenH^+$, in acidic media. Complex formation with iron is thus best described by the equation

$$Fe^{2+} + 3PhenH^+ \rightleftharpoons Fe(Phen)_3^{2+} + 3H^+$$

The structure of the complex is shown in Feature 16-1. The formation constant for this equilibrium is 2.5×10^6 at 25°C. Iron(II) is quantitatively complexed in the pH range between 3 and 9. A pH of about 3.5 is ordinarily recommended to prevent precipitation of iron salts, such as phosphates.

An excess of a reducing reagent, such as hydroxylamine or hydroquinone, is

---

[9] For further information, see J. O. Erickson and T. Surles, *Amer. Lab.*, **1976**, 8 (6), 50.

needed to maintain iron in the $+2$ state. The complex, once formed, is very stable.

This determination can be performed with a spectrophotometer set at 508 nm or with a photometer equipped with a green filter.

Preparation of Solutions

(a) *Standard iron solution, 0.01 mg/mL.* Weigh (to the nearest 0.2 mg) 0.0702 g of reagent-grade $Fe(NH_4)_2(SO_4)_2 \cdot 6H_2O$ into a 1-L volumetric flask. Dissolve in 50 mL of water that contains 1 to 2 mL of concentrated sulfuric acid; dilute to the mark, and mix well.

(b) *Hydroxylamine hydrochloride.* (Sufficient for 80 to 90 measurements.) Dissolve 10 g of $H_2NOH \cdot HCl$ in about 100 mL of distilled water.

(c) *1,10-phenanthroline solution.* (Sufficient for 80 to 90 measurements.) Dissolve 1.0 g of 1,10-phenanthroline monohydrate in about 1 L of water. Warm slightly if necessary. Each milliliter is sufficient for no more than about 0.09 mg of Fe. Prepare no more reagent than needed; it darkens on standing and must then be discarded.

(d) *Sodium acetate, 1.2 M.* (Sufficient for 80 to 90 measurements.) Dissolve 166 g of $NaOAc \cdot 3H_2O$ in 1 L of distilled water.

Procedure

Preparation of a Calibration Curve

Transfer 25.00 mL of the standard iron solution to a 100-mL volumetric flask and 25 mL of distilled water to a second 100-mL volumetric flask. Add 1 mL of hydroxylamine, 10 mL of sodium acetate, and 10 mL of 1,10-phenanthroline to each flask. Allow the mixtures to stand for 5 min; dilute to the mark and mix well.

Clean a pair of matched cells for the instrument. Rinse each cell with at least three portions of the solution it is to contain. Determine the absorbance of the standard with respect to the blank.

Repeat the above procedure with at least three other volumes of the standard iron solution; attempt to encompass an absorbance range between 0.1 and 1.0. Plot a calibration curve.

Determination of Iron

Transfer 10.00 mL of the unknown to a 100-mL volumetric flask; treat in exactly the same way as the standards, measuring the absorbance with respect to the blank. If necessary, alter the volume of unknown taken to obtain absorbance measurements for replicate samples that are within the range of the calibration curve.

Report the parts per million of iron in the unknown.

29K-3 The Determination of Manganese in Steel

Discussion

Small quantities of manganese are readily determined photometrically by the oxidation of Mn(II) to the intensely colored permanganate ion. Potassium periodate is an effective oxidizing reagent for this purpose. The reaction is

$$5IO_4^- + 2Mn^{2+} + 3H_2O \rightarrow 5IO_3^- + 2MnO_4^- + 6H^+$$

Permanganate solutions that contain an excess of periodate are quite stable.

Interferences to the method are few. The presence of most colored ions can be compensated for with a blank. Cerium(III) and chromium(III) are exceptions; these yield oxidation products with periodate that absorb to some extent at the wavelength used for the measurement of permanganate.

The method given here is applicable to steels that do not contain large amounts of chromium. The sample is dissolved in nitric acid. Any carbon in the steel is oxidized with peroxodisulfate. Iron(III) is eliminated as a source of interference by complexation with phosphoric acid. The standard-addition method (page 431) is used to establish the relationship between absorbance and amount of manganese in the sample.

A spectrophotometer set at 525 nm or a photometer with a green filter can be used for the absorbance measurements.

Preparation of Solutions

(a) *Standard manganese(II) solution.* (Sufficient for several hundred analyses.) Weigh 0.1 g (to the nearest 0.1 mg) of manganese into a 50-mL beaker, and dissolve in about 10 mL of 6 M HNO_3 (**use a hood**). Boil gently to eliminate oxides of nitrogen. Cool; then transfer the solution quantitatively to a 1-L volumetric flask. Dilute to the mark with water, and mix thoroughly. The manganese in 1 mL of the standard solution, after being converted to permanganate, causes a volume of 50 mL to increase in absorbance by about 0.09.

Procedure

The unknown does not require drying. If there is evidence of oil on the surface of the unknown, rinse with acetone and dry briefly. Weigh (to the nearest 0.1 mg) duplicate samples (Note 1) into 150-mL beakers. Add about 50 mL of 6 M HNO_3 and boil gently (**use a hood**); heating for 5 to 10 min should suffice. Cautiously add about 1 g of ammonium peroxodisulfate, and boil gently for an additional 10 to 15 min. If the solution is pink or has a deposit of MnO_2, add 1 mL of NH_4HSO_3 (or 0.1 g of $NaHSO_3$) and heat for 5 min. Cool; transfer quantitatively (Note 2) to 250.0-mL volumetric flasks. Dilute to the mark with water, and mix well. Use a 20.00-mL pipet to transfer three aliquots of each sample to individual beakers. Treat as follows:

Aliquot	Volume of 85% H_3PO_4, mL	Volume of Standard Mn, mL	Mass of KIO_4, g
1	5	0.00 (Note 3)	0.4
2	5	5.00 (Note 3)	0.4
3	5	0.00 (Note 3)	0.0

Boil each solution gently for 5 min, cool, and transfer it quantitatively to a 50-mL volumetric flask. Dilute to the mark, and mix well. Measure the absorbance of aliquots 1 and 2 using aliquot 3 as the blank (Note 4).

Report the percentage of manganese in the unknown.

Notes

1. The sample size depends on the manganese content of the unknown; consult with the instructor.
2. If there is evidence of turbidity, filter the solutions as they are transferred to the volumetric flasks.
3. The volume of the standard addition may be dictated by the absorbance of the sample. It is useful to obtain a rough estimate by generating permanganate in about 20 mL of sample, diluting to about 50 mL, and measuring the absorbance.
4. A single blank can be used for all measurements, provided the samples weigh within 50 mg of one another.

29K-4 The Spectrophotometric Determination of pH

Discussion

The pH of an unknown buffer is determined by addition of an acid/base indicator and spectrophotometric measurement of the absorbance of the resulting solution. Because overlap exists between the spectra for the acid and base forms of the indicator, it is necessary to evaluate individual molar absorptivities for each form at two wavelengths. See page 432 for further discussion.

The relationship between the two forms of bromocresol green in an aqueous solution is described by the equilibrium

$$HIn + H_2O \rightleftharpoons H_3O^+ + In^-$$

for which

$$K_a = \frac{[H_3O^+][In^-]}{[HIn]} = 1.6 \times 10^{-5}$$

The spectrophotometric evaluation of $[In^-]$ and $[HIn]$ permits the calculation of $[H_3O^+]$.

Preparation of Solutions

(a) *Bromocresol green*, 1.0×10^{-4} *M*. (Sufficient for about five determinations.) Dissolve 40.0 mg (to the nearest 0.1 mg) of the sodium salt of bromocresol green (720 g/mol) in water, and dilute to 500 mL in a volumetric flask.

(b) *HCl*, *0.5 M*. Dilute about 4 mL of concentrated HCl to approximately 100 mL with water.

(c) *NaOH*, *0.4 M*. Dilute about 7 mL of 6 M NaOH to about 100 mL with water.

Procedure

Determination of Individual Absorption Spectra

Transfer 25.00-mL aliquots of the bromocresol green indicator solution to two 100-mL volumetric flasks. To one add 25 mL of 0.5 M HCl; to the other add 25 mL of 0.4 M NaOH. Dilute to the mark and mix well.

Obtain the absorption spectra for the acid and conjugate-base forms of the indicator between 400 and 600 nm, using water as a blank. Record absorbance values at 10-nm intervals routinely and at closer intervals as needed to define maxima and minima. Evaluate the molar absorptivity for HIn and In$^-$ at wavelengths corresponding to their absorption maxima.

Determination of the pH of an Unknown Buffer

Transfer 25.00 mL of the stock bromocresol green indicator to a 100-mL volumetric flask. Add 50.0 mL of the unknown buffer, dilute to the mark, and mix well. Measure the absorbance of the diluted solution at the wavelengths for which absorptivity data were calculated.

Report the pH of the buffer.

29L   ATOMIC SPECTROSCOPY

Several methods of analysis based on atomic spectroscopy are discussed in Chapter 24. One such application is atomic absorption, which is demonstrated in the experiment that follows.

29L-1   The Determination of Lead in Brass by Atomic Absorption Spectroscopy

Discussion

Brasses and other copper-based alloys usually contain lead, zinc, and tin in various concentrations. Atomic absorption spectroscopy permits the quantitative determination of these elements. The accuracy of this procedure is not as great as that obtainable with gravimetric or volumetric measurements, but much less time is needed to acquire the analytical information.

A weighed sample is dissolved in a mixture of nitric and hydrochloric acids, the latter being needed to prevent the precipitation of tin as metastannic acid, $SnO_2 \cdot xH_2O$. After suitable dilution, the sample is aspirated into a flame, and the absorption of radiation from a hollow cathode lamp is measured.

Preparation of Solution

Standard lead solution, 100 mg/L. Dry a quantity of reagent-grade $Pb(NO_3)_2$ for about 1 h at 110°C. Cool; weigh (to the nearest 0.1 mg) 0.17 g into a 1-L volumetric flask. Dissolve in a solution of 5 mL water and 1 to 3 mL of concentrated HNO_3. Dilute to the mark with distilled water, and mix well.

Procedure

Weigh duplicate samples of the unknown (Note 1) into 150-mL beakers. Cover with watch glasses, and then dissolve (**use a hood**) in a mixture consisting of about 4 mL of concentrated HNO_3 and 4 mL of concentrated HCl (Note 2). Boil gently to remove oxides of nitrogen. Cool; transfer the solutions quantitatively to individual 250-mL volumetric flasks, dilute to the mark with water, and mix well.

Use a buret to deliver 0-, 5-, 10-, 15-, and 20-mL portions of the standard lead solution to individual 50-mL volumetric flasks. Add 4 mL of concentrated HNO_3 and 4 mL of concentrated HCl to each, and dilute to the mark with water.

Transfer 10.00-mL aliquots of each sample to 50-mL volumetric flasks, and dilute to the mark with water.

Set the monochromator at 283.3 nm, and measure the absorbance for each standard and the sample at that wavelength. Take at least three—and preferably more—readings for each measurement.

Plot the calibration data. Report the percentage of lead in the brass.

Notes

1. The weight of sample depends on the lead content of the brass and on the sensitivity of the instrument used for the absorption measurements. A sample containing 6 to 10 mg of lead is reasonable. Consult with the instructor.
2. Brasses that contain a large percentage of tin require additional HCl to prevent the formation of metastannic acid. The diluted samples may develop some turbidity on prolonged standing; a slight turbidity has no effect on the determination of lead.

29L-2 The Determination of Sodium, Potassium, and Calcium in Mineral Waters by Atomic Emission Spectroscopy

Discussion

A convenient method for the determination of alkali and alkaline earth metals in water and in blood serum is based on the characteristic spectra these elements emit when they are aspirated into a natural gas/air flame. The accompanying directions are suitable for the analysis of the three elements in water samples. Radiation buffers (page 405) are used to minimize the effect of each element on the emission intensity of the others.

Preparation of Solutions

(a) *Standard calcium solution, approximately 500 ppm.* Dry a quantity of $CaCO_3$ for about 1 h at 110°C. Cool in a desiccator; weigh (to the nearest milligram) 1.25 g into a 600-mL beaker. Add about 200 mL of distilled water and about 10 mL of concentrated HCl; cover the beaker with a watch glass during the addition of the acid to avoid loss due to spattering. After reaction is complete, transfer the solution quantitatively to a 1-L volumetric flask, dilute to the mark and mix well.

(b) *Standard potassium solution, approximately 500 ppm.* Dry a quantity of KCl for about 1 h at 110°C. Cool; weigh (to the nearest milligram) about 0.95 g into a 1-L volumetric flask. Dissolve in distilled water, and dilute to the mark.

(c) *Standard sodium solution, approximately 500 ppm.* Proceed as in (b), using 1.25 g (to the nearest milligram) of dried NaCl.

(d) *Radiation buffer for the determination of calcium.* Prepare about 100 mL of a solution that has been saturated with NaCl, KCl, and $MgCl_2$, in that order.

(e) *Radiation buffer for the determination of potassium.* Prepare about 100 mL of a solution that has been saturated with NaCl, $CaCl_2$, and $MgCl_2$, in that order.

(f) *Radiation buffer for the determination of sodium.* Prepare about 100 mL of a solution that has been saturated with $CaCl_2$, KCl, and $MgCl_2$, in that order.

Procedure

Preparation of Working Curves

Add 5.00 mL of the appropriate radiation buffer to each of a series of 100-mL volumetric flasks. Add volumes of standard that will produce solutions that range from 0 to 10 ppm in the cation to be determined. Dilute to the mark with water, and mix well.

Measure the emission intensity for each solution, taking at least three readings for each. Aspirate distilled water between each set of measurements. Correct the average values for background luminosity, and prepare a working curve from the data.

Repeat for the other two cations.

Analysis of a Water Sample

Prepare duplicate aliquots of the unknown as directed for preparation of working curves. If necessary, use a standard to calibrate the response of the instrument to the working curve; then measure the emission intensity of the unknown. Correct the data for background. Determine the cation concentration in the unknown by comparison with the working curve.

29M THE SEPARATION OF CATIONS BY ION EXCHANGE

The application of ion-exchange resins to the separation of ionic species of opposite charge is discussed in Section 25B. Directions follow for the ion-exchange separation of nickel(II) from zinc(II) based on converting the zinc ions to negatively charged chloro complexes. After separation, each of the cations is determined by EDTA titration.

Discussion

The separation of the two cations is based on differences in their tendency to form anionic complexes. Stable chlorozincate(II) complexes (such as $ZnCl_3^-$ and $ZnCl_4^{2-}$) are formed in 2 M hydrochloric acid and retained on an anion-exchange resin. In contrast, nickel(II) is not complexed appreciably in this medium and passes rapidly through such a column. After separation is complete, elution with water effectively decomposes the chloro complexes and permits removal of the zinc.

Both nickel and zinc are determined by titration with standard EDTA at pH 10. Eriochrome Black T is the indicator for the zinc titration. Bromopyrogallol Red or murexide is used for the nickel titration.

Preparation of Solutions

(a) *Standard EDTA.* See Section 29D-2.

(b) *pH-10 buffer.* See Section 29D-1.

(c) *Eriochrome Black T indicator.* See Section 29D-1.

(d) *Bromopyrogallol Red indicator.* (Sufficient for 100 titrations.) Dissolve 0.5 g of the solid indicator in 100 mL of 50% (v/v) ethanol.

(e) *Murexide indicator.* The solid is approximately 0.2% indicator by weight in NaCl. Approximately 0.2 g is needed for each titration. The solid preparation is used because solutions of the indicator are quite unstable.

Preparation of Ion-Exchange Columns

A typical ion-exchange column is a cylinder 25 to 40 cm in length and 1 to 1.5 cm in diameter. A stopcock at the lower end permits adjustment of liquid flow through the column. A buret makes a convenient column. It is recommended that two columns be prepared to permit the simultaneous treatment of duplicate samples.

Insert a plug of glass wool to retain the resin particles. Then introduce sufficient strong-base anion-exchange resin (Note) to give a 10- to 15-cm column. Wash the column with about 50 mL of 6 M NH_3, followed by 100 mL of water and 100 mL of 2 M HCl. At the end of this cycle, the flow should be stopped so that the liquid level remains about 1 cm above the resin column. *At no time should the liquid level be allowed to drop below the top of the resin.*

Note

Amberlite® CG 400 or its equivalent can be used.

Procedure

Obtain the unknown, which should contain between 2 and 4 mmol of Ni^{2+} and Zn^{2+}, in a clean 100-mL volumetric flask. Add 16 mL of 12 M HCl, dilute to the mark with distilled water, and mix well. The resulting solution is approximately 2 M in acid. Transfer 10.00 mL of the diluted unknown onto the column. Place a 250-mL conical flask beneath the column, and slowly drain until the liquid level is barely above the resin. Rinse the interior of the column with several 2 to 3 mL portions of the 2 M HCl; lower the liquid level to just above the resin surface after each washing. Elute the nickel with about 50 mL of 2 M HCl at a flow rate of 2 to 3 mL/min.

Elute the Zn(II) by passing about 100 mL of water through the column, using the same flow rate; collect the liquid in a 500-mL conical flask.

Titration of Nickel

Evaporate the solution containing the nickel to dryness to eliminate excess HCl. Avoid overheating; the residual $NiCl_2$ must not be permitted to decompose to NiO. Dissolve the residue in 100 mL of distilled water, and add 10 to 20 mL of pH-10 buffer. Add 15 drops of bromopyrogallol red indicator or 0.2 g of murexide. Titrate to the color change (blue to purple for bromopyrogallol red, yellow to purple for murexide).

Calculate the milligrams of nickel in the unknown.

Titration of Zinc

Add 10 to 20 mL of pH-10 buffer and 1 to 2 drops of Eriochrome Black T to the eluate. Titrate with standard EDTA solution to a color change from red to blue.

Calculate the milligrams of zinc in the unknown.

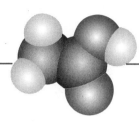

SOLUBILITY PRODUCT CONSTANTS AT 25°C

Compound	Formula	K_{sp}	Notes
Aluminum hydroxide	$Al(OH)_3$	3×10^{-34}	
Barium carbonate	$BaCO_3$	5.0×10^{-9}	
Barium chromate	$BaCrO_4$	2.1×10^{-10}	
Barium hydroxide	$Ba(OH)_2 \cdot 8H_2O$	3×10^{-4}	
Barium iodate	$Ba(IO_3)_2$	1.57×10^{-9}	
Barium oxalate	BaC_2O_4	1×10^{-6}	
Barium sulfate	$BaSO_4$	1.1×10^{-10}	
Cadmium carbonate	$CdCO_3$	1.8×10^{-14}	
Cadmium hydroxide	$Cd(OH)_2$	4.5×10^{-15}	
Cadmium oxalate	CdC_2O_4	9×10^{-8}	
Cadmium sulfide	CdS	1×10^{-27}	
Calcium carbonate	$CaCO_3$	4.5×10^{-9}	Calcite
	$CaCO_3$	6.0×10^{-9}	Aragonite
Calcium fluoride	CaF_2	3.9×10^{-11}	
Calcium hydroxide	$Ca(OH)_2$	6.5×10^{-6}	
Calcium oxalate	$CaC_2O_4 \cdot H_2O$	1.7×10^{-9}	
Calcium sulfate	$CaSO_4$	2.4×10^{-5}	
Cobalt(II) carbonate	$CoCO_3$	1.0×10^{-10}	
Cobalt(II) hydroxide	$Co(OH)_2$	1.3×10^{-15}	
Cobalt(II) sulfide	CoS	5×10^{-22}	α
	CoS	3×10^{-26}	β
Copper(I) bromide	$CuBr$	5×10^{-9}	
Copper(I) chloride	$CuCl$	1.9×10^{-7}	

Compound	Formula	K_{sp}	Notes
Copper(I) hydroxide*	Cu_2O	2×10^{-15}	
Copper(I) iodide	CuI	1×10^{-12}	
Copper(I) thiocyanate	$CuSCN$	4.0×10^{-14}	
Copper(II) hydroxide	$Cu(OH)_2$	4.8×10^{-20}	
Copper(II) sulfide	CuS	8×10^{-37}	
Iron(II) carbonate	$FeCO_3$	2.1×10^{-11}	
Iron(II) hydroxide	$Fe(OH)_2$	4.1×10^{-15}	
Iron(II) sulfide	FeS	8×10^{-19}	
Iron(III) hydroxide	$Fe(OH)_3$	2×10^{-39}	
Lanthanum iodate	$La(IO_3)_3$	1.0×10^{-11}	
Lead carbonate	$PbCO_3$	7.4×10^{-14}	
Lead chloride	$PbCl_2$	1.7×10^{-5}	
Lead chromate	$PbCrO_4$	3×10^{-13}	
Lead hydroxide	$PbO†$	8×10^{-16}	Yellow
	$PbO†$	5×10^{-16}	Red
Lead iodide	PbI_2	7.9×10^{-9}	
Lead oxalate	PbC_2O_4	8.5×10^{-9}	$\mu = 0.05$
Lead sulfate	$PbSO_4$	1.6×10^{-8}	
Lead sulfide	PbS	3×10^{-28}	
Magnesium ammonium phosphate	$MgNH_4PO_4$	3×10^{-13}	
Magnesium carbonate	$MgCO_3$	3.5×10^{-8}	
Magnesium hydroxide	$Mg(OH)_2$	7.1×10^{-12}	
Manganese carbonate	$MnCO_3$	5.0×10^{-10}	
Manganese hydroxide	$Mn(OH)_2$	2×10^{-13}	
Manganese sulfide	MnS	3×10^{-11}	Pink
	MnS	3×10^{-14}	Green
Mercury(I) bromide	Hg_2Br_2	5.6×10^{-23}	
Mercury(I) carbonate	Hg_2CO_3	8.9×10^{-17}	
Mercury(I) chloride	Hg_2Cl_2	1.2×10^{-18}	
Mercury(I) iodide	Hg_2I_2	4.7×10^{-29}	
Mercury(I) thiocyanate	$Hg_2(SCN)_2$	3.0×10^{-20}	
Mercury(II) hydroxide	$HgO‡$	3.6×10^{-26}	
Mercury(II) sulfide	HgS	2×10^{-53}	Black
	HgS	5×10^{-54}	Red
Nickel carbonate	$NiCO_3$	1.3×10^{-7}	
Nickel hydroxide	$Ni(OH)_2$	6×10^{-16}	
Nickel sulfide	NiS	4×10^{-20}	α
	NiS	1.3×10^{-25}	β
Silver arsenate	Ag_3AsO_4	6×10^{-23}	
Silver bromide	$AgBr$	5.0×10^{-13}	
Silver carbonate	Ag_2CO_3	8.1×10^{-12}	
Silver chloride	$AgCl$	1.82×10^{-10}	
Silver chromate	$AgCrO_4$	1.2×10^{-12}	
Silver cyanide	$AgCN$	2.2×10^{-16}	
Silver iodate	$AgIO_3$	3.1×10^{-8}	
Silver iodide	AgI	8.3×10^{-17}	

Compound	Formula	K_{sp}	Notes
Silver oxalate	$Ag_2C_2O_4$	3.5×10^{-11}	
Silver sulfide	Ag_2S	8×10^{-51}	
Silver thiocyanate	$AgSCN$	1.1×10^{-12}	
Strontium carbonate	$SrCO_3$	9.3×10^{-10}	
Strontium oxalate	SrC_2O_4	5×10^{-8}	
Strontium sulfate	$SrSO_4$	3.2×10^{-7}	
Thallium(I) chloride	$TlCl$	1.8×10^{-4}	
Thallium(I) sulfide	Tl_2S	6×10^{-22}	
Zinc carbonate	$ZnCO_3$	1.0×10^{-10}	
Zinc hydroxide	$Zn(OH)_2$	3.0×10^{-16}	Amorphous
Zinc oxalate	ZnC_2O_4	8×10^{-9}	
Zinc sulfide	ZnS	2×10^{-25}	α
	ZnS	3×10^{-23}	β

Most of these data were taken from A. E. Martell and R. M. Smith, *Critical Stability Constants*, Vol. 3–6. New York: Plenum, 1976–1989. In most cases, the ionic strength was 0.0 and the temperature 25°C.

* $Cu_2O(s) + H_2O \rightleftharpoons 2Cu^+ + 2OH^-$

† $PbO(s) + H_2O \rightleftharpoons Pb^{2+} + 2OH^-$

‡ $HgO(s) + H_2O \rightleftharpoons Hg^{2+} + 2OH^-$

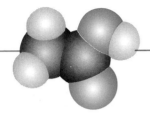

APPENDIX 2

ACID DISSOCIATION CONSTANTS AT 25°C

Acid	Formula	K_1	K_2	K_3
Acetic acid	CH_3COOH	1.75×10^{-5}		
Ammonium ion	NH_4^+	5.70×10^{-10}		
Anilinium ion	$C_6H_5NH_3^+$	2.51×10^{-5}		
Arsenic acid	H_3AsO_4	5.8×10^{-3}	1.1×10^{-7}	3.2×10^{-12}
Arsenous acid	H_3AsO_3	5.1×10^{-10}		
Benzoic acid	C_6H_5COOH	6.28×10^{-5}		
Boric acid	H_3BO_3	5.81×10^{-10}		
1-Butanoic acid	$CH_3CH_2CH_2COOH$	1.52×10^{-5}		
Carbonic acid	H_2CO_3	4.45×10^{-7}	4.69×10^{-11}	
Chloroacetic acid	$ClCH_2COOH$	1.36×10^{-3}		
Citric acid	$HOOC(OH)C(CH_2COOH)_2$	7.45×10^{-4}	1.73×10^{-5}	4.02×10^{-7}
Dimethyl ammonium ion	$(CH_3)_2NH_2^+$	1.68×10^{-11}		
Ethanol ammonium ion	$HOC_2H_4NH_3^+$	3.18×10^{-10}		
Ethyl ammonium ion	$C_2H_5NH_3^+$	2.31×10^{-11}		
Ethylene diammonium ion	$^+H_3NCH_2CH_2NH_3^+$	1.42×10^{-7}	1.18×10^{-10}	
Formic acid	$HCOOH$	1.80×10^{-4}		
Fumaric acid	*trans*-$HOOCCH:CHCOOH$	8.85×10^{-4}	3.21×10^{-5}	
Glycolic acid	$HOCH_2COOH$	1.47×10^{-4}		
Hydrazinium ion	$H_2NNH_3^+$	1.05×10^{-8}		
Hydrazoic acid	HN_3	2.2×10^{-5}		
Hydrogen cyanide	HCN	6.2×10^{-10}		
Hydrogen fluoride	HF	6.8×10^{-4}		
Hydrogen peroxide	H_2O_2	2.2×10^{-12}		
Hydrogen sulfide	H_2S	9.6×10^{-8}	1.3×10^{-14}	

Acid	Formula	K_1	K_2	K_3
Hydroxyl ammonium ion	$HONH_3^+$	1.10×10^{-6}		
Hypochlorous acid	$HOCl$	3.0×10^{-8}		
Iodic acid	HIO_3	1.7×10^{-1}		
Lactic acid	$CH_3CHOHCOOH$	1.38×10^{-4}		
Maleic acid	$cis\text{-}HOOCCH:CHCOOH$	1.3×10^{-2}	5.9×10^{-7}	
Malic acid	$HOOCCHOHCH_2COOH$	3.48×10^{-4}	8.00×10^{-6}	
Malonic acid	$HOOCCH_2COOH$	1.42×10^{-3}	2.01×10^{-6}	
Mandelic acid	$C_6H_5CHOHCOOH$	4.0×10^{-4}		
Methyl ammonium ion	$CH_3NH_3^+$	2.3×10^{-11}		
Nitrous acid	HNO_2	7.1×10^{-4}		
Oxalic acid	$HOOCCOOH$	5.60×10^{-2}	5.42×10^{-5}	
Periodic acid	H_5IO_6	2×10^{-2}	5×10^{-9}	
Phenol	C_6H_5OH	1.00×10^{-10}		
Phosphoric acid	H_3PO_4	7.11×10^{-3}	6.32×10^{-8}	4.5×10^{-13}
Phosphorous acid	H_3PO_3	3×10^{-2}	1.62×10^{-7}	
o-Phthalic acid	$C_6H_4(COOH)_2$	1.12×10^{-3}	3.91×10^{-6}	
Picric acid	$(NO_2)_3C_6H_2OH$	4.3×10^{-1}		
Piperidinium ion	$C_5H_{11}NH^+$	7.50×10^{-12}		
Propanoic acid	CH_3CH_2COOH	1.34×10^{-5}		
Pyridinium ion	$C_5H_5NH^+$	5.90×10^{-6}		
Pyruvic acid	$CH_3COCOOH$	3.2×10^{-3}		
Salicylic acid	$C_6H_4(OH)COOH$	1.06×10^{-3}		
Sulfamic acid	H_2NSO_3H	1.03×10^{-1}		
Succinic acid	$HOOCCH_2CH_2COOH$	6.21×10^{-5}	2.31×10^{-6}	
Sulfuric acid	H_2SO_4	Strong	1.02×10^{-2}	
Sulfurous acid	H_2SO_3	1.23×10^{-2}	6.6×10^{-8}	
Tartaric acid	$HOOC(CHOH)_2COOH$	9.20×10^{-4}	4.31×10^{-5}	
Thiocyanic acid	$HSCN$	0.13		
Thiosulfuric acid	$H_2S_2O_3$	0.3	2.5×10^{-2}	
Trichloroacetic acid	Cl_3CCOOH	3		
Trimethyl ammonium ion	$(CH_3)_3NH^+$	1.58×10^{-10}		

Most data are for zero ionic strength. (From A. E. Martell and R. M. Smith, *Critical Stability Constants*, Vol. 1–6. New York: Plenum Press, 1974–1989.)

APPENDIX 3

FORMATION CONSTANTS AT 25°C

Ligand	Cation	$\log K_1$	$\log K_2$	$\log K_3$	$\log K_4$	Ionic Strength
Acetate	Ag^+	0.73	−0.9			0.0
(CH_3COO^-)	Ca^{2+}	1.18				0.0
	Cd^{2+}	1.93	1.22			0.0
	Cu^{2+}	2.21	1.42			0.0
	Fe^{3+}	3.38*	3.1*	1.8*		0.1
	Hg^{2+}	$\log K_1K_2 = 8.45$				0.0
	Mg^{2+}	1.27				0.0
	Pb^{2+}	2.68	1.40			0.0
Ammonia	Ag^+	3.31	3.91			0.0
(NH_3)	Cd^{2+}	2.55	2.01	1.34	0.84	0.0
	Co^{2+}	1.99*	1.51	0.93	0.64	0.0
		$\log K_5 = 0.06$	$\log K_6 = -0.74$			0.0
	Cu^{2+}	4.04	3.43	2.80	1.48	0.0
	Hg^{2+}	8.8	8.6	1.0	0.7	0.5
	Ni^{2+}	2.72	2.17	1.66	1.12	0.0
		$\log K_5 = 0.67$	$\log K_6 = -0.03$			0.0
	Zn^{2+}	2.21	2.29	2.36	2.03	0.0
Bromide	Ag^+	$Ag^+ + 2Br^- \rightleftharpoons AgBr_2^-$	$\log K_1K_2 = 7.5$			0.0
(Br^-)	Hg^{2+}	9.00	8.1	2.3	1.6	0.5
	Pb^{2+}	1.77				0.0
Chloride	Ag^+	$Ag^+ + 2Cl^- \rightleftharpoons AgCl_2^-$	$\log K_1K_2 = 5.25$			0.0
(Cl^-)		$AgCl_2^- + Cl^- \rightleftharpoons AgCl_3^{2-}$	$\log K_3 = 0.37$			0.0
	Cu^+	$Cu^+ + 2Cl^- \rightleftharpoons CuCl_2^-$	$\log = 5.5*$			0.0
	Fe^{3+}	1.48	0.65			0.0

Ligand	Cation	$\log K_1$	$\log K_2$	$\log K_3$	$\log K_4$	Ionic Strength
	Hg^{2+}	7.30	6.70	1.0	0.6	0.0
	Pb^{2+}		$Pb^{2+} + 3Cl^- \rightleftharpoons PbCl_3^-$	$\log K_1K_2K_3 = 1.8$		0.0
	Sn^{2+}	1.51	0.74	−0.3	−0.5	0.0
Cyanide	Ag^+		$Ag^+ + 2CN^- \rightleftharpoons Ag(CN)_2^-$	$\log K_1K_2 = 20.48$		0.0
(CN^-)	Cd^{2+}	6.01	5.11	4.53	2.27	0.0
	Hg^{2+}	17.00	15.75	3.56	2.66	0.0
	Ni^{2+}		$Ni^{2+} + 4CN^- \rightleftharpoons Ni(CN)_4^-$	$\log K_1K_2K_3K_4 = 30.22$		0.0
	Zn^{2+}	$\log K_1K_2 = 11.07$		4.98	3.57	0.0
EDTA	See Table 14-1, page 239.					
Fluoride	Al^{3+}	7.0	5.6	4.1	2.4	0.0
(F^-)	Fe^{3+}	5.18	3.89	3.03		0.0
Hydroxide	Al^{3+}		$Al^{3+} + 4OH^- \rightleftharpoons Al(OH)_4^-$	$\log K_1K_2K_3K_4 = 33.4$		0.0
(OH^-)	Cd^{2+}	3.9	3.8			0.0
	Cu^{2+}	6.5				0.0
	Fe^{2+}	4.6				0.0
	Fe^{3+}	11.81	11.5			0.0
	Hg^{2+}	10.60	11.2			0.0
	Ni^{2+}	4.1	4.9	3		0.0
	Pb^{2+}	6.4	$Pb^{2+} + 3OH^- \rightleftharpoons Pb(OH)_3^-$	$\log K_1K_2K_3 = 13.9$		0.0
	Zn^{2+}	5.0	$Zn^{2+} + 4OH^- \rightleftharpoons Zn(OH)_4^{2-}$	$\log K_1K_2K_3K_4 = 15.5$		0.0
Iodide	Cd^{2+}	2.28	1.64	1.0	1.0	0.0
(I^-)	Cu^+		$Cu^+ + 2I^- \rightleftharpoons CuI_2^-$	$\log K_1K_2 = 8.9$		0.0
	Hg^{2+}	12.87	10.95	3.8	2.2	0.5
	Pb^{2+}		$Pb^{2+} + 3I^- \rightleftharpoons PbI_3^-$	$\log K_1K_2K_3 = 3.9$		0.0
			$Pb^{2+} + 4I^- \rightleftharpoons PbI_4^{2-}$	$\log K_1K_2K_3K_4 = 4.5$		0.0
Oxalate	Al^{3+}	5.97	4.96	5.04		0.1
$(C_2O_4^{2-})$	Ca^{2+}	3.19				0.0
	Cd^{2+}	2.73	1.4	1.0		1.0
	Fe^{3+}	7.58	6.23	4.8		1.0
	Mg^{2+}	3.42(18°C)				
	Pb^{2+}	4.20	2.11			1.0
Sulfate	Al^{3+}	3.89				0.0
(SO_4^{2-})	Ca^{2+}	2.13				0.0
	Cu^{2+}	2.34				0.0
	Fe^{3+}	4.04	1.34			0.0
	Mg^{2+}	2.23				0.0
Thiocyanate	Cd^{2+}	1.89	0.89	0.1		0.0
(SCN^-)	Cu^+		$Cu^+ + 3SCN^- \rightleftharpoons Cu(SCN)_3^{2-}$	$\log K_1K_2K_3 = 11.60$		0.0
	Fe^{3+}	3.02	0.62*			0.0
	Hg^{2+}	$\log K_1K_2 = 17.26$		2.7	1.8	0.0
	Ni^{2+}	1.76				0.0
Thiosulfate	Ag^+	8.82*	4.7	0.7		0.0
$(S_2O_3^{2-})$	Cu^{2+}	$\log K_1K_2 = 6.3$				0.0
	Hg^{2+}	$\log K_1K_2 = 29.23$		1.4		0.0

* 20°C

Data from A. E. Martell and R. M. Smith, *Critical Stability Constants*, Vol. 3–6. New York: Plenum Press, 1974–1989.

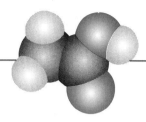

SOME STANDARD AND FORMAL ELECTRODE POTENTIALS

Half-Reaction	E^0, V*	Formal Potential, V†
Aluminum		
$Al^{3+} + 3e^- \rightleftharpoons Al(s)$	-1.662	
Antimony		
$Sb_2O_5(s) + 6H^+ + 4e^- \rightleftharpoons 2SbO^+ + 3H_2O$	$+0.581$	
Arsenic		
$H_3AsO_4 + 2H^+ + 2e^- \rightleftharpoons H_3AsO_3 + H_2O$	$+0.559$	0.577 in 1 M HCl, HClO$_4$
Barium		
$Ba^{2+} + 2e^- \rightleftharpoons Ba(s)$	-2.906	
Bismuth		
$BiO^+ + 2H^+ + 3e^- \rightleftharpoons Bi(s) + H_2O$	$+0.320$	
$BiCl_4^- + 3e^- \rightleftharpoons Bi(s) + 4Cl^-$	$+0.16$	
Bromine		
$Br_2(\ell) + 2e^- \rightleftharpoons 2Br^-$	$+1.065$	1.05 in 4 M HCl
$Br_2(aq) + 2e^- \rightleftharpoons 2Br^-$	$+1.087‡$	
$BrO_3^- + 6H^+ + 5e^- \rightleftharpoons \frac{1}{2}Br_2(\ell) + 3H_2O$	$+1.52$	
$BrO_3^- + 6H^+ + 6e^- \rightleftharpoons Br^- + 3H_2O$	$+1.44$	
Cadmium		
$Cd^{2+} + 2e^- \rightleftharpoons Cd(s)$	-0.403	
Calcium		
$Ca^{2+} + 2e^- \rightleftharpoons Ca(s)$	-2.866	
Carbon		
$C_6H_4O_2$ (quinone) $+ 2H^+ + 2e^- \rightleftharpoons C_6H_4(OH)_2$	$+0.699$	0.696 in 1 M HCl, HClO$_4$, H$_2$SO$_4$
$2CO_2(g) + 2H^+ + 2e^- \rightleftharpoons H_2C_2O_4$	-0.49	

Half-Reaction	E^0, V*	Formal Potential, V†
Cerium		
$Ce^{4+} + e^- \rightleftharpoons Ce^{3+}$		+1.70 in 1 M $HClO_4$; +1.61 in 1 M HNO_3; + 1.44 in 1 M H_2SO_4
Chlorine		
$Cl_2(g) + 2e^- \rightleftharpoons 2Cl^-$	+1.359	
$HClO + H^+ + e^- \rightleftharpoons \frac{1}{2}Cl_2(g) + H_2O$	+1.63	
$ClO_3^- + 6H^+ + 5e^- \rightleftharpoons \frac{1}{2}Cl_2(g) + 3H_2O$	+1.47	
Chromium		
$Cr^{3+} + e^- \rightleftharpoons Cr^{2+}$	−0.408	
$Cr^{3+} + 3e^- \rightleftharpoons Cr(s)$	−0.744	
$Cr_2O_7^{2-} + 14H^+ + 6e^- \rightleftharpoons 2Cr^{3+} + 7H_2O$	+1.33	
Cobalt		
$Co^{2+} + 2e^- \rightleftharpoons Co(s)$	−0.277	
$Co^{3+} + e^- \rightleftharpoons Co^{2+}$	+1.808	
Copper		
$Cu^{2+} + 2e^- \rightleftharpoons Cu(s)$	+0.337	
$Cu^{2+} + e^- \rightleftharpoons Cu^+$	+0.153	
$Cu^+ + e^- \rightleftharpoons Cu(s)$	+0.521	
$Cu^{2+} + I^- + e^- \rightleftharpoons CuI(s)$	+0.86	
$CuI(s) + e^- \rightleftharpoons Cu(s) + I^-$	−0.185	
Fluorine		
$F_2(g) + 2H^+ + 2e^- \rightleftharpoons 2HF(aq)$	+3.06	
Hydrogen		
$2H^+ + 2e^- \rightleftharpoons H_2(g)$	0.000	−0.005 in 1 M HCl, $HClO_4$
Iodine		
$I_2(s) + 2e^- \rightleftharpoons 2I^-$	+0.5355	
$I_2(aq) + 2e^- \rightleftharpoons 2I^-$	+0.615‡	
$I_3^- + 2e^- \rightleftharpoons 3I^-$	+0.536	
$ICl_2^- + e^- \rightleftharpoons \frac{1}{2}I_2(s) + 2Cl^-$	+1.056	
$IO_3^- + 6H^+ + 5e^- \rightleftharpoons \frac{1}{2}I_2(s) + 3H_2O$	+1.196	
$IO_3^- + 6H^+ + 5e^- \rightleftharpoons \frac{1}{2}I_2(aq) + 3H_2O$	+1.178‡	
$IO_3^- + 2Cl^- + 6H^+ + 4e^- \rightleftharpoons ICl_2^- + 3H_2O$	+1.24	
$H_5IO_6 + H^+ + 2e^- \rightleftharpoons IO_3^- + 3H_2O$	+1.601	
Iron		
$Fe^{2+} + 2e^- \rightleftharpoons Fe(s)$	−0.440	
$Fe^{3+} + e^- \rightleftharpoons Fe^{2+}$	+0.771	0.700 in 1 M HCl; 0.732 in 1 M $HClO_4$; 0.68 in 1 M H_2SO_4
$Fe(CN)_6^{3-} + e^- \rightleftharpoons Fe(CN)_6^{4-}$	+0.36	0.71 in 1 M HCl; 0.72 in 1 M $HClO_4$, H_2SO_4
Lead		
$Pb^{2+} + 2e^- \rightleftharpoons Pb(s)$	−0.126	−0.14 in 1 M $HClO_4$; −0.29 in 1 M H_2SO_4
$PbO_2(s) + 4H^+ + 2e^- \rightleftharpoons Pb^{2+} + 2H_2O$	+1.455	
$PbSO_4(s) + 2e^- \rightleftharpoons Pb(s) + SO_4^{2-}$	−0.350	
Lithium		
$Li^+ + e^- \rightleftharpoons Li(s)$	−3.045	

Half-Reaction	E^0, V*	Formal Potential, V†
Magnesium		
$Mg^{2+} + 2e^- \rightleftharpoons Mg(s)$	−2.363	
Manganese		
$Mn^{2+} + 2e^- \rightleftharpoons Mn(s)$	−1.180	
$Mn^{3+} + e^- \rightleftharpoons Mn^{2+}$		1.51 in 7.5 M H_2SO_4
$MnO_2(s) + 4H^+ + 2e^- \rightleftharpoons Mn^{2+} + 2H_2O$	+1.23	
$MnO_4^- + 8H^+ + 5e^- \rightleftharpoons Mn^{2+} + 4H_2O$	+1.51	
$MnO_4^- + 4H^+ + 3e^- \rightleftharpoons MnO_2(s) + 2H_2O$	+1.695	
$MnO_4^- + e^- \rightleftharpoons MnO_4^{2-}$	+0.564	
Mercury		
$Hg_2^{2+} + 2e^- \rightleftharpoons 2Hg(\ell)$	+0.788	0.274 in 1 M HCl; 0.776 in 1 M $HClO_4$; 0.674 in 1 M H_2SO_4
$2Hg^{2+} + 2e^- \rightleftharpoons Hg_2^{2+}$	+0.920	0.907 in 1 M $HClO_4$
$Hg^{2+} + 2e^- \rightleftharpoons Hg(\ell)$	+0.854	
$Hg_2Cl_2(s) + 2e^- \rightleftharpoons 2Hg(\ell) + 2Cl^-$	+0.268	0.244 in sat'd KCl; 0.282 in 1 M KCl; 0.334 in 0.1 M KCl
$Hg_2SO_4(s) + 2e^- \rightleftharpoons 2Hg(\ell) + SO_4^{2-}$	+0.615	
Nickel		
$Ni^{2+} + 2e^- \rightleftharpoons Ni(s)$	−0.250	
Nitrogen		
$N_2(g) + 5H^+ + 4e^- \rightleftharpoons N_2H_5^+$	−0.23	
$HNO_2 + H^+ + e^- \rightleftharpoons NO(g) + H_2O$	+1.00	
$NO_3^- + 3H^+ + 2e^- \rightleftharpoons HNO_2 + H_2O$	+0.94	0.92 in 1 M HNO_3
Oxygen		
$H_2O_2 + 2H^+ + 2e^- \rightleftharpoons 2H_2O$	+1.776	
$HO_2^- + H_2O + 2e^- \rightleftharpoons 3OH^-$	+0.88	
$O_2(g) + 4H^+ + 4e^- \rightleftharpoons 2H_2O$	+1.229	
$O_2(g) + 2H^+ + 2e^- \rightleftharpoons H_2O_2$	+0.682	
$O_3(g) + 2H^+ + 2e^- \rightleftharpoons O_2(g) + H_2O$	+2.07	
Palladium		
$Pd^{2+} + 2e^- \rightleftharpoons Pd(s)$	+0.987	
Platinum		
$PtCl_4^{2-} + 2e^- \rightleftharpoons Pt(s) + 4Cl^-$	+0.73	
$PtCl_6^{2-} + 2e^- \rightleftharpoons PtCl_4^{2-} + 2Cl^-$	+0.68	
Potassium		
$K^+ + e^- \rightleftharpoons K(s)$	−2.925	
Selenium		
$H_2SeO_3 + 4H^+ + 4e^- \rightleftharpoons Se(s) + 3H_2O$	+0.740	
$SeO_4^{2-} + 4H^+ + 2e^- \rightleftharpoons H_2SeO_3 + H_2O$	+1.15	
Silver		
$Ag^+ + e^- \rightleftharpoons Ag(s)$	+0.799	0.228 in 1 M HCl; 0.792 in 1 M $HClO_4$; 0.77 in 1 M H_2SO_4
$AgBr(s) + e^- \rightleftharpoons Ag(s) + Br^-$	+0.073	
$AgCl(s) + e^- \rightleftharpoons Ag(s) + Cl^-$	+0.222	0.228 in 1 M KCl
$Ag(CN)_2^- + e^- \rightleftharpoons Ag(s) + 2CN^-$	−0.31	
$Ag_2CrO_4(s) + 2e^- \rightleftharpoons 2Ag(s) + CrO_4^{2-}$	+0.446	
$AgI(s) + e^- \rightleftharpoons Ag(s) + I^-$	−0.151	

Half-Reaction	E^0, V*	Formal Potential, V†
$Ag(S_2O_3)_2^{3-} + e^- \rightleftharpoons Ag(s) + 2S_2O_3^{2-}$	+0.017	
Sodium		
$Na^+ + e^- \rightleftharpoons Na(s)$	−2.714	
Sulfur		
$S(s) + 2H^+ + 2e^- \rightleftharpoons H_2S(g)$	+0.141	
$H_2SO_3 + 4H^+ + 4e^- \rightleftharpoons S(s) + 3H_2O$	+0.450	
$SO_4^{2-} + 4H^+ + 2e^- \rightleftharpoons H_2SO_3 + H_2O$	+0.172	
$S_4O_6^{2-} + 2e^- \rightleftharpoons 2S_2O_3^{2-}$	+0.08	
$S_2O_8^{2-} + 2e^- \rightleftharpoons 2SO_4^{2-}$	+2.01	
Thallium		
$Tl^+ + e^- \rightleftharpoons Tl(s)$	−0.336	−0.551 in 1 M HCl; −0.33 in 1 M HClO$_4$, H$_2$SO$_4$
$Tl^{3+} + 2e^- \rightleftharpoons Tl^+$	+1.25	0.77 in 1 M HCl
Tin		
$Sn^{2+} + 2e^- \rightleftharpoons Sn(s)$	−0.136	−0.16 in 1 M HClO$_4$
$Sn^{4+} + 2e^- \rightleftharpoons Sn^{2+}$	+0.154	0.14 in 1 M HCl
Titanium		
$Ti^{3+} + e^- \rightleftharpoons Ti^{2+}$	−0.369	
$TiO^{2+} + 2H^+ + e^- \rightleftharpoons Ti^{3+} + H_2O$	+0.099	0.04 in 1 M H$_2$SO$_4$
Uranium		
$UO_2^{2+} + 4H^+ + 2e^- \rightleftharpoons U^{4+} + 2H_2O$	+0.334	
Vanadium		
$V^{3+} + e^- \rightleftharpoons V^{2+}$	−0.256	−0.21 in 1 M HClO$_4$
$VO^{2+} + 2H^+ + e^- \rightleftharpoons V^{3+} + H_2O$	+0.359	
$V(OH)_4^+ + 2H^+ + e^- \rightleftharpoons VO^{2+} + 3H_2O$	+1.00	1.02 in 1 M HCl, HClO$_4$
Zinc		
$Zn^{2+} + 2e^- \rightleftharpoons Zn(s)$	−0.763	

* G. Milazzo, S. Caroli, and V. K. Sharma, *Tables of Standard Electrode Potentials*. London: Wiley, 1978.
† E. H. Swift and E. A. Butler, *Quantitative Measurements and Chemical Equilibria*. New York: Freeman, 1972.
‡ These potentials are hypothetical because they correspond to solutions that are 1.00 M in Br_2 or I_2. The solubilities of these two compounds at 25°C are 0.18 M and 0.0020 M, respectively. In saturated solutions containing an excess of $Br_2(\ell)$ or $I_2(s)$, the standard potentials for the half-reaction $Br_2(\ell) + 2e^- \rightleftharpoons 2Br^-$ or $I_2(s) + 2e^- \rightleftharpoons 2I^-$ should be used. In contrast, at Br_2 and I_2 concentrations less than saturation, these hypothetical electrode potentials should be employed.

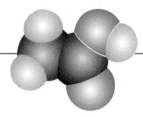

USE OF EXPONENTIAL NUMBERS AND LOGARITHMS

Scientists frequently find it necessary (or convenient) to use exponential notation to express numerical data. A brief review of this notation follows.

EXPONENTIAL NOTATION

An exponent is used to describe the process of repeated multiplication or division. For example, 3^5 means

$$3 \times 3 \times 3 \times 3 \times 3 = 3^5 = 243$$

The power 5 is the exponent of the number (or base) 3; thus, 3 raised to the fifth power is equal to 243.

A negative exponent represents repeated division. For example, 3^{-5} means

$$\frac{1}{3} \times \frac{1}{3} \times \frac{1}{3} \times \frac{1}{3} \times \frac{1}{3} = \frac{1}{3^5} = 3^{-5} = 0.00412$$

Note that changing the sign of the exponent yields the *reciprocal* of the number; that is,

$$3^{-5} = \frac{1}{3^5} = \frac{1}{243} = 0.00412$$

It is important to note that a number raised to the first power is the number itself, and any number raised to the zero power has a value of 1. For example,

$$4^1 = 4$$
$$4^0 = 1$$
$$67^0 = 1$$

Fractional Exponents

A fractional exponent symbolizes the process of extracting the root of a number. The fifth root of 243 is 3; this process is expressed exponentially as

$$(243)^{1/5} = 3$$

Other examples are

$$25^{1/2} = 5$$
$$25^{-1/2} = \frac{1}{25^{1/2}} = \frac{1}{5}$$

The Combination of Exponential Numbers in Multiplication and Division

Multiplication and division of exponential numbers that have the same base are accomplished by adding and subtracting the exponents. For example,

$$3^3 \times 3^2 = (3 \times 3 \times 3)(3 \times 3) = 3^{(3+2)} = 3^5 = 243$$

$$3^4 \times 3^{-2} \times 3^0 = (3 \times 3 \times 3 \times 3)\left(\frac{1}{3} \times \frac{1}{3}\right) \times 1 = 3^{(4-2+0)} = 3^2 = 9$$

$$\frac{5^4}{5^2} = \frac{5 \times 5 \times 5 \times 5}{5 \times 5} = 5^{(4-2)} = 5^2 = 25$$

$$\frac{2^3}{2^{-1}} = \frac{(2 \times 2 \times 2)}{1/2} = 2^4 = 16$$

Note that in the last equation the exponent is given by the relationship

$$3 - (-1) = 3 + 1 = 4$$

Extraction of the Root of an Exponential Number

To obtain the root of an exponential number, the exponent is divided by the desired root. Thus,

$$(5^4)^{1/2} = (5 \times 5 \times 5 \times 5)^{1/2} = 5^{(4/2)} = 5^2 = 25$$
$$(10^{-8})^{1/4} = 10^{(-8/4)} = 10^{-2}$$
$$(10^9)^{1/2} = 10^{(9/2)} = 10^{4.5}$$

THE USE OF EXPONENTS IN SCIENTIFIC NOTATION

Scientists and engineers are frequently called on to use very large or very small numbers for which ordinary decimal notation is either awkward or impossible. For example, to express Avogadro's number in decimal notation would require 21 zeros following the number 602. In scientific notation the number is written as a multiple of two numbers, the one number in decimal notation and the other expressed as a power of 10. Thus, Avogadro's number is written as 6.02×10^{23}. Other examples are

$$4.32 \times 10^3 = 4.32 \times 10 \times 10 \times 10 = 4320$$

$$4.32 \times 10^{-3} = 4.32 \times \frac{1}{10} \times \frac{1}{10} \times \frac{1}{10} = 0.00432$$

$$0.002002 = 2.002 \times \frac{1}{10} \times \frac{1}{10} \times \frac{1}{10} = 2.002 \times 10^{-3}$$

$$375 = 3.75 \times 10 \times 10 = 3.75 \times 10^2$$

Note that the scientific notation for a number can be expressed in any of several equivalent forms. Thus,

$$4.32 \times 10^3 = 43.2 \times 10^2 = 432 \times 10^1 = 0.432 \times 10^4 = 0.0432 \times 10^5$$

The number in the exponent is equal to the number of places the decimal must be shifted to convert a number from scientific to purely decimal notation. Shift the decimal point to the right if the exponent is positive and to the left if it is negative. Reverse the process when decimal numbers are converted to scientific notation.

ARITHMETIC OPERATIONS WITH SCIENTIFIC NOTATION

The use of scientific notation is helpful in preventing decimal errors in arithmetic calculations. Some examples follow.

Multiplication

Here, the decimal parts of the numbers are multiplied and the exponents are added; thus,

$$420,000 \times 0.0300 = (4.20 \times 10^5)(3.00 \times 10^{-2})$$
$$= 12.60 \times 10^3 = 1.26 \times 10^4$$
$$0.0060 \times 0.000020 = 6.0 \times 10^{-3} \times 2.0 \times 10^{-5}$$
$$= 12 \times 10^{-8} = 1.2 \times 10^{-7}$$

Division

Here, divide the decimal parts of the numbers; subtract the exponent in the denominator from that in the numerator. For example,

$$\frac{0.015}{5000} = \frac{15 \times 10^{-3}}{5 \times 10^3} = 3.0 \times 10^{-6}$$

Addition and Subtraction

To add or subtract in scientific notation, express all numbers to a common power of 10. Then add or subtract the decimal parts, as appropriate. Thus,

$$2.00 \times 10^{-11} + 4.00 \times 10^{-12} - 3.00 \times 10^{-10}$$
$$= 2.00 \times 10^{-11} + 0.400 \times 10^{-11} - 30.0 \times 10^{-11}$$
$$= -27.6 \times 10^{-11} = -2.76 \times 10^{-10}$$

Raising to a Power a Number Written in Exponential Notation

Here, raise each part of the number to the power separately. For example,

$$(2 \times 10^{-3})^4 = (2.0)^4 \times (10^{-3})^4 = 16 \times 10^{-(3 \times 4)}$$
$$= 16 \times 10^{-12} = 1.6 \times 10^{-11}$$

Extraction of the Root of a Number Written in Exponential Notation

Here, write the number in such a way that the exponent of 10 is evenly divisible by the root. Thus,

$$(4.0 \times 10^{-5})^{1/3} = \sqrt[3]{40 \times 10^{-6}} = \sqrt[3]{40} \times \sqrt[3]{10^{-6}}$$
$$= 3.4 \times 10^{-2}$$

LOGARITHMS

In this discussion, we will assume that you have available an electronic calculator for obtaining logarithms and antilogarithms of numbers. It is desirable, however, that you understand what a logarithm is as well as some of its properties. The discussion that follows provides this information.

A logarithm (or log) of a number is the power to which some base number (often 10) must be raised in order to give the desired number. Thus, a base-ten logarithm is an exponent of the base 10. From the discussion in the previous paragraphs about exponential numbers, we can draw the following conclusions with respect to logs:

1. The logarithm of a product is the sum of the logarithms of the individual numbers in the product.

$$\log (100 \times 1000) = \log 10^2 + \log 10^3 = 2 + 3 = 5$$

2. The logarithm of a quotient is the difference between the logarithms of the individual numbers.

$$\log (100/1000) = \log 10^2 - \log 10^3 = 2 - 3 = -1$$

3. The logarithm of a number raised to some power is the logarithm of the number multiplied by that power.

$$\log (1000)^2 = 2 \times \log 10^3 = 2 \times 3 = 6^-(-)-$$
$$\log (0.01)^6 = 6 \times \log 10^{-2} = 6 \times (-2) = -12$$

4. The logarithm of a root of a number is the logarithm of that number divided by the root.

$$\log (1000)^{1/3} = \frac{1}{3} \times \log 10^3 = \frac{1}{3} \times 3 = 1$$

The following two examples demonstrate that the logarithm of a number is the sum of two parts, a *characteristic* located to the left of the decimal point and a *mantissa* that lies to the right. The characteristic is the logarithm of 10 raised to a power and serves to indicate the location of the decimal point in the original number when that number is expressed in decimal notation. The mantissa is the logarithm of a number in the range between 0.00 and 9.99. . . .

$$\log 40 \times 10^{20} = \log 4.0 \times 10^{21} = \log 4.0 + \log 10^{21}$$
$$= 0.60 + 21 = 21.60$$
$$\log 2.0 \times 10^{-6} = \log 2.0 + \log 10^{-6} = 0.30 + (-6) = -5.70$$

THE LEAST-SQUARES METHOD FOR DERIVING CALIBRATION CURVES

When the method of least squares is used to generate a linear calibration curve, two assumptions are required. The first is that there is actually a linear relationship between the measured variable (y) and the analyte concentration (x). This relationship is stated mathematically as

$$y = mx + b$$

where b is the y intercept (the value of y when x is zero) and m is the slope of the line (see Figure 5-5). We also assume that any deviation of individual points from the straight line results from error in the *measurement*. That is, we must assume that there is no error in the x values of the points. For a calibration curve, for example, we assume that exact concentrations of the standards are known from the way they were prepared. Both of these assumptions are appropriate for most analytical methods.

As illustrated in Figure 5-4, the vertical deviation of each point from the straight line is called a *residual*. The line generated by the least-squares method is the one that minimizes the sum of the squares of the residuals from all of the points. In addition to providing the best fit between the experimental points and the straight line, the method gives the standard deviations for m and b.[1]

---

[1] R. L. Anderson, *Practical Statistics for Analytical Chemists*, pp. 89–121. New York: Van Nostrand-Reinhold, 1987.

For convenience, we define three quantities S_{xx}, S_{yy}, and S_{xy} as follows:

$$S_{xx} = \Sigma(x_i - \bar{x})^2 = \Sigma x_i^2 - \frac{(\Sigma x_i)^2}{N} \tag{A6-1}$$

$$S_{yy} = \Sigma(y_i - \bar{y})^2 = \Sigma y_i^2 - \frac{(\Sigma y_i)^2}{N} \tag{A6-2}$$

$$S_{xy} = \Sigma(x_i - \bar{x})(y_i - \bar{y}) = \Sigma x_i y_i - \frac{\Sigma x_i \Sigma y_i}{N} \tag{A6-3}$$

where x_i and y_i are individual pairs of data for x and y, N is the number of pairs of data used in preparing the calibration curve, and $\bar{x}$ and $\bar{y}$ are the average values for the variables, that is,

$$\bar{x} = \frac{\Sigma x_i}{N} \quad \text{and} \quad \bar{y} = \frac{\Sigma y_i}{N}$$

Note that S_{xx} and S_{yy} are the sum of the squares of the deviations from the mean for the individual values of x and y. The equivalent expressions shown to the far right in Equations A6-1 through A6-3 are more convenient when a hand-held calculator is being used.

Six useful quantities can be derived from S_{xx}, S_{yy}, S_{xy}, $\bar{x}$, and $\bar{y}$.

1. The slope of the line m:

$$m = S_{xy}/S_{xx} \tag{A6-4}$$

2. The intercept b:

$$b = \bar{y} - m\bar{x} \tag{A6-5}$$

3. The standard deviation about regression s_r:

$$s_r = \sqrt{\frac{S_{yy} - m^2 S_{xx}}{N - 2}} \tag{A6-6}$$

4. The standard deviation of the slope s_m:

$$s_m = \sqrt{s_r^2/S_{xx}} \tag{A6-7}$$

5. The standard deviation of the intercept s_b:

$$s_b = s_r \sqrt{\frac{\Sigma x_i^2}{N \Sigma x_i^2 - (\Sigma x_i)^2}} = s_r \sqrt{\frac{1}{N - (\Sigma x_i)^2/\Sigma x_i^2}} \tag{A6-8}$$

6. The standard deviation for results obtained from the calibration curve s_c:

$$s_c = \frac{s_r}{m} \sqrt{\frac{1}{M} + \frac{1}{N} + \frac{(\bar{y}_c - \bar{y})^2}{m^2 S_{xx}}} \tag{A6-9}$$

Equation A6-9 gives us a way to calculate the standard deviation for the mean $\bar{y}_c$ of a set of M replicate analyses of unknowns when a calibration curve that contains N points is used; recall that $\bar{y}$ is the mean value of y for the N calibration data.

The standard deviation about regression s_r (Equation A6-6) is the standard deviation for y when the deviations are measured not from the mean of y (as is usually the case) but from the derived straight line:

$$s_r = \sqrt{\frac{\sum_{i=1}^{N} [y_i - (b + mx_i)]^2}{N - 2}} \qquad \textbf{(A6-10)}$$

In this equation, the number of degrees of freedom is $N - 2$ since one degree is lost in calculating m and one in determining b.

Example A6-1

Carry out a least-squares analysis of the experimental data provided in the first two columns in Table A6-1 and plotted in Figure 5-4.

Columns 3, 4, and 5 of the table contain computed values for x_i^2, y_i^2, and $x_i y_i$, with their sums appearing as the last entry in each column. Note that the number of digits carried in the computed values should be the *maximum allowed by the calculator or computer*; that is, *rounding off should not be performed until the calculation is complete*.

We now substitute into Equations A6-1, A6-2, and A6-3 and obtain

$$S_{xx} = \Sigma x_i^2 - (\Sigma x_i^2)^2/N = 6.90201 - (5.365)^2/5 = 1.14536$$
$$S_{yy} = \Sigma y_i^2 - (\Sigma y_i^2)^2/N = 36.3775 - (12.51)^2/5 = 5.07748$$
$$S_{xy} = \Sigma x_i y_i - \Sigma x_i \Sigma y_i/N = 15.81992 - 5.365 \times 12.51/5 = 2.39669$$

Substitution of these quantities into Equations A6-4 and A6-5 yields

$$m = 2.39669/1.14536 = 2.0925 = 2.09$$
$$b = \frac{12.51}{5} - 2.0925 \times \frac{5.365}{5} = 0.2567 = 0.26$$

Thus, the equation for the least-squares line is

$$y = 2.09x + 0.26 \qquad \textbf{(A6-11)}$$

Substitution into Equation A6-6 yields the standard deviation about regression

$$s_r = \sqrt{\frac{S_{yy} - m^2 S_{xx}}{N - 2}} = \sqrt{\frac{5.07748 - (2.0925)^2 \times 1.14536}{5 - 2}}$$
$$= 0.144 = 0.14$$

Table A6-1

CALIBRATION DATA FOR A CHROMATOGRAPHIC METHOD
FOR THE DETERMINATION OF ISOOCTANE IN A
HYDROCARBON MIXTURE

Mole Percent Isooctane, x_i	Peak Area, y_i	x_i^2	y_i^2	$x_i y_i$
0.352	1.09	0.12390	1.1881	0.38368
0.803	1.78	0.64481	3.1684	1.42934
1.08	2.60	1.16640	6.7600	2.80800
1.38	3.03	1.90440	9.1809	4.18140
1.75	4.01	3.06250	16.0801	7.01750
5.365	12.51	6.90201	36.3775	15.81992

and substitution into Equation A6-7 gives the standard deviation of the slope:

$$s_m = \sqrt{s_r^2/S_{xx}} = \sqrt{(0.144)^2/1.14536} = 0.13$$

Finally, we find the standard deviation of the intercept from Equation A6-8.

$$s_b = 0.144 \sqrt{\frac{1}{5 - (5.365)^2/6.90201}} = 0.16$$

Example A6-2

The calibration curve derived in Example A6-1 was used for the chromatographic determination of isooctane in a hydrocarbon mixture. A peak area of 2.65 was obtained. Calculate the mole percent of isooctane and the standard deviation for the result if the area was (a) the result of a single measurement, and (b) the mean of four measurements.

In either case, substituting into Equation A6-11 and rearranging gives

$$x = \frac{y - 0.26}{2.09} = \frac{2.65 - 0.26}{2.09} = 1.14 \text{ mol } \%$$

(a) Substituting into Equation A6-9, we obtain

$$s_c = \frac{0.14}{2.09} \sqrt{\frac{1}{1} + \frac{1}{5} + \frac{(2.65 - 12.51/5)^2}{(2.09)^2 \times 1.145}} = 0.07 \text{ mol } \%$$

(b) For the mean of four measurements,

$$s_c = \frac{0.14}{2.09} \sqrt{\frac{1}{4} + \frac{1}{5} + \frac{(2.65 - 12.51/5)^2}{(2.09)^2 \times 1.145}} = 0.04 \text{ mol } \%$$

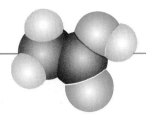

VOLUMETRIC CALCULATIONS USING NORMALITY AND EQUIVALENT WEIGHT

The *normality* of a solution expresses the number of equivalents of solute contained in 1 L of solution or the number of milliequivalents in 1 mL. The equivalent and milliequivalent, like the mole and millimole, are units for describing the amount of a chemical species. The former are defined, however, in such a way that it is possible to state that, at the equivalence point in *any* titration,

$$\text{no. meq analyte present} = \text{no. meq standard reagent added} \quad \textbf{(A7-1)}$$

or

$$\text{no. eq analyte present} = \text{no. eq standard reagent added} \quad \textbf{(A7-2)}$$

As a consequence, stoichiometric ratios such as those described in Section 9C-3 need not be derived every time a volumetric calculation is performed. Instead, the stoichiometry is taken into account by the way equivalent or milliequivalent weight is defined.

THE DEFINITION OF EQUIVALENT AND MILLIEQUIVALENT

In contrast to the mole, the amount of a substance contained in one equivalent can vary from reaction to reaction. Consequently, the weight of one equivalent of a compound can never be computed *without reference to a chemical reaction* in which that compound is, directly or indirectly, a participant. Similarly, the normality of a solution can never be specified *without knowledge about how the solution will be used.*

Equivalent Weights in Neutralization Reactions

One equivalent weight of a substance participating in a neutralization reaction is that amount of substance (molecule, ion, or paired ion such as NaOH) that either reacts with or supplies 1 mol of hydrogen ions *in that reaction.*[1] A milli-equivalent is simply 1/1000 of an equivalent.

The relationship between equivalent weight (eqw) and the molar mass ($\mathcal{M}$) is straightforward for strong acids or bases and for other acids or bases that contain a single reactive hydrogen or hydroxide ion. For example, the equivalent weights of potassium hydroxide, hydrochloric acid, and acetic acid are equal to their molar masses because each has but a single reactive hydrogen ion or hydroxide ion. Barium hydroxide, which contains two identical hydroxide ions, reacts with two hydrogen ions in any acid/base reaction, and so its equivalent weight is one half its molar mass:

$$\text{eqw } Ba(OH)_2 = \frac{\mathcal{M}_{Ba(OH)_2}}{2}$$

The situation becomes more complex for acids or bases that contain two or more reactive hydrogen or hydroxide ions with different tendencies to dissociate. With certain indicators, for example, only the first of the three protons in phosphoric acid is titrated:

$$H_3PO_4 + OH^- \rightarrow H_2PO_4^- + H_2O$$

With certain other indicators, a color change occurs only after two hydrogen ions have reacted:

$$H_3PO_4 + 2OH^- \rightarrow HPO_4^{2-} + 2H_2O$$

For a titration involving the first reaction, the equivalent weight of phosphoric acid is equal to the molar mass; for the second, the equivalent weight is one half the molar mass. (Because it is not practical to titrate the third proton, an equivalent weight that is one third the molar mass is not generally encountered for H_3PO_4.) If it is not known which of these reactions is involved, an unambiguous definition of the equivalent weight for phosphric acid *cannot be made.*

---

[1] An alternative definition, proposed by the International Union of Pure and Applied Chemistry, is as follows: An equivalent is "that amount of substance, which, in a specified reaction, releases or replaces that amount of hydrogen that is combined with 3 g of carbon-12 in methane $^{12}CH_4$" (see *Information Bulletin* No. 36, International Union of Pure and Applied Chemistry, August 1974). This definition applies to acids. For other types of reactions and reagents, the amount of hydrogen referred to may be replaced by the equivalent amount of hydroxide ions, electrons, or cations. The reaction to which the definition is applied must be specified.

Equivalent Weights in Oxidation/Reduction Reactions

The equivalent weight of a participant in an oxidation/reduction reaction is that amount that directly or indirectly produces or consumes 1 mol of electrons. The numerical value for the equivalent weight is conveniently established by dividing the molar mass of the substance of interest by the change in oxidation number associated with its reaction. As an example, consider the oxidation of oxalate ion by permanganate ion:

$$5C_2O_4^{2-} + 2MnO_4^- + 16H^+ \rightarrow 10CO_2 + 2Mn^+ + 8H_2O \qquad \textbf{(A7-3)}$$

In this reaction, the change in oxidation number of manganese is 5 because the element passes from the $+7$ to the $+2$ state; the equivalent weights for MnO_4^- and Mn^{2+} are therefore one fifth their molar masses. Each carbon atom in the oxalate ion is oxidized from the $+3$ to the $+4$ state, leading to the production of two electrons by that species. Therefore, the equivalent weight of sodium oxalate is one half of its molar mass. It is also possible to assign an equivalent weight to the carbon dioxide produced by the reaction. Since this molecule contains but a single carbon atom and since that carbon undergoes a change in oxidation number of 1, the molar mass and equivalent weight of the two are identical.

It is important to note that in evaluating the equivalent weight of a substance, *only* its change in oxidation number during the titration is considered. For example, suppose the manganese content of a sample containing Mn_2O_3 is to be determined by a titration based on the reaction given in Equation A7-3. The fact that each manganese in the Mn_2O_3 has an oxidation number of $+3$ plays no part in determining equivalent weight. That is, we must assume that by suitable treatment, all the manganese is oxidized to the $+7$ state before the titration is begun. Each manganese from the Mn_2O_3 is then reduced from the $+7$ to the $+2$ state in the titration step. The equivalent weight is thus the molar mass of Mn_2O_3 divided by $2 \times 5 = 10$.

As in neutralization reactions, the equivalent weight for a given oxidizing or reducing agent is not invariant. Potassium permanganate, for example, reacts under some conditions to give MnO_2:

$$MnO_4^- + 3e^- + 2H_2O \rightarrow MnO_2(s) + OH^-$$

The change in the oxidation state of manganese in this reaction is from $+7$ to $+4$, and the equivalent weight of potassium permanganate is now equal to its molar mass divided by 3 (instead of 5 as in the earlier example).

Equivalent Weights in Precipitation and Complex-Formation Reactions

The equivalent weight of a participant in a precipitation or a complex-formation reaction is the weight that reacts with or provides one mole of the *reacting* cation if it is univalent, one-half mole if it is divalent, one-third mole if it is trivalent, and so on. It is important to note that the cation referred to in this definition is always *the cation directly involved in the analytical reaction* and not necessarily the cation contained in the compound whose equivalent weight is being defined.

Example A7-1

Define equivalent weights for $AlCl_3$ and $BiOCl$ if the two compounds are determined by a precipitation titration with $AgNO_3$:

$$Ag^+ + Cl^- \rightarrow AgCl(s)$$

In this instance, the equivalent weight is based on the number of moles of *silver ions* involved in the titration of each compound. Since 1 mol of Ag^+ reacts with 1 mol of Cl^- provided by one-third mole of $AlCl_3$, we can write

Recall that $\mathcal{M}$ is the symbol for molar mass (Section 2A-2).

$$\text{eqw } AlCl_3 = \frac{\mathcal{M}_{AlCl_3}}{3}$$

Because each mole of $BiOCl$ reacts with only 1 Ag^+ ion,

$$\text{eqw } BiOCl = \frac{\mathcal{M}_{BiOCl}}{1}$$

Note that the fact that Bi^{3+} (or Al^{3+}) is trivalent has no bearing because the definition is based *on the cation involved in the titration*: Ag^+.

THE DEFINITION OF NORMALITY

The normality c_N of a solution expresses the number of milliequivalents of solute contained in 1 mL of solution or the number of equivalents contained in 1 L. Thus, a 0.20 N hydrochloric acid solution contains 0.20 meq of HCl in each milliliter of solution or 0.20 eq in each liter.

The normal concentration of a solution is defined by equations analogous to Equation 2-1. Thus, for a solution of the species A, the normality $c_{N(A)}$ is given by the equations

$$c_{N(A)} = \frac{\text{no. meq A}}{\text{no. mL solution}} \tag{A7-4}$$

$$c_{N(A)} = \frac{\text{no. eq A}}{\text{no. L solution}}\text{m} \tag{A7-5}$$

SOME USEFUL ALGEBRAIC RELATIONSHIPS

Two pairs of algebraic equations, analogous to Equations 9-1 and 9-2 as well as 9-3 and 9-4, apply when normal concentrations are used:

$$\text{amount A} = \text{no. meq A} = \frac{\text{mass A (g)}}{\text{meqw A (g/meq)}} \tag{A7-6}$$

$$\text{amount A} = \text{no. eq A} = \frac{\text{mass A (g)}}{\text{eqw A (g/eq)}} \qquad \textbf{(A7-7)}$$

$$\text{amount A} = \text{no. meq A} = V\,(\text{mL}) \times c_{N(A)}\,(\text{meq/mL}) \qquad \textbf{(A7-8)}$$

$$\text{amount A} = \text{no. eq A} = V\,(\text{L}) \times c_{N(A)}\,(\text{eq/L}) \qquad \textbf{(A7-9)}$$

CALCULATION OF THE NORMALITY OF STANDARD SOLUTIONS

Example A7-2 shows how the normality of a standard solution is computed from preparatory data. Note the similarity between this example and Example 9-1.

Example A7-2

Describe the preparation of 5.000 L of 0.1000 N Na_2CO_3 (105.99 g/mol) from the primary-standard solid, assuming the solution is to be used for titrations in which the reaction is

$$CO_3^{2-} + 2H^+ \rightarrow H_2O + CO_2$$

Applying Equation A7-9 gives

$$\text{amount Na}_2\,CO_3 = V \text{ soln (L)} \times c_{N(Na_2CO_3)}\,(\text{eq/L})$$
$$= 5.000 \text{ Ł} \times 0.1000 \text{ eq/Ł} = 0.5000 \text{ eq Na}_2CO_3$$

Rearranging Equation A7-7 gives

$$\text{mass Na}_2CO_3 = \text{no. eq Na}_2CO_3 \times \text{eqw Na}_2CO_3$$

But 2 eq of Na_2CO_3 are contained in each mole of the compound; therefore,

$$\text{mass Na}_2CO_3 = 0.5000 \text{ eq Na}_2CO_3 \times \frac{105.99 \text{ g Na}_2CO_3}{2 \text{ eq Na}_2CO_3} = 26.50 \text{ g}$$

Therefore, dissolve 26.50 g in water and dilute to 5.000 L.

It is worth noting that when the carbonate ion reacts with two protons, the weight of sodium carbonate required to prepare a 0.10 N solution is just one half that required to prepare a 0.10 M solution.

The Treatment of Titration Data with Normalities

Calculation of Normalities from Titration Data

Examples A7-3 and A7-4 illustrate how normality is computed from standardization data. Note that these examples are similar to Examples 9-5 and 9-6.

Example A7-3

Exactly 50.00 mL of an HCl solution required 29.71 mL of 0.03926 N $Ba(OH)_2$ to give an end point with bromocresol green indicator. Calculate the normality of the HCl.

Note that the molarity of $Ba(OH)_2$ is one half its normality. That is,

$$c_{Ba(OH)_2} = 0.03926 \, \frac{\text{meq}}{\text{mL}} \times \frac{1 \text{ mmol}}{2 \text{ meq}} = 0.01963 \text{ M}$$

Because we are basing our calculations on the milliequivalent, we write that at equivalence

$$\text{no. meq HCl} = \text{no. meq } Ba(OH)_2$$

The number of milliequivalents of standard is obtained by substituting into Equation A7-8:

$$\text{amount } Ba(OH)_2 = 29.71 \, \text{mL Ba(OH)}_2 \times 0.03926 \, \frac{\text{meq Ba(OH)}_2}{\text{mL Ba(OH)}_2}$$

To obtain the number of milliequivalents of HCl, we write

$$\text{amount HCl} = (29.71 \times 0.03926) \, \text{meq Ba(OH)}_2 \times \frac{1 \text{ meq HCl}}{1 \text{ meq Ba(OH)}_2}$$

Equating this result to Equation A7-8 yields

$$\text{amount HCl} = 50.00 \text{ mL} \times c_{N(HCl)}$$
$$= (29.71 \times 0.03926 \times 1) \text{ meq HCl}$$
$$c_{N(HCl)} = \frac{(29.71 \times 0.03926 \times 1) \text{ meq HCl}}{50.00 \text{ mL HCl}} = 0.02333 \text{ N}$$

Example A7-4

A 0.2121-g sample of pure $Na_2C_2O_4$ (134.00 g/mol) was titrated with 43.31 mL of $KMnO_4$. What is the normality of the $KMnO_4$ solution? The chemical reaction is

$$2MnO_4^- + 5C_2O_4^{2-} + 16H^+ \rightarrow 2Mn^{2+} + 10CO_2 + 8H_2O$$

By definition, at the equivalence point in the titration,

$$\text{no. meq } Na_2C_2O_4 = \text{no. meq } KMnO_4$$

Substituting Equations A7-8 and A7-6 into this relationship gives

$$V_{KMnO_4} \times c_{N(KMnO_4)} = \frac{\text{mass } Na_2C_2O_4 \text{ (g)}}{\text{meqw } Na_2C_2O_4 \text{ (g/meq)}}$$

$$43.31 \text{ mL KMnO}_4 \times c_{N(KMnO_4)} = \frac{0.2121 \text{ g } Na_2C_2O_4}{0.13400 \text{ g } Na_2C_2O_4/2 \text{ meq}}$$

$$c_{N(KMnO_4)} = \frac{0.2121 \text{ g } \cancel{Na_2C_2O_4}}{43.31 \text{ mL KMnO}_4 \times 0.1340 \text{ g } \cancel{Na_2C_2O_4}/2 \text{ meq}}$$

$$= 0.073093 \text{ meq/mL KMnO}_4 = 0.07309 \text{ N}$$

Note that the normality found here is five times the molarity computed in Example 9-6.

Calculation of the Quantity of Analyte from Titration Data

The examples that follow illustrate how analyte concentrations are computed when normalities are involved.

Example A7-5

A 0.8040-g sample of an iron ore was dissolved in acid. The iron was then reduced to Fe^{2+} and titrated with 47.22 mL of 0.1121 N (0.02242 M) $KMnO_4$ solution. Calculate the results of this analysis in terms of (a) percent Fe (55.847 g/mol) and (b) percent Fe_3O_4 (231.54 g/mol). The reaction of the analyte with the reagent is described by the equation

$$MnO_4^- + 5Fe^{2+} + 8H^+ \rightarrow Mn^{2+} + 5Fe^{3+} + 4H_2O$$

(a) At the equivalence point, we know that

$$\text{no. meq KMnO}_4 = \text{no. meq Fe}^{2+} = \text{no. meq Fe}_3O_4$$

Substituting Equations A7-8 and A7-6 leads to

$$V_{KMnO_4} \text{ (mL)} \times c_{N(KMnO_4)} \text{ (meq/mL)} = \frac{\text{mass Fe}^{2+} \text{ (g)}}{\text{meqw Fe}^{2+} \text{ (g/meq)}}$$

After substituting numerical data into this equation and rearranging, we have

$$\text{mass Fe}^{2+} = 47.22 \text{ mL } \cancel{KMnO_4} \times 0.1121 \frac{\text{meq}}{\text{mL } \cancel{KMnO_4}} \times \frac{0.055847 \text{ g}}{1 \text{ meq}}$$

Note that the milliequivalent weight of the Fe is equal to its millimolar mass.

The percentage of iron is

$$\text{percent Fe} = \frac{(47.22 \times 0.1121 \times 0.055847)\ \text{g Fe}^{2+}}{0.8040\ \text{g sample}} \times 100\%$$

$$= 36.77\%$$

(b) Here,

$$\text{no. meq KMnO}_4 = \text{no. meq Fe}_3\text{O}_4$$

and

$$V_{\text{KMnO}_4}\ (\text{mL}) \times c_{\text{N(KMnO}_4)}\ (\text{meq/mL}) = \frac{\text{mass Fe}_3\text{O}_4\ (\text{g})}{\text{meqw Fe}_3\text{O}_4\ (\text{g/meq})}$$

Substituting numerical data and rearranging gives

$$\text{mass Fe}_3\text{O}_4 = 47.22\ \text{mL} \times 0.1121\ \frac{\text{meq}}{\text{mL}} \times 0.23154\ \frac{\text{g Fe}_3\text{O}_4}{3\ \text{meq}}$$

Note that the milliequivalent weight of Fe_3O_4 is one third its millimolar mass because each Fe^{2+} undergoes a one-electron change and the compound is converted to $3Fe^{2+}$ before titration. The percentage of Fe_3O_4 is then

$$\text{percent Fe}_3\text{O}_4 = \frac{(47.22 \times 0.1121 \times 0.23154/3)\ \text{g Fe}_3\text{O}_4}{0.8040\ \text{g sample}} \times 100\%$$

$$= 50.81\%$$

Note that the answers to this example are identical to those in Example 9-7.

Example A7-6

A 0.4755-g sample containing $(NH_4)_2C_2O_4$ and inert compounds was dissolved in water and made alkaline with KOH. The liberated NH_3 was distilled into 50.00 mL of 0.1007 N (0.05035 M) H_2SO_4. The excess H_2SO_4 was back-titrated with 11.13 mL of 0.1214 N NaOH. Calculate the percentage of N (14.007 g/mol) and of $(NH_4)_2C_2O_4$ (124.10 g/mol) in the sample.

At the equivalence point, the number of milliequivalents of acid and base are equal. In this titration, however, two bases are involved: NaOH and NH_3. Thus,

$$\text{no. meq H}_2\text{SO}_4 = \text{no. meq NH}_3 + \text{no. meq NaOH}$$

After rearranging,

$$\text{no. meq NH}_3 = \text{no. meq N} = \text{no. meq H}_2\text{SO}_4 - \text{no. meq NaOH}$$

To obtain the number of milliequivalents of H_2SO_4 and NaOH, we write

$$\text{no. meq } H_2SO_4 = 50.00 \text{ mL } H_2SO_4 \times 0.1007 \frac{\text{meq}}{\text{mL } H_2SO_4} = 5.0350$$

$$\text{no. meq NaOH} = 11.13 \text{ mL NaOH} \times 0.1214 \frac{\text{meq}}{\text{mL NaOH}} = 1.3512$$

$$\text{no. meq } NH_3 = \text{no. meq N} = 5.0350 - 1.3512 = 3.6838$$

$$\text{mass N} = 3.6838 \text{ meq N} \times 0.014007 \text{ g N/meq} = 0.05160 \text{ g}$$

$$\text{percent N} = \frac{0.05160 \text{ g N}}{0.4755 \text{ g sample}} \times 100\% = 10.85\%$$

The number of milliequivalents of $(NH_4)_2C_2O_4$ is also equal to the number of milliequivalents of N, or 3.6838. The milliequivalent weight of $(NH_4)_2C_2O_4$ is, however, only one half of its molar mass. Thus,

$$\text{mass } (NH_4)_2C_2O_4 = 3.6838 \text{ meq } (NH_4)_2C_2O_4 \times \frac{0.12410 \text{ g } (NH_4)_2C_2O_4}{2 \text{ meq}}$$

$$= 0.22858 \text{ g}$$

$$\text{percent } (NH_4)_2C_2O_4 = \frac{0.22858 \text{ g } (NH_4)_2C_2O_4}{0.4755 \text{ g sample}} \times 100\% = 48.07\%$$

Note that the results obtained here are identical to those obtained in Example 9-9.

ANSWERS TO QUESTIONS AND PROBLEMS

Chapter 2

2-1. (a) The dalton, which is synonymous with the atomic mass unit, is a relative mass unit that is equal to 1/12 of the mass of one neutral ^{12}C atom.

    (c) The p-value for a chemical species X is the negative logarithm of the molar concentration of that species. That is, $pX = -\log[X]$, where $[X]$ is the molar concentration of X.

2-2. (a) Mass is an invariant measure of the amount of matter in an object. Weight is the force of attraction between that object and the earth.

    (c) The density of a substance is its mass per unit volume. Its specific gravity is the ratio of its mass to an equal volume of water usually measured at 0°C.

2-3. The atomic mass of iron is 55.85 daltons, or 55.85 amu. Its molar mass is 55.85 grams.

2-5. $1\ L = 10^{-3}\ m^3$

$$1\ M = \frac{1\ mol}{L} = \frac{1\ mol}{10^{-3}\ m^3}$$

$1\ \overset{\circ}{A} = 10^{-10}\ m$

2-6. (a) 1.5 MHz   (c) 62.3 mmol   (e) 96.495 kC

2-7. (a) 0.0401 mol Na^+ (b) 0.0616 mol Na^+

    (c) 0.0256 mol Na^+ (d) 0.0596 mol Na^+

2-9. (a) 0.0982 mol   (b) 7.76×10^{-4} mol

    (c) 0.0382 mol   (d) 5.26×10^{-4} mol

2-11. (a) 5.52 mmol   (b) 31.2 mmol

    (c) 6.58×10^{-3} mmol   (d) 966 mmol

2-13. (a) 4.20×10^4 mg   (b) 1.21×10^4 mg

    (c) 1.52×10^6 mg   (d) 2.92×10^6 mg

2-15. (a) 1.33×10^3 mg   (b) 520 mg

2-16. (a) 2.51 g   (b) 2.88×10^{-3} g

2-17. (a) pNa = 0.618, pCl = 0.936, pOH = 0.903

    (c) pH = −0.176, pZn = 0.921, pCl = −0.241

    (e) pK = 4.249, pOH = 4.385, pFe(CN)₆ = 5.421

2-18. (a) pNa = 2.000, pBr = 2.000, pH = pOH = 7.000

    (c) pBa = 2.46, pOH = 2.16, pH = 11.84

    (e) pCa = 2.28, pBa = 2.44, pCl = 1.75, pH = pOH = 7.00

2-19. (a) 6.5×10^{-10} M   (b) 1.5×10^{-5} M

    (c) 6.8×10^{-3} M   (d) 1.08 M

2-20. (a) 7.2×10^{-8} M   (b) 9.3×10^{-1} M

    (c) 2.5×10^{-10} M   (d) 8.5 M

2-21. (a) $[Na^+] = 4.79 \times 10^{-2}$, $[SO_4^{2-}] = 2.87 \times 10^{-3}$

    (b) pNa = 1.320, pSO₄ = 2.543

2-23. (a) 1.821×10^{-2} M   (b) 1.821×10^{-2} M

    (c) 5.463×10^{-2} M   (d) 0.506%

    (e) 1.366 mmol   (f) 712 ppm

    (g) 1.740   (h) 1.263

2-25. (a) 0.346 M   (b) 1.038 M   (c) 83.7 g/L

2-27. (a) Dilute 32.5 g C_2H_5OH to 500 mL with H_2O

    (b) Mix 32.5 g C_2H_5OH with 467.5 g H_2O

    (c) Dilute 32.5 mL C_2H_5OH to 500 mL with H_2O

2-29. Dilute 26 mL of concentrated $HClO_4$ to 2.0 L

2-31. (a) 1 mol AgCl/mol $AgNO_3$

    (c) 2 mol H_3O^+/mol Na_3PO_4

    (e) 2 mol Ag/mol Cu

    (g) 1 mol $NaAg(CN)_2$/2 mol NaCN

2-33. (a) Dissolve 6.37 g $AgNO_3$ in H_2O and dilute to 500 mL.

    (b) Dilute 52.5 mL of 6.00 M HCl to 1.00 L.

    (c) Dissolve 4.56 g $K_4Fe(CN)_6$ in H_2O and dilute to 600 mL.

    (d) Dilute 144 mL of 0.400 M $BaCl_2$ to 400 mL.

    (e) Dilute 25 mL of the reagent to 2.0 L.

    (f) Dissolve 1.67 g Na_2SO_4 in H_2O and dilute to 9.00 L.

2-35. 3.35 g $La(IO_3)_3$

2-37. (a) 0.04652 g CO_2 (b) 0.0286 M HCl

2-39. (a) 1.602 g SO_2 (b) 0.0386 M $HClO_4$

2-41. 318.4 mL $AgNO_3$

Chapter 3

3-1. (a) A weak electrolyte is a substance that ionizes only partially in a solvent.

    (c) The conjugate base of a Brønsted–Lowry acid is the species formed when the acid has donated a proton.

    (e) An amphiprotic solute is one that can act either as an acid or as a base when dissolved in a solvent.

(g) Autoprotolysis is self-ionization of a solvent to give a conjugate acid and a conjugate base.

(i) The Le Châtelier principle states that the position of equilibrium in a system always shifts in a direction that tends to relieve an applied stress to the system.

3-2. (a) An amphiprotic solvent is a solvent that acts as a base with acidic solutes and as an acid with basic solutes.

(c) A leveling solvent is one in which a series of acids (or bases) all dissociate completely.

3-3. For an aqueous equilibrium in which water is a reactant or a product, the concentration of water is normally so much larger than the concentrations of the reactants and products that its concentration can be assumed to be constant and independent of the position of the equilibrium. Thus its concentration is assumed to be constant and is included in the equilibrium constant. For a solid reactant or product, it is the concentration of that reactant in the solid phase that would influence the position of equilibrium. However, the concentration of a species in the solid phase is constant. Thus as long as some solid exists as a second phase, its effect on the equilibrium is constant, and its concentration is included in the equilibrium constant.

3-4.

	Acid	Conjugate Base
(a)	HCN	CN^-
(c)	NH_4^+	NH_3
(e)	$H_2PO_4^-$	HPO_4^{2-}

3-5.

	Base	Conjugate Acid
(a)	H_2O	H_3O^+
(c)	H_2O	H_3O^+
(e)	PO_4^{3-}	HPO_4^{2-}

3-6. (a) $2H_2O \rightleftharpoons H_3O^+ + OH^-$

(c) $2CH_3NH_2 \rightleftharpoons CH_3NH_3^+ + CH_3NH^-$

3-7. (a) $K_{sp} = [Ag^+][I^-] = 8.3 \times 10^{-17}$

(c) $K_{sp} = [Ag^+]^2[CrO_4^{2-}] = 1.2 \times 10^{-12}$

3-8. (a) $C_2H_5NH_2 + H_2O \rightleftharpoons C_2H_5NH_3^+ + OH^-$

$$K_b = \frac{K_w}{K_a} = \frac{1.00 \times 10^{-14}}{2.31 \times 10^{-11}}$$

$$= \frac{[C_2H_5NH_3^+][OH^-]}{[C_2H_5NH_2]} = 4.33 \times 10^{-4}$$

(c) $C_5H_5NH^+ + H_2O \rightleftharpoons C_5H_5NH_2 + H_3O^+$

$$K_a = \frac{[H_3O^+][C_5H_5NH_2]}{[C_5H_5NH^+]} = 5.90 \times 10^{-6}$$

(e) $H_3AsO_4 + 3H_2O \rightleftharpoons 3H_3O^+ + AsO_4^{3-}$

$$K_1K_2K_3 = 5.8 \times 10^{-3} \times 1.1 \times 10^{-7}$$
$$\times 3.2 \times 10^{-12} = 2.0 \times 10^{-21}$$

$$\frac{[H_3O^+]^3[AsO_4^{3-}]}{[H_3AsO_4]} = 2.0 \times 10^{-21}$$

3-9. (a) 4.0×10^{-16}　　(c) 3.1×10^{-6}　　(e) 3.5×10^{-10}

3-10. (a) 8.0×10^{-15} M　(c) 3.9×10^{-3} M　(e) 6.4×10^{-4} M

3-11. (a) 8.0×10^{-15} M　(c) 1.2×10^{-3} M　(e) 2.8×10^{-6} M

3-12. (a) 2.18×10^{-7} M　　(b) 0.98 M

3-14. (a) 0.0250 M　　(b) 1.7×10^{-2} M

(c) 1.9×10^{-3} M　　(d) 7.6×10^{-7} M

3-16. (a) PbI_2 (1.2×10^{-3} M) > TlI (2.5×10^{-4} M) > BiI_3 (1.3×10^{-5} M) > AgI (9.1×10^{-9} M)

(b) PbI_2 (7.1×10^{-7} M) > TlI (6.5×10^{-7} M) > AgI (8.3×10^{-16} M) > BiI_3 (8.1×10^{-16} M)

(c) PbI_2 (4.2×10^{-4} M) > TlI (6.5×10^{-6} M) > BiI_3 (1.4×10^{-6} M) > AgI (8.3×10^{-15} M)

3-19.

	$[H_3O^+]$	$[OH^-]$
(a)	2.4×10^{-5} M	4.1×10^{-10} M
(c)	1.1×10^{-12} M	9.3×10^{-3} M
(e)	5.0×10^{-11} M	2.0×10^{-4} M
(g)	3.32×10^{-4} M	3.02×10^{-11} M

3-20. (a) 1.10×10^{-2} M　　(b) 1.17×10^{-8} M

(e) 1.46×10^{-4} M

Chapter 4

4-1. (a) *Accuracy* is the agreement between an experimentally measured value and the true value. *Precision* describes the agreement among measurements that have been performed in exactly the same way.

(c) The *mean* is the sum of the measurements in a set divided by the number of measurements. The *median* is the central value for a set of data; half of the remaining measurements are larger and half are smaller than the median.

(e) The *sample variance, s^2,* is given by the expression

$$s^2 = \frac{\sum_{i=1}^{N} (x_i - \bar{x})^2}{N - 1}$$

where $\bar{x}$ is the sample mean. The *sample standard deviation* is given by

$$s = \sqrt{\frac{\sum_{i=1}^{N} (x_i - \bar{x})^2}{N - 1}}$$

4-2. (a) The *range* is the difference between the largest value and the smallest value in a set of two or more replicate data.

(c) *Significant figures* include all numbers that are known with certainty plus the first uncertain number.

4-3. (a) Random temperature fluctuations causing random changes in the length of the metal rule.

(b) Uncertainties resulting from moving and positioning the rule twice.

(c) Personal judgment in reading the rule.

(d) Vibrations in the table and/or rule.

(e) Uncertainty in locating the rule perpendicular to the edge of the table.

4-4.

	Set A	Set C	Set E
(a)	$\bar{x} = 2.08$	0.0918	69.53
(b)	median = 2.1	0.0894	69.635
(c)	$w = 0.9$	0.0116	0.44
(d)	$s = 0.35$	0.0055	0.22
(e)	CV = 17%	6.0%	0.31%
(f)	$kw = 0.39$	0.0057	0.22

4-5.

	Set A	Set C	Set E
(a)	+0.08	−0.0012	+0.48
(b)	+40 ppt	−13 ppt	+7.0 ppt

4-6.

	s_y	CV	y
(a)	0.030	5.2%	0.57 (±0.03)
(b)	0.089	0.42%	21.3 (±0.1)
(c)	0.14×10^{-16}	2.0%	$6.9 (\pm 0.1) \times 10^{-16}$
(d)	1.4×10^3	0.77%	$1.84 (\pm 0.01) \times 10^5$
(e)	5.1×10^{-3}	8.5%	$6.0 (\pm 0.5) \times 10^{-2}$
(f)	1.1×10^{-4}	1.3%	$8.1 (\pm 0.1) \times 10^{-3}$

4-8. (a) 0.238   (c) 23.7796
(e) 3.4×10^{-4} (g) 9.8

4-9. (a)

Sample	s, % K
1	0.095
2	0.12
3	0.11
4	0.10
5	0.10

(b) $s_{pooled} = 0.11\%$ K
4-11. $s_{pooled} = 0.29\%$ heroin
4-13. (a) 0.06%   (c) 0.12%
4-14. (a) −1.0%   (c) −0.10%

Chapter 5

5-1. The three types of systematic error are *instrumental error, method error,* and *personal error.*
5-3. Constant errors.
5-4. (a) −0.04%   (c) −0.3%
5-5. (a) 13 g   (c) 3 g
5-6. Set A: CL = 2.1 ± 0.4, Set C: CL = 0.092 ± 0.009, Set E: CL = 69.5 ± 0.3. The 95% confidence limit is an interval around a sample mean within which the population mean is expected to be with a 95% probability.
5-7. Set A: CL = 2.1 ± 0.2, Set C: CL = 0.092 ± 0.007, Set E: CL = 69.5 ± 0.2
5-8. Set A: retain, Set C: reject, E: reject

5-9.

	80% CL	95% CL
(a)	18 ± 3	18 ± 5
(b)	18 ± 2	18 ± 3
(c)	18 ± 2	18 ± 2

5-11. For 95% confidence interval, $N = 10$
For 99% confidence interval, $N = 17$
5-13. (a) 3.22 ± 0.15 mmol/mL   (b) 3.22 ± 0.06 mmol/mL
5-15. (a) 12 measurements
5-16. (a) Retain   (b) reject
5-18. (a) $R = 0.232 c_x + 0.162$
(b) 15.1 mg SO_4^{2-}/L, $s_c = 1.4$ mg SO_4^{2-}/L, CV = 9.3%
(c) $s_c = 0.81$ mg SO_4^{2-}/L, CV = 5.4%
5-20. (a) $A_X = 5.57 c_{MVK} + 0.902$, where A_X is the peak area and c_{MVK} is mmol MVK/L.
(b) $c_{MVK} = 0.97$ mmol MVK/L
(c) For M = 1, $s_c = 0.09$ mmol MVK/L, $(s_c)_r$ = CV = 8.8%; For M = 4, $s_c = 0.06$ and CV = 6.0%
(d) $c_{MVK} = 4.78$ mmol MVK/L; For M = 1, $s_c = 0.08$ and CV = 1.8%; For M = 4, $s_c = 0.056$ and CV = 1.2%

Chapter 6

6-1. (a) The individual particles of a *colloid* are smaller than about 10^{-5} mm in diameter, while those of a *crystalline precipitate* are larger. As a consequence, crystalline precipitates settle out of solution relatively rapidly, whereas colloidal particles do not unless they can be caused to agglomerate.
(c) *Precipitation* is the process by which a solid phase forms and is carried out of solution when the solubility product of a species is exceeded. *Coprecipitation* is the process in which a normally soluble species is carried out of solution during the formation of a precipitate.
(e) *Occlusion* is a type of coprecipitation in which an impurity is entrapped in a pocket formed by a rapidly growing crystal. *Mixed-crystal formation* is a type of coprecipitation in which a foreign ion is incorporated into a growing crystal in a lattice position that is ordinarily occupied by one of the ions of the precipitate.
6-2. (a) *Digestion* is a process for improving the purity and filterability of a precipitate by heating the solid in contact with the solution from which it is formed (the *mother liquor*).
(c) In *reprecipitation*, a precipitate is filtered, washed, redissolved, and then reformed from the new solution. Because the concentration of contaminant is lower in this new solution than in the original, the second precipitate contains less coprecipitated impurity.
(e) The *electric double layer* consists of lattice ions adsorbed on the surface of a solid (the primary adsorption layer) and a volume of solution surrounding the particle (the counter-ion layer) in which there is an excess of ions of opposite charge.
6-3. A *chelating agent* is an organic compound that contains two or more electron-donor groups located in such a con-

figuration that five- or six-membered rings are formed when the donor groups form a complex with a cation.

6-5. (a) Positive charge   (b) adsorbed Ag^+ (c) NO_3^-

6-7. *Peptization* is the process in which a coagulated colloid returns to its original dispersed state as a consequence of a decrease in the electrolyte concentration of the solution in contact with the precipitate. Peptization of a coagulated colloid can be avoided by washing with an electrolyte solution rather than with pure water.

6-9. (a) Generate hydroxide ions from urea.
(c) Generate hydrogen sulfide from thioacetamide.

6-10. (a) $\dfrac{1 \text{ mol } CO_2}{\text{mol } BaCO_3}$ (c) $\dfrac{1 \text{ mol } K_2O}{2 \times \text{mol } (C_6H_5)_4BK}$

(e) $\dfrac{1 \text{ mol } H_2S}{\text{mol } CdSO_4}$ (g) $\dfrac{1 \text{ mol } C_8H_6O_3Cl_2}{2 \times \text{mol } AgCl}$

(i) $\dfrac{1 \text{ mol } CoSiF_6 \cdot 6H_2O}{6 \times \text{mol } H_2O}$

6-12. (a) 1.172 g   (b) 2.050 g
6-14. (a) 0.318 g   (c) 0.438 g
6-15. $Ce(IO_3)_3$ is eleven times more soluble than $AgIO_3$.
6-18. (a) 0.369 g $Ba(IO_3)_2$ (b) 0.0149 g $BaCl_2 \cdot 2H_2O$
6-20. 95.35%
6-22. 18.99%
6-25. 46.40%
6-26. 0.03219 g/tablet
6-28. 1.867%
6-30. 96.12%
6-32. (a) 0.239 g sample   (b) 0.494 g AgCl
(c) 0.4065 g sample
6-34. 4.72% Cl^- and 27.05% I^-
6-36. 1.80% KI

6-37.

	% Ag	% Cu
(a)	80.00	20.00
(c)	90.00	10.00
(e)	50.00	50.00

Chapter 7

7-1. (a) Although their molar solubilities are nearly identical, K_{sp} for $Cd(OH)_2$ will be numerically smaller than that for $BaSO_4$ because K_{sp} for the former has units of mol^3/L^3. For $BaSO_4$, K_{sp} has units of mol^2/L^2.
(b) Barium sulfate has the larger molar mass and will thus have the larger solubility in g/L.

7-3. The calculations differ because of the contribution of water to the molar hydroxide ion concentration. For $Mg(OH)_2$ this source can be neglected as vanishingly small with respect to $[OH^-]$ from the solute. The opposite is true with $Pt(OH)_2$ which is very slightly soluble; here, water is the principal supplier of OH^- to the system.

7-5. The simplifications in equilibrium calculations involve assuming that the concentration of one or more species is 0.00 M. When 0.00 is inserted into equilibrium-constant expressions, the constant becomes equal to zero or infinity. Thus, the expression is meaningless.

7-7. (a) $0.10 = [H_3PO_4] + [H_2PO_4^-] + [HPO_4^{2-}] + [PO_4^{3-}]$

(c) $0.100 + 0.0500 = [HNO_2] + [NO_s^-]$
$[Na^+] = c_{NaNO_2} = 0.0500$
(e) $0.100 = [Na^+] = [OH^-] - 2[Zn(OH)_4^{2-}]$
(g) $[Ca^{2+}] = \frac{1}{2}([F^-] + [HF])$ and $[HF] = [OH^-]$

7-9. (a) 1.47×10^{-4} M   (c) 7.42×10^{-5} M
7-11. (a) 2.5×10^{-9} M   (b) 2.5×10^{-12} M
7-13. (a) 5×10^{-2} M   (b) 2×10^{-4} M
7-14. (a) 2.9×10^{-3} M   (c) 6.3×10^{-7} M
7-15. (a) 6.0×10^{-2} M   (c) 4.0×10^{-6} M
7-16. (a) 5.2×10^{-3} M   (c) 3.6×10^{-4} M
7-17. (a) 2×10^{-3} M   (c) 5×10^{-6} M
7-19. 3.6×10^{-15} M
7-20. (a) 5.7×10^{-2} M   (c) 5.9×10^{-4} M
7-22. (a) Separation not feasible   (b) separation feasible.
7-24. (a) 8.3×10^{-11} M   (b) 1.6×10^{-11} M
(c) 1.3×10^4 (d) 1.3×10^4

Chapter 8

8-1. (a) *Activity*, a, is the effective concentration of a species A in solution. The *activity coefficient*, γ_A, is the factor needed to convert a molar concentration to activity: $a_A = \gamma_A[A]$.
(b) The *thermodynamic equilibrium constant* refers to an ideal system within which each species is unaffected by any others. A *concentration equilibrium constant* takes account of the influence exerted by solute species upon one another. A thermodynamic constant is based upon activities of reactants and products; a concentration constant is based upon molar concentrations.

8-3. (a) Reaction: $MgCl_2 + 2NaOH \rightarrow Mg(OH)_2(s) + 2NaCl$. The addition of NaOH has the effect of replacing a divalent ion (Mg^{2+}) with a chemically equivalent quantity of a univalent ion (Na^+); μ should decrease.
(b) Reaction: $HCl + NaOH \rightarrow NaCl + H_2O$. Addition of NaOH has the effect of converting the HCl to an equivalent amount of NaOH. Thus, μ should remain unchanged.
(c) Reaction: $NaOH + HOAc \rightarrow NaOAc + H_2O$. Addition of NaOH has the effect of replacing a slightly ionized species (HOAc) with a chemically equivalent quantity of water and Na^+; μ should increase.

8-5. For a given ionic strength, activity coefficients for ions with multiple charge show greater departures from ideality.

8-7. (a) 0.16   (c) 1.2
8-8. (a) 0.20   (c) 0.073
8-9. (a) 0.21   (c) 0.079
8-10. (a) 1.7×10^{-12} (c) 7.6×10^{-11}
8-11. (a) 5.2×10^{-6} M   (b) 6.2×10^{-6} M
(c) 9.5×10^{-12} M   (d) 1.5×10^{-7} M

8-12.

	(1), M	(2), M
(a)	1.4×10^{-6}	1.0×10^{-6}
(b)	2.1×10^{-3}	1.3×10^{-3}
(c)	2.8×10^{-5}	1.0×10^{-5}
(d)	1.4×10^{-5}	2.0×10^{-6}

Chapter 9

9-2. (a) The *millimole* is the amount of an elementary species, such as an atom, an ion, a molecule, or an electron. A millimole contains

$$6.02 \times 10^{23} \frac{particles}{mole} \times 10^{-3} \frac{mole}{millimole} = 6.02 \times 10^{20} \text{ particles}$$

(c) The *stoichiometric factor* is the molar ratio of two species that appear in a balanced chemical equation.

9-3. (a) The *equivalence point* in a titration is the point at which sufficient titrant has been added so that stoichiometrically equivalent amounts of analyte and titrant are present. The *end point* in a titration is the point at which an observable physical change signals the equivalence point.

(c) A *primary standard* is a highly purified substance that serves as the basis for a titrimetric method. It is used either (1) to prepare a standard solution directly by mass or (2) to standardize a solution to be used in a titration.

A *secondary standard* is a material or solution whose concentration is determined from the stoichiometry of its reaction with a primary standard material. Secondary standards are employed when a reagent is not available in primary standard quality. For example, solid sodium hydroxide is hygroscopic and cannot be used to prepare a standard solution directly. A secondary standard solution of the reagent is easily prepared, however, by standardizing a solution of sodium hydroxide against a primary standard reagent such as potassium acid phthalate.

9-4. For a dilute aqueous solution, 1 L = 1000 mL = 1000 g, so

$$\frac{mg}{L} = \frac{10^{-3} \, g}{1000 \, g} = 10^{-6} = 1 \text{ ppm}$$

9-5. (a) $\dfrac{1 \text{ mol } H_2NNH_2}{2 \text{ mol } I_2}$ $\dfrac{1 \text{ mol } Na_2B_4O_7 \cdot 10H_2O}{2 \text{ mol } H^+}$

9-6. 19.0 M

9-8. (a) 6.161 M   (c) 4.669 M

9-9. (a) Dissolve 80 g C_2H_5OH and dilute to 500 mL with H_2O.
(b) Dilute 80.0 mL C_2H_5OH to 500 mL with H_2O.
(c) Dilute 80.0 g C_2H_5OH with 420.0 mL H_2O.

9-11. 9.151×10^{-2} M

9-13. 0.06114 M

9-14. 0.2970 M $HClO_4$ and 0.3259 M NaOH

9-16. 0.08411 M

9-17. 345.8 ppm

9-19. 5.471%

9-20. 7.317%

9-22. (a) 9.36×10^{-3} M   (b) 1.9×10^{-5} M   (c) -3 ppt

Chapter 10

10-1. (a) The initial pH of the NH_3 solution will be less than that for the solution containing NaOH. With the first addition of titrant, the pH of the NH_3 solution will decrease rapidly and then level off and become nearly constant throughout the middle part of the titration. In contrast, additions of standard acid to the NaOH solution will cause the pH of the NaOH solution to decrease gradually and nearly linearly until the equivalent point is approached. The equivalence point pH for the NH_3 solution will be well below 7, whereas for the NaOH solution it will be exactly 7.

(b) Beyond the equivalence point, the pH is determined by the excess titrant. Thus the curves become identical in this region.

10-3. The limited sensitivity of the eye to small color differences requires that there be a roughly tenfold excess of one or the other form of the indicator in order for the color change to be seen. This change corresponds to a pH range of ± 1 pH unit about the pK of the indicator.

10-5. The standard reagents in neutralization titrations are always strong acids or strong bases because the reactions with this type of reagent are more complete than with those of their weaker counterparts. Sharper end points are the consequence of this difference.

10-7. The *buffer capacity* of a solution is the number of moles of hydronium ion or hydroxide ion needed to cause 1.00 L of the buffer to undergo a unit change in pH.

10-9. The three solutions will have the same pH, but solution (a) will have the greatest buffer capacity and (c) the least.

10-10. (a) malic acid/sodium malate   (c) NH_4Cl/NH_3

10-11. (a) NaOCl   (c) CH_3NH_2

10-12. (a) HIO_3 (c) pyruvic acid

10-14. 3.25

10-16. (a) 14.94

10-17. (a) 12.94

10-18. -0.607

10-20. 7.04

10-22. (a) 1.05   (b) 1.05   (c) 1.81
(d) 1.81   (e) 12.60

10-24. (a) 1.30   (b) 1.37

10-26. (a) 4.26   (b) 4.76   (c) 5.76

10-28. (a) 11.12   (b) 10.62   (c) 9.62 (9.53 by quadratic)

10-30.

	Quadratic	Approximate
(a)	12.04	12.06
(b)	11.48	11.56
(c)	9.97	10.56

10-32. (a) 1.94   (b) 2.44   (c) 3.52

10-34. (a) 2.41   (b) 8.35   (c) 12.35   (d) 3.84

10-37. (a) 3.85   (b) 4.06   (c) 2.63   (d) 2.10

10-39. (a) $\Delta pH = 0.00$ (c) $\Delta pH = -1.000$
(e) $\Delta pH = -0.500$ (g) $\Delta pH = 0.000$

10-40. (a) $\Delta pH = -5.00$ (c) $\Delta pH = -0.097$
(e) $\Delta pH = -3.369$ (g) $\Delta pH = -0.017$

10-41. (a) $\Delta pH = 5.00$ (c) $\Delta pH = 0.079$
(e) $\Delta pH = 3.272$ (g) $\Delta pH = 0.017$

10-42. (b) $\Delta pH = -0.141$

10-43. 15.5 g sodium formate

10-45. 194 mL HCl

10-47.

V_{HCl} (mL)	pH	V_{HCl} (mL)	pH
0.00	13.00	49.00	11.00
10.00	12.82	50.00	7.00
25.00	12.52	51.00	3.00
40.00	12.05	55.00	2.32
45.00	11.72	60.00	2.04

10-48. 24.95 mL reagent, pH = 6.44
25.05 mL reagent, pH = 9.82
Use cresol purple indicator.

10-50.

	(a)	(c)
Vol, mL	pH	pH
0.00	2.09	3.12
5.00	2.38	4.28
15.00	2.82	4.86
25.00	3.17	5.23
40.00	3.76	5.83
45.00	4.11	6.18
49.00	4.85	6.92
50.00	7.92	8.96
51.00	11.00	11.00
55.00	11.68	11.68
60.00	11.96	11.96

10-51. (a)

Vol HCl, mL	pH	Vol HCl, mL	pH
0.00	11.12	49.00	7.55
5.00	10.20	50.00	5.27
15.00	9.61	51.00	3.00
25.00	9.24	55.00	2.32
40.00	8.64	60.00	2.04
45.00	8.29		

10-52.

	(a)	(c)
Vol, mL	pH	pH
0.00	2.80	4.26
5.00	3.65	6.57
15.00	4.23	7.15
25.00	4.60	7.52
40.00	5.20	8.12
49.00	6.29	9.21
50.00	8.65	10.11
51.00	11.00	11.00
55.00	11.68	11.68
60.00	11.96	11.96

10-53. (a) $\alpha_0 = 0.215$; $\alpha_1 = 0.785$
(c) $\alpha_0 = 0.769$; $\alpha_1 = 0.231$
(e) $\alpha_0 = 0.917$; $\alpha_1 = 0.083$
10-54. $[HCOOH] = 6.61 \times 10^{-2}$ M

Chapter 11

11-1. Not only is NaHA a proton donor, it is also the conjugate base of the parent acid H_2A

$$HA^- + H_2O \underset{\displaystyle H_2A + OH^-}{\overset{\displaystyle H_3O^+ + A^-}{\rightleftharpoons}}$$

Solutions of acid salts are acidic or alkaline, depending upon which of these equilibria predominate. In order to compute the pH of solutions of this type it is necessary to take both equilibria into account.

11-3. The HPO_4^{2-} ion is such a weak acid ($K_a = 4.5 \times 10^{-13}$) that the change in pH in the vicinity of the third end point is too small to be observable.

11-4. (a) ~neutral   (c) neutral   (e) basic   (g) acidic

11-6. Phenolphthalein

11-8. (a) Cresol purple   (c) cresol purple
(e) Bromocresol green   (g) phenolphthalein

11-9. (a) 1.86   (c) 1.64   (e) 4.21

11-10. (a) 4.71   (c) 4.28   (e) 9.80

11-11. (a) 12.32   (c) 9.70   (e) 12.58

11-12. (a) 2.07   (b) 7.18   (c) 10.63
(d) 2.55   (e) 2.06

11-14. (a) 1.54   (b) 1.99   (c) 12.07   (d) 12.01

11-16. (a) $[HSO_3^-]/[SO_3^{2-}] = 15.2$
(b) $[HCit^{2-}]/[Cit^{3-}] = 2.5$
(c) $[HM^-]/[M^{2-}] = 0.498$
(d) $[HT^-]/[T^{2-}] = 0.0232$

11-18. 50.2 g

11-20. (a) 2.11   (b) 7.38

11-22. Mix 442 mL of 0.300 M Na_2CO_3 with 558 mL of 0.200 M HCl.

11-24. Mix 704 mL of 0.400 M HCl with 296 mL of 0.500 M Na_3AsO_4.

11-26. (a) Titration with NaOH of a solution containing a mixture of two weak acids HA_1 and HA_2. HA_1 is present in a greater concentration and has a dissociation constant that is larger by a factor of about 10^4.
(b) Titration of a typical monoprotic weak acid.
(c) Titration of a mixture of a weak base, such as Na_2CO_3 and an acid salt, such as $NaHCO_3$.

11-28.

	(a)	(c)
Vol Reagent, mL	pH	pH
0.00	11.66	0.96
12.50	10.33	1.28
20.00	9.73	1.50
24.00	8.95	1.63
25.00	8.34	1.67
26.00	7.73	1.70
37.50	6.35	2.19
45.00	5.75	2.70
49.00	4.97	3.46
50.00	3.83	7.35
51.00	2.70	11.30
60.00	1.74	12.26

11-29.

mL HClO₄	pH	mL HClO₄	pH
0.00	13.00	35.00	7.98
10.00	12.70	44.00	6.70
20.00	12.15	45.00	4.68
24.00	11.43	46.00	2.68
25.00	10.35	50.00	2.00
26.00	9.26		

11-31. (a) $\dfrac{K_2}{K_1} = \dfrac{[H_3AsO_4][HAsO_4^{2-}]}{[H_2AsO_4^-]^2} = 1.9 \times 10^{-5}$

11-32. 3.26×10^{-5}

11-33.

	pH	D	α_0	α_1	α_2	α_3
(a)	2.00	1.112×10^{-4}	0.899	0.101	3.94×10^{-5}	
	6.00	5.500×10^{-9}	1.82×10^{-4}	0.204	0.796	
	10.00	4.379×10^{-9}	2.28×10^{-12}	2.56×10^{-5}	1.000	
(c)	2.00	1.075×10^{-6}	0.931	6.93×10^{-2}	1.20×10^{-4}	4.82×10^{-9}
	6.00	1.882×10^{-14}	5.31×10^{-5}	3.96×10^{-2}	0.685	0.275
	10.00	5.182×10^{-15}	1.93×10^{-16}	1.44×10^{-9}	2.49×10^{-4}	1.000
(e)	2.00	4.000×10^{-4}	0.250	0.750	1.22×10^{-5}	
	6.00	3.486×10^{-8}	2.87×10^{-5}	0.861	0.139	
	10.00	4.863×10^{-9}	2.06×10^{-12}	6.17×10^{-4}	0.999	

Chapter 12

12-1. Carbon dioxide is not strongly bonded by water molecules, and thus is readily volatilized from aqueous media. Gaseous HCl molecules, on the other hand, are fully dissociated into H_3O^+ and Cl^- when dissolved in water; neither of these species is volatile.

12-3. Primary standard Na_2CO_3 can be obtained by heating primary standard grade $NaHCO_3$ for about an hour at 270 to 300°C. The reaction is

$$2NaHCO_3(s) \rightarrow Na_2CO_3(s) + H_2O(g) + CO_2(g)$$

12-5. For, let us say, a 40-mL titration

mass $KH(IO_3)_2$ required

$$= 40 \text{ mL} \times 0.010 \frac{\text{mmol}}{\text{mL}} \times 0.390 \frac{\text{g}}{\text{mmol}} = 0.16 \text{ g}$$

mass HBz required

$$= 40 \text{ mL} \times 0.010 \frac{\text{mmol}}{\text{mL}} \times 0.122 \frac{\text{g}}{\text{mmol}} = 0.049 \text{ g}$$

The $KH(IO_3)_2$ is preferable because the relative weighing error would be less with a 0.16-g sample than with a 0.049-g sample.

12-8. (a) Dissolve 17 g KOH and dilute to 2.0 L.
(b) Dissolve 9.4 g of the solid and dilute to 2.0 L.
(c) Dilute about 120 mL of the reagent to 2.0 L.

12-10. (a) 0.1026 M   (b) $s = 0.00039$ and CV = 0.38%

12-12. (a) 0.1388 M   (b) 0.1500 M

12-14. (a) 0.08387 M   (b) 0.1007 M   (c) 0.1311 M

12-16. (a) 0.28 to 0.36 g Na_2CO_3 (c) 0.85 to 1.1 g HBz
(e) 0.17 to 0.22 g THAM

12-17. (a) 0.067%   (b) 0.16%   (c) 0.043%

12-19. 0.1217 g H_2T/100 mL

12-21. (a) 46.25% $Na_2B_4O_7$ (b) 87.67% $Na_2B_4O_7 \cdot 10H_2O$
(c) 32.01% B_2O_3 (d) 9.94% B

12-23. 24.4% HCHO

12-25. 7.079%

12-27. $MgCO_3$

12-29. 3.35×10^3 ppm CO_2

12-31. 6.333% P

12-32. 13.86% analyte

12-33. 22.08% analyte

12-35. 3.885% N

12-37. (a) 10.09% N   (c) 47.61% $(NH_4)_2SO_4$

12-39. 24.39% NH_4NO_3 and 15.23% $(NH_4)_2SO_4$

12-40. 69.84% KOH, 21.04% K_2CO_3, and 9.12% H_2O

12-42. (a) 18.15 mL   (b) 45.37 mL
(c) 38.28 mL   (d) 12.27 mL

12-44. (a) 4.314 mg NaOH/mL
(b) 7.985 mg Na_2CO_3/mL and 4.358 mg $NaHCO_3$/mL
(c) 3.455 mg Na_2CO_3/mL and 4.396 mg NaOH/mL
(d) 8.215 mg Na_2CO_3/mL
(e) 13.46 mg $NaHCO_3$/mL

Chapter 13

13-1. The Fajans determination of chloride involves a direct titration, while a Volhard approach requires two standard solutions and a filtration step to eliminate AgCl.

13-3. In contrast to Ag_2CO_3 and AgCN, the solubility of AgI is unaffected by the acidity and additionally is less soluble than AgSCN. The filtration step is thus unnecessary, whereas it is necessary with the other two compounds.

13-5. Potassium is determined by precipitation with an excess of a standard solution of sodium tetraphenylboron. An excess of standard $AgNO_3$ is then added, which precipitates the excess tetraphenylboron ion. Next, the excess $AgNO_3$ is titrated with a standard solution of SCN^-. The reactions are

$$K^+ + B(C_6H_5)_4^- \rightleftharpoons KB(C_6H_5)_4(s)$$
$$\text{[measured excess } B(C_6H_5)_4^-]$$
$$Ag^+ + B(C_6H_5)_4^- \rightleftharpoons AgB(C_6H_5)_4(s)$$
$$\text{[measured excess } AgNO_3]$$

The excess $AgNO_3$ is then determined by a Volhard titration with KSCN.

13-7. (a) 51.78 mL   (c) 10.64 mL   (e) 46.24 mL

13-9. (a) 44.70 mL   (c) 14.87 mL

13-11. 28.5%

13-13. 18.9 ppm H_2S

13-15. 1.472% P_2O_5

13-16. 116.7 mg analyte

13-19. 0.07052 M

13-20. Only one of the chlorine atoms in the heptachlor reacts with $AgNO_3$.

13-23. 15.60 mg/tablet

13-24. 21.5% CH_2O

13-25. 0.4348% $C_{19}H_{16}O_4$

13-28. 10.60% Cl^- and 55.65% ClO_4^-

13-29. (a)

V_{NH_4SCN} (mL)	$[Ag^+]$	$[SCN^-]$	pAg
30.00	9.09×10^{-3}	1.2×10^{-10}	2.04
40.00	3.85×10^{-3}	2.9×10^{-10}	2.42
49.00	3.38×10^{-4}	3.3×10^{-9}	3.47
50.00	1.05×10^{-6}	1.05×10^{-6}	5.98
51.00	3.3×10^{-9}	3.3×10^{-4}	8.48
60.00	3.7×10^{-10}	2.94×10^{-3}	9.43
70.00	2.1×10^{-10}	5.26×10^{-3}	9.68

(c)

V_{NaCl}	$[Ag^+]$	$[Cl^-]$	pAg
10.00	3.75×10^{-2}	4.85×10^{-8}	1.43
20.00	1.50×10^{-2}	1.21×10^{-8}	1.82
29.00	1.27×10^{-3}	1.43×10^{-7}	2.90
30.00	1.35×10^{-5}	1.35×10^{-5}	4.87
31.00	1.48×10^{-7}	1.23×10^{-3}	6.83
40.00	1.70×10^{-8}	1.07×10^{-2}	7.77
50.00	9.71×10^{-9}	1.88×10^{-2}	8.01

(e)

$V_{Na_2SO_4}$ (mL)	$[Ba^{2+}]$	$[SO_4^{2-}]$	pBa
0.00	2.50×10^{-2}	0.0	1.60
10.00	1.00×10^{-2}	1.1×10^{-8}	2.00
19.00	8.48×10^{-4}	1.3×10^{-7}	3.07
20.00	1.05×10^{-5}	1.05×10^{-5}	4.98
21.00	1.3×10^{-7}	8.20×10^{-4}	6.87
30.00	1.5×10^{-8}	7.14×10^{-3}	7.81
40.00	8.8×10^{-9}	1.25×10^{-2}	8.06

Chapter 14

14-1. (a) A *chelate* is a cyclic complex consisting of metal ion and a reagent that contains two or more electron donor groups located in such a position that they can bond with the metal ion to form a heterocyclic ring structure.

   (c) A *ligand* is a species that contains one or more electron pair donor groups that tend to form bonds with metal ions.

   (e) A *conditional formation constant* is an equilibrium constant for the reaction between a metal ion and a complexing agent that applies only when the pH and/

or the concentration of other complexing ions are carefully specified.

   (g) *Water hardness* is the concentration of calcium carbonate that is equivalent to the total concentration of all of the multivalent metal carbonates in the water.

14-2. Three general methods for performing EDTA titrations include (1) direct titration, (2) back titration, and (3) displacement titration. Method (1) is simple, rapid, and requires but one standard reagent. Method (2) is advantageous for those metals that react so slowly with EDTA as to make direct titration inconvenient. In addition, this procedure is useful for cations for which satisfactory indicators are not available. Finally, it is useful for analyzing samples that contain anions that form sparingly soluble precipitates with the analyte under the analytical conditions. Method (3) is particularly useful in situations where no satisfactory indicators are available for direct titration.

14-4. (a) $Ag^+ + S_2O_3^{2-} \rightleftharpoons Ag(S_2O_3)^-$

$$K_1 = \frac{[Ag(S_2O_3)^-]}{[Ag^+][S_2O_3^{2-}]}$$

$Ag(S_2O_3)^- + S_2O_3^{2-} \rightleftharpoons Ag(S_2O_3)_2^{3-}$

$$K_2 = \frac{[Ag(S_2O_3)_2^{3-}]}{[AgS_2O_3^-][S_2O_3^{2-}]}$$

   (c) $Cd^{2+} + NH_3 \rightleftharpoons Cd(NH_3)^{2+}$

$$K_1 = \frac{[Cd(NH_3)^{2+}]}{[Cd^{2+}][NH_3]}$$

$Cd(NH_3)^+ + NH_3 \rightleftharpoons Cd(NH_3)_2^{2+}$

$$K_2 = \frac{[Cd(NH_3)_2^{2+}]}{[Cd(NH_3)^{2+}][NH_3]}$$

$Cd(NH_3)_2^{2+} + NH_3 \rightleftharpoons Cd(NH_3)_3^{2+}$

$$K_3 = \frac{[Cd(NH_3)_3^{2+}]}{[Cd(NH_3)_2^{2+}][NH_3]}$$

$Cd(NH_3)_3^{2+} + NH_3 \rightleftharpoons Cd(NH_3)_4^{2+}$

$$K_4 = \frac{[Cd(NH_3)_4^{2+}]}{[Cd(NH_3)_3^{2+}][NH_3]}$$

14-5. The overall formation constant is equal to the product of the individual stepwise constants. Thus the overall constant for formation of $Cd(NH_3)_4^{2+}$ in Example 14-4(c) is

$$\beta_4 = K_1K_2K_3K_4 = \frac{[Cd(NH_3)_4^{2+}]}{[Cd^{2+}][NH_3]^4}$$

which is the equilibrium constant for the reaction $Cd^{2+} + 4NH_3 \rightleftharpoons Cd(NH_3)_4^{2+}$.

14-8. 0.01032 M EDTA

14-10. (a) 39.1 mL   (c) 41.6 mL   (e) 31.2 mL

14-12. 3.028% Zn

14-13. 0.998 mg Cr/cm^2

14-14. 1.228% Tl_2SO_4

14-16. 213.0 ppm Fe^{2+} and 184.0 ppm Fe^{3+}

14-18. 55.16% Pb and 44.86% Cd

14-20. 99.7% ZnO and 0.256% Fe_2O_3

14-22. 64.68 ppm K^+

14-24. 8.518% Pb, 24.86% Zn, 64.08% Cu, 2.54% Sn

14-25. (a) 4.6×10^9 (b) 1.1×10^{12} (c) 7.4×10^{13}

14-27.

mL EDTA	pSr	mL EDTA	pSr
0.00	2.00	25.00	5.37
10.00	2.30	25.10	6.16
24.00	3.57	26.00	7.16
24.90	4.57	30.00	7.86

Chapter 15

15-1. (a) *Oxidation* is a process in which a species loses electrons.

   (c) A *salt bridge* consists of a U-shaped tube that is filled with a concentrated solution of an electrolyte (most commonly a saturated solution of potassium chloride). It is used to provide electrical contact between two half-cells making up an electrochemical cell. The bridge prevents the contents of the two half-cells from coming into direct contact with one another.

   (e) The *Nernst equation* relates the electrical potential for a half-reaction to the concentrations (strictly, activities) of the participants in the reaction.

15-2. (a) The *electrode potential* is the potential of an electrochemical cell in which a standard hydrogen electrode acts as anode and the half-cell of interest is the cathode.

   (c) The *standard electrode potential* for a half-reaction is the potential of a *cell* consisting of a cathode at which that half-reaction is occurring and a standard hydrogen electrode behaving as the anode. The activities of all of the participants in the half-reaction are specified as having a value of unity. The additional specification that the standard hydrogen electrode is the anode implies that the standard potential for a half-reaction is always a *reduction potential*.

   (e) An *oxidation potential* is the potential of an electrochemical cell in which the cathode is a standard hydrogen electrode and the half-cell of interest acts as anode.

15-3. (a) *Reduction* is the process whereby a substance acquires electrons; a *reducing agent* is a supplier of electrons.

   (c) The *anode* of an electrochemical cell is the electrode at which oxidation occurs. The *cathode* is the electrode at which reduction occurs.

   (e) The *standard electrode potential* is the potential of an electrochemical cell in which the standard hydrogen electrode acts as an anode and all participants in the cathode process have unit activity. The formal potential differs in that the molar concentrations of the reactants and products are unity and the *concentration* of other species in the solution are carefully specified.

15-4. The first standard potential is for a solution that is satu-

rated with I_2, which has an $I_2(aq)$ activity significantly less than one. The second potential is for a *hypothetical* half-cell in which the $I_2(aq)$ activity is unity. Such a half-cell, if it existed, would have a greater potential since the driving force for the reduction would be greater at the higher I_2 concentration. The second half-cell potential, although hypothetical, is nevertheless useful for calculating electrode potentials for solutions that are undersaturated in I_2.

15-5. It is necessary to bubble hydrogen through the electrolyte in a hydrogen electrode in order to keep the solution saturated with the gas. Only under these circumstances is the hydrogen activity constant so that the electrode potential is constant and reproducible.

15-7. (a) $2Fe^{3+} + Sn^{2+} \rightarrow 2Fe^{2+} + Sn^{4+}$
   (c) $2NO_3^- + Cu(s) + 4H^+ \rightarrow 2NO_2(g) + 2H_2O + Cu^{2+}$
   (e) $Ti^{3+} + Fe(CN)_6^{3-} + H_2O \rightarrow$
       $TiO^{2+} + Fe(CN)_6^{4-} + 2H^+$
   (g) $2Ag(s) + 2I^- + Sn^{4+} \rightarrow 2AgI(s) + Sn^{2+}$
   (i) $5HNO_2 + 2MnO_4^- + H^+ \rightarrow 5NO_3^- + 2Mn^{2+} + 3H_2O$

15-8. (a) Oxidizing Agent Fe^{3+}; $Fe^{3+} + e^- \rightleftharpoons Fe^{2+}$
     Reducing Agent Sn^{2+}; $Sn^{2+} \rightleftharpoons Sn^{4+} + 2e^-$
   (c) Oxidizing Agent NO_3^-;
     $NO_3^- + 2H^+ + e^- \rightleftharpoons NO_2(g) + H_2O$
     Reducing Agent Cu; $Cu(s) \rightleftharpoons Cu^{2+} + 2e^-$
   (e) Oxidizing Agent $Fe(CN)_6^{3-}$;
     $Fe(CN)_6^{3-} + e^- \rightleftharpoons Fe(CN)_6^{4-}$
     Reducing Agent Ti^{3+};
     $Ti^{3+} + H_2O \rightleftharpoons TiO^{2+} + 2H^+ + e^-$
   (g) Oxidizing Agent Sn^{4+}; $Sn^{4+} + 2e^- \rightleftharpoons Sn^{2+}$
     Reducing Agent Ag; $Ag(s) + I^- \rightleftharpoons AgI(s) + e^-$
   (i) Oxidizing Agent MnO_4^-;
     $MnO_4^- + 8H^+ + 5e^- \rightleftharpoons Mn^{2+} + 4H_2O$
     Reducing Agent HNO_2;
     $HNO_2 + H_2O \rightleftharpoons NO_3^- + 3H^+ + 2e^-$

15-9. (a) $MnO_4^- + 5VO^{2+} + 11H_2O \rightarrow$
     $Mn^{2+} + 5V(OH)_4^+ + 2H^+$
   (c) $Cr_2O_7^{2-} + 3U^{4+} + 2H^+ \rightarrow 2Cr^{3+} + 3UO_2^{2+} + H_2O$
   (e) $IO_3^- + 5I^- + 6H^+ \rightarrow 3I_2 + H_2O$
   (g) $HPO_3^{2-} + 2MnO_4^- + 3OH^- \rightarrow$
     $PO_4^{3-} + 2MnO_4^{2-} + 2H_2O$
   (i) $V^{2+} + 2V(OH)_4^+ + 2H^+ \rightarrow 3VO^{2+} + 5H_2O$

15-10. (a) Oxidizing Agent MnO_4^-;
     $MnO_4^- + 8H^+ + 5e^- \rightleftharpoons Mn^{2+} + 4H_2O$
     Reducing Agent VO^{2+};
     $VO^{2+} + 3H_2O \rightleftharpoons V(OH)_4^+ + 2H^+ + e^-$
   (c) Oxidizing Agent $Cr_2O_7^{2-}$;
     $Cr_2O_7^{2-} + 14H^+ + 6e^- \rightleftharpoons 2Cr^{3+} + 7H_2O$
     Reducing Agent U^{4+};
     $U^{4+} + 2H_2O \rightleftharpoons UO_2^{2+} + 4H^+ + 2e^-$
   (e) Oxidizing Agent IO_3^-;
     $IO_3^- + 6H^+ + 5e^- \rightleftharpoons \frac{1}{2}I_2 + 3H_2O$
     Reducing Agent I^-; $I^- \rightleftharpoons \frac{1}{2}I_2 + e^-$
   (g) Oxidizing Agent MnO_4^-; $MnO_4^- + e^- \rightleftharpoons MnO_4^{2-}$
     Reducing Agent HPO_3^{2-};
     $HPO_3^{2-} + 3OH^- \rightleftharpoons PO_4^{3-} + 2H_2O + 2e^-$
   (i) Oxidizing Agent $V(OH)_4^+$;
     $V(OH)_4^+ + 4H^+ + 2e^- \rightleftharpoons V^{3+} + H_2O$
     Reducing Agent V^{2+}; $V^{2+} \rightleftharpoons V^{3+} + e^-$

15-11.

(a)
$AgBr(s) + e^- \rightleftharpoons Ag(s) + Br^-$
$V^{2+} \rightleftharpoons V^{3+} + e^-$
$Tl^{3+} + 2e^- \rightleftharpoons Tl^+$
$Fe(CN)_6^{4-} \rightleftharpoons Fe(CN)_6^{3-} + e^-$
$V^{3+} + e^- \rightleftharpoons V^{2+}$
$Zn \rightleftharpoons Zn^{2+} + 2e^-$
$Fe(CN)_6^{3-} + e^- \rightleftharpoons Fe(CN)_6^{4-}$
$Ag(s) + Br^- \rightleftharpoons AgBr(s) + e^-$
$S_2O_8^{2-} + 2e^- \rightleftharpoons 2SO_4^{2-}$
$Tl^+ \rightleftharpoons Tl^{3+} + 2e^-$

(b), (c)	E^0
$S_2O_8^{2-} + 2e^- \rightleftharpoons 2SO_4^{2-}$	2.01
$Tl^{3+} + 2e^- \rightleftharpoons Tl^+$	1.25
$Fe(CN)_6^{3-} + e^- \rightleftharpoons Fe(CN)_6^{4-}$	0.36
$AgBr(s) + e^- \rightleftharpoons Ag(s) + Br^-$	0.073
$V^{3+} + e^- \rightleftharpoons V^{2+}$	−0.256
$Zn^{2+} + 2e^- \rightleftharpoons Zn(s)$	−0.763

15-13. (a) 0.297 V (b) 0.190 V
(c) −0.152 V (d) 0.048 V
(e) 0.007 V

15-16. (a) 0.78 V (b) 0.198 V
(c) −0.355 V (d) 0.210 V
(e) 0.177 V (f) 0.86 V

15-18. (a) −0.280 V, anode (b) −0.090 V, anode
(c) 1.002 V, cathode (d) 0.171 V, cathode
(e) −0.009 V, anode

15-20. 0.390 V
15-22. −0.964 V
15-24. −1.25 V
15-25. 0.13 V

Chapter 16

16-1. The electrode potential of a system is the electrode potential of all half-cell processes at equilibrium in the system.

16-4. With the notable exception of the equivalence point, concentration data needed for the Nernst equation of a participant in an oxidation/reduction titration can be evaluated directly from the stoichiometry of the reactions. Short of the equivalence point, we have sufficient information for the analyte; beyond the equivalence point, we have sufficient information for the titrant.

16-6. An asymmetric titration curve will be encountered whenever the titrant and the analyte react in a ratio that is not 1 : 1.

16-7. (a) −0.290 V, electrolytic (b) 2.03 V, galvanic
(c) −0.062 V, electrolytic (d) 1.037 V, galvanic
(e) 0.551 V, galvanic (f) 0.286 V, galvanic

16-9. (a) 0.631 V (c) 0.331 V (e) −0.620 V

16-10. (a) $Zn|Zn^{2+}(0.1364\ M)||Pb^{2+}(0.0848\ M)|Pb$
(c) $Pt|TiO^{2+}(1.46 \times 10^{-3}\ M),Ti^{3+}(0.02723\ M),H^+$
$(1.00 \times 10^{-3}\ M)||SHE$
(e) $Ag|Ag^+(0.2058\ M)||KBr(0.0791\ M),AgBr(sat'd)|Ag$

16-11. (a) $2.2 \times 10^{17} = \dfrac{[Fe^{2+}][V^{3+}]}{[Fe^{3+}][V^{2+}]}$

(c) $3 \times 10^{22} = \dfrac{[VO^{2+}]^2[UO_2^{2+}]}{[V(OH)_4^+]^2[U^{4+}]}$

(e) $9 \times 10^{37} = \dfrac{[Ce^{3+}]^2[H_3AsO_4]}{[Ce^{4+}]^2[H_3AsO_3]}$

(g) $2.4 \times 10^{10} = \dfrac{[V^{3+}]^2}{[VO^{2+}][V^{2+}][H^+]^2}$

16-12. (a) 0.258 V (c) 0.444 V (e) 0.951 V
(g) −0.008 V

16-13. (a) Phenosafranine
(c) Indigo tetrasulfonate or Methylene blue
(e) Erioglaucin A
(g) None

16-14.

Vol, mL	E, V		
	(a)	(c)	(e)
10.00	−0.292	0.32	0.316
25.00	−0.256	0.36	0.334
49.00	−0.156	0.46	0.384
49.90	−0.096	0.52	0.414
50.00	−0.017	0.95	1.17
50.10	−0.074	1.17	1.48
51.00	−0.104	1.20	1.49
60.00	−0.133	1.23	1.50

Chapter 17

17-1. (a) $2Mn^{2+} + 5S_2O_8^{2-} + 8H_2O \rightarrow$
$10SO_4^{2-} + 2MnO_4^- + 16H^+$
(b) $NaBiO_3(s) + 2Ce^{3+} + 4H^+ \rightarrow$
$BiO^+ + 2Ce^{4+} + 2H_2O + Na^+$
(c) $H_2O_2 + U^{4+} \rightarrow UO_2^{2+} + 2H^+$
(d) $V(OH)_4^+ + Ag(s) + Cl^- + 2H^+ \rightarrow$
$VO^{2+} + AgCl(s) + 3H_2O$
(e) $2MnO_4^- + 5H_2O_2 + 6H^+ \rightarrow 5O_2 + 2Mn^{2+} + 8H_2O$
(f) $ClO_3^- + 6I^- + 6H^+ \rightarrow 3I_2 + Cl^- + 3H_2O$

17-3. Only in the presence of Cl^- ion is Ag a sufficiently good reducing agent to be very useful for prereductions. In the presence of Cl^- ion, the half-reaction occurring in the Walden reductor is $Ag(s) + Cl^- \rightarrow AgCl(s) + e^-$. The excess HCl increases the tendency of this reaction to occur by the common ion effect.

17-5. $UO_2^{2+} + 2Ag(s) + 4H^+ + 2Cl^- \rightarrow$
$U^{4+} + 2AgCl(s) + H_2O$

17-7. Standard solutions of reductants find somewhat limited use because of their susceptibility to air oxidation.

17-8. Standard $KMnO_4$ solutions are seldom used to titrate solutions containing HCl because of the tendency of MnO_4^- to oxidize Cl^- to Cl_2, thus causing an over-consumption of MnO_4^-.

17-10. $2MnO_4^- + 3Mn^{2+} + 2H_2O \rightarrow 5MnO_2(s) + 4H^+$

17-12. Standard permanganate and thiosulfate solutions are generally stored in the dark because their decomposition reactions are catalyzed by light.

17-13. $4MnO_4^- + 2H_2O \rightarrow 4MnO_2(s) + 3O_2 + 4OH^-$

          brown

17-15. Iodine is not sufficiently soluble in water to produce a useful standard reagent. It is quite soluble in solutions that contain an excess of iodide, however, as a consequence of the formation of the triiodide complex. The rate at which iodine dissolves in iodide solutions increases as the concentration of iodide ion becomes greater. For this reason, iodine is always dissolved in a very concentrated solution of potassium iodide and diluted only after solution is complete.

17-17. $S_2O_3^{2-} + H^+ \rightarrow HSO_3^- + S(s)$

17-19. $BrO_3^- + 6I^- + 6H^+ \rightarrow Br^- + 3I_2 + 3H_2O$

     excess

   $I_2 + 2S_2O_3^{2-} \rightarrow 2I^- + S_4O_6^{2-}$

17-21. $2I_2 + N_2H_4 \rightarrow N_2 + 4H^+ + 4I^-$

17-23. (a) 0.1122 M Ce^{4+} (c) 0.02245 M MnO_4^-

   (e) 0.02806 M IO_3^-

17-24. Dissolve 2.574 g $K_2Cr_2O_7$ and dilute to 250.0 mL.

17-26. Dissolve about 24 g $KMnO_4$ in 1.5 L H_2O.

17-28. 0.01518 M $KMnO_4$

17-30. 0.06711 M $Na_2S_2O_3$

17-32. (a) 14.72% Sb (b) 20.54% Sb_2S_3

17-34. 9.38% analyte

17-35. (a) 32.08% Fe (b) 45.86% Fe_2O_3

17-37. 0.03074 M H_2NOH

17-39. 56.53% $KClO_4$

17-41. 0.701% As_2O_3

17-43. 3.64% C_2H_5SH

17-45. 2.635% KI

17-46. 69.07% Fe and 21.07% Cr

17-48. 0.5554 g Tl

17-49. 10.4 ppm SO_2

17-51. 19.5 ppm H_2S

Chapter 18

18-1. (a) An indicator electrode is an electrode used in potentiometry that responds to variations in the activity of an analyte ion or molecule.

 (c) An electrode of the first type is a metal electrode that is used to determine the concentration of its cation in a solution.

18-2. (a) A liquid-junction potential is the potential that develops across the interface between two solutions having different electrolyte compositions.

18-3. (a) An electrode of the first kind for Hg(II) would take the form $\|Hg^{2+}(x\ M)|Hg$,

$$E_{Hg} = E_{Hg}^0 - \frac{0.0592}{2}\log\frac{1}{[Hg^{2+}]} = E_{Hg}^0 - \frac{0.0592}{2}pHg$$

 (b) An electrode of the second kind for EDTA would take the form $\|HgY^{2-}(y\ M),\ Y^{4-}(x\ M)|Hg$, where a small and fixed amount of HgY^{2-} is introduced into the analyte solution. Here the potential of the mercury electrode is given by

$$E_{Hg} = K - \frac{0.0592}{2}\log[Y^{4-}] = K + \frac{0.0592}{2}pY$$

where

$$K = E_{HgY^{2-}}^0 - \frac{0.0592}{2}\log\frac{1}{a_{HgY^{2-}}}$$

$$\cong 0.21 - \frac{0.0592}{2}\log\frac{1}{[HgY^{2-}]}$$

18-5. The pH-dependent potential that develops across a glass membrane arises from the difference in positions of dissociation equilibria on each of the two surfaces. These equilibria are described by the equation

$$H^+Gl^- \rightleftharpoons H^+ + Gl^-$$

   membrane soln membrane

The surface exposed to the solution having the higher hydrogen ion concentration then becomes positive with respect to the other surface. This charge difference, or potential, serves as the analytical parameter when the pH of the solution on one side of the membrane is held constant.

18-7. Uncertainties that may be encountered in pH measurements include (1) the acid error in highly acidic solutions, (2) the alkaline error in strongly basic solutions, (3) the error that arises when the ionic strength of the calibration standards differ from that of the analyte solution, (4) uncertainties in the pH of the standard buffers, (5) nonreproducible junction potentials when samples of low ionic strength are measured, and (6) dehydration of the working surface.

18-9. The alkaline error arises when a glass electrode is employed to measure the pH of solutions having pH values in the 10 to 12 range or greater. In the presence of alkali ions, the glass surface becomes responsive to not only hydrogen but alkali ions as well. Low pH values arise as a consequence.

18-11. The direct potentiometric measurement of pH is intensive. That is, the pH of one drop of the solution will be the same as five liters of that solution. The information provided by a potentiometric titration is extensive, being dependent on the amount of hydronium ions in the sample.

18-12. (a) 0.354 V

  (b) $SCE\|IO_3^-(x\ M),AgIO_3(sat'd)|Ag$

  (c) $pIO_3 = (E_{cell} - 0.110)/0.0592$

  (d) 3.11

18-14. (a) $SCE\|SCN^-(x\ M),AgSCN(sat'd)|Ag$

  (c) $SCE\|SO_3^{2-}(x\ M),Ag_2SO_3(sat'd)|Ag$

18-15. (a) $pSCN = (E_{cell} + 0.153)/0.0592$

  (c) $pSO_3 = 2(E_{cell} - 0.146)/0.0592$

18-16. (a) 4.65 (c) 5.20

18-17. 6.76

18-19. (a) 0.157 V (b) -0.026 V

18-21. (a) pH = 12.629 and $a_{H^+} = 2.35 \times 10^{-13}$

  (b) pH = 5.579 and $a_{H^+} = 2.64 \times 10^{-6}$

  (c) For part (a), $a_{H^+} = 2.17 \times 10^{-13}$ to 2.54×10^{-13}

   For part (b), $a_{H^+} = 2.44 \times 10^{-6}$ to 2.85×10^{-6}

18-23. 2.0×10^{-6}

18-25. 136 g HA/mol

18-27.

mL Reagent	E vs. SCE, V	mL Reagent	E vs. SCE, V
5.00	0.58	49.00	0.66
10.00	0.59	50.00	0.80
15.00	0.60	51.00	1.10
25.00	0.61	55.00	1.14
40.00	0.62	60.00	1.15

Chapter 19

19-1. (a) *Concentration polarization* is a condition in which the current in an electrochemical cell is limited by the rate at which reactants are brought to or removed from the surface of one or both electrodes. *Kinetic polarization* is a condition in which the current in an electrochemical cell is limited by the rate at which electrons are transferred between the electrode surfaces and reactants in solution. For either type of polarization, the current is no longer proportional to the cell potential.

(c) Both the *coulomb* and the *Faraday* are units of quantity of charge, or electricity. The former is the quantity transported by one ampere of current in one second; the latter is equal to 96,495 coulombs or one mole of electrons.

(e) The *electrolysis circuit* consists of a working electrode and a counter electrode. The *control circuit* regulates the applied potential such that the potential between the working electrode and a reference electrode in the control circuit is constant and at a desired level.

19-2. (a) *Current density* is the current at an electrode divided by the surface area of that electrode. Ordinarily, it has units of amperes per square centimeter.

(c) A *coulometric titration* is an electroanalytical method in which a constant current of known magnitude generates a reagent that reacts with the analyte. The time required to generate enough reagent to complete the reaction is measured.

(e) *Current efficiency* is a measure of agreement between the number of faradays of current and the number of moles of reactant oxidized or reduced at a working electrode.

19-3. Mass transport in an electrochemical cell results from one or more of the following: (1) *diffusion,* which arises from concentration differences between the electrode surface and the bulk of the solution; (2) *migration,* which results from electrostatic attraction or repulsion; and (3) *convection,* which results from stirring, vibration, or temperature difference.

19-5. Both kinetic and concentration polarization cause the potential of a cell to be more negative than the thermodynamic potential. Concentration polarization arises from the slow rate at which reactants or products are transported to or away from the electrode surfaces. Kinetic polarization arises from the slow rate of the electrochemical reactions at the electrode surfaces.

19-7. Kinetic polarization is often encountered when the product of a reaction is a gas, particularly when the electrode

is a soft metal such as mercury, zinc, or copper. It is likely to occur at low temperatures and high current densities.

19-9. Temperature, current density, complexation of the analyte, and codeposition of a gas influence the physical properties of an electrogravimetric deposit.

19-11. (a) An *amperostat* is an instrument that provides a constant current for an electrolysis cell.

(b) A *potentiostat* controls the applied potential to maintain a constant potential between the working electrode and a reference electrode.

19-13. The species produced at the counter electrode is a potential interference by reacting with the product at the working electrode. Isolation of the one from the other is ordinarily necessary.

19-15. (b) 6.2×10^{16} cations

19-16. (a) -0.738 V   (c) -0.337 V

19-17. -0.913 V

19-19. (a) -0.676 V   (b) -0.36 V   (c) -1.54 V
(d) -1.67 V

19-21. (a) -0.94 V   (b) -0.35 V   (c) -2.09 V
(d) -2.37 V

19-23. (a) $[BiO^+] = 5 \times 10^{-28}$ (b) 0.103 V

19-25. (a) Separation not possible.
(b) Separation feasible.
(c) 0.099 V to -0.014 V

19-26. (a) $\Delta E^0 = 0.237$ V
(c) $\Delta E^0 = 0.0395$ V
(e) $\Delta E^0 = 0.118$ V
(g) $\Delta E^0 = 0.276$ V
(i) $\Delta E^0 = 0.0789$ V

19-27. (a) 28.4 min   (b) 9.47 min

19-29. 112.6 g/eq

19-31. 79.5 ppm $CaCO_3$

19-33. 4.06% $C_6H_5NO_2$

19-35. 2.471% CCl_4 and 1.854% $CHCl_3$

19-37. 50.9 μg $C_6H_5NH_2$

Chapter 20

20-1. (b) A *phototube* is a vacuum tube equipped with a photoemissive cathode. It has a high electrical resistance and requires a potential of 90 V or more to produce a photocurrent. The currents are generally small enough to require considerable amplification before they can be measured. A *photovoltaic cell* consists of a photosensitive semiconductor sandwiched between two electrodes. A current is generated between the electrodes when radiation is absorbed by the semiconducting layer. The current is generally large enough to be measured directly with a microammeter. The advantages of a phototube are greater sensitivity and wavelength range as well as better reproducibility. The advantages of the photocell are its simplicity, low cost, and general ruggedness. In addition, it does not require an external power supply or elaborate electronic circuitry. Its use is limited to visible radiation, however, and in addition it suffers from fatigue whereby its electrical output decreases gradually with time.

(d) A *photon detector* produces an electrical signal when photons of radiation are absorbed by a photoemissive surface in the detector. A *heat detector* responds to the increase in temperature brought about by absorption of radiation by a blackened surface in the detector. Photon detectors are generally more sensitive than heat detectors and less subject to interference by changes in the ambient temperature.

20-3. Photons from the infrared region of the spectrum do not have sufficient energy to cause photoemission from the cathode of a photomultipler tube.

20-5. *Tungsten/halogen lamps* contain a small amount of iodine in the evacuated quartz envelope that contains the tungsten filament. The iodine prolongs the life of the lamp and permits it to operate at a higher temperature. The iodine combines with gaseous tungsten that sublimes from the filament and causes the metal to be redeposited, thus adding to the life of the lamp.

20-6. (a) 1.13×10^{18} Hz   (c) 4.318×10^{14} Hz
(e) 1.53×10^{13} Hz

20-7. (a) 252.8 cm   (c) 286 cm

20-9. 1.00×10^{14} Hz to 1.62×10^{15} Hz
3.33×10^3 cm^{-1} to 5.41×10^4 cm^{-1}

20-11. 136 cm and 1.46×10^{-25} J

20-12. (a) 436 nm

Chapter 21

21-1. (a) A *chromophore* is an organic functional group that absorbs radiation in the ultraviolet/visible regions.
(c) Radiation that consists of a single wavelength is said to be *monochromatic*.
(e) An *absorption spectrum* is a plot of a spectral property (absorbance, log absorbance, absorptivity, transmittance) as the ordinate and wavelength, wavenumber, or frequency as the abscissa.

21-2. (a) Absorbance A and transmittance T are related by the equation

$$A = -\log T = \log \frac{1}{T}$$

21-4. Departures from Beer's law occur as a result of failure to use monochromatic radiation, existence of stray radiation, experimental uncertainties in measurement of low absorbances, molecular interactions at high absorbance, concentration-dependent association/dissociation.

21-6. A solution of $Cu(NH_3)_4^{2+}$ is blue because this ion absorbs yellow radiation and transmits blue radiation unchanged.

21-8. (a) cm^{-1} ppm^{-1} (c) cm^{-1}%$^{-1}$

21-9. (a) $T = 67.3\%$ $c = 4.07 \times 10^{-5}$ M
$c_{ppm} = 8.13$ ppm $a = 2.11 \times 10^{-2}$ cm^{-1} ppm^{-1}
(c) $T = 30.2\%$ $c = 6.54 \times 10^{-5}$ M
$c_{ppm} = 13.1$ ppm $a = 0.0397$ cm^{-1} ppm^{-1}
(e) $A = 0.638$ $T = 23.0\%$
$c_{ppm} = 342$ ppm $a = 1.87 \times 10^{-2}$ cm^{-1} ppm^{-1}
(g) $T = 15.9\%$ $c = 1.68 \times 10^{-4}$ M
$\varepsilon = 3.17 \times 10^3$ L cm^{-1} mol^{-1}
$a = 0.158$ cm^{-1} mol^{-1}
(i) $c = 2.62 \times 10^{-5}$ M $A = 1.281$
$b = 5.00$ cm $a = 0.0489$ cm^{-1} ppm^{-1}

21-10. (a) 88.9%   (c) 41.8%   (e) 32.7%

21-11. (a) 0.593   (c) 0.484   (e) 1.07

21-12. (a) 79.1%   (c) 17.5%   (e) 10.7%

21-13. (a) 0.894   (c) 0.785   (e) 1.37

21-14. 1.80×10^4 L cm^{-1} mol^{-1}

21-16. (a) 0.175   (b) 0.350   (c) 66.8% and 44.7%
(d) 0.476

21-17. 0.0530

21-19. (a) 0.590   (b) 25.7%   (c) 1.70×10^{-5} M
(d) 2.50 cm

21-20. (a)

c_{ind}, M	A_{430}	A_{600}
3.00×10^{-4}	1.54	1.06
2.00×10^{-4}	0.935	0.777
1.00×10^{-4}	0.383	0.455
0.500×10^{-4}	0.149	0.261
0.250×10^{-4}	0.056	0.145

Chapter 22

22-2. As a minimum, the radiation emitted by the source of a single-beam instrument must be stable long enough to make the 0% T adjustment, the 100% T adjustment, and the measurement of T for the sample.

22-4. pH, electrolyte concentration, temperature.

22-6. 1.6×10^{-5} M to 8.6×10^{-5} M

22-8. 5.4×10^{-5} M to 1.2×10^{-3} M

22-10. (a) $T = 26.1\%$, $A = 0.583$
(c) $T = 6.82\%$, $A = 1.166$

22-11. (b) 0.471   (d) 11.4%

22-14. The absorbance will decrease linearly before the equivalence point and reach a constant value of approximately zero as equivalence is passed.

22-16. 0.0214% Co

22-18. (a) 5.48×10^{-5} M Co and 1.31×10^{-4} M Ni
(c) 2.20×10^{-4} M Co and 4.41×10^{-5} M Ni

22-21. (a) 0.492   (c) 0.190

22-22. (a) 0.301   (c) 0.491

22-23. For solution A, pH = 5.60
For solution C, pH = 4.80

22-24. (a)

λ	A	λ	A	λ	A
420	0.414	470	0.490	585	0.123
445	0.510	510	0.321	595	0.114
450	0.513	550	0.170	610	0.103
455	0.513	570	0.139	650	0.082

22-26.

	c_P, M	c_Q, M
(a)	2.08×10^{-4}	4.91×10^{-5}
(c)	8.37×10^{-5}	6.10×10^{-5}
(e)	2.11×10^{-4}	9.65×10^{-5}

22-27. (b) $A_{510} = 0.03949 \, c_{Fe} - 0.001008$
(c) $s_m = 1.1 \times 10^{-4}$ and $s_b = 2.7 \times 10^{-3}$

22-28.

	c_{Fe}, ppm	s_c, rel %	
		1 Result	3 Results
(a)	3.65	2.8	2.1
(c)	1.75	6.1	4.6
(e)	38.3	0.27	0.20

Chapter 23

23-1. (a) *Fluorescence* is a process by which an excited atom relaxes by emitting electromagnetic radiation.

(c) *Internal conversion* is the nonradiative relaxation of a molecule from the lowest vibrational level of an excited electronic state to the highest vibrational level of a lower electronic state.

(e) The *Stokes shift* is the difference in wavelength between the radiation used to excite fluorescence and the wavelength of the emitted radiation.

(g) *Self-quenching* occurs when the fluorescent radiation from an excited analyte molecule is absorbed by an unexcited analyte molecule. This process results in a decrease in fluorescence intensity.

23-3. Compounds that fluoresce have structures that slow the rate of nonradiative relaxation to the point where there is time for fluorescence to occur. Compounds that do not fluoresce have structures that permit rapid relaxation by nonradiative processes.

23-5. See Figure 23-8. A fluorometer usually consists of a light source, a filter or a monochromator for selecting the excitation wavelength, a sample container, an emission filter or monochromator, and a detector. There may also be a reference detector for monitoring and correcting for fluctuations in the light source intensity. Emission is usually detected at right angles to the incident radiation to maximize the fluorescence signal.

23-7. Filter fluorometers are usually more sensitive than spectrofluorometers because filters have a higher radiation throughput than do monochromators. They also allow positioning of the detector closer to the sample, thus increasing the magnitude of the signal.

23-9. (a) Excitation of fluorescence usually involves transfer of an electron to a high vibrational state of an upper electronic state. Relaxation to a lower vibrational state of this electronic state goes on much more rapidly than fluorescence relaxation. When fluorescence relaxation occurs it is to a high vibrational state of the ground state or to a high vibrational state of an electronic state that is above the ground state. Such transitions involve less energy than the excitation energy. Therefore, the emitted radiation is longer in wavelength than the excitation wavelength.

(b) For spectrofluorometry, the analytical signal F is given by $F = 2.3 K' \varepsilon bc P_0$. The magnitude of F, and thus sensitivity, can be enhanced by increasing the source intensity P_0 or the transducer sensitivity.

For spectrophotometry, the analytical signal A is given by $A = \log P_0/P$. Increasing P_0 or the detector's response to P_0 is accompanied by a corresponding increase in P. Thus, the ratio does not change nor does the analytical signal. Consequently, no improvement in sensitivity accompanies such changes.

23-10. 533 mg quinine/tablet

23-12. 0.0170% warfarin

23-13. (c) $I_r = 22.4 \, c_{NADH} + 0.0700$

(d) $s_m = 0.340$, $s_b = 0.186$

(e) 0.540 μmol NADH/mL

(f) 0.0052 μmol NaDH/mL

Chapter 24

24-1. In *atomic emission spectroscopy* the radiation source is the sample itself. The energy for excitation of analyte atoms comes from a flame, a furnace, or a plasma. The signal is the measured intensity of the flame at the wavelength of interest. In *atomic absorption spectroscopy* the radiation source is usually a line source such as a hollow cathode lamp, and the output signal is the absorbance calculated from the incident power of the source and the resulting power after the light has passed through the atomized sample in the heated source.

24-2. (a) *Atomization* is a process in which a sample, usually in solution, is volatilized and decomposed to produce an atomic vapor.

(c) *Doppler broadening* is an increase in the width of atomic absorption or emission lines caused by the Doppler effect in which atoms moving toward a detector absorb or emit wavelengths that are slightly shorter than those absorbed or emitted by atoms moving at right angles to the detector. The effect is reversed for atoms moving away from the detector.

(e) A *plasma* is a conducting gas that contains a large concentration of ions and/or electrons.

(g) *Sputtering* is a process in which atoms of an element are dislodged from the surface of a cathode by bombardment by a stream of inert gas ions that have been accelerated toward the cathode by a high electrical potential.

(i) A *spectral interference* in atomic spectroscopy occurs when an absorption or emission peak of an element in the sample matrix overlaps that of the analyte.

(k) A *radiation buffer* is a substance that is added in large excess to both standards and samples in atomic spectroscopy to prevent the presence of that substance in the sample matrix from having an appreciable effect on the results.

(m) A *protective agent* in atomic spectroscopy is a substance, such as EDTA or 8-hydroxyquinoline, that forms stable but volatile complexes with analyte cation and thus prevents interference by anions that form nonvolatile species with the analyte.

24-3. In atomic emission spectroscopy, the analytical signal is produced by *excited* atoms or ions, whereas in atomic absorption the signal results from absorption by *unexcited* species. Typically, the number of unexcited species exceeds the excited by several orders of magnitude. The ratio of unexcited to excited atoms in a hot medium varies exponentially with temperature. Thus a small change in temperature brings about a large change in the number of excited atoms. The number of unexcited atoms changes very little, however, because they are present in an enormous excess. Therefore, emission spectroscopy is more sensitive to temperature changes than is absorption spectroscopy.

24-5. (a) Sulfate ion forms complexes with Fe(III) that are not readily atomized. Thus the concentration of iron atoms in a flame is less in the presence of sulfate ions.

(b) Sulfate interference could be overcome by (1) adding a releasing agent that forms complexes with sulfate that are more stable than the iron complexes, (2) adding a protective agent, such as EDTA, that forms a highly stable but volatile complex with the Fe(III), and (3) by employing a high temperature flame (oxygen/acetylene or nitrous oxide/acetylene).

24-7. The lack of linearity is a consequence of ionization of the uranium, which decreases the relative concentration of uranium atoms. Alkali metal atoms ionize readily and thus give a high concentration of electrons that represses the uranium ionization.

24-9. 0.504 ppm Pb

24-11. (a) A plot of A vs. mL standard (V_s) is a straight line.

(b) Rearranging Beer's law gives

$$c_{Cr} = \frac{A}{ab} = \frac{A}{k}$$

The concentration of chromium for each solution is given by

$$c_{Cr} = \frac{c_x V_x + c_s V_s}{V_t} = \frac{A}{k}$$

where c_s and c_x are the concentrations of chromium in the unknown and the standard, and V_x and V_s are the volumes of the standard and the unknown. $V_t = V_x + V_s$. Rearranging the foregoing equation gives

$$A = kc_s \frac{V_s}{V_t} + kc_x \frac{V_x}{V_t}$$

(c) The equation derived in part (b) can be written in the form $A = mV_s + b$, where

$$m = \text{slope} = \frac{kc_s}{V_t}$$

$$b = \text{intercept} = \frac{kc_x V_x}{V_t}$$

(d) $\dfrac{b}{m} = \dfrac{kc_x V_x/V_t}{kc_s/V_t}$.   Solving for c_x gives

$$c_x = \frac{b}{m} \times \frac{c_s}{V_x}$$

(e) $m = 8.81 \times 10^{-3}$, $b = 0.202$,
    $A = 8.81 \times 10^{-3} V_s + 0.202$

(f) $s_m = 4.11 \times 10^{-5}$, $s_b = 1.01 \times 10^{-3}$
(g) 28.0 ppm Cr

Chapter 25

25-1. A *masking agent* is a complexing reagent that reacts selectively with one or more components of a solution to prevent them from interfering in an analysis.

25-2. Strong acid synthetic ion exchangers have sulfonic acid groups attached to the resin molecules. Weak acid exchangers have carboxylic acid functional groups rather than sulfonic acid groups.

25-3. (a) 1.73×10^{-2} M   (b) 6.40×10^{-3} M
     (c) 2.06×10^{-3} M   (d) 6.89×10^{-4} M

25-5. (a) 75 mL   (b) 40 mL   (c) 22 mL

25-7. (a) 18.0   (b) 7.56

25-9. (a) 1.53
     (b) $[HA]_{aq} = 0.0147$, $[A^-]_{aq} = 0.0378$
     (c) 9.7×10^{-2}

25-11. (a) 12.4 meq Ca^{2+}/L   (b) 6.19×10^2 mg $CaCO_3$/L

25-13. Dissolve 17.53 g of NaCl in about 100 mL of water and pass the solution through a column packed with a cation exchange resin in its acid form. Wash the column with several hundred milliliters of water, collecting the liquid from the original solution and the washings in a 2-L volumetric flask. Dilute to the mark and mix well.

Chapter 26

26-1. (a) *Elution* is a process in which species are washed through a chromatographic column by additions of fresh solvent.

(c) The *stationary phase* in a chromatographic column is a solid or liquid that is fixed in place. A mobile phase then passes over or through the stationary phase.

(e) The *retention time* for an analyte is the time interval between its injection onto a column and the appearance of its peak at the other end of the column.

(g) The *selectivity factor* α of a column toward two species is given by the equation $\alpha = K_B/K_A$ where K_B is the partition ratio of the more strongly held species B and K_A is the corresponding ratio for the less strongly held solute A.

26-3. In gas-liquid chromatography, the mobile phase is a gas, whereas in liquid-liquid chromatography it is a liquid.

26-5. The number of plates in a column can be determined by measuring the retention time t_R and width of a peak at its base W. The number of plates N is then given by the equation $N = 16(t_R/W)^2$.

26-6. (a)

Peak	N
A	2775
B	2472
C	2363
D	2523

(b) $\overline{N} = 2.5 \times 10^3$ plates $s_N = 0.2 \times 10^3$
(c) 0.0097 cm

26-8.

Peak	k' (a)	K (b)
A	0.74	6.2
B	3.3	27
C	3.5	30
D	6.0	50

26-9. (a) 0.72 (b) 1.1 (c) 108 cm (d) 62 min

26-10. (a) 5.2 (b) 2.0 cm

26-14. (a) $k'_1 = 4.3$ $k'_2 = 4.7$ $k'_3 = 6.1$
(b) $K_1 = 14$ $K_2 = 15$ $K_3 = 19$
(c) $\alpha_{2,1} = 1.11$

26-15. (a) $k'_M = 2.54$ $k'_N = 2.62$
(b) $\alpha = 1.03$
(c) 8.1×10^4 plates
(d) 1.8×10^2 cm
(e) 91 min

Chapter 27

27-2. *Slow sample injection* in gas chromatography leads to band broadening and lowered resolution.

27-4. In *temperature programming*, the temperature of a column is increased either linearly or in steps as an elution is carried out. Temperature programming often leads to better separations in shorter times, as shown in Figure 27-7c.

27-5. N increases with increases in column efficiency. H decreases with increases in column efficiency.

27-7. In *open tubular columns*, the stationary phase is held on the inner surface of a capillary, whereas in packed columns, the stationary phase is supported on particles that are contained in a glass or metal tube. Open tubular columns, which are only applicable in gas and supercritical fluid chromatography, contain an enormous number of plates that permit rapid separations of closely related species. They suffer from small sample capacities.

27-8. (a) *Sparging* is the process of removing a dissolved gas from a liquid by bubbling an inert gas through the liquid.
(c) In an *isocratic elution*, the solvent composition is held constant throughout the elution.
(g) *Bonded phase packings* have organic groups chemically bonded to silica particles.

27-9. (a) Diethyl ether, benzene, *n*-hexane

27-10. (a) Ethyl acetate, dimethylamine, acetic acid

27-11. In *adsorption chromatography*, separations are based upon *adsorption equilibria* between the components of the sample and a solid surface. In partition chromatography, separations are based upon *distribution equilibria* between two immiscible liquids.

27-13. *Gel filtration* is a type of size-exclusion chromatography in which the packings are hydrophilic, and eluents are aqueous. It is used for separating high-molecular-weight polar compounds. *Gel-permeation chromatography* is a type of size-exclusion chromatography in which the packings are hydrophobic, and eluents are nonaqueous. It is used for separating high-molecular-weight nonpolar species.

27-15.

Peak Area	% Present
16.4	22.9
45.2	48.5
30.2	28.7

INDEX

Entries in **bold face** refer to specific laboratory directions; *t* refers to a table

INTERNATIONAL ATOMIC MASSES

Element	Symbol	Atomic Number	Atomic Mass	Element	Symbol	Atomic Number	Atomic Mass
Actinium	Ac	89	227	Mercury	Hg	80	200.59
Aluminum	Al	13	26.981539	Molybdenum	Mo	42	95.94
Americium	Am	95	243	Neodymium	Nd	60	144.24
Antimony	Sb	51	121.757	Neon	Ne	10	20.1797
Argon	Ar	18	39.948	Neptunium	Np	93	237
Arsenic	As	33	74.92159	Nickel	Ni	28	58.6934
Astatine	At	85	210	Niobium	Nb	41	92.90638
Barium	Ba	56	137.327	Nitrogen	N	7	14.00674
Berkelium	Bk	97	247	Nobelium	No	102	259
Beryllium	Be	4	9.012182	Osmium	Os	76	190.2
Bismuth	Bi	83	208.98037	Oxygen	O	8	15.9994
Boron	B	5	10.811	Palladium	Pd	46	106.42
Bromine	Br	35	79.904	Phosphorus	P	15	30.973762
Cadmium	Cd	48	112.411	Platinum	Pt	78	195.08
Calcium	Ca	20	40.078	Plutonium	Pu	94	244
Californium	Cf	98	251	Polonium	Po	84	210
Carbon	C	6	12.011	Potassium	K	19	39.0983
Cerium	Ce	58	140.115	Praseodymium	Pr	59	140.90765
Cesium	Cs	55	132.90543	Promethium	Pm	61	145
Chlorine	Cl	17	35.4527	Protactinium	Pa	91	231.03588
Chromium	Cr	24	51.9961	Radium	Ra	88	226
Cobalt	Co	27	58.93320	Radon	Rn	86	221
Copper	Cu	29	63.546	Rhenium	Re	75	186.207
Curium	Cm	96	247	Rhodium	Rh	45	102.90550
Dysprosium	Dy	66	162.50	Rubidium	Rb	37	85.4678
Einsteinium	Es	99	252	Ruthenium	Ru	44	101.07
Erbium	Er	68	167.26	Samarium	Sm	62	150.36
Europium	Eu	63	151.965	Scandium	Sc	21	44.955910
Fermium	Fm	100	257	Selenium	Se	34	78.96
Fluorine	F	9	18.9984032	Silicon	Si	14	28.0855
Francium	Fr	87	223	Silver	Ag	47	107.8682
Gadolinium	Gd	64	157.25	Sodium	Na	11	22.989768
Gallium	Ga	31	69.723	Strontium	Sr	38	87.62
Germanium	Ge	32	72.61	Sulfur	S	16	32.066
Gold	Au	79	196.96654	Tantalum	Ta	73	180.9479
Hafnium	Hf	72	178.49	Technetium	Tc	43	98
Helium	He	2	4.002602	Tellurium	Te	52	127.60
Holmium	Ho	67	164.93032	Terbium	Tb	65	158.92534
Hydrogen	H	1	1.00794	Thallium	Tl	81	204.3833
Indium	In	49	114.82	Thorium	Th	90	232.0381
Iodine	I	53	126.90447	Thulium	Tm	69	168.93421
Iridium	Ir	77	192.22	Tin	Sn	50	118.710
Iron	Fe	26	55.847	Titanium	Ti	22	47.88
Krypton	Kr	36	83.80	Tungsten	W	74	183.85
Lanthanum	La	57	138.9055	Uranium	U	92	238.0289
Lawrencium	Lr	103	262	Vanadium	V	23	50.9415
Lead	Pb	82	207.2	Xenon	Xe	54	131.29
Lithium	Li	3	6.941	Ytterbium	Yb	70	173.04
Lutetium	Lu	71	174.967	Yttrium	Y	39	88.90585
Magnesium	Mg	12	24.305	Zinc	Zn	30	65.39
Manganese	Mn	25	54.93805	Zirconium	Zr	40	91.224
Mendelevium	Md	101	258				